BUSINESS/SCIENCE/TECHNOLOGY DIVISION
CHICAGO PUBLIC LIBRARY
400 SOUTH STATE STREET
CHICAGO, IL 60605

CHICAGO PUBLIC LIBRARY
HAROLD WASHINGTON LIBRARY CENTER

R0006354083

R537.5352
G858M
cop. 1

FORM 125 M

The Chicago Public Library

SCIENCE DIVISION
Received May 8, 1972

Mössbauer
Spectroscopy

Mössbauer Spectroscopy

N. N. Greenwood
*Professor of Inorganic and Structural Chemistry,
University of Leeds*

T. C. Gibb
*Lecturer in Inorganic and
Structural Chemistry,
University of Leeds*

Chapman and Hall Ltd · London

First published 1971
by Chapman and Hall Ltd,
11 New Fetter Lane, London EC4P 4EE
Printed in Great Britain by
Butler & Tanner Ltd, Frome and London

SBN 412 10710 4

© 1971 T. C. Gibb and N. N. Greenwood
All rights reserved. No part of this
publication may be produced, stored in
a retrieval system or transmitted in
any form or by any means, electronic,
mechanical, photocopying, recording,
or otherwise, without the prior written
permission of the publisher.

Distributed in the U.S.A. by
Barnes and Noble Inc.

Contents

		page	
Preface		*page*	xi
1.	**The Mössbauer Effect**		1
	Introduction		1
1.1	Energetics of free-atom recoil and thermal broadening		2
1.2	Heisenberg natural linewidth		5
1.3	Energy and momentum transfer to the lattice		6
1.4	Recoil-free fraction and Debye–Waller factor		9
1.5	Cross-section for resonant reabsorption		11
1.6	A Mössbauer spectrum		15
2.	**Experimental Techniques**		17
2.1	Velocity modulation of gamma-rays		17
2.2	Constant-velocity drives		19
2.3	Repetitive velocity-scan systems		21
	(a) Pulse-height analysis mode		23
	(b) Time-mode (multiscalar-mode) spectrometers		25
	(c) On-line computers		26
2.4	Derivative spectrometers		26
2.5	Scattering experiments		27
2.6	Source and absorber preparation		30
2.7	Detection equipment		35
2.8	Cryogenic equipment and ovens		38
2.9	Velocity calibration		39
2.10	Curve fitting by computer		41
3.	**Hyperfine Interactions**		46
	Introduction		46
3.1	Chemical isomer shift, δ		46
3.2	Second-order Doppler shift and zero-point motion		50
3.3	Effect of pressure on the chemical isomer shift		53
3.4	Electric quadrupole interactions		54

3.5	Magnetic hyperfine interactions	59
3.6	Combined magnetic and quadrupole interactions	63
3.7	Relative intensities of absorption lines	66
3.8	Relaxation phenomena	72
3.9	Anisotropy of the recoilless fraction	74
3.10	The pseudoquadrupole interaction	76

4. Applications of the Mössbauer Effect — 80
4.1	Relativity and general physics	80
4.2	Nuclear physics	82
4.3	Solid-state physics and chemistry	84

5. ^{57}Fe – Introduction — 87
5.1	The γ-decay scheme	87
5.2	Source preparation and calibration	89
5.3	Chemical isomer shifts	90
5.4	Quadrupole splittings	96
5.5	Magnetic interactions	102
5.6	Polarised radiation studies	104
5.7	Energetic nuclear reactions	109
5.8	The 136-keV transition	110

6. High-spin Iron Complexes — 112

A. HIGH-SPIN IRON(II) COMPLEXES — 112
6.1	Iron(II) halides	113
6.2	Iron(II) salts of oxyacids and other anions	130
6.3	Iron(II) complexes with nitrogen ligands	140

B. HIGH-SPIN IRON(III) COMPLEXES — 148
6.4	Iron(III) halides	148
6.5	Iron(III) salts of oxyacids	155
6.6	Iron(III) complexes with chelating ligands	159

7. Low-spin Iron(II) and Iron(III) Complexes — 169
7.1	Ferrocyanides	169
7.2	Ferricyanides	173
7.3	Prussian blue	178
7.4	Substituted cyanides	182
7.5	Chelating ligands	187

8. Unusual Electronic Configurations of Iron — 194
8.1	Iron(II) compounds showing $^5T_2$–$^1A_1$ crossover	194
8.2	Iron(III) compounds showing $^6A_1$–$^2T_2$ crossover	202

8.3	Iron(II) compounds with $S = 1$ spin state	205
8.4	Iron(III) compounds with $S = \frac{3}{2}$ spin state	206
8.5	Iron 1,2-dithiolate complexes	212
8.6	Systems containing iron(I), iron(IV), and iron(VI)	216
9.	**Covalent Iron Compounds**	**221**
9.1	Binary carbonyls, carbonyl anions, and hydride anions	222
9.2	Substituted iron carbonyls	226
9.3	Ferrocene and other π-cyclopentadienyl derivatives	233
10.	**Iron Oxides and Sulphides**	**239**
10.1	Binary oxides and hydroxides	240
10.2	Spinel oxides AB_2O_4	258
10.3	Other ternary oxides	269
10.4	Iron(IV) oxides	280
10.5	Iron chalcogenides	283
10.6	Silicate minerals	286
10.7	Lunar samples	294
11.	**Alloys and Intermetallic Compounds**	**304**
11.1	Metallic iron	305
11.2	Iron alloys	308
11.3	Intermetallic compounds	317
12.	**^{57}Fe – Impurity Studies**	**329**
12.1	Chemical compounds	330
12.2	Metals	340
12.3	Miscellaneous topics	344
13.	**Biological Compounds**	**352**
13.1	Haemeproteins	353
13.2	Metalloproteins	365
14.	**Tin-119**	**371**
14.1	γ-Decay scheme and sources	371
14.2	Hyperfine interactions	375
14.3	Tin(II) compounds	381
14.4	Inorganic tin(IV) compounds	390
14.5	Organotin(IV) compounds	399
14.6	Metals and alloys	417
15.	**Other Main Group Elements**	**433**
15.1	Potassium (^{40}K)	433

15.2	Germanium (^{73}Ge)	434
15.3	Krypton (^{83}Kr)	437
15.4	Antimony (^{121}Sb)	441
15.5	Tellurium (^{125}Te)	452
15.6	Iodine (^{127}I, ^{129}I)	462
15.7	Xenon (^{129}Xe, ^{131}Xe)	482
15.8	Caesium (^{133}Cs)	486
15.9	Barium (^{133}Ba)	488

16.	**Other Transition-metal Elements**	**493**
16.1	Nickel (^{61}Ni)	493
16.2	Zinc (^{67}Zn)	497
16.3	Technetium (^{99}Tc)	499
16.4	Ruthenium (^{99}Ru)	499
16.5	Silver (^{107}Ag)	504
16.6	Hafnium (^{176}Hf, ^{177}Hf, ^{178}Hf, ^{180}Hf)	504
16.7	Tantalum (^{181}Ta)	507
16.8	Tungsten (^{182}W, ^{183}W, ^{184}W, ^{186}W)	509
16.9	Rhenium (^{187}Re)	514
16.10	Osmium (^{186}Os, ^{188}Os, ^{189}Os)	514
16.11	Iridium (^{191}Ir, ^{193}Ir)	518
16.12	Platinum (^{195}Pt)	524
16.13	Gold (^{197}Au)	526
16.14	Mercury (^{201}Hg)	530

17.	**The Rare-earth Elements**	**536**
17.1	Praseodymium (^{141}Pr)	537
17.2	Neodymium (^{145}Nd)	537
17.3	Promethium (^{147}Pm)	539
17.4	Samarium (^{149}Sm, ^{152}Sm, ^{154}Sm)	539
17.5	Europium (^{151}Eu, ^{153}Eu)	543
17.6	Gadolinium (^{154}Gd, ^{155}Gd, ^{156}Gd, ^{157}Gd, ^{158}Gd, ^{160}Gd)	558
17.7	Terbium (^{159}Tb)	563
17.8	Dysprosium (^{160}Dy, ^{161}Dy, ^{162}Dy, ^{164}Dy)	563
17.9	Holmium (^{165}Ho)	573
17.10	Erbium (^{164}Er, ^{166}Er, ^{167}Er, ^{168}Er, ^{170}Er)	574
17.11	Thulium (^{169}Tm)	579
17.12	Ytterbium (^{170}Yb, ^{171}Yb, ^{172}Yb, ^{174}Yb, ^{176}Yb)	585

18.	**The Actinide Elements**	**596**
18.1	Thorium (^{232}Th)	596
18.2	Protactinium (^{231}Pa)	596
18.3	Uranium (^{238}U)	597

18.4 Neptunium (^{237}Np) — 600
18.5 Americium (^{243}Am) — 604

Appendix 1. Table of nuclear data for Mössbauer transitions — 607

Appendix 2. The relative intensities of hyperfine lines — 612

Notes on the International System of Units (SI) — 619

Author Index — 621

Subject Index — 645

Preface

Rudolph Mössbauer discovered the phenomenon of recoil-free nuclear resonance fluorescence in 1957–58 and the first indications of hyperfine interactions in a chemical compound were obtained by Kistner and Sunyar in 1960. From these beginnings the technique of Mössbauer spectroscopy rapidly emerged and the astonishing versatility of this new technique soon led to its extensive application to a wide variety of chemical and solid-state problems. This book reviews the results obtained by Mössbauer spectroscopy during the past ten years in the belief that this will provide a firm basis for the continued development and application of the technique to new problems in the future.

It has been our aim to write a unified and consistent treatment which firstly presents the basic principles underlying the phenomena involved, then outlines the experimental techniques used, and finally summarises the wealth of experimental and theoretical results which have been obtained. We have tried to give some feeling for the physical basis of the Mössbauer effect without extensive use of mathematical formalism, and some appreciation of the experimental methods employed without embarking on a detailed discussion of electronics and instrumentation. However, full references to the original literature are provided and particular points can readily be pursued in more detail if required.

The vast amount of work which has been published using the ^{57}Fe resonance is summarised in nine chapters and again full references to the relevant literature are given. The aim here has been to give a critical treatment which provides perspective without loss of detail. The text has been updated to the beginning of 1970 and numerous references to important results after that time have also been included.

A similar approach has been adopted for tin-119 and for each of the forty other elements for which Mössbauer resonances have been observed. Though the results here are much less extensive than for iron, many new points of theory emerge and numerous ingenious applications have been devised. Indeed, one of the great attractions of Mössbauer spectroscopy is its applicability to a wide range of very diverse problems. At all times we have tried to emphasise the chemical implications of the results and have indicated both the inherent advantages and the occasional limitations of the technique.

Our own interest in Mössbauer spectroscopy dates from 1962 and, from the outset, we were helped immeasurably by that small international band of enthusiasts who in the early days so generously shared their theoretical insight and instrumental expertise with us. As more chemists became aware of the technique, various international conferences were arranged and, in this country, the Mössbauer Discussion Group was formed. The wide range of topics discussed at these meetings testifies to the metamorphosis of Mössbauer spectroscopy from its initial 'technique oriented phase' to its present, more mature 'problem oriented phase'. Accordingly, Mössbauer spectroscopy is now widely employed as one of a variety of experimental techniques appropriate to the particular problem being studied. With this maturity comes the realisation that in future it will no longer be feasible, or indeed desirable, to review each individual application of Mössbauer spectroscopy or to delineate its distinctive contribution to the solution of a particular problem. However, the present book, in reviewing the first decade of Mössbauer spectroscopy, enables an overall view of the scope of the method to be appreciated. We hope that it will serve both as an introductory text for those wishing to become familiar with the technique, and as a detailed source-book of references and ideas for those actively working in this field.

The book was written whilst we were in the Department of Inorganic Chemistry at the University of Newcastle upon Tyne, and it is a pleasure to record our appreciation of the friendly and stimulating discussions we have had over the years with the other members of staff, graduate students, and visiting research fellows in the Mössbauer Group which grew up there. We are indebted to Mrs Mary Dalgliesh for her good-humoured persistence and technical skill in preparing the extensive and often complex typescript, and would also like to thank Mrs Linda Cook, Mrs Millie Fenwick, and Miss Anne Greenwood for their help in checking proofs and preparing the indexes.

T.C.G., N.N.G.

Department of Inorganic Chemistry,
University of Newcastle upon Tyne.
May 1971.

1 | The Mössbauer Effect

The phenomenon of the emission or absorption of a γ-ray photon without loss of energy due to recoil of the nucleus and without thermal broadening is known as the Mössbauer effect. It was discovered by Rudolph Mössbauer in 1957 [1] and has had an important influence in many branches of physics and chemistry, particularly during the last five years. Its unique feature is in the production of monochromatic electromagnetic radiation with a very narrowly defined energy spectrum, so that it can be used to resolve minute energy differences. The direct application of the Mössbauer effect to chemistry arises from its ability to detect the slight variations in the energy of interaction between the nucleus and the extra-nuclear electrons, variations which had previously been considered negligible.

The Mössbauer effect has been detected in a total of 88 γ-ray transitions in 72 isotopes of 42 different elements. Although in theory it is present for all excited-state–ground-state γ-ray transitions, its magnitude can be so low as to preclude detection with current techniques. As will be seen presently, there are several criteria which define a useful Mössbauer isotope for chemical applications, and such applications have so far been restricted to about a dozen elements, notably iron, tin, antimony, tellurium, iodine, xenon, europium, gold, and neptunium, and to a lesser extent nickel, ruthenium, tungsten, and iridium. The situation thus parallels experience in nuclear magnetic resonance spectroscopy where virtually all elements have at least one naturally occurring isotope with a nuclear magnetic moment, though less than a dozen can be run on a routine basis (e.g. hydrogen, boron, carbon-13, fluorine, phosphorus, etc.). However, advances in technique will undoubtedly allow the development of other nuclei for chemical Mössbauer spectroscopy.

To gain an insight into the physical basis of the Mössbauer effect and the importance of recoilless emission of γ-rays, we must consider the interplay of a variety of factors. These are best treated under five separate headings:

1.1 Energetics of free-atom recoil and thermal broadening.
1.2 Heisenberg natural linewidth.
1.3 Energy and momentum transfer to the lattice.

2 | THE MÖSSBAUER EFFECT

1.4 Recoil-free fraction and Debye–Waller factor.
1.5 Cross-section for resonant absorption.

1.1 Energetics of Free-atom Recoil and Thermal Broadening

Let us consider an isolated atom in the gas phase and define the energy difference between the ground state of the nucleus (E_g) and its excited state (E_e) as

$$E = E_e - E_g$$

The following treatment refers for simplicity to one dimension only, that in which a γ-ray photon is emitted and in which the atom recoils. This simplification causes no loss of generality as the components of motion in the other two dimensions remain unchanged. If the photon is emitted from a nucleus of mass M moving with an initial velocity V_x in the chosen direction x at the moment of emission (see Fig. 1.1), then its total energy above the ground

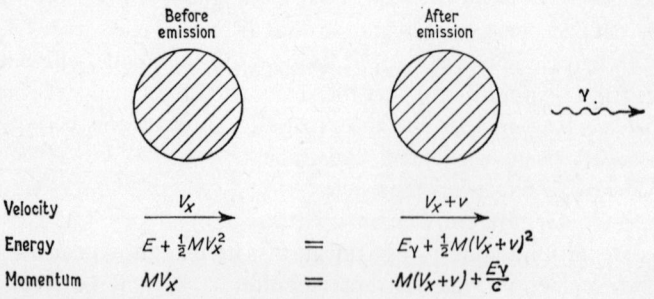

	Before emission		After emission
Velocity	V_x		$V_x + v$
Energy	$E + \tfrac{1}{2}MV_x^2$	=	$E_\gamma + \tfrac{1}{2}M(V_x+v)^2$
Momentum	MV_x	=	$M(V_x+v) + \tfrac{E_\gamma}{c}$

Fig. 1.1 The energy and momentum are conserved in the gamma emission process.

state nucleus *at rest* is $(E + \tfrac{1}{2}MV_x^2)$. After emission the γ-ray will have an energy E_γ and the nucleus a new velocity $(V_x + v)$ due to recoil (note that v is a vector so that its direction can be opposite to V_x). The total energy of the system is $E_\gamma + \tfrac{1}{2}M(V_x + v)^2$. By conservation of energy

$$E + \tfrac{1}{2}MV_x^2 = E_\gamma + \tfrac{1}{2}M(V_x + v)^2$$

The difference between the energy of the nuclear transition (E) and the energy of the emitted γ-ray photon (E_γ) is

$$\delta E = E - E_\gamma = \tfrac{1}{2}Mv^2 + MvV_x$$
$$\delta E = E_R + E_D \qquad 1.1$$

The γ-ray energy is thus seen to differ from the nuclear energy level separation by an amount which depends firstly on the recoil kinetic energy ($E_R = \tfrac{1}{2}Mv^2$) which is independent of the velocity V_x, and secondly on the term $E_D = MvV_x$ which is proportional to the atom velocity V_x and is a

[Refs. on p. 16]

Doppler-effect energy. Since V_x and v are both much smaller than the speed of light it is permissible to use non-relativistic mechanics.

The mean kinetic energy per translational degree of freedom of a free atom in a gas with random thermal motion is given by

$$\overline{E_K} = \tfrac{1}{2}M\overline{V_x^2} \simeq \tfrac{1}{2}kT$$

where $\overline{V_x^2}$ is the mean square velocity of the atoms, k is the Boltzmann constant and T the absolute temperature.

Hence $(\overline{V_x^2})^{\frac{1}{2}} = \sqrt{(2\overline{E_K}/M)}$ and the mean broadening

$$\overline{E_D} = Mv(\overline{V_x^2})^{\frac{1}{2}} = \sqrt{(2\overline{E_K}Mv^2)} = 2\sqrt{(\overline{E_K}E_R)} \qquad 1.2$$

Thus referring to equation 1.1, the γ-ray distribution is displaced by E_R and broadened by twice the geometric mean of the recoil energy and the average thermal energy. The distribution itself is Gaussian and is shown diagrammatically in Fig. 1.2. For E and E_γ having values of about 10^4 eV E_R and $\overline{E_D}$ are

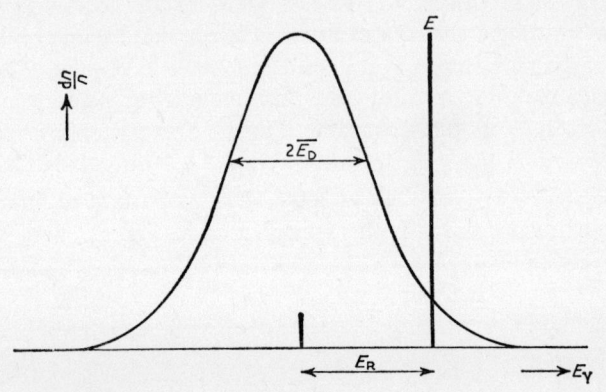

Fig. 1.2 The statistical energy distribution of the emitted γ-ray showing the interrelationship of E, E_R, and $\overline{E_D}$.

about 10^{-2} eV. (The 14·4-keV ^{57}Fe transition at 300 K has $E_R = 1·95 \times 10^{-3}$ eV and $\overline{E_D} \sim 1·0 \times 10^{-2}$ eV).

The values of E_R and $\overline{E_D}$ can be more conveniently expressed in terms of the γ-ray energy E_γ. Thus

$$E_R = \tfrac{1}{2}Mv^2 = \frac{(Mv)^2}{2M} = \frac{p^2}{2M}$$

where p is the recoil momentum of the atom. Since momentum must be conserved, this will be equal and opposite to the momentum of the γ-ray photon, p_γ,

$$p = -p_\gamma = -\frac{E_\gamma}{c}$$

4 | THE MÖSSBAUER EFFECT

hence
$$E_R = \frac{E_\gamma^2}{2Mc^2} \qquad 1.3$$

Expressing E_γ in eV, M in a.m.u., and with $c = 2\cdot998 \times 10^{11}$ mm s^{-1} gives

$$E_R \text{ (eV)} = 5\cdot369 \times 10^{-10} \frac{E_\gamma^2}{M} \qquad 1.4$$

Likewise from equations 1.2 and 1.3

$$\overline{E_D} = 2\sqrt{(\overline{E_K}E_R)} = E_\gamma \sqrt{\frac{2\overline{E_K}}{(Mc^2)}} \qquad 1.5$$

The relevance of E_R and $\overline{E_D}$ to gamma resonance emission and absorption can now be discussed. Fundamental radiation theory tells us that the proportion of absorption is determined by the overlap between the exciting and excited energy distributions. In this case the γ-ray has lost energy E_R due to recoil of the emitting nucleus. It can easily be seen that for the reverse process where a γ-ray is reabsorbed by a nucleus a further increment of energy E_R is required since the γ-ray must provide both the nuclear excitation energy and the recoil energy of the absorbing atom $(E + E_R)$. For example, for $E_\gamma = 10^4$ eV and a mass of $M = 100$ a.m.u., it is found that $E_R = 5\cdot4 \times 10^{-4}$ eV and $\overline{E_D} \sim 5 \times 10^{-3}$ eV at 300 K. The amount of resonance overlap is illustrated (not to scale) as the shaded area in Fig. 1.3 and is extremely small.

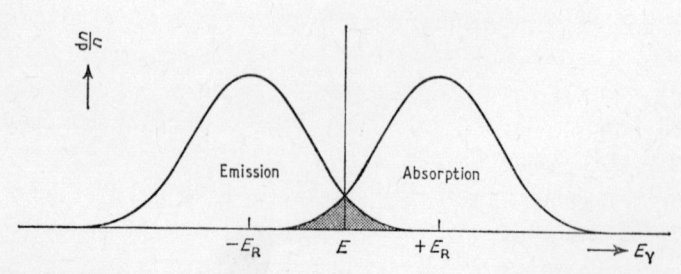

Fig. 1.3 The resonance overlap for free-atom nuclear gamma resonance is small and is shown shaded in black.

It is interesting to note that the principles outlined above are also relevant to the resonance absorption of ultraviolet radiation by atoms. However, for the typical values of $E = 6\cdot2$ eV (50,000 cm^{-1}) and $M = 100$ a.m.u. E_R is only $2\cdot1 \times 10^{-10}$ eV and E_D is $\sim 3 \times 10^{-6}$ eV. In this case strong resonance absorption is expected because the emission and absorption profiles overlap strongly. The problems associated with recoil phenomena are thus only important for very energetic transitions.

[*Refs. on p. 16*]

The earliest attempts to compensate for the energy disparity $2E_R$ involved the provision of a large closing Doppler velocity of about $2v$.

$$2v = \frac{2p}{M} = \frac{2p_\gamma}{M} = \frac{2E_\gamma}{Mc}$$

This was first successfully achieved by Moon [2] in 1950 using an ultracentrifuge. A mercury absorber was used, together with a ^{198}Au β-active source to generate an excited state of ^{198}Hg. The required velocity was about 7×10^5 mm s^{-1} (1600 mph). Since that time recoil has been compensated by a variety of techniques including prior radioactive decay, nuclear reactions, and the use of high temperatures to increase the Doppler broadening of the emission and absorption lines and so improve the resonant overlap. It is important to note that in all these techniques the recoil energy is being compensated for, whereas in the Mössbauer effect the recoil energy is eliminated and no compensation for recoil is required.

1.2 Heisenberg Natural Linewidth

One of the most important influences on a γ-ray energy distribution is the mean lifetime of the excited state. The uncertainties in energy and time are related to Planck's constant h ($= 2\pi\hbar$) by the Heisenberg uncertainty principle

$$\Delta E \, \Delta t \geqslant \hbar$$

The ground-state nuclear level has an infinite lifetime and hence a zero uncertainty in energy. However, the excited state of the source has a mean life τ of a microsecond or less, so that there will be a spread of γ-ray energies of width Γ_s at half height where

$$\Gamma_s \tau = \hbar$$

Whence, substituting numerical values and remembering that the mean life τ is related to the half-life $t_{\frac{1}{2}}$ by the relation $\tau = \ln 2 \times t_{\frac{1}{2}}$:

$$\Gamma_s \text{(eV)} = \frac{4 \cdot 562 \times 10^{-16}}{t_{\frac{1}{2}} \text{(s)}} \qquad 1.6$$

For a typical case (^{57}Fe), $t_{\frac{1}{2}}$ is 97·7 ns and Γ_s is $4 \cdot 67 \times 10^{-9}$ eV. This is some 10^6–10^7 times less than the values of E_R and $\overline{E_D}$ for a free atom and so can be neglected in that case; this was done implicitly in the preceding section where the γ-transition energy E was represented as a single value rather than as an energy profile. However, it can be seen that if the recoil and thermal broadening could be eliminated, radiation with a monochromaticity approaching 1 part in 10^{12} could be obtained.

[Refs. on p. 16]

1.3 Energy and Momentum Transfer to the Lattice

Chemical binding and lattice energies in solids are of the order of 1–10 eV, and are considerably greater than the free-atom recoil energies E_R. If the emitting atom is unable to recoil freely because of chemical binding, the recoiling mass can be considered to be the mass of the whole crystal rather than the mass of the single emitting atom. Equation 1.3 will still apply, but the mass M is now that of the whole crystallite which even in a fine powder contains at least 10^{15} atoms. This diminution of E_R by a factor of 10^{15} makes it completely negligible. From equation 1.5, $\overline{E_D}$ will also be negligible in these circumstances.

However, the foregoing treatment is somewhat of an over-simplification. The nucleus is not bound rigidly in the crystal as assumed above, but is free to vibrate. In these circumstances it is still true that the recoil momentum is transferred to the crystal as a whole, since the mean displacement of the vibrating atom about its lattice position averages essentially to zero during the time of the nuclear decay, and the only random translational motion involves the whole crystal, and is negligible. However, the recoil energy of a single nucleus can be taken up either by the whole crystal as envisaged above or it can be transferred to the lattice by increasing the vibrational energy of the crystal, particularly as the two energies are of the same order of magnitude. The vibrational energy levels of the crystal are quantised: only certain energy increments are allowed and unless the recoil energy corresponds closely with one of the allowed increments it cannot be transferred to the lattice, thus ensuring that the whole crystal recoils, leading to negligible recoil energy. It can be seen that the necessary condition for the Mössbauer effect to occur is that the nucleus emitting the γ-photon should be in an atom which has established vibrational integrity with the solid matrix. In practice this criterion can be relaxed slightly (see p. 346) and the effect can sometimes be observed in viscous liquids.

The vibrational energy of the lattice as a whole can only change by discrete amounts 0, $\pm\hbar\omega$, $\pm 2\hbar\omega$, etc. If $E_R < \hbar\omega$ then either zero or $\hbar\omega$ units of vibrational energy but nothing intermediate can be transferred. If a fraction, f, of γ-photons are emitted without transfer of recoil energy to the vibrational states of the lattice (zero-phonon transitions), then a fraction $(1 - f)$ will transfer one phonon $\hbar\omega$, neglecting all higher multiples of $\hbar\omega$ to a first approximation. Lipkin [3] has shown that if many emission processes are considered, the average energy transferred per event is exactly the free-atom recoil energy: or on our simple model

$$E_R = (1 - f)\hbar\omega \qquad 1.7$$

$$f = 1 - \frac{E_R}{\hbar\omega}$$

[Refs. on p. 16]

Since the preceding discussion is fundamental to an understanding of the Mössbauer effect, it is worthwhile restating the main lines of argument to emphasise their significance. The relevant orders of magnitude of the energy terms are given in Table 1.1. The Heisenberg widths of the Mössbauer γ-rays

Table 1.1 Some typical energies
(1 eV = 23·06 kcal mole^{-1} = 96·48 kJ mole^{-1})

Mössbauer γ-ray energies (E_γ)	10^4–10^5 eV
Chemical binding and lattice energies	1–10 eV
Free-atom recoil energies (E_R)	10^{-4}–10^{-1} eV
Lattice vibration phonon energies	10^{-3}–10^{-1} eV
Heisenberg natural linewidths (Γ_s)	10^{-9}–10^{-6} eV

permit an intrinsic resolving power of 10^{-10} to 10^{-14} of the γ-ray energies. When compared with other spectroscopic methods (e.g. u.v.–visible $\sim 10^{-1}$, gas-phase infrared $\sim 10^{-3}$, atomic line spectra $\sim 10^{-8}$), this emphasises the enormous potential for measuring minute proportionate energy differences. However, in the presence of free-atom recoil and thermal broadening effects the resolution drops to 10^{-6} to 10^{-9}.

It is important to emphasise that γ-ray energies cannot be measured with this accuracy on an absolute scale; indeed these energies are seldom known to better than 1 part in 10^4. In order to use the precision of the Heisenberg width, it is customary to use as a reference a radioactive source in which all the emitting atoms have an identical chemical environment. This is then compared to an absorbing chemical matrix of the same element and only the minute *difference* between the two transition energies is measured. The means whereby this is accomplished will be discussed shortly.

To summarise: in the Mössbauer experiment the emitting and absorbing nuclei are embedded in a solid lattice or matrix; this results in the recoil momentum being taken up by the crystal as a whole since the free-atom recoil energy ($\leq 10^{-1}$ eV) is insufficient to eject the atom from the lattice site (binding energy 1–10 eV) thus precluding momentum transfer to linear translational motion of the nucleus, and the lattice vibrations cannot take up the momentum because a time-average of zero is established within the decay time. Similarly, the total energy of the γ-transition E must be conserved and can only be shared between

(a) the energy of the γ-photon, E_γ;
(b) the lattice vibrations;
(c) the translational kinetic energy of the individual atom;
(d) the translational kinetic energy of the solid as a whole.

The third of these possibilities is eliminated by the high chemical binding energy and the fourth has been shown to be minute because of the large mass

8 | THE MÖSSBAUER EFFECT

involved. The γ-transition energy is thus shared between the γ-photons and the lattice vibration phonons. Because of quantisation conditions, a fraction of events, f, occur with no change in the lattice vibrations and the entire transition energy E is manifest in the γ-photon energy ($E = E_\gamma$).

It must be emphasised that the preceding discussion represents a considerable simplification of the actual situation. For example, a single lattice vibration frequency corresponds to an Einstein solid. A Debye model is more realistic, and includes frequencies ranging from zero to a maximum value ω_D. Fortunately, the very low frequencies are difficult to excite, otherwise the Mössbauer effect would not exist. Although it is not intended to discuss more advanced treatments here, several texts are readily available [4-5]. Since our main interest is in the events without recoil, the fact that γ-rays with an energy different to E are produced is often forgotten. A theoretical γ-photon energy spectrum calculated by Visscher [4] for iridium is shown in Fig. 1.4. The recoil-free line with zero energy shift should be contrasted with the very broad distribution of events with phonon excitations.

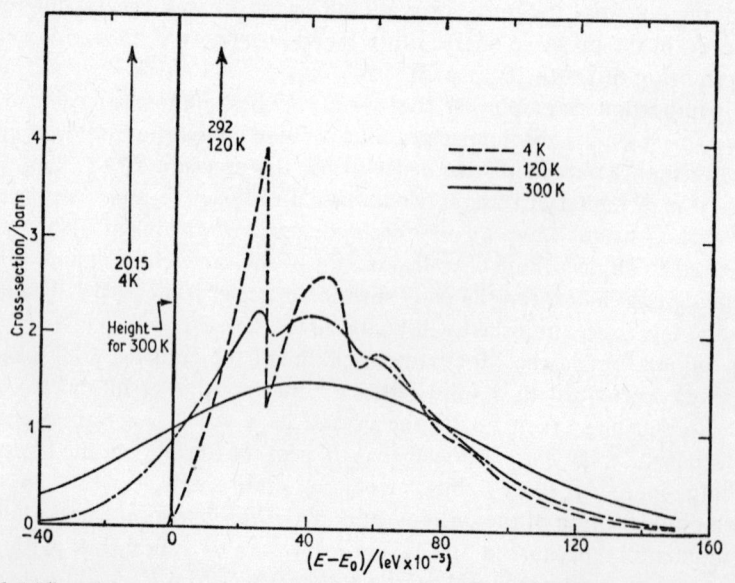

Fig. 1.4 The absorption cross-section for a nucleus in iridium metal calculated assuming a Debye model for the lattice vibrations at three different temperatures.

Looking back, physicists have been tempted to wonder why the recoilless emission of γ-rays was not discovered earlier. The full theory had been worked out by Lamb in 1939 for the analogous case of absorption of neutrons [6]. Similarly the diffraction of X-rays depends on the same effect. namely the elastic scattering which leaves the wavelength (energy) of the X-ray beam unchanged. Indeed, the fact that in X-ray diffraction thermal motion of the

[*Refs. on p. 16*]

atoms weakens the diffraction maxima but does not broaden them is precisely the Mössbauer effect: the probability of zero-phonon events diminishes but those observed are still unbroadened by Doppler or recoil effects.

We have now reached the stage of realising that, for a γ-emitter in a solid, there is a finite probability that the γ-ray can be emitted essentially without recoil or thermal broadening and that the width of the line derives from the Heisenberg uncertainty principle. We next seek to calculate what this probability will be.

1.4 Recoil-free Fraction and Debye–Waller Factor

We have already seen qualitatively that the recoil-free fraction or probability of zero-phonon events will depend on three things:

(a) the free-atom recoil energy, which is itself proportional to E_γ^2;
(b) the properties of the solid lattice;
(c) the ambient temperature.

Thus f will be greater the smaller the probability of exciting lattice vibrations, i.e. the smaller the γ-ray energy, the firmer the binding of the atom in the lattice and the lower the temperature.

From a quantitative viewpoint, the probability W of zero-phonon γ-emission from a nucleus embedded in a solid which simultaneously changes its vibration state can be calculated by dispersion theory [3, 5]; it is proportional to the square of the matrix element connecting the initial $|\text{i}\rangle$ and final $\langle\text{f}|$ states.

$$W = \text{const} \times |\langle\text{f}|\mathscr{H}|\text{i}\rangle|^2$$

where $\mathscr{H}$ is the interaction Hamiltonian operator and depends upon the positional coordinates of the atom (nucleus) and the momenta and spins of the particles within the nucleus. The forces acting within the nucleus are extremely short range whereas those holding the lattice together are of much longer range [6]. Hence the nuclear decay is independent of the vibrational state and *vice versa*, thus enabling the matrix element to be split into two parts of which the nuclear part will be a constant depending only on the properties of the particular nucleus. The matrix element can now be reduced to the one term for the transition from the initial vibrational state L_i to the final state L_f. The form of the operator is such that

$$W = \text{const} \times |\langle L_f|e^{i\mathbf{k}\cdot\mathbf{x}}|L_i\rangle|^2$$

where $\mathbf{k}$ ($= p/\hbar$) is the wave vector for the emitted γ-photon, i.e. the number of units of momentum it carries, and $\mathbf{x}$ is the coordinate vector of the centre of mass of the decaying nucleus. For zero-phonon emission the lattice modes are unchanged and hence the probability for recoilless emission is

$$f = \text{const} \times |\langle L_i|e^{i\mathbf{k}\cdot\mathbf{x}}|L_i\rangle|^2 \qquad 1.8$$

10 | THE MÖSSBAUER EFFECT

and since L_i is normalised

$$f = e^{-k^2 \cdot x^2} \qquad 1.9$$

Further, since x is a random vibration vector, x^2 can be replaced by $\langle x^2 \rangle$, the component of the mean square vibrational amplitude of the emitting atom in the direction of the γ-ray. Since $k^2 = 4\pi^2/\lambda^2 = E_\gamma^2/(\hbar c)^2$, where λ is the wavelength of the γ-ray, we obtain

$$f = \exp\left(\frac{-4\pi^2 \langle x^2 \rangle}{\lambda^2}\right) = \exp\left(\frac{-E_\gamma^2 \langle x^2 \rangle}{(\hbar c)^2}\right) \qquad 1.10$$

Equation (1.10) indicates that the probability of zero-phonon emission decreases exponentially with the square of the γ-ray energy. This places an upper limit on the usable values of E_γ, and the highest transition energy for which a measurable Mössbauer effect has been reported is 155 keV for ^{188}Os. Equation 1.10 also shows that f increases exponentially with decrease in $\langle x^2 \rangle$ which in turn depends on the firmness of binding and on the temperature. The displacement of the nucleus must be small compared to the wavelength λ of the γ-ray. This is why the Mössbauer effect is not detectable in gases and non-viscous liquids. Clearly, however, a study of the temperature dependence of the recoil-free fraction affords a valuable means of studying the lattice dynamics of crystals.

To proceed further it is necessary to assume some model for the vibrational modes of the crystal. The mathematics become more severe and the results are somewhat unrewarding, since very few of the solids which chemists may be interested in approximate even remotely to the assumptions made in setting up the models. However, the form of the expressions is instructive.

The simplest model is due to Einstein (1907) and assumes the solid to be composed of a large number of independent linear harmonic oscillators each vibrating at a frequency ω_E. The appropriate integration of equation 1.8 gives

$$f = \exp\left(\frac{-E_R}{\hbar \omega_E}\right) = \exp\left(\frac{-E_R}{k \theta_E}\right) \qquad 1.11$$

where θ_E is a characteristic temperature of the lattice given by $k\theta_E = \hbar\omega_E$. If $E_R \ll k\theta_E$ then $f \simeq 1 - E_R/k\theta_E$ as intuitively derived in equation 1.7.

The Debye model (1912) abandons the idea of a single vibration frequency and embodies a continuum of oscillator frequencies ranging from zero up to a maximum ω_D and following the distribution formula $N(\omega) = \text{const} \times \omega^2$. A characteristic temperature called the Debye temperature θ_D is defined as $\hbar\omega_D = k\theta_D$, and the average frequency is

$$\hbar\bar{\omega} = \tfrac{3}{4}\hbar\omega_D$$

The Debye temperature θ_D for metals varies from 88 K for lead to 1000 K for beryllium. Values of θ_D are often also assigned to chemical compounds,

[Refs. on p. 16]

but since the Debye model is grossly inadequate even for many of the pure metals, the values used for compounds are merely an indication of the approximate lattice properties and should not be considered too seriously.

The Debye model leads to

$$k^2 \cdot x^2 = \frac{\hbar}{2M} \int_0^{\omega_D} \frac{N(\omega)}{\omega} \coth\left(\frac{\hbar\omega}{2kT}\right) d\omega$$

which integrates to the more familiar equation

$$f = \exp\left[\frac{-6E_R}{k\theta_D}\left\{\frac{1}{4} + \left(\frac{T}{\theta_D}\right)^2 \int_0^{\theta_D/T} \frac{x\,dx}{e^x - 1}\right\}\right] \quad 1.12$$

This is often written as $f = e^{-2W}$. The factor W is sometimes loosely called the Debye–Waller factor though it might be better in this context to call it the Lamb–Mössbauer factor.

The Debye–Waller factor was originally derived during the development of Bragg X-ray scattering theory. The difference between the two effects is that X-ray scattering is fast when compared with the characteristic time for lattice vibrations whereas the mean lifetime of a Mössbauer nucleus is long when compared with the lattice time.

At low temperatures where $T \ll \theta_D$ we can reduce equation 1.12 to the approximation

$$f = \exp\left[\frac{-E_R}{k\theta_D}\left\{\frac{3}{2} + \frac{\pi^2 T^2}{\theta_D^2}\right\}\right] \quad T \ll \theta_D \quad 1.13$$

and at absolute zero

$$f = \exp\left[\frac{-3E_R}{2k\theta_D}\right] \quad 1.14$$

In the high temperature limit

$$f = \exp\left[\frac{-6E_R T}{k\theta_D^2}\right] \quad T \geqslant \tfrac{1}{2}\theta_D \quad 1.15$$

From equation 1.15 W is proportional to T at high temperatures. Experience shows that experimentally this is usually not the case because of anharmonicity of the lattice vibrations. The effects of anharmonicity have been discussed at length by Boyle and Hall [7] and will be met again in connection with the intensities of individual lines in hyperfine interactions.

1.5 Cross-section for Resonant Reabsorption

We have described the mechanism by which a γ-ray can be emitted without recoil and the same arguments apply to resonant reabsorption. Since Mössbauer experiments usually utilise the recoilless emission of γ-rays by a

12 | THE MÖSSBAUER EFFECT

radioactive source followed by their subsequent resonant recoilless reabsorption by a non-active absorber, we must now consider this process in detail.

The probability of recoilless emission from the source is f_s. This recoilless radiation has a Heisenberg width at half height of Γ_s and the distribution of energies about the energy E_γ is given by the Breit–Wigner formula [8]. This leads to a Lorentzian distribution, i.e. the number of transitions $N(E)$ with energy between $(E_\gamma - E)$ and $(E_\gamma - E + \mathrm{d}E)$ is given by

$$N(E)\,\mathrm{d}E = \frac{f_s \Gamma_s}{2\pi} \frac{\mathrm{d}E}{(E - E_\gamma)^2 + (\Gamma_s/2)^2} \qquad 1.16$$

This distribution is illustrated in Fig. 1.5.

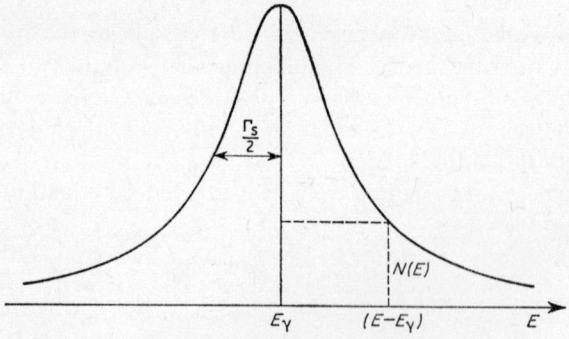

Fig. 1.5 The Lorentzian energy distribution of the source recoilless radiation.

The resonant absorption cross-section $\sigma(E)$ can be similarly expressed as

$$\sigma(E) = \sigma_0 \frac{(\Gamma_a/2)^2}{(E - E_\gamma)^2 + (\Gamma_a/2)^2} \qquad 1.17$$

where Γ_a is the Heisenberg width at half-height of the absorption profile and σ_0 is the effective cross-section given by

$$\sigma_0 = 2\pi\lambda^2 \frac{2I_e + 1}{2I_g + 1} \frac{1}{1 + \alpha} \qquad 1.18$$

where I_e and I_g are the nuclear spin quantum numbers of the excited and ground states and α is the internal conversion coefficient of the γ-ray of wavelength λ. (Not all nuclear γ transitions produce a physically detectable γ-ray; a proportion eject electrons from the atomic orbitals, giving X-rays and these internal conversion electrons instead. The internal conversion coefficient of a γ-transition is defined as the ratio of the number of conversion electrons to

[Refs. on p. 16]

the number of γ-ray photons emitted.) Using the relation between λ and E_γ in equation 1.10 we obtain the expression

$$\sigma_0(\text{cm}^2) = \frac{2\cdot446 \times 10^{-15}}{(E_\gamma, \text{keV})^2} \frac{2I_e + 1}{2I_g + 1} \frac{1}{1 + \alpha} \qquad 1.19$$

σ_0 values are sometimes expressed in units of barns ($= 10^{-24}$ cm^2). Equation 1.19 shows that the desirability of a high absorption cross-section requires that both E_γ and α have low values. The resonant absorption process will also be in competition with other absorption processes such as photoelectric absorption, and it is important that the cross-section for nuclear resonance absorption should be higher than that for any other method of γ-ray attenuation. A full tabulation of physical parameters for Mössbauer resonances is to be found in Appendix 1 on p. 607.

The preceding equations already indicate the numerous parameters which will influence the intensity of an emission-absorption resonance:

(a) nuclear properties: the cross-section of γ-ray absorption and hence E_γ, I_e, I_g, and α
(b) source properties: the recoil-free fraction f_s and the Heisenberg width Γ_s
(c) absorber properties: the recoil-free fraction for absorption f_a and the Heisenberg width Γ_a.

A completely general evaluation of the problem is impossible, but useful results are obtained if we assume that both source and absorber have the same linewidth ($\Gamma = \Gamma_s = \Gamma_a$). Margulies and Ehrman [9] showed that the γ-transmission through a uniform resonant absorber can be represented by

$$T(\varepsilon) = e^{-\mu_a t_a} \Bigg\{ (1 - f_s) \int_0^\infty \rho(x) e^{-\mu_s x} \, dx$$

$$+ \frac{f_s \Gamma}{2\pi} \int_{-\infty}^\infty \exp\left[-f_a n_a a_a \sigma_0 t_a \frac{(\Gamma/2)^2}{(E - E_\gamma)^2 + (\Gamma/2)^2}\right]$$

$$\times \int_0^\infty \frac{\rho(x)}{(E - E_\gamma + \varepsilon)^2 + (\Gamma/2)^2}$$

$$\times \exp\left[-\left(f_s n_s a_s \sigma_0 \frac{(\Gamma/2)^2}{(E - E_\gamma + \varepsilon)^2 + (\Gamma/2)^2} + \mu_s\right)x\right] dx \, dE \Bigg\} \qquad 1.20$$

where ε is an energy displacement between the source and absorber distribution maxima,

n_s, n_a = number of atoms of the element concerned per cm^3 in the source and the absorber respectively,

a_s, a_a = fractional abundance of the resonant isotope,

[Refs. on p. 16]

14 | THE MÖSSBAUER EFFECT

μ_s, μ_a = mass absorption coefficients (cm^{-1}) for the source and absorber evaluated at energy E_γ,
t_a = thickness of the absorber in cm,
x = depth of the emitting atom normal to the surface of the source as shown in Fig. 1.6.
$\rho(x)$ = distribution of emitting atoms along the x direction in the source.

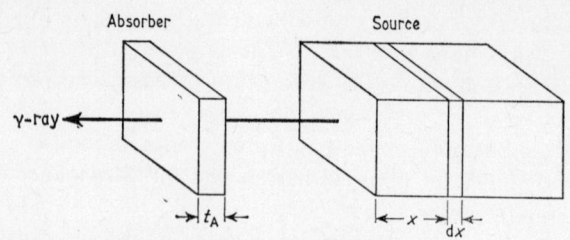

Fig. 1.6 Schematic representation of parameters used in calculating resonant transmission for source and absorber of finite thickness.

The first term in equation 1.20 is the transmission of the non-resonant radiation and is independent of ε. The first part of the second term is the resonant absorption in the absorber and the second part is the resonant absorption in the source. The factor $\varepsilon^{-\mu_a t_a}$ accounts for non-resonant scattering in the absorber.

Margulies and Ehrman solved this equation for certain interesting simplified cases. If the source and absorber have a thickness tending to zero (i.e. are ideally thin), the decrease in transmission with respect to ε is given by

$$I(\varepsilon) = \frac{\Gamma_r}{2\pi} \times \frac{1}{(\varepsilon - E_\gamma)^2 + (\Gamma_r/2)^2} \qquad 1.21$$

This is normalised so that $\int_0^\infty I(\varepsilon)\,d\varepsilon = 1$.

Γ_r is the sum of the emission and absorption half-widths, i.e. $\Gamma_r = 2\Gamma$ and the distribution is still Lorentzian. If the absorber has an effective thickness $T = f_a n_a a_a \sigma_0 t_a$ the shape is still basically Lorentzian but will be broadened[5] so that

$$\frac{\Gamma_r}{\Gamma} = 2 \cdot 00 + 0 \cdot 27 T \qquad 0 < T \leqslant 5 \qquad 1.22$$

$$\frac{\Gamma_r}{\Gamma} = 2 \cdot 02 + 0 \cdot 29 T - 0 \cdot 005 T^2 \qquad 4 \leqslant T \leqslant 10$$

If we can measure Γ_r for a series of thicknesses, T, it is possible to obtain the true value of $\Gamma = \tfrac{1}{2}\Gamma_r(T \to 0)$.

The numerical significance of the equations will be discussed more fully in connection with absorber preparation in Chapter 2.

[*Refs.* on p. 16]

1.6 A Mössbauer Spectrum

The preceding five sections have provided the basic concepts necessary for an understanding of the Mössbauer effect and we now describe the experimental observation of a Mössbauer spectrum. If γ-rays from a source which has a substantial recoil-free fraction are passed through an absorber of the same material the transmission of the γ-rays in the direction of the beam will be less than expected because of their resonant reabsorption and subsequent re-emission over a 4π solid angle. Paradoxically, this resonance can only be shown to be present by its destruction, e.g. by warming the absorber so that the fraction of recoil-free events will decrease and the transmission increase. The principal technique of Mössbauer spectroscopy is, however, rather more subtle. It was shown in the previous section in equation 1.21 that the decrease in transmission (i.e. decrease in the extent of resonant overlap) is affected by the difference in the relative values of E_γ for the source and the absorber. The *effective* E_γ value can be altered by moving the source and absorber relative to each other with a velocity v, i.e. by using an externally applied Doppler effect $[\varepsilon = (v/c)E_\gamma]$. If the effective E_γ values are exactly matched at a certain Doppler velocity, resonance will be at a maximum and the count-rate a minimum. At any higher or lower applied velocity, the resonance will

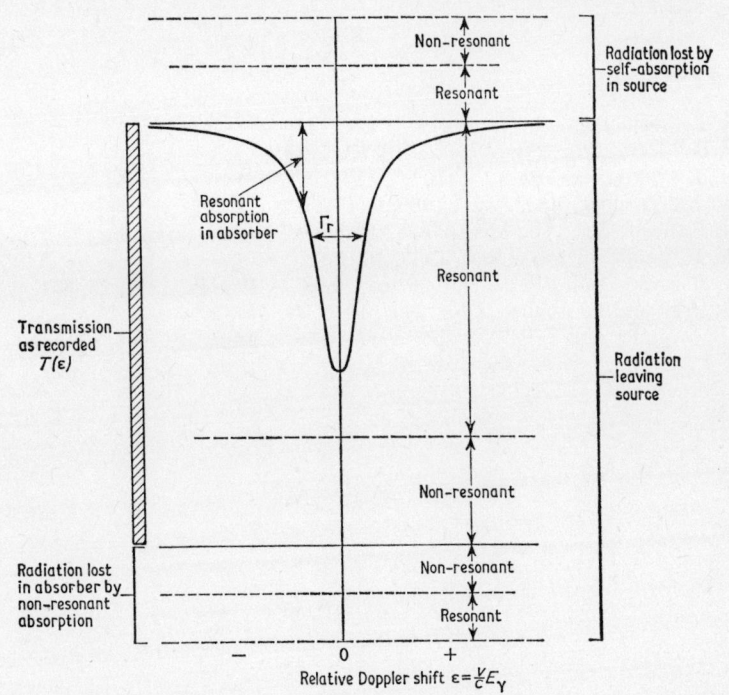

Fig. 1.7 A Mössbauer transmission spectrum produced by Doppler scanning, and the factors influencing it.

[*Refs. on p. 16*]

decrease until it is effectively zero at velocities well away from that defining the maximum resonance.

This is the basic form of a Mössbauer spectrum; a plot of transmission versus a series of Doppler velocities between source and absorber (i.e. versus the effective γ-ray energy), the absorption line being Lorentzian in shape with a width at half-height corresponding to 2Γ. The various factors influencing the transmission spectrum are illustrated schematically in Fig. 1.7, which is self-explanatory.

It should not be forgotten that the excited absorber nuclei re-emit the γ-ray within $\sim 10^{-7}$ s. However, if the internal conversion coefficient is high, correspondingly fewer γ-rays will be emitted. More important, however, the re-emission is not directional but takes place over the full 4π solid angle. Consequently the number of secondary events recorded at the detector in a collimated transmission experiment are few and are usually neglected.

We shall return to the theoretical discussion again in Chapter 3 to see how the spectrum can be influenced in detail by various properties of the resonant nucleus and by the extra-nuclear electrons. Before then, however, it will be convenient to outline the experimental techniques of Mössbauer spectroscopy and this forms the subject of the next chapter.

REFERENCES

[1] R. L. Mössbauer, *Z. Physik*, 1958, **151**, 124.
[2] P. B. Moon, *Proc. Phys. Soc.*, 1950, **63**, 1189.
[3] H. J. Lipkin, *Ann. Phys.*, 1960, **9**, 332.
[4] W. M. Visscher, *Ann. Phys.*, 1960, **9**, 194.
[5] H. Frauenfelder, 'The Mössbauer Effect', W. A. Benjamin, Inc., N.Y., 1962.
[6] W. E. Lamb, Jr, *Phys. Rev.*, 1939, **55**, 190.
[7] A. J. F. Boyle and H. E. Hall, *Repts. Progr. in Physics*, 1962, **25**, 441.
[8] G. Breit and E. Wigner, *Phys. Rev.*, 1936, **49**, 519.
[9] S. Margulies and J. R. Ehrman, *Nuclear Instr. Methods*, 1961, **12**, 131.

2 | Experimental Techniques

In this chapter we shall consider the various techniques which have been used for observation of the Mössbauer effect, together with methods of source and absorber preparation and computer techniques for data analysis. Some of the advantages and limitations of Mössbauer spectroscopy will become apparent during the discussion of these problems. References to more recent development will be found in the review by J. R. De Voe and J. J. Spijkerman in *Analytical Chemistry*, 1970, **42**, 366R, and in 'Spectroscopic Properties of Inorganic and Organometallic Compounds' published annually by the Chemical Society (London).

A few of the very earliest observations of the Mössbauer effect involved a static experiment in which a source and absorber of identical chemical form were maintained in a stationary transmission arrangement. Under these conditions one would expect some nuclear resonance absorption but this can only be verified by demonstrating its diminution or increase; thus an increase in the temperature of the system will increase the counting rate because of the decreasing fraction of resonant absorption, whereas a lowering of the temperature increases the recoil-free fraction and lowers the counting rate of transmitted γ-photons. This technique was the basis of Mössbauer's original observations [1]. It does not however provide much significant information and may fail completely if the source and absorber are not chemically identical.

The method of velocity modulation of the γ-ray energy by means of the Doppler effect was described by Mössbauer in 1958 [2] and provides the basis for all modern spectrometers.

2.1 Velocity Modulation of Gamma-rays

The recoil-free γ-ray energy of a typical Mössbauer transition is so precisely defined that its Heisenberg width corresponds to the energy change produced by an applied Doppler velocity of the order of 1 mm s^{-1}. It is therefore possible to imagine a particular relative velocity between source and absorber at which the γ-ray energy from the source will precisely match the nuclear

[*Refs. on p. 43*]

energy level gap in the absorber and resonant absorption will be at a maximum. For a source and absorber which are chemically identical this relative velocity will be zero. Application of an additional velocity increment will lower the resonant overlap and decrease the absorption. Application of a sufficiently large relative velocity will destroy the resonance completely.

A Mössbauer spectrum comprises a series of measurements at different velocities (that is energies) across the resonant region. The convention universally adopted is that a closing velocity between source and absorber (i.e. a higher energy) is defined as positive.

It has already been shown in Chapter 1 that the resonant absorption curve for an ideally thin source and absorber has a width at half-height Γ_r which is twice the Heisenberg width of the emitted γ-photon. The Doppler velocity v corresponding to this energy Γ_r is given by

$$\frac{v}{c} = \frac{\Gamma_r}{E_\gamma}$$

where c is the velocity of light. Typical numerical values for commonly used Mössbauer isotopes are

^{57}Fe (14 keV) 0·192 mm s^{-1} ^{119}Sn (24 keV) 0·626 mm s^{-1}
^{127}I (58 keV) 2·54 mm s^{-1} ^{125}Te (35 keV) 5·02 mm s^{-1}
^{195}Pt (99 keV) 16 mm s^{-1}

A full list of such values in Appendix I shows that the velocities range from $3\cdot1 \times 10^{-4}$ for ^{67}Zn up to 202 mm s^{-1} for ^{187}Re. They are all very small when compared with the tremendous velocities ($\sim 7 \times 10^5$ mm s^{-1}) used by Moon in 1950 to detect nuclear resonance fluorescence without recoilless emission, and show dramatically that the Mössbauer technique eliminates both recoil and thermal broadening. The Heisenberg relation means that an excited state with a shorter half-life has a greater uncertainty in the γ-transition energy and hence a broader resonance line.

Since the Doppler energy shift is relative to the source and absorber only, it is independent of the frame of reference. In practice one resonant matrix is maintained at rest. For transmission experiments it is usually more satisfactory mechanically to move the source. It is then much easier to change absorbers and to vary their temperature. The only major problem is that the source of the γ-rays and the detector are then not at rest relative to each other, and the solid angle subtended by the source varies during the motion. If the amplitude of the source motion is large, a correction term should be applied. There is also a limitation on the counting geometry since the Doppler shift $E_\gamma v/c$ is only accurate along the axis of source motion. A γ-photon travelling to the detector at an angle θ to this axis will have an effective Doppler shift of $E_\gamma v \cos \theta / c$. A large solid angle in the counting geometry will thus cause a distortion in the shape of the Mössbauer absorption line.

[Refs. on p. 43]

The difficulty can be overcome by maintaining an adequate separation between source and detector or by collimation of the γ-ray beams.

There are two general approaches to the measurement of γ-ray transmission at different Doppler velocities:

(a) measurement of the total number of transmitted γ-photons in a fixed time at a constant velocity, followed by subsequent counts at other velocities; in this way the spectrum is scanned stepwise one velocity at a time;

(b) rapid scanning through the whole velocity range and subsequent numerous repetitions of this scan, thereby accumulating all the data for the individual velocities essentially simultaneously.

The relatively low photon flux-density necessitates much longer counting times to achieve significant counting statistics than, for example, in optical spectroscopy. The statistical behaviour of the γ-emission results in a standard deviation of $\sqrt{N}$ for a total number N of registered γ-counts. Hence the standard deviation in 10,000 counts is 100 (1%); in 1,000,000 counts is 1000 (0·1%). The longer the counting time, the better the definition of the resonance line, but the improvement to be gained must be balanced against the experimental time required and the long-term stability or reproducibility of the apparatus.

2.2 Constant-velocity Drives

To move an object at a constant velocity with high reproducibility and stability, when it is restricted by both a relatively small amplitude of movement and the necessity for repetitive motion, is a difficult problem in applied mechanics. Several constant-velocity spectrometers have been described, and these can be briefly classified under eight different headings. A simple diagram of each type is shown in Fig. 2.1 and references to more detailed descriptions are given where available.

(i) lathe and gears: a lead screw parallel to the γ-ray axis is used to drive the source, the velocity being varied by use of gears or a variable-speed motor [3, 4].

(ii) lathe and inclined plane: the lead screw is perpendicular to the γ-ray axis and is driven by a constant-speed motor. The source is moved by a slide-plate on an inclined plane mounted on the lead screw carriage and the velocity is varied by altering the angle of inclination of this plane [5].

(iii) hydraulic devices: the driving force is provided by hydraulic pressure. A typical system embodies a double-action cylinder to enable continuous repetitive movement [6].

(iv) cam and gears: a wide variety of rotating cam drives has been used. The exact shape of the eccentric determines the motion of the follower [7]. One of the more recent instruments uses a 70-kg rotating disc mounted with its axis vertical [8]. It features a helicoidal surface on the outer circumference

[Refs. on p. 43]

transmitting a linear velocity to a piston aligned vertically and gliding on the surface.

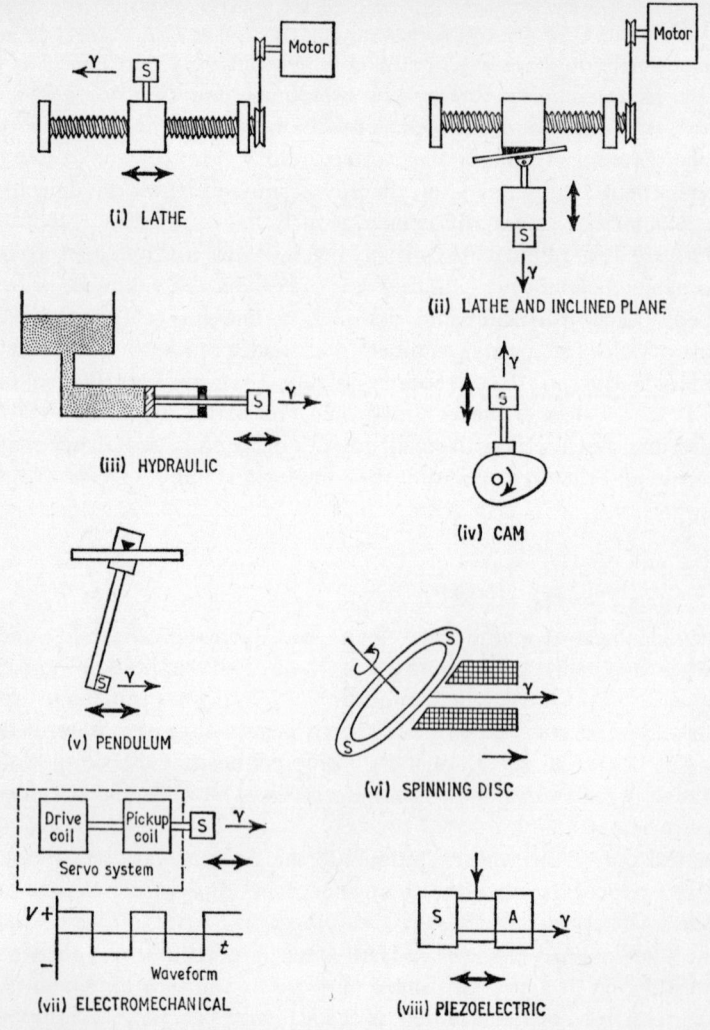

Fig. 2.1 Eight drive systems for producing constant velocities. The direction of motion is indicated by the heavy arrow. The source is labelled S.

(v) pendulum: a long pendulum device gives a good approximation to constant velocity at the bottom of its swing and can be activated by an electromagnetic drive coil [9, 10].

(vi) spinning disc: an accurately driven rotating disc can be used, the Doppler velocity along the γ-ray axis being varied by altering the speed of rotation, the diameter of the disc or its inclination.

[*Refs.* on p. 43]

(vii) electromechanical drive: an electromechanical servo-drive can be driven with a square wave to give a constant velocity [11, 12].

(viii) piezoelectric drive: a triangular voltage waveform applied to a piezoelectric crystal which is glued in between source and absorber produces alternate forward and backward constant velocity [13].

The only other essential equipment in these constant-velocity spectrometers comprises a γ-ray detection system, a timer, and a scalar for registering the accumulated count at each successive velocity.

The advantages offered by a constant-velocity spectrometer include the ability to examine a small velocity range not centred on zero velocity, and to calibrate the instrument directly in terms of absolute velocity. It is also often considered to be less expensive than the alternative velocity-sweep techniques considered in Section 2.3, but this probably applies only to simple demonstration equipment. A constant-velocity instrument of any precision must involve considerable expense in machine-shop and development time.

There are many disadvantages. Such an instrument is extremely tedious to operate unless it has been fully automated (with considerable increase in expense). It is difficult to machine cams and lead screws to the precision required to give an accurately linear drive. Mechanical wear and vibrations are inevitably a problem. The velocities attainable are limited to the range 0.1–10 mm s^{-1}. Cam and lead screw drives require a complicated automatic reversing system with rejection of all counts received at the end of the travel when the velocity is changing. The solid angle and cosine effects have already been referred to though they can be minimised by moving the absorber rather than the source. Rotating discs make very inefficient use of the resonant matrix because this must be distributed over a large annulus rather than in a small area. Finally, if the source activity is short-lived compared with the experimental time-scale, the observed count rates over successive constant intervals of time must be corrected for the decreasing rate of γ-emission.

The cumulative result of these disadvantages has been to restrict the continued development of constant-velocity spectrometers. One of the few major developments of recent years has been the description of an instrument which takes readings at a number of velocities pre-programmed on a length of punched tape [14]. The great majority of investigations are now made using repetitive velocity-scan systems.

2.3 Repetitive Velocity-scan Systems

Mechanical drives can be very tedious to operate (unless they are very fully automated) because of frequently having to alter the velocity setting. The availability of small, transistorised, multichannel analysers embodying typically 400 or 512 individual scalars in a computer-type memory store prompted de Bennedetti to suggest their use for Mössbauer data acquisition

22 | EXPERIMENTAL TECHNIQUES

using repetitive scanning techniques. The Doppler motion is provided by an electromechanical drive system which is controlled by a servo-amplifier. The amplifier is fed with a reference voltage waveform which repeats itself exactly with a frequency of between about 5 and 40 Hz (1 Hz = 1 cycle s^{-1}). The actual drive or transducer embodies two coils, one of which produces a voltage proportional to the *actual* velocity of the shaft. The servo-amplifier compares this signal to the reference waveform and applies corrections to the drive coil to minimise any differences. In this way, the centre shaft which is usually rigidly connected to the Mössbauer source executes an accurate periodic motion. A considerable volume of literature is available on the design of suitable transducers [15–22]. Velocities of over 600 mm s^{-1} can be achieved, or as low at 10^{-2} mm s^{-1} [23]. Even smaller velocities can be obtained by a piezoelectric crystal glued between source and absorber, and in one instance has been used at speeds of up to about 8 mm s^{-1} [24].

Several types of command waveform have found favour, and three of these

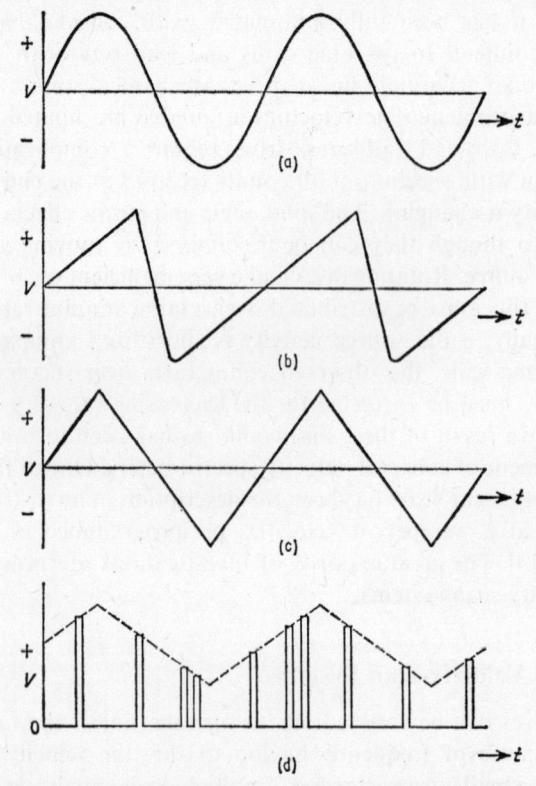

Fig. 2.2 (a), (b), and (c) show three of the most popular voltage waveforms for electromechanical drive systems. (d) illustrates how each detected γ-ray can be used to produce a pulse with amplitude characteristic of the instantaneous velocity.

[*Refs. on p. 43*]

are illustrated in Fig. 2.2. (Fig. 2.2d is discussed on pp. 23–24.) The sine wave (2.2a) is less demanding on the mechanical adjustment of the transducer, but since the velocity is directly proportional to the voltage it produces a markedly non-linear scale on the final spectrum. The asymmetric double ramp (2.2b) executes about 80% of the motion with a constant acceleration and therefore a linear velocity scale, but the high-frequency components cause some difficulties. The symmetrical double-ramp (2.2c) scans the spectrum alternately with constant acceleration in opposite directions, and in some equipment cuts by half the number of data units available and produces a double (mirror-image) spectrum; this can be folded to give an additional check on linearity.

Many spectrometers have been described, but most fall into one of three basic classes as follows:

(a) Pulse-height Analysis Mode

Many multichannel analysers have a facility known as pulse-height analysis, and a typical system employing an analyser in this mode is shown in Fig. 2.3. Several detailed descriptions are available [25–27].

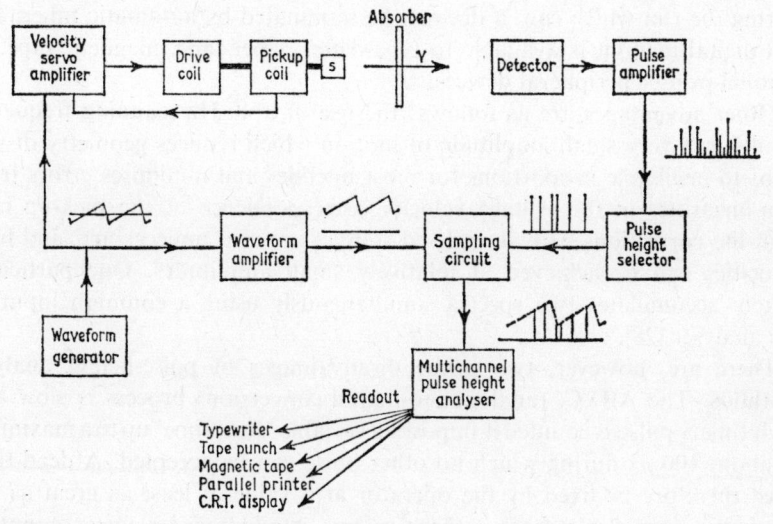

Fig. 2.3 Schematic arrangement for the pulse-height analysis spectrometer described in the text.

A reference waveform identical to that supplying the drive is given a d.c. shift such that it is always positive. When a γ-ray which has passed through the absorber is detected, the waveform is sampled so that a pulse is produced which has a voltage maximum characteristic of the instantaneous velocity of the source at the time of emission (see Fig. 2.2d). An analogue-to-digital converter (A.D.C.) transforms this pulse into a train of pulses from a constant-frequency clock such that the number of pulses is proportional to the voltage. These pulses then sequentially step through the channel addresses of the analyser until the train is exhausted, at which point the number stored in the open channel is increased by one.

In this way each channel in the analyser will receive those γ counts which are registered in the narrow velocity range assigned to it, and in the cases of constant-acceleration waveforms such as 2.2b and 2.2c the channel number is directly proportional to the velocity. Many pulses are detected and stored during each cycle of the motion, and successive cycles over a long period of time allow the spectrum to build up as a whole. At velocities where resonance absorption occurs the accumulation rate will be slower. The analyser has a cathode-ray tube (C.R.T.) display in which a voltage proportional to the accumulated count of a channel is used as the vertical deflection, and a voltage proportional to the channel address number is used as the horizontal deflection. At the frequency of scan used in the experiments this gives a continuous visual display of all the channels. The spectrum can thus be inspected at any time during the course of a run. This is a significant advantage over constant-velocity instrumentation. No operator attention is required during the run which can, if desired, be terminated by automatic timer, and full digital readout is available to typewriter, paper tape, magnetic tape, or parallel-printer peripheral devices.

Other advantages are as follows: the use of a 40-Hz scanning frequency results in a very small amplitude of motion which reduces geometry distortions to negligible proportions for most nuclides and minimises errors from non-linearities in the voltage/velocity correspondence of the pick-up coil. Half-life corrections with short-lived sources are also unnecessary, and high velocities can be achieved at relatively small amplitudes. One particular system accumulates two spectra simultaneously using a common input to the analyser [28].

There are, however, two prime disadvantages of pulse-height analysis methods. The A.D.C. (analogue-to-digital-conversion) process is slow and each time a pulse is counted it imposes a variable 'dead time' up to a maximum of about 100 μs during which no other pulse can be accepted. A dead-time must therefore be fixed by the operator at a value at least as great as the maximum time for storage, otherwise one would register faster counting rates at the lower channel address numbers. The absolute counting rate of the analyser is thereby limited to 10^4 pulses per second for normal 1μs pulses. The second difficulty concerns the relatively poor linearity and stability of an A.D.C. unit. A differential linearity of $\pm 1\%$ is normal and $\pm 0.2\%$ can be achieved with some difficulty. However, this means a potential variation of 1% of the baseline across the final spectrum on which may be superimposed a Mössbauer resonance of the same order of magnitude.

[Refs. on p. 43]

This presents a difficulty if the data are to be analysed by computer and it reduces the feasible accuracy of measurements.

Both these objections can be overcome by using a time-mode system and spectrometers based on this latter principle have now virtually replaced those operating in the pulse-height analysis mode.

(b) Time-mode (Multiscalar-mode) Spectrometers

The time-mode system dispenses entirely with the A.D.C. processing. The channel address is advanced sequentially under the control of a very accurate crystal oscillator with a dwell time in each channel of say 50 or 100 μs. While a channel is open it accepts all input pulses from the detection system. The only dead-time is that incurred during the channel advance, and this is only 5–10% of the total time. In this way count-rates approaching 10^6 s^{-1} can be achieved. The baseline constancy is directly dependent on the short-term (one period of the drive) stability of the crystal oscillator. Usually it is good enough not to present any problems, and the biggest potential non-linearity is from geometric effects when using high-velocity scans.

The servo-controlled transducer system is identical to that used in the pulse-height analysis mode, but it must be coupled to the channel advance frequency. Several variations have been proposed for this. Some workers have built their own waveform and channel advance circuitry. Others have taken an existing analogue voltage, proportional to the channel address, from an analyser using its own internal crystal oscillator, and have used this as the drive reference signal for a waveform of type shown in Fig. 2.2b. A schematic layout is given in Fig. 2.4. The 'staircase' nature of the signal is smoothed by the servo system to produce a constant acceleration over 80% of

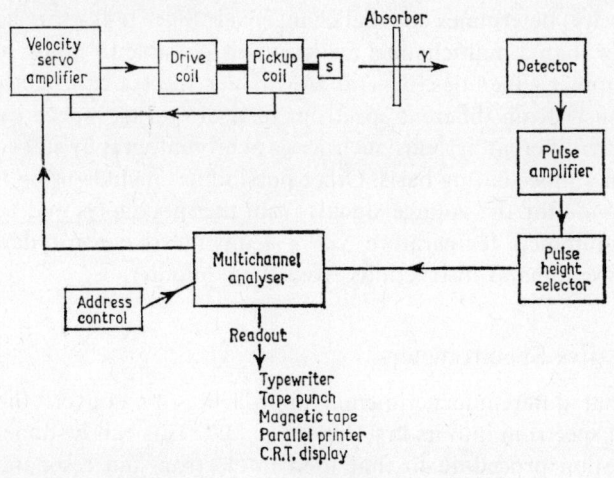

Fig. 2.4 Schematic arrangement for a time-mode spectrometer.

[*Refs.* on p. 43]

26 | EXPERIMENTAL TECHNIQUES

the scan. A third method has been to take a signal from an internal bi-stable of the analyser and produce a double ramp as in Fig. 2.2c. Several detailed descriptions of time-mode operation are available [12, 29–32].

A more recent innovation which is only completely satisfactory if the transducer unit is of high quality is to use forward–backward address scaling [33, 34]. While the source executes one period of a symmetrical double-ramp waveform as in Fig. 2.2c the channel addresses are stepped once in a forward and once in a backward direction. The result is a superposition of forward and backward scans and the accumulation of only one spectrum. This method would suffer if the return scan were not the mirror-image of the forward, but otherwise it has the advantages of allowing more efficient use of the channel storage and a reduction in geometry effects. An alternative approach [35] is to activate the analyser for $\frac{1}{2}$ or $\frac{1}{4}$ of the double-ramp cycle thereby doubling or quadrupling the effective resolution. The prime disadvantage of these modifications is a considerable increase in 'dead-time', necessitating longer counting times.

(c) On-line Computers

A recent development which should see more widespread application in the next decade is the use of small on-line computers. Typically these contain a memory of 4096 (4k) 12-bit words and a typical arrangement has already been described in some detail [36, 37]. The pick-up voltage which is a direct measure of velocity is digitised by an A.D.C. unit. Since this voltage varies slowly, a continuous and accurate value of the voltage can be stored by the computer. When the digital equivalent of the voltage changes by a predefined increment, the channel address for data storage is altered. This differs from the time-mode system where the input waveform (and not the actual velocity) determines channel changeover. Since a 4k-store computer is much larger than a multichannel analyser it is possible to utilise much more of the computer's facilities. Several Mössbauer spectra can be accumulated simultaneously from different spectrometers, or storage space can be used with entirely different systems such as gas-chromatography data collection, possibly on a time-sharing basis. Other possibilities include using the A.D.C. facility to monitor the voltage signals from thermocouples and to maintain a pre-programmed temperature via a temperature-control device which receives its command instructions from the computer.

2.4 Derivative Spectrometers

A somewhat different experimental approach is to convert the resonant absorption spectrum into its first derivative [38]. This can be done by a similar modulation procedure to that used in electron-spin resonance spectrometers. The major complication is that two modulating velocity terms are required, a constant velocity v_0 and a high-frequency, low-amplitude, periodic velocity of sinuisoidal type. The resultant velocity is then

$$v = v_0 + k \sin \omega t$$

The transmission $\tau(v)$ is now given by

$$\tau(v) = \tau(v_0 + k \sin \omega t)$$

which for appropriate values of the amplitude k and frequency ω approximates to

$$\tau(v_0, \omega) = k \frac{d\tau}{dv_0} \sin \omega t$$

Thus $\tau(v_0, \omega)$ is proportional to $d\tau/dv_0$, the derivative of the normal transmission curve. The amplitude of the secondary modulation must be much less than the width of the Mössbauer line so that $d\tau/dv_0$ is nearly constant over the amplitude.

Bressani, Brevetto, and Chiavassa have described an instrument using this feature [39]. The absorber is moved with a constant velocity v_0 and the source with varying velocity $k \sin \omega t$. The detected γ pulses are not distributed statistically but are periodically bunched by the extra modulation. A lock-in amplifier can be made to respond to the bunching frequency and with a suitably long integration time will produce a voltage proportional to $d\tau/dv_0$.

A resultant spectrum is shown in Fig. 2.5 where it is compared with the normal transmission spectrum. The method has the advantage of being more selective in the presence of strong background radiation since the latter will not have a time distribution containing the modulation frequency. However, strong sources are required to obtain the best results, and it is unlikely that the technique will ever replace the very popular velocity-scan systems.

2.5 Scattering Experiments

Although the most usual method of registering a Mössbauer spectrum is by a transmission experiment, it is also possible to observe the resonance by a scattering method [40–43]. γ-rays can be scattered by several mechanisms. The electrons in the atoms can scatter the incident γ-rays without change in the wavelength (Rayleigh scattering) or with an increase in the wavelength (Compton scattering). The corresponding processes in which the nucleus is the scattering agent are usually weak enough to be neglected. There are also the γ-rays produced by recoilless and non-recoilless nuclear resonance scattering. The Rayleigh and resonant scattered radiation are indistinguishable except that the latter will be affected by applying a Doppler shift to the system. The scattered intensity will be higher when Mössbauer resonance absorption occurs.

28 | EXPERIMENTAL TECHNIQUES

The scattering method uses very similar equipment to that employed in the transmission technique. The only major difference is in the counting geometry. Two examples are shown in Fig. 2.6. The detector must be com-

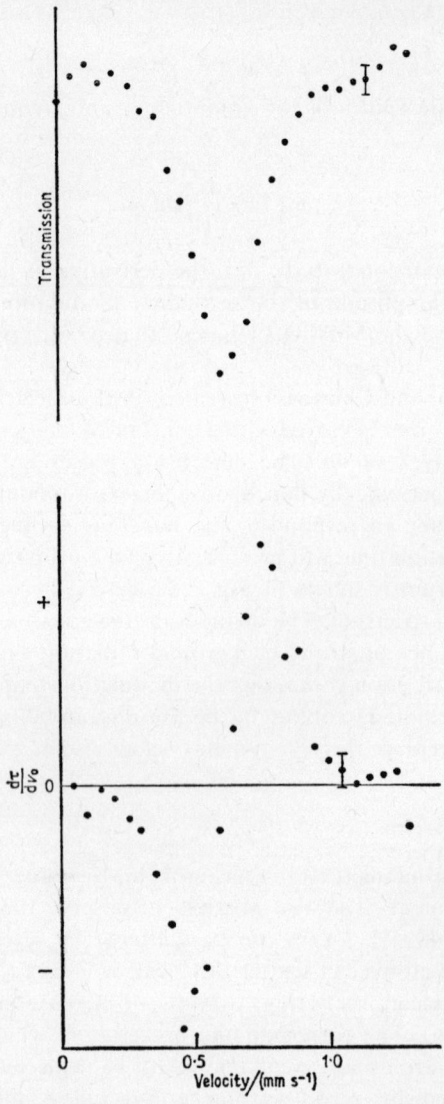

Fig. 2.5 The Mössbauer transmission spectrum of a resonance line and the corresponding derivative spectrum.

pletely shielded from the primary source radiation. The cylindrical and conical scatterers illustrated have been used because of the greater solid

[*Refs. on p. 43*]

angle which they present without a corresponding increase in the velocity spread. Computation of the effective Doppler shift from the known drive motion can be quite complicated.

The intensity of the scattered radiation is far weaker than the transmitted

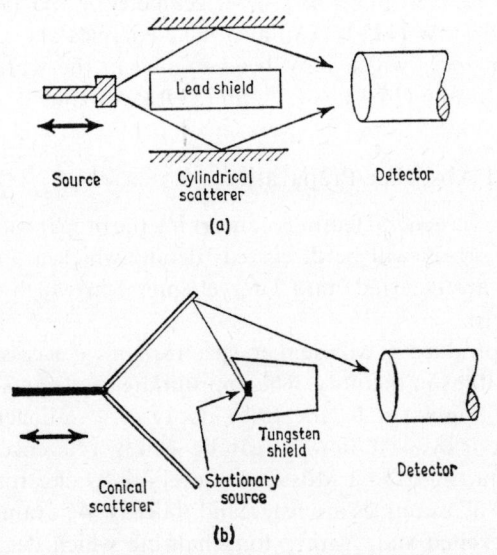

Fig. 2.6 Two types of scattering geometry: (a) using a cylindrical scatterer with a moving source; (b) using a conical moving back-scatterer.

portion and this entails the disadvantage of high-intensity sources. However, the cross-section for resonant scattering can easily be several orders of magnitude higher than that for Rayleigh scattering, giving potentially considerable improvement in signal to noise ratio. This is particularly significant when the recoil-free fraction is very low, and some isotopes such as ^{188}Os (155 keV) have only been observed to give a resonance by scattering methods (see Chapter 16). The technique is also potentially of use for samples which are necessarily too thick for transmission experiments, and when surface states are of particular interest.

One complication which can arise is interference of the recoilless γ-rays with the Rayleigh scattering. Since the interference term shows a phase change at the resonance maximum, a dispersion curve is added to the Mössbauer scattering spectrum causing an asymmetric distortion of the spectrum shape. This can be partially overcome by suitable choice of the scattering angles.

Historically, resonant scattering was first recorded by Barloutaud, Picou, and Tzara [44] in 1960 using ^{119}Sn. Considerable effort has been expended in the study of scattering from the aspect of basic phenomena using metal foils as the scattering materials, but since no chemical application has yet

30 | EXPERIMENTAL TECHNIQUES

been made it is only appropriate to mention here a bibliography included in ref. 42. The interference between the Rayleigh and resonant radiation has been observed in ^{57}Fe by Black *et al.* [45, 46]. Of the isotopes where scattering experiments have been reported, ^{186}Os (137·2 keV) and ^{188}Os (155·0 keV) have not been observed by transmission geometry: ^{195}Pt (98·8 keV), ^{182}W (100·1 keV), and ^{186}W (122·6 keV) also have energies above 90 keV. Other Mössbauer resonances which have been studied in the scattering mode are ^{169}Tm (8·4 keV), ^{57}Fe (14·4 keV), ^{119}Sn (23·9 keV), and ^{191}Ir (82·3 keV).

2.6 Source and Absorber Preparation

In this section, some general features concerning the preparation of Mössbauer sources and absorbers will be discussed; details which are specific to individual nuclides are deferred until later chapters, in which each element is considered in turn.

In order to produce a Mössbauer spectrum it is necessary to produce recoil-free γ-photons in quantity. The appropriate excited state of a resonant nucleus can be populated by the prior decay of a radioactive isotope, by nuclear reaction or by excitation. As can be seen by reference to Appendix 1, the most frequent routes to a Mössbauer level are by electron-capture decay (E.C.) for which 37 examples are listed and β-decay (46 examples). Some isotopes have an excited state with a long half-life which decays by isomeric transition (I.T.) with γ-ray emission (6 examples). α-emission from ^{241}Am has been used to populate ^{237}Np. Coulombic excitation, that is the bombardment of a target material with very high-energy particles such as oxygen ions has been used for 24 Mössbauer transitions, but has the disadvantage of necessitating *in situ* experimentation because of the effectively instantaneous decay. Nuclear reactions such as (n, γ) and (d, p) which likewise fail to generate a long-lived intermediate (5 transitions) also fall into this category.

The β^-, I.T., and E.C. routes are most conveniently illustrated by the ^{83}Kr decay scheme shown in Fig. 2.7. The details are taken from ref. 47. The isomeric state ^{83m}Kr is produced by the ^{83}Kr$(n, \gamma)^{83m}$Kr reaction and has a 1·86-hour half-life. ^{83}Br gives a β^- decay to ^{83m}Kr with a 2·41-hour half-life. ^{83}Rb undergoes E.C. direct to the higher excited states of ^{83}Kr with an 83-day half-life.

The likelihood of success in observing a Mössbauer resonance will depend on numerous factors as follows:
(1) The energy of the γ-ray must be ideally between 10 and 100 keV. This is because γ-rays with energies less than 10 keV are very strongly absorbed in solid matter, and for those above 100 keV the recoil-free fraction which is proportional to $\exp(-E_\gamma^2)$ falls to a very low value (equations 1.3 and 1.12). The absorption cross-section σ_0 is proportional to E_γ^{-2} and also decreases rapidly as E increases. These are the main reasons why the Mössbauer effect

[*Refs. on p. 43*]

has not yet been recorded for any element lighter than ^{40}K. The energy level separations in light nuclei are usually quite large, and the γ-rays emitted are too energetic to produce a detectable recoil-free fraction. However, as will be seen in later chapters, it is possible to study the chemical bonding effects of light elements in the compounds of those heavier nuclides which do give a Mössbauer resonance.

(2) The half-life of the excited state $t_{\frac{1}{2}}$ should preferably be between about

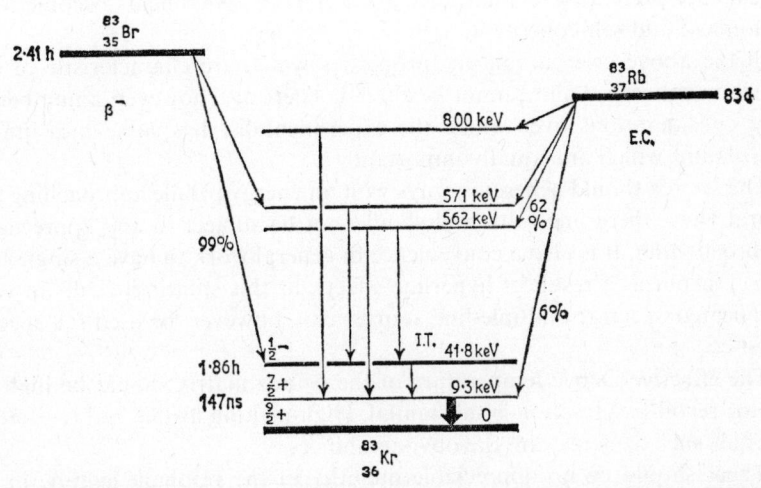

Fig. 2.7 The decay scheme of ^{83}Kr showing how the 9·3-keV Mössbauer level is populated by β^-, I.T., and E.C. decay. The levels are not drawn to scale, and the details are taken from ref. 47.

1 and 100 ns. As we saw in Section 1.3 it is this time which controls the Heisenberg linewidth of the γ-ray energy. If the relative linewidth Γ/E_γ is too narrowly defined (i.e. a long half-life) there are considerable problems in damping out mechanical vibrations in the spectrometer. Conversely, a short lifetime gives a broad line which is difficult to observe and which usually also obscures any chemical hyperfine effects.

(3) The internal conversion coefficient α must be small so that the γ-transition has a high probability of producing a γ-photon rather than a conversion electron. This will also increase the absorption cross-section σ_0 (equation 1.18). Most values listed in Appendix 1 fall in the range 0–10.

(4) The absorption cross-section σ_0 should be large and the free-atom recoil energy E_R should be small. Both of these factors have already been mentioned in connection with other quantities above. Appendix 1 lists 5 transitions for which $\sigma_0 > 10^{-18}$ cm^2 and few Mössbauer resonances have been observed for transitions in which $\sigma_0 < 0.06 \times 10^{-18}$ cm^2. Likewise only 5 transitions have a free-atom recoil $E_R > 6 \times 10^{-2}$ eV and values are normally in the range $(0.1–5) \times 10^{-2}$ eV.

[*Refs. on p. 43*]

(5) The method of generation of the source γ-rays should ideally be such that a source can be encapsulated and then used for a long period with only the minimum of handling precautions. This implies a long-lived precursor which is easily obtained in high activity, and explains the popularity of those β^-, E.C., and I.T. sources, which can be purchased commercially. The Coulomb, (n, γ), and (d, p) reactions require access to very expensive equipment. For routine chemical applications particularly, sources having lifetimes of months or years are preferable so that prolonged series of experiments become both economical and self-consistent.

All the above criteria concern properties which are characteristic of the transition concerned and cannot be altered. There are, however, a number of other considerations over which the experimentalist has some measure of control, and which are equally important:

(6) The source should generate γ-rays with an energy profile approaching the natural Heisenberg linewidth and should not be subject to any appreciable line broadening. It is also a convenience in general work to have a single-line source unsplit as a result of hyperfine effects, as this splitting results in very complicated spectra; multiple-line sources can, however, be used for special purposes.

(7) The effective Debye temperature of the source matrix should be high so that the recoil-free fraction is substantial. High-melting metals and refractory materials such as oxides are the obvious choices.

(8) There should be no appreciable quantity of the resonant isotope in its ground state in the source, otherwise the self-resonant source term in equation 1.20 becomes important. This will result in an effective source linewidth which is greater than the Heisenberg natural linewidth. In this respect it is interesting to note that the linewidth of a ^{57}Co/(iron) source is greater than that for ^{57}Co/(iron enriched in ^{56}Fe) because of the greater self-resonance in the former case [48].

(9) Non-resonant scattering inside the source can be reduced, either by careful choice of any other elements in the matrix, or, in the case of metal foils doped with a radioisotope which is the Mössbauer precursor, by controlling the depth to which this generating impurity is diffused.

(10) The ground-state isotope should ideally be stable and have a high natural abundance, otherwise it may become necessary to use artificially enriched compounds at greatly increased cost and inconvenience. Isotopic enrichment is, however, a very important method of improving resolution of spectra and may become essential in work with biological materials or in the study of doped solids where the actual concentration of the element of interest is extremely small.

An important consideration is the chemical effects of the nuclear reactions preceding the occurrence of a Mössbauer event. If, for example, the source preparation involves a long neutron irradiation of the intended source matrix at high flux as in ^{118}Sn$(n, \gamma)^{119m}$Sn, considerable radiation damage may occur with the formation of lattice defects. Since ideally all the excited atoms should be in identical chemical environments, such defects will result in unwanted line-broadening effects. For this reason it may be desirable either to anneal the source so as to restore the regularity of the crystal lattice, or to extract the radioisotope chemically and then incorporate it in a new matrix.

Very high-energy processes immediately preceding the Mössbauer γ-emission must also be considered. The α-decay of ^{241}Am to give ^{237}Np is sufficiently energetic to displace the daughter nucleus from its initial lattice site. At the same time a considerable number of neighbours are thermally excited. The fact that a Mössbauer spectrum is recorded at all in this case shows that the Np atom comes to rest on a normal lattice site and establishes vibrational integrity with the lattice as a whole in a time which is short compared to the lifetime of the 59·54-keV level, i.e. 63 ns. Complications can occur if the displaced atom is capable of residing on more than one type of chemical site in the crystal.

Similar situations arise, for example, in Coulomb excitation reactions. In the ^{73}Ge case, the low Debye temperature of the Ge metal produces a very low recoil-free fraction. As mentioned in more detail later (p. 109), it is possible to displace the excited atoms completely out of the target material and implant them into a new matrix with a high Debye temperature, thereby obtaining a considerable improvement in the quality of the spectra.

Electron capture involves the incorporation of an inner-shell electron (usually the K-shell) into the nucleus (thereby converting a proton into a neutron and emitting a neutrino). This event is followed by an Auger cascade which results in production of momentary charge states on the atom of up to +7. Many papers on ^{57}Fe experiments have reported the apparent detection of decaying higher charge states resulting from E.C. in ^{57}Co, but as detailed in Chapter 12 it has since been proven that although such states are indeed generated by the after-effects of electron capture, they have already reached stable equilibrium with the solid before the subsequent Mössbauer transition occurs. It would appear that all transient nuclear decay after-effects are over within 10^{-8} of a second. It is important, however, to avoid the production of multiple charge states if the source is not intended to be the subject of a special study designed to detect such states but is to be used only as a source of monochromatic radiation. Some of these topics will be considered in more detail under specific isotopes in later chapters.

Although the preparation of a good source may in some cases be difficult, an adequate absorber can usually be made quite easily. A few general remarks may be useful to illustrate the balance of factors involved, and several texts with full mathematical equations are available [49–52]. Integration of equation 1.20 for simplified cases shows that the measured linewidth will increase

34 | EXPERIMENTAL TECHNIQUES

above 2Γ as the absorber thickness increases. This will decrease the resolution of a multiple-line spectrum or the precision with which a single line can be located. Furthermore, an increase in absorber thickness also diminishes the transmission of resonant radiation as a result of non-resonant scattering. However, it is equally clear that the absorber must have a finite thickness for the resonance to be observed at all, hence it follows that there must be an optimum absorber thickness for transmission geometry.

Integration of equation 1.20 for a uniform resonant absorber and an ideally thin source gives the transmission at the resonance maximum as

$$T(0) = e^{-\mu_a t_a}\{(1 - f_s) + f_s e^{-\frac{1}{2}T_a}J_0(\tfrac{1}{2}iT_a)\} \qquad 2.1$$

where J_0 is the zero-order Bessel function

$$J_0(ix) = 1 + \left(\frac{x}{2}\right)^2 + \frac{\left(\frac{x}{2}\right)^4}{1^2 \cdot 2^2} + \frac{\left(\frac{x}{2}\right)^6}{1^2 \cdot 2^2 \cdot 3^2} \cdots \qquad 2.2$$

and $T_a = f_a n_a a_a \sigma_0 t_a$. The function $T(0)$ is plotted in Fig. 2.8 for ^{57}Fe using

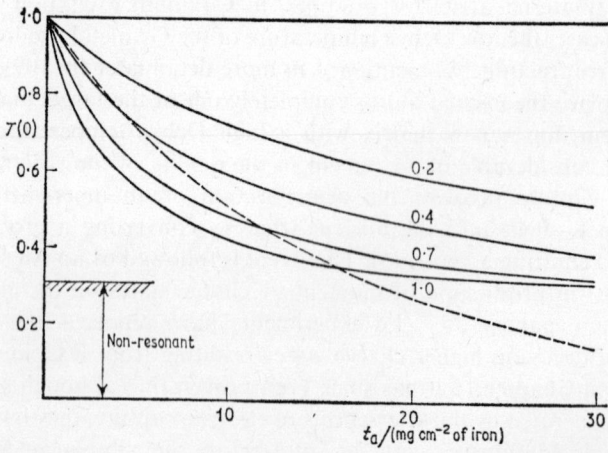

Fig. 2.8 The solid lines are the function $T(0)$ plotted for a zero value of μ_a and four different values of f_a, together with parameters appropriate to ^{57}Fe (14·4 keV). The dashed curve represents the non-resonant attentuation with $\mu_a = 0.067$.

the parameters $\mu_a = 0$ (i.e. no non-resonant scattering), $a_a = 0.0219$, $\sigma_0 = 2.57 \times 10^{-18}$ cm^2, $f_s = 0.7$ and for f_a values of 0·2, 0·4, 0·7, and 1·0 (the thickness has been converted to mg cm^{-2} of natural iron). The resonant absorption shows a saturation behaviour with increasing thickness. The dashed curve shows the non-resonant scattering attenuation. The quantity to optimise is obviously the absorption in the final transmitted radiation which is $e^{-\mu_a t_a} - T(0) = \Delta T(0)$.

$$\Delta T(0) = e^{-\mu_a t_a} f_s[1 - e^{-\frac{1}{2}T_a}J_0(\tfrac{1}{2}iT_a)] \qquad 2.3$$

[Refs. on p. 43]

A series of curves for ^{57}Fe with $\mu_a = 0.067$ (the value for natural iron) are shown in Fig. 2.9. The optimum value is seen to be about 10 mg cm^{-2} of

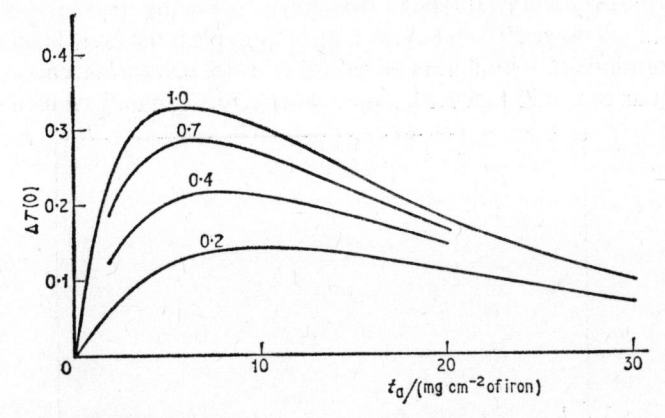

Fig. 2.9 The function $\Delta T(0)$ for ^{57}Fe (14·4 keV) plotted for four values of f_a to show the optimum absorber thickness.

total iron. Fortunately, this value is virtually independent of the recoil-free fraction, a parameter which cannot always be determined in advance. Compounds of iron can be visualised using the same thickness scale, but with a higher non-resonant attenuation coefficient, so that the maximum in the curves moves to lower t_a values.

This type of calculation is only of limited value. The major problem is the possibility of additional non-resonant intensity at the counter produced by the Compton scattering of higher-energy γ-rays. Another factor concerns particle size. The presence of large granules in the absorber can cause a significant reduction in the observed absorption. Calculations [53] show that reduction in particle size eventually results in a reversion to the uniform absorber model.

A discussion of the effects of orientation of the granules and the use of single-crystal absorbers will be deferred until Chapter 3 when the intensities of hyperfine components are discussed.

In general, absorbers can be metal foils, compacted powders, mixtures with inert solid diluents, mixtures with inert greases, frozen liquids, or frozen solutions. The only limitation is on the material used for the windows of the sample container; this must be free of the resonant isotope and have a low mass attenuation coefficient for the γ-ray being studied. Organic plastics and aluminium foil are most commonly used.

2.7 Detection Equipment

Of the four principal types of γ-ray detector used in Mössbauer spectroscopy,

36 | EXPERIMENTAL TECHNIQUES

three are conventional instruments and require only brief mention. Further descriptive material and detailed references are readily available [54, 55].

The scintillation-crystal type of detector is frequently used for γ-rays with energies in the range 50–100 keV. A typical example is the NaI/Tl scintillator. The resolution of scintillators deteriorates with decreasing energy of the γ-photon as shown in Fig. 2.10, and such detectors can only be used for very

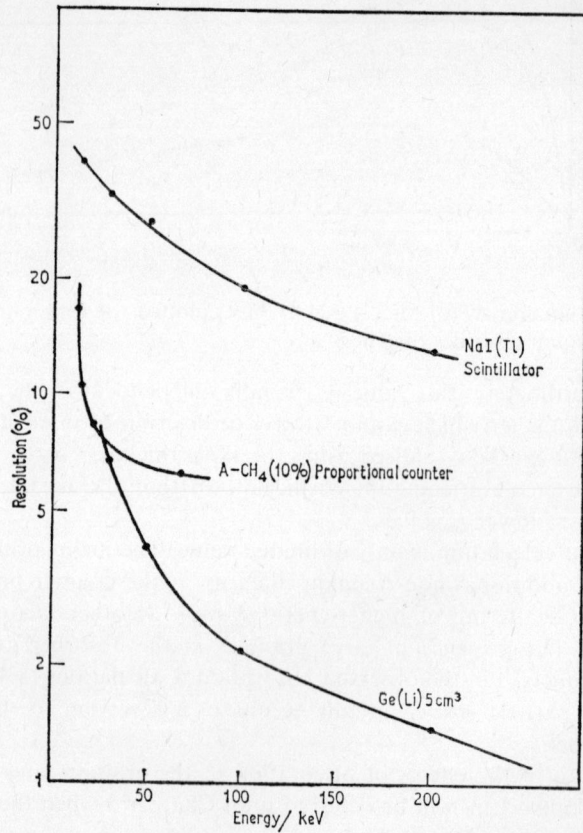

Fig. 2.10 Typical resolution of commonly used γ-ray detectors.

soft γ-rays if the radiation background is low and there are no other X-ray or γ-ray lines with energies near that of the Mössbauer transition. The scintillation unit has a prime advantage of a high efficiency.

Below 40 keV, the gas-filled proportional counter gives better resolution but at the expense of a low efficiency and generally lower reliability. It is possible to lower the radiation background in some cases by the choice of a suitable 'filter'; for example a copper foil will absorb the 27·5- and 31·0-keV ^{125}Te X-rays strongly and transmit most of the 35·5-keV γ-rays because of the higher mass attenuation coefficient for the former.

[*Refs.* on p. 43]

A more recent development has been the introduction of lithium-drifted germanium detectors. As shown in Fig. 2.10 these give a very highly resolved energy spectrum, but at the expense of low sensitivity, and some inconvenience in use. They are of obvious application where there are several γ-rays of similar energy as often found in the complicated decay schemes of the heavier isotopes. Again the resolution drops drastically with decreasing energy, and they are only of use at the higher end of the Mössbauer energy range. Other drawbacks include a high cost and the necessity for maintaining them indefinitely at liquid nitrogen temperature; irreparable damage results if they are allowed to warm to room temperature.

Comparative examples [56] of the resolution for a source of ^{125}I populating the ^{125}Te 35·5-keV level are given in Fig. 2.11. Although only the Li-drifted Ge crystal gives good resolution of the 35·5 keV γ-ray from the

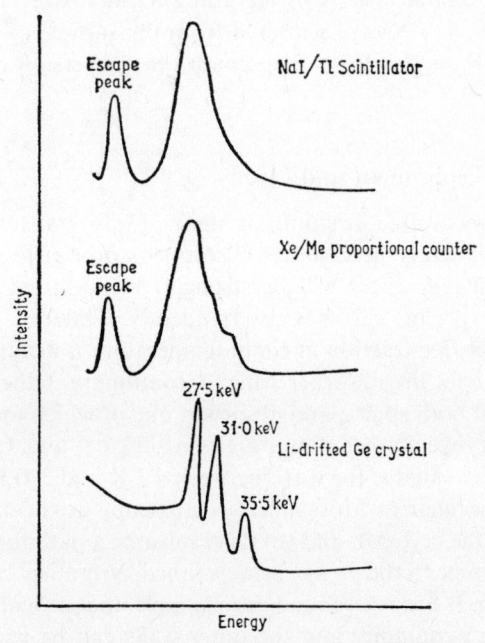

Fig. 2.11 The energy spectrum of an ^{125}I source as recorded with three different detection systems.

27·5- and 31·0-keV X-rays, in this particular instance the other two systems can make use of the 'escape peak' which is produced by loss of an iodine (or xenon) K X-ray from the capturing medium. The tellurium X-rays are too low in energy to generate an escape peak.

The fourth detector system is to use a resonance scintillation counter [57]. A standard type of plastic scintillator for β-detection is doped with the resonant absorber. It is insensitive to the non-resonant background of primary

[*Refs. on p. 43*]

γ- and X-photons, but will detect the secondary conversion electrons after γ-capture by the resonant nuclei. Although effective, the main difficulty is one of inconvenience, because a new plastic scintillator must be made for each compound used as an absorber. Alternatively, one can use an ordinary Geiger or proportional counter with its inner surface thinly coated with a compound of the isotope being studied [58] or with the proper absorber itself located inside the counter [59]. Again, the recoilless γ-rays are resonantly absorbed by the coating on the internal absorber, and the internal conversion electrons or low-energy X-rays emitted in the subsequent decay of the excited state are then counted with almost 100% efficiency in 4π-geometry. The counter has been used for ^{57}Fe [59], ^{119}Sn [60], and ^{169}Tm [61], and can be used in principle for any Mössbauer isotope in a compound which has a large recoil-free fraction at room temperature.

Other closely related means of recording a Mössbauer resonance are to record the conversion X-rays scattered from the surface of an absorber [62] or transmitted through it [63], or to count the conversion electrons emitted [64].

2.8 Cryogenic Equipment and Ovens

The very low recoil-free fraction of many γ-ray transitions at ambient temperatures frequently necessitates experimentation at lower temperatures where the Mössbauer effect will be stronger. In addition, the temperature dependence of hyperfine effects is also frequently of interest. If a good source with a high recoil-free fraction at room temperature is available, it may only be necessary to cool the absorber which is stationary. Otherwise, it may be necessary to cool both source and absorber, one of which must also be moving. Standard cryogenic techniques are used [55, 65] and vacuum cryostats are commercially available for work between 4·2 K and 300 K. The two main considerations peculiar to Mössbauer spectroscopy are (a) there must be no vibration inside the cryostat, and (b) there must be a path through the system which is transparent to the γ-rays being studied. Vibration can be eliminated by careful design. It has been found, for example, that a rigid interconnection between a helium container and the outer walls can be made by means of two stacks of several hundred aluminised Mylar washers loosely threaded together, between two nylon washers for support, the large number of contact surfaces lowering the heat transmission to acceptable limits [66]. Cooling the vibrating source as well as the absorber can only be accomplished by inserting the entire transducer unit inside the vacuum space.

Although work below liquid nitrogen temperature necessitates a precision-engineered vacuum cryostat, it is quite easy to work between 78 K and 300 K without a vacuum by using styrofoam insulation. Standard coolants are liquid nitrogen (b.p. 77·3 K), liquid hydrogen (b.p. 20·4 K), and liquid helium (b.p. 4·2 K). Temperatures below these boiling points can also be maintained by pumping on the liquid coolant. Some work calls for more flexible variable-temperature control and in such cases it is possible to use either a temperature gradient along a conducting metal rod regulated by a heating element, or a cold-gas flow technique. Many systems are briefly described in the literature, and there are also some detailed texts available [55, 65, 67–69].

The window material used in vacuum cryostats is usually beryllium, aluminium, or aluminised mylar. It should be noted, however, that commercial beryllium and aluminium often contain sufficient iron to give a detectable ^{57}Fe (14 keV) resonance, and this has been known to cause problems when working with this isotope.

Temperatures above 300 K are sometimes desirable for ^{57}Fe work, and a controlled-temperature furnace is then required [68, 69]. The sample is frequently sandwiched between thin discs of beryllium, graphite, or aluminium (m.p. 660°C) attached to an electric heating coil in a vacuum. Alternatively [70], the sample can be placed in a furnace with thin entrance and exit windows and containing an atmosphere of hydrogen to prevent oxidation. A vacuum furnace capable of providing temperatures up to 1000°C has been described [71] and temperatures of 1700°C have been reached with a helium-filled oven with beryllium windows [72].

High-pressure work has been published by relatively few laboratories because pressure cells require special experience and engineering [73–75].

If a large external magnetic field is to be applied to an absorber, this is usually done with a superconducting magnet installation which also requires a liquid helium cryostat for its operation [76]. Again, several commercial instruments are available.

2.9 Velocity Calibration

One of the more difficult experimental aspects of Mössbauer spectroscopy is the accurate determination of the absolute velocity of the drive. The calibration is comparatively easy for constant-velocity instruments, but most spectrometers now use constant-acceleration drives. The least expensive method, and therefore that commonly used, is to utilise the spectrum of a compound which has been calibrated as a reference. Unfortunately, suitable international standards and criteria for calibration have yet to be decided. As a result, major discrepancies sometimes appear in the results from different laboratories. The problem is accentuated by having figures quoted with respect to several different standards, necessitating conversion of data before comparison can be made. However, calibration of data from an arbitrary standard spectrum will at least give self-consistency within each laboratory.

Several systems for a more direct absolute-velocity calibration have been developed. In one such method [31], the output of the monitoring pick-up

40 | EXPERIMENTAL TECHNIQUES

coil is fed to a voltage-to-frequency converter and thence to the time-mode multichannel analyser as a train of pulses with the same frequency. The counting rate in a channel is thus a function of the instantaneous velocity, provided of course that the linearities of the pick-up coil and the voltage-to-frequency converter are both accurate. This method gives a good check on

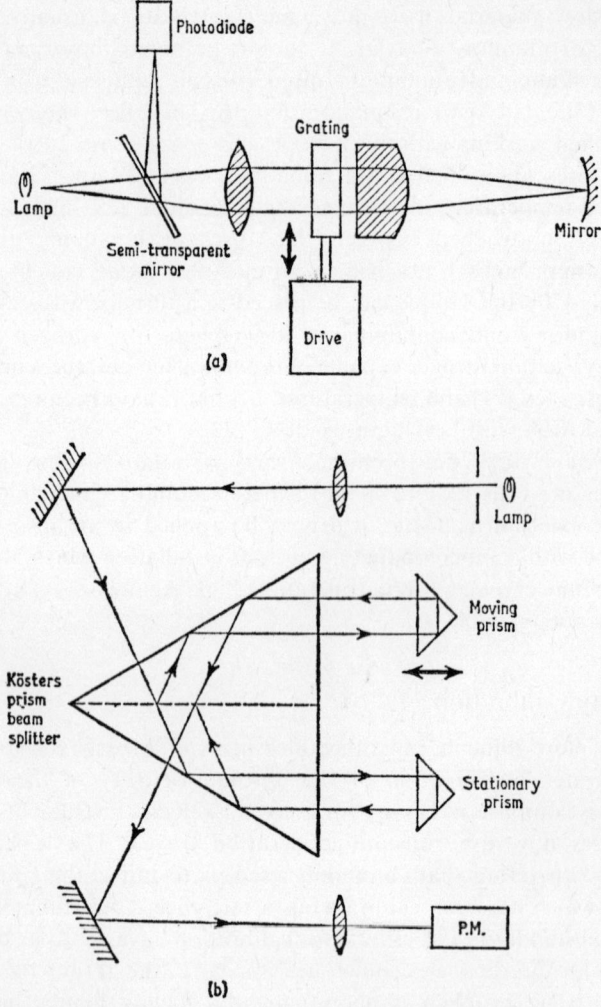

Fig. 2.12 Absolute calibration of constant-acceleration drives using (a) diffraction grating, (b) optical interferometer.

the day-to-day linearity and stability of the drive, but is not easily calibrated in absolute terms from a known hyperfine spectrum.

An absolute calibration can be obtained by mounting a diffraction grating

[*Refs. on p. 43*]

(typically 8 μm spacing) on the reverse end of the source drive-shaft [77]. Reflection of the diffraction image will produce a light intensity which is periodic by interference because of the varying velocity of the grating. The fluctuations can be detected and converted to pulses which after accumulation in the time-mode analyser give a calibration similar to that mentioned above. The basic components are illustrated schematically in Fig. 2.12a. The absolute accuracy is claimed to be as good as that of the best mechanical drives.

A similar method uses an optical interferometer as illustrated in Fig. 2.12b [78]. The motion of the small prism, which is attached to the drive, causes a time-dependent interference which is again converted to pulses and registered in the analyser. Neither method is widely applicable because of the prohibitive cost of the precision optical equipment.

Calibration in terms of a known frequency has also been accomplished by mounting the absorber on a quartz crystal and calculating the velocity scale from the spectrum sidebands produced by frequency modulation [79].

2.10 Curve Fitting by Computer

Much of the Mössbauer spectroscopic data which is published comes from institutions which also have large computer facilities. Since the raw data of a spectrum are already digitised, it is in a very convenient form for automated analysis by digital computer. The majority of multichannel analysers have output signals which are either directly compatible with or easily adapted for appropriate types of punched paper tape or magnetic tape units, so that accumulated spectra can be printed out in a form which is compatible with the computer installation.

The information which may be required from the computer analysis of the data are the parameters of a selected function which are the best possible fit to the observed spectrum. In the simplest cases this function will be one or more Lorentzian curves. However, more complex types of function may sometimes be required, for example when fitting multi-line spectra which are subject to motional narrowing (see p. 72).

The general problem can be described as follows. The data comprise N digitised values Y_i. The function which is to be fitted contains n variables, represented by the vector V, and will give a calculated value at data point i of $A(V)_i$. The error thereby incurred is

$$e_i = Y_i - A(V)_i \qquad 2.4$$

The best possible fit to the N data points, which is the answer required, must be such that the overall error is a minimum. The statistics of radioactive counting processes [80] tell us that the values of Y_i will be distributed as a Poisson distribution with a variance equal to the mean, and hence the standard deviation in Y_k is $\sqrt{Y_k}$. The quantity to be minimised for a best fit is

[*Refs. on p. 43*]

therefore the 'goodness of fit' function

$$F(V) = \sum_{i=1}^{N} \left\{ \frac{[Y_i - A(V)_i]^2}{Y_i} \right\} \qquad 2.5$$

which at the minimum is a chi-squared function and can be used to determine the probability of the fit being a valid one.

Equation 2.5 is non-linear in the variables V and represents a complex mathematical problem even for high-speed digital computers. Considerable effort has been expended since 1959 in developing appropriate methods, and their application to Mössbauer spectroscopy has been specifically detailed in several papers [80–82].

The chi-squared (χ^2) function, $F(V)$, has $N - n$ degrees of freedom, and the validity of the result can be determined from standard tables [83]. As a general rule, the statistical significance of the fit decreases rapidly as $F(V)$ rises in value. For example, a computed fit with 400 degrees of freedom is within the 25–75% confidence limits of a χ^2 distribution if $380 < F(V) < 419$, and within the 5–95% limits if $355 < F(V) < 448$.

The χ^2 values obtained in Mössbauer spectroscopy are frequently higher than one would expect because the function used is not absolutely correct. Thus, peaks may not be strictly Lorentzian in shape, as in cases of order–disorder phenomena where they may be distorted by a large number of small variations in the electronic state of the absorbing atoms due to environment fluctuations. Again, if the counting rate is greater than about 10^5 counts per second in the detection system the 'dead-time' effects which occur in the instrumentation will distort the Poisson distribution. In this case the total registered count-rate could still be low if the rejection fraction at the single-channel discriminator were high. Other factors which affect the χ^2 values include instrumental drift, counting geometry, cosine effects, and the proportion of the spectrum being computed which is zero-absorption baseline.

Introducing additional variables will invariably lower the value of $F(V)$, but the decrease may not be large enough to justify the new fit. The final decision regarding the valid interpretation of the data must and should remain with the experimenter, using his full experience and chemical knowledge.

One useful feature of some of the mathematical methods of solution is that they also provide information on the precision of the final values of the variables [80], but the effects of instrument drift are likely to be underestimated as it is assumed that the velocity of any given channel is not a function of time.

Although data are usually analysed by specifying an exact mathematical function, it is possible (with slightly different methods) to use an experimentally defined peak-shape if the former is not known with precision [84].

One obvious development of these various methods is the application of direct on-line time-sharing computer techniques. In principle, this enables the analyser output to be transmitted directly to the central processing unit from the laboratory, and the results of the calculation can then be returned at least in abbreviated form to a remote terminal for rapid preliminary assessment of the experiment.

REFERENCES

[1] R. L. Mössbauer, *Z. Physik*, 1958, **151**, 124.
[2] R. L. Mössbauer, *Naturwiss.*, 1958, **45**, 538.
[3] K. Cassell and A. H. Jiggins, *J. Sci. Instr.*, 1967, **44**, 212.
[4] R. Booth and C. E. Violet, *Nuclear Instr. Methods*, 1963, **25**, 1.
[5] C. E. Johnson, W. Marshall, and G. J. Perlow, *Phys. Rev.*, 1962, **126**, 1503.
[6] C. W. Kocher, *Rev. Sci. Instr.*, 1965, **36**, 1018.
[7] A. J. Bearden, M. G. Hauser, and P. L. Mattern, 'Mössbauer Effect Methodology Vol. 1', Ed. I. J. Gruverman, Plenum Press, N.Y., 1965, p. 67.
[8] H. M. Kappler, A. Trautwein, A. Mayer, and H. Vogel, *Nuclear Instr. Methods*, 1967, **53**, 157.
[9] P. A. Flinn, *Rev. Sci. Instr.*, 1963, **34**, 1422.
[10] P. Flinn, 'Mössbauer Effect Methodology Vol. 1', Ed. I. J. Gruverman, Plenum Press, N.Y., 1965, p. 75.
[11] J. Lipkin, B. Schechter, S. Shtrikman, and D. Treves, *Rev. Sci. Instr.*, 1964, **35**, 1336.
[12] F. W. D. Woodhams, *J. Sci. Instr.*, 1967, **44**, 285.
[13] V. P. Alfimenkov, Yu. M. Ostanevich, T. Ruskov, A. V. Strelkov, F. L. Shapiro, and W. K. Yen, *Soviet Physics – J.E.T.P.*, 1962, **15**, 713 (*Zhur. eksp. teor. Fiz.*, 1962, **42**, 1029).
[14] R. C. Knauer and J. G. Mullen, *Rev. Sci. Instr.*, 1967, **38**, 1624.
[15] P. E. Clark, A. W. Nichol, and J. S. Carlow, *J. Sci. Instr.*, 1967, **44**, 1001.
[16] J. Pahor, D. Kelsin, A. Kodre, D. Hanzel, and A. Moljk, *Nuclear Instr. Methods*, 1967, **46**, 289.
[17] Y. Hazony, *Rev. Sci. Instr.*, 1967, **38**, 1760.
[18] D. St P. Bunbury, *J. Sci. Instr.*, 1966, **43**, 783.
[19] R. Zane, *Nuclear Instr. Methods*, 1966, **43**, 333.
[20] D. Rubin, *Rev. Sci. Instr.*, 1962, **33**, 1358.
[21] R. L. Cohen, *Rev. Sci. Instr.*, 1966, **37**, 957.
[22] R. L. Cohen, P. G. McMullin, and G. K. Wertheim, *Rev. Sci. Instr.*, 1963, **34**, 671.
[23] V. Vali and T. W. Nybakken, *Rev. Sci. Instr.*, 1964, **35**, 1085.
[24] R. Gerson and W. S. Denno, *Rev. Sci. Instr.*, 1965, **36**, 1344.
[25] L. Lovborg, *Nuclear Instr. Methods*, 1965, **34**, 307.
[26] M. Bornaz, G. Filoti, A. Gelberg, V. Grabari, and C. Nistor, *Nuclear Instr. Methods*, 1966, **40**, 61.
[27] J. D. Cooper, T. C. Gibb, N. N. Greenwood, and R. V. Parish, *Trans. Faraday Soc.*, 1964, **60**, 2097.
[28] J. D. Cooper and N. N. Greenwood, *J. Sci. Instr.*, 1966, **43**, 71.
[29] C. A. Miller and J. H. Broadhurst, *Nuclear Instr. Methods*, 1965, **36**, 283.
[30] T. E. Cranshaw, *Nuclear Instr. Methods*, 1964, **30**, 101.

[31] F. C. Ruegg, J. J. Spijkerman, and J. R. de Voe, *Rev. Sci. Instr.*, 1965, **36**, 356.
[32] E. Kankeleit, *Rev. Sci. Instr.*, 1964, **35**, 194.
[33] Y. Reggev, S. Bukshpan, M. Pasternak, and D. A. Segal, *Nuclear Instr. Methods*, 1967, **52**, 193.
[34] E. Nadav and M. Palmai, *Nuclear Instr. Methods*, 1967, **56**, 165.
[35] M. Michalski, J. Piekoszewskii, and A. Sawicki, *Nuclear Instr. Methods*, 1967, **48**, 349.
[36] R. H. Goodman and J. E. Richardson, *Rev. Sci. Instr.*, 1966, **37**, 283.
[37] R. H. Goodman, 'Mössbauer Effect Methodology Vol. 3', Ed. I. J. Gruverman, Plenum Press, N.Y., 1967, p. 163.
[38] J. E. S. Bradley and J. Marks, *Nature*, 1961, **192**, 1176.
[39] T. Bressani, P. Brovetto, and E. Chiavassa, *Phys. Letters*, 1966, **21**, 299; *Nuclear Instr. Methods*, 1967, **47**, 164.
[40] H. Frauenfelder, 'The Mössbauer Effect', W. A. Benjamin, Inc., N.Y., 1962.
[41] A. J. F. Boyle and H. E. Hall, *Repts. Progr. in Physics*, 1962, **25**, 441.
[42] J. K. Major, 'Mössbauer Effect Methodology Vol. 1', Ed. I. J. Gruverman, Plenum Press, N.Y., 1965, p. 89.
[43] P. Debrunner, 'Mössbauer Effect Methodology Vol. 1', Ed. I. J. Gruverman, Plenum Press, N.Y., 1965, p. 97.
[44] R. Barloutaud, J-L. Picou, and C. Tzara, *Compt. rend.*, 1960, **250**, 2705.
[45] P. J. Black, D. E. Evans, and D. A. O'Connor, *Proc. Roy. Soc.*, 1962, **A270**, 168.
[46] P. J. Black, G. Longworth, and D. A. O'Connor, *Proc. Phys. Soc.*, 1964, **83**, 925.
[47] C. M. Lederer, J. M. Hollander, and I. Perlman, 'Table of Isotopes – 6th Ed.', John Wiley & Sons, Inc., N.Y., 1967.
[48] G. A. Chackett, K. F. Chackett, and B. Singh, *J. Inorg. Nuclear Chem.*, 1960, **14**, 138.
[49] S. L. Ruby and J. M. Hicks, *Rev. Sci. Instr.*, 1962, **33**, 27.
[50] D. A. O'Connor, *Nuclear Instr. Methods.*, 1963, **21**, 318.
[51] S. Margulies and J. R. Ehrman, *Nuclear Instr. Methods*, 1961, **12**, 131.
[52] S. Margulies, P. Debrunner, and H. Frauenfelder, *Nuclear Instr. Methods*, 1963, **21**, 217.
[53] J. D. Bowman, E. Kankeleit, E. N. Kaufmann, and B. Persson, *Nuclear Instr. Methods*, 1967, **50**, 13.
[54] K. Siegbahn (Ed.), 'Alpha, Beta, and Gamma Ray Spectroscopy', North Holland Publ., Amsterdam, 1965.
[55] N. Benczer-Koller and R. H. Herber, Chap. 2, 'Experimental Methods' in V. I. Goldanskii and R. H. Herber (Eds.), 'Chemical Applications of Mössbauer Spectroscopy', Academic Press, New York, 1968.
[56] C. E. Violet, 'The Mössbauer Effect and its Appl. to Chem.', *Adv. Chem. Ser.*, 1967, **68**, Chap. 10.
[57] L. Levy, L. Mitrani, and S. Ormandjiev, *Nuclear Instr. Methods*, 1964, **31**, 233.
[58] K. P. Mitrofanov and V. S. Shpinel, *Zhur. eksp. teor. Fiz.*, 1961, **40**, 983.
[59] D. A. O'Connor, N. M. Butt, and A. S. Chohan, *Rev. Mod. Phys.*, 1964, **36**, A 361.
[60] K. P. Mitrofanov, N. V. Illarionova, and V. S. Shpinel, *Pribory i Tekhn. Eksperim.*, 1963, **8**, 49; English Transl. in *Instr. Exptl. Tech. USSR*, 1963, **3**, 415.
[61] J. S. Eck, N. Hershkowitz, and J. C. Walker, *Bull. Am. Phys. Soc.*, 1965, **10**, 577.

[62] H. Frauenfelder, D. R. F. Cochran, D. E. Nagle, and R. D. Taylor, *Nuovo Cimento*, 1961, **19**, 183.
[63] N. Hershkowitz and J. C. Walker, *Nuclear Instr. Methods*, 1967, **53**, 273.
[64] E. Kankeleit, *Z. Physik*, 1961, **164**, 442.
[65] M. Kalvius, 'Mössbauer Effect Methodology Vol. 1', Ed. I. J. Gruverman, Plenum Press, N.Y., 1965, p. 163.
[66] D. P. Johnson, G. A. Erickson, and J. G. Dash, *Rev. Sci. Instr.*, 1968, **39**, 420.
[67] A. J. Nozik and M. Kaplan, *Anal. Chem.*, 1967, **39**, 854.
[68] F. Van der Woude and G. Boom, *Rev. Sci. Instr.*, 1965, **36**, 800.
[69] B. Sharon and D. Treves, *Rev. Sci. Instr.*, 1966, **37**, 1252.
[70] D. E. Nagle, H. Frauenfelder, R. D. Taylor, D. R. F. Cochran, and B. T. Matthias, *Phys. Rev. Letters*, 1960, **5**, 364.
[71] R. S. Preston, S. S. Hanna, and J. Heberle, *Phys. Rev.*, 1962, **128**, 2207.
[72] R. G. Barnes, R. L. Mössbauer, E. Kankeleit, and J. M. Poindexter, *Phys. Rev.*, 1964, **136**, A 175.
[73] R. Ingalls, 'Mössbauer Effect Methodology Vol. 1', Ed. I. J. Gruverman, Plenum Press, N.Y., 1965, p. 185.
[74] D. Pipkorn, C. K. Edge, P. Debrunner, G. de Pasquali, H. G. Drickamer, and H. Frauenfelder, *Phys. Rev.*, 1964, **135**, A 1604.
[75] P. Debrunner, R. W. Vaughan, A. R. Champion, J. Cohen, J. A. Moyzis, and H. G. Drickamer, *Rev. Sci. Instr.*, 1966, **37**, 1310.
[76] J. Heberle, 'Mössbauer Effect Methodology Vol. 2', Ed. I. J. Gruverman, Plenum Press, N.Y., 1966, p. 95.
[77] H. de Waard, *Rev. Sci. Instr.*, 1965, **36**, 1728.
[78] J. J. Spijkerman, F. C. Ruegg, and J. R. De Voe, International Atomic Energy Agency Technical Reports Series No. 50, Vienna, 1966, p. 53.
[79] T. E. Cranshaw and P. Reivari, *Proc. Phys. Soc.*, 1967, **90**, 1059.
[80] B. J. Duke and T. C. Gibb, *J. Chem. Soc. (A)*, 1967, 1478.
[81] S. W. Marshall, J. A. Nelson, and R. M. Wilenzick, *Comm. Assoc. Computing Machinery*, 1965, **8**, 313.
[82] G. M. Bancroft, A. G. Maddock, W. K. Ong, R. H. Prince, and A. J. Stone, *J. Chem. Soc. (A)*, 1967, 1966.
[83] D. B. Owen, 'Handbook of Statistical Tables', Pergamon Press, 1962.
[84] A. Gavron, D. Kedem, and T. Rothem, *Nuclear Instr. Methods*, 1968, **61**, 213.

3 | Hyperfine Interactions

It was shown in Chapter 1 that the Mössbauer effect produced monochromatic γ-radiation with a definition of the order of 1 part in 10^{12} and we now seek ways to use this extremely high precision to obtain chemical information. The key to the problem lies in the total interaction Hamiltonian for the atom, which contains terms relating to interactions between the nucleus on the one hand and the electrons (and hence the chemical environment) on the other. The Hamiltonian can be written as

$$\mathcal{H} = \mathcal{H}_0 + E_0 + M_1 + E_2 + \ldots \qquad 3.1$$

where $\mathcal{H}_0$ represents all terms in the Hamiltonian for the atom except the hyperfine interactions being considered; E_0 refers to electric monopole (i.e. Coulombic) interactions between the nucleus and the electrons; M_1 refers to magnetic dipole hyperfine interactions; and E_2 refers to electric quadrupole interactions. Higher terms are usually negligible.

The E_0 Coulombic interaction alters the energy separation between the ground state and the excited state of the nucleus, thereby causing a slight shift in the position of the observed resonance line. The shift will be different in various chemical compounds, and for this reason is generally known as the chemical isomer shift. It is also frequently referred to as the isomer shift or chemical shift, but in view of the earlier use of these terms in optical spectroscopy and nuclear magnetic resonance spectroscopy respectively, the longer expression is preferred. A less frequently used synonym is centre shift.

The electric quadrupole and magnetic dipole interactions both generate multiple-line spectra, and consequently can give a great deal of information. All three interactions can be expressed as the product of a nuclear term which is a constant for a given Mössbauer γ-ray transition and an electronic term which can be varied and related to the chemistry of the resonant absorber being studied.

3.1 Chemical Isomer Shift, δ

For many purposes it is adequate to consider the nucleus as a point charge which influences the electrons via the Coulombic potential. However, the nucleus has a finite volume, and this must be taken into account when considering nucleus–electron interactions because an *s*-electron wavefunction implies a non-zero electron charge density within the nuclear volume. During the course of a nuclear γ-transition, it is usual for the effective nuclear size to alter, thereby changing the nucleus–electron interaction energy. This change is only a minute fraction of the total Coulombic interaction but is dependent on chemical environment. Although we cannot measure this energy change directly, it is possible to *compare* values by means of a suitable reference which can be either the γ-ray source used in recording the Mössbauer spectrum or another absorber. The observed range of chemical isomer shifts for a given nuclide is frequently within an order of magnitude of the Heisenberg natural linewidth of the transition (i.e. $0 \cdot 1\Gamma$–10Γ). The Mössbauer resonance line recorded by velocity scanning may thus be measurably displaced from zero velocity if the chemical environment of the nuclide in the source and absorber differ. For convenience, the chemical isomer shift δ is usually quoted in mm s^{-1} rather than in energy units, but it is important to remember that, because of the relation $\delta = (v/c) \times E_\gamma$, a given displacement velocity, v, represents a different energy for each particular Mössbauer transition.

The mathematical expression of the above concepts is virtually identical to that for the optical isomeric shift in electronic spectra, and as such is well documented [1–5]. One form of the theory is given in the following paragraphs.

Electrons of charge $-e$ in the field of a point nucleus of charge $+Ze$ experience a normal Coulombic potential and the integrated electrostatic energy will be

$$E_0 = \frac{-Ze^2}{\kappa} \int \rho_e \frac{d\tau}{r} \qquad 3.2$$

where κ is the dielectric constant of a vacuum, r is the radial distance from the nucleus, and $-e\rho_e$ is the charge density of the orbital electrons in the volume element $d\tau$. For a spherical nucleus of finite radius R, equation 3.2 is still correct for $r > R$, but is invalid if $r < R$. The electrostatic energy in this case is

$$E = -e \int \rho_n \phi_e \, d\tau \qquad 3.3$$

where $e\rho_n$ is the charge density of the nucleus and ϕ_e is the electrostatic potential of the electrons. The integrand becomes equivalent to that in 3.2 for $r > R$. The change W in the electrostatic energy caused by the finite radius of the nucleus can therefore be expressed as

$$W = e \int_0^R \rho_n \phi_e \, d\tau - \frac{Ze^2}{\kappa} \int_0^R \rho_e \frac{d\tau}{r} \qquad 3.4$$

[*Refs. on p. 78*]

48 | HYPERFINE INTERACTIONS.

The probability density of an s-electron in the neighbourhood of a point charge is given in the Dirac theory by

$$\rho_e = \frac{2(\rho + 1)|\psi_s(0)|^2}{\Gamma^2(2\rho + 1)}\left(\frac{2Z}{a_H}\right)^{2\rho-2} r^{2\rho-2} \qquad 3.5$$

where $\rho = \sqrt{(1 - \alpha^2 Z^2)}$, $\alpha = e^2/\hbar c$, a_H is the first Bohr radius, $\psi_s(0)$ is the non-relativistic Schrödinger wavefunction at $r = 0$, and $\Gamma(n)$ is the gamma function

$$\Gamma(n) = \int_0^\infty e^{-x} x^{n-1}\,dx \quad n > 0 \qquad 3.6$$

Equation 3.5 ignores distortion of $\psi_s(0)$ by the finite nuclear size which will cause an overestimation of the shift. It should also be mentioned that there is a second possible contribution to electron density at the nucleus if $p_{\frac{1}{2}}$ electronic states of the atom are occupied. However, this term will always be much less than the equivalent s-term. Furthermore, the $p_{\frac{1}{2}}$ electronic state is not usually encountered in applications of Mössbauer spectroscopy and will therefore not be considered further.

The lack of knowledge of the precise charge distribution and potential inside the nucleus necessitates further approximations to equation 3.4. The customary method is to assume a potential equation which has the value of the point charge Coulombic potential at $r = R$ and also has the same gradient. This ensures that the potential is continuous. If the nuclear charge density is taken to be a uniform sphere, the change in the electrostatic energy between the point-charge and finite-radius models is given by

$$W = \frac{24\pi(\rho + 1)|\psi_s(0)|^2 Z e^2}{\kappa 2\rho(2\rho + 1)(2\rho + 3)\Gamma^2(2\rho + 1)}\left(\frac{2Z}{a_H}\right)^{2\rho-2} R^{2\rho} \qquad 3.7$$

The difference in energy caused by the nuclear radius change δR will then be

$$\Delta W = \frac{24\pi(\rho + 1)|\psi_s(0)|^2 Z e^2}{\kappa(2\rho + 1)(2\rho + 3)\Gamma^2(2\rho + 1)}\left(\frac{2Z}{a_H}\right)^{2\rho-2} R^{2\rho} \frac{\delta R}{R} \qquad 3.8$$

The appropriate potential curves are illustrated in Fig. 3.1.

The chemical isomer shift, δ, as measured in a Mössbauer experiment is a difference in energy between two chemical environments A and B and from 3.8 is seen to be

$$\delta = \frac{24\pi(\rho + 1)\{|\psi_s(0)_A|^2 - |\psi_s(0)_B|^2\}Z e^2}{\kappa(2\rho + 1)(2\rho + 3)\Gamma^2(2\rho + 1)}\left(\frac{2Z}{a_H}\right)^{2\rho-2} R^{2\rho} \frac{\delta R}{R} \qquad 3.9$$

This equation is rather intractable, and is commonly simplified by assuming

[Refs. on p. 78]

that $\rho = 1$. This is not strictly valid for the heavier elements (for $Z = 26$(Fe) $\rho = 0.98$ and for $Z = 80$(Hg) $\rho = 0.80$). With $\kappa = 1$ and $\rho = 1$

$$\delta = \frac{4}{5}\pi Ze^2\{|\psi_s(0)_A|^2 - |\psi_s(0)_B|^2\}R^2\frac{\delta R}{R} \qquad 3.10$$

The nucleus will usually be slightly non-spherical, and the chemical isomer

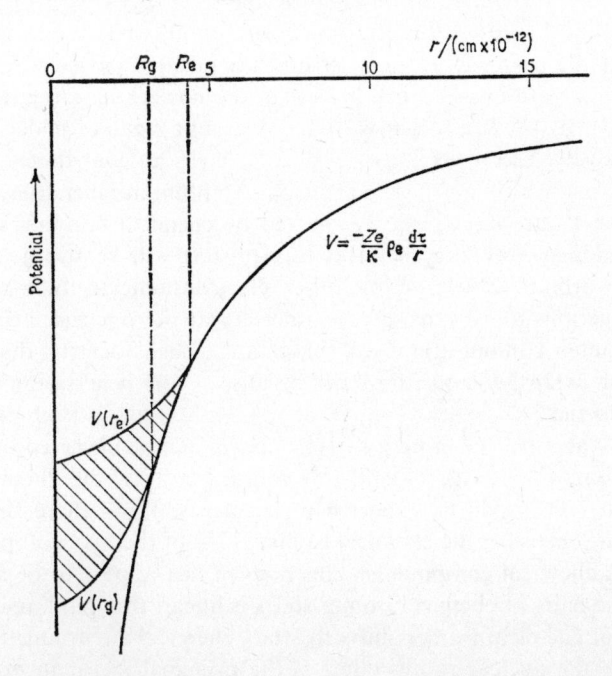

Fig. 3.1 The electrostatic potential of an electric charge of $-e\rho_e \, d\tau$ at distance r from a point nucleus is given by V, but when the nucleus has a finite radius, the potential curve within the sphere is different. The shaded area indicates the effect of a change in the nuclear radius from R_e to R_g.

shift is therefore frequently designated in terms of the differences in the root mean square radii of the excited and ground states.

$$\delta = \tfrac{2}{3}\pi Ze^2\{|\psi_s(0)_A|^2 - |\psi_s(0)_B|^2\}\{\langle R_e^2\rangle - \langle R_g^2\rangle\} \qquad 3.11$$

where $\quad \langle R_e^2\rangle - \langle R_g^2\rangle \equiv \dfrac{6}{5}R^2\dfrac{\delta R}{R} \quad$ and $\quad \delta\langle R^2\rangle/\langle R^2\rangle = \dfrac{2\delta R}{R}.$

$R = 1.2 \times A^{\frac{1}{3}}$ fm to a close approximation.

Equations 3.10 and 3.11 can be seen to be the product of a chemical term and a nuclear term. If the electron densities are known, the latter can be calculated or vice versa. In practice the nuclear term is a constant for a given transition, and for chemical applications the important equation is

$$\delta = \text{const} \times \{|\psi_s(0)_A|^2 - |\psi_s(0)_B|^2\} \qquad 3.12$$

A and B are normally the absorber and source respectively. $|\psi_s(0)|^2$ should not be confused with the number of s-electrons in the atomic environment. It is the s-electron density *at the nucleus*, and as such will be affected not only by the s-electron population but also by the screening effects of p-, d-, and f-electrons and by covalency and bond formation, that is by the chemical bonding of the atom. If $\delta R/R$ is positive, a positive chemical isomer shift implies an increase in s-electron density at the nucleus in going from source to absorber. If $\delta R/R$ is negative, the same shift signifies a decrease in s-electron density. Electrons in $1s$, $2s$, $3s$ . . ., shells all contribute to $|\psi_s(0)|^2$ but in decreasing amounts as the principal quantum number rises. However, the inner shells are not markedly affected by chemical bonding so that the principal influence on the chemical isomer shift will be by the outermost occupied s-orbital. Shielding by other electrons effectively increases the s-radial functions and decreases the s-density at the nucleus. For example, a $3d^64s^1$ outer configuration will have a higher s-density than $3d^74s^1$; likewise for $3s^23p^63d^5$ and $3s^23p^63d^6$ because of the penetration of the $3d$-orbitals into the $3s$.

In cases where two or more γ-ray resonances can be observed in the same chemical compounds, e.g. ^{127,129}I, ^{151,153}Eu, ^{191,193}Ir, the chemical terms in equation 3.11 would be expected to be identical and there should be a constant ratio between the chemical isomer shifts of the two isotopes in pairs of identical chemical compounds. This does in fact appear to be so because a plot of the pairs of chemical isomer shifts is linear, thus lending confidence to the use of chemical isomer shifts to study chemical environment.

Although the nuclear radius effect is the principal factor in producing a shift of the resonance line, there are two other factors, namely temperature and pressure, which are also acting, and it is often forgotten that the term chemical isomer shift is generally applied to the sum effect of all three. The temperature effect can be very important when measuring small differences in s-electron density at the nucleus and is considered in the next section.

3.2 Second-order Doppler Shift and Zero-point Motion

The existence of a relativistic temperature-dependent contribution to the chemical isomer shift was pointed out independently by Pound and Rebka [6] and by Josephson [7]. The emitting or absorbing atom is vibrating on its lattice site in the crystal. The frequency of oscillation about the mean position is of the order of 10^{12} per second, so that the average displacement during the Mössbauer event is zero. However, there is a term in the Doppler shift which depends on v^2, so that the mean value $\langle v^2 \rangle$ is non-zero.

[*Refs. on p. 78*]

SECOND-ORDER DOPPLER SHIFT AND ZERO-POINT MOTION | 51

The relativistic equation [8] for the Doppler effect on an emitted photon gives the observed frequency v' for a closing velocity v as

$$v' = v\left(1 - \frac{v}{c}\right)\left(1 - \frac{v^2}{c^2}\right)^{-\frac{1}{2}}$$

Hence

$$v' \simeq v\left(1 - \frac{v}{c}\right)\left(1 + \frac{v^2}{2c^2}\right) \qquad 3.13$$

where v is the frequency for a stationary system. The first-order term in velocity is a function of the velocity of the atom vibrating on its lattice site in the direction of the γ-ray and will average to zero over the lifetime of the state. The second-order term will not average to zero as it is a term in v^2 and is therefore independent of direction. It is usually referred to as a second-order Doppler shift, and for a Mössbauer resonance

$$v' = v\left(1 + \frac{\langle v^2 \rangle}{2c^2}\right) \qquad 3.14$$

Accordingly, there is a shift in the Mössbauer line given by

$$\frac{\delta E_\gamma}{E_\gamma} = \frac{\delta v}{v} = -\frac{\langle v^2 \rangle}{2c^2} \qquad 3.15$$

The kinetic energy per mole of the solid, $\frac{1}{2}M\langle v^2 \rangle$, can be related to the total energy of the solid per unit mass, U, ($\frac{1}{2}M\langle v^2 \rangle = \frac{1}{2}MU$) so that

$$\frac{\delta v}{v} = -\frac{U}{2c^2} \qquad 3.16$$

The temperature dependence of this quantity is thus related to the molar heat capacity at constant pressure, C_p,

$$\frac{1}{v}\left(\frac{\partial v}{\partial T}\right)_P = -\frac{C_p}{2Mc^2} \qquad 3.17$$

Equation 3.17 is not very convenient in this form as it is far easier to measure the molar heat capacity at constant volume, C_v.

$$\frac{1}{v}\left(\frac{\partial v}{\partial T}\right)_P = \frac{1}{v}\left(\frac{\partial v}{\partial T}\right)_V + \frac{1}{v}\left(\frac{\partial v}{\partial \ln V}\right)_T \left(\frac{\partial \ln V}{\partial T}\right)_P$$

$$\frac{1}{v}\left(\frac{\partial v}{\partial T}\right)_P = -\frac{C_v}{2Mc^2} + \frac{1}{v}\left(\frac{\partial \ln v}{\partial \ln V}\right)_T \left(\frac{\partial \ln V}{\partial T}\right)_P \qquad 3.18$$

To a first approximation, C_p and C_v are similar in value.

It is instructive to consider the description of $\delta v/v$ in terms of the lattice dynamics of the solid. The following treatment is a simplification of that by

[Refs. on p. 78]

Hazony [9]. An harmonic approximation is used, and the average energy associated with each atom is

$$\tfrac{1}{2}M\langle v^2\rangle = 3(n_j + \tfrac{1}{2})\hbar\omega_j \qquad 3.19$$

where $n_j = [\exp(\hbar\omega_j/kT) - 1]^{-1}$ and ω_j is the oscillation frequency. We wish to sum over all possible frequencies and modes of vibration, so that

$$\frac{\delta v}{v} = -\frac{3}{2Mc^2}\sum_j A_j{}^2\hbar\omega_j(\tfrac{1}{2} + n_j) \qquad 3.20$$

where the $A_j{}^2$ terms are weighting factors such that $\sum_j A_j{}^2 = 1$. M is the atomic mass of the Mössbauer nuclide. The classical high-temperature limit of this expression is

$$\frac{\delta v}{v} = -\frac{3}{2}\frac{RT}{Mc^2} \qquad 3.21$$

From the point of view of intercomparison of chemical shifts, it is useful to consider the general equation as $T \to 0$. There is a zero-point motion term given by

$$\frac{\delta v_0}{v} = -\frac{3}{4}\frac{1}{Mc^2}\sum_j A_j{}^2\hbar\omega_j \qquad 3.22$$

The magnitude of the zero-point motion will be dependent on the exact mode of vibration in the crystal, so that $\delta v_0/v$ will not in general be the same in all compounds.

If we adopt the simple picture of an Einstein solid such that there is one frequency only given by ω_E then

$$\frac{\delta v_0}{v} = -\frac{3}{4}\frac{\hbar\omega_E}{Mc^2} \qquad 3.23$$

In the Debye model ω_j can have any frequency between 0 and ω_D with a probability of $9N\omega_j{}^2/\omega_D{}^3$, the average value of $\hbar\omega_j$ being given by $\tfrac{3}{4}\hbar\omega_D$. Hence

$$\frac{\delta v_0}{v} = -\frac{9}{16}\frac{\hbar\omega_D}{Mc^2} \qquad 3.24$$

or using the Debye temperature defined as $\hbar\omega_D = k\theta_D$

$$\frac{\delta v_0}{v} = -\frac{9}{16}\frac{k\theta_D}{Mc^2} \qquad 3.25$$

The zero-point motion term is proportional to the Debye temperature of the solid if this vibrational model is valid. In practice, of course, the vibrational

[Refs. on p. 78]

modes of chemical compounds are very complex, and Hazony [9] and Epstein [10] have both pointed out cases where both the temperature dependence and zero-point contributions are very different in apparently similar compounds. The second-order Doppler shift for the iron-resonance in a solid of Debye temperature 200 K is illustrated in Fig. 3.2. These considerations are very

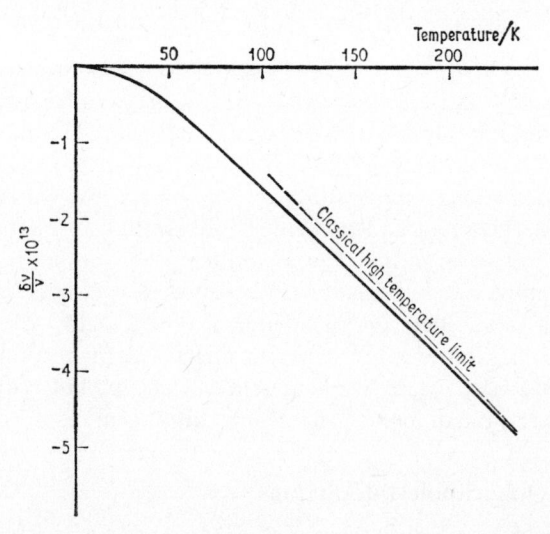

Fig. 3.2 The second-order Doppler shift calculated for a Debye model with $\theta_D = 200$ K and an atomic mass of 55·9.

relevant to the interpretation of small differences in chemical isomer shift. As a general rule, however, if data to be compared are all obtained at the same low temperature, the respective zero-point and temperature-dependent terms will largely cancel each other out.

3.3 Effect of Pressure on the Chemical Isomer Shift

The chemical isomer shift is the sum of two terms, the one being proportional to the s-electron density at the nucleus, and the other being the second-order Doppler shift. Both of these will be influenced by a change in pressure, but it is the overall effect on the former which is the more significant.

The measured pressure coefficient at constant temperature can be represented [11] by

$$\frac{1}{E_\gamma}\left(\frac{\partial \delta}{\partial P}\right)_T = \frac{1}{\nu}\left(\frac{\partial \nu}{\partial P}\right)_T$$

$$\left(\frac{\partial \nu}{\partial P}\right)_T = K\left(\frac{\partial |\psi_s(0)|^2}{\partial \ln V}\right)_T\left(\frac{\partial \ln V}{\partial P}\right)_T + \left(\frac{\partial \nu_{\text{rel}}}{\partial \ln V}\right)_T\left(\frac{\partial \ln V}{\partial P}\right)_T \quad 3.26$$

The first term is the effect of pressure on the electron density at the nucleus and the second term describes the relativistic effect of altering the mean vibrational energy. The latter can be simply expressed, using the Debye model for the vibrational energy of $E_{\text{vib}} = 3kT[1 + \frac{1}{20}(\theta_D/T)^2 + \ldots]$, as

$$\frac{1}{\nu}\left(\frac{\partial \nu_{\text{rel}}}{\partial \ln V}\right)_T\left(\frac{\partial \ln V}{\partial P}\right)_T \approx -\frac{3}{20}\frac{k\theta_D}{Mc^2}\left(\frac{\theta_D}{T}\right)\left(\frac{\partial \ln \theta_D}{\partial \ln V}\right)_T\left(\frac{\partial \ln V}{\partial P}\right)_T \quad 3.27$$

The quantity $-(\partial \ln \theta_D/\partial \ln V)_T$ can be calculated from Gruneisen's constant γ and $(\partial \ln V/\partial P)_T$ is the compressibility. The magnitude of the expression will be small for a low Debye temperature of the lattice, and is generally less than the electronic term [11, 12].

The effect of pressure on $|\psi_s(0)|^2$ will depend on the detailed electronic structure of the material, and is therefore of use in investigating the latter. Examples can be found in the chapters on the ^{57}Fe isotope.

The total spin-density imbalance is also affected by change in volume, so that there is a small effect on an internal magnetic field with pressure [13] (see Section 3.5). Compression can also affect the quadrupole splitting by decreasing the radial extent of the valence electrons, but redistribution of electrons between the different orbitals may also occur.

3.4 Electric Quadrupole Interactions

The existence of an electric quadrupole interaction is one of the most useful features of Mössbauer spectroscopy. The theory is closely related to that used in nuclear quadrupole resonance spectroscopy [14, 15]. Any nucleus with a spin quantum number of greater than $I = \frac{1}{2}$ has a non-spherical charge distribution, which if expanded as a series of multipoles contains a quadrupole term. The magnitude of the charge deformation is described as the nuclear quadrupole moment Q, given by

$$eQ = \int \rho r^2 (3\cos^2\theta - 1)\,d\tau \quad 3.28$$

where e is the charge of the proton, ρ is the charge density in a volume element $d\tau$, which is at a distance r from the centre of the nucleus and making an included angle θ to the nuclear spin quantisation axis. The sign of Q depends on the shape of the deformation. A negative quadrupole moment indicates that the nucleus is oblate or flattened along the spin axis, whereas for a positive moment it is prolate or elongated.

In a chemically bonded atom, the electronic charge distribution is usually not spherically symmetric. The electric field gradient at the nucleus is defined as the tensor $E_{ij} = -V_{ij} = -(\partial^2 V/\partial x_i x_j)(x_i, x_j = x, y, z)$ where V is the electrostatic potential. It is customary to define the axis system of the resonant atom so that $V_{zz} = eq$ is the maximum value of the field gradient. The orientation of the nuclear axis with respect to the principal axis, z, is quantised. There is an interaction energy between Q and eq which is different for each possible orientation of the nucleus.

The Laplace equation requires that the electric field gradient be a traceless tensor, i.e. the sum of the second derivatives of the electrostatic potential vanish:

$$V_{zz} + V_{xx} + V_{yy} = 0 \qquad 3.29$$

Consequently, only two independent parameters are needed to specify the electric field gradient completely, and the two which are usually chosen are V_{zz} and an asymmetry parameter η defined as

$$\eta = \frac{(V_{xx} - V_{yy})}{V_{zz}} \qquad 3.30$$

Using the convention that $|V_{zz}| > |V_{yy}| \geqslant |V_{xx}|$ ensures that $0 \leqslant \eta \leqslant 1$.

The Hamiltonian describing the interaction can be written as

$$\mathcal{H} = \frac{eQ}{2I(2I-1)}(V_{zz}\hat{I}_z^2 + V_{xx}\hat{I}_x^2 + V_{yy}\hat{I}_y^2) \qquad 3.31$$

where I is the nuclear spin and $\hat{I}_z$, $\hat{I}_x$, and $\hat{I}_y$ are the conventional spin operators. Using equation 3.30 the Hamiltonian therefore becomes

$$\mathcal{H} = \frac{e^2qQ}{4I(2I-1)}[3\hat{I}_z^2 - I(I+1) + \eta(\hat{I}_x^2 - \hat{I}_y^2)]$$

or

$$\mathcal{H} = \frac{e^2qQ}{4I(2I-1)}[3\hat{I}_z^2 - I(I+1) + \frac{\eta}{2}(\hat{I}_+^2 + \hat{I}_-^2)] \qquad 3.32$$

where $\hat{I}_+$ and $\hat{I}_-$ are shift operators.

The simplest case to consider is when the electric field gradient has axial symmetry, i.e. $V_{xx} = V_{yy}$ and $\eta = 0$. The matrix elements for the nucleus of spin I are given by

$$\langle I_z' | \mathcal{H} | I_z \rangle = \frac{e^2qQ}{4I(2I-1)}[3I_z^2 - I(I+1)]\delta_{I_z'I_z} \qquad 3.33$$

$|I_z\rangle$ and $|I_z'\rangle$ are two different quantum states of the orientation and $\delta_{I_z'I_z}$ is a Krönecker delta such that $\delta_{I_z'I_z} = 0$ unless $I_z' = I_z$. In other words the matrix is diagonal in form and the energy levels are given directly by

$$E_Q = \frac{e^2qQ}{4I(2I-1)}[3I_z^2 - I(I+1)] \qquad 3.34$$

Instead of a single energy level we now have a series of Kramers' doublets identified by the $|I_z|$ quantum number. For $I = \frac{1}{2}$ there is only one level, but for $I = \frac{3}{2}$ there are two distinct eigenvalues of energy $+(e^2qQ/4)$ (for $I_z = \pm\frac{3}{2}$) and $-(e^2qQ/4)$ (for $I_z = \pm\frac{1}{2}$).

In general, a Mössbauer transition occurs between two nuclear levels, each of which may have a nuclear spin and quadrupole moment. This means that both the ground-state and excited-state levels may show a quadrupole interaction. Because the energy separations due to these quadrupole interactions are very small, all the levels connected with a given nuclear spin are equally populated at temperatures above 1 K. A change in the I_z quantum number is allowed during the γ-ray transition, and we shall discuss the relative probabilities of transitions between sublevels more fully in Section 3.7. Basically the laws of conservation of angular momentum and of parity lead to the formulation of definite selection rules which characterise the transition between the two states and these ensure that there is a high probability for transitions in which the change in the z quantum numbers $[(I_z)_e - (I_z)_g] = m$ is 0 or ± 1. The very important example of a γ-transition between states with $I = \frac{3}{2}$ and $I = \frac{1}{2}$ results in two transitions from the $\{I = \frac{3}{2} | \pm \frac{3}{2}\rangle\}$ and $\{I = \frac{3}{2} | \pm \frac{1}{2}\rangle\}$ sublevels to the unsplit $\{I = \frac{1}{2} | \pm \frac{1}{2}\rangle\}$ level. The resultant spectrum will comprise two lines, which are in fact of equal intensity if the sample is an isotropic powder, and the energy separation between the two lines is $|(e^2qQ/2)|$. The centroid of the doublet corresponds to the energy of the γ-transition without a quadrupole interaction so that we can still measure the chemical isomer shift. The energy level schemes and observed spectra for an $I = \frac{3}{2} \to I = \frac{1}{2}$ transition and an $I = \frac{5}{2} \to I = \frac{7}{2}$ transition are represented in Fig. 3.3.

A lack of axial symmetry in the electric field gradient introduces matrix elements which are off-diagonal with $[(I_z)_e - (I_z)_g] = \pm 2$. Exact solutions of the secular equations for the eigenvalues can only be given for $I = \frac{3}{2}$. These are

$$E_Q = \frac{e^2qQ}{4I(2I-1)}[3I_z^2 - I(I+1)]\left(1 + \frac{\eta^2}{3}\right)^{\frac{1}{2}} \qquad 3.35$$

For higher spin states, convergent series can be calculated by perturbation procedures, or else the energies can be solved accurately by matrix diagonalisation on a computer. It is interesting to note, however, that as η can never be greater than unity (equation 3.30), the maximum difference between the quadrupole splittings derived from equations 3.34 and 3.35 is about 16%. A method for analysing spectra with a non-zero value of η has been described [16].

The magnitude of the quadrupole interaction is a product of two factors, eQ is a nuclear constant for the resonant isotope, while eq is a function of chemical environment. For a $\frac{3}{2} \to \frac{1}{2}$ transition it is not possible to determine

[Refs. on p. 78]

the sign of e^2qQ or the magnitude of η from the line positions alone. This is not the case for higher spin states where the sign of e^2qQ can be uniquely determined from the unequally spaced lines. If both I_e and I_g are greater than $\frac{1}{2}$, there is little difficulty in estimating eq as the ground-state quadrupole

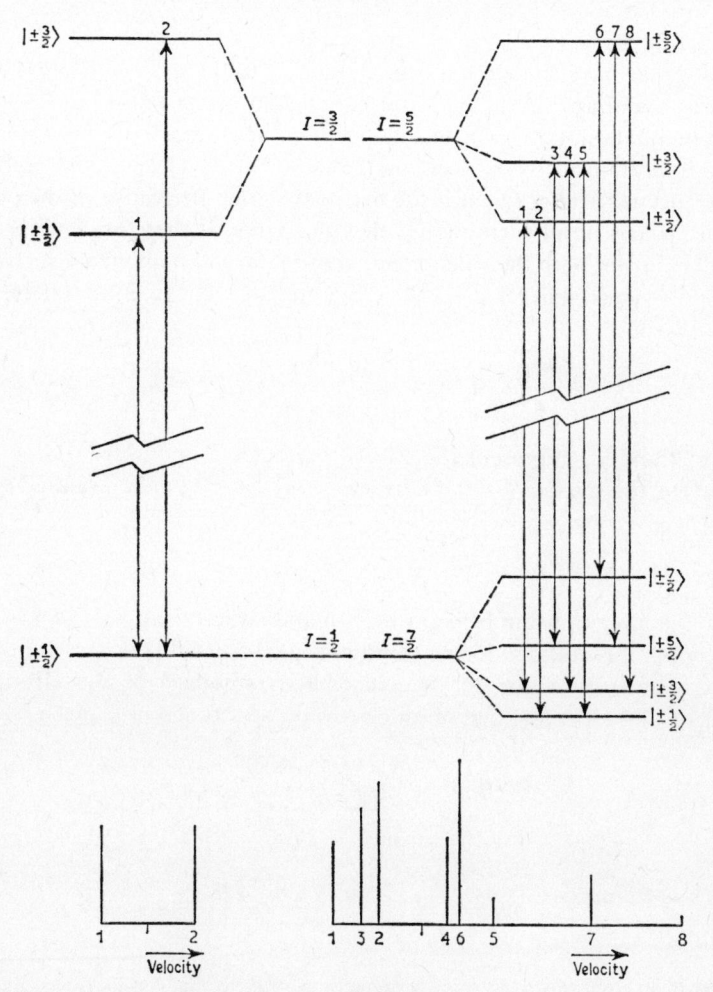

Fig. 3.3 Typical energy level schemes and observed spectra for $\frac{3}{2} \to \frac{1}{2}$ and $\frac{5}{2} \to \frac{7}{2}$ transitions with quadrupole hyperfine interactions.

moment will be accurately known from other measurements. If $I_g = 0$ or $\frac{1}{2}$ then Q for the excited state is usually calculated from estimated values of eq in chemical compounds and will not be known with any great accuracy.

Although the quadrupole coupling constant e^2qQ and asymmetry parameter η can easily be evaluated from a Mössbauer spectrum, it is much more

[*Refs. on p. 78*]

difficult to relate these parameters to the electronic structure which generates them. Although frequent reference is made to the sign of the electric field gradient, this is misleading because the latter is a tensor quantity. What is actually meant is the sign of the principal component, V_{zz}. Confusion over the sign convention used can be avoided by considering the coupling constant e^2qQ. If the sign of Q is known, that of q is immediately implied. Throughout this work we have deliberately chosen to refer to the sign of e^2qQ rather than of eq to avoid any ambiguity, and have taken considerable care in quoting published data not given in this convention. The observed sign of e^2qQ may be an important factor in deciding the origin of the electric field gradient.

The electric field gradient is the negative second derivative of the potential at the nucleus of all surrounding electric charge. It therefore embraces contributions from both the valence electrons of the atom and from surrounding ions. In ionic complexes it is customary to consider these separately, and to write q as

$$\frac{V_{zz}}{e} = q = (1 - R)q_{\text{ion}} + (1 - \gamma_\infty)q_{\text{latt}} \qquad 3.36$$

where R and γ_∞ represent the effects of shielding and antishielding respectively* of the nucleus by the core electrons [17, 18]. The asymmetry parameter is given by

$$(V_{xx} - V_{yy})/e = \eta q = (1 - R)\eta_{\text{ion}}q_{\text{ion}} + (1 - \gamma_\infty)\eta_{\text{latt}}q_{\text{latt}} \qquad 3.37$$

Although a relationship between the ion and lattice terms has been suggested [17], opinion is still divided as to whether this is valid [19].

The numerical value of the principal component of the electric field gradient ($V_{zz} = -E_{zz}$) due to an electronic wavefunction is given by

$$V_{zz} = eq = -e\left\langle \psi \left| \frac{3\cos^2\theta - 1}{r^3} \right| \psi \right\rangle \qquad 3.38$$

and

$$\eta q = \left\langle \psi \left| \frac{3\sin^2\theta \cos 2\phi}{r^3} \right| \psi \right\rangle \qquad 3.39$$

* When an atom which has some asymmetric p- or d-electron density is placed in a non-homogeneous electric field, the unbalanced electron charge generates an electric field gradient q_{ion}. Simultaneously, this spacial asymmetry of the electron charge results in a polarisation of the inner, spherically symmetrical electron shells which in turn induces an electric field gradient of opposite sign on the nucleus. This is represented by the Sternheimer shielding factor $(1 - R)$ where R is frequently in the range $0 < R < 1$. Likewise, if in the absence of the atomic electrons, an electric field gradient of q_{latt} is set up by the interaction of the crystal electric field on the nucleus, then this same crystal field will induce a further gradient by polarisation of the electrons in the inner core of the atom, give a resultant effective electric field gradient of $q_{\text{latt}}(1 - \gamma_\infty)$. The term γ_∞ is called the Sternheimer antishielding factor and can take on positive or negative values which, in some cases, are calculated to be as large as 10^2.

[Refs. on p. 78]

A table of the appropriate values for some standard wavefunctions is given in Table 3.1. The $\langle r^{-3} \rangle$ value is the expectation value of $1/r^3$ for the appro-

Table 3.1 Magnitude of q and η for various atomic orbitals

Orbital	q	η
p_z	$-\frac{4}{5}\langle r^{-3}\rangle$	0
p_x	$+\frac{2}{5}\langle r^{-3}\rangle$	-3
p_y	$+\frac{2}{5}\langle r^{-3}\rangle$	$+3$
$d_{x^2-y^2}$	$+\frac{4}{7}\langle r^{-3}\rangle$	0
d_{z^2}	$-\frac{4}{7}\langle r^{-3}\rangle$	0
d_{xy}	$+\frac{4}{7}\langle r^{-3}\rangle$	0
d_{xz}	$-\frac{2}{7}\langle r^{-3}\rangle$	$+3$
d_{yz}	$-\frac{2}{7}\langle r^{-3}\rangle$	-3

priate function, and it follows that a 4p-electron for example will give a smaller value of q than a 3p-electron and that a 3d-electron will also give a smaller value than for the corresponding 3p-electron. An s-electron has spherical symmetry and gives no electric field gradient.

The total electric field gradient from the valency electrons of the ion can be obtained by summing the appropriate wavefunction contributions. If the orbital populations alter with temperature because of excitation to low-lying higher states, then the electric field gradient will probably show a strong temperature dependence. This point is developed more fully on p. 99.

The lattice term q_{latt} can be similarly evaluated as the sum of contributions from individual charges Z_i and can be written

$$q_{\text{latt}} = \sum_i Z_i \frac{3\cos^2\theta_i - 1}{r_i^3} \qquad 3.40$$

Generally the valence term is the major contribution to the electric field gradient unless the ion has the high intrinsic symmetry of an S-state ion such as high-spin $Fe^{3+}(d^5)$. In the latter case the lattice term will be dominant.

In molecular complexes the same arguments apply except that the components contributing to the electric field gradient are now molecular orbitals, and unless the mathematical forms of these are available, it is rarely possible to do more than generalise by semi-empirical chemical reasoning. The problems associated with the interpretation of the quadrupole coupling constant will be discussed in more detail in later chapters dealing with specific chemical applications.

3.5 Magnetic Hyperfine Interactions

The third important hyperfine interaction is the nuclear Zeeman effect [20]. This will occur if there is a magnetic field at the nucleus. The magnetic field can originate either within the atom itself, within the crystal via exchange interactions, or as a result of placing the compound in an externally applied magnetic field; for the moment, however, it is only necessary to consider that there is a magnetic field with a flux density of H gauss ($= H \times 10^{-4}$ tesla) and that its direction defines the principal z axis.

The Hamiltonian describing the magnetic dipole hyperfine interaction is

$$\mathcal{H} = -\boldsymbol{\mu}\cdot\boldsymbol{H} = -g\mu_N \boldsymbol{I}\cdot\boldsymbol{H} \qquad 3.41$$

where μ_N is the nuclear Bohr magneton ($e\hbar/2Mc$), $\boldsymbol{\mu}$ is the nuclear magnetic moment, $\boldsymbol{I}$ is the nuclear spin, and g is the nuclear g-factor [$g = \mu/(I\mu_N)$]. μ_N has the value of $5{\cdot}04929 \times 10^{-24}$ ergs per gauss or $5{\cdot}04929 \times 10^{-27}$ J T^{-1}. The matrix elements are easy to evaluate using the spin operator form of $\mathcal{H} = -g\mu_N \hat{I}_z H$ and result in eigenvalues of

$$E_m = \frac{-\mu H m_I}{I} = -g\mu_N H m_I \qquad 3.42$$

where m_I is the magnetic quantum number representing the z component of I (i.e. $m_I = I, I-1 \ldots -I$). The magnetic field splits the nuclear level of spin I into $(2I+1)$ equi-spaced non-degenerate substates. As already seen for quadrupole spectra, the Mössbauer transition can take place between different nuclear levels if the change in the m_I value is 0, ± 1, or additionally in some cases ± 2. The allowed transitions for a $\tfrac{3}{2} \to \tfrac{1}{2}$ Mössbauer γ-ray are illustrated in Fig. 3.4.

In a similar way to the chemical isomer shift and quadrupole splitting, the magnetic hyperfine effect is the product of a nuclear term which is a constant for a given Mössbauer transition, and a magnetic field which, as will now be shown, can be produced by the electronic structure.

The magnetic field at the nucleus can originate in several ways [21–23]. A general expression would be

$$H = H_0 - DM + \tfrac{4}{3}\pi M + H_S + H_L + H_D \qquad 3.43$$

H_0 is the value of the magnetic field at the nucleus generated by an external magnet and is effectively zero away from a large magnet. The next term, $-DM$, is the demagnetising field and $\tfrac{4}{3}\pi M$ is the Lorentz field (the coefficient being strictly applicable for cubic symmetry only) but both are small. H_S arises as a result of the interaction of the nucleus with an imbalance in the s-electron spin density at the nucleus, and can be written as

$$H_S = -\left(\frac{16\pi}{3}\right)\mu_B \left\langle \sum_i S_i^z \,\delta(r_i) \right\rangle \qquad 3.44$$

The expression contained in the angular brackets is the expectation value of the spin density, r_i being the radial coordinate of the ith electron. μ_B is the

[Refs. on p. 78]

Bohr magneton. H_S is usually referred to as the Fermi contact term. Its actual origin may be from intrinsic impairing of the actual s-electrons, or indirectly as a result of polarisation effects on filled s-orbitals. These can occur if the atom has unpaired electrons in d- or f-orbitals, or if it is chemically bonded to such an atom. Intuitively one can see that the interaction of an unpaired

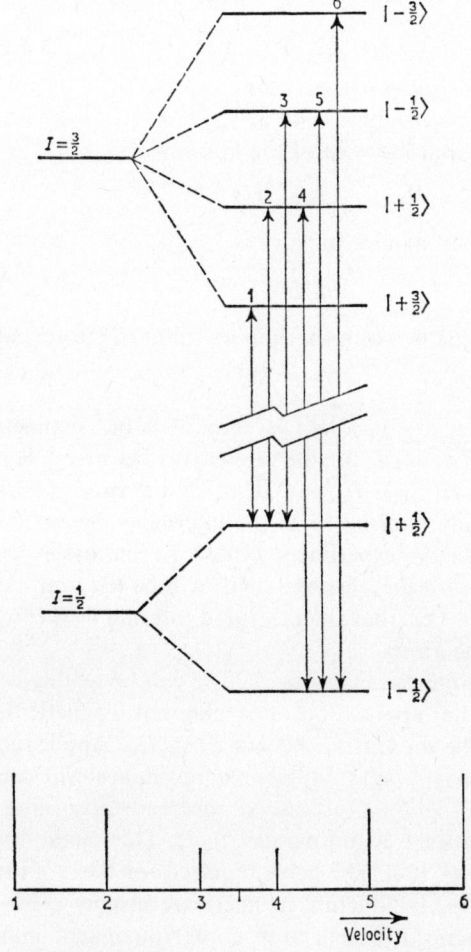

Fig. 3.4 Magnetic splitting for a $\frac{3}{2} \to \frac{1}{2}$ transition. The relative inversion of the $\frac{3}{2}$ and $\frac{1}{2}$ multiplets signifies a change in sign of the nuclear magnetic moment. Only transitions for a change in m_I of 0, ± 1 are shown.

d-electron with the s-electrons of parallel spin will be different to that with the s-electrons of opposed spin. The result is a slight imbalance of spin density at the nucleus. In the case of metals, direct conduction-electron polarisation as well as indirect core-polarisation effects may be important.

[*Refs. on p. 78*]

As mentioned at the end of Section 3.3, application of pressure can also cause small changes in the imbalance of the spin density and thus in H.

If the orbital magnetic moment of the parent atom is non-zero, there is a further term in equation 3.43 given by either

$$H_\text{L} = -2\mu_\text{B}\langle r^{-3}\rangle\langle L\rangle$$

or
$$H_\text{L} = -2\mu_\text{B}\langle r^{-3}\rangle(g-2)\langle S\rangle \qquad 3.45$$

$\langle L \rangle$ and $\langle S \rangle$ are the appropriate expectation values of the orbital and spin angular momenta, and g is the usual Landé splitting factor.

The final term in equation 3.43 arises from the dipolar interaction of the nucleus with the spin moment of the atom

$$H_\text{D} = -2\mu_\text{B}\langle 3r(s.r)r^{-5} - sr^{-3}\rangle \qquad 3.46$$

which in the case of axial symmetry is

$$H_\text{D} = -2\mu_\text{B}\langle S\rangle\langle r^{-3}\rangle\langle 3\cos^2\theta - 1\rangle \qquad 3.47$$

where θ is the angle between the spin axis and the principal axis. Since

$$e^2qQ = 2e^2Q\langle r^{-3}\rangle\langle 3\cos^2\theta - 1\rangle \,|\,\langle S\rangle\,|$$

we have that $|H_\text{D}| = \mu_\text{B} q$. H_D is zero in cubic symmetry for transition elements, but can be large for the rare earths because L is not quenched.

The terms H_S, H_L, and H_D can all be of the order of 10^4–10^5 gauss and their sum is usually referred to as the internal magnetic field. Considerable data are available from experiments with ^{57}Fe and other nuclides but a more detailed discussion of the observed fields will be left until later chapters. It is already clear, however, that the measured internal field can be related to the orbital state of the atom.

The sign of an internal magnetic field H can be readily determined. Equation 3.43 shows that application of an external magnetic field H_0 alters the effective field at the nucleus to the sum $H + H_0$. Application of an external field to a magnetically split Mössbauer resonance will cause the apparent internal magnetic field to increase or decrease according to whether the applied field is parallel or antiparallel to H. This method was described by Hanna *et al.* in 1960 [24]. The fields required for this are rather large, about 30–50 kG, and superconducting magnets are usually used. If there are two or more field directions present as in antiferromagnetic materials, it may be possible to distinguish the sublattice resonance lines by the opposite effects of the applied field. This method will fail if the magnetic interaction is highly anisotropic so as to prevent complete polarisation of the ordered spins by the external field.

From the foregoing description of the origins of the internal magnetic field, it might be assumed that all compounds containing unpaired valence electrons would show a hyperfine magnetic splitting effect. There is, however, another factor which has not been considered, namely that the Hamiltonian in equation 3.41 contains I and H as a vector product, and that the observation time-scale is of the order of 10^{-8} s. The electronic spins which generate H are subject to changes of direction, known as electronic spin relaxation. In paramagnetic compounds, the spin relaxation is usually rapid and results in H having a time-average of zero so that no magnetic splitting is seen. The major exceptions are found with the rare earths or with magnetically dilute solid solutions. When cooperative phenomena such as ferromagnetism or antiferromagnetism operate, the relaxation rates are effectively slower and a splitting will be recorded. There are, however, intermediate possibilities where the electronic spins are relaxing on a time-scale comparable with that of the nuclear transition, and these result in more complicated spectra (see Section 3.8). Such systems embrace both ordered materials with unusually fast relaxation and also paramagnetic compounds with slow relaxation.

Because the internal magnetic field of a magnetically ordered material is generally proportional to the magnetisation, its temperature dependence will reflect the latter and follows a Brillouin function, becoming zero at the Curie or Néel temperature. In cases where two or more distinct magnetic lattices are present, the Mössbauer spectrum will give the internal field at each individual site, whereas the bulk magnetisation is an average effect. This differentiation is particularly significant for antiferromagnetic compounds where the Mössbauer spectrum can conclusively confirm that magnetic ordering is present.

3.6 Combined Magnetic and Quadrupole Interactions

The chemical isomer shift with magnetic or quadrupole split spectra merely causes a uniform shift of all the resonance lines without altering their separation. Both the magnetic and quadrupole interactions, however, are direction-dependent effects, and consequently when both are present the general interpretation of the spectrum can be quite complex.

If the electric field gradient tensor is axially symmetric and its principal axis makes an angle θ with the magnetic axis, then a relatively simple solution exists providing that $e^2qQ \ll \mu H$; in this case the quadrupole interaction can be treated as a first-order perturbation to the magnetic interaction. The eigenvalues are

$$E = -g\mu_N H m_I + (-1)^{|m_I|+\frac{1}{2}} \frac{e^2 q Q}{4}\left(\frac{3\cos^2\theta - 1}{2}\right) \qquad 3.48$$

and the level splitting is illustrated for a $\frac{3}{2} \to \frac{1}{2}$ decay in Fig. 3.5. The angle θ is not determinable from the spectrum in this case, so that e^2qQ cannot be calculated unless the direction of magnetisation relative to the symmetry axis

[Refs. on p. 78]

64 | HYPERFINE INTERACTIONS

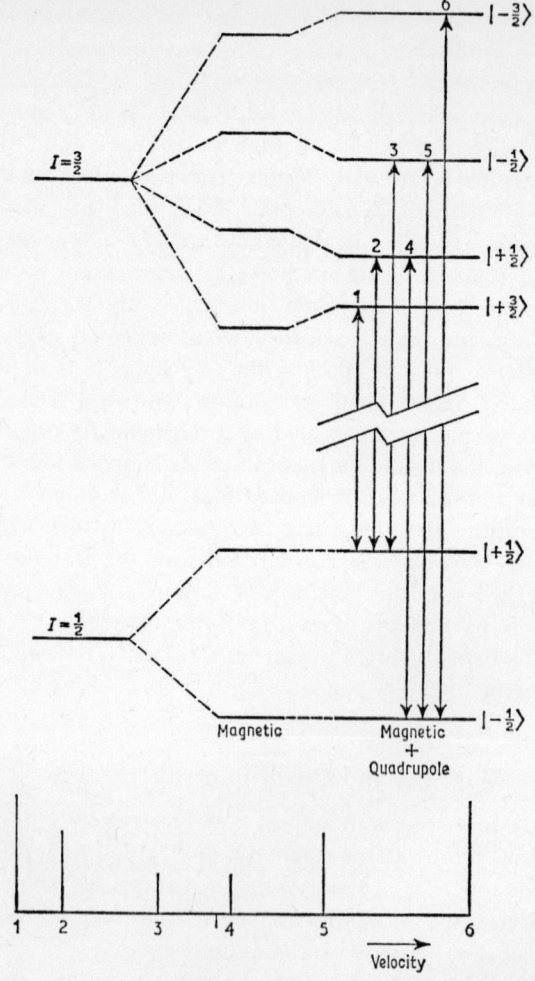

Fig. 3.5 The effect of a first-order quadrupole perturbation on a magnetic hyperfine spectrum for a $\frac{3}{2} \to \frac{1}{2}$ transition. Lines 1,2 and 5,6 have equal separations only when there is no quadrupole effect acting, or when $\cos \theta = 1/\sqrt{3}$.

is measured by other means. If $\cos \theta = 1/\sqrt{3}$ then the quadrupole interaction fortuitously appears to be absent. The expression

$$\frac{e^2qQ}{4}\left(\frac{3\cos^2\theta - 1}{2}\right)$$

is often denoted as ε if θ is unknown.

If the electric field gradient tensor is not axially symmetric but the mag-

[*Refs.* on p. 78]

COMBINED MAGNETIC AND QUADRUPOLE INTERACTIONS | 65

netic axis lies along one of its principal axes, then the excited-state splitting for $I = \frac{3}{2}$ gives the four energies [25]

$$\tfrac{1}{2}g\mu_N H \pm \frac{e^2qQ}{4}\left[\left(1 + \frac{4g\mu_N H}{e^2qQ}\right)^2 + \frac{\eta^2}{3}\right]^{\frac{1}{2}}$$

and

$$-\tfrac{1}{2}g\mu_N H \pm \frac{e^2qQ}{4}\left[\left(1 - \frac{4g\mu_N H}{e^2qQ}\right)^2 + \frac{\eta^2}{3}\right]^{\frac{1}{2}} \qquad 3.49$$

In many of the observed spectra referred to in later chapters, it is not possible to use one of the simplifications described above. Solutions are not obtained analytically but by full mathematical analysis using a digital computer. A method of computing the spectrum appropriate to any given symmetry has been given by Gabriel et al. [26, 27]. Similar texts have been specifically directed to the solution of $\frac{3}{2} \to \frac{1}{2}$ ^{57}Fe spectra [28–30].

One must return to the complete Hamiltonian of

$$\mathscr{H} = -g\mu_N \hat{I}_z' H + \frac{e^2qQ}{4I(2I-1)}[3\hat{I}_z^2 - I(I+1) + \eta(\hat{I}_x^2 - \hat{I}_y^2)] \qquad 3.50$$

$\hat{I}_z'$ and $\hat{I}_z$ are defined with respect to two different axis systems. If we define a new coordinate system in which the principal z axis is now the γ-ray direction, it is possible to generate new spin operators $\hat{S}_x$, $\hat{S}_y$, and $\hat{S}_z$ defined in the new system. The magnetic interaction can be related to it by using direction cosines:

$$\mathscr{H}_{\text{Zeeman}} = -g\mu_N H(l\hat{S}_x + m\hat{S}_y + n\hat{S}_z)$$
$$l = \sin\theta\cos\phi, \quad m = \sin\theta\sin\phi, \quad n = \cos\theta \qquad 3.51$$

Similarly the quadrupole interaction coordinate system can be related to the new one by an Eulerian angle transformation which will enable $\hat{I}_z$, $\hat{I}_x$, $\hat{I}_y$ to be expressed in terms of $\hat{S}_z$, $\hat{S}_x$, $\hat{S}_y$. In total, the angles θ and ϕ and the Eulerian angles α, β, and γ are required to define the geometry uniquely.

A simulated Mössbauer spectrum can be calculated by digital computer for assumed values of H, e^2qQ, η, θ, ϕ, α, β, and γ. The eigenstates are no longer pure $|I_z\rangle$ functions, but are linear combinations. As we shall see in the next section this affects the intensities of the lines. Although the calculations are difficult, a combined magnetic–quadrupole interaction does potentially provide a large amount of data regarding the symmetry of the atomic environment.

One aspect of combined interactions which has proved useful is the application of an external magnetic field to non-oriented polycrystalline absorbers with $\frac{3}{2} \to \frac{1}{2}$ γ-transitions. In the zero-field case the quadrupole doublet is symmetrical and the sign of e^2qQ is not determined. Collins [31] used a first-order perturbation treatment to show that an applied field would cause the transition from the $\pm\frac{3}{2}$ state of ^{57}Fe to split into a doublet, and the

[Refs. on p. 78]

transition from the $\pm\frac{1}{2}$ component of the excited state to split into an apparent triplet (unresolved quartet). The sign of e^2qQ could then be determined. The calculations have since been verified by computer methods [26], and the technique has been experimentally used with success. The method will fail if the asymmetry parameter is substantially different from zero. The doublet/triplet observation is then not applicable. Similar experiments using ^{119}Sn are described in Chapter 14.

3.7 Relative Intensities of Absorption Lines

The interpretation of a complex Mössbauer spectrum will obviously be simplified if the relative intensities of the various components are known. Once the energy levels of the Zeeman/quadrupole Hamiltonian have been calculated, and the spin quantum numbers for each state assigned (or appropriate linear combinations if the states are mixed), it is possible to calculate the intensities from the theory of the coupling of two angular momentum states [32, 33].

The final results contain two terms which are respectively angular dependent and angular independent. We shall discuss the latter first, as it corresponds to the situation where there is no preferred orientation in the laboratory frame for the appropriate principal axis, i.e. the absorber is a non-oriented polycrystalline material.

If the γ-ray transition is between two levels of nuclear spin I_1 and I_2, and furthermore between the two substates with I_z values of m_1 and m_2 respectively, then the angular-independent probability term is given by the square of the appropriate Clebsch–Gordan coefficients [32].

$$\text{Intensity} \propto \langle I_1 J -m_1 m \mid I_2 m_2 \rangle^2$$

J is the vector sum $J = I_1 + I_2$, and m is the vector sum $m = m_1 - m_2$. J is also known as the multipolarity of the transition, and the smaller values of J give the larger intensities. $J = 1$ is a dipole transition and $J = 2$ is a quadrupole transition, etc. If there is no change in parity during the decay it is classified as magnetic dipole (M1) or electric quadrupole (E2). Electric dipole (E1) transitions with a change in parity also come within our scope. In some cases the γ decay is a mixed dipole–quadrupole radiation, so that both must be included in the calculations.

Most Mössbauer isotopes decay primarily by a dipole transition. This puts an effective bar on the number of values which m can adopt. All values of m other than 0, ± 1 cause the coefficient to be zero. For quadrupole transitions, m can be 0, ± 1, ± 2. The majority of the Mössbauer isotopes have ground/excited state spins which are one of the following: $0,2$; $\frac{1}{2},\frac{3}{2}$; $\frac{3}{2},\frac{5}{2}$; $\frac{5}{2},\frac{5}{2}$; or $\frac{5}{2},\frac{7}{2}$. The first of these $0 \rightarrow 2$, which is always an E2 decay, is trivial in that all five possible transitions have an equal intensity. The values of the

Clebsch–Gordan coefficients and other useful data for the $\frac{1}{2} \to \frac{3}{2}$ dipole transitions are given in Table 3.2. Similar tables for the $\frac{1}{2} \to \frac{3}{2}$ quadrupole transitions and for higher spin states are given in Appendix 2. It is immaterial whether I_1 is the ground or excited state.

We shall now consider the $\frac{1}{2}$, $\frac{3}{2}$ isotopes in more detail. In a magnetically split spectrum there are eight possible transitions between levels, and initially we shall assume these levels to be the pure states described by the m_1 and m_2 quantum numbers. The coefficients are given in Table 3.2 for a dipole transition ($J = 1$). The $+\frac{3}{2} \to -\frac{1}{2}$ and $-\frac{3}{2} \to +\frac{1}{2}$ transitions have zero probability

Table 3.2 Relative probabilities for a dipole $\frac{3}{2}$, $\frac{1}{2}$ transition

Magnetic spectra (M1)

m_2	$-m_1$	m	C (1)	C^2 (2)	Θ (2)	$\theta = 90°$ (3)	$\theta = 0°$ (3)
$+\frac{3}{2}$	$+\frac{1}{2}$	$+1$	1	3	$1 + \cos^2 \theta$	3	6
$+\frac{1}{2}$	$+\frac{1}{2}$	0	$\sqrt{\frac{2}{3}}$	2	$2 \sin^2 \theta$	4	0
$-\frac{1}{2}$	$+\frac{1}{2}$	-1	$\sqrt{\frac{1}{3}}$	1	$1 + \cos^2 \theta$	1	2
$-\frac{3}{2}$	$+\frac{1}{2}$	-2	0	0	0	0	0
$+\frac{3}{2}$	$-\frac{1}{2}$	$+2$	0	0	0	0	0
$+\frac{1}{2}$	$-\frac{1}{2}$	$+1$	$\sqrt{\frac{1}{3}}$	1	$1 + \cos^2 \theta$	1	2
$-\frac{1}{2}$	$-\frac{1}{2}$	0	$\sqrt{\frac{2}{3}}$	2	$2 \sin^2 \theta$	4	0
$-\frac{3}{2}$	$-\frac{1}{2}$	-1	1	3	$1 + \cos^2 \theta$	3	6

Quadrupole spectra (M1)

Transitions	C^2 (2)	Θ (2)	$\theta = 90°$ (3)	$\theta = 0°$ (3)
$\pm\frac{1}{2}, \pm\frac{1}{2}$	1	$2 + 3 \sin^2 \theta$	5	2
$\pm\frac{3}{2}, \pm\frac{1}{2}$	1	$3(1 + \cos^2 \theta)$	3	6

(1) The Clebsch–Gordan coefficients $\langle \frac{1}{2} 1 -m_1 m | \frac{3}{2} m_2 \rangle$ calculated using the formulae in ref. 32 for $\langle \frac{3}{2} \frac{1}{2} m_2 m_1 | 1 m \rangle$ and converted using the relationship $\langle \frac{1}{2} 1 -m_1 m | \frac{3}{2} m_2 \rangle = (-)^{\frac{1}{2} + m_1} \sqrt{(\frac{4}{3})} \langle \frac{3}{2} \frac{1}{2} m_2 m_1 | 1 m \rangle$
(2) C^2 and Θ are the angular independent and dependent terms arbitrarily normalised.
(3) Relative intensities observed at 90° and 0° to the principal axis. Normalisation arbitrary.

as $m = +2$ and -2 respectively. There are thus six finite coefficients. The squares of these normalised to give integral values are also given in Table 3.2 and are seen to be $3 : 2 : 1 : 1 : 2 : 3$. These are the theoretical intensity ratios for a [57]Fe (14·4-keV) Zeeman spectrum, for example, and have already been illustrated in Fig. 3.4. Appendix 2 contains data for other spin states.

The angular terms are formulated as the radiation probability in a direction θ to the principal axis (z axis) of the magnetic field or the electric field gradient tensor. The appropriate functions $\Theta(J, m)$ are listed in Table 3.3. The

Table 3.3 Direction-dependent probability term $\Theta\,(J, m)$

	$J = 1$	$J = 2$
$m = 0$	$\frac{1}{2}\sin^2\theta$	$\frac{3}{8}\sin^2 2\theta$
$m = \pm 1$	$\frac{1}{4}(1 + \cos^2\theta)$	$\frac{1}{4}(\cos^2\theta + \cos^2 2\theta)$
$m = \pm 2$	—	$\frac{1}{4}\left(\sin^2\theta + \dfrac{\sin^2 2\theta}{4}\right)$

probability in a given direction is then $\langle I_1 J -m_1 m \mid I_2 m_2 \rangle^2\, \Theta\,(J, m)$. For a polycrystalline absorber we can integrate $\Theta\,(J, m)$ over all orientations to give the average $\overline{\Theta\,(J, m)}$. For example the average of $\sin^2\theta$ is given by

$$\overline{\sin^2\theta} = \frac{1}{4\pi}\int_0^{2\pi}\int_0^{\pi}(\sin^2\theta)\sin\theta\,d\theta\,d\phi = \tfrac{2}{3}$$

The total emitted radiation given by

$$\sum_{m_1 m_2}\langle I_1 J -m_1 m \mid I_2 m_2 \rangle^2\, \overline{\Theta\,(J, m)} \qquad 3.52$$

is independent of θ, i.e. the γ-ray emission is isotropic, and may be normalised to a total probability of unity.

The above discussion is important in Mössbauer experimentation. A randomly packed polycrystalline sample gives a spectrum which is not affected by orientation of the whole sample. However, a single-crystal absorber does have angular properties, and the intensity of the hyperfine lines will vary according to orientation. The relative intensities for the γ-ray direction parallel ($\theta = 0°$) and perpendicular ($\theta = 90°$) to the principal axis are given in the Tables. Thus an oriented absorber with a $\tfrac{3}{2} \to \tfrac{1}{2}$ decay will have zero intensity in the $m = 0$ lines at $\theta = 0°$. The effects of rotation on a magnetic splitting are illustrated in Fig. 3.6. From this, and the expressions for $C^2\Theta$ in Table 3.2, it is clear that the ratios of the intensities of the six transitions in a single-crystal absorber are $3:x:1:1:x:3$ where $x = 4\sin^2\theta/(1 + \cos^2\theta)$. Thus the relative intensities of the $\mid \pm\tfrac{3}{2}\rangle \to \mid \pm\tfrac{1}{2}\rangle$ transitions to those of the $\mid \mp\tfrac{1}{2}\rangle \to \mid \pm\tfrac{1}{2}\rangle$ transitions is always $3:1$ whereas the relative intensity of the $\mid \pm\tfrac{1}{2}\rangle \to \mid \pm\tfrac{1}{2}\rangle$ transitions for which $\Delta m = 0$ are a function of θ and fall in the range 0–4.

The quadrupole spectra will have several degenerate transitions, and the appropriate summations are listed in Table 3.2 and Appendix 2. A dipole $\tfrac{3}{2} \to \tfrac{1}{2}$ transition shows $1:1$ intensity for a polycrystalline sample, and $5:3$ and $1:3$ intensity ratios for a single crystal with the γ-ray axis perpendicular to and parallel to the principal axis of a symmetric electric field gradient tensor.

We can now see that an angular dependence study of the quadrupole-split spectrum of a single crystal absorber will define which of the two lines is the $|\pm\frac{3}{2}\rangle \to |\pm\frac{1}{2}\rangle$ transition and thence give the sign of the principal component of the electric field gradient tensor, a parameter which cannot be determined from the spectrum of a polycrystalline sample in the absence of indigenous

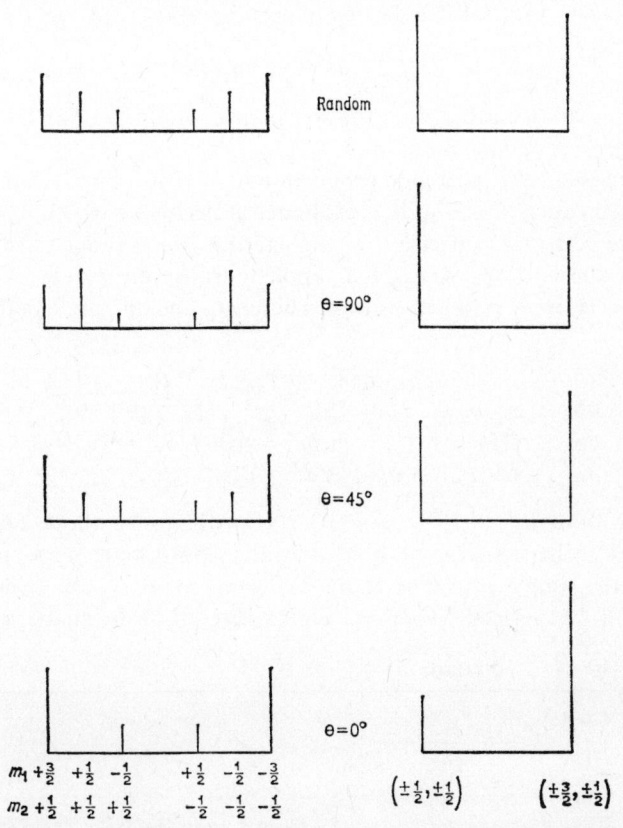

Fig. 3.6 The effect of orientation upon the intensities of magnetic and quadrupole spectra for a $\frac{3}{2} \to \frac{1}{2}$ transition.

or applied magnetic interactions. Magnetic metals and alloys can be polarised by an external field and behave in a similar manner to single crystals since the magnetic direction is uniquely defined.

Thus far we have only discussed transitions between 'pure' states. In the event that the energy levels are formed from a linear combination of the basis set, it is not possible to use our earlier arguments, and a more fundamental approach is required. We shall assume that the linear combination of the $2I_1 + 1$ terms in one state have the coefficients a_1, and that the $2I_2 + 1$ terms in the other state have the coefficients b_2. The intensity cannot be calculated directly because of interference between the various components.

[*Refs.* on p. 78]

It is therefore necessary to sum the amplitudes, taking into account the possible presence of a quadrupole/dipole mixing ratio of δ^2 and a phase factor between these of $e^{i\xi}$. Discussions of various aspects of the theory are available [34–38]. The amplitude $\vec{a}$ is proportional to

$$\sum_{m_1}\sum_{m_2} \{a_1^* b_2 [\sqrt{(3)}i\langle I_1 1 -m_1 m | I_2 m_2\rangle(D^1_{+1,m} + D^1_{-1,m})$$
$$- \sqrt{(5)} |\delta| e^{i\xi}\langle I_1 2 -m_1 m | I_2 m_2\rangle(iD^2_{+1,m} - iD^2_{-1,m})]\} \qquad 3.53$$

The $D^J_{\pm 1,m}$ expressions are the rotation matrices for right-handed (+1) and left-handed (−1) polarisation.

The intensity now becomes proportional to $\vec{a}^*\vec{a}$ which is an extremely complicated expression, and best evaluated if necessary by a computer technique. One important effect of 'mixing' states is to introduce a weak probability for the $m = \pm 2$ lines of a dipole γ-transition which are normally 'forbidden'. For a 'pure' transition where m_1 and m_2 are good quantum numbers

$$\vec{a}^*\vec{a} \propto \{\tfrac{3}{2}\langle I_1 1 -m_1 m | I_2 m_2\rangle^2 [(d^1_{+1,m})^2 + (d^1_{-1,m})^2]$$
$$+ \tfrac{5}{2}\delta^2 \langle I_1 2 -m_1 m | I_2 m_2\rangle^2 [(d^2_{+1,m})^2 + (d^2_{-1,m})^2]$$
$$- \cos\xi\sqrt{(15)} |\delta| \langle I_1 1 -m_1 m | I_2 m_2\rangle\langle I_1 2 -m_1 m | I_2 m_2\rangle$$
$$\times [(d^1_{+1,m})(d^2_{+1,m}) - (d^1_{-1,m})(d^2_{-1,m})]\} \qquad 3.54$$

The expressions $[(d^1_{+1,m})^2 + (d^1_{-1,m})^2]$ etc. prove to be the $\Theta(J, m)$ terms introduced earlier, so that with $\delta^2 = 0$ and appropriate normalisation we return to the simple case. The analytical forms for $d^J_{\pm 1,m}$ are given in Table 3.4. It will be noticed, however, that where there is dipole–quadrupole

Table 3.4 The $d^J_{\pm 1,m}$ polarisation coefficients

	$J = 1$		$J = 2$	
	+1	−1	+1	−1
$m = +2$	—	—	$-(2\sin\theta + \sin 2\theta)/4$	$-(2\sin\theta - \sin 2\theta)/4$
$m = +1$	$\cos^2\left(\tfrac{\theta}{2}\right)$	$\sin^2\left(\tfrac{\theta}{2}\right)$	$(\cos\theta + \cos 2\theta)/2$	$(\cos\theta - \cos 2\theta)/2$
$m = 0$	$(\sin\theta)/\sqrt{2}$	$-(\sin\theta)/\sqrt{2}$	$(\tfrac{3}{8})^{\frac{1}{2}}\sin 2\theta$	$-(\tfrac{3}{8})^{\frac{1}{2}}\sin 2\theta$
$m = -1$	$\sin^2\left(\tfrac{\theta}{2}\right)$	$\cos^2\left(\tfrac{\theta}{2}\right)$	$(\cos\theta - \cos 2\theta)/2$	$(\cos\theta + \cos 2\theta)/2$
$m = -2$	—	—	$(2\sin\theta - \sin 2\theta)/4$	$(2\sin\theta + \sin 2\theta)/4$

mixing there is an interference term with a phase angle $\cos\xi$ which must be taken into account. Such a treatment has only proved necessary for a few isotopes such as ^{99}Ru (90 keV) and ^{193}Ir (73 keV) where δ^2 is significantly large, and polarised magnetic spectra show the effects of the interference term [39, 40].

[Refs. on p. 78]

RELATIVE INTENSITIES OF ABSORPTION LINES | 71

All that remains now is to relate the relative line intensities to the total absorption intensity. The equations for the resonant cross-section of a single peak in Section 1 are still basically valid except that the term for each hyperfine line must additionally contain the appropriate

$$\langle I_1 J - m_1 m \mid I_2 m_2 \rangle^2 \Theta (J, m)$$

coefficient. The total cross-section is then the same, but that for each component is less than that for an unsplit resonance. Complicated hyperfine

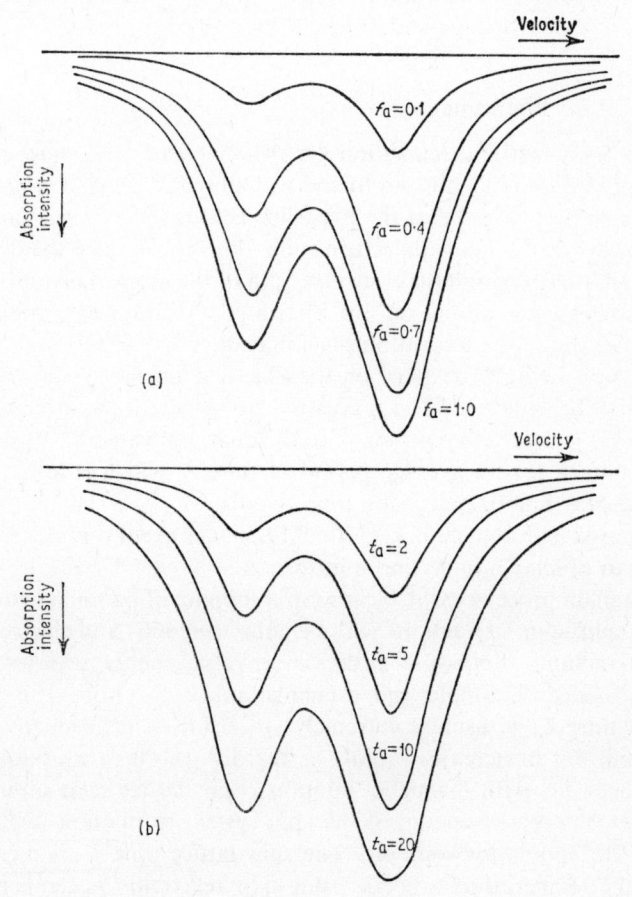

Fig. 3.7 Curves illustrating the changes in the relative intensity of two hyperfine lines caused by saturation effects: (a) with change in recoilless fraction, (b) with change in absorber thickness.

spectra will therefore require longer counting times to obtain good counting statistics and resolution.

It is also important to note that components with differing relative probabilities will show a difference in saturation behaviour. This will cause an

[Refs. on p. 78]

apparent accentuation of the nominally weaker component with increasing cross-section, i.e. with increasing absorber thickness and with decreasing temperature because of the increase in recoilless fraction. The true relative line intensities can only be obtained by extrapolation to zero cross-section. This phenomenon has been discussed in detail [41], and is illustrated by the calculated saturation effects on an asymmetric quadrupole doublet in Fig. 3.7. Thus the apparent intensity of the stronger component relative to the weaker one diminishes from 2·0 to 1·3 as f_a increases from 0·1 to 1·0 and from 2·0 to 1·2 as the absorber thickness increases from 2 to 20.

3.8 Relaxation Phenomena

The effects of electronic relaxation on the magnetic hyperfine interaction have already been briefly alluded to and will now be discussed in more detail. It was seen in Section 3.5 that the hyperfine field is usually generated by the polarising effects of unpaired electron spins. The direction of the field will be related to that of the resultant electronic spin of the atom. This spin direction is not invariant but can alter or 'flip' after a period of time by one of several mechanisms; this is the relaxation phenomenon.

Influence of the nuclear spin I on the time dependence of the atomic spin direction can be neglected. The precession frequency of the atomic spin S is such that the nuclear spin reacts only to the quantum value S_z. If the value of S_z is maintained for an average period of time τ_S which is long compared with the nuclear Lamor precession time $1/\omega_L$ (i.e. $\omega_L \tau_S \gg 1$) and the latter is long compared with the lifetime of the Mössbauer event (i.e. $\omega_L t_{\frac{1}{2}} \gg 1$) then we expect to observe a hyperfine splitting.

The two main processes which cause spin-flipping of paramagnetic ions are electronic spin–spin interactions with neighbouring ions and electronic spin–lattice interactions. Spin–spin processes involve energy transfer between interacting spins via dipole and exchange spin relaxation. The spin–spin relaxation time T_1 is usually extremely small for a magnetically 'concentrated' solid, but it increases rapidly as the distance between paramagnetic ions increases, i.e. with magnetic 'dilution'. Spin–lattice relaxation involves the transfer of Zeeman energy of the spin system to phonon modes of the lattice via the spin–orbit coupling. The spin–lattice time T_2 is increased by lowering the temperature, whereas spin–spin relaxation is temperature independent. In the case of a paramagnetic impurity in a metal, spin interactions with the conduction electrons are important. A third type of spin relaxation known as cross-relaxation involves mutual spin-flips so that energy is effectively conserved.

If we have a paramagnetic solid containing ions with $S = \frac{1}{2}$, then there are two orientations of S_z, namely $+\frac{1}{2}$ and $-\frac{1}{2}$. These form a degenerate Kramers' doublet. Both configurations generate a field H at the nucleus but with opposing sign. There is no preferred direction for S_z in contrast to the ordered systems. If the spin relaxation is very slow, we expect to see a magnetic splitting corresponding to the field H. Other spin-states behave similarly. Such fields have been reported for $Fe^{3+} 3d^5$ ions (^6S state), and for several of the rare-earth transitions [42], and many examples will be found in later chapters. In the case of Fe^{3+}, the $S_z = \pm\frac{5}{2}, \pm\frac{3}{2}, \pm\frac{1}{2}$ Kramers' doublets generated by the influence of the crystal field on the free-ion ground term all lie very close together. In consequence, all are populated at temperatures above 1 K, and can in principle generate three hyperfine spectra. However, the spin–lattice relaxation times are not identical for the three doublets. The $\pm\frac{5}{2}$ states are the longest-lived, while the $\pm\frac{1}{2}$ and $\pm\frac{3}{2}$ states are usually relaxed in a Mössbauer experiment.

For cooperative phenomena such as ferromagnetism or antiferromagnetism, interactions between spins are so strong as to create a preferred direction for S_z. It has been shown by van der Woude and Dekker [43] that the observed magnetic hyperfine spectra are best interpreted using a spin-wave treatment. The angular frequency of the spin waves ω_s is far greater than ω_L so that the nucleus sees a time-average of S_z which behaves as a function of the magnetisation. Even if $S > \frac{1}{2}$ only a single hyperfine spectrum without line broadening is seen when all the sites are chemically equivalent. If the spin relaxation frequency is of the order of the Lamor frequency ($\omega_L \tau_s \approx 1$), i.e. the spin-wave time-average is itself time dependent, the static viewpoint is invalid, and an inward collapse (motional narrowing) of the hyperfine spectrum is seen. If $\omega_L \tau_s \ll 1$ no hyperfine splitting will be recorded.

The collapse of a six-line hyperfine spectrum with decreasing relaxation time is illustrated by the calculated spectra in Fig. 3.8 [44] for a ^{57}Fe nucleus in a fluctuating magnetic field and a fixed electric field gradient for different values of the electronic relaxation time. If the fluctuation rate is very slow compared to the precession frequency of the nucleus in the field H, the full six-line hyperfine pattern is observed. If the fluctuation rate is extremely rapid the nucleus will see only the time-averaged field which is zero and a symmetric quadrupolar pattern will be seen. At intermediate frequencies the spectra reflect the fact that the $|\pm\frac{3}{2}\rangle \to |\pm\frac{1}{2}\rangle$ transitions which make up the low-velocity component of the quadrupole doublet relax at higher frequencies than do the $|\pm\frac{1}{2}\rangle \to |\pm\frac{1}{2}\rangle$ and $|\pm\frac{1}{2}\rangle \to |\mp\frac{1}{2}\rangle$ transitions which make up the high-velocity component because of their differing Zeeman precession frequencies. Thus at an electronic relaxation time of 3×10^{-9} s the high-velocity component has already collapsed to a sharp line whilst the low-velocity component is still broad and only begins to resolve as the electronic relaxation time approaches 10^{-10} s. Conversely, it is to be noted that in the presence of magnetic interactions the $|\pm\frac{3}{2}\rangle$ component of a quadrupole doublet broadens considerably before the other, and this is a potential cause of apparent asymmetry in quadrupole spectra.

[Refs. on p. 78]

74 | HYPERFINE INTERACTIONS

It will be obvious that there are close similarities between the ordered magnetic and disordered paramagnetic hyperfine spectra. The hyperfine fields originate in similar ways, and the angular properties of the transitions

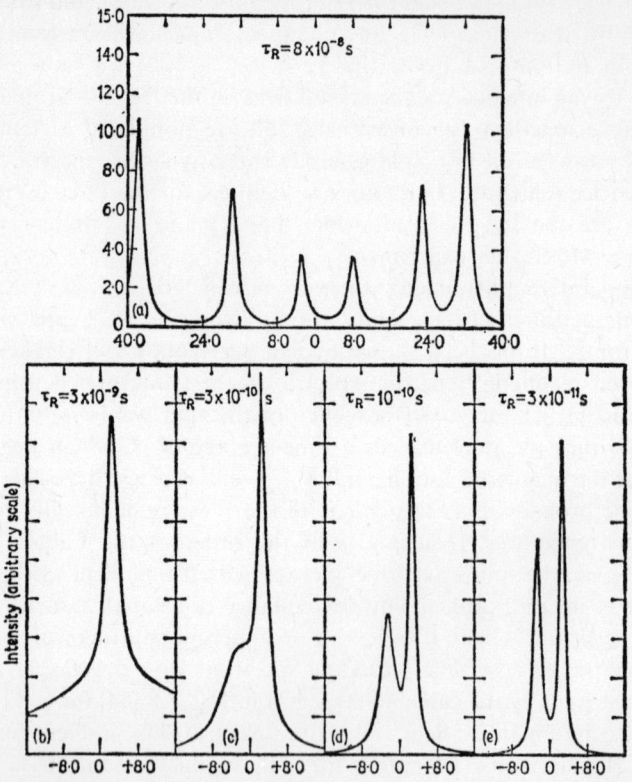

Fig. 3.8 Line shapes for an ^{57}Fe nucleus in a fluctuating magnetic field and a fixed electric field gradient, for different values of the electronic relaxation time τ_R. [Ref. 44, Fig. 2]

are identical. The main difference is caused by the spin-wave time-averaging effect in the ordered case.

The detailed mathematical theories of the intermediate spectra ($\omega_L \tau_s \approx 1$) are complex and will not be discussed here. Much of the work has been excellently reviewed by Wickman and Wertheim [42] and Wickman [45], and other useful texts are by Blume [44], Boyle and Gabriel [46], Nowik [47], van Zorge et al. [48], Wegener [49], Bradford and Marshall [50], and Wickman [51].

3.9 Anisotropy of the Recoilless Fraction

Although, as was seen in Section 3.7, a quadrupole split spectrum for a $\frac{3}{2} \rightarrow \frac{1}{2}$ transition in a powdered absorber with random orientation should be

[*Refs. on p. 78*]

two lines of equal intensity, several early measurements showed unexpected asymmetry. A possible explanation for this was put forward by Goldanskii *et al.* in 1962 [52] and developed mathematically by Karyagin [53]. Since that time the effect has been variously described as the 'Goldanskii effect' or the 'Karyagin effect'.

The relative intensity $I_{\frac{3}{2}}/I_{\frac{1}{2}}$ of the lines involving the $|\pm\frac{3}{2}\rangle$ and $|\pm\frac{1}{2}\rangle$ levels of the excited state (also frequently written as I_π/I_σ) depends on the angle θ between the γ-ray direction and the principal axis of the electric field gradient: from Table 3.2 this can be written as

$$\frac{I_{\frac{3}{2}}(\theta)}{I_{\frac{1}{2}}(\theta)} = \frac{1 + \cos^2 \theta}{\frac{2}{3} + \sin^2 \theta} \qquad 3.55$$

For polycrystalline samples

$$\frac{\int_0^\pi (1 + \cos^2 \theta) \sin \theta \, d\theta}{\int_0^\pi (\frac{2}{3} + \sin^2 \theta) \sin \theta \, d\theta} = 1 \qquad 3.56$$

However, if the recoilless radiation is anisotropic and the recoil-free fraction f is itself a function of θ, then

$$\frac{\int_0^\pi (1 + \cos^2 \theta) f(\theta) \sin \theta \, d\theta}{\int_0^\pi (\frac{2}{3} + \sin^2 \theta) f(\theta) \sin \theta \, d\theta} = F[f(\theta)] \neq 1 \qquad 3.5$$

In other words an anisotropic recoilless fraction should produce an asymmetry in the $\frac{3}{2} \rightarrow \frac{1}{2}$ quadrupole doublet which is not angular dependent.

Experimental observation of this phenomenon was claimed in Ph_3SnCl by Goldanskii *et al.* [54], but was contested by Shpinel *et al.* [55]. Temperature dependence of the asymmetry has been reported by Stöckler and Sano [56] in polymeric organotin compounds. The precise form of the anisotropy of f cannot be studied from the polycrystalline data alone. Goldanskii *et al.* [57] measured the asymmetry angular dependence in the spectrum of a single crystal of the mineral siderite, $FeCO_3$, and claimed to have proved the polycrystalline asymmetry to be due to vibrational anisotropy. This work has been strongly contested by Housley *et al.* [58] on the grounds that the nuclear absorption cross-sections are polarisation dependent, a factor omitted in the earlier work. If the recoilless fraction is anisotropic, the transmitted radiation will be partially polarised. Publishing new data, they showed no evidence for an appreciable anisotropy in $FeCO_3$. Furthermore, they forecast difficulty in verifying the effect without extremely precise determination of the area ratios. This has since been confirmed in $FeCO_3$ by Goldanskii [59].

[*Refs. on p. 78*]

More recently a detailed study of Me_2SnF_2 by Herber and Chandra has shown substantial anisotropy in the recoilless fraction [60]. The ellipsoids of vibration of the tin atom have been measured by X-ray diffraction methods, and the ratio $F[f(\theta)]$ defined in equation 3.57 was calculated by a lattice dynamical treatment to be 0·72. The corresponding experimental value of 0·78 ± 0·06 is in good agreement, and the data probably constitute the best authenticated test of the Goldanskii–Karyagin effect at the present time.

Anisotropy effects have also been calculated to be detectable as an alteration of the intensity ratios in magnetic hyperfine spectra [61]. For quadrupole spectra with higher spin states it is sometimes possible to observe intensity discrepancies between the $m = 0$, $m = \pm 1$, and the $\pm\frac{1}{2} \rightarrow \pm\frac{1}{2}$ groups of transitions, each of which has different angular properties. Examples are to be found in I_2 [62] and IBr, ICl [63] where vibrational anisotropy is assumed.

One word of caution must be given. The main features of asymmetry due to vibrational anisotropy in polycrystalline materials are identical to the effects of partial orientation of crystallites [41]. Many papers claim anisotropy effects in polycrystals without giving details of absorber preparation, or stating whether the possibility of orientation has been eliminated by checking for angular dependence of the spectra. Certain types of microcrystallite are particularly prone to partial orientation on compacting, even after grinding vigorously, and much of the data might not stand re-inspection. The safest method of preparing unoriented absorbers appears to be by grinding with a large bulk of an abrasive powder such as Al_2O_3 or quartz glass.

3.10 The Pseudoquadrupole Interaction

A hyperfine effect has been described which can generate a temperature-dependent shift of the centroid of the resonance lines only in rare circumstances. This is the pseudoquadrupole hyperfine interaction [64].

We have already seen that the magnetic–dipole and electric–quadrupole interactions do not normally cause any shift in the weighted mean position of the spectrum. This is because there is no first-order magnetic hyperfine interaction acting on a non-degenerate electronic state. However, if there is an excited electronic state $|E\rangle$ which is only of the order of 1 cm^{-1} above the electronic ground state $|G\rangle$ there will be a second-order hyperfine interaction known as the 'pseudoquadrupole' interaction which causes mutual repulsion of $|E\rangle$ and $|G\rangle$. The magnitude of this effect will be different for the ground- and excited-state nuclear spins, E_g and E_e respectively. Assuming that $|E\rangle$ and $|G\rangle$ are so similar as to generate the same magnitude of quadrupole effect, the nuclear energy levels will be as shown in Fig. 3.9. The theory has been detailed [64, 65]. The γ-transition energies for the excited electronic state are different from those for the ground state. At ambient temperatures the ground and excited electronic states are equally populated, but at very low temperatures (< 4 K) the proportion of decays in the excited states is decreased.

TmCl$_3$.6H$_2$O and Tm$_2$(SO$_4$)$_3$.8H$_2$O both give $\frac{3}{2} \to \frac{1}{2}$ decays and have electronic levels separated by $\sim$1 K. Decreasing the temperature from 4·2 K

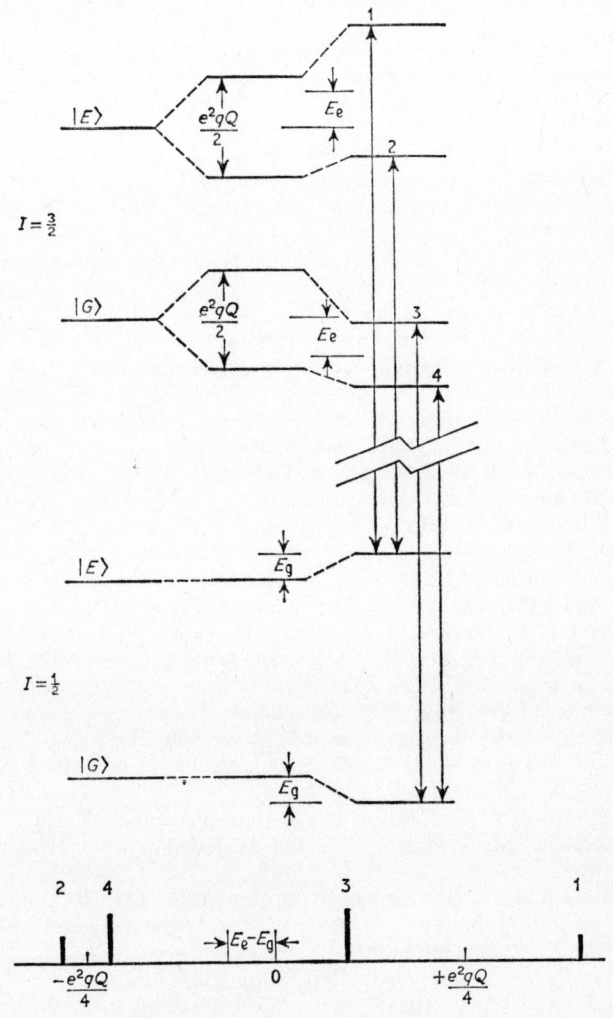

Fig. 3.9 Schematic representation of the pseudoquadrupole effect on two electronic levels | $E\rangle$ and | $G\rangle$ (not drawn to scale). Transitions 1 and 2 are weak at very low temperatures because of depopulation of the | $E\rangle$ levels

to 1·1 K gives a small apparent movement of the centroid of the quadrupole doublet because of the depopulation effect on the excited electronic state and the pseudoquadrupole interaction. The latter is less than the Heisenberg

[*Refs. on p. 78*]

width so that only two resonance lines are seen. Transitions from an $|E\rangle$ to a $|G\rangle$ state or vice versa during the γ-event are too weak to detect.

Although not widely applicable, the observation of the pseudoquadrupole effect can be used to measure separations between electronic states of less than 1 cm^{-1}.

REFERENCES

[1] J. E. Rosenthal and G. Breit, *Phys. Rev.*, 1932, **41**, 459.
[2] G. Breit, *Rev. Mod. Phys.*, 1958, **30**, 507.
[3] A. R. Bodmer, *Proc. Phys. Soc.*, 1953, **A66**, 1041.
[4] L. Wilets, 'Isotope Shifts', Encyclopedia of Physics, Vol. 38/1, p. 96, Springer-Verlag, 1958.
[5] H. Kopfermann (translated by E. E. Schneider), 'Nuclear Moments', Academic Press, N.Y., 1958.
[6] R. V. Pound and G. A. Rebka, Jr, *Phys. Rev. Letters*, 1960, **4**, 274.
[7] B. D. Josephson, *Phys. Rev. Letters*, 1960, **4**, 341.
[8] W. G. V. Rosser, 'An Introduction to the Theory of Relativity', Butterworths, 1964, p. 114.
[9] Y. Hazony, *J. Chem. Phys.*, 1966, **45**, 2664.
[10] L. M. Epstein, *J. Chem. Phys.* 1964, **40**, 435.
[11] R. V. Pound, G. B. Benedek, and R. Drever, *Phys. Rev. Letters*, 1961, **7**, 405.
[12] R. W. Vaughan and H. G. Drickamer, *J. Chem. Phys.*, 1967, **47**, 468.
[13] D. N. Pipkorn, C. K. Edge, P. Debrunner, G. de Pasquali, H. G. Drickamer, and H. Frauenfelder, *Phys. Rev.*, 1964, **135**, A1604.
[14] T. P. Das and E. L. Hahn, 'Nuclear Quadrupole Resonance Spectroscopy', Solid State Physics, Supp. No. 1, Academic Press, 1958.
[15] M. Kubo and D. Nakamura, *Adv. Inorg. Chem. and Radiochem.*, 1966, **8**, 257.
[16] G. K. Shenoy and B. D. Dunlap, *Nuclear Instr. Methods*, 1969, **71**, 285.
[17] R. Ingalls, *Phys. Rev.*, 1964, **133**, A787.
[18] C. E. Johnson, *Proc. Phys. Soc.*, 1966, **88**, 943.
[19] A. J. Nozik and M. Kaplan, *Phys. Rev.*, 1967, **159**, 273.
[20] A. Abragam, 'The Principles of Nuclear Magnetism', Clarendon Press, Oxford, 1961.
[21] W. Marshall and C. E. Johnson, *J. Phys. Radium*, 1962, **23**, 733.
[22] C. E. Johnson, M. S. Ridout, and T. E. Cranshaw, *Proc. Phys. Soc.*, 1963, **81**, 1079.
[23] R. E. Watson and A. J. Freeman, *Phys. Rev.*, 1961, **123**, 2027.
[24] S. S. Hanna, J. Heberle, G. J. Perlow, R. S. Preston, and D. H. Vincent, *Phys. Rev. Letters*, 1960, **4**, 513.
[25] G. K. Wertheim, *J. Appl. Phys.*, 1961, Suppl. 32, 110S.
[26] J. R. Gabriel and S. L. Ruby, *Nuclear Instr. Methods*, 1965, **36**, 23.
[27] J. R. Gabriel and D. Olsen, *Nuclear Instr. Methods*, 1969, **70**, 209.
[28] W. Kundig, *Nuclear Instr. Methods*, 1967, **48**, 219.
[29] G. R. Hoy and S. Chandra, *J. Chem. Physics*, 1967, **47**, 961.
[30] P. G. L. Williams and G. M. Bancroft, *Chem. Phys. Letters*, 1969, **3**, 110.
[31] R. L. Collins, *J. Chem. Phys.*, 1965, **42**, 1072.
[32] E. U. Condon and G. H. Shortley, 'The Theory of Atomic Spectra', Cambridge Univ. Press, 1935.

[33] M. E. Rose, 'Multipole Fields', Chapman & Hall, 1955.
[34] H. Frauenfelder, D. E. Nagle, R. D. Taylor, D. R. F. Cochran, and W. M. Visscher, *Phys. Rev.*, 1962, **126**, 1065.
[35] M. K. F. Wong, *Phys. Rev.*, 1966, **149**, 378.
[36] J. T. Dehn, J. G. Marzolf, and J. F. Salmon, *Phys. Rev.*, 1964, **135**, B1307.
[37] S. K. Misra, *Nuovo Cimento*, 1969, **59**, B, 152.
[38] C. E. Violet, *J. Appl. Phys.*, 1969, **40**, 4959.
[39] O. C. Kistner, *Phys. Rev. Letters*, 1967, **19**, 872.
[40] M. Atac, B. Chrisman, P. Debrunner, and H. Frauenfelder, *Phys. Rev. Letters*, 1968, **20**, 691.
[41] T. C. Gibb, R. Greatrex, and N. N. Greenwood, *J. Chem. Soc. (A)*, 1968, 890.
[42] H. H. Wickman and G. K. Wertheim, 'Spin Relaxation in Solids', Chap. 11, in 'Chemical Applications of Mössbauer Spectroscopy' by V. I. Goldanskii and R. H. Herber, Eds., Academic Press, New York, 1968.
[43] F. van der Woude and A. J. Dekker, *Phys. Stat. Sol.*, 1965, **9**, 775.
[44] M. Blume, *Phys. Rev. Letters*, 1965, **14**, 96.
[45] H. H. Wickman, 'Mössbauer Effect Methodology, Vol. 2', Ed. I. J. Gruverman, Plenum Press, N.Y., 1966, p. 39.
[46] A. J. F. Boyle and J. R. Gabriel, *Phys. Letters*, 1965, **19**, 451.
[47] I. Nowik, *Phys. Letters*, 1967, **24A**, 487.
[48] B. C. van Zorge, W. J. Caspers, and A. J. Dekker, *Phys. Stat. Sol.*, 1966, **18**, 761.
[49] H. Wegener, *Z. Physik*, 1965, **186**, 498.
[50] E. Bradford and W. Marshall, *Proc. Phys. Soc.*, 1966, **87**, 731.
[51] H. H. Wickman, 'Hyperfine Structure and Nuclear Radiations', Ed. E. Matthias and D. A. Shirley, North Holland, Amsterdam, 1968, p. 928.
[52] V. I. Goldanskii, G. M. Gorodinskii, S. V. Karyagin, L. A. Korytko, L. M. Krizhanskii, E. F. Makarov, I. P. Suzdalev, and V. V. Khrapov, *Doklady Akad. Nauk S.S.S.R.*, 1962, **147**, 127.
[53] S. V. Karyagin, *Doklady Akad. Nauk S.S.S.R.*, 1963, **148**, 1102.
[54] V. I. Goldanskii, E. F. Makarov, and V. V. Khrapov, *Phys. Letters*, 1963, **3**, 344.
[55] V. S. Shpinel, A. Yu Aleksandrov, G. K. Ryasnin, and O. Yu. Okhlobystin, *Soviet Physics – J.E.T.P.*, 1965, **21**, 47.
[56] H. A. Stöckler and H. Sano, *Phys. Letters*, 1967, **25A**, 550.
[57] V. I. Goldanskii, E. F. Makarov, I. P. Suzdalev, and I. A. Vinogradov, *Phys. Rev. Letters*, 1968, **20**, 137.
[58] R. M. Housley, U. Gonser, and R. Grant, *Phys. Rev. Letters*, 1968, **20**, 1279.
[59] V. I. Goldanskii, E. F. Makarov, I. P. Suzdalev, and I. A. Vinogradov, *Zhur. eksp. teor. Fiz.*, 1970, **58**, 760.
[60] R. H. Herber and S. Chandra, *J. Chem. Phys.*, 1970, **52**, 6045.
[61] S. G. Cohen, P. Gielen, and R. Kaplow, *Phys. Rev.*, 1966, **141**, 423.
[62] M. Pasternak, A. Simopoulos, and Y. Hazony, *Phys. Rev.* 1965, **140**, A1892.
[63] M. Pasternak and T. Sonnino, *J. Chem. Phys.*, 1968, **48**, 2004.
[64] M. J. Clauser, E. Kankeleit, and R. L. Mössbauer, *Phys. Rev. Letters*, 1966, **17**, 5.
[65] M. J. Clauser and R. L. Mössbauer, *Phys. Rev.*, 1969, **178**, 559.

4 Applications of the Mössbauer Effect

Although a number of possible applications of the Mössbauer effect have been suggested in the previous chapters, it is worthwhile to consider its use in general terms before developing the theme of more specific chemical application. Generally speaking, three broad areas can be defined in addition to that of the basic phenomenon of resonant absorption itself; namely relativity and general physics, nuclear physics, and solid-state physics and chemistry.

4.1 Relativity and General Physics

The advent of a radiation defined to 1 part in 10^{12} to 10^{14} prompted many physicists to find means of using it to verify some of the predictions of relativistic mechanics experimentally. In fact, many of the early Mössbauer experiments were directed towards such ends. After an early flood of papers, it was realised that many of the experiments could not provide an unambiguous proof of the validity of the physical laws with which they were predicted and interpreted, and consequently interest waned again. However, many of the experiments are extremely interesting in their own right, and some of them will now be outlined.

(a) Gravitational Red Shift

Let us consider a Mössbauer experiment in which the source is mounted a distance d above the absorber so that the photon travels through a gravitational field which would introduce an acceleration in the atom of g if it were free to fall. The photon interacts with the gravitational field as though it had a mass of E_γ/c^2 and we find a change in the γ-ray energy, i.e. a gravitational red shift of $\partial E/E_\gamma = gd/c^2$ and thus a shift in the position of the Mössbauer resonance line. The effect was first predicted by Einstein in 1907, but in view of the very small magnitude of the effect (1·1 parts in 10^{-16} m^{-1}) it is difficult to demonstrate inside the laboratory frame.

The first published attempt to measure the shift was not conclusive [1], but in 1960 Pound and Rebka [2], using the ^{57}Fe (14·4-keV) resonance in iron foils at a height separation of 22·5 m, showed that the experimental shift was

[*Refs. on p. 86*]

0·94 ± 0·10 of that predicted. Later experiments have given values of 0·859 ± 0·085 [3] and probably the most accurate value, 0·9970 ± 0·0076 [4, 5]. In this work scrupulous attention must be paid to the elimination of small temperature differentials between source and absorber because of the possibility of spurious results due to the second-order Doppler shift (see p. 50). Although Mössbauer lines with a definition better than that of ^{57}Fe are known, the technical difficulties involved have prevented their use in gravitational red shift experiments.

(b) Accelerational Shift

Similar experiments have been devised to show a red shift in a non-gravitational accelerated system. If a source is mounted at the centre of a high-speed rotor, a shift due to the angular acceleration along the photon path will be detected at an absorber mounted on the periphery. The shift becomes larger as the angular velocity increases in agreement with relativity predictions [6, 7]. Mounting the source and absorber at opposite ends of a diameter of the rotor gives a null result as expected [8].

(c) The 'Clock-paradox'

One of the more intriguing consequences of relativity theory is the so-called 'clock-paradox'. A space traveller goes on an accelerational journey of some magnitude, and carries with him one of two identical clocks, the other being left at his point of departure. Upon his return, he is expected to find that his own clock appears to have been running slowly when compared with the other. In other words, he could become younger than a twin brother who did not travel with him. Sherwin has claimed [9] that the rotor experiment [6] provides experimental proof of this concept.

(d) Aether Drift

In 1960 Ruderfer [10] proposed a uni-directional photon experiment to verify the existence or otherwise of an aether drift. Mounting the Mössbauer source at the centre of a revolving turntable with the absorber and detector fixed on its circumference would cause a line-shift of a diurnal nature as the orientation of the laboratory frame in the aether alters with the earth's revolution. Champeney et al. [11] observed count-rates through absorbers at opposite ends of a rotor diameter with time of day, and found no evidence of a steady drift past the earth resolved parallel to the equatorial plane down to an upper limit of 1·9 m s^{-1}.

(e) Anisotropy of Inertia

It has been postulated that the inertial mass of a body may alter depending on whether it is accelerating towards the centre of the galaxy or in a direction perpendicular to this. The result is a possible splitting or broadening of the Zeeman component of a Mössbauer spectrum [12]. Experiments [13] have shown that such an anisotropy is less than 5×10^{-16}.

(f) Time-reversal Invariance

The suggestion that the electromagnetic interaction involving M1 and E2 multipoles might not be invariant under time-reversal has led to experiments with ^{99}Ru [14, 15] and ^{193}Ir [16] to verify this. In both cases time-reversal invariance was found to hold to within the experimental error.

(g) Refractive-index Measurements

The γ-ray beam between a Mössbauer source and absorber can be frequency-modulated by passage through a γ-transparent medium with an optical path length which varies with time. In this way the refractive index of the medium for the γ-radiation can be calculated [17].

(h) Scattering and Interference

A considerable volume of work has been published on the scattering and interference of the Mössbauer radiation. Some reference to this has already been made in Chapter 2, but as the experiments are not particularly relevant to the theme of this book, it is not intended to enlarge further on that discussion.

4.2 Nuclear Physics

Mössbauer hyperfine spectra are useful in the determination of nuclear parameters, especially those of the excited states. Their significance stems from the fact that the structure of the nucleus is still poorly understood. Comparison of the parameters as measured with the values estimated from theory is used to discover the validity or inadequacy of the nuclear model. The rare-earth elements are popular for this type of work because of the proliferation of Mössbauer resonances, making it feasible to study the effects of successive proton or neutron addition over a range of nuclei. Although theory and experiment are sometimes in accord, gross differences are not unusual.

The following nuclear parameters may be estimated from a Mössbauer hyperfine spectrum. Specific examples will not be given here, but many will be found in Chapter 5 and thereafter. The subject has also been extensively reviewed recently [18].

(a) Lifetime of the Excited State

The observed width of the resonance line when extrapolated to zero absorber thickness can be used to determine the lifetime of the excited state (see

Chapter 1.2 and equation 1.22) as $\Gamma_r(eV) = 9.124 \times 10^{-16}/t_{\frac{1}{2}}$ (s). This only provides a lower limit to the lifetime as the possibility of intrinsic broadening from unresolved hyperfine effects cannot be eliminated. For this reason the lifetime is more usually determined by other methods, but in a few cases the first estimate has been derived from a Mössbauer experiment.

(b) Nuclear Spin

The excited-state nuclear spin quantum number can be determined from the hyperfine spectrum, but this parameter is usually known before a Mössbauer experiment is attempted with an isotope.

(c) Nuclear Quadrupole Moment

If both ground and excited states of the nucleus have a spin of $I > \frac{1}{2}$, then the ratio of the quadrupole moments Q_e/Q_g can be determined. Q_g is usually known from other experiments so that Q_e is derived. If Q_g is zero, Q_e can be calculated using an estimated value of the principal component of the electric field gradient tensor, but in this case the inherent accuracy of the result is usually low.

(d) g-Factors and Nuclear Magnetic Moments

Excited-state g-factors and nuclear magnetic moments are usually determinable with high accuracy because the ground-state values or alternatively the internal magnetic fields are known from other measurements. The excited-state moments have been determined for nearly all Mössbauer transitions by this method. Both the quadrupole and the magnetic moments provide highly significant tests of nuclear models.

(e) Nuclear-radius Changes

The fractional change in the nuclear radius $\delta R/R$ can also be compared with theory. For some of the heavier elements it is also common to describe a deformation parameter $\delta\beta/\beta$ which is given by $[4\pi/(5\beta^2)]\,\delta R/R$ [19]. The difficulties of estimating the electronic charge density at the nucleus result in experimental $\delta R/R$ values being approximate only. Numerical values are listed under individual Mössbauer nuclides in later chapters.

(f) Internal Conversion Coefficient

Careful comparative measurements of the intensity of resonance in the absorption and scattering modes can be used to determine the effective cross-section and hence from equation 1.18 the internal conversion coefficient.

(g) M1/E2 Mixing Ratios

As seen in Chapter 3, the relative proportions of M1 and E2 radiation in a γ-decay can be determined from the relative intensities of hyperfine components.

[Refs. on p. 86]

(h) Decay After-effects

An upper limit can be set to the duration time of several nuclear processes and their consequences such as Auger cascades in the case of electron capture reactions, or the thermal displacement of atoms from one site to another in α-emitting or Coulombic-excitation processes. The subject has recently been reviewed [20].

4.3 Solid-state Physics and Chemistry

The third area of application is also the most extensive and the most important, and by virtue of the diversity of the information obtainable is particularly difficult to summarise concisely. This is partly because the type of information obtainable is very dependent on the nature of the chemical system under study. Some of the more striking applications will be found to be effective only in specific examples and are not general. The insight and ingenuity of the researcher are then paramount.

Generally speaking, the following parameters may be observable:

(1) chemical isomer shift δ
(2) quadrupole splitting Δ
(3) internal magnetic field H
(4) shape of the resonance lines, or their width at half-height Γ
(5) intensity of the resonance lines (and hence the recoilless fraction f)

Additionally it may be advantageous to consider

(6) temperature dependence of (1)–(5)
(7) pressure dependence of (1)–(5)
(8) effect of an externally applied magnetic field on the spectrum
(9) the angular dependence of (5) in magnetically polarised alloys or single crystals
(10) the spatial relationship between H and the electric field gradient tensor.

Mössbauer spectroscopy has the two important advantages that it is non-destructive, and that it probes the solid state microscopically at the atomic dimensions rather than in bulk, so that one observes the statistical sum of the individual isotope environments rather than a bulk average. The following synopses indicate the general properties which can be determined in suitable cases.

(a) Molecular Compounds and Ionic Complexes

The formal oxidation state of the resonant atom can be determined, together with the coordination number, and the site symmetry including the parameters describing the electric field gradient tensor and its relation to the internal magnetic field if one is present. Temperature-dependence data give ligand-field parameters such as energy level separations, tetragonal or trigonal distortions, degree of covalency and electron delocalisation, and spin–orbit coupling. Ligand-field strengths and π-bonding characters of a series of ligands in related complexes can be compared. Distortions due to nearest-neighbour interactions or Jahn–Teller effects may be seen. Where two chemically different positions of the resonant isotope occur in the compound, they can be distinguished unambiguously if the compound is prepared with selective isotopic enrichment at one of them.

The detailed electronic configuration of a transition-metal ion (e.g. $S = \frac{5}{2}, \frac{3}{2}$, or $\frac{1}{2}$ for a d^5 ion), orbital populations, and mixed ligand-field crossover complexes can be studied in detail. Electronic spin relaxation is known in paramagnetic ions. The pressure dependence may give additional information on the bonding-type and anisotropy of the crystal.

The recoilless fraction and the second-order Doppler shift reflect the vibrational properties of the crystal. Different atoms in the same molecule may have dissimilar vibrational properties, but interpretation of recoilless fractions can only be rudimentary.

(b) Oxides and Chalcogenides

The formal oxidation states of metal cations can be deduced, and fast 'electron hopping' processes detected. It is possible to study site-occupancy and disorder in ferrite type oxides, as well as other forms of nonstoichiometry and lattice defects. In magnetic ferrites the magnetisation of distinct sites can be followed as opposed to the bulk magnetisation properties, and the number and directions of the individual magnetic axes obtained. This feature is particularly useful when the ordering is antiferromagnetic. Particle-size and electronic-relaxation effects are known. Oxides of non-Mössbauer elements can be studied indirectly by impurity doping with a suitable radio-isotope. If the latter is a paramagnetic ion the magnetic ordering in the host material can be traced by the interaction with the impurity.

(c) Metals and Alloys

The magnetic properties of an alloy can be studied as a function of the temperature and concentration of the constituent metals. Order–disorder phenomena can be measured directly and compared with the thermal history and mechanical properties of the alloy. The statistical occupation of the first coordination sphere round the resonant atoms in binary alloys can be derived from magnetic spectra.

Core-polarisation and conduction-electron polarisation effects can be studied as can exchange polarisation of diamagnetic atoms in magnetic hosts. The lattice dynamics of the metal lattice are examined via the temperature dependence of the f-factor. Many metals approximate closely to the Debye model, and a Debye temperature has some significance. Impurity doping can also be used to study the magnetisation and lattice dynamics of non-resonant metals.

(d) Miscellaneous Topics

The chemical bonding of adsorbed surface states can be studied using radioactive doping source experiments, or scattering experiments. Although non-viscous solutions do not give a Mössbauer effect, frozen solutions usually do; however, it is not easy to relate the data obtained to the solution phase. It is possible to identify and analyse minerals containing the resonant isotope. Phase changes and solid-state reactions can be observed, and kinetic experiments are feasible if the rate of reaction is slow. Non-destructive testing of materials such as ferrites is applicable to quality control techniques.

Examples of nearly all the above-mentioned applications are to be found in the following chapters on the 14·4-keV decay of ^{57}Fe. The massive volume of data available for this resonance far exceeds the combined data for all other isotopes.

REFERENCES

[1] T. E. Cranshaw, J. P. Schiffer, and A. B. Whitehead, *Phys. Rev. Letters*, 1960, **4**, 163.
[2] R. V. Pound and G. A. Rebka, Jr, *Phys. Rev. Letters*, 1960, **4**, 337.
[3] T. E. Cranshaw and J. P. Schiffer, *Proc. Phys. Soc.*, 1964, **84**, 245.
[4] R. V. Pound and J. L. Snider, *Phys. Rev. Letters*, 1964, **13**, 539.
[5] R. V. Pound and J. L. Snider, *Phys. Rev.*, 1965, **140**, B788.
[6] H. J. Hay, J. P. Schiffer, T. E. Cranshaw, and P. A. Egelstaff, *Phys. Rev. Letters*, 1960, **4**, 165.
[7] W. Kundig, *Phys. Rev.*, 1963, **129**, 2371.
[8] D. C. Champeney and P. B. Moon, *Proc. Phys. Soc.*, 1961, **77**, 350.
[9] C. W. Sherwin, *Phys. Rev.*, 1960, **120**, 17.
[10] M. Ruderfer, *Phys. Rev. Letters*, 1960, **5**, 191.
[11] D. C. Champeney, G. R. Isaak, and A. M. Khan, *Phys. Letters*, 1963, **7**, 241.
[12] C. Cocconi and E. E. Salpeter, *Phys. Rev. Letters*, 1960, **4**, 176.
[13] C. W. Sherwin, J. Frauenfelder, E. L. Garwin, E. Lüscher, S. Margulies, and R. N. Peacock, *Phys. Rev. Letters*, 1960, **4**, 399.
[14] O. C. Kistner, *Phys. Rev.*, 1966, **144**, 1022.
[15] O. C Kistner, *Phys. Rev. Letters*, 1967, **19**, 872.
[16] M. Atac, B. Chrisman, P. Debrunner, and H. Frauenfelder, *Phys. Rev. Letters*, 1968, **20**, 691.
[17] L. Grodzins and E. A. Phillips, *Phys. Rev.*, 1961, **124**, 774.
[18] N. A. Burgov and A. V. Davydov, 'Applications of the Mössbauer Effect in Nuclear Physics', Chap. 12 in 'Chemical Applications of Mössbauer Spectroscopy' by V. I. Goldanskii and R. H. Herber, Eds., Academic Press, New York, 1968.
[19] J. A. Stone and W. L. Pillinger, *Symposia Faraday Soc.*, No. 1, 1967, p. 77.
[20] H. H. Wickman and G. K. Wertheim, 'Spin Relaxation in Solids and After-effects of Nuclear Transformations', Chap. 11 in 'Chemical Applications of Mössbauer Spectroscopy' by V. I. Goldanskii and R. H. Herber, Eds., Academic Press, New York, 1968.

5 | ⁵⁷Fe – Introduction

Seventy-five per cent of all papers on experimental Mössbauer spectroscopy are concerned with the first excited-state decay of ^{57}Fe; indeed, because of this, ^{57}Fe and Mössbauer spectroscopy are synonymous to many people. The sheer volume of published work makes a completely exhaustive survey impossible, but in the following chapters a series of critical reviews will be given of specific areas defined by chemical classification. In this way a comprehensive view of the field can be obtained. The nuclear parameters of the ^{57}Fe γ-decay have also received more attention than usual. This chapter summarises the currently available data on the ^{57}Fe nuclear parameters, and then applies the general theory of hyperfine interactions given in Chapter 3 to the specific case of this isotope.

5.1 The γ-Decay Scheme

It is fortunate, though quite fortuitous, that an element of great commercial importance should have an isotope with an almost ideal combination of physical properties for Mössbauer spectroscopy. The energy levels of the excited states are shown in Fig. 5.1. The excited state at 136·32 keV is populated by electron capture with 99·84% efficiency from ^{57}Co whose half-life of 270 days is conveniently large; 11% of the decays from the 136·32-keV state result in a 136·32-keV γ-ray, and 85% a 121·9-keV γ-ray. The first excited state of energy 14·412 keV [1] is thus efficiently populated by ^{57}Co decay. A Mössbauer resonance from the 136·32-keV level has also been reported, but as this is of minor importance, it will be left until the last section in this chapter. The 14·41-keV transition occurs, without change in parity, between an excited state of nuclear spin quantum number $\frac{3}{2}$ and a ground state of spin $\frac{1}{2}$, and the transition is therefore primarily of M1 type, the amount of E2 admixture being only 0·0006%, which is negligible. The lifetime of this level has been variously given in recent years as 98 ± 1 ns [2], 98 ± 4 ns [3], 97·7 ± 0·2 ns [4], and 99·3 ± 0·5 ns [5]. Taking the last value, one obtains a Heisenberg width Γ_r of 0·192 mm s^{-1} for the Mössbauer

[*Refs. on p. 110*]

88 | ⁵⁷Fe – INTRODUCTION

resonance, which is large enough to prevent major problems in instrumentation.

The 14·41-keV γ-transition is internally converted to a high degree during the decay. Many early papers used a value for the internal conversion

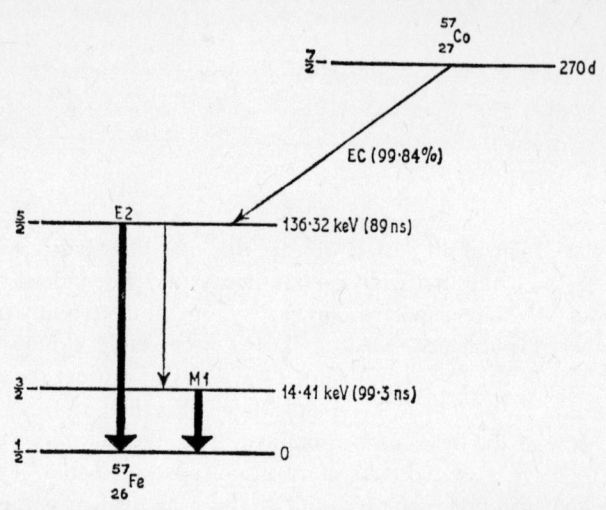

Fig. 5.1 The γ-decay scheme of ⁵⁷Co showing the 14·41-keV and 136·32-keV Mössbauer transitions.

coefficient, α, of 15, but this was found to be obviously too large when used for calculating absolute recoil-free fractions. Some of the recently available values for α are 9·0 ± 0·4 from magnetic field polarisation measurements on an iron foil [6], 9·3 ± 0·5 from scattering measurements [7], 9·5 ± 0·5 from K X-ray/γ-ray intensity ratios [8], 9·0 ± 0·5 from γ–γ coincidence measurements [2], and 8·9 ± 0·6 from Mössbauer absorption cross-sections [2]. The most recent value of 8·17 ± 0·25 obtained from X–X and X–γ coincidence counting is considerably below this range but may prove to be the most accurate yet obtained [9]. The important consequence of the large value of α is that only about 10% of the nuclear decays via the first excited level actually emit a γ-ray.

The relative abundance of ⁵⁷Fe in natural iron is rather low (2·19%) but the isotope has a large cross-section for capture of the 14·41-keV γ-ray (theoretically $2·57 \times 10^{-18}$ cm²), and isotopic enrichment is only needed when the iron content of the sample is very low, as for example in proteins or very dilute iron alloys. Enrichment with 90% ⁵⁷Fe in the latter cases generally gives an adequate cross-section. The free-atom recoil energy of $1·95 \times 10^{-3}$ eV is low and ensures a high proportion of recoilless events.

[*Refs.* on p. *110*]

5.2 Source Preparation and Calibration

The universal source for ^{57}Fe 14·41-keV work is the ^{57}Co isotope. This is produced by the (d, n) reaction on ^{56}Fe (91·52% abundant in natural iron). The process is most effective when the deuteron energies are about 9·5 meV. The cyclotron target is dissolved in mixed mineral acids, extracted with isopropyl ether and then passed down an anion-exchange column to give nearly carrier-free activity. Two different experimental situations can now be envisaged. For examination of a series of compounds rich in iron it is convenient to have a source with a single emission line of the minimum possible linewidth so that any hyperfine effects observed are produced in the absorbers only. Alternatively, when the content is very small, it is better to introduce the ^{57}Co into each compound or matrix to be studied, thereby generating ^{57}Fe impurity nuclei, which can be compared with an appropriate single-line absorber. The former technique requires a narrow-line source with a high recoilless fraction at room temperature, and much effort has been expended in the search for such sources. Metallic matrices are the obvious choice because of the ease of manufacturing homogeneous alloys with the cobalt. Furthermore, such alloys avoid decay after-effects which sometimes produce the resonant nucleus in an anomalous charge-state in insulating materials. Additional advantages are the high f-factors, and the involatility and chemical stability of such sources in the normal laboratory atmosphere. Early experiments used ^{57}Co diffused into metallic iron, but magnetically ordered metals give hyperfine magnetic splittings resulting in complex spectra. Certain stainless steels give a single line, but these are broader than the Heisenberg width because of self-resonant absorption. Chromium, copper, platinum, and palladium have all been used with success, the last mentioned possibly being the best, as the matrices are not self-resonant and give a high recoilless fraction with no appreciable line broadening.

The most difficult part of the source preparation is the incorporation of the ^{57}Co activity; this is usually achieved by electrolytic deposition and subsequent diffusion by annealing. Many papers give brief details, but the precise conditions have often been jealously guarded at the time of their development. However, most of the commonly used sources are now available commercially with guaranteed performance. Some useful accounts have been published [10–13].

It is not uncommon to use the energy of the source transition as the arbitrary reference point when defining the chemical isomer shift (see equation 3.12). However, the chemical isomer shift observed is different for each type of source matrix. This means that although the data from one particular laboratory may be self-consistent, intercomparison of data from different institutions may be difficult. Chemical isomer shifts for the more common matrices are given in Table 5.1.

[*Refs. on p. 110*]

90 | ^{57}Fe – INTRODUCTION

Some attempt has been made to establish a reference standard for calibration of chemical isomer shifts. The metallic iron spectrum is popular because there are six hyperfine components giving a linearity check as well as defining a zero. α-Fe$_2$O$_3$ has been used similarly. The spectrum of sodium

Table 5.1 Chemical isomer shifts of commonly used ^{57}Co sources with respect to iron metal at room temperature

Source	Shift/(mm s^{-1})	Abbreviation
^{57}Co/Au	+0·620	Au
^{57}Co/Ag	+0·514	Ag
^{57}Co/Pt	+0·347	Pt
^{57}Co/Cu	+0·226	Cu
^{57}Co/Ir	+0·215	Ir
^{57}Co/Pd	+0·185	Pd
^{57}Co/Fe	0·000	Fe
^{57}Co/SS*	~−0·09	SS
^{57}Co/Cr	−0·152	Cr
Na$_2$[Fe(CN)$_5$NO].2H$_2$O	−0·257	NP

* SS = Stainless steel. The precise shifts depend upon the type of steel, e.g. Vacromium −0·077; 310 SS −0·096. All figures contain an uncertainty of about ±0·01 mm s^{-1}.

nitroprusside, Na$_2$[Fe(CN)$_5$NO].2H$_2$O, was adopted by the National Bureau of Standards, U.S.A., but, because it has only two peaks with a large separation, it provides no check on linearity and may cause comparatively large errors in recorded shifts. It does, however, have the advantage that nearly all ^{57}Fe chemical isomer shifts are then positive in sign. Because it seems probable that metallic iron foil will eventually be adopted as the international standard, all the ^{57}Fe data in this volume are quoted with respect to it; where this is different from the reference zero adopted in the publication being discussed the original reference zero compound is also given. The appropriate addition factors and the abbreviations used are in Table 5.1. A detailed discussion of the iron-foil spectrum is given in Chapter 11. The most accurate determination of the parameters for sodium nitroprusside gives [14] the quadrupole splitting as 1·7048 ± 0·0025 mm s^{-1} and a chemical isomer shift relative to the ^{57}Co/Cu source of −0·4844 ± 0·0010 mm s^{-1}.

5.3 Chemical Isomer Shifts

The existence of a chemical isomer shift produced by the change in nuclear radius and differing chemical environments was first demonstrated by Kistner and Sunyar [15] in 1960 for the case of a stainless steel source and an absorber of α-Fe$_2$O$_3$. Early work quickly showed that the chemical isomer shift measured was related to the formal oxidation state of the iron. This is illus-

[*Refs. on p. 110*]

trated in Fig. 5.2 in which the range of shifts observed in practice is approximately indicated for each principal oxidation state or electronic configuration. The first attempted theoretical correlation was by Walker, Wertheim, and Jaccarino [16] in 1961; this will be referred to as the WWJ model. The WWJ model applies to high-spin compounds of iron and was based on some

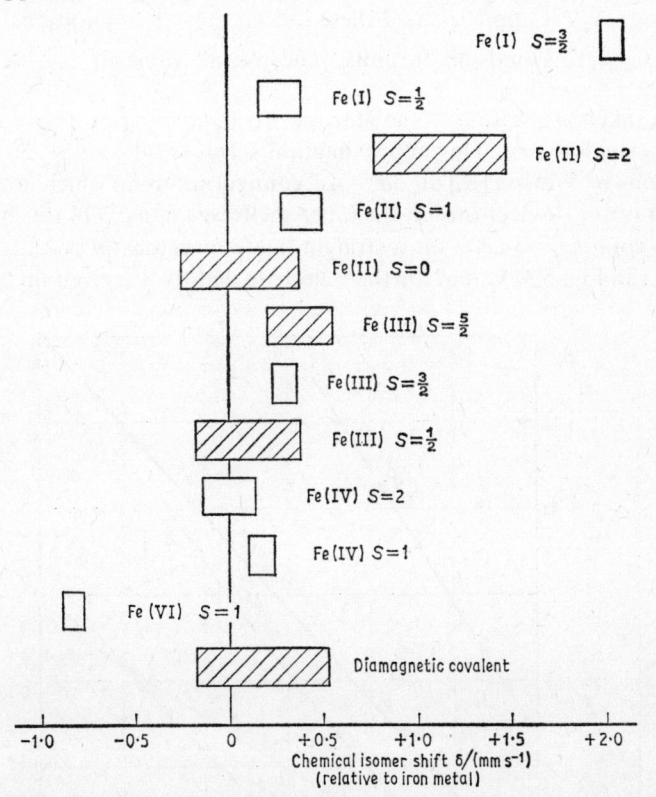

Fig. 5.2 Approximate representation of the ranges of chemical isomer shifts found in iron complexes. The values are related to iron metal at room temperature, but do allow for temperature variations in the absorber. The width of the individual distributions may be attributed to covalency and as a general rule the more ionic compounds have the higher shift. The most common configurations are cross-hatched, while the diamagnetic covalent classification includes carbonyls and ferrocenes etc. In most series the shift decreases as spin pairing takes place. Some of the ranges may in fact extend beyond the limits shown, and in particular it is difficult to predict where the lower limit of the Fe(II) $S = 2$ group lies. Note that the Fe(IV) $S = 2$ subgroup refers to oxides with collective-electron bonding and for which the electron configuration is not simply defined. The Fe(IV) $S = 1$ compounds have a distinct low-spin configuration.

restricted Hartree–Fock calculations by Watson of $|\psi_{3s}(0)|^2$ and the two inner s-shell values for various outer configurations $3d^n$ [17, 18]. Unrestricted

^{57}Fe – INTRODUCTION

Hartree–Fock calculations give exactly the same total density at the nucleus. The 1s- and 2s-electron densities are not appreciably affected by the change in the number of 3d-electrons, but the variations in $|\psi_{3s}(0)|^2$ can be attributed directly to 3d-shielding.

Faced with the problem of estimating the contributions from the outer 4s-electrons, WWJ approximated these for $3d^{n-1}4s^1$ configurations by using the Fermi–Segrè–Goudsmit formula. The overall value of $\sum_{n=1-3} |\psi_{ns}(0)|^2$ was then taken as the sum of the Hartree–Fock values appropriate to $3d^{n-1}$ plus the extra 4s-term. This approximation seems valid in view of the later calculations by Watson [18] on $3d^{n-2}4s^2$ configurations, in which no substantial effect of the 4s-electrons on the inner shells was noted. On this basis, the $3d^{n-x}4s^x$ configuration lies on a straight line connecting $3d^n$ and $3d^{n-1}4s$ as illustrated in Fig. 5.3. Values for the s-electron density are given in Table 5.2.

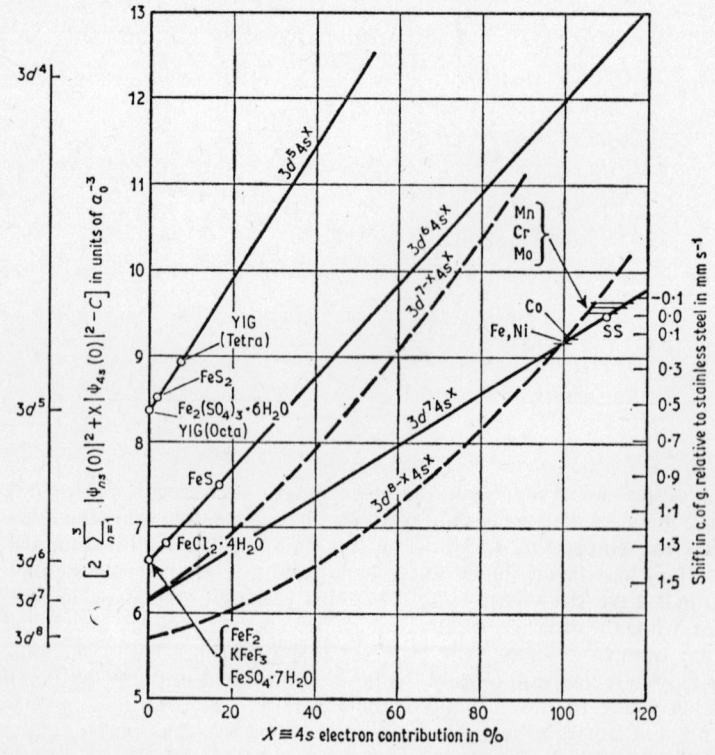

Fig. 5.3 The Walker–Wertheim–Jaccarino interpretation of the ^{57}Fe chemical isomer shift. The values of the total s-electron density are probably realistic, but the difficulty arises in relating these to actual chemical compounds. In particular the assumption of a $3d^5$ configuration for $Fe_2(SO_4)_3.6H_2O$ is suspect. The constant $C = 11{,}873 a_0^{-3}$. [Ref. 16, Fig. 1]

[*Refs. on p. 110*]

CHEMICAL ISOMER SHIFTS | 93

There is no reason to doubt the general validity of these assumptions for free-atom configurations. The main problem is in applying these free-atom values to a practical scale of Mössbauer chemical shifts in real compounds.

The WWJ model assumed that the 'typical' ferrous and ferric salts have $3d^64s^0$ and $3d^54s^0$ atomic configurations respectively, and a scale of chemical

Table 5.2 Electron densities at $r = 0$ for different configurations of an Fe atom [17–18]

Electrons per cubic Bohr radius	$3d^8$	$3d^7$ Fe$^+$	$3d^6$ Fe^{2+}	$3d^5$ Fe^{3+}	$3d^64s^2$ free atom
$\psi^2_{1s}(0)$	5378·005	5377·973	5377·840	5377·625	5377·873
$\psi^2_{2s}(0)$	493·953	493·873	493·796	493·793	493·968
$\psi^2_{3s}(0)$	67·524	67·764	68·274	69·433	68·028
$\psi^2_{4s}(0)$					3·042
$2 \sum_n \psi^2_{ns}(0)$	11,878·9	11,879·2	11,879·8	11,881·7	11,885·8

isomer shifts was derived accordingly. The compounds $Fe_2(SO_4)_3 \cdot 6H_2O$ and FeF_2, $KFeF_3$, and $FeSO_4 \cdot 7H_2O$ were those used, and the lower chemical isomer shifts in FeS for example were attributed to some additional 4s-density. The derived difference in shift for $3d^5$ and $3d^6$ configurations was then used with the s-electron density values to derive a value of $-1 \cdot 8 \times 10^{-3}$, for $\delta R/R$ (or $\delta \langle R^2 \rangle / \langle R^2 \rangle = -3 \cdot 6 \times 10^{-3}$), i.e. the nucleus expands after emission of the γ-ray and contracts on absorption of a γ-ray by the ground state. Addition of the 3d-electron decreases the total s-density at the nucleus and increases the chemical isomer shift.

The WWJ treatment was formulated at a time when little chemical isomer shift data had been measured for iron compounds, and it also tended to neglect the effects of chemical bonding, a point which has been mentioned by several authors. Danon [19] attempted to improve the WWJ treatment by allowing for a change in configuration of the ferric ion on bonding. The high charge of the nominal ferric ion is effectively reduced in complexes by displacement of ligand charge towards the cation, bringing the 4s-shell into the bonding scheme. Danon adopted the configuration $3d^54s^{0 \cdot 32}$ for the complex ferric anion $[Fe^{III}F_6]^{3-}$, a value which had been deduced from M.O. calculations and is significantly different from the free-ion value of $3d^54s^0$. This value can then be used with the $3d^64s^0$ configuration still assumed for ionic ferrous complexes to re-calibrate the chemical isomer shift scale as shown in Fig. 5.4. One implication that follows is that $\delta R/R$ was overestimated by WWJ, and is closer to -7×10^{-4}. Typical configurations derived from Danon's calibration are $3d^54s^{0 \cdot 40}$ for $[Fe^{III}Cl_4]^-$, $3d^44s^{0 \cdot 08}$ for

[Refs. on p. 110]

94 | ^{57}Fe – INTRODUCTION

SrFe^{IV}O$_3$, and 3d$^{>3}$ for K$_2$Fe^{VI}O$_4$. These figures are chemically more satisfactory than any interpretation using the WWJ model.

Neither model allows for covalent bonding to the ligands and consequently low-spin complexes and covalent diamagnetic complexes with low formal

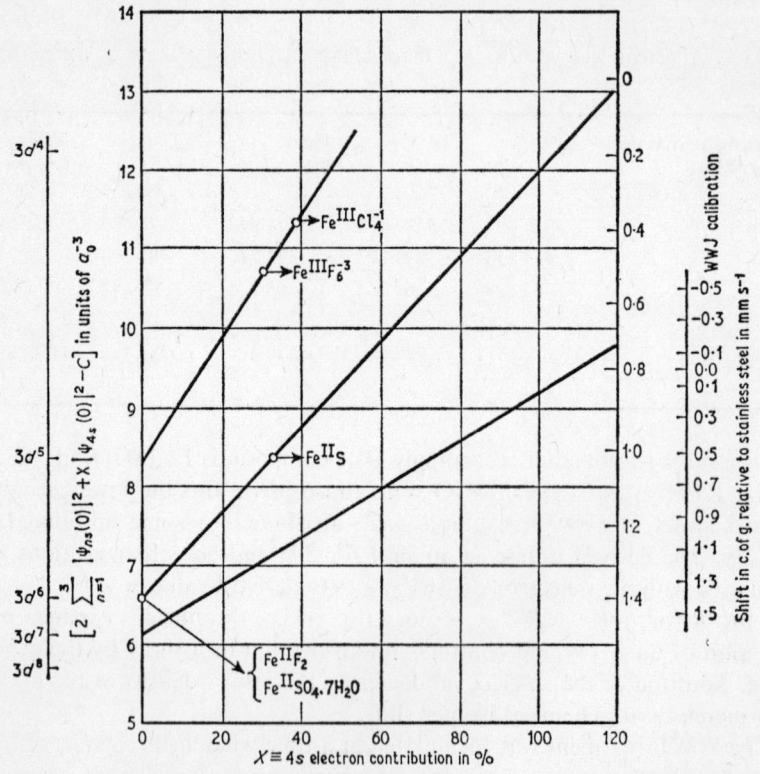

Fig. 5.4 A reinterpretation of the WWJ figures by Danon allowing for substantial 4s density in ferric compounds. [Ref. 19, Fig. 6]

oxidation states lie outside the scope of the theory. Difficulties are also encountered in the quantitative interpretation of the chemical isomer shift in metals and alloys. One of the more recent examples of this is in iron metal itself [20]. A further consideration which was neglected is electron correlation interactions in non-spherical symmetry which could produce shielding effects.

A more recent approach by Šimánek and Šroubek [21] has been to interpret the pressure dependence of the chemical isomer shift for a single compound. The first attempt involved the calculation of the overlap distortion effect on $|\psi_s(0)|^2$ due to a decreasing Fe–O bond distance for an ^{57}Fe divalent impurity atom in CoO. Incorporation of 4s-bonding into the orbitals used

[Refs. on p. 110]

showed that a large decrease in shift could originate from a small increase in 4s-electron occupancy. The net result is a value of $\delta R/R$ of $\sim -4 \times 10^{-4}$, and the configurations of Fe^{2+} and Fe^{3+} in their most ionic compounds in this system are $3d^6 4s^0$ and $3d^5 4s^{0.2}$ respectively. The difference in chemical isomer shift of ferric iron at the two sites of garnets and spinel oxides can be interpreted as the result of overlap effects on configurations of $3d^5 4s^{0.2}$ for the octahedral B site and $3d^5 4s^{0.32}$ for the tetrahedral A site. The increase in $|\psi_s(0)|^2$ at the tetrahedral site arises mainly from the direct 4s-covalency effect.

A later analysis by Šimánek and Wong [22] of pressure data for $KFeF_3$ is more accurate because of the better wavefunctions available for the fluorine ion and gives a value for $\delta R/R$ of -5.2×10^{-4}. A possible objection to the method is that the orbital populations may change with pressure, a factor not easily verified, but the results are at least more consistent with chemical considerations.

The low-spin and diamagnetic complexes such as carbonyls show very little correlation of chemical isomer shift with formal oxidation state. This is because of the high degree of covalent overlap between the central metal atom and the ligands. Many of the main problems which apply to these compounds have been set out by Shulman and Sugano [23] in a detailed molecular-orbital analysis of the complex iron cyanides. The cyanide ligand represents one of the more complicated cases to consider, as not only does it have filled π-orbitals of energy suitable for bonding, but it also has unfilled π^*-antibonding orbitals which can accept electrons from the central ion.

The various electronic terms in the structure can be classified as

  a) ligand to metal σ donation from filled ligand σ-orbitals
  b) metal to ligand σ donation into unfilled ligand σ^*-orbitals
  c) ligand to metal π donation from filled ligand π-orbitals
  d) metal to ligand π donation into unfilled ligand π^*-orbitals.

a), c), and d), are all likely to be significant in the cyanides. The important metal orbitals concerned will be the 3d- and 4s-orbitals. Mechanisms a) and c) will increase the effective number of 3d-electrons at the metal, thereby increasing the screening of the 3s-electrons, decreasing the s-electron density at the nucleus, and increasing the value of the chemical shift. This contribution is opposite in sign to the effect of any concurrent increase by a) in the 4s-population. Mechanisms b) and d) have the reverse effect, but since the major term, d) can only involve the 3d-orbitals and not the 4s-, it will usually result in a decrease in the chemical shift.

The calculations for the ferrocyanide and ferricyanide complex anions used parameters from electron-spin resonance and optical absorption spectra [23]. The net result is that in $[Fe^{II}(CN)_6]^{4-}$ about one more electron charge resides on the ligands than in $[Fe^{III}(CN)_6]^{3-}$. In other words, when the paramagnetic

[*Refs. on p. 110*]

ferricyanide ion is reduced to the diamagnetic ferrocyanide ion one electron is added to the iron t_{2g} system, but simultaneously the equivalent of almost one electron charge is delocalised via d_π–p_π back bonding from the central atom on to the peripheral cyanide ligands, thus leaving the overall charge density on the iron atom unaltered. This is fully consistent with the observed small difference in chemical isomer shift of 0·08 mm s^{-1} between ferro- and ferri-cyanides (cf. the difference of ~0·8 mm s^{-1} between Fe^{2+} and Fe^{3+}). As Danon has pointed out [19], the high-spin and low-spin complexes are essentially different in other ways. The series of chemical isomer shifts for the high-spin compounds FeF_3 > $FeCl_3$ > $FeBr_3$ decrease with increasing tendency of the ligands to donate electrons to the metal ion, whereas the series for low-spin complexes $[Fe(CN)_5(NO)]^{2-}$ < $[Fe(CN)_5CO]^{3-}$ < $[Fe(CN)_6]^{4-}$ reflect the increasing tendency of the ligand to accept electrons from back donation by the metals.

Further correlations of s-density with environment have been made purely on an *ad hoc* empirical basis for particular groups of compounds, and at present it is rarely possible to discuss more than relative trends as opposed to detailed quantitative values.

The second-order Doppler shift (see Chapter 3.2) is usually not negligible in iron compounds, and because it is a function of the lattice dynamics of the compound, which are rarely understood, it is difficult to eliminate prior to making comparisons of s-electron densities. Consequently there is usually a limit to the significance of small differences in chemical isomer shift in related compounds and care should be taken not to over-analyse data in terms of very small changes in electron density.

5.4 Quadrupole Splittings

As detailed in Chapter 3.4 the quadrupole hyperfine spectrum of a $\frac{3}{2} \to \frac{1}{2}$ Mössbauer transition in a polycrystalline absorber will be a simple doublet in which the separation of the lines is $\Delta = (e^2qQ/2) \times (1 + \eta^2/3)^{\frac{1}{2}}$. Many iron compounds have a finite electric field gradient, $-eq$, and in practice Δ is found to have values in the range 0–4 mm s^{-1}, the two lines being easily resolved for values greater than 0·2 mm s^{-1}. Since the electric field gradient is generated predominantly by the electronic orbitals in the environment of the resonant nucleus, and to a lesser extent by the distribution of lattice charges, estimation of these effects is of fundamental importance in studying the electronic configuration of iron complexes. In the absence of an internal magnetic field, the asymmetry parameter η and the sign of e^2qQ can only be determined from spectra of single crystals or from powders in high applied magnetic fields, so that for the majority of compounds, only $|\Delta|$ has been measured. As with chemical isomer shift data, difficulties arise in constructing an absolute scale of values for the chemical term. This is because the value of the nuclear quadrupole moment Q cannot be determined independently for iron.

Five basic approaches have been adopted for the estimation of Q and the results are summarised in Table 5.3. The first method involves the observed quadrupole spectrum of a high-spin ferric complex. The $3d^5$-ion is spherically symmetric in the free state so that it has no intrinsic electric field gradient. One can therefore evaluate eq in a crystal lattice as a summation of all the external (lattice) charges of other ions. Although the concept is simple, small errors in X-ray structural data can cause large errors in eq, and the ionic charge is not concentrated exclusively at the lattice positions (as assumed in the point-charge model). An even greater difficulty is in allowing for the antishielding term $(1 - \gamma_\infty)$ in equation 3.36, some of the problems of which have been outlined by Freeman and Watson [24].

The second method of evaluating Q is from the calculated electric field gradient of a high-spin ferrous complex in which the electronic ground state is known so that the term q_{ion} in equation 3.36 can be evaluated from the expressions in Table 3.1. Here again, the value of $(1 - R)\langle r^{-3} \rangle$ is also difficult to evaluate as the degree of ligand-ferrous ion interaction will affect the radial expansion and shielding of the electrons. For this reason one of the most ionic ferrous complexes, $FeSiF_6.6H_2O$, has normally been used.

A third method uses data for a neutral molecular complex. Under the assumption that ferrocene and cobalticinium perchlorate have similar chemical bonding, the pure nuclear quadrupole resonance data for the latter together with the independent determination of Q_g for ^{59}Co from atom-beam spectroscopy data have been used to estimate $Q_e(^{57}Fe)$. The method has the advantage that the lattice sums are less than 1% of the total electric field gradient.

The fourth method for evaluating Q measures the quadrupole splitting induced at 4.2 K by an applied field of 50 kG acting on an oriented single crystal of MgO doped with Fe^{2+}. The magnetic field induces a quadrupole hyperfine interaction in the following way: the ^5D electronic term of Fe^{2+} is split in the cubic field into an upper orbital doublet and a lower triplet, the latter being further split linearly by the applied magnetic field; at low temperature the three states are not equally populated, thus inducing an electric field gradient which in turn produces the electrical quadrupole hyperfine interaction. Using the accepted values of $(1 - R)\langle r^{-3} \rangle = 3.3$ au and an orbital reduction factor of $k = 0.8$, the observed quadrupole splitting of 0.32 mm s^{-1} leads to a value of $Q = +0.21 \pm 0.03$ barn.

The fifth approach involves a direct comparison of ^{27}Al and ^{57}Fe quadrupole coupling constants from nuclear magnetic resonance and Mössbauer data respectively in $CoAl_2O_4$ and $ZnAl_2O_4$ doped with 5% ^{57}Fe. Ferric iron substitutes at aluminium sites and it is assumed that there is no additional distortion induced upon replacement of one spherical ion by another. The

[Refs. on p. 110]

^{57}Fe – INTRODUCTION

Table 5.3 Estimates of the quadrupole moment, Q, of the 14·41-keV level of ^{57}Fe

Q/barn	Method	Reference
−0·19	α-Fe$_2$O$_3$ lattice sums	1
+0·1	FeF$_2$	2
+0·4	YIG and α-Fe$_2$O$_3$ lattice sums	3
+0·12	Δ in FeSiF$_6$.6H$_2$O	4
+0·15	Δ in FeSiF$_6$.6H$_2$O, FeSO$_4$.7H$_2$O	5
+0.15	Δ in FeSiF$_6$.6H$_2$O	6
+0·28	Recalculation of Burns' data	7
+0·29	Δ in FeSiF$_6$.6H$_2$O	8
+0·2	lattice sums in ferric garnets	9
+0·277	α-Fe$_2$O$_3$ lattice sums	10
+0·41	α-Fe$_2$O$_3$ lattice sums	11
+0·28	α-FeOOH lattice sums	12
+0·25	lattice sum of 4CaO.Al$_2$O$_3$.Fe$_2$O$_3$	13
+0·20	Δ and lattice sum in FeSiF$_6$.6H$_2$O	14
+0·17	Fe^{2+} and Fe^{3+} haemoglobins	15
+0·18	anisotropy of H in FeSiF$_6$.6H$_2$O	16
+0·283	α-Fe$_2$O$_3$ lattice sums	17
+0·17	ferrocene	18
+0·21	applied H on Fe^{2+}-doped MgO	19
+0·21	Δ and covalency in FeSiF$_6$.6H$_2$O	20
+0·16 < Q < 0·29	α-Fe$_2$O$_3$ lattice sums	21
+0·20	NMR and Δ in CoAl$_2$O$_4$ (^{57}Fe)	22

References
1. R. Bersohn, *Phys. Rev. Letters*, 1960, **4**, 690.
2. A. Abragam and F. Boutron, *Compt. rend.*, 1961, **252**, 2404.
3. G. Burns, *Phys. Rev.*, 1961, **124**, 524.
4. C. E. Johnson, W. Marshall, and G. J. Perlow, *Phys. Rev.*, 1962, **126**, 1503.
5. R. Ingalls, *Phys. Rev.*, 1962, **128**, 1155.
6. H. Eicher, *Z. Physik*, 1963, **171**, 582.
7. R. M. Sternheimer, *Phys. Rev.*, 1963, **130**, 1423.
8. R. Ingalls, *Phys. Rev.*, 1964, **133**, A787.
9. W. J. Nicholson and G. Burns, *Phys. Rev.*, 1964, **133**, A1568.
10. R. R. Sharma and T. P. Das, *J. Chem. Phys.*, 1964, **41**, 3581.
11. J. O. Artman, *Phys. Rev.*, 1966, **143**, 541.
12. F. van der Woude and A. J. Dekker, *Phys. Stat. Sol.*, 1966, **13**, 181.
13. F. Wittmann, *Phys. Letters*, 1967, **24A**, 252.
14. A. J. Nozik and M. Kaplan, *Phys. Rev.*, 1967, **159**, 273.
15. M. Weissbluth and J. E. Maling, *J. Chem. Phys.*, 1967, **47**, 4166.
16. C. E. Johnson, *Proc. Phys. Soc.*, 1967, **92**, 748.
17. J. O. Artman, A. H. Muir, Jr, and H. Wiedersich, *Phys. Rev.*, 1968, **173**, 337.
18. C. B. Harris, *J. Chem. Phys.*, 1968, **49**, 1648.
19. J. Chappert, R. B. Frankel, A. Misetich, and N. A. Blum, *Phys. Letters*, 1969, **28B**, 406.
20. R. Ingalls, *Phys. Rev.*, 1969, **188**, 1045.
21. M. Raymond and S. S. Hafner, *Phys. Rev. B*, 1970, **1**, 979.
22. M. Rosenberg, S. Mandache, H. Niculescu-Majewska, G. Filotti, and V. Gomolea, *Phys. Letters*, 1970, **31A**, 84.

^{27}Al quadrupole moment is known to be 0·149 barn, and by assuming appropriate values for the Sternheimer antishielding factors a value of 0·20(1) barn is derived directly. The accuracy of the method is determined by the accuracy with which the coefficients can be estimated.

A large number of estimates of Q are available, and these are listed in Table 5.3 in approximate chronological order. Assuming that the more recent determinations are the less inaccurate, it appears that Q has a value of between 0·18 and 0·28 barns.

Fortunately, the preceding difficulties of separating the nuclear and chemical contributions to e^2qQ do not prevent considerable use being made of quadrupole splitting data on a comparative basis. The type of behaviour anticipated is strongly dependent on the oxidation state and electronic spin multiplicity of the iron. Thus, the high-spin Fe^{3+} cation has a $t_{2g}^3 e_g^2(^6S)$ electronic ground state and is spherically symmetric. Low-spin Fe^{III} behaves similarly as a result of its $t_{2g}^6(^1A)$ configuration in the ground state. Any electric field gradient will arise solely from charges external to the central atom either by direct contribution or by indirect polarisation. It is unlikely that there will be any low-lying excited levels in either of these configurations, so that thermal alterations of electron configuration are small, and it is predicted (and observed) that the electric field gradient in both Fe^{3+} and Fe^{II} compounds will be essentially independent of temperature.

The effects of anisotropic covalency on the quadrupole splitting of high-spin ferric compounds has been considered theoretically for compounds of D_{4h} symmetry [25]. Using appropriate molecular orbitals and the spectroscopic g-tensor to give a physical significance to the results, it can be shown that an anisotropy of the g-tensor of $\Delta g_\parallel - \Delta g_\perp$ of the order of 0·0001 is sufficient to cause a predicted Mössbauer splitting of 0·01 mm s^{-1}. Anisotropy of covalency in ferric complexes may therefore be highly significant.

The situation is somewhat more complex for high-spin Fe^{2+} and low-spin Fe^{III} systems and considerable temperature dependence may be observed. The $t_{2g}^4 e_g^2(^5D)$ free-ion configuration for Fe^{2+} signifies a single 3d-electron in addition to the spherical half-filled shell, whereas the $t_{2g}^5(^2T)$ state of Fe^{III} corresponds to a single electron hole in an otherwise cubic triplet level. In a ligand field of cubic symmetry, the 3d-orbitals in such compounds retain sufficient degeneracy for there to be no electric field gradient. For example, no electric field gradient would be expected in a complex with perfect octahedral symmetry. However, in complexes of lower symmetry such as trigonal or tetragonal further degeneracy is removed, thereby producing a non-zero electric field gradient. Thermal population of the low-lying excited states then causes a variation in the electron configuration with temperature, and since as seen in Table 3.1 the different 3d-orbitals have differing electric field gradients, there is a strong temperature dependence of Δ. As the level separations need only be very small to observe this effect, it is possible

100 | ⁵⁷Fe – INTRODUCTION

to detect and measure distortions which are too small to detect by optical spectroscopy.

The Fe^{2+} case in octahedral symmetry has been discussed by Ingalls [26], octahedral Fe^{III} by Golding [27, 28], and tetrahedral Fe^{2+} by Gibb and Greenwood [29]. All three have been re-examined more specifically for the determination of ligand-field parameters [30].

The basic theoretical treatment devolves largely on the crystal-field approach, although as will be seen shortly this can also approximate to a molecular-orbital method. The effect of the environment on the energies of the $3d$-shell can be treated as a perturbation Hamiltonian:

$$\mathscr{H} = v_0 + v_T + v_R + \lambda \hat{\mathbf{L}}.\hat{\mathbf{S}}$$

where v_0 is the cubic field term, v_T is the tetragonal distortion, v_R is the rhombic distortion, and $\lambda\hat{\mathbf{L}}.\hat{\mathbf{S}}$ is the spin–orbital coupling. The last term must be included as it has a large effect on the value of $|\Delta|$ at low temperatures. When the spin degeneracy is included in the calculations, the ferrous case with five d-orbitals and a spin of $S = 2$ assumes a dimension of 25 wavefunctions, while the Fe^{III} calculations with $S = \frac{1}{2}$ and three t_{2g} orbitals, have 6. Accurate solutions can be obtained by digital computer for the energy levels and the appropriate linear combinations of the orbital set which

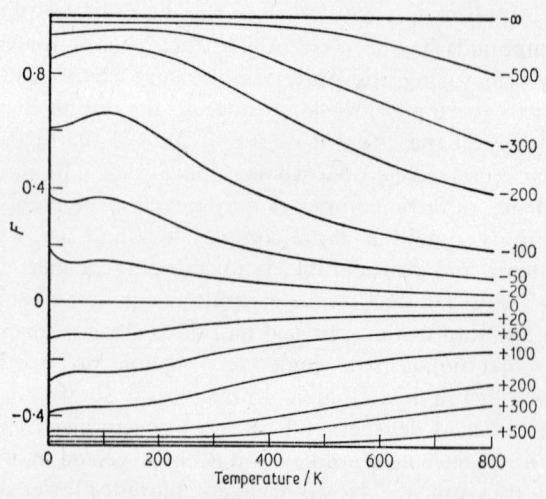

Fig. 5.5 Calculated values of the reduction factor F for different degrees of tetragonal distortion of an octahedral Fe^{2+} environment. The 10 Dq $t_{2g} - e_g$ separation = 10,000 cm⁻¹ and the spin-orbit coupling $\lambda = -80$ cm⁻¹. The numerical figures refer to a convenient splitting factor Ds derived from the operator form of $v_T = Ds(\hat{l}_z^2 - 2)$, splitting the t_{2g} levels by $3Ds$. The low values of F at low temperatures and small distortions are caused by the spin-orbit coupling. [Ref. 30, Fig. 1]

[*Refs. on p. 110*]

comprise them. The value of Δ at any given temperature Δ_T is then obtained by Boltzmann summation.

Δ_T is always less than that equivalent to the electric field gradient produced by a single unpaired 3d-electron, i.e. $\frac{4}{7}(1-R)\langle r^{-3}\rangle$ (about 4·0 mm s^{-1}), and the effect of temperature is best expressed as a reduction factor F such that $1 \geqslant |F| \geqslant 0$. Values of F for octahedral Fe^{2+} and Fe^{III} with tetragonal distortions and spin–orbit coupling included are given in Figs. 5.5 and 5.6.

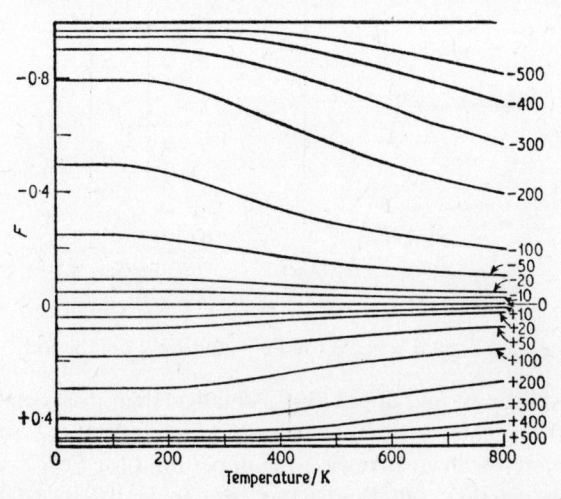

Fig. 5.6 Calculated values of the reduction factor F for low-spin Fe^{III}, derived in a similar way to Fig. 5.5 but using a spin–orbit coupling parameter of $\lambda = -300$ cm^{-1}. [Ref. 30, Fig. 6]

A much simplified model neglecting spin–orbit coupling will give reasonable approximations provided that $v_T > \lambda \hat{\mathbf{L}} \cdot \hat{\mathbf{S}}$, and is illustrated in Fig. 5.7. The reduction factor is then given by

$$F = \frac{(1 + e^{-2E_1/kT} + e^{-2E_2/kT} - e^{-E_1/kT} - e^{-E_2/kT} - e^{-(E_1+E_2)/kT})^{\frac{1}{2}}}{(1 + e^{-E_1/kT} + e^{-E_2/kT})}$$

For an axial field only (i.e. the level splittings $E_1 = E_2 = E_0$) the expression is

$$F = \frac{1 - e^{-E_0/kT}}{1 + 2e^{-E_0/kT}}$$

$|\Delta|$ will generally decrease with increasing temperature, particularly if the level separations are small. Tetragonal or trigonal distortions can be estimated easily and may be distinguished by the opposing signs of the coupling constant e^2qQ, but rhombic distortions and spin–orbit coupling are harder to assign unambiguously.

[*Refs. on p. 110*]

Tetrahedral Fe^{2+} is similar to octahedral Fe^{2+} but has little spin–orbit interaction. Admixture of metal 4p-electrons is allowed into the T_2 levels, but not into the E levels which produce the electric field gradient. The observed

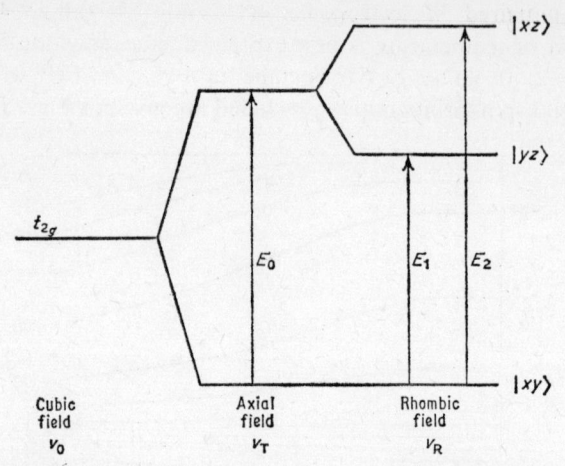

Fig. 5.7 A simplified orbital scheme for Fe^{2+} and Fe^{III} octahedral complexes.

quadrupole splittings are about 30% smaller than for octahedral Fe^{2+} because of the greater covalency effects which affect the screening and radial function terms in Δ. Likewise, the upper limit for Fe^{III} Δ values is also expected to be smaller, but this is less easy to verify. The 'electron hole' reverses the sign of the coupling constant e^2qQ. The introduction of appreciable covalency does not alter the overall validity of the results, provided that it is realised that the ligand-field parameters used do not now apply to the free ion, but are 'effective' parameters.

Theoretical estimations of the quadrupole splitting in covalent diamagnetic complexes of iron with low formal oxidation states (e.g. $Fe(CO)_5$) are difficult. The electric field gradient is now generated by the molecular orbitals of the complex, and in view of the large number of orbitals involved in bonding is difficult to interpret. Consequently in this type of compound the quadrupole splitting is used semi-empirically as a 'distortion parameter'. Low-lying excited levels are generally absent so that there is little temperature dependence of Δ in such complexes.

5.5 Magnetic Interactions

The appropriate theory for a $\frac{3}{2} \to \frac{1}{2}$ magnetic hyperfine spectrum was given in Chapter 3.5. The internal fields found in iron compounds and alloys are of the order of several hundred kG and can generate clearly resolved six-line spectra. The ground state magnetic moment, μ_g, has been accurately determined to be $+0.0903 \pm 0.0007$ nuclear magnetons (n.m.) [31]. Recent accurate determinations of the ratio μ_e/μ_g and thence μ_e are given in Table 5.4. As the nuclear term in equation 3.41 is thus known to high precision, the absolute magnitude of the internal magnetic fields can be found.

Table 5.4 The magnetic moment, μ_e, of the 14·41-keV level of ^{57}Fe

μ_e/μ_g	$\mu_e/$(n.m.)	Source of data	Reference
−1·715 ± 0·004	−0·1549 ± 0·0013	iron metal	1
−1·7135 ± 0·0015	−0·1547 ± 0·0013	^{57}Co in single-crystal cobalt	2
−1·715 ± 0·003	−0·1549 ± 0·0013	ultrasonic splitting of iron	3
−1·712 ± 0·013	−0·1545 ± 0·0024	rare-earth orthoferrites	4

1. R. S. Preston, S. S. Hanna, and J. Heberle, *Phys. Rev.*, 1962, **128**, 2207.
2. G. J. Perlow, C. E. Johnson, and W. Marshall, *Phys. Rev.*, 1965, **140**, A875.
3. T. E. Cranshaw and P. Reivari, *Proc. Phys. Soc.*, 1967, **90**, 1059.
4. M. Eibschutz, S. Shtrikman, and D. Treves, *Phys. Rev.*, 1967, **156**, 562.

Diagnostic application of the measured values of the hyperfine magnetic field, H, requires a knowledge of the origins of the field. The various terms which contribute were described in Chapter 3.5, but the particular importance of each for ^{57}Fe has not yet been outlined. The most significant is the Fermi contact interaction. In ionic complexes where the spin density is produced by 3d-polarisation effects on the internal s-shells, the field observed at low temperatures where it has reached the limiting value of the Brillouin function is of the order of 110 kG for each unpaired 3d-electron. This is sometimes referred to as the $220\langle S_z\rangle$ rule. For the Fe^{3+} ion ($3d^5$) the Fermi term is about 550 kG, and is 440 kG for the Fe^{2+} ion ($3d^6$). It should be noted that the sign of H_S is negative.

The orbital term H_L and spin moment H_D are more difficult to estimate. For an Fe^{3+} ion in cubic symmetry, both are zero, and even in distorted environments they are small. Calculations of these contributions in α-Fe_2O_3 have been made [32]. The limiting value of ferric hyperfine fields is therefore about 550 kG, and it is difficult to distinguish between two cation sites in the same sample unless they show significant differences in chemical isomer shift or quadrupole splitting. By contrast, for ferrous iron, the magnitudes of H_L and H_D can be of the order of H_S, and the resultant field $|H|$ can be anywhere between 0 and 450 kG. As a consequence, compounds where cation randomisation can occur may have large local variations in the ferrous-ion environment; the observed electric field gradient has no unique value and gross broadening of the spectrum is observed. This occurs, for example, in iron–titanium spinel oxides [33].

Interpretation of magnetic effects in metals and alloys is difficult because a definite electronic structure cannot be easily assigned to individual atoms, and there may be some unpaired 4s-spin density in the conduction band. The

104 | ^{57}Fe – INTRODUCTION

field is usually attributed to two effects, core polarisation and conduction-electron polarisation. Core polarisation is the induction of a resultant spin density in the inner s-shells by unpaired electrons in the outer electron shells. The 3d-contribution will be similar to that for the discrete cations. Conduction-electron polarisation involves the polarisation of the 4s-conduction electrons by both exchange interaction with the 3d-electrons and mixing with the 3d-band. The core-polarisation term is believed to be the major one in the 3d-magnetic elements, but it is not proposed to discuss the problem further here. Examples will be found in Chapter 13.

The temperature dependence of a hyperfine field from an Fe^{3+} spin configuration follows the appropriate spin-only $S = \frac{5}{2}$ Brillouin function as applied to cooperative magnetic phenomena [34, 35]. Small deviations sometimes occur, but in general H decreases with rising temperature, falling increasingly rapidly towards zero as the critical temperature (the Curie point) is approached. The magnetic behaviour of Fe^{2+} is less regular because of the influence of orbital repopulation as outlined in the previous section. S and L are not easily defined for metals, but as L is generally quenched and the Brillouin curves are not very sensitive to the total spin value, the behaviour is basically similar.

Motional narrowing of magnetic hyperfine interactions by spin relaxation (superparamagnetism) has also been recorded; it results from a reduction in particle size or from magnetic dilution. Conversely, long spin relaxation times can be induced in paramagnetic ferric compounds by appropriate lowering of the temperature (spin-lattice) or magnetic dilution (spin–spin), resulting in a complex magnetic spectrum. The $|\pm\frac{5}{2}\rangle$ Kramers' doublet is often fully resolved, but the $|\pm\frac{3}{2}\rangle$ and $|\pm\frac{1}{2}\rangle$ states tend to contribute to a broad unresolved background. Examples of this behaviour will be found in the detailed chapters on experimental results which follow.

5.6 Polarised Radiation Studies

A very useful technique which has recently been developed for ^{57}Fe work, although its application is more general, is the use of polarised radiation. It is convenient to discuss the subject here, but the reader is advised to refer to the appropriate sections in later chapters for more detailed discussion of the spectra of some of the materials mentioned. Polarisation of the emitted γ-ray was first shown in 1960 [36], and can take place by the Stark and Zeeman effects already familiar in optical polarisation studies [37].

There are two basic approaches, the simplest of which is to use a magnetically split source matrix, such as an iron foil, and to magnetise it in a fixed direction. With iron foils it is convenient to magnetise perpendicular to the direction of emission. The emitted radiation is then not monochromatic but consists of six individual components with intensity ratios 3 : 4 : 1 : 1 : 4 : 3,

[Refs. on p. 110]

each of which has a definite polarisation. In the case we are considering the $|\pm\frac{3}{2}\rangle \to |\pm\frac{1}{2}\rangle$ and $|\pm\frac{1}{2}\rangle \to |\mp\frac{1}{2}\rangle$ transitions have a linear polarisation with the electric vector vibration parallel with respect to the magnetisation, and the $|\pm\frac{1}{2}\rangle \to |\pm\frac{1}{2}\rangle$ transitions are similar but perpendicularly polarised.

In the event that the Mössbauer absorber is unsplit or is a random polycrystal, the observed spectrum shows no new features. However, if both source and absorber are directionally polarised by a magnetic field which can be internal in origin or by an electric field such as is associated with the electric field gradient tensor in a single-crystal absorber the polarisation of each emission line becomes important.

In principle each source emission line will come into resonance with each line in the absorber, but where the emission spectrum has a high symmetry such as in iron foil there is generally a reduction in the number of non-coincident components. One important effect in this type of experiment is a dramatic variation of line intensity with change in the relative orientation of the principal directions in the source and absorber. Calculation of the line intensities is difficult, but the detailed theory has been outlined [38, 39]. Equation 3.53 can be simplified to give an equation for the proportional amplitude probability of a source (M1 only) emission line as

$$\varepsilon_{fi} = \langle I_i 1 -m_i M \mid I_f m_f \rangle (\hat{\eta}_1 D_{+1,M} + \hat{\eta}_{-1} D_{-1,M})$$

where i and f denote the initial and final states, M is the change in the m quantum number $= m_i + m_f$, and $\hat{\eta}_1$ and $\hat{\eta}_{-1}$ are complex unit vectors such that $\hat{\eta}_a^* \cdot \hat{\eta}_b = \delta_{ab}$. Similarly the amplitude probability for absorption is

$$\varepsilon_{f'i'} = \langle I_{i'} 1 -m_{i'} m \mid I_{f'} m_{f'} \rangle (\hat{\eta}_1 D_{+1,m} + \hat{\eta}_{-1} D_{-1,m})$$

The line intensity is then given by

$$|\varepsilon_{f'i'} \varepsilon_{fi}|^2 = \langle I_{i'} 1 -m_{i'} m \mid I_{f'} m_{f'} \rangle^2$$
$$\times \langle I_i 1 -m_i M \mid I_f m_f \rangle^2 [d^2_{+1,M} d^2_{+1,m} + d^2_{-1,M} d^2_{-1,m}$$
$$+ 2\cos\{2(\gamma - \gamma')\} d_{+1,M} d_{-1,M} d_{+1,m} d_{-1,m}]$$

The $d_{\pm 1,M}$ functions are those tabulated in Table 3.4 for $J = 1$, and $(\gamma - \gamma')$ is the relative angle between the source and absorber polarisation directions projected onto the plane perpendicular to the γ-ray axis. In the event that the absorber eigenstates are not 'pure' eigenfunctions, e.g. for an electric field gradient tensor with an asymmetry parameter, the more complex expression

$$\varepsilon_{f'i'} = \sum_{m_i' m_f'} a_{f'}^* b_{i'} \langle I_{i'} 1 -m_i' m \mid I_{f'} m_{f'} \rangle (\hat{\eta} D_{+1,m} + \hat{\eta}_{-1} D_{-1,m})$$

must be used, and complex calculations of this nature are best treated by computer analysis. The addition of M1/E2 mixing would considerably complicate the calculations.

Such polarisation effects were shown in iron-foil sources and absorbers in

1960 [40], and one of the first chemical applications of an iron-foil source was in the determination of the sign of the quadrupole coupling constant in a single crystal of FeSiF$_6$.6H$_2$O [41]. As an illustration of the types of behaviour to be expected we have calculated the theoretical line intensities for two typical experimental situations and present the results in Fig. 5.8. In all cases

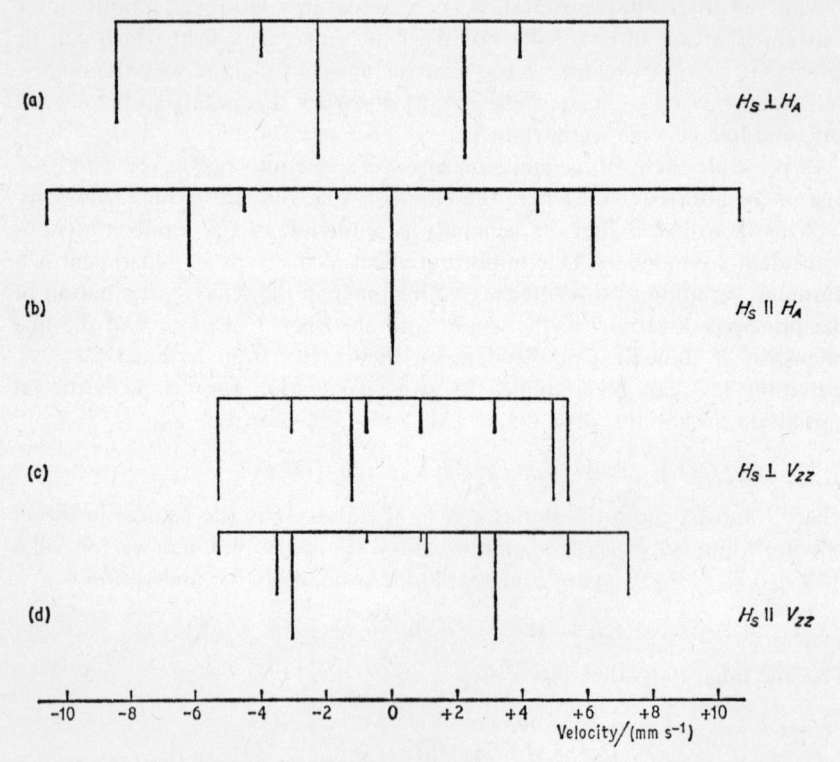

Fig. 5.8 Predicted spectra using an iron metal source at room temperature magnetised perpendicular to the axis of propagation.
(a) an iron foil magnetised perpendicular to the axis of propagation but with $H_S \perp H_A$
(b) as (a) but $H_S \parallel H_A$
(c) an FeCO$_3$ single crystal with V_{zz} perpendicular to the axis of propagation and $H_S \perp V_{zz}$
(d) as (c) but $H_S \parallel V_{zz}$.

the source is assumed to be iron metal at room temperature magnetised perpendicular to the direction of propagation. In (a) and (b) are seen the spectra for an iron metal absorber, also magnetised in the perpendicular plane, but aligned perpendicular and parallel to the source magnetisation.

In (c) and (d) are shown the predicted spectra for an FeCO$_3$ single crystal with the principal axis of the electric field gradient tensor aligned perpendicu-

[Refs. on p. 110]

lar to the direction of propagation but perpendicular and parallel to the source magnetisation. Experimental data for these systems confirm the analysis [42].

The dramatic change in the observed spectrum with orientation opens up many possibilities for the study of directional properties in single crystal materials, and several examples will be found in the following chapters.

In the preceding discussion we have been considering polarised sources which are *not* monochromatic. The assignment of hyperfine components in a complex spectrum would obviously be easier if the source were effectively polarised and monochromatic so that resonant absorption would only be observed in those transitions with the same polarisation. This method was first demonstrated in 1968 [43]. The spectrum of a ^{57}Fe quadrupole splitting observed in a direction perpendicular to the principal value of an axially symmetric electric field gradient consists of two lines in the intensity ratio 5 : 3. The weaker line is linearly polarised in the plane parallel to V_{zz}, while the other is 40% unpolarised and 60% linearly polarised perpendicular to V_{zz}. If the latter can be filtered out a linearly polarised source will result.

This was achieved using a ^{57}Co/Be metal source which has a quadrupole splitting of 0·56 mm s^{-1}, and single crystals of which can be oriented with V_{zz} perpendicular to the direction of γ-ray propagation. The more intense line from $|\pm\frac{1}{2}\rangle \rightarrow |\pm\frac{1}{2}\rangle$ is then selectively filtered out using a layer of Fe(NH$_4$)$_2$(SO$_4$)$_2$.6H$_2$O clamped to the source. One line of the quadrupole splitting in this material coincides in energy with the unwanted line in the Be metal. The result is an apparently monochromatic source with a linearly polarised radiation. The absorption spectrum of a crystal of FeSiF$_6$.6H$_2$O oriented with V_{zz} perpendicular to the γ-ray axis obtained using this source was simply a single line because only the $|\pm\frac{1}{2}\rangle \rightarrow |\pm\frac{3}{2}\rangle$ component can resonantly absorb the radiation.

An equivalent method which does not require a split source is to use a polariser, and this has been demonstrated, firstly in 1967 using iron foils magnetised perpendicular to the propagation axis [44], and more recently using a polycrystalline sample of yttrium iron garnet (YIG) [45]. If the latter material is magnetised in an external magnetic field of 3 kG applied parallel to the axis of propagation the $\Delta m = 0$ components in its spectrum become forbidden, while the others become right or left circularly polarised. The garnet is completely polarised by comparatively small fields which is an advantage for these experiments. This is illustrated schematically in Fig. 5.9(ii). Note that there are two sites (signified by *a* and *d* as detailed in Chapter 10), each magnetically split to give a total of 12 lines in the spectrum. The monochromatic unpolarised source is kept stationary while the YIG polariser (enriched in ^{57}Fe) is moved at a *constant* velocity chosen so that one of the resonance lines coincides with the emission line. If this is line 'A' the polariser can absorb nearly all the right-circularly polarised radiation, but is transparent to the left-circularly polarised radiation. It therefore transmits a monochromatic circularly polarised radiation, the direction of polarisation being reversible by using the line 'B', for example. If this combination is now used with an absorber of YIG which is Doppler scanning conventionally and

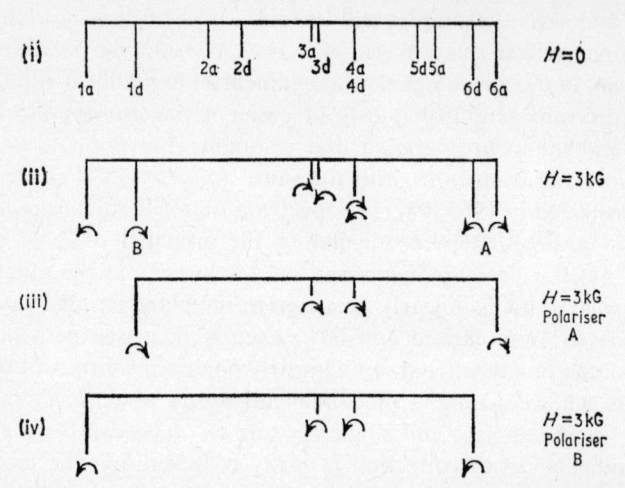

Fig. 5.9 Schematic representation of the use of an yttrium iron garnet polariser:
(i) the zero-field conventional absorption spectrum of YIG showing the two 6-line patterns from the *a* and *d* sites.
(ii) as (i) but with a field of 3 kG applied parallel to the γ-ray axis. Note the disappearance of the $\Delta m = 0$ transitions because of complete magnetisation. The polarisations are shown.
(iii) the absorption spectrum of YIG in a 3-kG field using radiation passing through a YIG polariser in which the incident line is coincident with line A. Note the absence of left circularly polarised lines.
(iv) as (iii) but with the emission line coincident with line B in the polariser.

is magnetised along the same axis, the resultant spectra will be as shown in (iii) and (iv) depending on the polarisation used. Note that the opposite polarisations found for lines 6*a* and 6*d* is consistent with antiparallel hyperfine fields at the two sublattices in YIG.

There is an apparent similarity between these polarisation experiments and some work on the selective excitation of nuclear sublevels [46]. However, the latter involves population of single sublevels of a hyperfine splitting, i.e. the polarisation of excited-state *nuclei*, and not the polarisation of the γ-radiation.

Although both experiment and theory of polarisation experiments are complex, the information obtainable far exceeds that from a simpler single-crystal experiment, and one can expect more intensive application to be made in the immediate future.

[*Refs. on p. 110*]

5.7 Energetic Nuclear Reactions

Several experiments have been performed to observe a Mössbauer resonance in the immediate time interval following an energetic nuclear reaction. Since the experimental conditions for these are severe compared to the ^{57}Co source method, they are discussed here more as a curiosity than as a recommended means of attempting chemical studies.

Coulombic excitation of ^{57}Fe to the 136-keV level was first achieved by Lee et al. in 1965 using a 1·5-μA beam of 3-MeV α-particles [47, 48]. The target foil of α-iron gave a recoil-free fraction and internal magnetic field in the 14·41-keV resonance essentially identical to those recorded in the normal ^{57}Co decay process. An Fe$_2$O$_3$ target showed a 50% reduction in recoil-free fraction but otherwise behaved normally. The high energy of the Coulomb excitation process displaces the excited atom from its original site, but the observation of a normal spectrum indicates that the recoiling atom must come to rest on a normal lattice site by a replacement collision in a time less than the 14·4-keV excited-state lifetime ($\tau_{\frac{1}{2}} = 10^{-7}$ s). In the case of iron metal, nearly all the atoms do this, but in the oxide it appears that there is an appreciable probability of the displaced atom coming to rest on an oxygen site, in which case its contribution to the resonant spectrum is either slight or broadened into the background. The problem of replacement collisions has been treated theoretically [49, 50].

An improved technique, using recoil implantation through vacuum, involves excitation of an ^{57}Fe target by 36-MeV ^{16}O ions [51]. Excited atoms displaced from the target travel through high vacuum to a catcher material which is shielded from the direct ion beam. Copper, aluminium, gold, and iron catcher foils gave good spectra, but the silicate mineral olivine gave no resonance. The interpretation parallels that given in the preceding paragraph.

Deuteron capture using the ^{56}Fe$(d, p)^{57}$Fe reaction gives similar results. A 0·025-μA beam of 2·8-MeV deuterons incident on an enriched ^{56}Fe iron foil produces the normal Fe metal spectrum [52], as does a 4·8-MeV pulsed beam of deuterons with a stainless steel target [53].

In situ neutron capture in an ^{56}Fe metal target results in a spectrum identical to that obtained from ^{57}Co decay [54]. An Fe$_2$O$_3$ target shows a slight reduction in internal magnetic field, whereas FeSO$_4$.7H$_2$O appears to suffer considerable oxidation to ferric during the neutron capture process [55]. Ordered Fe$_3$Al alloy targets show considerable differences from the conventional spectra because of the variations in position in the lattice of the excited atoms after they have been displaced from their initial sites by the nuclear reaction.

5.8 The 136·4-keV Transition

The 136·4-keV resonance in ^{57}Fe (i.e. the second excited state to ground state $\frac{5}{2}- \to \frac{1}{2}-$ transition) is extremely difficult to detect because of the low recoil-free fraction and resonant-absorption cross-section associated with this γ-ray. In addition, the latter is comparable to the Compton and Rayleigh scattering cross-sections. A claim to have detected hyperfine lines from an iron metal scatterer using a ^{57}Co source has been made [56]. Both were maintained at liquid helium temperature. The spectra produced were of very poor resolution, and it was not possible to analyse the hyperfine patterns with confidence. There is little hope that better data can be obtained with present techniques.

REFERENCES

[1] J. A. Bearden, *Phys. Rev.*, 1965, **137**, B455.
[2] O. C. Kistner and A. W. Sunyar, *Phys. Rev.*, 1965, **139**, B295.
[3] S. L. Ruby, Y. Hazony, and M. Pasternak, *Phys. Rev.*, 1963, **129**, 826.
[4] M. Eckhause, R. J. Harris, Jr, W. B. Shuler, and R. E. Welsh, *Proc. Phys. Soc.*, 1966, **89**, 187.
[5] C. Hohenemser, R. Reno, H. C. Benski, and J. Lehr, *Phys. Rev.*, 1969, **184**, 298.
[6] R. H. Nussbaum and R. M. Housley, *Nuclear Phys.*, 1965, **68**, 145.
[7] G. R. Isaak and U. Isaak, *Phys. Letters*, 1965, **17**, 51.
[8] A. H. Muir, Jr, E. Kankeleit, and F. Boehm. *Phys. Letters*. 1963, **5**, 161.
[9] W. Rubinson and K. P. Gopinathan, *Phys. Rev.*, 1968, **170**, 969.
[10] G. A. Chackett, K. F. Chackett, and B. Singh, *J. Inorg. Nuclear Chem.*, 1960, **14**, 138.
[11] A. Mustachi, *Nuclear Instr. Methods*, 1964, **26**, 219.
[12] P. M. Dryburgh, *J. Sci. Instr.*, 1964, **41**, 640.
[13] J. Stephen, *Nuclear Instr. Methods*, 1964, **26**, 269.
[14] R. W. Grant, R. M. Housley, and U. Gonser, *Phys. Rev.*, 1969, **178**, 523.
[15] O. C. Kistner and A. W. Sunyar, *Phys. Rev. Letters*, 1960, **4**, 229.
[16] L. R. Walker, G. K. Wertheim, and V. Jaccarino, *Phys. Rev. Letters*, 1961, **6**, 98.
[17] R. E. Watson, *Phys. Rev.*, 1960, **118**, 1036.
[18] R. E. Watson, *Phys. Rev.*, 1960, **119**, 1934.
[19] J. Danon, 'Applications of the Mössbauer Effect in Chemistry and Solid-state Physics', Int. Atomic Energy Agency, Vienna, 1966, p. 89.
[20] R. Ingalls, *Phys. Rev.*, 1967, **155**, 157.
[21] E. Šimánek and Z. Šroubek, *Phys. Rev.*, 1967, **163**, 275.
[22] E. Šimánek and A. Y. C. Wong, *Phys. Rev.*, 1968, **166**, 348.
[23] R. G. Shulman and S. Sugano, *J. Chem. Phys.*, 1965, **42**, 39.
[24] A. J. Freeman and R. E. Watson, *Phys. Rev.*, 1963, **131**, 2566.
[25] Sh. Sh. Bashkirov and E. K. Sadykov, *Doklady Akad. Nauk S.S.S.R.*, 1968, **180**, 137.
[26] R. Ingalls, *Phys. Rev.*, 1964, **133**, A787.

[27] R. M. Golding and H. J. Whitfield, *Trans. Faraday Soc.*, 1966, **62,** 1713.
[28] R. M. Golding, *Mol. Phys.*, 1967, **12,** 13.
[29] T. C. Gibb and N. N. Greenwood, *J. Chem. Soc.*, 1965, 6989.
[30] T. C. Gibb, *J. Chem. Soc. (A)*, 1968, 1439.
[31] G. W. Ludwig and H. H. Woodburg, *Phys. Rev.*, 1960, **117,** 1286.
[32] F. van der Woude, *Phys. Stat. Sol.*, 1966, **17,** 417.
[33] S. K. Banerjee, W. O'Reilly, T. C. Gibb, and N. N. Greenwood, *J. Phys. and Chem. Solids*, 1967, **28,** 1323.
[34] A. H. Morrish, 'The Physical Principles of Magnetism', John Wiley & Son, Inc., New York, 1965.
[35] J. B. Goodenough, 'Magnetism and the Chemical Bond', Interscience Publishers, New York, 1963.
[36] G. J. Perlow, S. S. Hanna, M. Hamermesh, C. Littlejohn, D. H. Vincent, R. S. Preston, and S. Heberle, *Phys. Rev. Letters*, 1960, **4,** 74.
[37] W. A. Shurcliff, 'Polarized Light: production and use', Oxford Univ. Press, London, 1962.
[38] H. Frauenfelder, D. E. Nagle, R. D. Taylor, D. R. F. Cochran, and W. M. Visscher, *Phys. Rev.*, 1962, **126,** 1065.
[39] J. T. Dehn, J. G. Marzolf, and J. F. Salmon, *Phys. Rev.*, 1964, **135,** B1307.
[40] S. S. Hanna, J. Heberle, C. Littlejohn, G. J. Perlow, R. S. Preston, and D. H. Vincent, *Phys. Rev. Letters*, 1960, **4,** 177.
[41] C. E. Johnson, W. Marshall, and G. J. Perlow, *Phys. Rev.*, 1962, **126,** 1503.
[42] U. Gonser, p. 343 of 'Hyperfine Structure and Nuclear Radiations', Ed. E. Matthias and D. A. Shirley, North-Holland Publishing Co., Amsterdam, 1968.
[43] R. M. Housley, *Nuclear Instr. Methods*, 1968, **62,** 321.
[44] S. Shtrikman, *Solid State Commun.*, 1967, **5,** 701.
[45] S. Shtrikman and S. Somekh, *Rev. Sci. Instr.*, 1969, **40,** 1151.
[46] N. D. Heiman, J. C. Walker, and L. Pfeiffer, *Phys. Rev.*, 1969, **184,** 281.
[47] Y. K. Lee, P. W. Keaton, Jr, E. T. Ritter, and J. C. Walker, *Phys. Rev. Letters*, 1965, **14,** 957.
[48] E. T. Ritter, P. W. Keaton, Jr, Y. K. Lee, R. R. Stevens, Jr, and J. C. Walker, *Phys. Rev.*, 1967, **154,** 287.
[49] P. H. Dederichs, C. Lehmann, and H. Wegener, *Phys. Stat. Sol.*, 1965, **8,** 213.
[50] D. A. Goldberg, Y. K. Lee, E. T. Ritter, R. R. Stevens, Jr, and J. C. Walker, *Phys. Letters*, 1966, **20,** 571.
[51] G. D. Sprouse, G. M. Kalvius, and S. S. Hanna, *Phys. Rev. Letters*, 1967, **18,** 1041.
[52] Y. K. Lee, P. W. Keaton, Jr, E. T. Ritter, and J. C. Walker, *Phys. Rev. Letters*, 1965, **14,** 957.
[53] J. Christiansen, E. Recknagel, and G. Weiger, *Phys. Letters*, 1966, **20,** 46.
[54] W. G. Berger, *Z. Physik*, 1969, **225,** 139.
[55] W. G. Berger, J. Fink, and F. E. Obenshain, *Phys. Letters*, 1967, **25,** A466.
[56] N. Hershkowitz and J. C. Walker, *Phys. Rev.*, 1967, **156,** 391.

6 | High-spin Iron Complexes

The facility with which ^{57}Fe 14·41-keV Mössbauer spectra can be measured has resulted in considerable use of the technique for probing iron solid-state systems; so much so that it would be impossible to summarise the data adequately in a single chapter. The general field of iron chemistry has therefore been subdivided into a number of subsections and, although this division is to some extent arbitrary it simplifies the task of presenting the results considerably. The subsections are as follows,

(i) iron(II) and iron(III) complexes with high-spin electronic configurations (Chapter 6);
(ii) iron(II) and iron(III) complexes with low-spin electronic configurations (Chapter 7);
(iii) ligand-field crossover situations and uncommon paramagnetic oxidation states (Chapter 8);
(iv) covalent, diamagnetic iron complexes (Chapter 9);
(v) iron oxides, chalcogenides, and minerals (Chapter 10);
(vi) intermetallic systems and alloys (Chapter 11);
(vii) decay after-effects and impurity studies, together with some miscellaneous topics (Chapter 12);
(viii) biological systems (Chapter 13).

The present chapter will be restricted to the first of these subjects. Since high-spin ferrous (Fe^{2+}) and ferric (Fe^{3+}) cations have somewhat different orbital properties, as outlined in Chapter 5, we shall consider each in turn. Many complexes have been examined by several workers in the past few years, and the data currently available for each will be presented. Early references which do not contribute significant information beyond that given in the later papers will be omitted where appropriate, but references to such papers will in many cases be found in the papers cited.

Great difficulty has been found in comparing chemical isomer shifts from different laboratories. Comparatively large discrepancies have frequently been found, and it has been necessary to restrict comment to examples where differences in chemical isomer shift can confidently be held to be significant.

[*Refs.* on p. 164]

IRON(II) HALIDES | 113

Ferrous ions usually have a non-cubic environment and give a large quadrupole splitting in paramagnetic complexes. Many ferrous compounds are antiferromagnetic at low temperatures and the spin and electric field gradient axes are not necessarily collinear. This causes considerable difficulty in interpretation, and computer methods must be used as already outlined in Chapter 3.6. Accordingly the results will be quoted without detailed proof in the discussion of actual spectra.

By contrast, high-spin ferric complexes show only small quadrupole splittings due to the lattice terms, and in the magnetically ordered state the field is basically the Fermi term. Consequently less information is obtainable, with the notable exception of electronic relaxation times which can be deduced from the broadening which often occurs in ferric complexes.

Appropriate illustrations are given of the principal features of high-spin ferrous and ferric spectra, and examples of the Mössbauer parameters observed are tabulated for many of the complexes. The tables are not intended to be completely comprehensive, but virtually all the significant and reliable results have been incorporated. Where possible, data are quoted for room temperature (RT), liquid nitrogen temperature (LN), and liquid helium temperature (HE).

A. HIGH-SPIN IRON(II) COMPLEXES
6.1 Iron(II) Halides
FeF$_2$

Ferrous fluoride was one of the first ferrous compounds to be investigated by Mössbauer spectroscopy [1]. It has the rutile structure with the iron coordinated by six fluorines, and extensive investigations have now been made [2, 3]. The compound gives a large quadrupole splitting in the paramagnetic state. It is antiferromagnetic at low temperatures, and the extrapolation of the magnetic field data gives the Néel temperature accurately as $T_N = 78 \cdot 12$ K [3]. The effects of the combined quadrupole and magnetic interactions are illustrated in Figs. 6.1 and 6.2.

At 78·21 K, which is 0·09° above the Néel temperature, a sharp quadrupole doublet is seen (Fig. 6.1). At 78·11 K, only 0·01° below the critical point, considerable line broadening has appeared, and the subsequent splitting of the two lines into a triplet and a doublet is very similar to the results of application of an external magnetic field for determination of the sign of the coupling constant e^2qQ. If the triplet is found at the lower Doppler velocities the sign of e^2qQ is positive. The internal magnetic field in FeF$_2$ increases very rapidly at first with decreasing temperature (see Fig. 6.3) but reaches a saturation value at absolute zero. In the neighbourhood of the Néel temperature the hyperfine effective magnetic field is given by the usual function

$$H_T = H_0 D(1 - T/T_N)^\beta$$

[*Refs. on p. 164*]

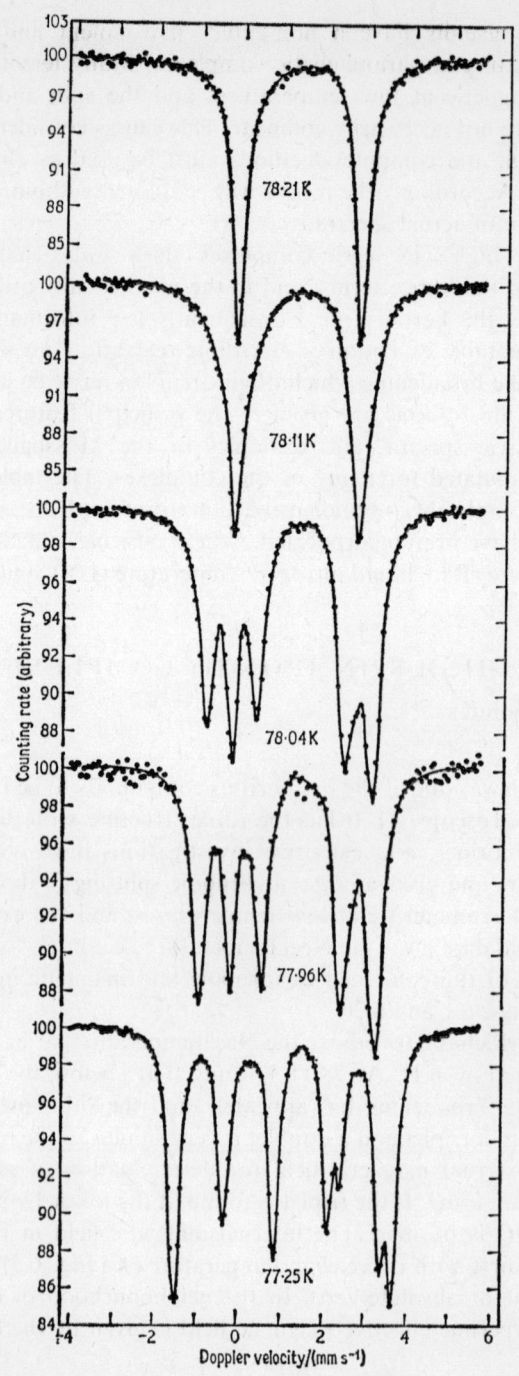

Fig. 6.1 Hyperfine structure of ^{57}Fe in FeF$_2$ within 1 K of the Néel temperature. [Ref. 3, Fig. 1]

IRON(II) HALIDES | 115

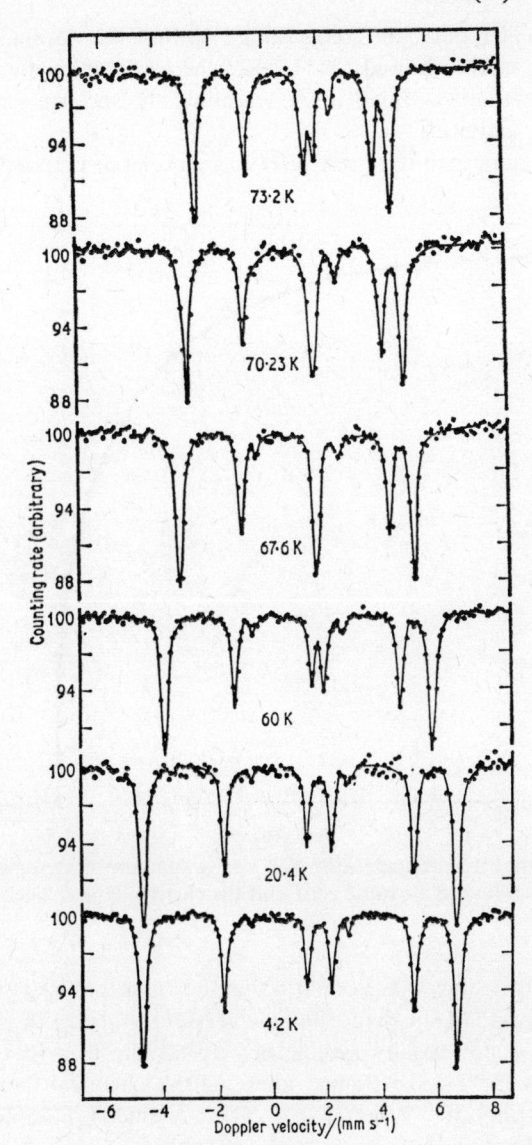

Fig. 6.2 Hyperfine structure of ^{57}Fe in FeF$_2$ when the magnetic coupling is stronger than the quadrupole coupling. [Ref. 3, Fig. 2]

For FeF$_2$ it was found that $D = 1\cdot36 \pm 0\cdot03$, $\beta = 0\cdot325 \pm 0\cdot005$, and the hyperfine field at the absolute zero $H_0 = 329 \pm 2$ kG. This behaviour is remarkably similar to that of MnF$_2$ despite the appreciably lower anisotropy in this latter compound.

The appearance of more than six Zeeman lines indicates that the Zeeman

[*Refs. on p. 164*]

116 | HIGH-SPIN IRON COMPLEXES

levels are no longer pure $|m_z\rangle$ eigenstates so that the nominal $\Delta m = \pm 2$ transitions are weakly allowed. At 4·2 K, where the magnetic field is considerably greater than the quadrupole coupling, the spectrum approximates best to a six-line pattern.

Single-crystal data with the γ-ray direction parallel or perpendicular to the

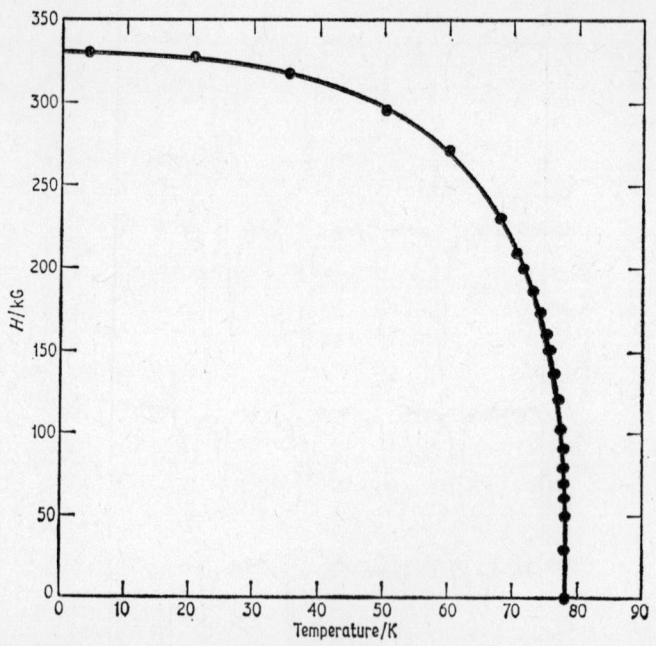

Fig. 6.3 The internal magnetic field in FeF_2 as a function of temperature. Note the saturation behaviour at absolute zero and the clearly defined Néel temperature. [Ref. 3, Fig. 3]

c axis show line intensities which confirm that the c axis is indeed the magnetic-spin axis [1]. Because of the orthorhombic crystal symmetry the electric field gradient tensor is not axially symmetric; it has the principal value, V_{zz}, perpendicular to the c axis. Some values of the chemical isomer shift δ, quadrupole splitting $\Delta (= \frac{1}{2}e^2qQ[1 + \frac{1}{3}\eta^2]^{\frac{1}{2}})$, asymmetry parameter η, and hyperfine field H are given as appropriate in Table 6.1.

The effect of the orbital state of the Fe^{2+} cation has been considered by several groups, but the very low symmetry of the environment causes difficulties [4–7]. If a rotated coordinate system is used, based approximately on the distorted fluorine octahedron, then the electronic ground state is $0.99 |x^2 - y^2\rangle + 0.10 |z^2\rangle$. The quadrupole coupling constant, e^2qQ, is therefore positive (see Table 3.1). Ingalls [5] proposed that the very weak temperature dependence of the quadrupole interaction was compatible with

[Refs. on p. 164]

Table 6.1 Mössbauer parameters for high-spin iron(II) halides

Compound	T/K	Δ /(mm s^{-1})	η	H/kG	δ /(mm s^{-1})	δ (Fe) /(mm s^{-1})	Reference
FeF$_2$	692	+1·70	—	—	—	—	7
	298	+2·79	—	—	1·18 (Pd)	1·37	
	78·21	+2·92	—	—	1·29 (Pd)	1·48	3
	4·2	+2·85	0·40	329	1·29 (Pd)	1·48	
RbFeF$_3$	102·3	0	—	0	—	—	
	98·4	+0·064	—	94·2	—	—	11
	88·2	+0·147	—	153·0	—	—	
KFeF$_3$	RT	—	—	—	1·14 (Pd)	1·33	15
CsFeF$_3$	298	{1·55	—	—	1·17 (Pd)	1·36	
		{0·47	—	—	1·17 (Pd)	1·36	16
	4·2	{−2·85	0·1	318	1·30 (Pd)	1·49	
		{−1·90	0·7	274	1·30 (Pd)	1·49	
Rb$_2$FeF$_4$	298	2·04	—	—	1·15 (Pd)	1·34	
	78	2·80	—	—	1·26 (Pd)	1·45	17
	4·2	2·77	0·065	364	1·27 (Pd)	1·46	
FeCl$_2$	78	+0·895	—	—	0·750 (Pt)	1·097	
	4·2	+1·210	—	4	0·746 (Pt)	1·093	20
	5	+0·98	—	∼3	0·97 (Pd)	1·16	25
FeCl$_2$.H$_2$O	RT	+2·03	—	—	1·13 (Fe)	1·13	26
	(0)	+2·50	—	251	—	—	27
FeCl$_2$.2H$_2$O	298	+2·50	—	—	0·80 (Cu)	1·03	
	78	+2·70	—	—	0·85 (Cu)	1·08	29
	4·2	+2·30	0·3	250	0·90 (Cu)	1·13	
FeCl$_2$.4H$_2$O	RT	+2·984	—	—	0·874 (Pt)	1·221	
	5	+3·117	—	—	1·014 (Pt)	1·361	32
	0·15	+3·10	−0·2	266	1·47 (?)	—	34
(Me$_4$N)$_2$FeCl$_4$	293	−0·72	—	—	0·91 (Cr)	0·76	
	77	−2·61	—	—	1·01 (Cr)	0·86	36
	4·2	−3·27	—	—	1·05 (Cr)	0·90	
FeBr$_2$	78	1·043	—	—	0·99 (Pt)	1·34	
	4·2	1·132	—	28·4	1·005 (Pt)	1·352	20
	5	+1·09	—	+29·6	0·94 (Pd)	1·12	37
FeBr$_2$.2H$_2$O	RT	2·49	—	—	1·14 (Fe)	1·14	26
FeBr$_2$.4H$_2$O	RT	2·83	—	—	1·22 (Fe)	1·22	26
(Et$_4$N)$_2$FeBr$_4$	293	0·52	—	—	0·88 (Cr)	0·73	
	77	2·29	—	—	1·02 (Cr)	0·87	36
	4·2	3·23	—	—	1·12 (Cr)	0·97	
FeI$_2$	78	0·802	—	—	0·646 (Pt)	0·993	
	4·2	0·962	—	74	0·697 (Pt)	1·044	20
K$_4$[Fe(NCS)$_6$].4H$_2$O	RT	—	—	—	1·24 (SS)	1·15	38
(Me$_4$N)$_2$Fe(NCS)$_4$	77	2·10	—	—	0·97 (Cr)	0·82	
	4·2	2·83	—	—	0·97 (Cr)	0·82	36
(Cat)Fe(NCSe)$_4$*	77	2·69	—	—	0·99 (Cr)	0·84	
	4·2	2·73	—	—	1·04 (Cr)	0·89	36

* 'Cat' is α,α'-(bis-triphenylphosphonium)p-xylene.

excited-state levels at 750 cm^{-1} and 2900 cm^{-1}, whereas Okiji and Kanamori [6] deduced values of 1028 cm^{-1} and 1484 cm^{-1}. A more recent analysis of data between 97 and 692 K gives values of 800 cm^{-1} and 900 cm^{-1} [7], while study of the second-order Doppler shift, recoil-free fraction and quadrupole splitting between 80 and 300 K has led to values of 740 and 930 cm^{-1} [8].

High pressures acting on FeF$_2$ produce a reversible decrease in the chemical isomer shift which is regular in behaviour [9]. This corresponds to an increase in electron density at the iron nucleus. Evidence from optical studies suggests that this results from a decrease in the $3d$-shielding effects with increasing pressure.

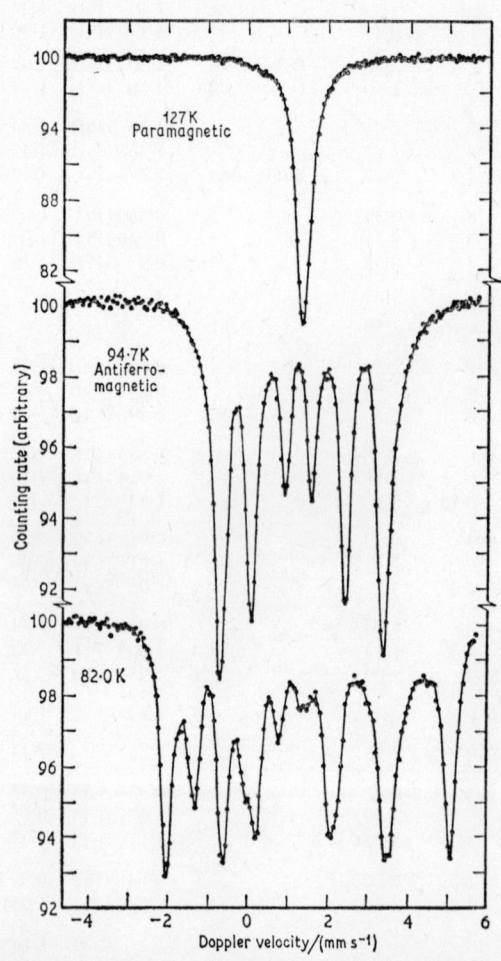

Fig. 6.4 Mössbauer spectra of RbFeF$_3$ in the paramagnetic, antiferromagnetic, and ferrimagnetic regions. [Ref. 11, Fig. 3]

[*Refs. on p. 164*]

RbFeF$_3$

One of the more unusual ferrous compounds is RbFeF$_3$; unusual in that it possesses an ideal cubic perovskite structure at ambient temperatures with the ferrous ion octahedrally coordinated by fluorine. The high degeneracy in the electronic ground state ensures the absence of an electric field gradient and only a single absorption is seen [10, 11]. This is illustrated in Fig. 6.4 by the single-line paramagnetic spectrum. Below 103 K antiferromagnetic ordering occurs and a six-line spectrum is found (Fig. 6.4). Simultaneously a small quadrupole interaction appears. It was originally thought that the magnetic axis lowered the symmetry via spin–orbit coupling, thereby inducing an electric field gradient, and attempts to interpret the spectra theoretically using this assumption have been made [12, 13]; however, the positive sign of the quadrupole splitting dictates that any distortion should be trigonal. In fact a tetragonal distortion in the antiferromagnetic region has since been detected which is at variance with the simple theory [14].

Below 87 K there is a further transition to a ferrimagnetic state with two inequivalent ferrous sites [11] and a complex spectrum from the two superimposed six-line hyperfine patterns is seen in Fig. 6.4.

KFeF$_3$

KFeF$_3$, like RbFeF$_3$, has a cubic perovskite lattice above the Néel temperature of 115 K, but below this temperature there is a trigonal distortion along the [111] axis which is expected to be the spin-axis. A quadrupole interaction is seen in the ordered phase [15] which is positive as required by a trigonal distortion. Relaxation narrowing effects are seen close to the critical point.

As with FeF$_2$ an increase in pressure decreases the chemical isomer shift [9]. It also produces a substantial quadrupole splitting as seen in Fig. 6.5. This implies that a substantial crystallographic distortion is induced. The rather broad room-temperature spectrum was believed to be the result of stresses created during formation of the absorber pellet. These high-pressure data for KFeF$_3$ were used by Šimánek and Wong as already discussed in Chapter 5 (p. 95) to calculate a value of $\delta R/R$ for the 14·41-keV transition.

CsFeF$_3$

CsFeF$_3$ has hexagonal symmetry at room temperature with two inequivalent octahedral Fe^{2+} sites in the ratio 2:1 which give two quadrupole splittings with the corresponding intensity ratios [16]. Although there is a crystallographic distortion at 83 K this has no effect on the Mössbauer spectrum, but ferrimagnetic ordering occurs below 62 K to give two superimposed hyperfine patterns. Full analysis has not been attempted.

Rb$_2$FeF$_4$

This compound, which has the K$_2$NiF$_4$ structure, has very unusual magnetic properties. The local iron environment is an octahedron of fluorines, but the

120 | HIGH-SPIN IRON COMPLEXES

iron atoms themselves are in well-separated layers containing square planar nets of Fe^{2+} ions. The crystal symmetry forbids magnetic interaction between layers, and it has been shown theoretically that an isotropic two-dimensional

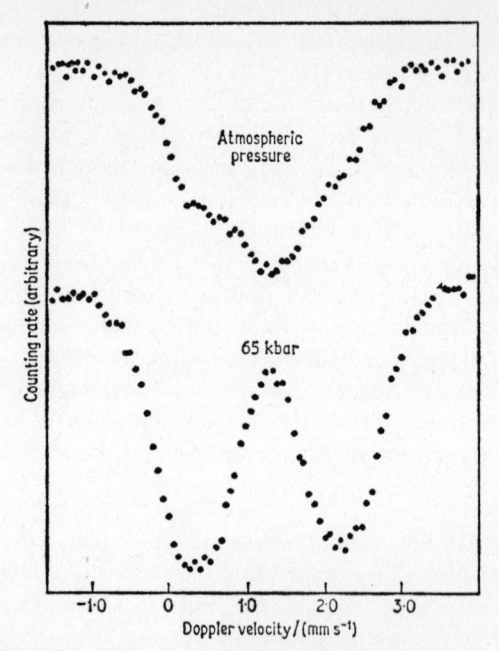

Fig. 6.5 The effect of pressure on the Mössbauer spectrum of $KFeF_3$. [Ref. 9, Fig. 5]

interaction can be neither ferromagnetic nor antiferromagnetic at non-zero temperature.

The Mössbauer spectra [17] show only a sharp quadrupole doublet between 60 K and the suspected ordering transition at 90 K. This confirms that there is no long-range order and that the spin relaxation is very fast. Long-range order appears as the temperature is lowered to 50 K (see Fig. 6.6). The change is accompanied by a crystallographic distortion which lowers the symmetry of the crystal. The presence of an asymmetry parameter shows that the fourfold axis has been destroyed. It is probable that the lowered symmetry allows magnetic interaction between Fe^{2+} layers so that the ordering becomes three-dimensional. The quadrupole splitting and chemical isomer shift are similar to those of FeF_2.

$FeCl_2$

Anhydrous iron(II) chloride has the $CdCl_2$ structure and has been studied in detail in three independent investigations. Ono *et al.* [18] found a strong temperature dependence in Δ and no significant internal magnetic field

[*Refs. on p. 164*]

(<10 kG) down to 4 K. This is unusual in view of the known antiferromagnetic transition at 24 K, but could be explained as a chance cancellation in the $H_S + H_L + H_D$ summation. Later work gave a value of 4 kG for the magnetic

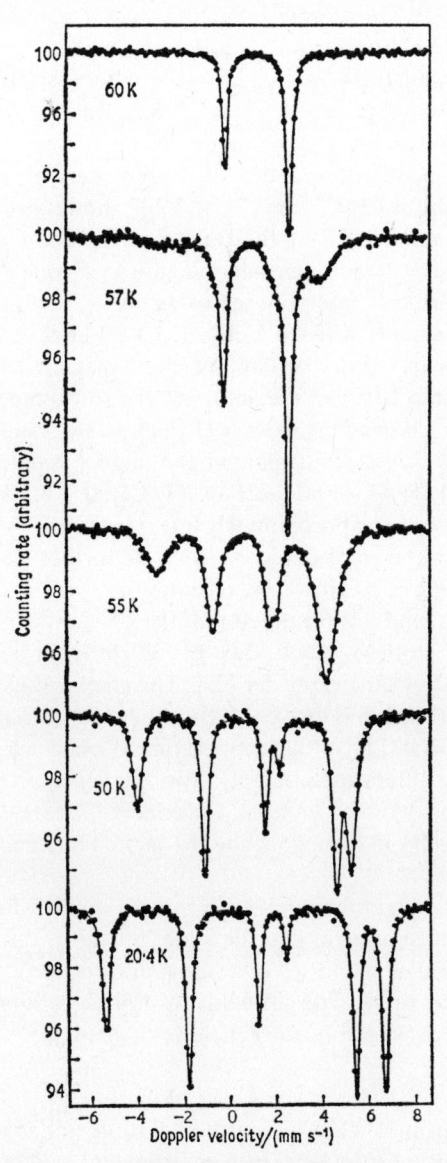

Fig. 6.6 Mössbauer spectra of a single crystal of Rb_2FeF_4 oriented with the c axis parallel to the γ-ray beam. Note the onset of long-range magnetic ordering below 60 K. [Ref. 17, Fig. 2]

[Refs. on p. 164]

field at 4·2 K. A single-crystal spectrum at room temperature gave an asymmetric doublet intensity and confirmed that e^2qQ was positive. The Fe^{2+} local symmetry is close to an octahedron of chlorine atoms with a trigonal distortion parallel to the crystal c axis. The very small trigonal field splits the t_{2g} triplet and results in a large reduction in Δ from the limiting value of $\frac{2}{7}(1-R)\langle r^{-3}\rangle$. The ground state is a doublet and is 145 ± 35 cm^{-1} below the singlet level [20] (also quoted as 119 cm^{-1} [18] and 150 cm^{-1} [19]). Below the Néel temperature there is a small increase in the quadrupole splitting due to an exchange coupling term of $-2Jx\langle S_z\rangle S_z$ where x is the number of Fe^{2+} neighbours and $\langle S_z\rangle$ is the thermal average of S_z. The chemical isomer shift of 1·093 mm s^{-1} at 4·2 K should be compared with that of 1·48 mm s^{-1} for FeF_2. Using the Danon interpretation (p. 93), this large decrease signifies [20] a considerable increase in covalency and an effective configuration of $3d^64s^{0\cdot 12}$ as opposed to $3d^64s^0$.

The chemical isomer shifts in FeF_2, $FeCl_2$, $FeBr_2$, and FeI_2 have been found to vary linearly with the Pauling electronegativity of the halide [21]. A linear relationship between the shift and the low-temperature quadrupole splitting was also claimed to exist and to indicate a high degree of covalent bonding and $3d$-delocalisation in the higher halides. Similar studies on $FeCl_2$, $FeCl_2.H_2O$, $FeCl_2.2H_2O$, $FeCl_2.4H_2O$, $FeSO_4.7H_2O$, and $FeSiF_6.6H_2O$ showed an approximately linear increase in shift with increasing coordination to water molecules, as well as further correlation with the quadrupole splitting [22]. However, recent work has questioned the validity of such arguments, and has proposed that the substantial changes in chemical isomer shift are not so much due to $3d$-electron delocalisation as to ligand-$p \rightarrow$ iron $4s$-electron transfer [23]. The chemical isomer shift–quadrupole splitting relationship is also believed to be fortuitous rather than general. A more recent analysis [24] of the temperature dependence of the quadrupole splitting in $FeCl_2$ differs considerably from earlier approaches but uses the argument of considerable $3d$-delocalisation. Controversies such as these illustrate the difficulty in distinguishing the factors influencing the Mössbauer spectrum.

Although $FeCl_2$ is antiferromagnetic in zero applied field, it undergoes a spin-flip transition to a ferromagnetic state in fields of greater than 10·5 kG (the entire antiparallel sublattice reverses spin direction). Spectra of a single crystal of $FeCl_2$ in small external fields do not show any separation of the two sublattices because the intrinsic field is so small [25].

$FeCl_2.H_2O$

The structure of iron(II) chloride monohydrate is unknown but, by analogy with the other hydrates, the local iron environment is likely to be one oxygen atom and five bridging chloride ions which are unlikely to be equivalent. This increased distortion accounts for the much larger quadrupole splitting

[Refs. on p. 164]

in FeCl$_2$.H$_2$O than in FeCl$_2$ [26]. The compound is antiferromagnetic at low temperatures and the saturation value of the internal magnetic field ($T \to 0$ K) is about 251 kG [27].

A related complex with stilbene PhCH:CHPh(FeCl$_2$.H$_2$O)$_4$, is also antiferromagnetic below 23 K and shows a very similar behaviour to FeCl$_2$.H$_2$O and FeCl$_2$.2H$_2$O [27]. Temperature-dependence data give excited-state levels at 309 cm^{-1} and 618 cm^{-1}.

FeCl$_2$.2H$_2$O

The crystal structure of FeCl$_2$.2H$_2$O is known, and the water molecules occupy the *trans*-positions in the near-octahedral environment. The a,b,c,

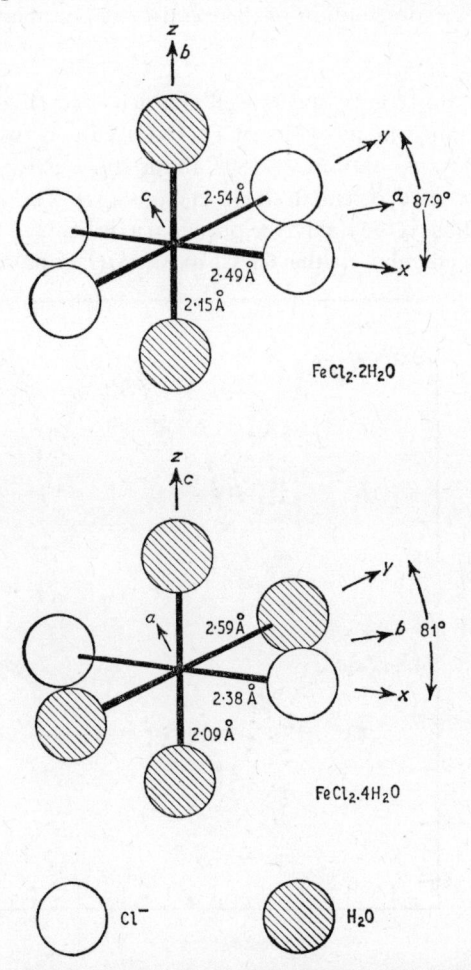

Fig. 6.7 The iron environments in FeCl$_2$.2H$_2$O and FeCl$_2$.4H$_2$O. The two sites in the latter are related by a 180° rotation about the b axis.

[*Refs. on p. 164*]

crystal axes are shown in relation to the bonds in Fig. 6.7. The compound becomes antiferromagnetic below 23 K. Comprehensive measurements on oriented single crystals [28] show that e^2qQ is positive and that the principal axis corresponds closely to the z axis (crystal b axis). The quadrupole temperature dependence is consistent with an electronic ground state which is predominantly $|xy\rangle$ with an $|xz\rangle$ state at $\sim$520 cm^{-1}. The combined quadrupole–magnetic interactions show that the spin axis is in the ac plane along the direction of the short Fe–Cl bond (x axis), and that the asymmetry parameter $\eta = 0.2$. Independent data obtained on polycrystalline powders [29] confirm the perpendicular relation and directions of the electric field gradient tensor and H. Direct lattice-sum calculations have been used to estimate the lattice contribution to the electric field gradient [30].

FeCl$_2$.4H$_2$O

The structure of this compound is well characterised (Fig. 6.7); it contains discrete *trans*-octahedral sub-units of Fe(H$_2$O)$_4$Cl$_2$ on two positions in the unit cell related by a rotation of 180° about the b axis. The first detailed Mössbauer study used oriented single crystals [31]. The minor axis of the electric field gradient is the c axis, the principal axis being in the ab plane either parallel or perpendicular to the Cl–Cl axis. e^2qQ is positive and $\eta = 0.1$.

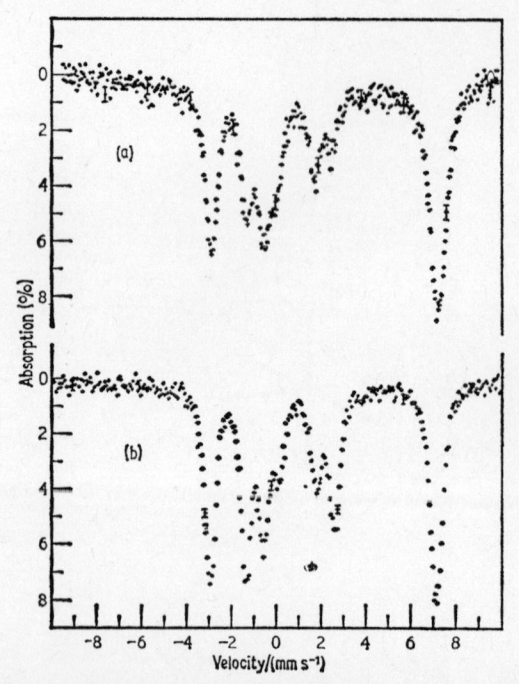

Fig. 6.8 The Mössbauer spectrum of FeCl$_2$.4H$_2$O at 0.4 K in (a) a single crystal with the γ-ray beam parallel to the b axis, and (b) in polycrystals. [Ref. 33, Fig. 1]

[*Refs. on p. 164*]

The sign of e^2qQ was also shown to be positive by using the applied magnetic field method [32]. The electronic ground state is similar to FeF$_2$, i.e. $0.99\,|\,x^2-y^2\rangle + 0.10\,|\,z^2\rangle$, and Ingalls [5] has estimated the electronic excited-state levels to lie at 750 cm^{-1} and 2900 cm^{-1}. However, these figures may not be accurate in view of direct lattice-sum calculations on FeCl$_2$.4H$_2$O which suggest that the lattice term contributes a large proportion of the electric field gradient [30].

The compound is antiferromagnetic below 1·1 K, and single-crystal and polycrystal spectra at 0·4 K [33] show that the spin axis and the principal electric field gradient axis are close to each other, the angle being estimated to be 15°. The magnetic axis is at 30° to the b axis. Line broadening is seen immediately below the Néel temperature because the electron spin relaxation times are comparable to the nuclear precession time. The effect of state mixing on the line intensities of a quadrupole/magnetic spectrum is clearly shown by this compound (see Fig. 6.8). Seven hyperfine components are visible instead of the usual six because the ±2 transition contain an appreciable admixture of other states.

Independent data [34] confirm these conclusions, and give a more extensive interpretation of the spin ordering. The field at the two iron sites makes an angle of ±15° to the b axis of the crystal, the angle between the two fields being 30°. The principal direction of the electric field gradient is at 40° to the b axis, compared with a value of 45° determined by Zory.

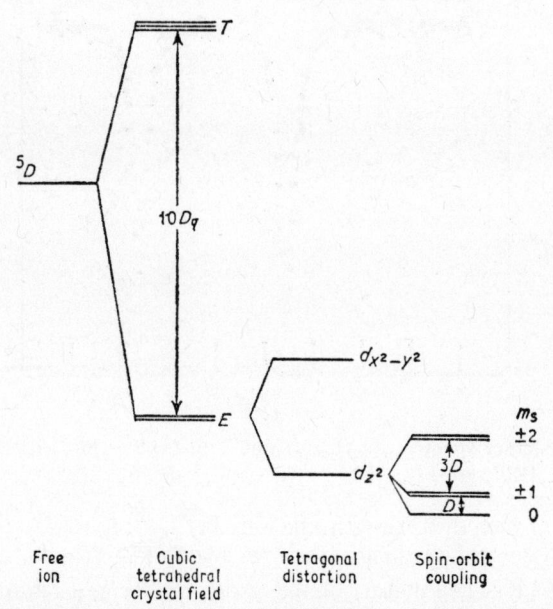

Fig. 6.9 Splittings of the electronic energy levels of a high-spin d^6 ion by the crystal-field and spin–orbit interaction in FeCl$_4^{2-}$. [Ref. 36, Fig. 5]

FeCl$_4{}^{2-}$, FeBr$_4{}^{2-}$, Fe(NCSe)$_4{}^{2-}$, and Fe(NCS)$_4{}^{2-}$

The measurement of a small distortion in the tetrahedral FeCl$_4{}^{2-}$ ion was first reported by Gibb and Greenwood [35]. The E level is split by a tetragonal distortion in the crystal field and this results in a markedly temperature-dependent quadrupole splitting as the thermal population of the $d_{x^2-y^2}$ and d_{z^2} levels changes (see Fig. 6.9). A more extensive investigation [36] has shown

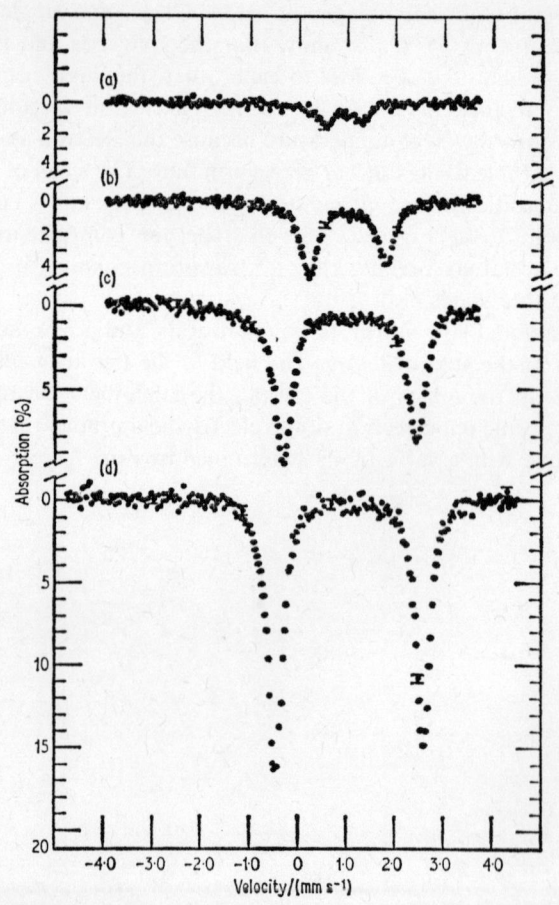

Fig. 6.10 Mössbauer spectra of (Me$_4$N)$_2$FeCl$_4$ at (a) 290 K, (b) 195 K, (c) 77 K, and (d) 4·2 K. [Ref. 36, Fig. 2]

the existence of such distortions in the ions FeCl$_4{}^{2-}$, FeBr$_4{}^{2-}$, Fe(NCSe)$_4{}^{2-}$, and Fe(NCS)$_4{}^{2-}$. In particular, (Me$_4$N)$_2$FeCl$_4$ has been studied in detail from 1·5 to 363 K. The intensity of the resonance decreases dramatically with temperature rise as seen in Fig. 6.10. The effective Debye temperature is only of the order of 200 K and reflects the increase in covalency over the octa-

[*Refs. on p. 164*]

hedral Fe²⁺ ions which give much more intense spectra. The chemical isomer shift shows the typical second-order Doppler shift behaviour as shown in Table 6.1. Covalency also leads to a much lower chemical isomer shift in tetrahedral complexes in general. This is illustrated in Fig. 6.11 in which

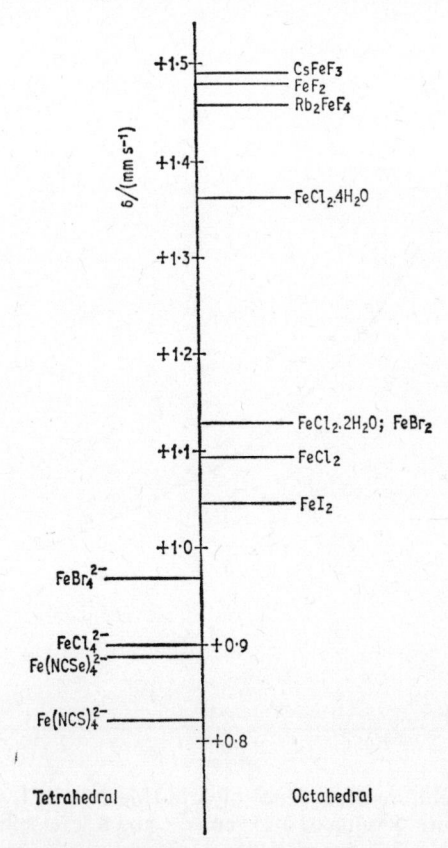

Fig. 6.11 The chemical isomer shifts at 4·2 K for a number of ferrous halides with octahedral and tetrahedral coordination.

chemical isomer shifts relative to iron metal are represented for a number of iron(II) complexes with the halogens and pseudohalogens. Data are compared at 4·2 K where available to reduce contributions from the second-order Doppler effect.

Application of an external magnetic field to $(Me_4N)_2FeCl_4$ gives a negative sign for e^2qQ and shows that the electronic ground state is $|z^2\rangle$, indicating the distortion to be a compression along the z axis (see Chapter 3.6).

The low-temperature limit of the quadrupole splitting for $(Me_4N)_2FeCl_4$ (3·30 mm s⁻¹ at 1·5 K) corresponds to the effective value of $\frac{4}{7}(1-R)\langle r^{-3}\rangle$ (see Section 5.4) and shows the effect of covalency in diminishing $\langle r^{-3}\rangle$ when

[*Refs. on p. 164*]

128 | HIGH-SPIN IRON COMPLEXES

compared to octahedral complexes such as [Fe(H$_2$O)$_6$]SiF$_6$ for which $\Delta = 3.6$ mm s^{-1} at 29 K. The temperature dependence shown in Fig. 6.12

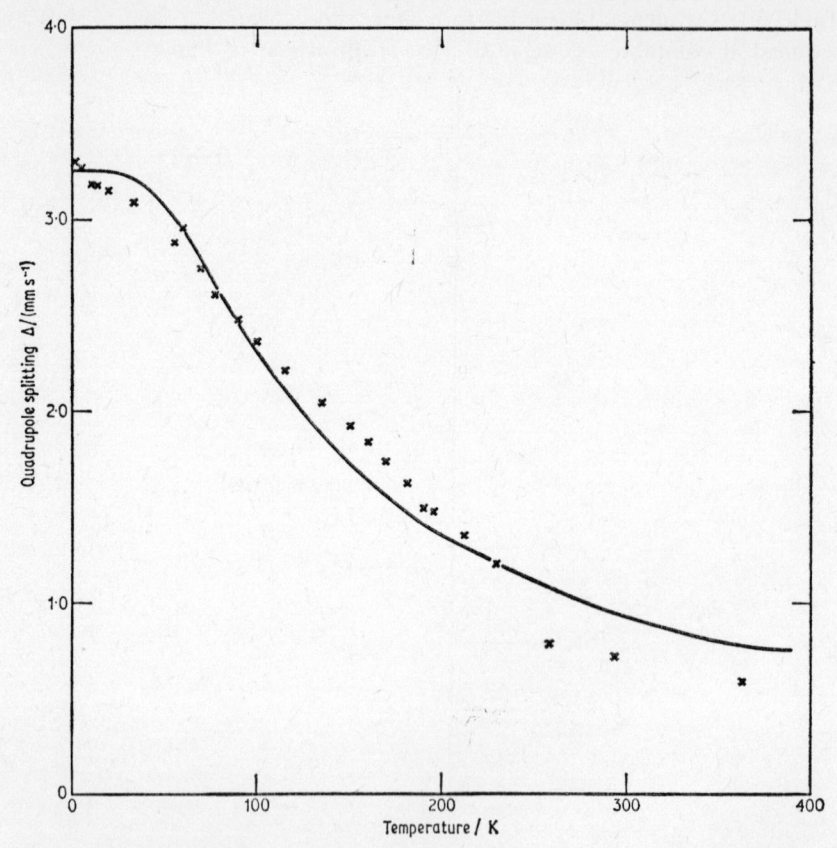

Fig. 6.12 The temperature dependence of Δ in (Me$_4$N)$_2$FeCl$_4$. The solid curve is for a zero temperature Δ value of 3·25 mm s^{-1} and a level splitting of 125 cm^{-1}. [Ref. 36, Fig. 7]

is approximately appropriate to a splitting of 125 cm^{-1} in the E doublet level, but, as shown by the poor agreement of experiment with theory, there appear to be other unknown factors also involved. The splittings of the E doublet for the other compounds were

(Et$_4$N)$_2$FeCl$_4$	135 cm^{-1}
(N,N'-dimethyl-4,4'-dipyridyl)FeCl$_4$	470 cm^{-1}
(Et$_4$N)$_2$FeBr$_4$	96 cm^{-1}
[α,α'-(*bis*-triphenylphosphonium)*p*-xylene]Fe(NCSe)$_4$	292 cm^{-1}
(Me$_4$N)$_2$Fe(NCS)$_4$	101 cm^{-1}

The tetrahedral Fe(NCS)$_4^{2-}$ ion has a much lower chemical isomer shift (0·82 mm s^{-1} at 77 K) than the octahedral Fe(NCS)$_6^{4-}$ ion (1·15 mm s^{-1}

[*Refs. on p. 164*]

at room temperature) [38], again because of increased covalency. An asymmetry was observed in the intensities of the two lines in the $(Me_4N)_2FeCl_4$ spectrum, but this has since been shown to be due to the presence of an Fe(III) impurity [37].

FeBr₂

Anhydrous $FeBr_2$ has a trigonally distorted octahedron of bromine atoms around the iron atom; it has the CdI_2 structure but is otherwise very similar to $FeCl_2$ which has the $CdCl_2$ structure. It too has a very small internal magnetic field below the Néel point of 11 K [20], and the doublet ground state is 175 ± 40 cm^{-1} below the singlet excited state. The value of the hyperfine magnetic field at 4·2 K is 28·4 kG. Like $FeCl_2$ it shows a spin-flip in fields above 31·5 kG and becomes ferromagnetic. The analysis of spectra in large external fields confirms that the normal internal field at 4·2 K is $+28$ kG [25]. The external field values for $FeCl_2 < FeBr_2 < FeI_2$ are consistent with increasing bond covalency in that order.

FeBr₂.2H₂O and FeBr₂.4H₂O

Only one spectrum has been reported for each of these compounds and the parameters are given in Table 6.1. They are probably similar to the chlorides.

FeI₂

Like ferrous bromide, FeI_2 has the CdI_2 structure, and is antiferromagnetic below 10 K. Partial orientation of powdered samples gave a line asymmetry

Table 6.2 Contributions to the internal magnetic field. Errors are at least of the order of 10%

Compound	T/K	H_s/kG	H_L/kG	H_D/kG	H/kG	Reference
FeF₂	4·2	−558	+276	−52	−334	6
FeCl₂	4·2	−411	+387	+24	0	18
	4·2	−520	+490	+30	∼0	20
	5	−419	—	—	0	25
FeBr₂	5	−390	—	—	+30	25
FeI₂	4·2	−515	{+560 / +410}	+30	{+74 / −74}	20
	5	−345	—	—	+74	25
RbFeF₃	(0)	−412	+240	+5	(172)	11
FeSO₄	4·2	−550	+217	+148	(−)185	42
FeCO₃	4·2	−467	+550	+88	(+)172	42
	4·2	−500	+583	+87	(+)170	6
FeC₂O₄.2H₂O	4·2	−550	+383	+13	(−)154	42

[Refs. on p. 164]

130 | HIGH-SPIN IRON COMPLEXES

which suggests that e^2qQ is positive [39]. The magnetic field is somewhat larger in this compound (74 kG at 4·2 K [20]), because of differences in the orbital term H_L (see Table 6.2) and is assumed to be positive [25]. The exchange coupling term recorded in the chloride is again significant in evaluation of the quadrupole splitting at low temperatures and the doublet–singlet separation has been given as 183 cm^{-1} [39] and 135 cm^{-1} [20]. The low chemical isomer shift reflects increased covalency along the halide series and leads to a configuration of $3d^6 4s^{0·15}$ on the Danon model.

6.2 Iron(II) Salts of Oxyacids and Other Anions

[Fe(H$_2$O)$_6$]SiF$_6$

Hydrated iron(II) fluorosilicate contains the SiF$_6^{2-}$ anion and the octahedral cation [Fe(H$_2$O)$_6$]$^{2+}$. The octahedron is trigonally distorted by compression along the [111] axis, and the ground-state configuration is well known and is a $|z^2\rangle$ state. This simple electronic configuration with no asymmetry parameter or close-lying excited states has made [Fe(H$_2$O)$_6$]SiF$_6$ a popular choice for the estimation of the valence-electron contribution to the electric field gradient tensor and hence to the quadrupole coupling constant e^2qQ; the many attempts at such calculations have been listed in Table 5.2.

The first experiments [40] used a single-crystal absorber and a polarised magnetic foil source. The intensities of the spectral components show that the coupling constant e^2qQ is negative as required for a $|z^2\rangle$ state, and this sign has also been checked by the magnetic perturbation technique [32]. The quadrupole splitting is large (3·612 mm s^{-1} at 29 K [32]) because of the lack of appreciable state mixing. Although an attempt has been made to derive excited-state electronic level separations from the temperature dependence of the quadrupole splitting [5] this parameter is not very sensitive to a level separation of the order of 1000 cm^{-1}. Direct lattice-sum calculations give only a very small value for the lattice term contribution to the electric field gradient tensor [30], a result not predicted in earlier work [5].

Hydrated iron(II) fluorosilicate is still paramagnetic at 1·8 K, and the fast ionic relaxation times prevent inherent magnetic splitting in the Mössbauer spectrum. However, the ^5D configuration of the high-spin iron(II) ion gives it anisotropic magnetic properties. Application of an external magnetic field H_{ext} then results in an internal field at the iron nucleus of

$$H = H_{\text{ext}} + H_n$$

or

$$H = H_{\text{ext}} + \frac{\langle S \rangle}{S} H_n^{(0)}$$

where $H_n^{(0)}$ is the saturation value of the tensor quantity H_n of the intrinsic magnetic field at the nucleus, and $\langle S \rangle$ is a thermal average of the total

[Refs. on p. 164]

electronic spin ($S = 2$) [41]. In terms of the susceptibility tensor $\chi_i (i = x, y, z)$

$$H = H_{\text{ext}}\left(1 + \frac{\chi_i H_{ni}^{(0)}}{N g_i \mu_N S}\right)$$

for $\langle S \rangle / S < 0.25$, i.e. for small magnetisation. For a paramagnetic Fe^{2+} ion in a zero field H_{ext}, it is found that $\langle S \rangle$ is zero because of the fast relaxation and $H = 0$.

At high temperatures the susceptibility χ is small and at fields up to 30 kG the value of $(\langle S \rangle / S) H_n^{(0)}$ is very small. The effective field is close to H_{ext} as assumed in the magnetic perturbation method for determining the sign of

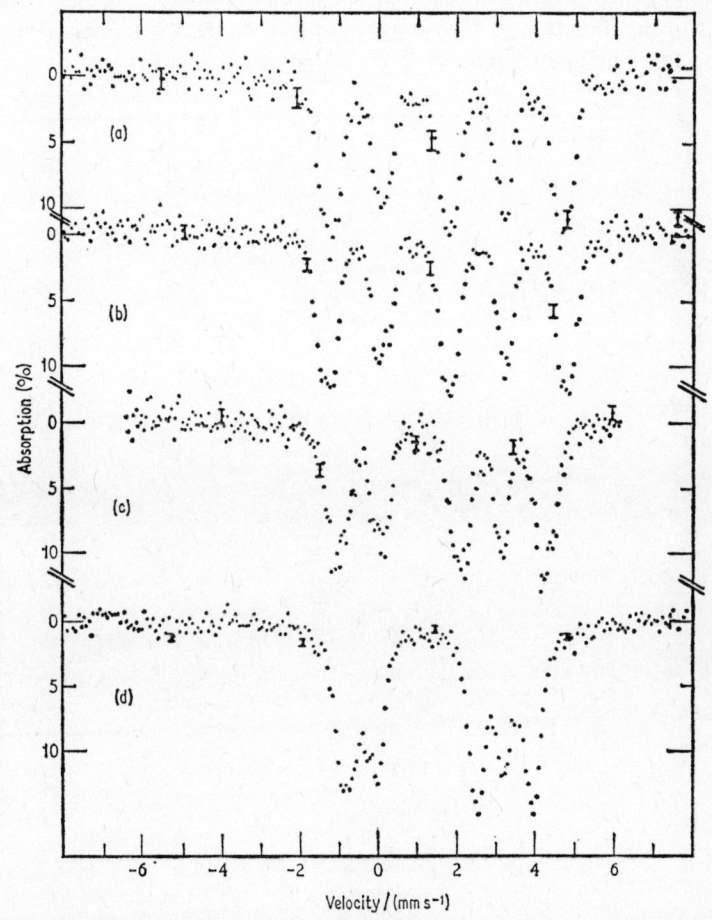

Fig. 6.13 Mössbauer spectra of an [Fe(H$_2$O)$_6$]SiF$_6$ single crystal in an external field of 30 kG applied at angles of (a) 90°, (b) 70°, (c) 45° and (d) 20° to the trigonal axis. [Ref. 41, Fig. 2]

[*Refs. on p. 164*]

132 | HIGH-SPIN IRON COMPLEXES

e^2qQ. At liquid helium temperatures where the susceptibility is large, H_n will be significant and, because it is a tensor quantity, the resultant value of H is dependent on the anisotropy of the magnetisation.

This behaviour is clearly illustrated by the behaviour of $[Fe(H_2O)_6]SiF_6$ in an external field at 4·2 K. Fig. 6.13 shows spectra for a single crystal with a 30-kG field applied at different angles to the trigonal axis [41]. The susceptibility is small along the trigonal axis so that at angles close to zero the field H is nearly equal to H_{ext}. At directions perpendicular to the trigonal axis the magnetisation is large so that a very substantial augmentation of H and a large magnetic splitting is found. Polycrystalline samples in an external field show a similar behaviour to the single crystals, because although the crystal orientation is now averaged, the high anisotropy tends to make the resultant field act in the direction of high magnetisation. As H_{ext} is increased in magnitude, a large magnetic splitting is generated (see Fig. 6.14).

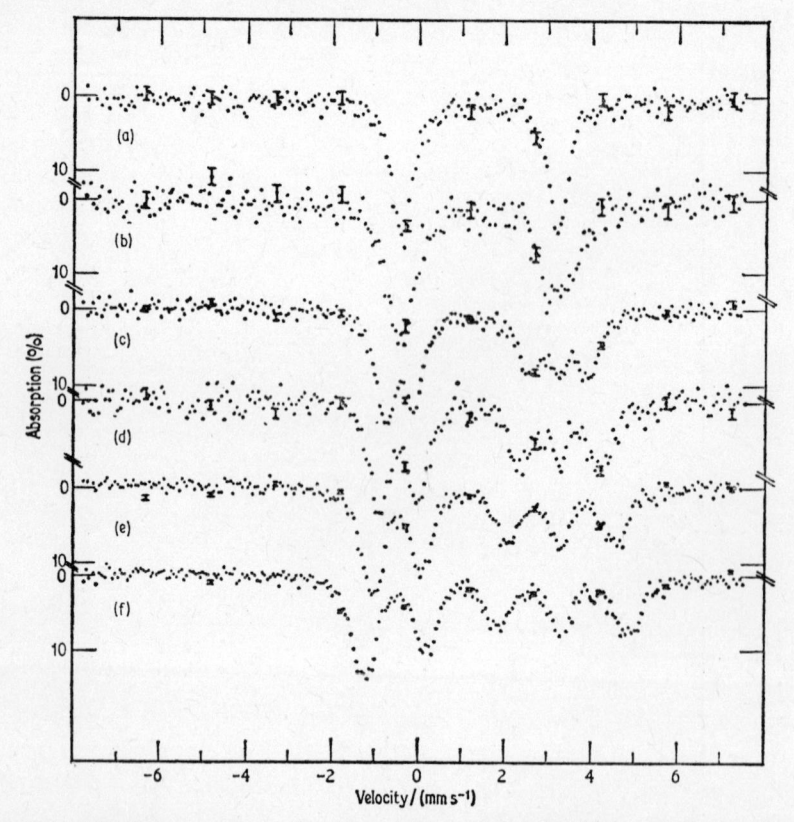

Fig. 6.14 Mössbauer spectra of polycrystalline $[Fe(H_2O)_6]SiF_6$ at 4·2 K with an external field applied perpendicular to the direction of the γ-rays. (a) 0 kG, (b) 4·5 kG, (c) 10·5 kG, (d) 15 kG, (e) 21 kG, (f) 30 kG. [Ref. 41, Fig. 3]

[*Refs. on p. 164*]

FeSO$_4$

FeSO$_4$ has a complicated orthorhombic crystal structure with four FeSO$_4$ per unit cell. Each Fe^{2+} ion is surrounded by a distorted octahedron of oxygen atoms and each FeO$_6$ group has the same geometry and one axis nearly parallel to the crystal b axis. The low-temperature antiferromagnetic Mössbauer spectrum [42] is consistent with a positive value of e^2qQ, an asymmetry parameter of $\eta = 0.4$, and a ground-state configuration of $0.99 \mid x^2 - y^2 \rangle + 0.09 \mid z^2 \rangle$. The contributions to H were estimated assuming that the observed field of 185 kG at 4·2 K was negative in sign (see Table 6.2). Excited-state electronic levels have been estimated to lie at 360 cm^{-1} and 1680 cm^{-1} [5, 19], and the positive sign of e^2qQ has been verified by the magnetic perturbation method [32].

[Fe(H$_2$O)$_6$]SO$_4$.H$_2$O

Iron(II) sulphate heptahydrate contains four formula weights per unit cell on two distinct sites A and B. The immediate iron environment in both cases is a distorted octahedron of oxygens so that the compound contains [Fe(H$_2$O)$_6$]$^{2+}$. The two types of sites are so similar as to generate only one Mössbauer quadrupole doublet, but the relative orientations of the four iron positions are important in interpreting the asymmetry of the quadrupole doublet in oriented single crystals. Such measurements have been made [43], and indicate an $\mid xy \rangle$ ground state (positive e^2qQ) and an asymmetry parameter of $\eta = 0.1$ for both the A and B sites. A model for the parameters of the electric field gradient tensors was proposed and found to agree with the experimental line intensities.

Direct lattice-sum calculations show that the lattice contributions to e^2qQ of the two sites are of opposite sign [30]. The signs of the coupling constants have been confirmed as positive by the magnetic perturbation method [32], and the electronic excited states estimated assuming only one type of site [5, 19].

[Fe(H$_2$O)$_6$](NH$_4$SO$_4$)$_2$

The interpretation of the Mössbauer spectrum of hydrated iron(II) ammonium sulphate presents considerable difficulty. The crystal unit cell is monoclinic with two iron sites which are equivalent but with different relative orientations. The immediate iron environment is a slightly distorted [Fe(H$_2$O)$_6$]$^{2+}$ octahedron. Early work [5] using the temperature dependence of the quadrupole splitting proposed an $\mid xy \rangle$ ground state and a rhombic symmetry giving excited levels at 240 cm^{-1} and 320 cm^{-1}. Values of 230 cm^{-1} and 300 cm^{-1} have also been quoted [44].

Two later studies have measured the line asymmetry in oriented single crystals and compared the observed values with those predicted by an adopted

134 | HIGH-SPIN IRON COMPLEXES

model of the electric field gradient tensor [43, 45]. The more detailed paper [45] included data at 300 K and 4·2 K. The asymmetry parameter is 0·7 at 300 K and 0·3 at 4·2 K. The distortion appears to be close to rhombic with a $|xy\rangle$ ground state and $|y^2 - z^2\rangle$ and $|xz\rangle$ excited states, but the electric field gradient tensor and susceptibility axes do not coincide, suggesting that the actual symmetry is lower than rhombic. An attempt to measure the sign of e^2qQ by the magnetic perturbation technique [32] failed because of the large asymmetry parameter.

The temperature dependence of the quadrupole splitting of the iron(II) potassium, rubidium, and caesium sulphates has also been examined but in less detail. As can be seen from Table 6.3 they appear to resemble the ammonium salt closely [44].

Table 6.3 Mössbauer parameters for high-spin iron(II) cations bonding to oxygen

Compound	T/K	Δ/(mm s^{-1})	η	H/kG	δ/(mm s^{-1})	δ (Fe)/(mm s^{-1})	Reference
[Fe(H$_2$O)$_6$]SiF$_6$	RT	−3·385	0	—	0·933 (Pt)	1·280	32
	29	−3·612	0	—	1·069 (Pt)	1·416	
FeSO$_4$	RT	+2·726	—	—	0·923 (Pt)	1·270 ⎫	32
	80	+3·104	—	—	1·043 (Pt)	1·390 ⎭	
	HE	+3·89	0·4	185	—	—	42
[Fe(H$_2$O)$_6$]SO$_4$.H$_2$O	RT	+3·217	—	—	0·914 (Pt)	1·261	32
	5	+3·384	—	—	1·044 (Pt)	1·391	
[Fe(H$_2$O)$_6$](NH$_4$SO$_4$)$_2$	RT	+1·717	—	—	0·897 (Pt)	1·244	46
FeCO$_3$	RT	+1·798	—	—	0·892 (Pt)	1·239 ⎫	42
	80	+2·043	—	—	1·014 (Pt)	1·361 ⎭	
	(0)	+2·06	—	+184	1·36 (Fe)	1·36	51
Fe$_3$(PO$_4$)$_2$.8H$_2$O	295 {1	+2·50	—	—	0·86 (Pt)	1·21 ⎫	
	{2	+2·98	0·2	—	0·88 (Pt)	1·23	
	80 {1	+2·59	—	—	0·95 (Pt)	1·30	54
	{2	+3·16	0·2	—	0·99 (Pt)	1·34	
	5 {1	+2·59	—	268	0·96 (Pt)	1·31	
	{2	+3·18	0·3	135	0·99 (Pt)	1·34 ⎭	
Fe$_3$(PO$_4$)$_2$.4H$_2$O	4·2 {1	2·50	0·25	203	0·98 (Cu)	1·21	56
	{2	2·64	0·23	362	0·98 (Cu)	1·21	
Fe(HCOO)$_2$.2H$_2$O	298 {A	0·59	—	—	1·34 (Cr)	1·19 ⎫	
	{B	2·97	—	—	1·39 (Cr)	1·24	
	77 {A	1·37	—	—	1·56 (Cr)	1·41	59
	{B	3·42	—	—	1·56 (Cr)	1·41	
	6 {A	1·48	—	—	1·48 (Cr)	1·33	
	{B	3·36	—	—	1·53 (Cr)	1·38 ⎭	
FeC$_2$O$_4$.2H$_2$O	RT	1·721	—	—	—	—	60
	4	−1·89	0·68	148	—	—	

[Refs. on p. 164]

FeCO₃

FeCO₃ which occurs naturally as the mineral siderite has a rhombohedral structure. The octahedron of oxygens around the iron is trigonally distorted along the c axis which is the magnetic axis below the Néel temperature. The ground state is a doublet and the high symmetry eases interpretation of the data.

Early data on the antiferromagnetic state ($T_N = 38$ K) showed a positive value for e^2qQ as expected and allowed estimates of the orbital and dipolar terms [6, 42]. The sign of e^2qQ was confirmed as positive by the magnetic perturbation method [32], and polarisation measurements using magnetically split iron metal sources and absorbers in external fields verify the positive sign of the internal field [47]. Goldanskii *et al.* [48] used oriented single crystals of FeCO₃ in experiments to measure quantitatively any anisotropy of the recoil-free fraction. A substantial anisotropy was claimed to be derived from the angular variation of the line intensities of the quadrupole doublet. However, a counter-claim by Housley *et al.* [49] refutes their results on the grounds that no allowance was made for the polarisation dependence of the molecular absorption cross-sections, and therefore the calculated areas for γ-ray directions not parallel to the c axis are incorrect.

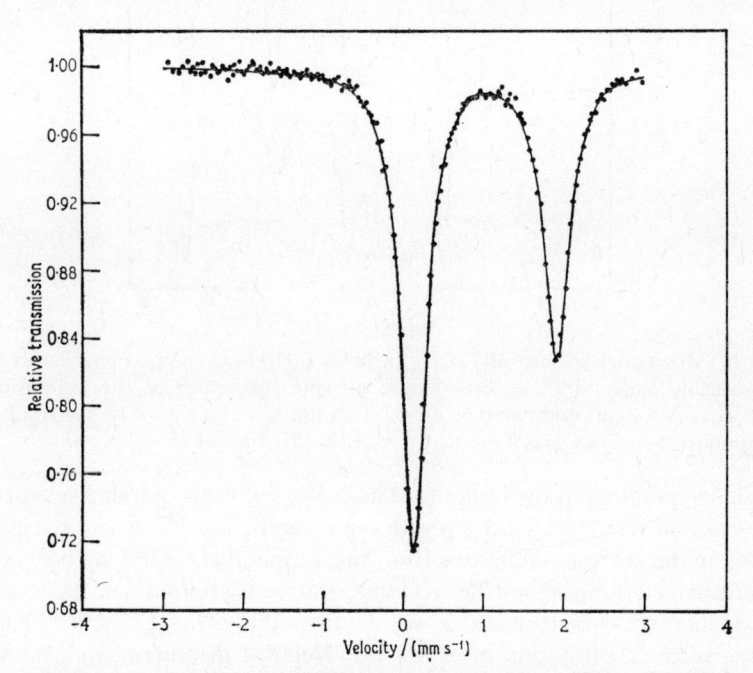

Fig. 6.15 Room-temperature Mössbauer spectrum of a siderite (FeCO₃) single crystal with the γ-ray beam normal to the c axis. [Ref. 49, Fig. 2]

The *total* cross-section in any direction *is* independent of polarisation, but the individual line cross-sections are not. A typical oriented crystal Mössbauer spectrum of $FeCO_3$ is shown in Fig. 6.15. One of the consequences of the modified theory is that the theoretical area ratio is comparatively insensitive to the recoil-free fraction and that it is unrealistic to determine it by an area ratio method. The presence of impurities and imperfections in the mineral specimens was held to explain some of the experimental deviations from prediction. No direct evidence for anisotropy of the recoil-free fraction was obtained, a conclusion since verified by Goldanskii [50], who made new polarisation measurements using single crystals of $FeCO_3$ as polariser and analyser.

$FeCO_3$ can show a transition from antiferromagnetic to ferromagnetic ordering in very large applied fields (metamagnetism) [51] in the same way as already described for $FeCl_2$. In a field of 100 kG the sublattice with spins normally antiparallel to the field have a faster relaxation time than the parallel sublattice. Consequently as the temperature is raised the resonance lines from the former show line broadening as shown in Fig. 6.16. The outer line on the

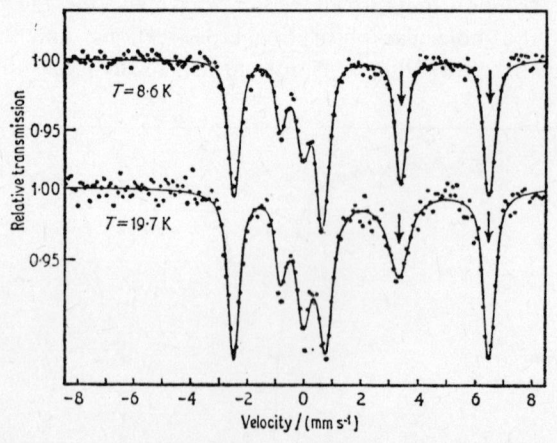

Fig. 6.16 Mössbauer spectra of $FeCO_3$ in fields of 100 kG applied parallel to the spin ordering axis. Note the broadening by spin relaxation of the antiparallel sublattice components compared to the parallel sublattice as shown by the resolved pair of equivalent lines which are arrowed. [Ref. 51, Fig. 3]

right, corresponding to the larger field associated with the parallel sublattice in the external field, remains sharp with rise in temperature; whereas the inner line, being the corresponding one from the antiparallel sublattice, broadens and effectively disappears as the Néel temperature is approached. Being able to distinguish the sublattice in this way confirms that the magnetic field has a positive value at saturation of +184 kG. Detailed measurements near the Néel temperature without an applied field also show evidence for the relaxation effects [52].

[*Refs. on p. 164*]

[Fe(H$_2$O)$_6$](ClO$_4$)$_2$

Although little is known about the structure of hydrated iron(II) perchlorate it is likely that it contains [Fe(H$_2$O)$_6$]$^{2+}$ units like the isomorphic [Mg(H$_2$O)$_6$](ClO$_4$)$_2$. The typical Fe^{2+} quadrupole doublet is seen [53], but an unusual behaviour is observed between about 220 and 250 K. In this region one pair of lines is gradually replaced by another, showing that a change has taken place in the symmetry of the Fe^{2+} ion. The same phenomenon is found in the isomorphic cobalt(II), nickel, and magnesium compounds when crystallised with 1·5 atom % of ^{57}Fe although the change occurs at different temperatures. The quadrupole splitting decreases from 3·4 mm s^{-1} at 108 K to 1·4 mm s^{-1} at 298 K and a change in the distortion of the octahedron is suggested. No other data are available.

Fe$_3$(PO$_4$)$_2$.8H$_2$O

Vivianite, Fe$_3$(PO$_4$)$_2$.8H$_2$O, has a monoclinic structure with two crystallographically inequivalent Fe^{2+} positions, Fe(1) and Fe(2), in the relative proportions 1 : 2. Fe(1) is octahedrally coordinated by four water molecules and two oxygens of (PO$_4$)$^{3-}$ ions, whereas Fe(2) sites occur in pairs as edge-shared octahedra and are each coordinated by two water molecules and four oxygens of (PO$_4$)$_3^-$ ions. Vivianite becomes antiferromagnetic at 8·8 K.

Single crystals and polycrystalline samples of naturally occurring vivianite have been examined in detail [54]. The paramagnetic state gives a four-line Mössbauer spectrum at 80 K in which the weaker quadrupole doublet, which also has the smaller splitting, can be assigned to the Fe(1) site. Both sites lie on the crystal b axis and have axial symmetry so that the b axis must represent one of their respective electric field gradient axes, and the other two lie in the ac plane. The area ratios of the spectra of single crystals show that V_{zz} is positive ($e^2qQ > 0$) and is in the ac plane. At 5 K a complex magnetic spectrum with at least 9 resolved components is seen. Gonser and Grant reject an earlier interpretation by Ono et al. [42] in which equal magnetic fields were assumed for the two sites. The new values quoted are 268 kG for the Fe(1) site and 135 kG for the Fe(2) site. Several components of the two spectra overlap fortuitously. Orientation experiments verify that the spin axes of both sites are parallel to the ac plane.

Determination of the spin directions within the ac plane was accomplished by using a α-Fe source and a single crystal of vivianite at room temperature. The spectra contained many lines because both absorber and source were split, but could be analysed successfully to show that V_{zz} bisects the 104·5° ac angle for the Fe(1) site and that V_{zz} for the Fe(2) site is perpendicular to this. Although the spin axes for both sites at low temperature were thought to be parallel to the V_{zz} direction for Fe(1) an independent investigation by neutron diffraction and Mössbauer methods did not confirm this [55]. The spin axes are not collinear but are angled at 42° to each other; V_{zz} for Fe(1) is approximately parallel to its spin axis and makes an angle of 48° with V_{zz} for Fe(2). The considerable depth of the data analysis gives an unusually detailed insight into the magnetic ordering processes in the vivianite.

$Fe_3(PO_4)_2.4H_2O$

$Fe_3(PO_4)_2.4H_2O$, known as ludlamite, also contains two iron sites, Fe(1) and Fe(2), in the ratios 1 : 2 and is antiferromagnetic below 15 K. The ordered phase gives a Mössbauer spectrum with two hyperfine fields [56]. The relation between the electric field gradient tensors and the spin axes were determined and the symmetry was shown to be lower than in $Fe_3(PO_4)_2.8H_2O$; however, the magnetic ordering has not yet been studied in detail.

Li(Fe,Mn)PO$_4$

The lithium orthophosphate $LiFe_{0.8}Mn_{0.2}PO_4$ is a naturally occurring mineral. It orders antiferromagnetically at low temperature, and the spectrum at 4·2 K indicates a magnetic field of 129 kG, $e^2qQ/2 = +2.71$ mm s^{-1}, and $\eta = 0.8$, with the magnetic and electric field gradient axes collinear [57]. Temperature-dependence data have not been published.

$Fe(HCOO)_2.2H_2O$

Iron(II) formate contains two basically different types of Fe^{2+} ion, one of which (A) is coordinated to six formate oxygens and the other (B) to two formate oxygens in a *trans*-configuration and four water molecules. Four equally intense lines are observed in the Mössbauer spectrum [58, 59]. The greater sensitivity of Δ than δ to environment allows a confident assignment to two quadrupole doublets on a BAAB basis. The smaller quadrupole splitting is probably associated with the more symmetrical site, and the orbital states have been considered [59].

$FeC_2O_4.2H_2O$

Although hydrated iron(II) oxalate is a well-known complex, its crystal structure is unknown; it has been classified as possessing rhombic symmetry. The spectrum of the low-temperature antiferromagnetic phase [42, 60] shows that the asymmetry parameter is approximately 0·7. The major axis of the electric field gradient tensor appears to be perpendicular to the magnetic axis. e^2qQ is negative, but this cannot be confirmed by the magnetic perturbation technique because of the effect of the large asymmetry parameter [30]. Only two broad lines are seen at 20 K and presumably the Néel point is close to this temperature. The contributions to the internal field were estimated (see Table 6.2), as were the excited-state electronic levels [5].

$Fe_3B_7O_{13}X$

The iron boracites have the formula $Fe_3B_7O_{13}X$ where X = Cl, Br or I. The iron is coordinated by two halogens as well as by four oxygen ions, and so the

IRON(II) SALTS OF OXYACIDS AND OTHER ANIONS | 139

compounds belong equally to this section and to the preceding one on halogen complexes of iron(II). They have a well-known orthorhombic cubic high-temperature phase transition. The cubic phase has a very high symmetry at the iron site, but in the orthorhombic phase two different site distortions occur. The resultant spectra [61, 62] are illustrated in Fig. 6.17, and clearly

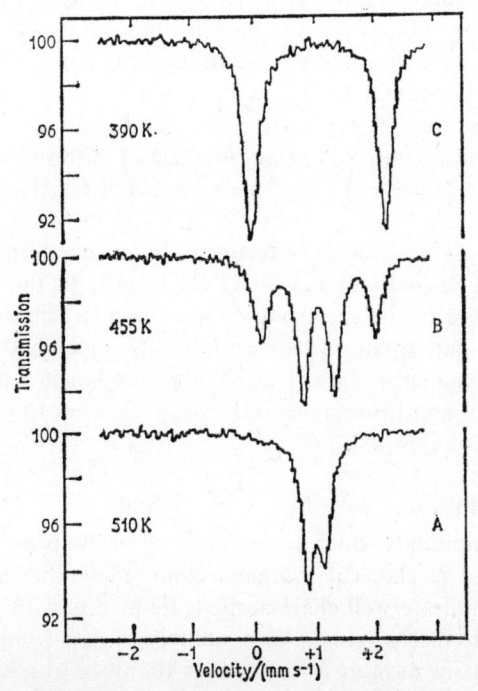

Fig. 6.17 Mössbauer spectra of the three phases of $Fe_3B_7O_{13}Br$. (A) cubic, (B) orthorhombic, (C) trigonal. [Ref. 61, Fig. 1]

show the two types of iron in the latter phase. Unexpectedly, a third type of spectrum was seen at lower temperatures in which there is only one type of iron and this gave the first evidence of a second phase transition from orthorhombic to a trigonal phase. The chemical isomer shifts showed no discontinuities at the phase transitions and at 300 K were as follows: $Fe_3B_7O_{13}Cl$, 1·14; $Fe_3B_7O_{13}Br$, 1·14; $Fe_3B_7O_{13}I$, 1·12 mm s^{-1}.

Frozen Solutions

Some initial experiments on frozen solutions of the iron(II) salts $FeCl_2 \cdot 4H_2O$ $[Fe(H_2O)_6](ClO_4)_2$, $[Fe(H_2O)_6]SO_4 \cdot H_2O$, and $[Fe(H_2O)_6](NH_4)_2(SO_4)_2$, showed that the iron species in each matrix was identical and unaffected by the anion [63]. Very unusual behaviour was detected at about 190 K due to a phase transition in the ice [64], and this was later confirmed [65] in a detailed study of the frozen solutions of iron(II) chloride. Quenching from the liquid state to 78 K produces $[Fe(H_2O_6)]^{2+}$ ions trapped in a cubic ice lattice rather than in the stable hexagonal ice form. Warming to 193 K induces an irreversible transformation to hexagonal ice which is clearly detected by the discontinuous change in the quadrupole splitting. In the immediate region of the transition temperature the Mössbauer spectrum disappears completely because the increase in diffusion reduces the f-factor. Subsequent slow cooling does not regenerate the cubic form although quenching does. The $[Fe(H_2O)_6]^{2+}$ in hexagonal ice has axial symmetry with an $|xy\rangle$ ground state and its orbital properties have been discussed. Above 233 K the spectrum again disappears because of eutectic formation. It also proved possible to generate the unstable hydrate $FeCl_2.6H_2O$ in a precipitated form [65, 66].

Other frozen solution studies of ferrous salts include ternary systems such as $Fe^{2+}-Sn^{4+}-H_2O$ [66] and $FeCl_2-KF-H_2O$ [67]. In the latter case there was no dependence of the spectrum on the anion, which may be presumed therefore not to participate in formation of the inner hydrate layer. Non-aqueous solvent studies are also feasible, e.g. $FeCl_2$ when dissolved in mixtures of methanol and formamide and frozen, appears to form at least two distinct solvated species [68].

6.3 Iron(II) Complexes with Nitrogen Ligands

Many of the compounds which were discussed in the previous two sections could be described as 'classical' inorganic compounds; they have been known for a long time and are well characterised, frequently with an X-ray crystal structure. In more recent years a large number of new complexes have been prepared in which one or more of the ligands about the iron is a nitrogen base, e.g. pyridine. Detailed characterisation of these complexes is still in progress, and one of the physical techniques which has been applied is Mössbauer spectroscopy, although the lack of structural data frequently limits its use to a more qualitative approach. However, the spectrum obtained can usually establish the oxidation state and spin state unequivocally, and might also show the presence of any major impurity in the preparation.

The ligand-field strengths of nitrogen bases are generally greater than those of oxygen and the halogens, and are sometimes high enough to overcome the electron repulsion energies and cause spin pairing. In a number of examples an equilibrium is found between the high-spin $^5T_2$ and low-spin $^1A_1$ states, and this ligand-field 'crossover' situation is considered separately in more detail in Chapter 8.

$[Fe(NH_3)_6]^{2+}$

Atypical results have been found in ferrous complexes containing the $[Fe(NH_3)_6]^{2+}$ cation [69]. $[Fe(NH_3)_6][Fe_2(CO)_8]$ and $[Fe(NH_3)_6]Cl_2$ both give a very sharp single-line resonance from the cation at 295 K, despite the fact that the electronic spectra show clear splitting of the $^5T_{1g} \to {}^5E_g$ bands which indicates distortion of the octahedron. In order to explain the lack of a quadrupole splitting a dynamic Jahn–Teller effect was proposed. This results in time-averaging of the electric field gradient within the Mössbauer lifetime, but not in the optical measurement time-scale. A small doublet contribution appears in the spectra at 80 K, the explanation for which is not clear. The similar octahedral cations $[Fe(MeCN)_6]^{2+}$ and $[Fe(PhCN)_6]^{2+}$ both differ in that a small quadrupole splitting is seen despite a smaller ligand-field splitting in these compounds. Parameters for the cations only are [69]

	T/K	Δ/(mm s^{-1})	δ (Fe)/(mm s^{-1})
$[Fe(NH_3)_6][Fe_2(CO)_8]$	295	0	1·04
	80	0	1·15
$[Fe(NH_3)_6]Cl_2$	295	0	1·02
	80	{0 1·50	1·10 1·12
$[Fe(MeCN)_6][FeCl_4]_2$	295	0·27	1·14
	80	0·48	1·23
$[Fe(PhCN)_6][FeCl_4]_2$	295	0·75	1·02
	80	1·31	1·22

Independent work on $[Fe(NH_3)_6]Cl_2$ has found that the appearance of a doublet spectrum at low temperatures is accompanied by hysteresis effects [70]. The lack of splitting at room temperature was attributed to the onset of free rotation about the Fe–N bonds of the NH_3 ligands, thereby generating a higher effective site symmetry at the iron, although there appears to be less evidence for this than for the dynamic Jahn–Teller effect.

Fe(py)$_4$X$_2$

A number of spectra have been measured for pyridine complexes of the general type Fe(py)$_4$X$_2$ where X = Cl, Br, I, NCS, or NCO [71–74]. The quadrupole splittings, chemical isomer shifts, and magnetic moments (μ) are listed in Table 6.4. The shift is generally lower than in complexes with co-ordination to oxygen by about 0·1–0·2 mm s^{-1}.

The magnetic moment and the quadrupole splitting of a complex are both dependent on its orbital state because of the anisotropy of the Fe^{2+} ion, and it is possible to formulate a mathematical relationship between the two using a simplified model with a *trans*-geometry and axial symmetry [71]. The quadrupole splitting is more sensitive to small distortions from octahedral symmetry than the magnetic moment and the converse is true for large distortions. Although experiment is approximately in agreement with theory, many approximations have to be made and the degree of error in magnetic moment measurements may be quite large [72].

Yellow and violet forms of Fe(py)$_4$(NCS)$_2$ are known and were originally

[Refs. on p. 164]

142 | HIGH-SPIN IRON COMPLEXES

Table 6.4 Mössbauer parameters for high-spin iron(II) complexes with pyridine

Compound	T/K	Δ /(mm s^{-1})	δ /(mm s^{-1})	δ (Fe) /(mm s^{-1})	μ/BM	Reference
Fe(py)$_4$Cl$_2$	RT	3·14	1·13 (SS)	1·04	5·34	71
	295	3·08	0·86 (Pd)	1·04	5·35	72
Fe(py)$_4$Br$_2$	RT	2·52	1·16 (SS)	1·07	5·28	71
Fe(py)$_4$I$_2$	RT	0·65	1·25 (SS)	1·16	5·90	71
Fe(py)$_4$I$_2$.2py	RT	0·83	1·17 (SS)	1·08	5·64	71
Fe(py)$_4$(NCS)$_2$	RT	1·52	1·21 (SS)	1·12	5·47	71
	RT	1·53	1·16 (Fe)	1·16		73
	295	1·56	0·90 (Pd)	1·08	5·41	72
	300	1·52	1·33 (NP)	1·07		74
	80	1·97	1·42 (NP)	1·16		
Fe(py)$_4$(NCO)$_2$	RT	2·52	1·24 (SS)	1·13	5·13	71
Fe(py)$_4$(NCO)$_2$py	RT	2·59	1·24 (SS)	1·15	5·20	71

thought to be *cis*- and *trans*-isomers respectively. Their Mössbauer spectra are identical and this initiated further work which indicated that the violet form was identical to the yellow form except for a small amount of Fe^{3+} impurity [73, 74]. The possibility of linkage isomerism was also ruled out and comparison of the single-crystal X-ray patterns with those of the isomorphous *trans*-nickel and cobalt complexes established that both the yellow and violet forms were, in fact, *trans*.

If all the pyridine complexes in Table 6.4 are considered to have a *trans*-geometry then the quadrupole splitting will provide a measure of the elongation or compression along this axis. It is interesting to note that in the pairs Fe(py)$_4$I$_2$/Fe(py)$_4$I$_2$.2py and Fe(py)$_4$(NCO)$_2$/Fe(py)$_4$(NCO)$_2$.2py where there is no change in the primary coordination sphere there is little change in the Mössbauer parameters.

The thermal decomposition of Fe(py)$_4$Cl$_2$ gives a series of intermediate products [75]. The stepwise process is

Fe(py)$_4$Cl$_2$ → Fe(py)$_2$Cl$_2$ → Fe(py)Cl$_2$ → Fe$_3$(py)$_2$Cl$_6$ → FeCl$_2$

All but the first of these show a small quadrupole splitting and may have an octahedral iron environment with bridging chlorines.

Fe(phen)$_2$X$_2$

Parameters for the 1,10-phenanthroline complexes, Fe(phen)$_2$X$_2$, with X = Cl$^-$, Br$^-$, I$^-$, NCS$^-$, NCSe$^-$, N$_3^-$ are listed in Table 6.5. The local symmetry is lower than in the Fe(py)$_4$X$_2$ complexes because of the bidentate phenanthroline group, and this results in a larger Fe^{2+} quadrupole splitting and smaller magnetic moment. Of particular interest are Fe(phen)$_2$(NCS)$_2$

[*Refs. on p. 164*]

Table 6.5 Mössbauer parameters for high-spin iron(II) complexes with 1,10-phenanthroline

Compound	T/K	Δ /(mm s⁻¹)	δ /(mm s⁻¹)	δ (Fe) /(mm s⁻¹)	μ/BM	Reference
Fe(phen)₂Cl₂	RT	3·28	1·15 (SS)	1·06	5·27	71
	77	3·15	1·025 (Fe)	1·025		76
Fe(phen)₂Br₂	RT	3·26	1·16 (SS)	1·07	5·15	72
	77	2·81	1·009 (Fe)	1·009		76
Fe(phen)₂I₂	77	2·80	1·052 (Fe)	1·052		76
Fe(phen)₂(NCS)₂*	RT	3·09	1·13 (SS)	1·04	5·22	71
	RT	2·67	0·98 (Fe)	0·98	5·20	77
	298	2·7	1·17 (Cr)	1·02		78
Fe(phen)₂(NCSe)₂*	RT	2·52	1·03 (Fe)	1·03	5·20	77
Fe(phen)₂(N₃)₂	77	2·88	1·000 (Fe)	1·000		76
Fe(2-Cl-phen)₃(ClO₄)₂	77	1·50	1·017 (Fe)	1·017		76

* Ligand-field crossover phenomena recorded at lower temperatures (see **Chapter 8**).

and Fe(phen)₂(NCSe)₂ which show a ligand-field crossover below ambient temperature. These are discussed more fully in Chapter 8.

Fe²⁺ o-Phenylenediamine Complexes

The new complexes of *o*-phenylenediamine with iron are all high-spin iron(II), but feature several structural types: [79]

(i) [Fe(opda)₂X₂] (X = Cl, Br, and I): 6-coordinated with two chelating (opda) ligands and probably with a *trans*-geometry.

(ii) [Fe(opda)₃Cl₂]: two iron environments.

(iii) [Fe(opda)₃X₂] (X = Br, I): 6-coordinated with three chelating (opda) ligands.

(iv) [Fe(opda)₄X₂] (X = Cl, Br): 6-coordinated with 4 monodentate (opda) ligands.

(v) [Fe(opda)₄I₂]: probably 6-coordinated with two chelate and 2 monodentate (opda) ligands.

(vi) [Fe(opda)₆Cl₂]: 6 monodentate ligands.

The Mössbauer parameters, which are in Table 6.6, have not been fully analysed, but of particular interest is the discovery of two different iron environments in [Fe(opda)₃Cl₂].

Other Fe²⁺ Octahedral Complexes

Several other monodentate and bidentate nitrogen ligands have been studied and parallel the pyridine and 1,10-phenanthroline systems closely. The isoquinoline and γ-picoline derivatives are believed to have axial symmetry and an $|xy\rangle$ ground state [72]. The complex Fe(di-2-pyridylamine)₂(NCS)₂

Table 6.6 Mössbauer parameters for other octahedral high-spin iron (II) complexes

Compound	T/K	Δ/(mm s^{-1})	δ/(mm s^{-1})	δ (Fe)/(mm s^{-1})	μ/BM	Reference
[Fe(opda)$_2$Cl$_2$]	290	2·89	1·02 (NP)	0·76		79
	80	3·17	1·14 (NP)	0·88		
[Fe(opda)$_2$Br$_2$]	290	2·35	0·96 (NP)	0·70		79
	80	2·94	1·09 (NP)	0·83		
[Fe(opda)$_2$I$_2$]	290	0·87	0·92 (NP)	0·66		79
	80	1·19	1·09 (NP)	0·83		
[Fe(opda)$_3$Cl$_2$]	290	2·89, 0·86	1·02, 1·00 (NP)	0·76, 0·74		79
	80	3·16, 1·27	1·15, 1·10 (NP)	0·89, 0·84		
[Fe(opda)$_3$Br$_2$]	80	1·12	1·09 (NP)	0·83		79
[Fe(opda)$_3$I$_2$]	290	0·84	0·98 (NP)	0·72		79
	80	1·34	1·07 (NP)	0·81		
[Fe(opda)$_4$Cl$_2$]	290	0·87	0·99 (NP)	0·73		79
	80	1·20	1·08 (NP)	0·82		
[Fe(opda)$_4$Br$_2$]	80	1·56	1·11 (NP)	0·85		79
[Fe(opda)$_4$I$_2$]	290	1·42	0·98 (NP)	0·72		79
	80	2·06	1·05 (NP)	0·79		
[Fe(opda)$_6$Cl$_2$]	290	1·30	0·96 (NP)	0·70		79
	80	2·00	1·09 (NP)	0·83		
Fe(IQ)$_4$Cl$_2$	295	3·18	0·85 (Pd)	1·03	5·30	72
Fe(IQ)$_4$Br$_2$	295	2·21	0·82 (Pd)	1·00	5·51	72
Fe(IQ)$_4$I$_2$	295	0·40	0·79 (Pd)	0·97	5·61	72
Fe(IQ)$_4$(NCS)$_2$	295	1·50	0·93 (Pd)	1·11	5·44	72
Fe(γ-pic)$_4$Cl$_2$	295	2·98	0·87 (Pd)	1·05	5·38	72
Fe(γ-pic)$_4$Br$_2$	295	1·28	0·81 (Pd)	0·99	5·60	72
Fe(γ-pic)$_4$I$_2$	295	0·19	0·75 (Pd)	0·93	5·67	72
Fe(γ-pic)$_4$(NCS)$_2$	295	1·67	0·90 (Pd)	1·08	5·35	72
Fe(dipyam)$_3$(ClO$_4$)$_2$	RT	2·37	0·85 (Pd)	1·03	5·52	80
	RT	2·51	1·27 (NP)	1·01	5·36	81
Fe(dipyam)$_2$Cl$_2$	RT	2·23	0·84 (Pd)	1·02	5·41	80
Fe(dipyam)$_2$Cl$_2$.2H$_2$O	RT	2·35	1·22 (NP)	0·96		81
Fe(dipyam)$_2$Br$_2$	RT	2·12	0·83 (Pd)	1·01	5·52	80
Fe(dipyam)$_2$(NCS)$_2$	RT	2·62, 1·66	0·85 0·89, (Pd)	1·03, 1·07	5·48	80
*Fe(bipy)$_2$(NCS)$_2$ (I)	RT	2·18	1·06 (Fe)	1·06		82
* (II)	RT	2·31	1·06 (Fe)	1·06		82
* (III)	RT	2·13	1·06 (Fe)	1·06		82

IQ = isoquinoline; γ-pic = γ-picoline; dipyam = di-2-pyridylamine; bipy = 2,2'-bipyridyl.

* Spin 'crossover' seen at lower temperatures (see Chapter 8).

IRON(II) COMPLEXES WITH NITROGEN LIGANDS | 145

is anomalous in that it contains two iron sites, [80] although it does not show the 'crossover' found for the 1,10-phenanthroline and 2,2'-bipyridyl isothiocyanates.

Relevant Mössbauer parameters are listed in Table 6.6. The 2,2'-bipyridyl complex with isothiocyanate exists in three polymorphs [82], all of which show spin 'crossover' (see Chapter 8).

Several compounds of the type $FeL_6(ClO_4)_2$ where e.g. L = pyridine-N-oxide or dimethyl sulphoxide have been characterised [83].

Other Fe^{2+} Complexes

Parameters have been measured [80, 84] for a number of 4-coordinate iron(II) complexes which can be thought of as derived from the corresponding FeX_4^{2-} ions (see Table 6.7). The $Fe(dipyam)X_2$ and $Fe(quin)_2X_2$ complexes show a larger quadrupole splitting than the $[Fe(quin)X_3]^-$ anions implying a greater distortion from T_d symmetry in the former cases as might be expected. The increased covalency in tetrahedral environment causes a decrease in chemical isomer shift of about $0.1-0.2$ mm s^{-1} from the corresponding octahedral type of complex. Triphenylphosphine oxide has a ligand-field strength similar to that of chlorine and only a small quadrupole splitting is seen in $Fe(Ph_3PO)_2Cl_2$, similar to that in the $FeCl_4^{2-}$ ion.

The nominally similar complexes $FeCl_2(acetamide)_2$, $FeCl_2(formamide)_2$, $FeCl_2(benzamide,)$ and $FeCl_2(aniline)$ have all been characterised as high-spin Fe(II) compounds [85], but are believed to have halogen bridged structures with a higher coordination. However, the parameters are listed in Table 6.7 for convenience. $Fe(terpy)Br_2$, $Fe(terpy)I_2$, and $Fe(terpy)(NCS)_2$ are all believed to be 5-coordinate complexes [86], but $Fe(terpy)Cl_2$ on the other hand shows two resonances corresponding to high and low-spin Fe(II) and is obviously formulated as $[Fe(terpy)_2][FeCl_4]$.

The high-spin Fe^{2+} complexes which have been discussed so far in this chapter have chemical isomer shifts at room temperature in the range $0.7-1.5$ mm s^{-1} relative to iron metal. Although there appears to be little possibility of finding shifts above this range, there is some evidence of values much lower than this. The compounds

di[ω,ω'-(N,N'-dimethylethylene diamine)]-o-tolyliron(II), structure I and di[ω,ω'-ethylenedioxy]-o-tolyliron(II), structure II

may fall into this category [87].

[Refs. on p. 164]

Table 6.7 Mössbauer parameters for other high-spin iron(II) complexes

Compound	T/K	Δ /(mm s⁻¹)	δ /(mm s⁻¹)	δ (Fe) /(mm s⁻¹)	μ/BM	Reference
Fe(dipyam)Cl₂	RT	2·71	0·64 (Pd)	0·82	5·34	80
Fe(dipyam)Br₂	RT	2·42	0·64 (Pd)	0·82	5·40	80
Fe(dipyam)I₂	RT	1·83	0·63 (Pd)	0·81	5·57	80
Fe(quin)₂Cl₂	295	2·72	0·87 (Fe)	0·87	5·20	84
Fe(quin)₂Br₂	295	2·71	0·83 (Fe)	0·83	5·19	84
Fe(quin)₂I₂	295	2·22	0·78 (Fe)	0·78	5·28	84
(Et₄N)[Fe(quin)Cl₃]	295	2·08	0·88 (Fe)	0·88	5·30	84
(Et₄N)[Fe(quin)Br₃]	295	2·1	0·86 (Fe)	0·86	5·35	84
Fe(3-methylIQ)₂Cl₂	295	2·21	0·86 (Fe)	0·86	5·26	84
Fe(3-methylIQ)₂Br₂	295	1·81	0·85 (Fe)	0·85	5·29	84
Fe(Ph₃PO)₂Cl₂	295	0·80	0·97 (Fe)	0·97		84
Fe(Ph₃PO)₂Br₂	295	2·21	1·01 (Fe)	1·01		84
FeCl₂(acetamide)₂	295	2·64	1·39 (NP)	1·13		85
	77	2·80	1·55 (NP)	1·29		85
FeCl₂(formamide)₂	295	2·65	1·42 (NP)	1·16		85
	77	3·00	1·59 (NP)	1·33		85
FeCl₂(benzamide)	295	2·57	1·41 (NP)	1·15		85
	77	2·87	1·54 (NP)	1·28		85
FeCl₂(aniline)	295	1·69	1·37 (NP	1·11		85
	77	2·18	1·49 (NP)	1·23		85
Fe(terpy)Br₂	RT	2·76	0·89 (Fe)	0·89		86
	78	3·16	1·01 (Fe)	1·01		86
Fe(terpy)I₂	RT	2·51	0·83 (Fe)	0·83		86
	78	3·18	0·98 (Fe)	0·98		86
Fe(terpy)(NCS)₂	RT	2·46	0·88 (Fe)	0·88		86
	78	2·78	1·03 (Fe)	1·03		86
Fe(terpy)Cl₂	RT	0·98, 1·69	0·21, 0·91 (Fe)	0·21, 0·91		86
	78	0·99, 3·05	0·27, 1·06 (Fe)	0·27, 1·06		86

quin = quinoline; 3-methylIQ = 3-methylisoquinoline; dipyam = di-2-pyridylamine; terpy = 2,2′,2″-terpyridyl.

The iron–carbon bonds can be presumed to be strongly covalent. The appropriate Mössbauer parameters are

Compound	T/K	Δ/(mm s⁻¹)	δ (Fe)/(mm s⁻¹)	μ/BM
I	298	2·16	0·32	4·45
	77	1·28	0·44	—
II	298	2·08	0·52	—
	77	2·40	0·68	—

[*Refs.* on p. 164]

Two disturbing features about the data on compound I are that its magnetic moment of 4·45 BM is significantly lower than the spin-only value of 4·9 BM for a d^6 high-spin complex with an assumed distorted tetrahedral geometry, and that its quadrupole splitting *decreases* drastically with decrease in temperature. It may be that detailed characterisation will reveal some tendency to an $S = 0$ (or intermediate $S = 1$ state if the geometry is considerably distorted) at low temperatures which could account both for the low moment and for the reduced splitting. The complex II appears to be more normal in behaviour.

Some early work on chelate complexes of high-spin iron(II) with salicylaldoxime (III) and salicylaldehyde (IV) derivatives gave abnormally low chemical isomer shifts in the range 0·40–0·56 mm s^{-1} [88–90]. It has recently been shown [91], however, that oxidation takes place unless the preparations are conducted under strict anaerobic conditions. The true high-spin iron(II) chelates show shifts in the range 0·94–1·30 mm s^{-1} with large quadrupole splittings and are in every way typical of the oxidation state (the values were measured at liquid nitrogen temperature with respect to sodium nitroprusside and have been converted).

Compound	before oxidation		after oxidation	
	Δ/(mm s^{-1})	δ(Fe)/(mm s^{-1})	Δ/(mm s^{-1})	δ(Fe)/(mm s^{-1})
III (X = OH)	2·53	1·30	0·86	0·36
IV	2·51	1·17	0·82	0·42
III (X = H)	2·53	1·21	0·91	0·43
V (Z = (CH$_2$)$_2$)	2·50	0·98	0·82	0·39
V (Z = o-C$_6$H$_4$)	2·42	0·94	0·81	0·37

(III)   (IV)

(V)

[*Refs. on p. 164*]

B. HIGH-SPIN IRON(III) COMPLEXES

6.4 Iron(III) Halides

FeF$_3$

Iron(III) fluoride has a unique iron environment comprising an almost regular octahedron of fluorines. The compound is antiferromagnetic below 363 K [92, 93]. Above this temperature only a single resonance line is seen, while below it the spin relaxation time is sufficiently short that the internal field is a true time-average and a six-line spectrum results. The chemical isomer shift at 296·5 K is 0·489 mm s^{-1}, and at 4·2 K the internal magnetic field is 618·1 kG with an estimated quadrupole coupling constant $e^2qQ/2$ of $-0·044$ mm s^{-1}. The latter is too small to cause other than a slight broadening of the paramagnetic resonance, but is clearly seen as a first-order perturbation of the six-line spectra (see section 3.6). Some typical spectra are shown in Fig. 6.18 and illustrate the higher symmetry (compared with Fe^{2+}) inherent in an Fe^{3+}(d^5) hyperfine pattern because of the low magnitude of quadrupole effects. One group [95] reported partial collapse of the hyperfine splitting in the 10° immediately below T_N, but this has been proven to be a result of slight impurity in the FeF$_3$ and not a property of the pure compound [93]. Detailed study of the sublattice magnetisation close to the Néel temperature [93] shows that it follows the equation given for $M(T)$ under iron(II) fluoride on p. 113 with $\beta = 0·352$ and $D = 1·21$. Essentially the same results were obtained independently [96]. The data on FeF$_3$ are compared with those obtained for other high-spin iron(III) halogen complexes in Table 6.8.

FeF$_3$.3H$_2$O

FeF$_3$.3H$_2$O gives a smaller chemical isomer shift than FeF$_3$ at room temperature but shows a substantial quadrupole splitting (see Table 6.8). The shift decreases regularly and reversibly with increase in pressure up to 175 kbar, and at the same time the quadrupole splitting increases from 0·48 mm s^{-1} to 0·722 mm s^{-1} [9]. The electron density at the nucleus and the electric field gradient thus both increase with increase in pressure.

FeCl$_3$

X-ray studies on FeCl$_3$ at room temperature indicate a hexagonal unit cell. The chlorine atoms are hexagonally close-packed with the iron atoms occupying two-thirds of the possible octahedral sites in a layered array. The iron environment is not perfectly octahedral although no quadrupole splitting is resolved in the paramagnetic Mössbauer spectrum. The chemical isomer shift at room temperature [94] is 0·436 mm s^{-1} which is to be compared with 0·442 mm s^{-1} for FeCl$_3$.6H$_2$O and 0·489 mm s^{-1} for FeF$_3$;

[Refs. on p. 164]

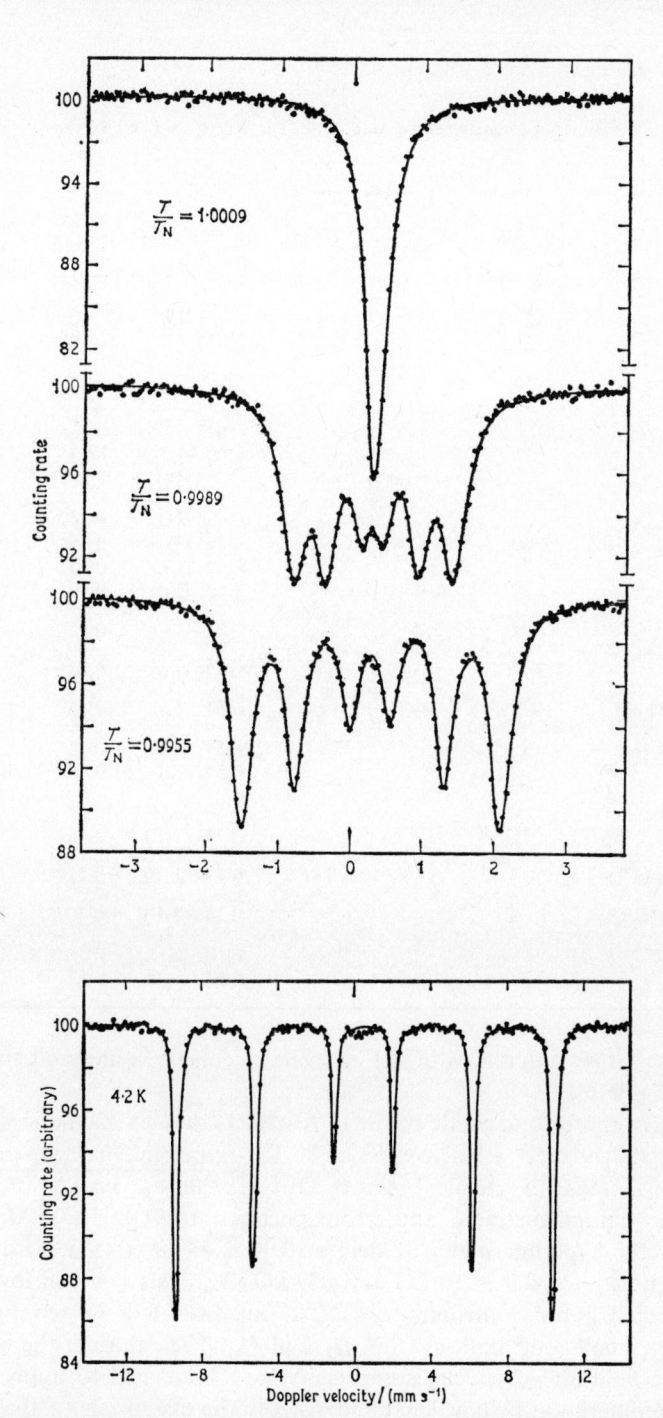

Fig. 6.18 Mössbauer spectrum of FeF$_3$ at 4·2 K and in the vicinity of the Néel temperature. [Ref. 93, Fig. 1 and Fig. 2]

Table 6.8 Mössbauer parameters for octahedral halogen complexes of high-spin iron(III)

Compound	T/K	Δ/(mm s^{-1})	H/kG	δ/(mm s^{-1})	δ (Fe)/(mm s^{-1})	Reference
FeF$_3$	364	—	—	0·265 (Pd)	0·450	93
	296·5	—	—	0·304 (Pd)	0·489	93
	4·2	−0·044	618·1			93
FeF$_3$.3H$_2$O	RT	0·48	—	0·445 (Fe)	0·445	9
FeCl$_3$	RT	—	—	0·693 (NP)	0·436	94
	78	—	—	0·69 (Cr)	0·53	103
	4·2	0·04	445	0·114 (Pt)	0·461	20
FeCl$_3$.6H$_2$O	RT	—	—	0·699 (NP)	0·442	94
	300	0·97	—	0·27 (Pd)	0·45	100
	78	—	—	0·69 (Cr)	0·53	103
	78	0·91	—	0·42 (Pd)	0·60	100
FeBr$_3$	78	—	—	0·71 (Cr)	0·55	103
K$_3$FeF$_6$	RT	0·38	—	0·42 (Fe)	0·42	9
K$_2$[FeF$_5$.H$_2$O]	RT	0·40	—	0·54 (SS)	0·45	105
[Co(NH$_3$)$_6$]FeCl$_6$	77	—	—	0·54 (Fe)	0·54	106
	78	—	—	0·59 (SS)	0·50	109
[Co(en)$_3$]FeCl$_6$	78	—	—	0·60 (SS)	0·51	109
	78	—	—	0·744 (NP)	0·487	104
[Rh(pn)$_3$]FeCl$_6$	78	—	—	0·777 (NP)	0·520	104
(MeNH$_3$)$_4$FeCl$_7$	78	—	—	0·746 (NP)	0·489	104
(NH$_4$)$_2$[FeCl$_5$.H$_2$O]	RT	—	—	0·60 (SS)	0·51	105
	LN	0·23	—	0·64 (SS)	0·55	105

the much larger differences in the equivalent iron(II) compound should be noted (Table 6.1).

FeCl$_3$ is antiferromagnetic below 15 K with a complex magnetic structure. The Mössbauer spectra between 6 and 80 K confirm the ordering point and show an increase in chemical isomer shift of +0·040 mm s^{-1} on passing from the paramagnetic to antiferromagnetic state [95]. The extrapolated value of the hyperfine magnetic field at 0 K is 487 ± 15 kG. More recent values quoted are 468 ± 10 kG and 458 kG [20]. This is much lower than the expected Fermi contribution of $H_S = 550$ kG for a $3d^5$ configuration with only small contributions from H_L and H_D. The values of the hyperfine magnetic field and chemical isomer shift have been held to imply a $3d^{5·6} 4s^{0·45}$ configuration by covalent bonding [20]. The magnitude of the internal field in an iron(III) complex is a good indication of the extent of covalency,

[*Refs.* on p. 164]

and the available data are listed in approximate order of increasing covalency in Table 6.9.

Table 6.9 The saturation internal magnetic field in iron(III) complexes. Available data are averaged when necessary and given to the nearest 10 kG

Compound	H/kG	Nearest neighbours	Reference
FeF_3	620	$6F^-$	93
$NH_4Fe(SO_4)_2.12H_2O$	~580*	$6H_2O$	112, 116
$NH_4(Fe,Al)(SO_4)_2.12H_2O$	570*	$6H_2O$	119, 120
$Fe_2(SO_4)_3$	550		98
$MFe_3(OH)_6(SO_4)_2$†	470	$4OH^-, 2O^{2-}$	123
$FeCl_3$	470	$6Cl^-$	20, 27, 95
$FeCl_4^-$	470	$4Cl^-$	106
$FeBr_4^-$	420	$4Br^-$	106
$Fe(NCO)_4^-$	390*	$4(NCO)^-$	106

* From relaxation data.
† $M = Na^+, K^+, NH_4^+, H_3O^+, \frac{1}{2}Pb^{2+}$.

A recent study has revealed a previously unknown phase transition in $FeCl_3$ at about 250 K [94]. There is a small discontinuous decrease in the chemical isomer shift with decreasing temperature of 0·003 mm s^{-1} which has been confirmed by extremely careful measurement. A change in the close-packed chlorine lattice to a face-centred cubic one by a shearing motion of adjacent chlorine layers was postulated and subsequently verified by X-ray diffraction. There is also the suggestion of a return to the hcp structure at 165 K so the cubic phase may be metastable.

$FeCl_3.6H_2O$

Hydrated iron(III) chloride has a *trans*-octahedral arrangement of $2Cl^-$ and $4H_2O$ around the iron atom and can be formulated as the complex $[FeCl_2(H_2O)_4]Cl.2H_2O$ [97]. As shown in Table 6.8 its chemical isomer shift is very similar to that of anhydrous $FeCl_3$ itself. Wignall [98] has recorded a symmetrical quadrupole-split doublet at 1·8 K, one component of which broadens considerably as the temperature is raised. The iron-iron distance is estimated to be about 6 Å and it is at this separation that the spin-spin relaxation times in high-spin iron(III) complexes become sufficiently large to cause incomplete averaging during the Mössbauer event. A slight narrowing of the resonance lines is seen when a small (2kG) external magnetic field is applied at room temperature, because of relaxation effects [99]. This narrowing is even more pronounced in a field of 8 kG, and this is illustrated in Fig. 6.19 [100]. The lowest-lying Kramers' doublet of the $^6S_{\frac{5}{2}}$ ion is the $|\pm\frac{1}{2}\rangle$ state with a comparatively small separation from the higher levels.

[*Refs. on p. 164*]

152 | HIGH-SPIN IRON COMPLEXES

The zero-field spectra with gross asymmetry are a common feature of high-spin iron(III) compounds. A difference has been noted between the crystalline

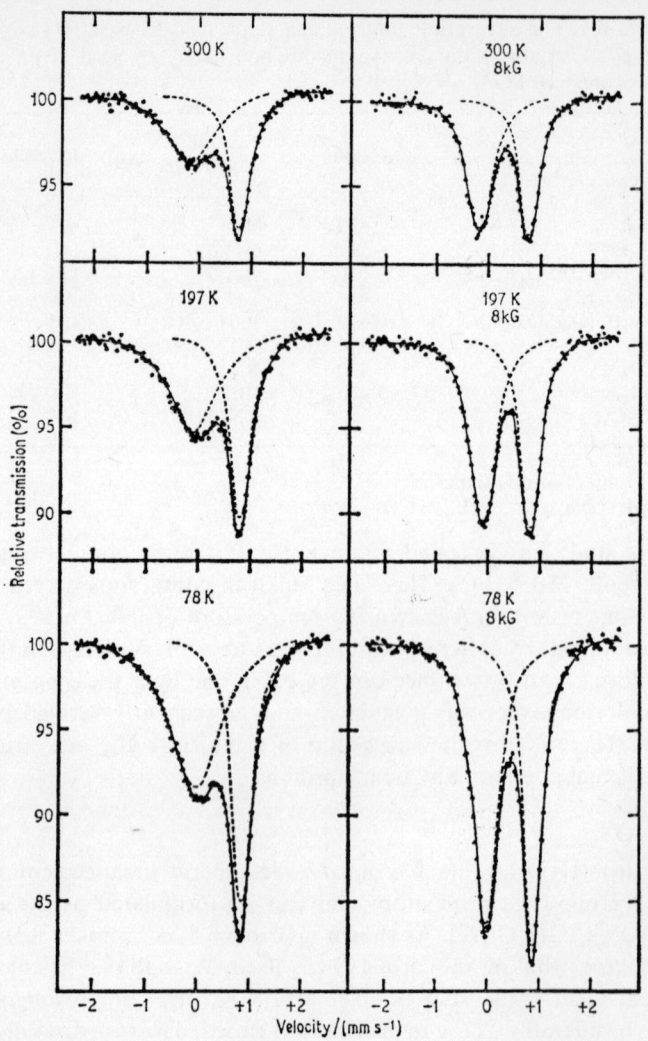

Fig. 6.19 Mössbauer spectra of $FeCl_3.6H_2O$ at 78, 197, and 300 K in an applied magnetic field of 0 and 8 kG. [Ref. 100, Fig. 1]

and vitreous forms of $FeCl_3.6H_2O$, the latter exhibiting only a single resonance line [101].

Application of pressure causes an apparent reversible reduction to an Fe^{2+} electronic configuration by charge transfer of an electron from a non-bonding ligand orbital on to the central iron atom. The related $[Fe(NH_3)_6]Cl_3$ complex shows a different pressure dependence between 15–25 kbar and above

[*Refs. on p. 164*]

25 kbar, and it is suggested that two different electron-transfer processes are involved, such as from Cl$^-$ to iron–ammonia complex, and from NH$_3$ to Fe^{3+} [102].

FeBr$_3$

Only one chemical isomer shift value has been published for the bromide [103] (see Table 6.8).

FeX$_6^{3-}$ Complexes

Few data are available on FeF$_6^{3-}$ and FeCl$_6^{3-}$ complexes other than those listed in Table 6.8. The pressure dependence of the chemical isomer shift and quadrupole splitting of K$_3$FeF$_6$ are quite regular and similar to those of FeF$_3$.3H$_2$O [9]. The small quadrupole splitting indicates a site symmetry lower than cubic although the crystal itself has cubic symmetry.

Although FeCl$_6^{3-}$ was previously thought to be stable only in the presence of large cations such as [Co(NH$_3$)$_6$]$^{3+}$ and [Rh(1,2-propanediamine)$_3$]$^{3+}$, the spectrum of the complex (MeNH$_3$)$_4$FeCl$_7$ has been used [104] to show that it is in fact a double salt which should be formulated as the FeCl$_6^{3-}$ complex [MeNH$_3$]$_3$[FeCl$_6$].MeNH$_3$Cl.

The complexes K$_2$[FeF$_5$.H$_2$O] and (NH$_4$)$_2$[FeCl$_5$.H$_2$O] show a quadrupole splitting as expected from their lower symmetry [105].

FeX$_4^-$ Complexes

Available data on tetrahedral halogen complexes of Fe^{3+} are summarised in Table 6.10, the most comprehensive study of such anions being that by Edwards and Johnson [106]. The complexes Ph$_4$AsFeCl$_4$, Me$_4$NFeCl$_4$, and Et$_4$NFe(NCO)$_4$ remain paramagnetic down to at least 1·1 K, but Et$_4$NFeCl$_4$ and Et$_4$NFeBr$_4$ become antiferromagnetic at 3·0 and 3·9 K respectively. Et$_4$NFe(NCO)$_4$ is an exception in that it shows a comparatively large quadrupole splitting (+0·86 mm s^{-1} at 77 K) indicating a large deviation from T_d local symmetry in this anion (see Fig. 6.20). A small splitting of 0·2 mm s^{-1} is found in Ph$_4$AsFeCl$_4$. The others show no measurable electric field gradient. The chemical isomer shift for Et$_4$NFeCl$_4$ (0·29 mm s^{-1}) is much less than those of FeCl$_3$ (0·53 mm s^{-1}) and [Co(NH$_3$)$_6$]FeCl$_6$ (0·54 mm s^{-1}) because of strong covalency and the short Fe–Cl bond (2·19 Å compared to 2·48 Å in FeCl$_3$), a similar situation to that already mentioned for the FeX$_4^{2-}$ complexes.

The zero-field spectrum of Ph$_4$AsFeCl$_4$ is a very broad line because of spin relaxation, no doubt due to a large Fe–Fe separation. The magnitude of the internal field at the nucleus at 1·6 K was measured in all the compounds (except the Et$_4$N$^+$ salts of FeCl$_4^-$ and FeBr$_4^-$ which are already antiferromagnetically ordered) by applying a large external field to orient the spins, and ranged from 470 kG for FeCl$_4^-$ to 420 kG for FeBr$_4^-$ and 394 kG for Fe(NCO)$_4^-$, being independent of the cation for the FeCl$_4^-$ anions. The

Table 6.10 Mössbauer parameters for tetrahedral halogen complexes of high-spin iron(III)

Compound	T/K	Δ /(mm s⁻¹)	H/kG	δ/(mm s⁻¹)	δ (Fe) /(mm s⁻¹)	Reference
Me₄NFeCl₄	77	0·00	479*	0·30 (Fe)	0·30	106
Ph₄AsFeCl₄	77	0·2	462*	0·3 (Fe)	0·3	106
Et₄NFeCl₄	77	0·00	465, 475*	0·29 (Fe)	0·29	106
Et₄NFeBr₄	77	0·00	420*	0·36 (Fe)	0·36	106
Et₄NFe(NCO)₄	77	+0·86	394*	0·34 (Fe)	0·34	106
[Ti(acac)₃]FeCl₄	78	—	—	0·45 (Cr)	0·30	108
[Si(acac)₃]FeCl₄	78	—	—	0·39 (SS)	0·30	109
[Ge(acac)₃]FeCl₄	78	—	—	0·39 (SS)	0·30	109
[Zr(bzbz)₃]FeCl₄	78	—	—	0·40 (SS)	0·31	109
(pyH)₃Fe₂Cl₉	78	∼0	—	0·47 (Cr)	0·32	107
(pyH)₃Fe₂Br₉	78	∼0	—	0·51 (Cr)	0·36	107
β-Cs₃Fe₂Cl₉	20	∼0	—	0·52 (Cr)	0·37	107
α-Cs₃Fe₂Cl₉†	78	0·32	—	0·65 (Cr)	0·50	107
Et₄NFeCl₄	77	—	—	0·539 (NP)	0·282	110
Et₄NFeCl₃Br	77	—	—	0·542 (NP)	0·285	110
Et₄NFeCl₂Br₂	77	—	—	0·578 (NP)	0·321	110
Et₄NFeClBr₃	77	—	—	0·590 (NP)	0·333	110
Et₄NFeBr₄	77	—	—	0·607 (NP)	0·350	110

* Field measured at 1·6 K in an external magnetic field.
† α-Cs₃Fe₂Cl₉ has a distorted octahedral coordination.

reduction in H when compared to the hyperfine field in FeF_3 for example reflects the diminished spin density at the nucleus when the electrons of the ion are delocalised onto the ligands by covalent bonding.

Suggestions that the pyridinium complexes $(pyH)_3Fe_2Cl_9$ and $(pyH)_3Fe_2Br_9$ contained a μ-trihalo dimeric anion, $(Fe_2X_9)^{3-}$, were countered [107] by the observation of a single-line Mössbauer spectrum compatible only with tetrahedral $FeCl_4^-$ anions. The complex $Cs_3Fe_2Cl_9$ exists in two forms. The β-form evidently contains $FeCl_4^-$ anions, but the α-form gives a quadrupole-split spectrum with a higher chemical isomer shift and therefore does contain the $Fe_2Cl_9^{3-}$ anion with 6-coordination to iron.

The complex $[Fe(acac)_3]TiCl_4$ was shown by Mössbauer spectroscopy to have been incorrectly formulated; it is in fact $[Ti(acac)_3]FeCl_4$ [108]. The complexes $[Ge(acac)_3]FeCl_4$, $[Si(acac)_3]FeCl_4$, and $[Zr\{(PhCO)_2CH_2\}_3]FeCl_4$ are similar [109].

The mixed halides in the series $Et_4N[FeCl_{4-n}Br_n]$ ($n = 0, 1, 2, 3, 4$) all show a single-line resonance at 77 K despite the non-cubic symmetry of $FeCl_3Br^-$, $FeCl_2Br_2^-$, and $FeClBr_3^-$ [110]. Presumably there is little difference in the bonding between the chloride and bromide ligands, although there is a small regular increase in the chemical isomer shift along the series (see Table 6.10).

[*Refs. on p. 164*]

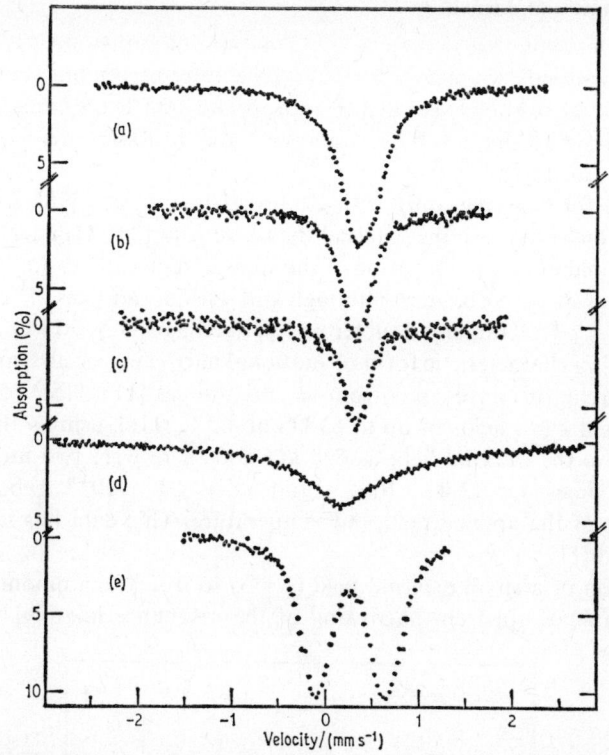

Fig. 6.20 Mössbauer spectra of (a) Me$_4$NFeCl$_4$ at 4·2 K, (b) Et$_4$NFeCl$_4$ at 77 K, (c) Et$_4$NFeBr at 77 K, (d) Ph$_4$AsFeCl$_4$ at 77 K, and (e) Et$_4$NFe(NCO)$_4$ at 77 K. [Ref. 106, Fig. 1]

The compounds Ph$_3$PFeCl$_3$ and Ph$_3$AsFeCl$_3$ have been briefly characterised and are probably tetrahedrally coordinated [85].

6.5 Iron(III) Salts of Oxyacids

Fe$_2$(SO$_4$)$_3$

Only brief details are available for anhydrous iron(III) sulphate; it gives a paramagnetic spectrum at room temperature [111] with a chemical isomer shift of 0·39 mm s^{-1} and a quadrupole splitting of 0·60 mm s^{-1} [9]. A sharp magnetic hyperfine spectrum is seen at 1·8 K with a field of 550 kG [98].

Fe$_2$(SO$_4$)$_3$ is one of a number of iron(III) compounds which show a partial reversible change to iron(II) ions under high pressure [9]. The change is 15% complete at 25 kbar and 49% complete at 150 kbar. Presumably the energy barrier for ligand-to-Fe^{3+} electron transfer is sufficiently small for the pressure to lower the barrier to the point where thermal excitation can effect transfer. An analogy can be drawn with the better-known photochemical reactions of iron.

[Refs. on p. 164]

Iron(III) Alums and Jarosites

Although relaxation broadening is very common in paramagnetic high-spin iron(III) compounds, the theory for a complete interpretation is complex and requires a detailed knowledge of the appropriate spin Hamiltonian. One of the few cases where this has been available is for iron(III) ammonium alum, $NH_4Fe(SO_4)_2.12H_2O$.

At 1·8 K, for example, only a single broad line is seen [98]. Obenshain et al. [112] found that a large external magnetic field (24 kG) parallel to the γ-ray axis could induce polarisation of the spins and give a four-line spectrum ($\Delta m = 0$ transitions forbidden) although still with broad lines. The effective field for the $|\pm\frac{5}{2}\rangle$ Kramers' doublet was derived as 563 ± 30 kG. At intermediate fields a characteristic form of motional narrowing occurs, and theoretical interpretations have come from several sources [113–115]. Later work extended the data to fields of up to 53 kG at 4·2 K [116], using a theoretical fit which gave the internal field as 598 kG and there were two atomic spin correlation times $\tau_c = 2\cdot 4 \times 10^{-9}$ s and $\tau_c' = 2\cdot 1 \times 10^{-9}$ s, both being independent of the applied field and temperature. This data has since been reinterpreted [117].

Application of a small external field (1 kG) to iron(III) ammonium alum causes substantial apparent narrowing of the resonance line [99], although

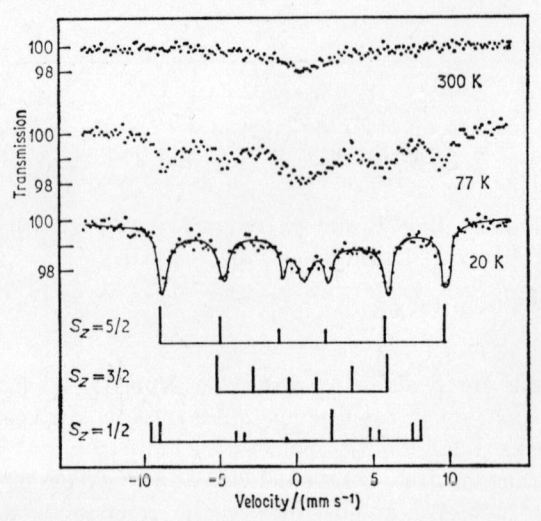

Fig. 6.21 Mössbauer spectra of 1·0% Fe^{3+} in $NH_4Al(SO_4)_2.12D_2O$ at three temperatures. The spectra predicted by the spin Hamiltonian for the three Kramers' doublets are shown. [Ref. 119, Fig. 1]

the total spectrum now comprises a narrow component superimposed on a very diffuse background [118]. No satisfactory explanation of the low-field data is yet available.

[Refs. on p. 164]

The doping of aluminium ammonium alum, with up to 2% of the corresponding iron(III) salt, NH$_4$(Fe,Al)(SO$_4$)$_2$.12H$_2$O, enabled a detailed study to be made of relaxation processes as a function of iron concentration, temperature, deuteration, and externally applied magnetic field [119, 120]. At high temperatures and zero field only a very broad resonance was seen, but at low temperatures and low iron concentration substantial resolution of magnetic hyperfine patterns can be obtained. At 20 K the magnetic lines are comparatively sharp below an aluminium replacement concentration of 2 atom-per cent of Fe. The hyperfine pattern of the $|\pm\frac{5}{2}\rangle$ Kramers' doublet gives a field of -572 kG, with a small quadrupole splitting of $\frac{1}{2}e^2qQ = 0.11$ mm s^{-1} and a chemical isomer shift of 0·37 mm s^{-1}. The $|\pm\frac{3}{2}\rangle$ pattern could be seen by 'stripping' the $|\pm\frac{5}{2}\rangle$ components from the observed spectrum, but was broadened, while the $|\pm\frac{1}{2}\rangle$ components were never resolved because of their high relaxation rate. Some typical spectra are shown in Fig. 6.21.

Other alums have been cursorily examined [121].

A number of jarosites of general formula MFe$_3$(OH)$_6$(SO$_4$)$_2$(M = Na$^+$, K$^+$, NH$_4^+$, H$_3$O$^+$, $\frac{1}{2}$Pb^{2+}) have been examined [122, 123]. They all give a chemical isomer shift of about 0·43 mm s^{-1} and a quadrupole splitting of 1·2 mm s^{-1} at 300 K; they are antiferromagnetically ordered below about 60 K with internal fields of 470–480 kG at 4·2 K.

Fe^{3+} Salts with Other Oxyanions

The nitrate, Fe(NO$_3$)$_3$.9H$_2$O, gives a broad resonance at 1·8 K due to relaxation [98]. The application at room temperature of an external magnetic field of up to 5 kG produces a central narrow component on a broad background similar to that observed with iron(III) alum, but the detailed relaxation processes have not been studied [118].

Chemical shifts for iron(III) chromate, basic chromate, and arsenate have been reported [121]. Ferric phosphate, FePO$_4$, is like Fe$_2$(SO$_4$)$_3$ in that it is partially converted to an Fe^{2+} species by application of high pressure [9]. Typical spectra are shown in Fig. 6.22. Iron(III) dichromate, Fe$_2$(Cr$_2$O$_7$)$_3$, is also converted to an Fe^{2+} species by pressure.

Iron(III) Oxalates and Acetates

Mössbauer spectroscopy has been used [124] to study the thermal decomposition of the compounds Fe$_2$(C$_2$O$_4$)$_3$.6H$_2$O, Ba$_3$[Fe(C$_2$O$_4$)$_3$]$_2$.8H$_2$O, and Sr$_3$[Fe(C$_2$O$_4$)$_3$]$_2$.2H$_2$O. In air, all three show an initial decomposition process involving a reduction of the iron(III) to iron(II) at about 250°C. The next step produces small-particle Fe$_2$O$_3$ which is superparamagnetic. The crystallites grow at higher temperatures and produce the normal hyperfine splitting. In the case of the strontium and barium compounds, the alkaline-earth carbonate produced with the Fe$_2$O$_3$ reacts at higher temperature to

158 | HIGH-SPIN IRON COMPLEXES

form successively oxides in the $MFeO_{3-x}$ and $M_3Fe_2O_{7-x}$ series both of which contain iron(IV) (for further details on oxide systems see Chapter 10).

The γ-radiolysis of hydrated iron(III) oxalate has been shown to give new components in the Mössbauer spectrum identical to those of iron(II) oxalate [125]. A more detailed study of radiolytic decomposition has used as the starting materials $K_3Fe(C_2O_4)_3 \cdot 3H_2O$ and $K_3Fe(C_2O_4)_3$, both of which

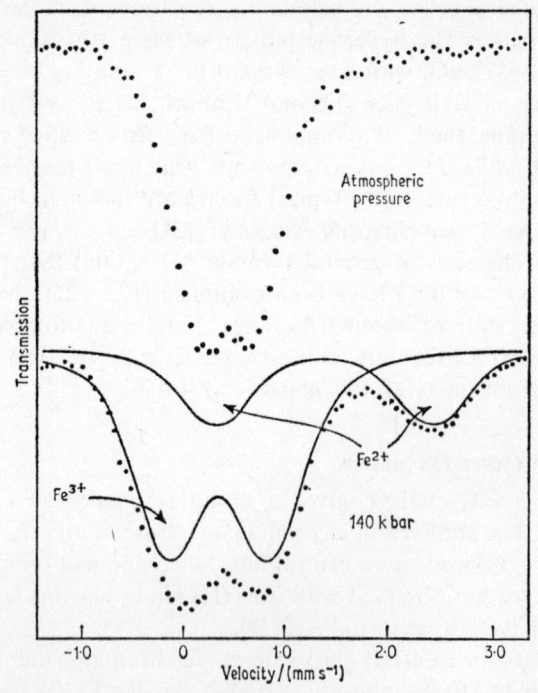

Fig. 6.22 Mössbauer spectra of iron(III) phosphate showing the electron-transfer process which takes place at high pressure. [Ref. 9, Fig. 6]

show relaxation broadening [126]. The irradiated materials were examined before and after subsequent thermal annealing. The primary product of the irradiation is an iron(II) compound with an octahedral environment as shown by the high chemical isomer shift of the quadrupole doublet (1·16 mm s^{-1}). There are also signs of an unstable iron(II) intermediate which is 4-coordinate in the hydrated but 6-coordinate in the anhydrous oxalate. Thermal annealing causes breakdown to the starting material and the final iron(II) product.

Parameters for the stable oxalates are given in Table 6.11.

The spectra of five trinuclear iron(III) carboxylate complexes of the type $[Fe_3(RCO_2)_6(OH)_2]X \cdot xH_2O$ are consistent with high-spin iron(III) in a dis-

[*Refs. on p. 164*]

torted environment but give little information about the structures of the clusters [127].

Table 6.11 Mössbauer parameters for iron(III) oxalates

Compound	T/K	Δ /(mm s^{-1})	δ/(mm s^{-1})	δ (Fe) /(mm s^{-1})	Reference
Fe$_2$(C$_2$O$_4$)$_3$.6H$_2$O	300	0·57	0·15 (Cu)	0·37	124
	4	0·65	0·22 (Cu)	0·44	124
Sr$_3$[Fe(C$_2$O$_4$)$_3$]$_2$.2H$_2$O	300	0·44	0·16 (Cu)	0·38	124
	78	—	0·23 (Cu)	0·45	124
	4	—	0·23 (Cu)	0·45	124
Ba$_3$[Fe(C$_2$O$_4$)$_3$]$_2$.8H$_2$O	300	0·32	0·03 (Cu)	0·25	124
	4	—	0·24 (Cu)	0·46	124
K$_3$Fe(C$_2$O$_4$)$_3$.3H$_2$O	RT	—	0·15 (Pd)	0·33	126
K$_3$Fe(C$_2$O$_4$)$_3$	RT	—	0·15 (Pd)	0·33	126

Frozen Solutions

The study of frozen aqueous solutions of iron(III) compounds is only of limited value because of the tendency to hydrolyse, and the relaxation broadening [63, 128], and little work has been done. A potentially more profitable application is the study of solvent extracts. Frozen nitrobenzene extracts of iron(III) in HCl, HBr, and NaSCN solutions contain respectively the FeCl$_4^-$ and FeBr$_4^-$ anions and a 6-coordinate complex Fe(NCS)$_4$X$_2^-$ (X is either water or nitrobenzene) [129, 130]. The spin–spin relaxation behaviour is dependent on the solvent, and trioctylphosphine oxide extracts show partially resolved hyperfine structure.

6.6 Iron(III) Complexes with Chelating Ligands

Fe(acac)$_3$ and Related Complexes

The spectrum of trisacetylacetonatoiron(III) was reported by several groups during early Mössbauer investigations, but their data for the chemical isomer shift were not self-consistent, mainly because the single-line spectrum is extremely broad (about 2·0 mm s^{-1} at 4 K, decreasing to 1·4 mm s^{-1} at 300 K) and slightly asymmetrical. The width is largely attributable to relaxation processes [98], although the possibility of a quadrupole interaction is not entirely excluded [131]. The chemical isomer shift is about 0·53 mm s^{-1} at 78 K. Dilute frozen solutions of Fe(acac)$_3$ in a mixture of 5 parts (by volume) ethyl ether, 5 parts isopropane, and 2 parts ethyl alcohol show resolved magnetic hyperfine structure as a result of a change in the spin–spin relaxation time on increasing the iron–iron separation [98].

[Refs. on p. 164]

160 | HIGH-SPIN IRON COMPLEXES

Seven substituted-acetylacetonate complexes show very similar behaviour to the parent species, except that in some cases the relaxation broadening at 78 K is not as great and a quadrupole effect of up to 0·75 mm s^{-1} is in evidence [131]. Further interpretation would require detailed knowledge of the spin interactions.

The iron(III) compound Fe(acac)$_2$Cl shows temperature-dependent spin-relaxation properties similar to those already found in iron(III) chloro-haemin, and the compound probably has the same square pyramidal structure [132]. The $|\pm\frac{1}{2}\rangle$ Kramers' doublet lies lowest, and at low temperatures results in a fast relaxation and a narrow spectrum. Increasing temperature causes population of the $|\pm\frac{3}{2}\rangle$ level, a slowing of the spin relaxation, and considerable line broadening. The compound is definitely in an $S = \frac{5}{2}$ spin state ($\mu = 5\cdot98$ BM at 301 K) and makes an interesting contrast with the square pyramidal $S = \frac{3}{2}$ compounds such as (R$_2$NCS$_2$)$_2$FeX discussed in Chapter 8.

EDTA and DTPA Chelates of Fe^{3+}

A number of chelates derived from ethylenediaminetetraacetic acid, H$_4$EDTA, and diethylenetriaminepentaacetic acid, H$_5$DTPA, have been examined

Table 6.12 Mössbauer parameters* for iron(III) complexes with ethylenediaminetetraacetic acid (H$_4$EDTA) and diethylenetriaminepentaacetic acid (H$_5$DTPA) [133–134]

Compound	T/K	Δ /(mm s^{-1})	δ (Fe) /(mm s^{-1})	μ/BM	Coordination number
HFe(H$_2$O)EDTA	298	0·498	0·371	6·16	6
	78	0·422	0·457	—	
	4·2	0·358	0·466	—	
LiFe(H$_2$O)EDTA.2H$_2$O	298	0·700	0·456	6·16	7
NaFe(H$_2$O)EDTA.2H$_2$O	298	0·695	0·472	6·22	7
KFe(H$_2$O)EDTA.H$_2$O	298	0·830	0·473	6·06	7
	78	0·813	0·568	—	
	4·2	0·760	0·603	—	
RbFe(H$_2$O)EDTA.H$_2$O	298	0·521	0·455	5·82	7
CsFe(H$_2$O)EDTA.1½H$_2$O	298	0·557	0·447	5·99	7
NH$_4$Fe(H$_2$O)EDTA.H$_2$O	298	0·726	0·458	5·96	7
Me$_4$NFe(H$_2$O)EDTA.2H$_2$O	298	0·655	0·447	6·06	7
H$_2$FeDTPA.2H$_2$O	298	0·844	0·375	6·16	6
Li$_2$FeDTPA.3H$_2$O	298	0·829	0·386	5·96	8
Na$_2$FeDTPA.H$_2$O	298	0·953	0·396	5·99	8
K$_2$FeDTPA.2½H$_2$O	298	0·987	0·405	6·20	8
NH$_4$HFeDTPA.H$_2$O	298	1·095	0·358	6·21	7

* Standard deviation in Δ and δ is 0·005 mm s^{-1}; δ originally given with respect to sodium nitroprusside.

[Refs. on p. 164]

[133, 134]. The Mössbauer parameters are not very sensitive as to whether the coordination number of the iron is six, seven, or eight, and the spectra are broadened considerably by spin relaxation. Numerical data are in Table 6.12.

Hydroxamates

Spectra of several tris(hydroxamato) iron(III) complexes such as tris(salicylhydroxamato) iron(III) trihydrate show chemical isomer shifts typical of high-spin iron(III) compounds, but gross relaxation broadening complicates any attempt at detailed interpretation [135].

Nitrogen Chelates

A particularly interesting class of high-spin iron(III) compounds are those with the ligand 1,10-phenanthroline (phen), 2,2'-bipyridine (bipy), 2,2',2''-terpyridine (terpy), and N,N'-ethylenebis(salicylideneiminato) (salen). All four form monomeric complexes which may be 5-coordinate in the case of (salen), and also dimeric complexes with an Fe–O–Fe bridge. The intermolecular bonding in the binuclear species is weak, but does allow an antiferromagnetic coupling between pairs of Fe^{3+} ions where the Fe–O–Fe bridge is linear (but not if it is non-linear). The result of the spin coupling is an anomalously low magnetic moment which is strongly temperature dependent. The spin state is clearly shown to be $S = \frac{5}{2}$ from the Mössbauer chemical isomer shift, and the latter is therefore more clearly characteristic of the iron than the magnetic moment (for examples appropriate to an $S = \frac{1}{2}$ state see Chapter 7 and for $S = \frac{3}{2}$ see Chapter 8). The available data summarised in Table 6.13 are from several sources [136–141]. The classification according to molecularity is made from a combination of infrared, magnetic, and Mössbauer data. Typical examples only are given in cases where a large family of complexes has been studied.

The ligand salen can use one of its oxygen atoms to form a bent Fe–O–Fe bridge, and the [Fe(salen)Cl] complexes contain the organic solvent as solvent of crystallisation. The dimeric [Fe(salen)Cl]$_2$ compounds are 6-coordinate, and as such appear to have a slightly higher chemical isomer shift than those which are monomeric and 5-coordinate. The weakness of the Fe–Fe and Fe–solvent interactions is clearly shown by the very small variation in the parameters concerned. The quadrupole splittings are independent of temperature because of the lack of significant valence-orbital contribution, but in some cases are still very large (e.g. $\Delta = 2.35$ mm s^{-1} at 77 K for the complex [Fe(terpy)$_2$]O(NO$_3$)$_4$.H$_2$O) and reflect the low symmetry of the iron environment.

The magnetic moment in [Fe(salen)Cl]$_2$ decreases with decreasing temperature, and the two $S = \frac{5}{2}$ spins couple to produce multiplets with $S' = 0$, 1, ..., 5. The ground state with $S' = 0$ is populated at 4·2 K and a sharp

[Refs. on p. 164]

Table 6.13 Mössbauer parameters for iron(III) complexes with nitrogen as ligand

Compound	T/K	Δ /(mm s^{-1})	δ /(mm s^{-1})	δ (Fe) /(mm s^{-1})	μ/BM	Reference
Monomeric with six-coordination						
[Et$_4$N][Fe(phen)Cl$_4$]	300	0·00	0·65 (NP)	0·39	—	136
	80	0·05	0·65 (NP)	0·39	—	136
[Fe(phen)$_2$Cl$_2$]ClO$_4$	300	0·00	0·63 (NP)	0·37	—	136
	80	0·05	0·65 (NP)	0·39	—	136
[Fe(phen)$_2$Cl$_2$][Fe(phen)Cl$_4$]	300	0·00	0·63 (NP)	0·37	—	136
	80	0·05	0·63 (NP)	0·37	—	136
[Fe(terpy)Cl$_3$]	RT	0·55	0·34 (Fe)	0·34	5·78	137
	77	0·54	0·46 (Fe)	0·46	—	137
Believed dimeric with six-coordination						
α-Fe(phen)Cl$_3$	300	0·80	0·66 (NP)	0·40	—	136
	80	0·85	0·68 (NP)	0·42	—	136
β-Fe(phen)Cl$_3$	300	0·79	0·56 (NP)	0·30	—	136
	80	0·80	0·70 (NP)	0·44	—	136
Monomeric with five-coordination						
Fe(acac)$_2$Cl	80	∼1	0·60			132
[Fe(salen)Cl].2MeNO$_2$	295	1·30	0·42 (SS)	0·33	5·76	138
	80	1·33	0·49 (SS)	0·40	—	138
[Fe(salen)Cl].MeCN	295	0·83	0·41 (SS)	0·32	5·57	138
	80	0·85	0·47 (SS)	0·38	—	138
[Fe(salen)Br].MeNO$_2$	295	1·13	0·51 (SS)	0·42	5·87	138
	80	1·12	0·56 (SS)	0·47	—	138
[Fe(salen)Br].MeCN	295	0·51	0·52 (SS)	0·43	5·12	138
	80	0·54	0·58 (SS)	0·49	—	138
Dimeric with bent Fe–O–Fe bond and 'normal' magnetic moment						
[Fe(salen)Cl]$_2$	295	1·45	0·49 (SS)	0·40	5·30	138
	80	1·43	0·57 (SS)	0·48	—	138
[Fe(salen)Br]$_2$	295	1·63	0·51 (SS)	0·42	5·43	138
	80	1·64	0·58 (SS)	0·49	—	138
[Fe(salen)Cl].2MeOH	295	1·45	0·48 (SS)	0·39	5·72	138
	80	1·44	0·55 (SS)	0·46	—	138
[Fe(salen)Cl].2CHCl$_3$	295	1·40	0·47 (SS)	0·38	5·81	138
	80	1·39	0·54 (SS)	0·45	—·	138
[Fe(salen)Br].2MeOH	295	1·25	0·51 (SS)	0·42	5·69	138
	80	1·26	0·59 (SS)	0·50	—	138
[Fe(salen)Br].2CHCl$_3$	295	1·65	0·49 (SS)	0·40	5·79	138
	80	1·64	0·56 (SS)	0·47	—	138
Dimeric with linear Fe–O–Fe bond and 'low' magnetic moment						
[Fe$_2$(phen)$_4$(OH)$_2$Cl$_4$]	295	0·80	0·48 (SS)	0·39	1·92	138
	80	0·83	0·55 (SS)	0·46	—	138
[Fe(phen)$_2$]$_2$O(NO$_3$)$_4$.3H$_2$O	RT	1·39	0·37 (Fe)	0·37	1·74	137
	77	1·49	0·46 (Fe)	0·46	0·49	137

[*Refs.* on p. 164] *Continued*

IRON(III) COMPLEXES WITH CHELATING LIGANDS | 163

Table 6.13 continued

Compound	T/K	Δ /(mm s^{-1})	δ /(mm s^{-1})	δ (Fe) /(mm s^{-1})	μ/BM	Reference
[Fe(bipy)$_2$]$_2$O(SO$_4$)$_2$.3.5H$_2$O	RT	1·33	0·38 (Fe)	0·38	1·86	137
	77	1·51	0·48 (Fe)	0·48	0·51	137
[Fe(terpy)]$_2$O(NO$_3$)$_4$.H$_2$O	RT	1·93	0·44 (Fe)	0·44	1·83	137
	77	2·35	0·59 (Fe)	0·59	0·65	137
[Fe(salen)]$_2$O	RT	0·72	0·32 (Fe)	0·32	1·94	137
	77	0·78	0·46 (Fe)	0·46	0·48	137
[Fe(salen)]$_2$O.MeNO$_2$	295	0·91	0·45 (SS)	0·36	2·14	138
	80	0·90	0·52 (SS)	0·43	—	138
[Fe(salen)]$_2$O.2py	RT	0·92	0·36 (Fe)	0·36	2·04	137
	77	0·88	0·44 (Fe)	0·44	—	137

quadrupole doublet is seen [142], but considerable broadening occurs as the temperature is raised because population of the higher multiplets is accompanied by a decrease in the spin relaxation time. This is illustrated in Fig. 6.23 [141]. The very fast relaxation found in [Fe(salen)]$_2$O for example has been held to be due to transmission of spin–spin interactions through the

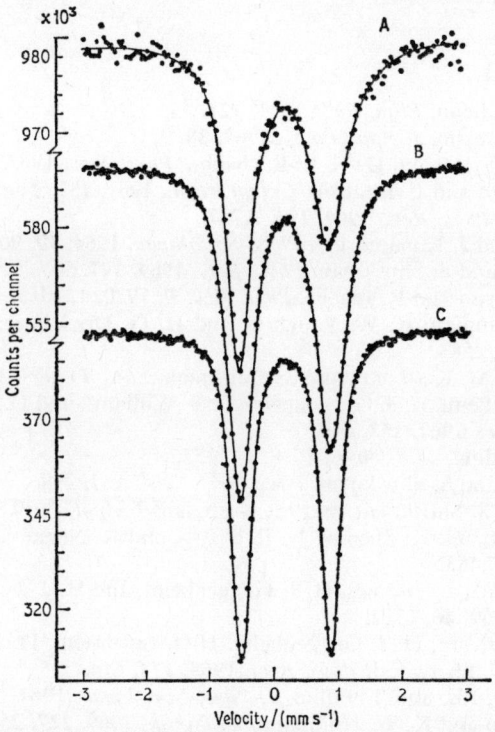

Fig. 6.23 Mössbauer spectra of [Fe(salen)Cl]$_2$ at A, 298 K; B, 78 K; C, 4·2 K. [Ref. 141, Fig. 3]

[*Refs. on p. 164*]

oxygen bridge, i.e. the relaxation is intramolecular [132], but counter-proposals favour intermolecular spin–spin relaxation in both [Fe(salen)]$_2$O and [Fe(salen)Cl]$_2$ [143].

[Fe(NH$_3$)$_5$NO]Cl$_2$

The complex [Fe(NH$_3$)$_5$NO]Cl$_2$ does not fall into any of the preceding sections and is included at this point for convenience. Its chemical isomer shift and quadrupole splitting are 0·54 and 1·40 mm s^{-1} respectively at 298 K, and have been held to favour an Fe^{3+} configuration rather than Fe$^{(I)}$ as suggested earlier [144]. Both parameters are rather larger than is characteristic for high-spin iron(III), showing that there is considerable π-bonding between the NO$^-$ group and the iron, with a resultant increase in 3d-population and considerable asymmetry in the bonding.

Dithiocarbamates

Some of these show spin crossover phenomena and they are therefore discussed in Chapter 8.

REFERENCES

[1] G. K. Wertheim, *Phys. Rev.*, 1961, **121**, 63.
[2] G. K. Wertheim, *J. Appl. Phys.*, 1967, **38**, 971.
[3] G. K. Wertheim and D. N. E. Buchanan, *Phys. Rev.*, 1967, **161**, 478.
[4] A. Abragam and F. Boutron, *Compt. rend.*, 1961, **252**, 2404.
[5] R. Ingalls, *Phys. Rev.*, 1964, **133**, A787.
[6] A. Okiji and J. Kanamori, *J. Phys. Soc. Japan*, 1964, **19**, 908.
[7] U. Ganiel and S. Shtrikman, *Phys. Rev.*, 1969, **177**, 503.
[8] D. P. Johnson and R. Ingalls, *Phys. Rev. B*, 1970, **1**, 1013.
[9] A. R. Champion, R. W. Vaughan, and H. G. Drickamer, *J. Chem. Phys.*, 1967, **47**, 2583.
[10] U. Ganiel, M. Kestigian, and S. Shtrikman, *Phys. Letters*, 1967, **24A**, 577.
[11] G. K. Wertheim, H. J. Guggenheim, H. J. Williams, and D. N. E. Buchanan, *Phys. Rev.*, 1967, **158**, 446.
[12] R. M. Golding, *Mol. Phys.*, 1968, **14**, 457.
[13] U. Ganiel and S. Shtrikman, *Phys. Rev.*, 1968, **167**, 258.
[14] U. Ganiel, S. Shtrikman, and M. Kestigian, *J. Appl. Phys.*, 1968, **39**, 1254.
[15] R. Fatchally, G. K. Shenoy, N. P. Sastry, and R. Nagaraian, *Phys. Letters*, 1967, **25A**, 453.
[16] M. Eibschutz, L. Holmes, H. J. Guggenheim, and H. J. Levinstein, *J. Appl. Phys.*, 1969, **40**, 1312.
[17] G. K. Wertheim, H. J. Guggenheim, H. J. Levinstein, D. N. E. Buchanan, and R. C. Sherwood, *Phys. Rev.*, 1968, **173**, 614.
[18] K. Ono, A. Ito, and T. Fujita, *J. Phys. Soc. Japan*, 1964, **19**, 2119.
[19] W. Klumpp and K. W. Hoffmann, *Z. Physik*, 1969, **227**, 254.
[20] E. Pfletschinger, *Z. Physik*, 1968, **209**, 119.

[21] R. C. Axtmann, Y. Hazony, and J. W. Hurley, *Chem. Phys. Letters*, 1968, **2**, 673.
[22] Y. Hazony, R. C. Axtmann, and J. W. Hurley, *Chem. Phys. Letters*, 1968, **2**, 440.
[23] G. A. Sawatzky and F. van der Woude, *Chem. Phys. Letters*, 1969, **4**, 335.
[24] Y. Hazony and H. N. Ok, *Phys. Rev.*, 1969, **188**, 591.
[25] D. J. Simkin, *Phys. Rev.*, 1969, **177**, 1008.
[26] C. D. Burbridge and D. M. L. Goodgame, *J. Chem. Soc. (A)*, 1968, 1410.
[27] G. Ziebarth, *Z. Physik*, 1968, **212**, 330.
[28] C. E. Johnson, *Proc. Phys. Soc.*, 1966, **88**, 943.
[29] S. Chandra and G. R. Hoy, *Phys. Letters*, 1966, **22**, 254.
[30] A. J. Nozik and M. Kaplan, *Phys. Rev.*, 1967, **159**, 273.
[31] P. Zory, *Phys. Rev.*, 1965, **140**, A1401.
[32] R. W. Grant, H. Wiedersich, A. H. Muir, Jr, U. Gonser, and W. N. Delgass, *J. Chem. Phys.*, 1966, **45**, 1015.
[33] C. E. Johnson and M. S. Ridout, *J. Appl. Phys.*, 1967, **38**, 1272.
[34] K. Ono, M. Shinohara, A. Ito, T. Fujita, and A. Ishigaki, *J. Appl. Phys.*, 1968, **39**, 1126.
[35] T. C. Gibb and N. N. Greenwood, *J. Chem. Soc.*, 1965, 6989.
[36] P. R. Edwards, C. E. Johnson, and R. J. P. Williams, *J. Chem. Phys.*, 1967, **47**, 2074.
[37] G. M. Bancroft and J. M. Dubery, *J. Chem. Phys.*, 1969, **50**, 2264.
[38] H. Sano and H. Kono, *Bull. Chem. Soc. Japan*, 1965, **38**, 1228.
[39] T. Fujita, A. Ito, and K. Ono, *J. Phys. Soc. Japan*, 1966, **21**, 1734.
[40] C. E. Johnson, W. Marshall, and G. J. Perlow, *Phys. Rev.*, 1962, **126**, 1503.
[41] C. E. Johnson, *Proc. Phys. Soc.*, 1967, **92**, 748.
[42] K. Ono and A. Ito, *J. Phys. Soc. Japan*, 1964, **19**, 899.
[43] K. Chandra and S. Puri, *Phys. Rev.*, 1968, **169**, 272.
[44] D. Raj, K. Chandra, and S. P. Puri, *J. Phys. Soc. Japan*, 1968, **24**, 35.
[45] R. Ingalls, K. Ono, and L. Chandler, *Phys. Rev.*, 1968, **172**, 295.
[46] W. Kerler, *Z. Physik*, 1962, **167**, 194.
[47] U. Gonser, R. M. Housley, and R. W. Grant, *Phys. Letters*, 1969, **29A**, 36.
[48] V. I. Goldanskii, E. F. Makarov, I. P. Suzdalev, and I. A. Vinogradov, *Phys. Rev. Letters*, 1968, **20**, 137.
[49] R. M. Housley, U. Gonser, and R. W. Grant, *Phys. Rev. Letters*, 1968, **20**, 1279.
[50] V. I. Goldanskii, E. F. Makarov, I. P. Suzdalev, and I. A. Vinogradov, *Zhur. eksp. teor. Fiz.*, 1970, **58**, 760.
[51] D. W. Forester and N. C. Koon, *J. Appl. Phys.*, 1969, **40**, 1316.
[52] H. N. Ok, *Phys. Rev.*, 1969, **185**, 472.
[53] I. Dezsi and L. Keszthelyi, *Solid State Commun.*, 1966, **4**, 511.
[54] U. Gonser and R. W. Grant, *Phys. Stat. Sol.*, 1967, **21**, 381.
[55] J. B. Forsyth, C. E. Johnson, and C. Wilkinson, *J. Phys. C*, 1970, **3**, 1127.
[56] S. Chandra and G. R. Hoy, *Phys. Letters*, 1967, **24A**, 377.
[57] J. A. Schideler and C. Terry, *Phys. Letters*, 1969, **28A**, 759.
[58] G. R. Hoy, S. de S. Barros, F. de S. Barros, and S. A. Friedberg, *J. Appl. Phys.*, 1965, **36**, 936.
[59] G. R. Hoy and F. de S. Barros, *Phys. Rev.*, 1965, **139**, A929.
[60] F. de S. Barros, P. Zory, and L. E. Campbell, *Phys. Letters*, 1963, **7**, 135.
[61] H. Schmid and J. M. Trooster, *Solid State Commun.*, 1967, **5**, 31.
[62] J. M. Trooster, *Phys. Stat. Sol.*, 1969, **32**, 179.

[63] I. Dezsi, L. Keszthelyi, L. Pocs, and L. Korecz, *Phys. Letters*, 1965, **14**, 14.
[64] I. Dezsi, L. Keszthelyi, B. Molnar, and L. Pocs, *Phys. Letters*, 1965, **18**, 28.
[65] A. J. Nozik and M. Kaplan, *J. Chem. Phys.*, 1967, **47**, 2960.
[66] J. V. DiLorenzo and M. Kaplan, *Chem. Phys. Letters*, 1969, **3**, 216.
[67] A. Vertes, *Magyar Kém. Folyóirat*, 1969, **75**, 175.
[68] A. Vertes, K. Burger, and L. Suba, *Magyar Kém. Folyóirat*, 1969, **75**, 317.
[69] G. M. Bancroft, M. J. Mays, and B. E. Prater, *Chem. Phys. Letters*, 1969, **4**, 248.
[70] L. Asch, J. P. Adloff, J. M. Friedt, and J. Danon, *Chem. Phys. Letters*, 1970, **5**, 105.
[71] R. M. Golding, K. F. Mok, and J. F. Duncan, *Inorg. Chem.*, 1966, **5**, 774.
[72] C. D. Burbridge, D. M. L. Goodgame, and M. Goodgame, *J. Chem. Soc. (A)*, 1967, 349.
[73] N. E. Erickson and N. Sutin, *Inorg. Chem.*, 1966, **5**, 1834.
[74] A. V. Ablov, V. I. Goldanskii, E. F. Makarov, and R. A. Stukan, *Doklady Akad. Nauk S.S.S.R.*, 1967, **173**, 595.
[75] T. Tominaga, T. Morimoto, M. Takeda, and N. Saito, *Inorg. Nuclear Chem. Letters*, 1966, **2**, 193.
[76] R. L. Collins, R. Pettit, and W. A. Baker, Jr, *J. Inorg. Nuclear Chem.*, 1966, **28**, 1001.
[77] E. König and K. Madeja, *Inorg. Chem.*, 1967, **6**, 48.
[78] I. Dezsi, B. Molnar, T. Tarnoczi, and K. Tompa, *J. Inorg. Nuclear Chem.*, 1967, **29**, 2486.
[79] G. A. Renovitch and W. A. Baker, Jr, *J. Chem. Soc. (A)*, 1969, 75.
[80] C. D. Burbridge and D. M. L. Goodgame, *J. Chem. Soc. (A)*, 1967, 694.
[81] W. R. McWhinnie, R. C. Poller, and M. Thevarasa, *J. Chem. Soc. (A)*, 1967, 1671.
[82] E. König, K. Madeja, and K. J. Watson, *J. Amer. Chem. Soc.*, 1968, **90**, 1146.
[83] J. Reedijk and A. M. van der Kraan, *Rec. Trav. chim.*, 1969, **88**, 828.
[84] C. D. Burbridge and D. M. L. Goodgame, *J. Chem. Soc. (A)*, 1968, 1074.
[85] T. Birchall, *Canad. J. Chem.*, 1969, **47**, 1351.
[86] W. M. Reiff, N. E. Erickson, and W. A. Baker, Jr, *Inorg. Chem.*, 1969, **8**, 2019.
[87] F. W. Kupper and U. Erich, *Z. Naturforsch.*, 1968, **23A**, 613.
[88] L. Korecz and K. Burger, *J. Inorg. Nuclear Chem.*, 1968, **30**, 781.
[89] R. A. Stukan, V. I. Goldanskii, E. F. Makarov, and E. G. Rukhadze, *Zhur. strukt. Khim.*, 1965, **8**, 239.
[90] A. Vertes, T. Tarnoczi, C. L. Egyed, E. Papp-Molnar, and K. Burger, *Magyar Kém. Folyóirat*, 1969, **75**, 17.
[91] J. L. K. F. de Vries, J. M. Trooster, and E. de Boer, *Chem. Comm.*, 1970, 604.
[92] U. Bertelsen, J. M. Knudsen, and H. Krogh, *Phys. Stat. Sol.*, 1967, **22**, 59.
[93] G. K. Wertheim, H. J. Guggenheim, and D. N. E. Buchanan, *Phys. Rev.*, 1968, **169**, 465.
[94] D. E. Earls, R. C. Axtmann, Y. Hazony, and I. Lefkowitz, *J. Phys. and Chem. Solids*, 1968, **29**, 1859.
[95] C. W. Kocher, *Phys. Letters*, 1967, 24A, 93.
[96] J. W. Pebler and F. W. Richter, *Z. Physik*, 1969, **221**, 480.
[97] M. D. Lind, *J. Chem. Phys.*, 1967, **47**, 990.
[98] J. W. G. Wignall, *J. Chem. Phys.*, 1966, **44**, 2462.
[99] R. M. Housley and H. de Waard, *Phys. Letters*, 1966, **21**, 90.
[100] N. Thrane and G. Trumpy, *Phys. Rev. B*, 1970, **1**, 153.

[101] H. Bernas and M. Langevin, *J. Phys. Radium*, 1963, **24,** 1034.
[102] W. Holzapfel and H. G. Drickamer, *J. Chem. Phys.*, 1969, **50,** 1480.
[103] A. P. Ginsberg and M. B. Robin, *Inorg. Chem.*, 1963, **2,** 817.
[104] C. A. Clausen and M. L. Good, *Inorg. Chem.*, 1968, **7,** 2662.
[105] N. L. Costa, J. Danon, and R. M. Xavier, *J. Phys. and Chem. Solids*, 1962, **23,** 1783.
[106] P. R. Edwards and C. E. Johnson, *J. Chem. Phys.*, 1968, **49,** 211.
[107] A. P. Ginsberg and M. B. Robin, *Inorg. Chem.*, 1963, **2,** 817.
[108] R. J. Woodruff, J. L. Marini, and J. P. Fackler, Jr, *Inorg. Chem.*, 1964, **3,** 687.
[109] G. M. Bancroft, A. G. Maddock, W. K. Ong, and R. H. Prince, *J. Chem. Soc. (A)*, 1966, 723.
[110] C. A. Clausen and M. L. Good, *Inorg. Chem.*, 1970, **9,** 220.
[111] S. DeBenedetti, G. Lang, and R. Ingalls, *Phys. Rev. Letters*, 1961, **6,** 60.
[112] F. E. Obenshain, L. D. Roberts, C. F. Coleman, and D. W. Forester, *Phys. Rev. Letters*, 1968, **14,** 692.
[113] A. J. F. Boyle and J. R. Gabriel, *Phys. Letters*, 1965, **19,** 451.
[114] F. van der Woude and A. J. Dekker, *Solid State Commun.*, 1965, **3,** 319.
[115] H. Wegener, *Z. Physik*, 1965, **186,** 498.
[116] W. Bruckner, G. Ritter, and H. Wegener, *Z. Physik*, 1967, **200,** 421.
[117] B. C. Van Zorge, F. van der Woude, and W. J. Caspers, *Z. Physik*, 1969, **221,** 113.
[118] R. M. Housley, *J. Appl. Phys.*, 1967, **38,** 1287.
[119] L. E. Campbell and S. DeBenedetti, *Phys. Letters*, 1966, **20,** 102.
[120] L. E. Campbell and S. DeBenedetti, *Phys. Rev.*, 1968, **167,** 556.
[121] Y. Takashima and S. Ohashi, *Bull. Chem. Soc. Japan*, 1965, **38,** 1684.
[122] A. Z. Hrynkiewicz, J. Kubisz, and D. S. Kulgawczuk, *J. Inorg. Nuclear Chem.*, 1965, **27,** 2513.
[123] M. Takano, T. Shinjo, M. Kiyama, and T. Takada, *J. Phys. Soc. Japan*, 1968, **25,** 902.
[124] P. K. Gallagher and C. R. Kurkjian, *Inorg. Chem.*, 1966, **5,** 214.
[125] N. Saito, H. Sano, T. Tominaga, and F. Ambe, *Bull. Chem. Soc. Japan*, 1965, **38,** 681.
[126] K. G. Dharmawardena and G. M. Bancroft, *J. Chem. Soc. (A)*, 1968, 2655.
[127] J. F. Duncan, C. R. Kanekar, and K. F. Mok, *J. Chem. Soc. (A)*, 1969, 480.
[128] I. Dezsi, A Vertes, and M. Komor, *Inorg. Nuclear Chem. Letters*, 1968, **4,** 649.
[129] A. G. Maddock, L. O. Medeiros, and G. M. Bancroft, *Chem. Comm.*, 1967, 1067.
[130] A. G. Maddock and L. O. Medeiros, *J. Chem. Soc. (A)*, 1969, 1946.
[131] G. M. Bancroft, A. G. Maddock, W. K. Ong, R. H. Prince, and A. J. Stone, *J. Chem. Soc. (A)*, 1967, 1966.
[132] M. Cox, B. W. Fitzsimmons, A. W. Smith, L. F. Larkworthy, and K. A. Rogers, *Chem. Comm.*, 1969, 183.
[133] J. J. Spijkerman, L. H. Hall, and J. L. Lambert, *J. Amer. Chem. Soc.*, 1968, **90,** 2039.
[134] L. H. Hall, J. J. Spijkerman, and J. L. Lambert, *J. Amer. Chem. Soc.*, 1968, **90,** 2044.
[135] L. M. Epstein and D. K. Straub, *Inorg. Chem.*, 1969, **8,** 453.
[136] R. R. Berrett, B. W. Fitzsimmons, and A. Owusu, *J. Chem. Soc. (A)*, 1968, 1575.
[137] W. M. Reiff, W. A. Baker, Jr, and N. E. Erickson, *J. Amer. Chem. Soc.*, 1968, **90,** 4794.

[138] G. M. Bancroft, A. G. Maddock, and R. P. Randl, *J. Chem. Soc. (A)*, 1968, 2939.
[139] W. M. Reiff, G. J. Long, and W. A. Baker, Jr, *J. Amer. Chem. Soc.*, 1968, **90**, 6347.
[140] K. S. Venkateswarlu, P. K. Mathur, and V. Ramshesh, *Indian J. Chem.* 1969, **7**, 915.
[141] A. van den Bergen, K. S. Murray, B. O. West, and A. N. Buckley, *J. Chem. Soc. (A)*, 1969, 2051.
[142] A. N. Buckley, G. V. H. Wilson, and K. S. Murray, *Solid State Commun.*, 1969, **7**, 471.
[143] A. N. Buckley, G. V. H. Wilson, and K. S. Murray, *Chem. Comm.*, 1969, 718.
[144] H. Mosbaek and K. G. Poulson, *Chem. Comm.*, 1969, 479.

7 | Low-spin Iron(II) and Iron(III) Complexes

It was seen in the preceding chapter that the differing electronic configurations of iron(II) and iron(III) in high-spin complexes caused substantial and characteristic differences in the chemical isomer shift, quadrupole splitting, and magnetic field parameters of their Mössbauer spectra. The corresponding strong-ligand-field or low-spin complexes have $^1A_{1g}$ and $^2T_{2g}$ configurations under octahedral symmetry. The low-spin iron(II) $^1A_{1g}(t_{2g}^6)$ configuration has no unpaired electrons and is diamagnetic, whereas low-spin iron(III) $^2T_{2g}(t_{2g}^5)$ embodies an 'electron hole' in the t_{2g} manifold and is paramagnetic, although magnetic ordering is sometimes observed at very low temperatures.

The increased involvement of the iron 4s-orbitals in covalent bonding with the ligands, and the π-acceptor properties of ligands such as CN$^-$ combine to increase the s-electron density at the iron nucleus, and the observed chemical isomer shifts are therefore less positive than for the high-spin complexes (see Fig. 5.2). The difference in chemical isomer shift between low-spin iron(II) and iron(III) is also much less marked than between high-spin iron(II) and iron(III) and this reduces the diagnostic usefulness of this parameter; however, the temperature dependence of the quadrupole splitting is a more reliable criterion of oxidation state. As seen in Chapter 5, the $^2T_{2g}$ configuration of Fe(III) shows appreciable thermal population of close-lying excited electronic levels when slightly distorted and thus has a strong temperature dependence of the quadrupole splitting whereas the $^1A_{1g}$ configuration of Fe(II), having no intrinsic valence contribution to the electric field gradient, shows only a temperature-independent quadrupole splitting from the lattice terms.

The comparative lack of magnetic exchange behaviour reduces the amount of information available from the Mössbauer spectra, and since the differences between the oxidation states are less marked, it is convenient to consider both in parallel.

7.1 Ferrocyanides

The best known of the low-spin iron complexes are undoubtedly the cyanides, and considerable Mössbauer data have now been collected. The diamagnetic

[*Refs. on p. 191*]

ferrocyanides, $[Fe^{II}(CN)_6]^{4-}$, give single-line resonances because of the octahedral ligand symmetry and the lack of a non-bonding 3d-electron contribution to the electric field gradient. The chemical isomer shift of the typical ferrocyanide $K_4Fe(CN)_6.3H_2O$ is -0.04 mm s^{-1} at room temperature, which may be compared to typical values in octahedral high-spin iron(II) complexes of 1·0–1·3 mm s^{-1}. The importance of the effect of π-bonding between the 3d-orbitals of the iron and the ligands was pointed out by Danon [1], and a more substantive molecular orbital analysis has been given by Shulman and Sugano [2].

π-bonding can occur between the filled t_{2g} orbitals of the iron and the π-bonding and π^*-antibonding orbitals of the CN$^-$ ligands. The chemical isomer shift will be influenced by several factors:

(1) any direct contribution from 4s-electrons;
(2) indirect contributions from 3d-electrons shielding the s-electrons, which may be considered in terms of
   (a) the purely ionic or non-bonding effect of a $3d^n$ configuration:
   (b) covalency with the ligands in which filled ligand orbitals donate an effective total of n_2 electrons to the 3d-orbitals;
   (c) covalency with the ligands in which the metal 3d-orbitals donate an effective total of n_3 3d-electrons to the π^*-ligand orbitals.

The effective number of resultant 3d-electrons is then

$$n_{\text{eff}} = n + n_2 - n_3$$

The measured orbital reduction factors from e.s.r. spectra of $[Fe^{III}(CN)_6]^{3-}$ and $[Mn^{II}(CN)_6]^{4-}$ were used to derive some of the molecular orbital coefficients for π-bonding and these show that

$$n_3[Fe(II)] - n_3[Fe(III)] \simeq 1.$$

We know that $n[Fe(II)] - n[Fe(III)] = 1$, and it also seems likely that the values of n_2 are very similar for the two oxidation states. The overall conclusion is that

$$n_{\text{eff}}[Fe(II)] \simeq n_{\text{eff}}[Fe(III)]$$

This implies that, in going from $[Fe^{III}(CN)_6]^{3-}$ to $[Fe^{II}(CN)_6]^{4-}$, although one electron is added to the iron atom, thus converting it from a paramagnetic to a diamagnetic configuration, nevertheless there is also a simultaneous delocalisation of approximately one electron on to the cyanide ligands, thereby leaving the effective charge density at the iron unaltered. This gives a plausible explanation of the very small chemical isomer shifts actually observed between ferricyanides and ferrocyanides.

The ferrocyanides $K_4Fe(CN)_6$ and $K_4Fe(CN)_6.3H_2O$ show a difference in chemical isomer shift of 0·018 mm s^{-1} at $-130°C$ [3]. Detailed measurements of the second-order Doppler shifts show that this difference is maintained upon extrapolation to zero temperature (Fig. 7.1). Theory shows

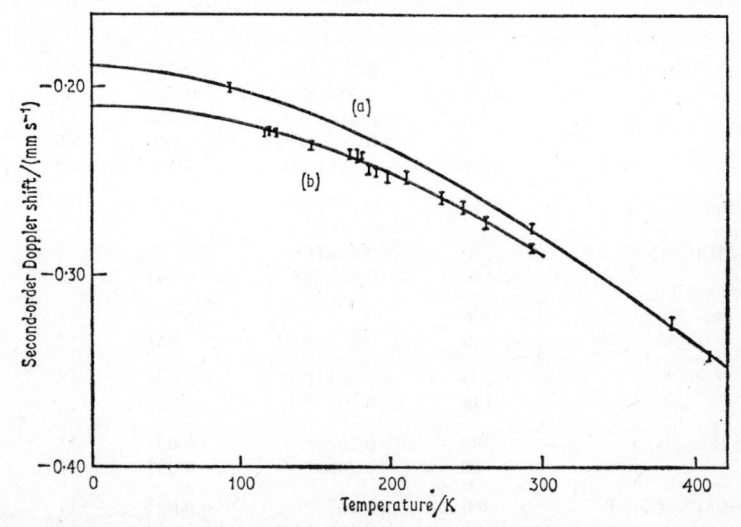

Fig. 7.1 The second-order Doppler shift of $K_4[Fe(CN)_6]$ (a) and $K_4[Fe(CN)_6].3H_2O$ (b). [Ref. 3, Fig. 4]

that this is due to the temperature-independent (zero-point motion) term in the second-order Doppler shift, and that there is no difference in the actual chemical isomer shift (see Chapter 3 for detailed equations). This result emphasises the futility of interpreting measured chemical isomer shift differences of less than 0·02 mm s^{-1} without a full lattice-dynamical study to correct for the zero-point motion.

The first measurements of the recoil-free fraction in an oriented single crystal of $K_4Fe(CN)_6.3H_2O$ were claimed to have shown that there is a change in f in the vicinity of its ferroelectric transition temperature perpendicular to the [010] direction, but not in the parallel direction [4]. However, more detailed study has failed to confirm any discontinuity in the recoil-free fraction or the chemical isomer shift [5, 6].

Application of pressure to $K_4[Fe(CN)_6].3H_2O$ causes a substantial decrease in the chemical isomer shift and a noticeable broadening of the line, indicating an increase in the s-electron density at the nucleus and a possible distortion of the octahedron [7]. High pressures at above 100°C on $Cu_2Fe(CN)_6$, $Ni_2Fe(CN)_6$, or $Zn_2Fe(CN)_6$ cause a partial change to a high-spin Fe^{2+} species [8]. The process is reversible but with considerable hysteresis.

Several papers have considered the effects of changing the cation on the chemical isomer shift of ferrocyanides [9–11, 15]. Selected values are given in Table 7.1. In the series of aquated $M_4[Fe^{II}(CN)_6]$ (M = H$^+$, Li$^+$, Na$^+$, K$^+$,

[*Refs. on p. 191*]

Table 7.1 Mössbauer parameters of some ferrocyanides

Compound	T/K	δ/(mm s^{-1})	δ (Fe) /(mm s^{-1})	Reference
K$_4$[Fe(CN)$_6$].3H$_2$O	298	−0·394 (Pt)	−0·047	12
	168	−0·344 (Pt)	+0·003	12
	77	−0·016 (Fe)	−0·016	13
	298	+0·217 (NP)	−0·040	14
	143	+0·115 (SS)	+0·02	3
K$_4$[Fe(CN)$_6$]	143	+0·133 (SS)	+0·04	3
Ag$_4$[Fe(CN)$_6$]	298	−0·469 (Pt)	−0·122	15
	169	−0·409 (Pt)	−0·062	15
H$_4$[Fe(CN)$_6$]	300	+0·115 (NP)	−0·142	11
	80	+0·173 (NP)	−0·084	11
Cu$_2$[Fe(CN)$_6$]	298	−0·445 (Pt)	−0·098	15
	148	−0·391 (Pt)	−0·044	15
Mg$_2$[Fe(CN)$_6$]	300	+0·160 (NP)	−0·097	11
	80	+0·206 (NP)	−0·051	11
Rb$_2$Ca[Fe(CN)$_6$]	300	+0·165 (NP)	−0·092	11
	80	+0·177 (NP)	−0·080	11
Al$_4$[Fe(CN)$_6$]$_3$	300	+0·100 (NP)	−0·157	11
	80	+0·177 (NP)	−0·080	11
KCe[Fe(CN)$_6$]	300	+0·135 (NP)	−0·122	11
	80	+0·185 (NP)	−0·072	11
Zr[Fe(CN)$_6$]	300	+0·125 (NP)	−0·132	11
	80	+0·176 (NP)	−0·081	11
H$_4$[Fe(CN)$_6$] etherate	300	+0·086 (NP)	−0·171	11
	80	+0·115 (NP)	−0·142	11
H$_4$[Fe(CN)$_6$].5H$_2$SO$_4$	300	+0·115 (NP)	−0·142	11
	80	+0·130 (NP)	−0·127	11

Rb$^+$, Cs$^+$) the chemical isomer shift increases systematically [9]. The implication is that either the central atom electron configuration changes or that the second-order Doppler shift contribution alters, or a combination of both effects. More precise interpretation is not possible as the degree of equation of the various salts was not stated. In a more detailed study of 29 different ferrocyanides [11], it was concluded that it is the effect of the changing polarising power of the cation on the π-bonding in the anion which is important. The ferrocyanide complexes with transition metals differ from the simple salts of the alkali metals in that they contain a polymer lattice with the metal cation octahedrally coordinated to nitrogen and any excess cations being accommodated in the large interstitial sites. Thus, they feature a second coordination sphere of metal cations which is of cubic symmetry

FERRICYANIDES | 173

and a third coordination sphere which may have lower symmetry (see Fig. 7.2). In the series $M_4[Fe(CN)_6]_3$ (M = Al^{3+}, Ga^{3+}, Sc^{3+}, In^{3+}, Y^{3+}) the

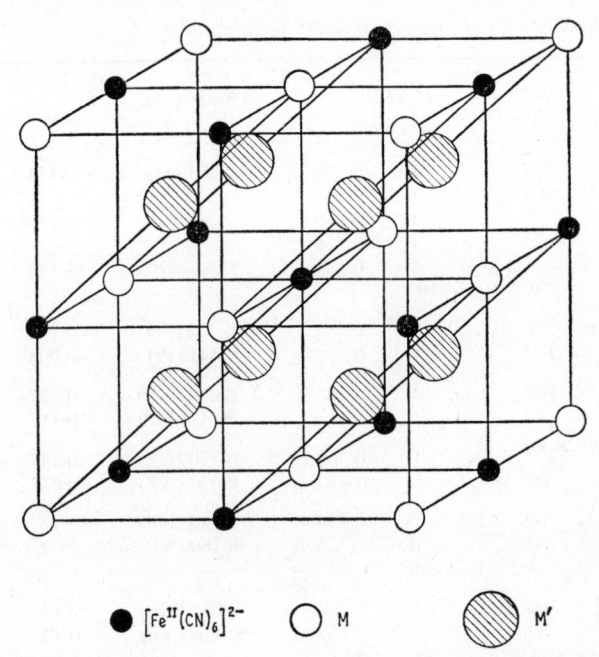

● $[Fe^{II}(CN)_6]^{2-}$ ○ M   ◎ M′

Fig. 7.2 The basic crystal structure of the transition-metal ferrocyanides.

decreasing polarising power of the cation is paralleled by an increase in the chemical isomer shift. A strongly polarising ligand will withdraw electrons from the iron-ligand π-bonds, causing a reduction in the screening of the s-electrons of the iron and the decrease in the chemical isomer shift observed.

A small quadrupole splitting of 0·28 mm s^{-1} has been recorded in anhydrous $H_4Fe(CN)_6$ at 292 K, and is taken to be consistent with the proposed hydrogen-bonded structure which has a symmetry lower than octahedral [16].

7.2 Ferricyanides

As already stated, the chemical isomer shifts of the ferricyanides differ little from those of the ferrocyanides, although they are generally smaller by about 0·1 mm s^{-1} for a given cation. Typical values are given in Table 7.2. They are, however, easily distinguished by the small quadrupole splittings displayed by the ferricyanides. The only example which has been extensively studied is $K_3[Fe^{III}(CN)_6]$, in which the quadrupole splitting is determined by the thermal population of three Kramers' doublets. The latter are linear combinations of the $|xy\rangle$, $|xz\rangle$ and $|yz\rangle$ basis wavefunctions, and the appropriate

[Refs. on p. 191]

coefficients have been determined from e.s.r. measurements [17]. As can be seen from Table 7.2 the quadrupole splitting varies considerably between

Table 7.2 Mössbauer parameters of some ferricyanides

Compound	T/K	H/kG	Δ /(mm s^{-1})	δ/(mm s^{-1})	δ (Fe) /(mm s^{-1})	Reference
K$_3$[Fe(CN)$_6$]	298	0	0·280	−0·471 (Pt)	−0·124	12
	300	0	0·262	—	—	18
	77	0	0·469	—	—	18
	20	0	0·526	—	—	18
	4·2	0	0·524	—	—	18
	0·025–0·04	193	—	—	—	20
Ag$_3$[Fe(CN)$_6$]	298	0	0·767	−0·494 (Pt)	−0·147	15
	147	0	0·858	−0·441 (Pt)	−0·094	15
H$_3$[Fe(CN)$_6$]	300	0	1·26	0·080 (NP)	−0·177	21
	80	0	1·61	0·140 (NP)	−0·117	21
Cs$_3$[Fe(CN)$_6$]	300	0	0·360	0·170 (NP)	−0·137	21
	80	0	0·470	0·220 (NP)	−0·037	21
Mn$_3$[Fe(CN)$_6$]$_2$	300	0	0·370	0·110 (NP)	−0·147	21
	80	0	0·755	0·160 (NP)	−0·097	21
	4·2	195	0·95	—	—	22
Cu$_3$[Fe(CN)$_6$]$_2$	300	0	0·400	0·08 (NP)	−0·18	21
	80	0	0·750	0·120 (NP)	−0·137	21
	4·2	266	0·81	—	—	21
Al[Fe(CN)$_6$]	300	0	0·240	0·13 (NP)	−0·13	21
	80	0	0·280	0·250 (NP)	−0·07	21
Sn$_3$[Fe(CN)$_6$]$_4$	300	0	0·480	0·160 (NP)	−0·097	21
	80	0	0·510	0·270 (NP)	+0·013	21

4·2 and 300 K, but an attempted theoretical fit gave very poor agreement in the low-temperature region [18]. The spectra are illustrated in Fig. 7.3. The small quadrupole splittings imply only a small distortion from octahedral symmetry.

Careful examination of the temperature dependence of the quadrupole splitting of K$_3$Fe(CN)$_6$ in the vicinity of 130 K has shown strong evidence for a sudden change in the gradient of the curve which cannot be derived solely by the Boltzmann population of the energy levels [19]. This is clear evidence in favour of the suspected change in the crystalline state of the salt at this temperature.

K$_3$[Fe(CN)$_6$] becomes antiferromagnetic below 0·13 K. The Mössbauer spectrum recorded using a demagnetisation cryostat was obtained between 0·025 and 0·04 K [20] and shows a six-line pattern with an internal magnetic field of 193 kG (see Fig. 7.4). This value is much higher than predicted by the

[Refs. on p. 191]

$-220\langle S_z \rangle$ rule for a single unpaired electron, but presumably there will also be orbital and dipolar terms, and these have not been estimated.

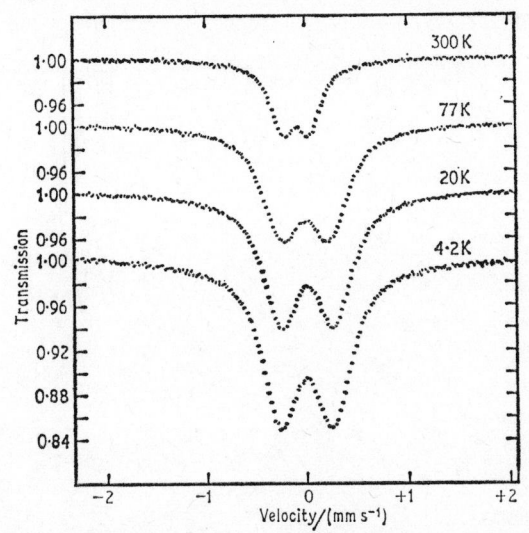

Fig. 7.3 Mössbauer spectra of powdered $K_3Fe(CN)_6$ at 300, 77, 20, and 4·2 K. [Ref. 24, Fig. 2]

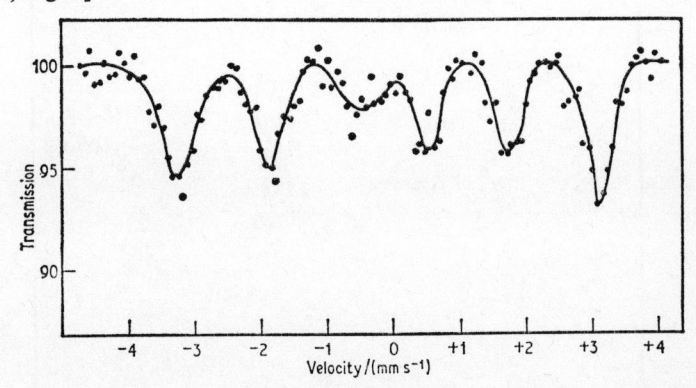

Fig. 7.4 Mössbauer spectrum of $K_3[Fe(CN)_6]$ obtained below the Néel point at 0·025–0·04 K. [Ref. 20, Fig. 3]

An extensive series of data on $K_3Fe(CN)_6$, both pure and magnetically diluted with the diamagnetic isomorphic $K_3Co(CN)_6$, have been obtained [23, 24]. When the iron concentration is less than 5% the spectra show magnetic hyperfine structure at low temperatures due to a slowing of the spin relaxation processes. The concentration dependence and the behaviour in external magnetic fields were compared with two theoretical models based on the e.s.r. and susceptibility results. Typical spectra are shown in Fig. 7.5. A frozen solution of $K_3Fe(CN)_6$ in glycerol shows basically similar behaviour

[*Refs. on p. 191*]

to the magnetically diluted salt, although with broader lines. The theory which generates the solid curve in Fig. 7.5 for instance is rather involved, and this

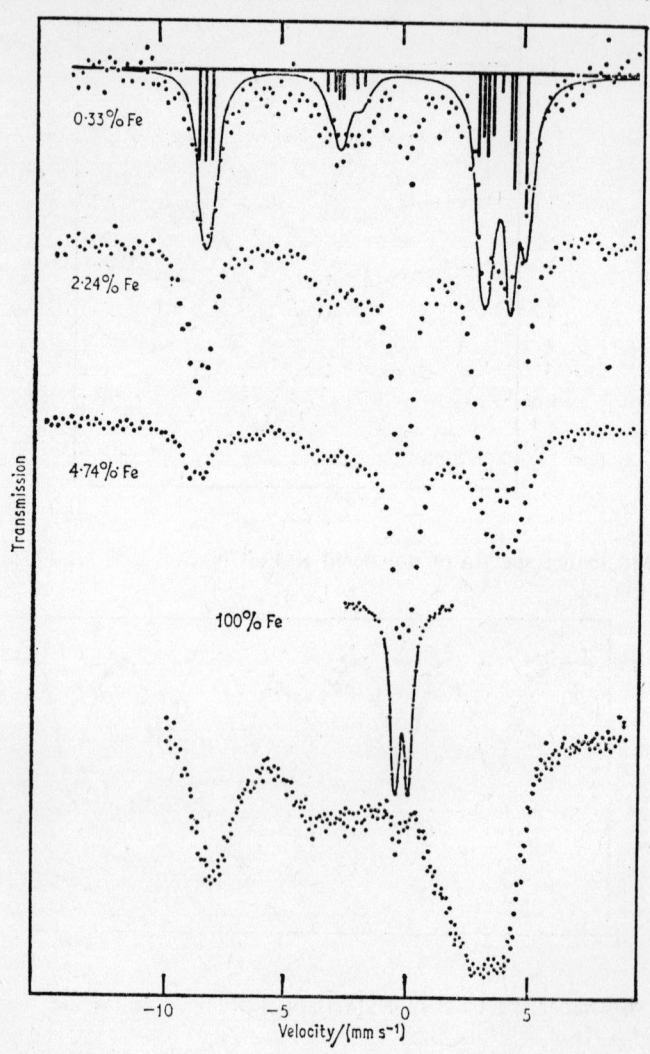

Fig. 7.5 The Mössbauer spectra at 4·2 K of four different concentrations of $K_3Fe(CN)_6$ in $K_3Co(CN)_6$. The solid curve is a theoretical spectrum, and the bottom spectrum is for $K_3Fe(CN)_6$ dissolved in glycerol. [Ref. 24, Fig. 8]

particular problem of an $S = \frac{1}{2}$ state Fe(III) cation with a long relaxation time has been discussed at length [25] for cubic, axial, and rhombic symmetries in zero and non-zero applied fields. Care must be exercised, however, in applying the very successful analysis of the doped salt to the undiluted ferricyanide.

[*Refs. on p. 191*]

The ferricyanide $Mn_3[Fe(CN)_6]_2 \cdot 3H_2O$ and the corresponding Cu and Ni compounds also show magnetic ordering below 15·8, 16·2, and 18·9 K respectively [22]. The magnetic fields at 4·2 K are 195, 266, and 269 kG respectively, which may be compared with the previously mentioned value of 193 kG for $K_3[Fe(CN)_6]$. Presumably the paramagnetic nature of the cation is important because the zinc and cadmium ferricyanides are still non-magnetic at 4·2 K. Spin relaxation has been found in the nickel ferricyanide [26].

$K_3[Fe(CN)_6]$ shows very unusual pressure effects [7]. Increase in pressure causes a substantial increase in the quadrupole splitting, together with the appearance of a third peak in the spectrum which has a chemical isomer shift similar to that of a ferrocyanide. The chemical isomer shift shows twice the pressure coefficient of the ferrocyanide, implying that the mechanism involves more than $3d$-shielding effects alone. It is proposed that an increase in pressure causes either an increase in the degree of back donation from the $3d$-orbitals to the empty ligand π^*-antibonding orbitals, or a change in the $4s$-admixture in the binding, or both. Near 50 kbar there is a phase transition which causes a discontinuity in the chemical isomer shift–pressure curves for both the ferricyanide and 'ferrocyanide' components, and at the same time produces a partial reversion to the oxidised species. All these changes are fully reversible on pressure release, and the apparent change in oxidation state is probably related to the known photochemical reduction properties of potassium ferricyanide (compare with high-spin iron(III) compounds).

$Cu_2Fe(CN)_6$, $Ni_2Fe(CN)_6$, and $Zn_2Fe(CN)_6$ also show a reduction to ferrocyanides with increasing pressure [8]. At elevated temperatures a second change to high-spin iron(II) is also seen.

As with the ferrocyanides, the effect of change in the metal cations has been studied [10, 21], and selected values are given in Table 7.2. The largest self-consistent set of data refers to 20 compounds [21]. In all cases quadrupole splittings are seen. Once again an increased polarising power of the cation causes a decrease in the chemical isomer shift for mono- and di-valent cations, but the opposite is found for trivalent cations. It is postulated that the very strongly polarising cations withdraw electron density from the σ-orbitals in the ferricyanides (but not in the ferrocyanides). The quadrupole splitting is also greater for strongly polarising cations. In the idealised polymer lattice structure (Fig. 7.2) the second coordination sphere has cubic symmetry, and the lattice term in the electric field gradient will only be generated by cations or water of crystallisation in the interstitial positions. One might expect that $M^{3+}[Fe^{III}(CN)_6]^{3-}$ complexes would show smaller quadrupole splittings because of the lack of ions on these sites, and that $Sn^{IV}{}_3[Fe^{III}(CN)_6]_4{}^{3-}$ would show a large splitting due to cation vacancies, but there is no corroboration of this, and the dominant factor appears to be the polarising power of the cation.

Preliminary temperature-dependence data have been given for the quadrupole splittings in the ferricyanides $M_3[Fe(CN)_6]_2$ (M = Mn, Co, Ni, Cu, Cd, or Ca) [27]. The values extrapolated to 0 K range from 0·49 mm s^{-1} in $Ca_3[Fe(CN)_6]_2$ to 0·96 mm s^{-1} in $Ni_3[Fe(CN)_6]_2$, and estimations of the T_{2g} level splittings were made using a simple model.

7.3 Prussian Blue

The dark blue substances formed in various reactions between solutions of iron high-spin salts and ferri- or ferrocyanides have long been an enigma to chemists. It is generally conceded that they have the infinite lattice structure common to the heavy metal ferrocyanides (Fig. 7.2) but it seems likely that there is no closely defined stoichiometry for any of them. 'Insoluble Prussian blue' is formed from an excess of an iron(III) salt and potassium ferrocyanide, and 'Turnbull's blue' is formed from an iron(II) salt and potassium ferricyanide. Both are approximately $Fe_7(CN)_{18}$, and can be written as $Fe_4[Fe(CN)_6]_3$. One can in principle assign electronic structures of $Fe_4{}^{3+}[Fe^{II}(CN)_6]_3$ or $Fe^{3+}Fe_3{}^{2+}[Fe^{III}(CN)_6]_3$, or in view of the unusual colour of the complexes assume that the oxidation state of the iron is indeterminate due to a fast electron-transfer process. The 'soluble Prussian blue' has the formula $KFe[Fe(CN)_6]$, and Robin [28] showed that the electronic spectra could be interpreted in terms of a $KFe^{3+}[Fe^{II}(CN)_6]$ formulation with a charge transfer excitation to the unstable $KFe^{2+}[Fe^{III}(CN)_6]$ alternative. The Mössbauer spectrum confirms that it is an iron(III) ferrocyanide, and that all three complexes contain high-spin iron(III) and low-spin iron(II) [15, 29]. The actual spectrum (a recent measurement is shown in Fig. 7.6

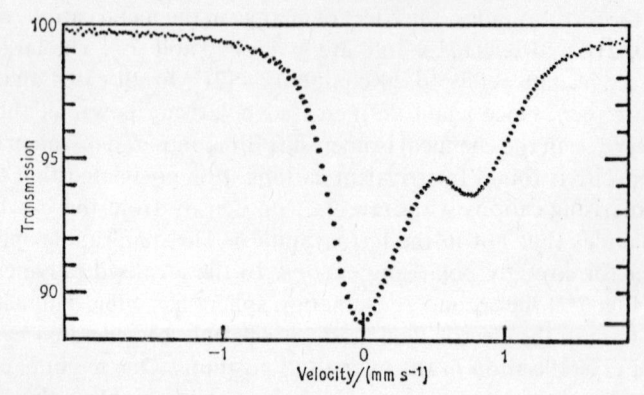

Fig. 7.6 Mössbauer spectrum at 77 K for 'Prussian blue' made from $Fe_2(SO_4)_3$ and $K_4Fe(CN)_6$ containing unenriched iron. [Ref. 30, Fig. 5]

[30]) is an asymmetric doublet due to the superposition of a singlet from the ferrocyanide and a quadrupole doublet from the high-spin iron(III) ions. Typical parameters are given in Table 7.3.

[*Refs. on p. 191*]

Table 7.3 Mössbauer parameters for iron ferro/ferricyanides

Compound		T/K	H/kG	Δ /(mm s^{-1})	δ /(mm s^{-1})	δ (Fe) /(mm s^{-1})	Reference
Fe$_4^{3+}$[Fe(CN)$_6$]$^{4-}$	(Fe^{3+})	77	0	0·57	+0·49 (Fe)	+0·49	30
insoluble Prussian	(Fe^{II})	77	0	0	−0·08 (Fe)	−0·08	30
blue	(Fe^{3+})	1·6	541	0·48*	+0·66 (SS)	+0·57	31
	(Fe^{II})	1·6	0	0	+0·13 (SS)	+0·04	31
Fe$_4^{3+}$[Fe(CN)$_6$]$^{4-}$	(Fe^{3+})	77	0	0·51	+0·49 (Fe)	+0·49	30
Turnbull's blue	(Fe^{II})	77	0	0	−0·07 (Fe)	−0·07	30
	(Fe^{3+})	1·6	543	0·52*	+0·65 (SS)	+0·56	31
	(Fe^{II})	1·6	0	0	+0·09 (SS)	+0·00	31
KFe^{3+}[Fe(CN)$_6$]$^{4-}$	(Fe^{3+})	1·6	536	0·37*	+0·66 (SS)	+0·57	31
soluble Prussian blue	(Fe^{II})	1·6	0	0	+0·15 (SS)	+0·06	31
Fe$_2^{2+}$[Fe(CN)$_6$]$^{4-}$	(Fe^{2+})	77	0	—	+1·42 (Fe)	+1·42	30
	(Fe^{II})	77	0	0	−0·09 (Fe)	−0·09	30
Fe^{3+}[Fe(CN)$_6$]$^{3-}$	(Fe^{3+})	77	0	0·52	+0·50 (Fe)	+0·50	30
	(Fe^{II})	77	0	0·43	−0·06 (Fe)	−0·06	30

* $\tfrac{1}{2}e^2qQ(3\cos^2\theta - 1)$ from magnetic splitting.

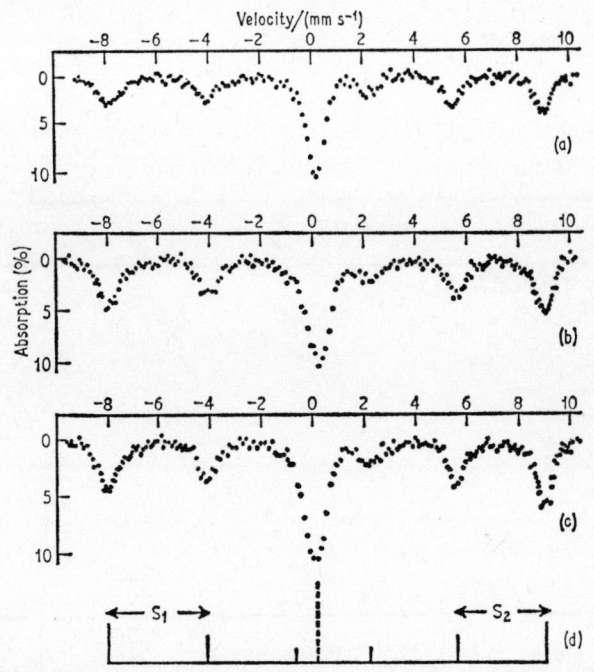

Fig. 7.7 The Mössbauer spectra at 1·6 K for (a) soluble Prussian blue, (b) insoluble Prussian blue, (c) Turnbull's blue. The ferric six-line spectrum shows a small quadrupole splitting, $S_1 - S_2 = \tfrac{1}{2}e^2qQ(3\cos^2\theta - 1)$. [Ref. 31, Fig. 1]

180 | LOW-SPIN IRON(II) AND IRON(III) COMPLEXES

The oxidation states concerned have since been confirmed by two different methods which identify the two species distinctly:

(a) All three complexes become ferromagnetic below 5·5 K, and the Mössbauer spectra at 1·6 K [31] show internal magnetic fields of 540 kG typical of ferric ion ($S = \frac{5}{2}$), and a superimposed singlet due to the diamagnetic ferrocyanides (Fig. 7.7). The chemical isomer shifts and quadrupole splittings are also compatible with this assignment. The blue colour is still retained at 1·6 K, which confirms the charge transfer model and refutes the oscillation of valence concept because the charge states are stable for at least 10^{-7} s. The preparation of Turnbull's blue therefore involves a mutual redox reaction.

(b) The other experiments introduce the technique of selective isotopic enrichment [30]. The natural abundance of ^{57}Fe is only 2·17%. Preparation of Prussian blue using enriched ^{57}Fe as iron(III) sulphate and unenriched

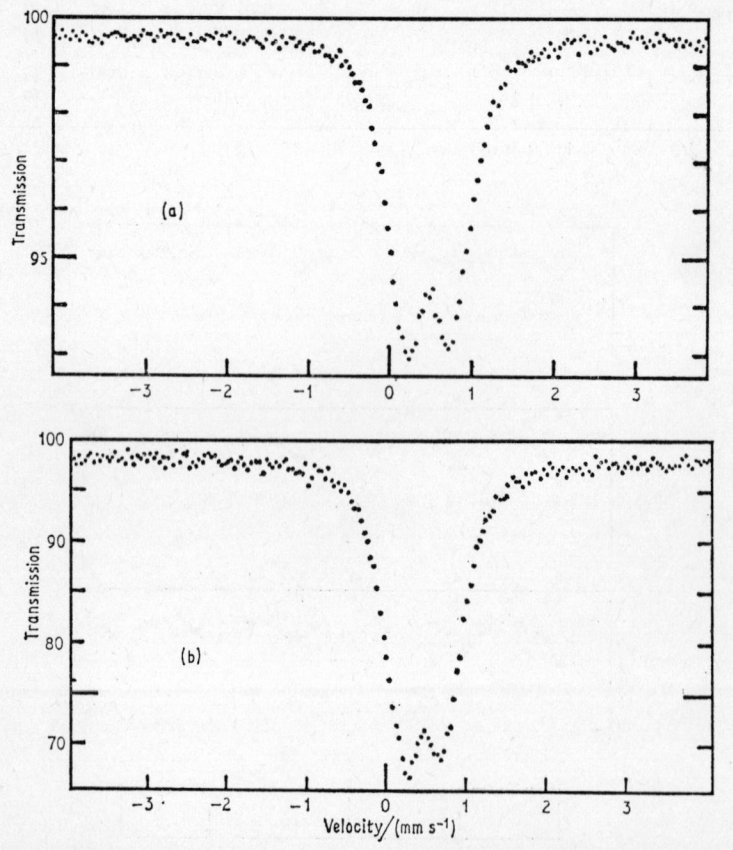

Fig. 7.8 Mössbauer spectrum at 77 K for 'Prussian blue' made from: (a) ^{57}Fe$_2$(SO$_4$)$_3$ plus unenriched K$_4$[Fe(CN)$_6$]; and (b) ^{57}FeCl$_2$ plus unenriched K$_3$[Fe(CN)$_6$]. [Ref. 30, Figs. 6 and 7]

[*Refs. on p. 191*]

potassium ferrocyanide gives a spectrum in which the dominant feature is an iron(III) quadrupole doublet (Fig. 7.8). There is little or no transfer of enriched iron to the low-spin species. The equivalent 'Turnbull's blue' preparation ^{57}FeCl$_2$ and unenriched ferricyanide also gives the enhanced iron(III) spectrum, thereby confirming the redox reaction. In both spectra the weaker singlet component arising from the unenriched ferrocyanide is obscured within the unresolved central portion of the spectrum.

Mössbauer studies also confirm the following reaction products [30]:

$$Fe^{2+}SO_4^{2-} + K_4[Fe^{II}(CN)_6] \rightarrow (Fe^{2+})_2[Fe^{II}(CN)_6]^{4-}$$
$$(Fe^{3+})_2(SO_4^{2-})_3 + K_3[Fe^{III}(CN)_6] \rightarrow Fe^{3+}[Fe^{III}(CN)_6]^{3-}$$
$$Fe^{2+}SO_4^{2-} + K_3[Co^{III}(CN)_6] \rightarrow (Fe^{2+})_3[Co^{III}(CN)_6]_2^{3-}$$
$$Co^{2+}(Cl^-)_2 + K_4[Fe^{II}(CN)_6] \rightarrow (Co^{2+})_2[Fe^{II}(CN)_6]^{4-}$$
$$Co^{2+}(Cl^-)_2 + K_3[Fe^{III}(CN)_6] \rightarrow (Co^{2+})_3[Fe^{III}(CN)_6]_2^{3-}$$
$$Ti^{3+}(Cl^-)_3 + H_4[Fe^{II}(CN)_6] \rightarrow (Ti^{3+})_4[Fe^{II}(CN)_6]_3^{4-}$$
$$Ti^{3+}(Cl^-)_3 + H_3[Fe^{III}(CN)_6] \rightarrow Ti^{4+}[Fe^{II}(CN)_6]^{4-}$$
$$Ti^{4+}(Cl^-)_4 + H_4[Fe^{II}(CN)_6] \rightarrow Ti^{4+}[Fe^{II}(CN)_6]^{4-}$$

It can be seen that only in the case of the titanium(III)–ferricyanide couple is a mutual redox reaction observed. The spectrum of $Fe_3^{2+}[Co^{III}(CN)_6]_2^{3-}$ is also interesting in that the Fe^{2+} spectrum shows signs of two superimposed quadrupole doublets, one from the nitrogen-coordinated position, and the other from the interstitial sites.

Application of pressure to insoluble Prussian blue causes considerable reversible reduction of high-spin ferric iron to high-spin ferrous iron [7].

Of particular interest is the observation of cyanide linkage isomerism in an iron(II) hexacyanochromate(III) [32]. The starting material is difficult to define but appears to have the composition $Fe_{1.6}[Cr(CN)_6](OH)_{0.2}$. The effects of isomerisation on long standing or heating were followed by infrared spectra, X-ray powder diffraction, magnetic susceptibility, and Mössbauer spectra. The unisomerised complex shows the presence of high-spin Fe^{2+} only, with some suggestion of resolution of the cations in interstitial sites and nitrogen-coordinated sites. The first stage of the isomerisation is the exchange of interstitial Fe^{2+} cations with Cr^{3+} so that the carbon-coordinated sites now contain 60% Fe^{II} (low-spin) and 40% Cr^{3+}. Oxidation by heating in air gives a complex which could not be positively identified, although high-spin Fe^{2+} is absent. Reduction of this material in an alcohol mixture with hydrazine hydrate gives a further product which appears to be the true isomerisation complex with Cr^{3+} in nitrogen coordination, Fe^{II} in carbon-coordination, and the original amount of Fe^{2+} in the interstitial sites. It is interesting to note that although the overall result is a reversal of the cyanide orientation, this is actually achieved by migration of the cations.

The reaction of $FeSO_4$ and $K_3Mn(CN)_6$ produces a product formulated as

Fe$_2$Mn(CN)$_6$ which from the Mössbauer spectrum contains [Fe^{II}(CN)$_6$]$^{4-}$ units [33]. Linkage isomerisation of the initial product is implied, although no intermediates were actually detected. Another solid state reaction, involving FeSO$_4$.7H$_2$O and KCN, has been shown to involve [Fe(CN)$_5$H$_2$O]$^{3-}$ as an intermediate, and gives [Fe(CN)$_6$]$^{4-}$ as the final product [34]. FeSO$_4$.2$\frac{1}{2}$H$_2$O does not show any reaction with KCN.

7.4 Substituted Cyanides

Less data are available for substituted iron cyanides. The exception is sodium nitroprusside which is one of the commonly adopted calibration standards. Single-crystal data for Na$_2$[Fe(CN)$_5$NO].2H$_2$O come from several sources [35–38]. The solid state contains diamagnetic [Fe(CN)$_5$NO]$^{2-}$ anions which have C_{4v} symmetry, the Fe–N–O bonds being strictly collinear. The electric field gradient tensor is found to be axially symmetric ($\eta < 0.01$ [37]) with the principal axis along the Fe–NO direction, and is not very sensitive to temperature [39]. The quadrupole splitting of a frozen solution is practically identical with that of the solid, showing that the electric field gradient is generated entirely by asymmetric covalent bonding of the 3d-orbitals. Molecular-orbital analysis has shown that the total 4p-population is only 0.25 electron, which can be neglected.

The ground state of the iron in the nitroprusside molecule is $(d_{xz}, d_{yz})^4(d_{xy})^2$ with the electrons in the lower (d_{xz}, d_{yz}) doublet strongly delocalised by back donation to the 2p-orbitals of the nitrosyl ligand with some slight delocalisation from the d_{xy} orbital to the equatorial cyanides. Using the known electric field gradient contribution for the 3d-orbitals (Table 3.1) and 'effective' numbers of electrons in each orbital we can write [36]

$$q = +\tfrac{4}{7}\langle r^{-3}\rangle n_{xy} - \tfrac{2}{7}\langle r^{-3}\rangle(n_{xz} + n_{yz})$$

The e.s.r. orbital reduction factor for [Fe^{III}(CN)$_6$]$^{3-}$ is $k = 0.87$, so that we can estimate n_{xy} as being ~1.74 electrons. Using a value for $\tfrac{4}{7}\langle r^{-3}\rangle$ of ~4.6 mm s^{-1} gives $(n_{xz} + n_{yz})$ as 2.8, and thence a molecular orbital expressing delocalisation to the nitrosyl as

$$\psi_{xz} = 0.84\phi_{xz} + 0.54\pi^* \text{ (NO)}$$

This gives 29% population of the π^* (NO) orbitals, in good agreement with other estimates. Since the 3d-electron population is effectively only 4.5, and there is a 4s-population of about 0.5, the electron density at the nucleus is much higher than usual and confers on sodium nitroprusside its unusually low chemical isomer shift. The electric field gradient and mean square displacement tensors have been completely determined in sodium nitroprusside single-crystal absorbers from the polarisation dependence of the absorption cross-section [37, 38]. The principal axes of these tensors do not coincide.

[*Refs. on p. 191*]

From 34 measurements using an absolutely calibrated mechanical drive the value for Δ was estimated as $1\cdot7048 \pm 0\cdot0025$ mm s^{-1}, and the chemical isomer shift relative to the ^{57}Fe/Cu source was $-0\cdot4844 \pm 0\cdot0010$ mm s^{-1}. This conflicts with an earlier value of $1\cdot726 \pm 0\cdot002$ mm s^{-1} at 23°C established by the National Bureau of Standards in the U.S.A., but a second reinvestigation [40] with many measurements obtained values in the range $1\cdot7078$ to $1\cdot7087$ mm s^{-1} with an estimated error of about $0\cdot0040$ mm s^{-1}.

The quadrupole splitting of sodium nitroprusside in frozen aqueous or alcoholic solution is $1\cdot86$ mm s^{-1}, the increase above the value in the solid being due to the change in the polarisation effect of the cations on the cyanide ligands [41].

Spectra have also been measured for a number of other pentacyanides of iron(II) and iron(III) (see Table 7.4). An increase in the π back-donation to

Table 7.4 Mössbauer parameters for iron pentacyanides

Compound	T/K	Δ /(mm s^{-1})	δ/(mm s^{-1})	δ (Fe) /(mm s^{-1})	Reference
Iron(II)					
Na$_2$[Fe(CN)$_5$NO].2H$_2$O	298	+1·705	−0·484 (Cu)	−0·258	37
K$_3$[Fe(CN)$_5$CO]	RT	?	−0·01 (SS)	−0·10	42
Na$_3$[Fe(CN)$_5$Ph$_3$P]	298	0·616	+0·231 (NP)	−0·026	14
Na$_5$[Fe(CN)$_5$SO$_3$].9H$_2$O	RT	0·80	+0·10 (SS)	+0·01	44
Na$_4$[Fe(CN)$_5$NO$_2$]	298	0·854	−0·342 (Pt)	+0·005	15
	146	0·855	−0·284 (Pt)	+0·063	15
Na$_3$[Fe(CN)$_5$NH$_3$].H$_2$O	298	0·671	−0·340 (Pt)	+0·007	15
	146	0·663	−0·285 (Pt)	+0·062	15
Na$_3$[Fe(CN)$_5$Ph$_3$As]	298	0·916	+0·29 (NP)	+0·003	14
Na$_3$[Fe(CN)$_5$Ph$_3$Sb]	298	0·94	+0·26 (NP)	0·00	14
Iron(III)					
Na$_2$[Fe(CN)$_5$Ph$_3$P]	298	1·038	+0·137 (NP)	−0·120	14
	117	1·516	+0·219 (NP)	−0·038	14
Na$_3$[Fe(CN)$_5$NO$_2$]	RT	1·78	0·00 (SS)	−0·09	44
Na$_2$[Fe(CN)$_5$NH$_3$].H$_2$O	RT	1·78	0·00 (SS)	−0·09	44
Na$_2$[Fe(CN)$_5$Ph$_3$As]	220	1·006	+0·205 (NP)	−0·052	14
	113	1·237	+0·251 (NP)	−0·006	14
Na$_2$[Fe(CN)$_5$Ph$_3$Sb]	220	0·936	+0·263 (NP)	+0·006	14
	112	0·943	+0·312 (NP)	+0·055	14
Na(Me$_4$N)$_2$[Fe(N$_3$)$_6$]	298	—	+0·183 (NP)	−0·074	14
	163	—	+0·103 (NP)	−0·154	14

the ligand antibonding orbitals causes a decrease in the chemical isomer shift in the order NO$^+$ > CO > CN$^-$ > Ph$_3$P > SO$_3^{2-}$ > NO$_2^-$ $\simeq$ NH$_3$ $\simeq$ Ph$_3$As $\simeq$ Ph$_3$Sb. The quadrupole splittings of the iron(II) pentacyanides also mirror the π back-donation effect, but this is not so in the equivalent iron(III) complexes because of the non-bonding electron contribution to the field gradient in the latter case.

[*Refs. on p. 191*]

184 | LOW-SPIN IRON(II) AND IRON(III) COMPLEXES

The $[Fe(CN)_5NO]^{2-}$ and $[Fe(CN)_5NH_3]^{3-}$ anions are converted to high-spin iron(II) configurations at high temperature and pressure in a similar manner to the ferro- and ferricyanides [45].

The sodium bis-(tetramethylammonium) hexaazidoferrate(III) complex, $Na(Me_4N)_2[Fe(N_3)_6]$, which is included in Table 7.4 for convenience, appears to be similar to the cyanide in its bonding [14].

The alkyl and aryl isocyanides form many complexes with iron which feature 6-coordination and may contain a variety of other ligands. Those of the general type $FeL_2(CNR)_4$ may occur as *cis*- and *trans*-geometric isomers. These can easily be distinguished in the Mössbauer spectrum from the

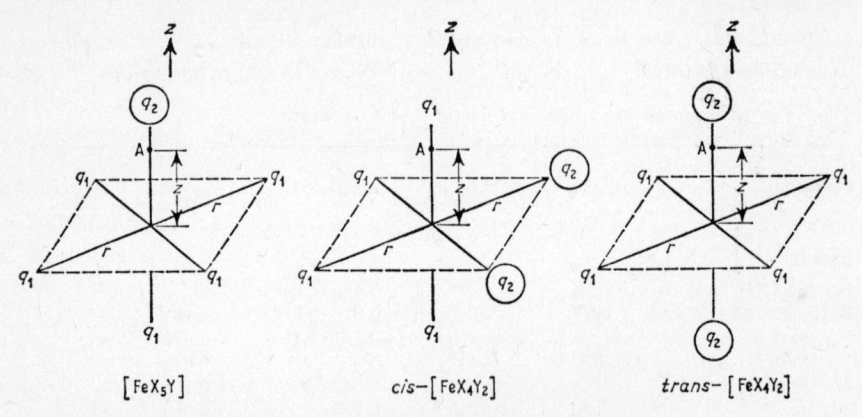

Fig. 7.9 Charge distribution in some six-coordinate complexes using the point charge approximation.

quadrupole splitting. Berrett and Fitzsimmons showed [46, 47] that if two similar ligands are represented as point charges q_1 and q_2 at an identical bond distance r from the central atom, then the electric field gradient and hence the quadrupole splitting for the *trans*-compound should be twice that of the *cis*-isomer in magnitude and opposite in sign. Similarly the mono-substituted complex should have the same magnitude of electric field gradient as the *cis*-disubstituted complex but be opposite in sign. These relations can be deduced as follows. For the monosubstituted complex $[FeX_5Y]$ reference to Fig. 7.9 shows that the potential, V_A at point A is given by

$$V_A = \frac{q_2}{(r-z)} + \frac{q_1}{(r+z)} + \frac{4q_1}{(r^2+z^2)^{\frac{1}{2}}}$$

$$\frac{d^2V_A}{dz^2} = \frac{2q_2}{(r-z)^3} + \frac{2q_1}{(r+z)^3} - \frac{4q_1(r^2-2z^2)}{(r^2+z^2)^{\frac{5}{2}}}$$

As the point A approaches the origin (i.e. $z \to 0$) this becomes

$$V_{zz}(mono) = \frac{2q_2}{r^3} - \frac{2q_1}{r^3} = 2\xi$$

[*Refs. on p. 191*]

For cis-[FeX$_4$Y$_2$] reference to Fig. 7.9 shows that the potential V_A is given by

$$V_A = \frac{2q_2}{(r^2+z^2)^{\frac{1}{2}}} + \frac{2q_1}{(r^2+z^2)^{\frac{1}{2}}} + \frac{q_1}{(r+z)} + \frac{q_1}{(r-z)}$$

Hence $\quad V_{zz}(cis) = \dfrac{-2q_2}{r^3} + \dfrac{2q_1}{r^3} = -2\xi$

$\qquad\qquad\qquad = -V_{zz}(mono)$

Likewise for trans-[FeX$_4$Y$_2$], V_A is given by

$$V_A = \frac{q_2}{(r+z)} + \frac{q_2}{(r-z)} + \frac{4q_1}{(r^2+z^2)^{\frac{1}{2}}}$$

Hence $\quad V_{zz}(trans) = \dfrac{4q_2}{r^3} - \dfrac{4q_1}{r^3} = 4\xi$

$\qquad\qquad\qquad = -2V_{zz}(cis)$

These results, together with those of similar calculations for other octahedral complexes, are summarised in Fig. 7.10.

These theoretical considerations have been verified experimentally though usually for the magnitude of the electric field gradient only. For example [47]

	Δ (mm s^{-1})
[Fe(CN)(CNEt)$_5$]ClO$_4$	0.17
cis-Fe(CN)$_2$(CNEt)$_4$	0.29
trans-Fe(CN)$_2$(CNEt)$_4$	0.59

Table 7.5 Mössbauer parameters for isocyanide complexes at room temperature

Compound	Δ /(mm s^{-1})	δ/(mm s^{-1})	δ (Fe) /(mm s^{-1})	Reference
[Fe(CNMe)$_6$](HSO$_4$)$_2$	0.00	−0.02 (SS)	−0.11	47
[Fe(CNEt)$_6$](ClO$_4$)$_2$	0.00	0.00 (SS)	−0.09	47
[Fe(CNCH$_2$Ph)$_6$](ClO$_4$)$_2$	0.00	+0.04 (SS)	−0.05	47
cis-Fe(CN)$_2$(CNMe)$_4$	0.24	0.00 (SS)	−0.09	47
trans-Fe(CN)$_2$(CNMe)$_4$	0.44	0.00 (SS)	−0.09	47
[Fe(CN)(CNEt)$_5$](ClO$_4$)	0.17	+0.04 (SS)	−0.05	47
cis-Fe(CN)$_2$(CNEt)$_4$	0.29	+0.05 (SS)	−0.04	47
trans-Fe(CN)$_2$(CNEt)$_4$	0.59	+0.05 (SS)	−0.04	47
[Fe(CN)(CNCH$_2$Ph)$_5$](ClO$_4$)	0.28	−0.02 (SS)	−0.11	47
trans-Fe(CN)$_2$(CNCH$_2$Ph)$_4$	0.56	−0.01 (SS)	−0.10	47
[FeCl(ArNC)$_5$](ClO$_4$)	0.70	+0.06 (SS)	−0.03	42, 49
cis-FeCl$_2$(ArNC)$_4$	−0.83	+0.12 (SS)	+0.03	42, 49
trans-FeCl$_2$(ArNC)$_4$	+1.59	+0.20 (SS)	+0.11	42, 49
[Fe(SnCl$_3$)(ArNC)$_5$](ClO$_4$)	—	+0.02 (SS)	−0.07	42
cis-Fe(SnCl$_3$)$_2$(ArNC)$_4$	0.54	+0.11 (SS)	+0.02	42, 49
trans-Fe(SnCl$_3$)$_2$(ArNC)$_4$	1.06	+0.08 (SS)	−0.01	42, 49
cis-FeCl(SnCl$_3$)(ArNC)$_4$	0.67	+0.09 (SS)	0.00	42, 49

ArNC = p-methoxyphenyl isocyanide

186 | LOW-SPIN IRON(II) AND IRON(III) COMPLEXES

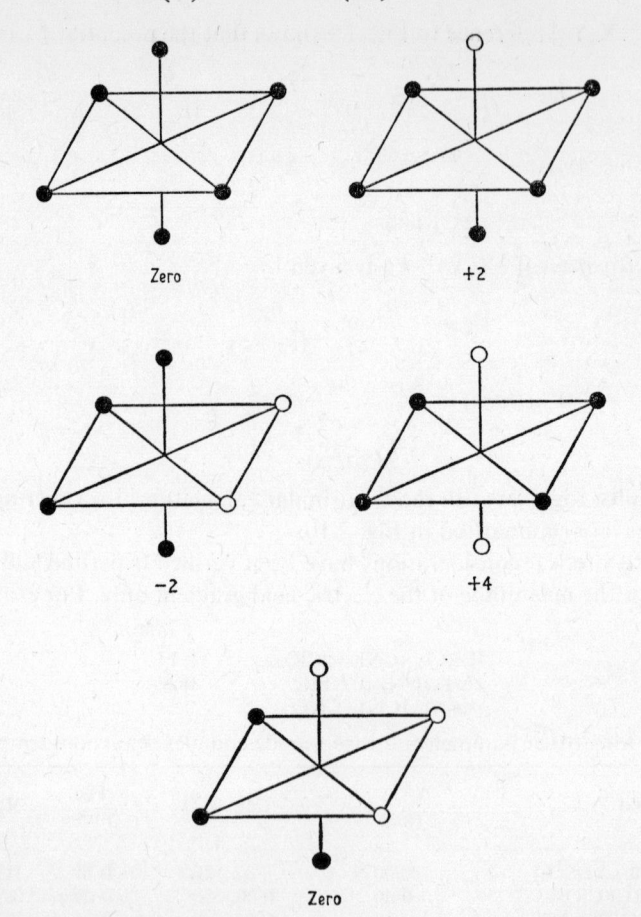

Fig. 7.10 Relative magnitude and sign of the electric field gradient in a series of octahedral complexes $MX_{6-n}Y_n$ where $n = 0, 1, 2, 3$. The ligand X is represented by a shaded circle and the ligand Y by an open circle.

Further examples are bracketed together in Table 7.5. The relationship holds for the CN^-, Cl^-, and $SnCl_3^-$ ligands. The observed splitting is relatively insensitive to the nature of the neutral ligand. Although the rule is only an approximation, it has been found useful for the assignment of the detailed stereochemistry in several series of *cis–trans*-isomers of iron and tin complexes where this had not previously been established by other means. In the case of *cis*- and *trans*-$FeCl_2(p\text{-MeO.}C_6H_4.NC)_4$, the magnetic perturbation method has been used to establish that they do have values of e^2qQ with the opposite sign, being negative for the *cis*- and positive for the *trans*-isomer [48].

Another semi-empirical observation is that the chemical isomer shift can be calculated as the sum of 'partial chemical isomer shifts' for each ligand

[42, 43]. These might be expected to be a function of the spectrochemical series, i.e. the ligand field strength, which in turn is related to the σ- and π-bonding effects on the s-density at the iron nucleus. Although some correlations have been claimed, gross anomalies occur if the ligands are widely different in their bonding, or if the stereochemistry is not the same, and this limits the usefulness of the concept. A further discussion of partial chemical isomer shifts is given in Chapter 9.

7.5 Chelating Ligands

It was pointed out in Chapter 6 that Mössbauer spectroscopy has been used as part of the general characterisation of a large number of relatively new 6-coordinate high-spin iron complexes with chelating ligands. Similar data have been obtained for low-spin complexes, many of which contain the same or similar ligands as those discussed in the high-spin examples but with more of the coordination positions being occupied by nitrogen donors to give the required increase in ligand field splitting [13, 42, 49–54]. The Mössbauer spectrum recorded at two temperatures can give a rapid identification of the oxidation state involved and also provides a check on purity.

The quadrupole splittings of the tris-(1,10-phenanthroline)-, tris-(2,2'-bipyridyl)-, and bis-(2,2',2''-terpyridine)-iron(II) perchlorates of composition [Fe(phen)$_3$](ClO$_4$)$_2$, [Fe(bipy)$_3$](ClO$_4$)$_2$, and [Fe(terpy)$_2$](ClO$_4$)$_2$, are 0.23, 0.39, and 1.14 mm s^{-1} at 77 K and show that the degree of distortion from octahedral (O_h) symmetry increases in the order phen < bipy < terpy. Further details and references are in Table 7.6. Corresponding iron(III) complexes are [Fe(en)$_3$]Cl$_3$ (Δ 1.09 mm s^{-1}), [Fe(phen)$_3$](ClO$_4$)$_3$H$_2$O (Δ 1.71 mm s^{-1}), [Fe(bipy)$_3$](ClO$_4$)$_3$ (Δ 1.80 mm s^{-1}), and [Fe(terpy)$_2$](ClO$_4$)$_3$ (Δ 3.43 mm s^{-1}) which also indicate the order en < phen < bipy < terpy. The last-mentioned compound gives one of the largest quadrupole splittings known for a low-spin complex. In general, as can be seen from Table 7.6, the iron(III) complexes show a much larger quadrupole splitting than the equivalent iron(II) complexes. The chemical isomer shift is also lower by $\sim$0.1–0.2 mm s^{-1}.

Data from a large series of substituted phenanthroline derivatives [13, 51] (see Table 7.6) show that there is little variation in the chemical isomer shift, and that its temperature dependence is often anomalously small [51].

The complex [Fe(phen)$_2$(CN)$_2$] shows a smaller chemical isomer shift (0.18 mm s^{-1} at 300 K) than does [Fe(phen)$_3$](ClO$_4$)$_2$ (0.31 mm s^{-1}) because of the greater π-bonding capabilities of the cyanide. The spectrum of so-called '*trans*'-[Fe(phen)$_2$(CN)$_2$] shows Mössbauer parameters identical to those of the *cis*-complex and from the arguments in Section 7.4 it can be concluded that the difference is not one of *cis–trans*-isomerism [47].

A rather unusual tridentate ligand is tripyridylamine (tripyam) which

188 | LOW-SPIN IRON(II) AND IRON(III) COMPLEXES

Table 7.6 Mössbauer parameters of complexes with nitrogen donor ligands

Compound	T/K	Δ /(mm s^{-1})	δ/(mm s^{-1})	δ (Fe) /(mm s^{-1})	Reference
Iron(II)					
[Fe(phen)$_3$](ClO$_4$)$_2$	300	0·15	0·57 (NP)	0·31	50
	80	0·17	0·62 (NP)	0·36	50
	77	0·23	0·341 (Fe)	0·341	13
cis-Fe(phen)$_2$(CN)$_2$	RT	0·58	0·27 (SS)	0·18	47
'*trans*'-Fe(phen)$_2$(CN)$_2$	RT	0·60	0·33 (SS)	0·24	47
Fe(phen)$_2$(NO$_2$)$_2$	293	0·38	0·28 (Fe)	0·28	54
	77	0·41	0·25 (Fe)	0·25	54
[Fe(4,7-diPh-phen)$_3$](ClO$_4$)$_2$	77	0·19	0·334 (Fe)	0·334	13
[Fe(5-Me-phen)$_3$](ClO$_4$)$_2$	77	0·19	0·334 (Fe)	0·334	13
[Fe(5-Cl-phen)$_3$](ClO$_4$)$_2$	77	0·24	0·346 (Fe)	0·346	13
[Fe(4,7-diOH-phen)$_3$](ClO$_4$)$_2$	77	0·61	0·373 (Fe)	0·373	13
[Fe(bipy)$_3$](ClO$_4$)$_2$	77	0·39	0·325 (Fe)	0·325	13
[Fe(bipy)$_2$(CN)$_2$].3H$_2$O	77	0·61	0·191 (Fe)	0·191	13
[Fe(bipy)$_2$(NCS)py](NCS)	293	0·52	0·36 (Fe)	0·36	54
	77	0·61	0·32 (Fe)	0·32	54
[Fe(bipy)$_2$(NCSe)py](NCSe)	293	0·47	0·35 (Fe)	0·35	54
	77	0·56	0·27 (Fe)	0·27	54
K$_2$[Fe(bipy)(CN)$_4$]	300	0·64	0·29 (NP)	0·03	50
[Fe(4,4'-diPh-bipy)$_3$](ClO$_4$)$_2$	77	0·00	0·311 (Fe)	0·311	13
[Fe(terpy)$_2$](ClO$_4$)$_2$	77	1·14	0·390 (Cr)	0·238	51
[Fe(tripyam)$_2$](ClO$_4$)$_2$	300	0·00	0·63 (NP)	0·37	50
[Fe(tripyam)$_2$](FeCl$_4$) {Fe^{II}	80	0·00	0·60 (NP)	0·34	50
{Fe^{2+}	80	2·63	1·28 (NP)	1·02	50
[Fe(bdh)$_3$](FeCl$_4$)* {Fe^{II}	80	0·42	0·52 (NP)	0·26	50
{Fe^{2+}	80	2·63	1·28 (NP)	1·02	50
Iron(III)					
[Fe(en)$_3$]Cl$_3$	RT	1·09	0·14 (Fe)	0·14	52
[Fe(phen)$_3$](ClO$_4$)$_3$.H$_2$O	300	1·62	0·31 (NP)	0·05	50
	80	1·71	0·36 (NP)	0·10	50
[Fe(bipy)$_3$](ClO$_4$)$_3$.3.H$_2$O	300	1·69	0·29 (NP)	0·03	50
	80	1·80	0·32 (NP)	0·06	50
[Fe(bipy)$_2$(CN)$_2$](ClO$_4$)	300	1·63	0·24 (NP)	−0·02	50
[Fe(terpy)$_2$](ClO$_4$)$_3$	RT	3·09	−0·01 (Fe)	−0·01	53
	77	3·43	0·07 (Fe)	0·07	53

* bdh = diacetyldihydrazone.

forms complexes such as [Fe(tripyam)$_2$](ClO$_4$)$_2$; these show no detectable quadrupole splitting despite the fact that the overall symmetry at the iron must be lower than cubic [50]. Presumably there is an accidental cancellation of the contributions to the electric field gradient tensor.

Brief details are also available for some complexes of the phosphine ligands bis-(diethylphosphino)ethane (I) and *o*-phenylene bis-(diethylphosphine) (II). Parameters are given in Table 7.7. The quadrupole splittings are larger than

[*Refs. on p. 191*]

CHELATING LIGANDS | 189

Table 7.7 Mössbauer parameters of phosphorus complexes

Compound	T/K	Δ /(mm s^{-1})	δ/(mm s^{-1})	δ (Fe) /(mm s^{-1})	Reference
trans-[FeBr$_2$(depe)$_2$]	80	1·45	0·50 (SS)*	0·41	42, 49
trans-[FeCl$_2$(depe)$_2$]	80	1·42	0·43 (SS)*	0·34	42, 49
trans-[FeI$_2$(depe)$_2$]	RT	—	0·49 (SS)	0·40	42
trans-[FeCl(SnCl$_3$)(depe)$_2$]	80	1·34	0·39 (SS)*	0·30	42, 49
trans-[FeHCl(depe)$_2$]	80	<0·13	0·23 (SS)*	0·14	42, 49
trans-[FeHI(depe)$_2$]	RT	—	0·23 (SS)	0·14	42
trans-[FeH(N$_2$)(depe)$_2$]BPh$_4$	RT	0·33	0·16 (SS)	0·05	55
trans[FeH(CO)(depe)$_2$]BPh$_4$	RT	1·00	−0·04 (SS)	−0·13	55
trans-[FeBr$_2$(depb)$_2$]	RT	—	0·45 (SS)	0·36	42
trans-[FeH$_2$(depb)$_2$]	RT	—	0·05 (SS)	−0·04	42
trans-[FeCl$_2$(depb)$_2$]	RT	—	0·43 (SS)	0·34	42

* room temperature value.

(depe) (depb)

(I) (II)

for the nitrogen complexes because the configurations are all *trans*. A hydride ligand, as in *trans*-[FeH$_2$(depb)$_2$], causes a large reduction in chemical isomer shift, showing that σ-donation effects can also be important. The new molecular nitrogen complex [*trans*-FeH(N$_2$)(depe)$_2$$^+$]BPh$_4$$^-$ and the corresponding carbonyl complexes show substantially different parameters (Table 7.7) because of the very different bonding characteristics of the N$_2$ and CO ligands [55].

(III) (IV)

[*Refs. on p. 191*]

190 | LOW-SPIN IRON(II) AND IRON(III) COMPLEXES

Iron(II) complexes derived from diacetylsemicarbazoneoxime (III) feature 6-coordination [56]; their stereochemistry is shown in structure IV. The complexes are formulated as [Fe(DTOH$_2$)$_2$]Cl$_2$ and [Fe(DTOH)$_2$], in which the proton is removed from the oximide group, and [Fe(DTOMe)$_2$] in which the proton is removed from the thiosemicarbazone group. The Mössbauer parameters (Table 7.8) of such complexes are particularly interesting because

Table 7.8 Mössbauer parameters of iron(II) oxime and phthalocyanine complexes

Compound	T/K	Δ/(mm s^{-1})	δ/(mm s^{-1})	δ (Fe)/(mm s^{-1})	Reference
[Fe(DTOH$_2$)$_2$]Cl$_2$	300	0·66	0·42 (Cr)	0·27	56
	78	0·65	0·47 (Cr)	0·32	56
[Fe(DTOH)$_2$]	300	2·02	0·22 (Cr)	0·07	56
	78	2·02	0·29 (Cr)	0·14	56
[Fe(DTOMe)$_2$]	300	0·88	0·45 (Cr)	0·30	56
	78	0·88	0·50 (Cr)	0·35	56
[Fe(niox)$_2$(py)$_2$]	293	1·79	0·21 (Fe)	0·21	58
	77	1·79	0·28 (Fe)	0·28	58
Fe(niox)$_2$(NH$_3$)$_2$]	293	1·75	0·21 (Fe)	0·21	58
	77	1·75	0·28 (Fe)	0·28	58
K$_2$[Fe(niox)$_2$(CN)$_2$]	77	0·80	0·16 (Fe)	0·16	58
[(Pc)Fe(py)$_2$]	293	2·02	0·26 (Fe)	0·26	58
	77	1·97	0·33 (Fe)	0·33	58
	4·2	+1·96	0·32 (Fe)	0·32	58
[(Pc)Fe(but)$_2$]	77	1·94	0·34 (Fe)	0·34	58
K$_2$[PcFe(CN)$_2$]	77	0·56	0·19 (Fe)	0·19	58

although the structural changes are nominally some distance from the iron atom, the values for Fe(DTOH)$_2$ are characteristically different. A semi-empirical molecular orbital interpretation has been formulated [57], and it is suggested that in Fe(DTOH)$_2$ the two pairs of Fe–N bonds are less alike so that the symmetry is effectively lowered from D_{4h} to D_{2h} with a simultaneous increase in quadrupole splitting.

A number of low-spin iron(II) complexes of the types bis-(1,2-cyclohexanedionedioxime)iron(II)XY and phthalocyanineiron(II)(Z)$_2$ (see Fig. 7.11) show an apparent relationship between the chemical isomer shift and quadrupole splitting in the respective series related to the donor properties of the ligands (Table 7.8) [58]. Application of a 30-kG external field to [PcFe(py)$_2$] shows that e^2qQ is positive in sign, and that the metal bonds more strongly to the planar ligands than to the axial ligands. The oxime complexes, however, have a very large asymmetry parameter. The difference in the ligand field strength of a nitrogen atom double-bonded to oxygen as opposed to a nitrogen atom single-bonded to a hydroxyl group is probably more important than the lack of fourfold rotation symmetry.

Other compounds of iron(II) and iron(III) such as dithiocarbamates and

CHELATING LIGANDS | 191

dithiolenes exhibit unusual phenomena such as intermediate spin states or spin-crossover equilibria, and these will be considered in detail in the next chapter.

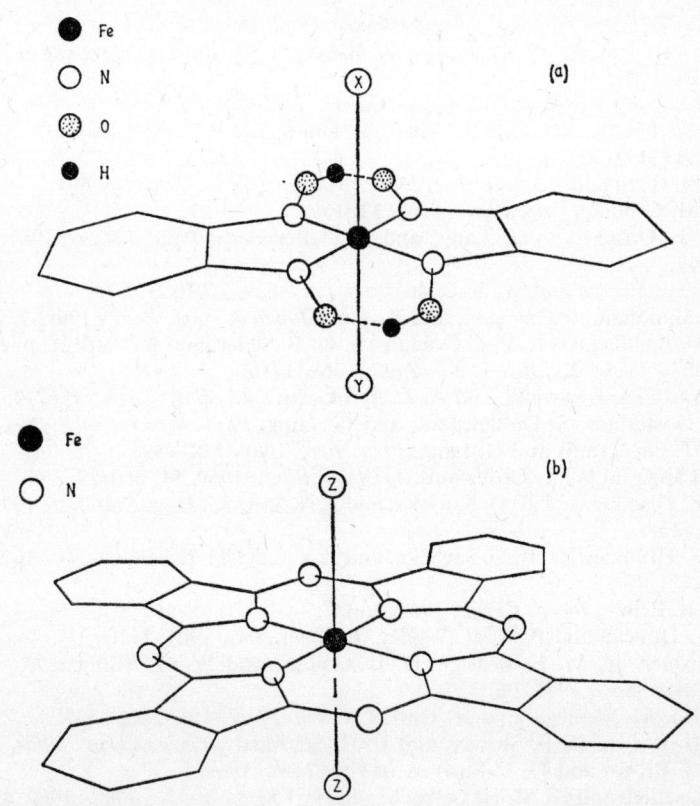

Fig. 7.11 (a) The structure of the complex $Fe^{II}(niox)_2XY$. Unlabelled atoms are carbon. Conjugation is confined to the two O—N—C—C—N—O fragments.
(b) The structure of the complex $PcFe(Z)_2$. The planar ligand is completely conjugated. [Ref. 58, Figs. 1 and 2]

REFERENCES

[1] J. Danon, *J. Chem. Phys.*, 1963, **39**, 236.
[2] R. C. Shulman and S. Sugano, *J. Chem. Phys.*, 1965, **42**, 39.
[3] Y. Hazony, *J. Chem. Phys.*, 1966, **45**, 2664.
[4] Y. Hazony, D. E. Earls, and I. Lefkowitz, *Phys. Rev.*, 1968, **166**, 507.
[5] T. G. Gleason and J. C. Walker, *Phys. Rev.*, 1969, **188**, 893.
[6] M. J. Clauser, *Phys. Rev. B*, 1970, **1**, 357.
[7] A. R. Champion and H. G. Drickamer, *J. Chem. Phys.*, 1967, **47**, 2591.

[8] S. C. Fung and H. G. Drickamer, *J. Chem. Phys.*, 1969, **51**, 4353.
[9] J. Matas and T. Zemcik, *Phys. Letters*, 1965, **19**, 111.
[10] K. Chandra, D. Raj, and S. P. Puri, *J. Chem. Phys.*, 1967, **46**, 1466.
[11] B. V. Borshagovskii, V. I. Goldanskii, G. B. Seifer, and R. A. Stukan, *Izvest. Akad. Nauk S.S.S.R., Ser. Khim.*, 1968, 87.
[12] W. Kerler and W. Neuwirth, *Z. Physik*, 1962, **167**, 176.
[13] R. L. Collins, R. Pettit, and W. A. Baker, Jr, *J. Inorg. Nuclear Chem.*, 1966, **28**, 1001.
[14] E. Fluck and P. Kuhn, *Z. anorg. Chem.*, 1967, **350**, 263.
[15] W. Kerler, W. Neuwirth, E. Fluck, P. Kuhn, and B. Zimmermann, *Z. Physik*, 1963, **173**, 321.
[16] A. N. Garg and P. S. Goel, *J. Inorg. Nuclear Chem.*, 1969, **31**, 697.
[17] R. M. Golding, *Mol. Phys.*, 1967, **12**, 13.
[18] W. T. Oosterhuis, G. Lang, and S. DeBenedetti, *Phys. Letters*, 1967, **24A**, 346.
[19] F. de S. Barros and W. T. Oosterhuis, *J. Phys. C*, 1970, **3**, L79.
[20] M. Shinohara, A. Ishigaki, and K. Ono, *Japan J. Appl. Phys.*, 1968, **7**, 170.
[21] B. V. Borshagovskii, V. I. Goldanskii, G. B. Seifer, and R. A. Stukan, *Izvest. Akad. Nauk S.S.S.R., Ser. Khim.*, 1968, 1716.
[22] B. Sawicka, J. Sawicki, and A. Z. Hyrnkiewicz, *Fiz. Tverd. Tela*, 1967, **9**, 1410.
[23] W. Oosterhuis, S. DeBenedetti, and G. Lang, *Phys. Letters*, 1968, **26A**, 214.
[24] W. T. Oosterhuis and G. Lang, *Phys. Rev.*, 1969, **178**, 439.
[25] G. Lang and W. T. Oosterhuis, *J. Chem. Phys.*, 1969, **51**, 3608.
[26] A. Z. Hrynkiewicz, B. D. Sawicka, and J. A. Sawicki, *Phys. Stat. Sol.*, 1970, **38**, K115.
[27] A. Z. Hrynkiewicz, B. D. Sawicka, and J. A. Sawicki, *Phys. Stat. Sol.*, 1970, **38**, K111.
[28] M. B. Robin, *Inorg. Chem.*, 1962, **1**, 337.
[29] J. F. Duncan and P. W. R. Wigley, *J. Chem. Soc.*, 1963, 1120.
[30] K. Maer, Jr, M. L. Beasley, R. L. Collins, and W. O. Milligan, *J. Amer. Chem. Soc.*, 1968, **90**, 3201.
[31] A. Ito, M. Suenega, and K. Ono, *J. Chem. Phys.*, 1968, **48**, 3597.
[32] D. B. Brown, D. F. Shriver, and L. H. Schwartz., *Inorg. Chem.*, 1968, **7**, 77.
[33] D. B. Brown and D. F. Shriver, *Inorg. Chem.*, 1969, **8**, 37.
[34] P. Guetlich and K. M. Hasselbach, *Angew. Chem., Int. Ed. Engl.*, 1969, **8**, 600.
[35] J. J. Spijkerman, F. C. Ruegg and J. R. DeVoe, p. 255 in 'Applications of the Mössbauer Effect in Chemistry and Solid-State Physics' ,Int. Atomic Energy Agency, Vienna, 1966.
[36] J. Danon and L. Iannarella, *J. Chem. Phys.*, 1967, **47**, 382.
[37] R. W. Grant, R. M. Housley, and U. Gonser, *Phys. Rev.*, 1969, **178**, 523.
[38] R. M. Housley, R. W. Grant, and U. Gonser, *Phys. Rev.*, 1969, **178**, 514.
[39] W. Kerler, *Z. Physik*, 1962, **167**, 194.
[40] R. Fritz, *Phys. Letters*, 1970, **31A**, 226.
[41] W. A. Mundt and T. Sonnino, *J. Chem. Phys.*, 1969, **50**, 3127.
[42] G. M. Bancroft, M. J. Mays, and B. E. Prater, *Chem. Comm.*, 1969, 39.
[43] G. M. Bancroft, M. J. Mays, and B. E. Prater, *Discuss. Faraday Soc.*, No. 47, 1969, p. 136.
[44] J. Danon, *J. Chem. Phys.*, 1964, **41**, 3378.
[45] S. C. Fung and H. G. Drickamer, *J. Chem. Phys.*, 1969, **51**, 4360.
[46] R. R. Berrett and B. W. Fitzsimmons, *Chem. Comm.*, 1966, 91.
[47] R. R. Berrett and B. W. Fitzsimmons, *J. Chem. Soc. (A)*, 1967, 525.

[48] G. M. Bancroft, R. E. B. Garrod, A. G. Maddock, M. J. Mays, and B. E. Prater, *Chem. Comm.*, 1970, 200.
[49] G. M. Bancroft, M. J. Mays, and B. E. Prater, *Chem. Comm.*, 1968, 1374.
[50] R. R. Berrett, B. W. Fitzsimmons, and A. Owusu, *J. Chem. Soc.* (*A*), 1968, 1575.
[51] L. M. Epstein, *J. Chem. Phys.*, 1964, **40,** 435.
[52] G. A. Renovitch and W. A. Baker, Jr, *J. Amer. Chem. Soc.*, 1968, **90,** 3585.
[53] W. M. Reiff, W. A. Baker, Jr, and N. E. Erickson, *J. Amer. Chem. Soc.*, 1968, **90,** 4794.
[54] E. Konig, S. Hufner, E. Steichele, and K. Madeja, *Z. Naturforsch.*, 1967, **22A,** 1543; 1968, **23A,** 632.
[55] G. M. Bancroft, M. J. Mays, and B. E. Prater, *Chem. Comm.*, 1969, 585.
[56] A. B. Ablov, G. N. Belozerskii, V. I. Gol'danskii, E. F. Makarov, V. A. Trukhtanov, and V. V. Khrapov, *Doklady Akad. Nauk S.S.S.R.*, 1963, **151,** 1352.
[57] A. V. Ablov, I. B. Bersuker, and V. I. Gol'danskii, *Doklady Akad. Nauk S.S.S.R.*, 1963, **152,** 1391.
[58] B. W. Dale, R. J. P. Williams, P. R. Edwards, and C. E. Johnson, *Trans. Faraday Soc.*, 1968, **64,** 620; 1968, **64,** 3011.

8 Unusual Electronic Configurations of Iron

In Chapters 6 and 7 the data for both high-spin and low-spin iron(II) and iron(III) complexes were discussed. Although these are the common electronic configurations for these oxidation states, there is also a minority group of compounds which show either ligand field crossover from high-spin to low-spin configuration or intermediate spin states, and in addition there are some uncommon oxidation states.

The following classification can be made:

(1) iron(II) showing $^5T_2$–$^1A_1$ crossover;
(2) iron(III) showing $^6A_1$–$^2T_2$ crossover;
(3) iron(II) with $S = 1$ spin state;
(4) iron(III) with $S = \frac{3}{2}$ spin state.

It is also convenient to include here the dithiolene group of complexes, as well as the oxidation states iron(I), iron(IV), and iron(VI). The iron(IV) oxidation state also occurs in oxide systems such as $BaFeO_3$, but these will be discussed separately in Chapter 10. Covalent diamagnetic complexes such as carbonyls are included in Chapter 9; the oxidation state of iron in these complexes spans both positive and negative values and includes iron(0) as in the binary carbonyls themselves.

8.1 Iron(II) Compounds Showing $^5T_2$–$^1A_1$ Crossover

Iron(II) complexes usually adopt either the $^5T_2$ high-spin ($S = 2$, weak ligand field) configuration with four unpaired electrons, or the $^1A_1$ low-spin ($S = 0$, strong ligand field) diamagnetic configuration. If the energies of the two possible states are very similar, then both forms should be capable of coexistence; this criterion for ligand field spin crossover is satisfied only infrequently, but the phenomenon is now well documented in several complexes. A comprehensive discussion of the factors involved has been given by Martin and White [1]. The Mössbauer spectrum has proved extremely useful in the study of such systems, and the results obtained are therefore discussed in detail.

[Refs. on p. 219]

1,10-Phenanthroline Complexes

Parameters for many $S = 2$ and $S = 0$ 1,10-phenanthroline iron(II) complexes are tabulated in Chapters 6 and 7. In only two cases, Fe(phen)$_2$(NCS)$_2$ and Fe(phen)$_2$(NCSe)$_2$, has a crossover been found [2]. Detailed magnetic measurements reveal that the $^5T_2$–$^1A_1$ transformation is a sudden one [3]. At high temperatures the magnetic moments are approximately 5·2 BM, indicating the $^5T_2$ state. The change to $^1A_1$ takes place at 174 K and 232 K respectively, although the low-temperature magnetisation data usually indicates a residual moment which is dependent on the particular sample preparation. The fact that the changeover is nearly discontinuous precludes any interpretation involving thermal population of close-lying $^5T_2$ and $^1A_1$ manifolds. Although no crystallographic phase change is involved, it seems likely that there is an alteration in molecular dimensions with a not inconsiderable potential barrier between the two states.

The Mössbauer parameters at 293 K and 77 K are clearly those of $^5T_2$ and $^1A_1$ configurations (Table 8.1). More extensive measurements [4] over this temperature range show that the spectrum of Fe(phen)$_2$(NCS)$_2$ exhibits lines due to both spin states in the vicinity of the transition region as shown in Fig. 8.1. This shows that the interconversion takes place on a time-scale of greater than 10^{-7} s, a factor also common to the other $^5T_2$–$^1A_1$ transitions, but contrasting with the $^6A_1$–$^2T_2$ transition in iron(III) where the spectrum is a time-average.

Tris-(2-aminomethylpyridine)iron(II) Halides

The tris-(2-aminomethylpyridine)iron(II) halides, [Fe(2-NH$_2$pic)$_3$]X$_2$ (X=Cl, Br, I), also show a spin crossover [5]. In this instance the halide anions have a very strong effect on the equilibrium, as shown by the magnetic moment versus temperature plot in Fig. 8.2. No interpretation can be formulated using a simple thermal excitation model. The chloride gives Mössbauer spectra indicating $^5T_2$ at room temperature, $^1A_1$ at 4·2 K, and both states in coexistence at 77 K, with no detectable interconversion within the observation time-scale.

Dithiocyanato-bis-(2,2'-bipyridyl)iron(II)

Three polymorphs (designated by A, B, and C in Table 8.1) of the compound [Fe(bipy)$_2$(NCS)$_2$] all show the same $^5T_2$–$^1A_1$ crossover behaviour [6]. Once again there is a precipitous change in the magnetic moment with temperature rise, presumably associated with significant changes of the molecular dimensions. At low temperatures the change to the $^1A_1$ state is not complete, a factor also noted in the 1,10-phenanthroline derivatives, and this may well be a consequence of accommodating the two different spatial arrangements into a single lattice to the best advantage.

[Refs. on p. 219]

Table 8.1 Mössbauer parameters for some iron(II) $^5T_2$–$^1A_1$ crossover complexes

Compound	T/K	Δ /(mm s^{-1})	$S=0$ δ /(mm s^{-1})	δ (Fe) /(mm s^{-1})	Δ /(mm s^{-1})	$S=2$ δ /(mm s^{-1})	δ (Fe) /(mm s^{-1})	Reference
Fe(phen)$_2$(NCS)$_2$	293	—	—	—	2·67	0·98(Fe)	0·98	3
	77	0·34	0·37(Fe)	0·37	—	—	—	3
Fe(phen)$_2$(NCSe)$_2$	293	—	—	—	2·52	1·03(Fe)	1·03	3
	77	0·18	0·35(Fe)	0·35	—	—	—	3
[Fe(2-NH$_2$pic)$_3$]Cl$_2$	RT	—	—	—	2·04	0·98(Fe)	0·98	5
	77	0·55	0·52(Fe)	0·52	—	—	—	5
[Fe(bipy)$_2$(NCS)$_2$] (A)	293	—	—	—	2·18	1·06(Fe)	1·06	6
	77	0·50	0·36(Fe)	0·36	—	—	—	6
[Fe(bipy)$_2$(NCS)$_2$] (B)	293	—	—	—	2·31	1·06(Fe)	1·06	6
	77	0·47	0·36(Fe)	0·36	—	—	—	6
[Fe(bipy)$_2$(NCS)$_2$] (C)	293	—	—	—	2·13	1·06(Fe)	1·06	6
	77	0·50	0·34(Fe)	0·34	—	—	—	6
[Fe(pyim)$_3$]Cl$_2$·2½H$_2$O	295	0·44	0·35(Fe)	0·35	2·30	0·90(Fe)	0·90	8
	77	0·48	0·44(Fe)	0·44	—	—	—	8
	4	0·49	0·47(Fe)	0·47	—	—	—	8
[Fe(pyim)$_3$](NCS)$_2$·2H$_2$O	295	0·34	0·35(Fe)	0·35	2·22	0·90(Fe)	0·90	8
	77	0·42	0·46(Fe)	0·46	3·05	1·15(Fe)	1·15	8

[Refs. on p. 219]

196

Table 8.1 continued

Compound	T/K	S = 0 Δ /(mm s⁻¹)	S = 0 δ /(mm s⁻¹)	S = 0 δ(Fe) /(mm s⁻¹)	S = 2 Δ /(mm s⁻¹)	S = 2 δ /(mm s⁻¹)	S = 2 δ(Fe) /(mm s⁻¹)	Reference
[Fe(pyim)₃](SO₄).3H₂O	295	0.46	0.40 (Fe)	0.40	2.30	0.90 (Fe)	0.90	8
	77	0.53	0.44 (Fe)	0.44	2.63	1.05 (Fe)	1.05	8
	4	0.50	0.45 (Fe)	0.45	2.60	1.10 (Fe)	1.10	8
[Fe(pyim)₃](ClO₄)₂.H₂O	RT	0.37	0.63 (NP)	0.37	2.0	1.25 (NP)	0.99	9
	80	0.54	0.71 (NP)	0.45	—	—	—	9
[Fe(pyiH)₃](ClO₄)₂ (A)	294	0.53	0.58 (NP)	0.32	—	—	—	10
	80	0.60	0.66 (NP)	0.40	—	—	—	10
[Fe(pyiH)₃](ClO₄)₂ (B)	294	—	—	—	2.29	1.28 (NP)	1.02	10
	80	0.75	0.71 (NP)	0.45	2.69	1.36 (NP)	1.10	10
[Fe(bipy)₂.₃₃(NCSe)₂]	293	0.32	0.30 (Fe)	0.30	1.23	1.02 (Fe)	1.02	7
I*	298	—	—	—	3.74	0.814 (Cu)	1.040	12
	230	0	0.20 (Cu)	0.43	3.82	0.818 (Cu)	1.044	12
	194	0	0.255 (Cu)	0.481	3.84	0.813 (Cu)	1.039	12
	4.2	0	0.240 (Cu)	0.466	—	—	—	12
II*	298	0	0.182 (Cu)	0.408	—	—	—	12
	4.2	0	0.130 (Cu)	0.356	—	—	—	12
III*	298	—	—	—	3.81	0.826	1.052	12
	4.2	—	—	—	3.84	0.821	1.047	12

* Formulae given in Fig. 8.3.

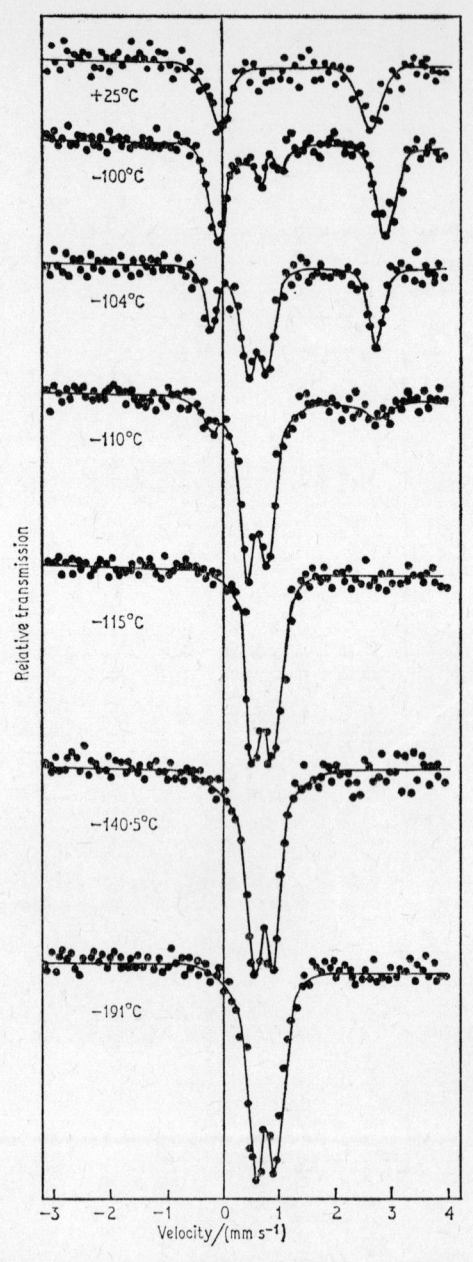

Fig. 8.1 The Mössbauer spectrum of Fe(phen)$_2$(NCS)$_2$ as a function of temperature. [Ref. 4, Fig. 2]

Attempts to prepare the selenocyanato derivative have been unsuccessful, but a compound nominally Fe(bipy)$_{2\cdot33}$(NCSe)$_2$ can be obtained which contains both high-spin and low-spin Fe(II) in the ratio 1 : 2 [7]. It shows no

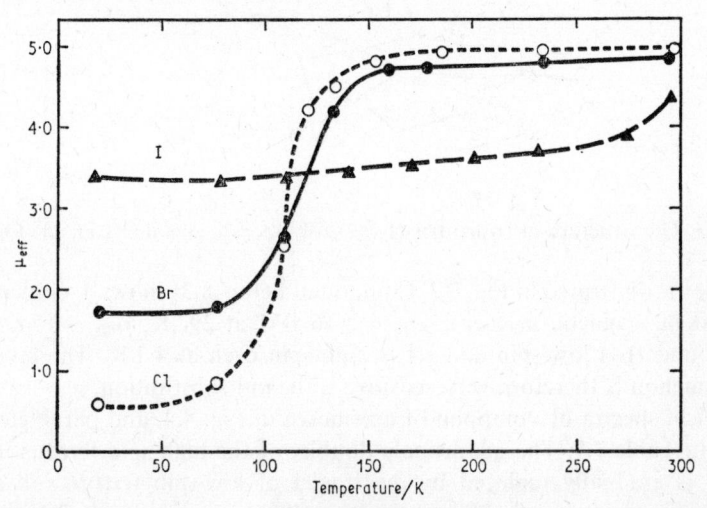

Fig. 8.2 The temperature dependence of the magnetic moment, μ_{eff}, of compounds of the type Fe(2-NH$_2$pic)$_3$X$_2$, showing the strong influence of the halide anion. [Ref. 5, Fig. 1]

evidence for a $^5T_2$–$^1A_1$ crossover in the range 77–293 K and is tentatively formulated as [Fe^{II}(bipy)$_2$(NCSe)$_2$]$_2$[Fe^{2+}(bipy)(NCSe)$_2$](bipy).

2-(2′-Pyridyl)imidazole Complexes

Several iron(II) complexes of 2-(2′-pyridyl)imidazole (pyim) also show crossover, but less data can be obtained because the proportion of high-spin iron is always small [8, 9]. However, the latter component is clearly seen in the Mössbauer spectra (Table 8.1), indicating that the interconversion is slow.

The closely related complexes with the ligand 2-(2′-pyridyl)imidazoline (pyiH) also feature spin crossover [10]. The perchlorate [Fe(pyiH)$_3$](ClO$_4$)$_2$ exists in a dark blue (A) form and a dark purple (B) form. Form A is in the low-spin $^1A_1$ configuration between 80 K and 294 K, but the B form shows an abrupt partial but reversible change from high-spin to low-spin as it is cooled below 100 K. A possible explanation for the existence of the magnetic isomers lies in the asymmetry of the chelating ligand, which allows in principle the formation of *vicinal* and *meridial* geometric isomers.

The Poly(1-pyrazolyl)borate System

$^5T_2$–$^1A_1$ crossover has also been reported in a ferrous complex with the chelating ligand hydro-tris-(1-pyrazolyl)borate [11, 12]. The geometry of the

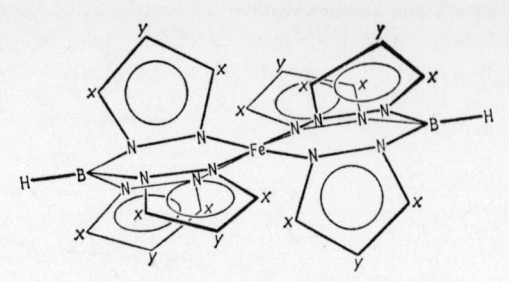

I. $x=CH_3, y=H$ II. $x=y=H$ III. $x=y=CH_3$

Fig. 8.3 The structure of hydro-tris-(1-pyrazolyl)borate chelate. [Ref. 11, Fig. 1]

complex is illustrated in Fig. 8.3. Compound I (Fig. 8.3) shows a temperature-dependent magnetic moment, $\mu_{eff} = 5.16$ BM at 297 K, $\mu_{eff} \sim 0$ at 4.2 K; compound II is low-spin and III is high-spin even at 4.2 K. The crossover phenomenon is therefore very sensitive to ligand substitution.

Typical spectra of compound I are shown in Fig. 8.4, and parameters are given in Table 8.1. The quadrupole doublet of the high-spin ferrous iron at 269 K is gradually replaced by the singlet of low-spin ferrous ion as the temperature is lowered. Phenomenologically it is possible to describe the high-temperature data in terms of an orbital splitting of the octahedral $^5T_{2g}$ state by ~ 1000 cm^{-1}, with a singlet $^1A_{1g}$ ground state. This accounts for the large quadrupole splitting observed. At low temperatures the occupied level is the $^1A_{1g}$ state. The intermediate crossover region can be simulated assuming that the singlet–quintet separation is an approximately linear function of temperature and giving a reversal in their relative positions. However, it seems likely that the agreement with experiment is fortuitous, and that the actual situation is more complex than that of a simple thermal equilibrium.

Examination of a single crystal of the compound suggests that the high- and low-spin forms adopt different crystal structures. Slow thermal cycling of the crystal through the transition region pulverises it completely. There is probably a large potential barrier between the two forms, resulting in a slow interconversion with cooperative collapse of the lattice. There are suggestions of both particle-size effects and hysteresis behaviour in the temperature dependence of the spin equilibrium, thus accounting for the change taking place over a range of temperature.

In summary, it can be seen that none of the $^5T_2$–$^1A_1$ conversions of iron(II) complexes so far investigated involve a simple thermal equilibrium between two energy levels. Interconversion is slow ($\gg 10^{-7}$ s) and appears to involve either a change in dimensions or geometric configuration of the complex. The change may occur over a temperature range of a few degrees (e.g. [Fe(phen)$_2$(NCS)$_2$]) or over more than one hundred degrees (e.g. [Fe(2-NH$_2$pic)$_3$]Cl$_2$).

[*Refs. on p. 219*]

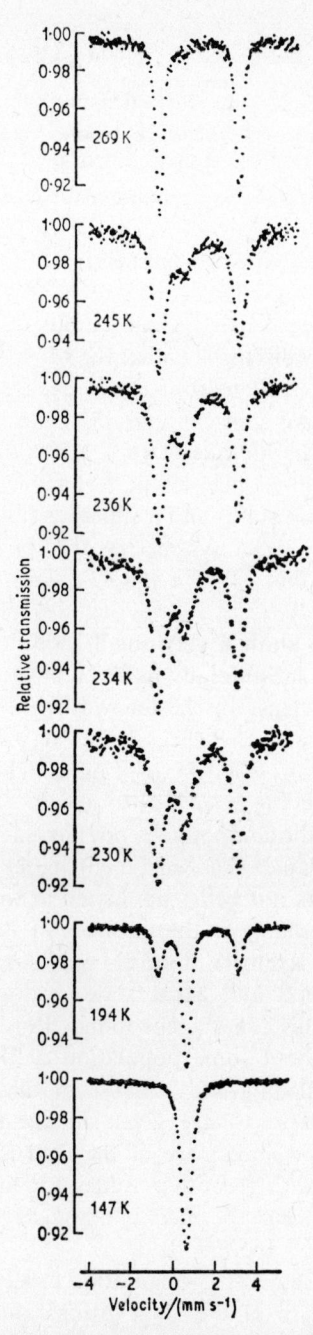

Fig. 8.4 Typical spectra of compound I showing the $^5T_2$–$^1A_1$ crossover. [Ref. 12, Fig. 5]

8.2 Iron(III) Compounds Showing $^6A_1$–$^2T_2$ Crossover

Many of the iron(III) N,N-dialkyldithiocarbamates show an anomalous magnetic moment which is intermediate between the values $\mu = 2 \cdot 0$ BM ($^2T_2$ state of low-spin iron(III)) and $\mu = 5 \cdot 9$ BM ($^6A_1$ state of high-spin iron(III)) and which is temperature dependent. The data are consistent with the existence of an equilibrium between these two configurations. Several groups have reported their Mössbauer spectra [13, 14], the definitive work being by Rickards et al. [15].

An important difference between these complexes and the examples of $^5T_2$–$^1A_1$ crossover already discussed is that the Mössbauer spectra only show evidence for one distinct iron environment in each case. The implication is that the lifetimes of the two spin configurations are small in comparison to the Mössbauer excited-state lifetime of 10^{-7} s. In other words the spectrum is time-averaged. In a ligand field of perfect cubic symmetry there is no interaction between the two states and a superposition of the spectra of each of these is predicted. For lower symmetries the spin–orbit coupling causes state mixing of the $^6A_1$ and $^2T_2$ via the $^4T_1$ state, which causes a rapid exchange.

Most of the complexes show a very small room-temperature quadrupole splitting which increases substantially as the temperature is lowered. There is also a concomitant increase in the linewidths due to an increase in the spin-lattice relaxation times, the effect being very marked at 4·2 K. The pyrrolidyl complex, $[(CH_2)_4NCS_2]_3Fe$, is a pure high-spin iron(III) complex and the relaxation broadening is such as to generate a well-resolved six-line spectrum at 4·2 K. Since the compound is not ferromagnetic, it would appear that only the $|\pm\frac{5}{2}\rangle$ Kramers' doublet contributes to the spectrum, the $|\pm\frac{1}{2}\rangle$ and $|\pm\frac{3}{2}\rangle$ doublets not being populated at this temperature. This is confirmed by the lack of change in the spectrum in an applied magnetic field at 4·2 K [16]. If the three Kramers' doublets were close in energy the applied field would split the levels and cause state mixing, thereby resulting in relaxation broadening. This effect is not found. Raising the temperature on the other hand brings about some population of the $|\pm\frac{3}{2}\rangle$ doublet with consequent inward broadening and collapse of the hyperfine components [16, 17]. Detailed calculations have given an accurate simulation of the observed spectra [17]. The magnitude of the field is 460 kG at 1·3 K [16] which, when compared with the other values for high-spin iron(III) compounds in Table 6.9, illustrates the effect of a high degree of covalency on the Fermi contact term of the hyperfine field.

Admixture of the low-spin $^2T_2$ configuration causes a much faster relaxation, and changing the size of the alkyl groups and thence the interionic distance also appears to cause significant changes in behaviour. Application of external magnetic fields decouples the electron and nuclear spins and causes magnetic six-line patterns to partially emerge. Typical spectra for the iso-butyl derivative are shown in Fig. 8.5.

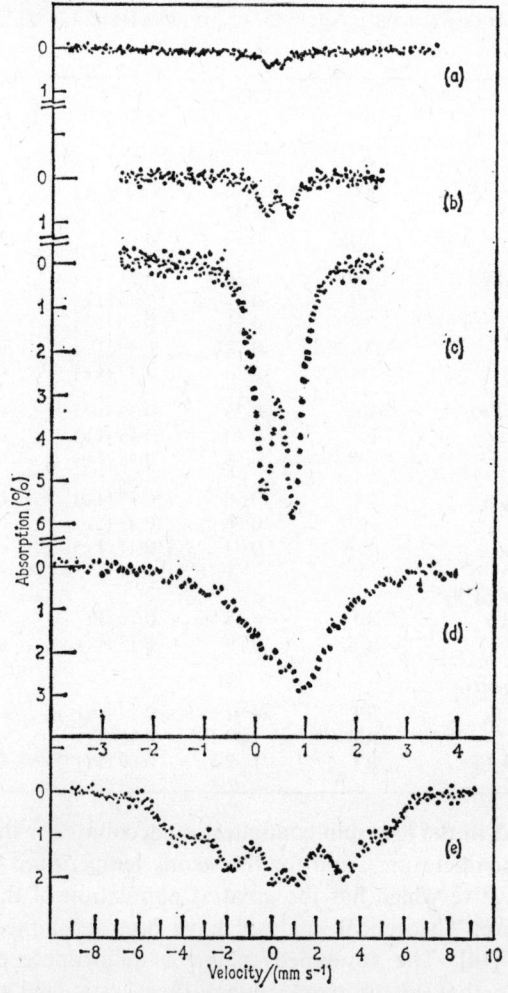

Fig. 8.5 Mössbauer spectra of iron(III) di-isobutyldithiocarbamate at (a) 300 K, (b) 195 K, (c) 77 K, (d) 4·2 K, (e) 4·2 K in 5-kG field. [Ref. 15, Fig. 3]

The chemical isomer shifts of all the crossover complexes are very close to those for the high-spin pyrollidyl derivative (Table 8.2), and in all cases are more typical of high-spin complexes even when there is considerable low-spin admixture. The highly polarisable sulphur ligands presumably donate strongly in σ-bonds with the metal d-orbitals, and thereby decrease the s-electron density at the nucleus, producing a higher chemical isomer shift than might otherwise be expected.

[*Refs. on p. 219*]

Both electronic configurations appear to show small quadrupole splittings because of distortions from true octahedral symmetry. The large orbital

Table 8.2 Mössbauer parameters for some iron(III) $^6A_1-^2T_2$ crossover complexes [15]

Compound	T/K	Δ /(mm s^{-1})	δ /(mm s^{-1})	δ (Fe) /(mm s^{-1})
[(CH$_2$)$_4$NCS$_2$]$_3$Fe	300	0·36	0·40 (Fe)	0·40
	77	0·42	0·51 (Fe)	0·51
	4·2	0·20	0·53 (Fe)	0·53
[Me$_2$NCS$_2$]$_3$Fe	300	0·26	0·41 (Fe)	0·41
	195	0·47	0·43 (Fe)	0·43
	77	0·71	0·42 (Fe)	0·42
	4·2	0·78	0·49 (Fe)	0·49
	1·5	0·78	0·49 (Fe)	0·49
[Bu$^i_2$NCS$_2$]$_3$Fe	300	0·33	0·39 (Fe)	0·39
	195	0·49	0·43 (Fe)	0·43
	77	0·59	0·46 (Fe)	0·46
[(C$_6$H$_{11}$)$_2$NCS$_2$]$_3$Fe	300	0·56	0·37 (Fe)	0·37
	195	0·66	0·43 (Fe)	0·43
	77	0·74	0·47 (Fe)	0·47
High-spin only [19]				
[Me$_2$CPS$_2$]$_3$Fe	RT	0·18	0·16 (Pt)	0·51
[Pr$^i_2$CPS$_2$]$_3$Fe	RT	0·29	0·13 (Pt)	0·48
Low-spin only [19]				
[Et$_2$NCSe$_2$]$_3$Fe	RT	0·91	−0·12 (Pt)	0·23
[PhCS$_2$]$_3$Fe	RT	1·87	−0·10 (Pt)	0·25
[PhCH$_2$CS$_2$]$_3$Fe	RT	1·72	0·00 (Pt)	0·35

contribution to Δ in the low-spin configuration accounts for the much larger splitting at low temperatures, and for the strong temperature dependence in the methyl derivative which has the greatest population of the $^6A_1$ state at room temperature. Theoretical analyses have been attempted but are not very successful [14]. The combined magnetic–quadrupole effects at low temperature show that the principal value of the electric field gradient tensor is positive and is perpendicular to the spin axis.

A more cursory examination of 24 of these tris-(N,N-disubstituted dithiocarbamato)iron(III) complexes showed no significant variation in the chemical isomer shift which was recorded as 0·38 mm s^{-1} at room temperature [18]. A significant variation was found in the quadrupole splitting despite the distance of the substituent groups from the central iron atom, and correlation with the structural type of ligand was attempted. In general the experimental parameters were similar to those listed in Table 8.2 from other workers, and are therefore not given.

[Refs. on p. 219]

A number of related sulphur and selenium complexes are either high-spin or low-spin iron(III) with no sign of a crossover equilibrium; they are included in Table 8.2 for comparison [19].

8.3 Iron(II) Compounds with $S = 1$ Spin State

It is a well-known direct consequence of ligand field theory that truly regular octahedral complexes cannot possess a ground state of intermediate spin. The majority of iron(II) complexes have a symmetry which is close to octahedral and as a result they show either $S = 2$ or $S = 0$ electronic configurations. There are, however, a few complexes in which the symmetry is very low, and the intermediate $S = 1$ state is the adopted ground level.

Phthalocyanine Iron(II)

Although the bis-substituted derivatives of phthalocyanine iron(II) are diamagnetic (see Chapter 7), the parent phthalocyanine iron(II) is paramagnetic, and detailed magnetic susceptibility measurements confirm that it has an $S = 1$ configuration [20]. The Mössbauer spectrum has been recorded by several workers, but the definitive work has included a detailed discussion [21]. A large quadrupole splitting with little dependence on temperature is found (Table 8.3). Application of a 30-kG magnetic field at 4·2 K causes splitting

Table 8.3 Mössbauer parameters for iron(II) $S = 1$ complexes

Compound	T/K	Δ /(mm s^{-1})	δ (Fe) /(mm s^{-1})	Reference
phthalocyanine iron(II)	293	2·62	0·40	21
	77	2·69	0·51	21
	4	2·70	0·49	21
[Fe(phen)$_2$ox].5H$_2$O	293	0·21	0·33	24
[Fe(phen)$_2$mal].7H$_2$O	293	0·18	0·34	24
	77	0·18	0·27	24
[Fe(phen)$_2$F$_2$].4H$_2$O	293	0·21	0·33	24
	77	0·16	0·30	24
[Fe(bipy)$_2$ox].3H$_2$O	293	0·26	0·34	24
	77	0·26	0·27	24
[Fe(bipy)$_2$mal].3H$_2$O	293	0·31	0·30	24
[Fe(4,7-dmph)$_2$ox].4H$_2$O	293	0·23	0·36	24
	77	0·21	0·29	24
[Fe(4,7-dmph)$_2$mal].7H$_2$O	293	0·27	0·33	24
	77	0·21	0·27	24

into a high-energy doublet and a low-energy triplet, showing that the sign of e^2qQ is positive and that the asymmetry parameter η is small. The precise

[Refs. on p. 21]

shape of the spectrum and the conventional magnetisation data show that the magnetic axis does not correspond to the molecular symmetry axis. This proves that there is no true fourfold symmetry axis, possibly the result of a Jahn–Teller distortion, for example. It seems likely that the magnetic z axis lies in the molecular plane. Approximate calculations of the contributions to the magnetisations suggest that the orbital and dipolar contributions to the hyperfine magnetic field are $H_L = +380$ kG and $H_D = +100$ kG; these together with the measured hyperfine field in the plane of maximum magnetisation of $+270$ kG give a Fermi contact term of $H_S = -210$ kG. This agrees well with the predictions of the $-220\langle S_z \rangle$ rule if $S = 1$.

The quadrupole splitting does not allow a unique electronic level scheme to be derived, although the temperature independence confirms lack of orbital degeneracy and low-lying excited states. It seems likely that covalent bonding is more important than the contributions from the $3d$-non-bonding electrons. The chemical isomer shift is intermediate between those of the $S = 2$ and $S = 0$ states, but is much closer to the latter.

Two superimposed quadrupole patterns with large and small Δ respectively have been observed in pressure-dependence studies [22]. Interpretation was in terms of spin excitation from $S = 1$ to $S = 2$ states, but the possibility of a phase change has not been eliminated.

The iron(II) phthalocyanine complexes $[FePc]^{n-}$ ($n = -1, 0, +1, 2, 3, 4$) have been studied over a range of temperature, and the observed spectra discussed in terms of the molecular orbital treatment of these complexes [23]. Progressive reduction has little effect on the spectrum.

Iron(II)-bis-(α-diimine) Complexes

A number of 1,10-phenanthroline, 2,2'-bipyridyl, and 4,7-dimethyl-1,10-phenanthroline complexes (see Table 8.3) have also been observed to have an $S = 1$ ground state and show magnetic moments of between 3·8 and 4·0 BM at 293 K [24, 25]. The chemical isomer shifts are very similar to those for corresponding $S = 0$ complexes. The very small quadrupole splittings observed are held to be compatible with a $|yz\rangle^2 |xz\rangle^2 |xy\rangle^1 |x^2 - y^2\rangle^1$ ground state in which the contributions to the electric field gradient effectively cancel.

8.4 Iron(III) Compounds with $S = \frac{3}{2}$ Spin State

Bis-(N,N-dialkyldithiocarbamato)iron(III) Halides

An interesting class of iron(III) complex is the type $(R_2NCS_2)FeX$ where R is an alkyl group and X is a halogen. These compounds are penta-coordinate with an approximately square pyramidal geometry around the iron atom and with the halide ligand at the apex, although the full symmetry is no higher than C_{2v} (Fig. 8.6). It is, however, a geometry which favours stabilisation

[Refs. on p. 219]

of the intermediate $S = \frac{3}{2}$ spin configuration. Four complexes have been studied in detail by e.s.r., Mössbauer, and magnetic susceptibility methods

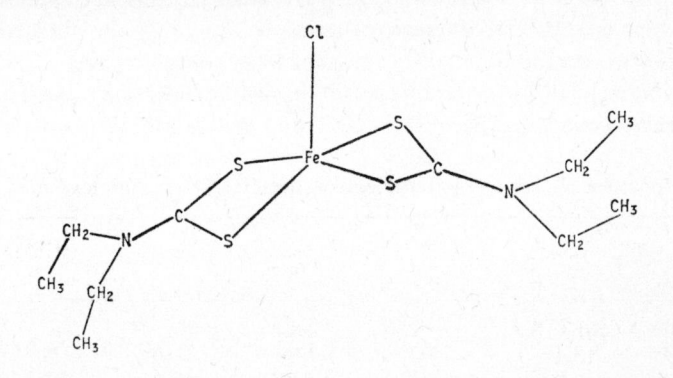

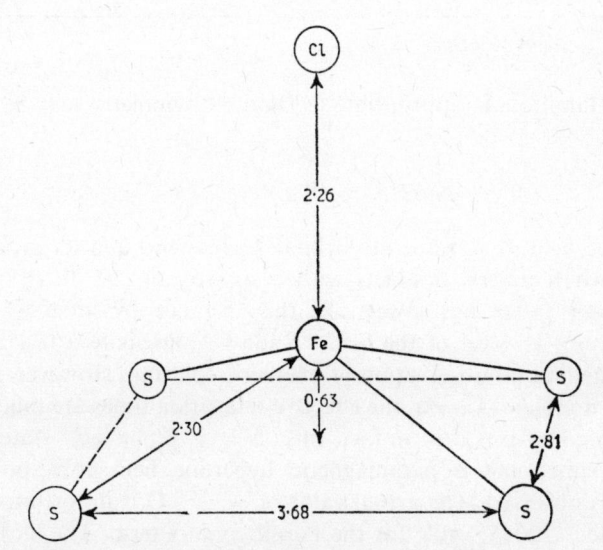

Fig. 8.6. Structural formula and local iron symmetry of [Et$_2$NCS$_2$]$_2$FeCl. [Ref. 26, Fig. 1]

[26–28], and in addition to possessing the unusual spin state, they have also been found to show unexpectedly complex magnetic properties.

Typical spectra at 1·2 K are shown in Fig. 8.7, and values are in Table 8.4. Prominent features of the Mössbauer spectra are as follows:

(a) [Et$_2$NCS$_2$]$_2$FeBr—a simple quadrupole splitting between 1·2 K and 300 K with no appreciable temperature dependence.
(b) [Pr$^i{}_2$NCS$_2$]$_2$FeCl—a sharp paramagnetic hyperfine splitting is shown at

1·2 K with the major axis of the electric field gradient tensor perpendicular to the spin axis.
(c) $[Et_2NCS_2]_2FeCl$—a spectrum which is virtually identical to spectrum (b) but arising from ferromagnetic ordering below 2·5 K; the distinction was established on the basis of d.c. susceptibility measurements.
(d) $[Me_2NCS_2]_2FeCl$—a paramagnetic hyperfine splitting which is considerably broadened because of electronic relaxation.

Table 8.4 Mössbauer parameters for iron(III) $S = \frac{3}{2}$ complexes [26]

Compound	T/K	$\frac{1}{2}e^2qQ$ /(mm s^{-1})	η	H/kG
$[Et_2NCS_2]_2FeBr$	1·2	2·88*	—	—
$[Pr^i{}_2NCS_2]_2FeCl$	1·2	2·68	0·16	334
$[Et_2NCS_2]_2FeCl$	1·2	2·68	0·15	333
$[Me_2NCS_2]_2FeCl$	1·2	2·66	0·15	338

* Actually $\frac{1}{2}e^2qQ(1 + \frac{1}{3}\eta^2)^{\frac{1}{2}}$

The spin Hamiltonian appropriate to rhombic symmetry may be approximated by

$$\mathcal{H} = g\mu_B H \cdot S + D\{S_z^2 - \tfrac{1}{3}S(S+1)\}$$

and when the field H is zero, the orbital singlet-spin quartet ground term splits into two Kramers' doublets with a spacing of $2D$. If D is positive the $|M_s = \pm\tfrac{1}{2}\rangle$ level lies lowest. In this instance by analogy with the $|S = \tfrac{5}{2}, M_s = \pm\tfrac{1}{2}\rangle$ level of the Fe^{3+} ^6S ion we anticipate a fast electronic relaxation rate and no observation of magnetic splitting. However, when the $|M_s = \pm\tfrac{3}{2}\rangle$ doublet is lowest, the effective relaxation times are much longer. In the chloro-complexes it is, in fact, this $|S = \tfrac{3}{2}, M_s = \pm\tfrac{3}{2}\rangle$ state which is the ground state, and a paramagnetic hyperfine field corresponding to $\langle S_z\rangle = \pm\tfrac{3}{2}$ is observed. The actual value of $\sim$335 kG is in fortuitous agreement with the $-220\langle S_z\rangle$ rule for the Fermi contact term. The lack of magnetic splitting in the bromo-complex may indicate that the $|M_s = \pm\tfrac{1}{2}\rangle$ level lies lowest in this case, and this has been confirmed by e.s.r. data.

The ferromagnetic ordering in $[Et_2NCS_2]_2FeCl$ merely removes the degeneracy of the $|M_s = \pm\tfrac{3}{2}\rangle$ level, but since the $|M_s = +\tfrac{3}{2}\rangle$ and $|M_s = -\tfrac{3}{2}\rangle$ states give identical spectra, there is no apparent effect on the Mössbauer spectrum as long as the relaxation times between the two states is long. The subtle differences in magnetic behaviour pose the interesting question as to how the magnetic order is achieved, and why it is only found in the one case.

Although these initial experiments did not show any distinction between

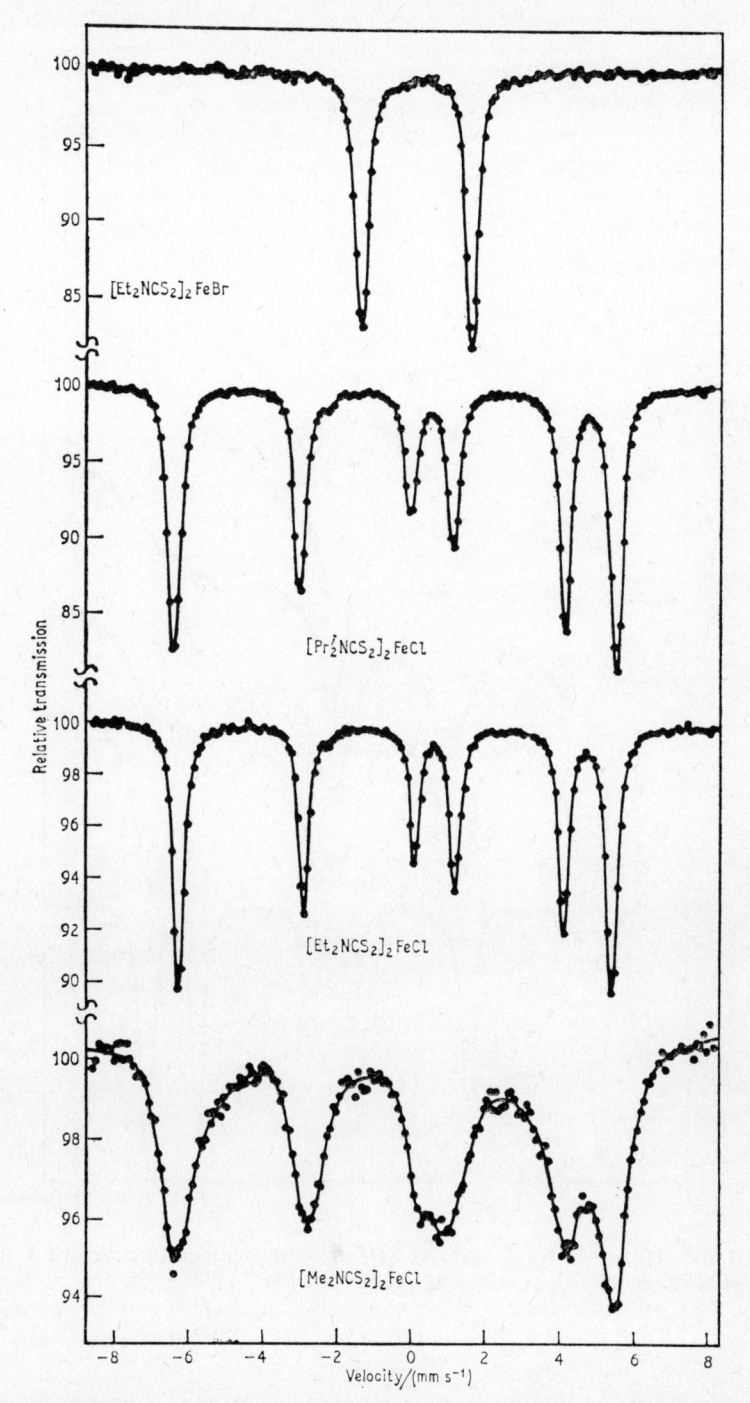

Fig. 8.7 Mössbauer spectra at 1·2 K in polycrystalline absorbers of four [R₂NCS₂]FeX compounds. [Ref. 26, Fig. 2]

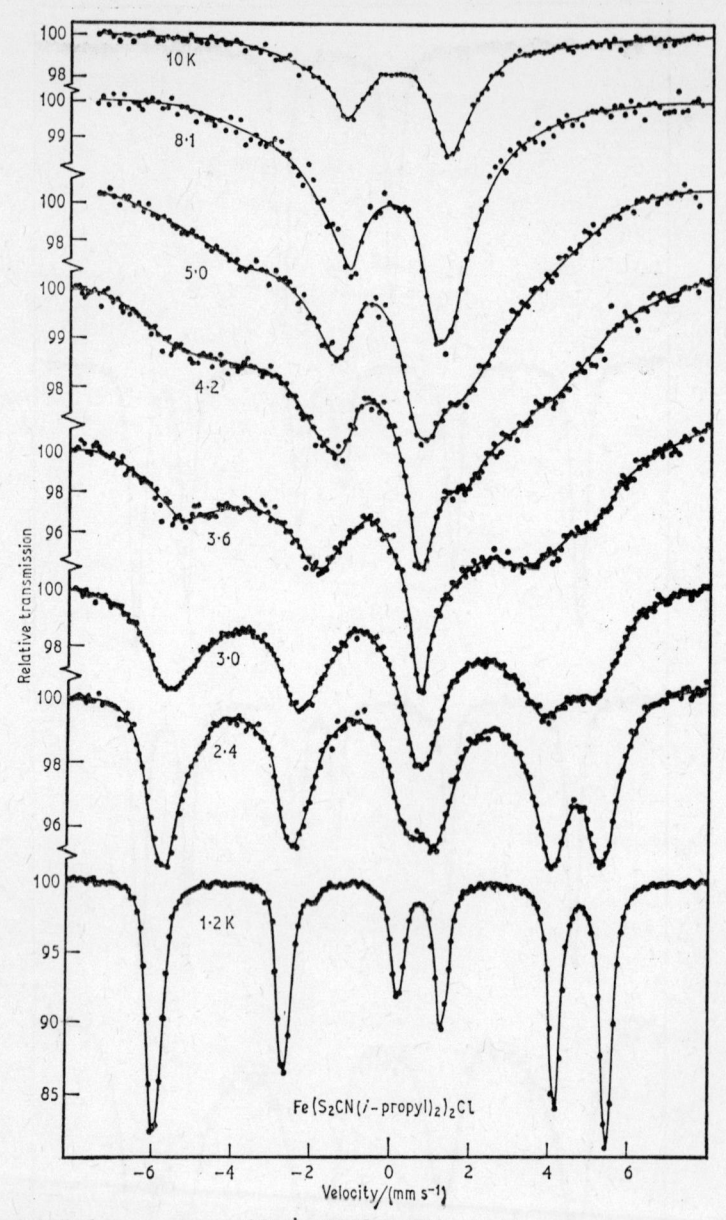

Fig. 8.8 Mössbauer spectra of [Pr$^i_2$ NCS]$_2$FeCl in the temperature range 1·2–10 K together with curves from theoretical prediction. [Ref. 17, Fig. 3]

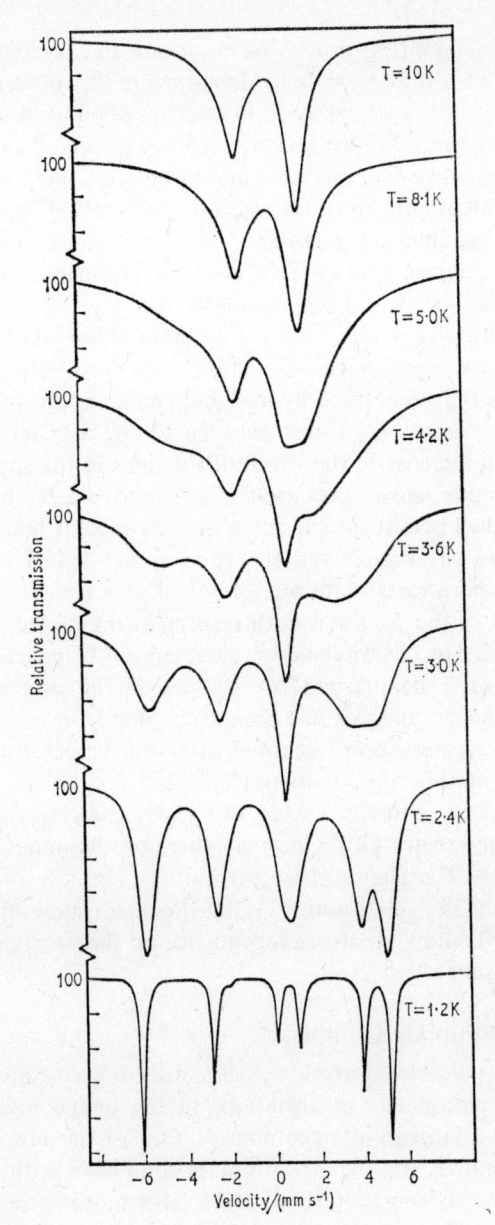

Fig. 8.8 continued

the paramagnetic splitting in [Pr$^i{}_2$NCS$_2$]$_2$FeCl and ferromagnetic ordering in [Et$_2$NCS$_2$]$_2$FeCl, it is possible to demonstrate this difference by applying an external magnetic field [29]. A 30-kG field applied to the latter compound at 1·6 K reduces the observed internal field by 22 kG, establishing that the sign of H_{eff} is negative (i.e. the spins are partially aligning antiparallel to the applied field; full alignment of the spins to give a 30-kG reduction does not occur because the internal magnetic field is not saturated even at 1·6 K). The temperature dependence of the field is also consistent with ferromagnetic ordering, and the spectrum lines remain sharp.

In the [Pr$^i{}_2$NCS$_2$]$_2$FeCl derivative a different behaviour is found. A 5-kG field at 1·3 K is sufficient to split the $|M_s = \pm\frac{3}{2}\rangle$ and $|M_s = \pm\frac{1}{2}\rangle$ Kramers' levels which are only separated by 4 K and causes appreciable state mixing. This allows additional transitions between levels, a faster relaxation time, and an observed increase in the width of the lines in the spectrum.

Both compounds show relaxation effects above 2 K, but in the paramagnetic case they persist over a much larger range of temperature. A very detailed analysis has been given [17], and some typical computed curves together with experimental data are given in Fig. 8.8.

The direction of the spin axis with respect to the molecular axes has not been determined. The electric field gradient cannot be generated by the non-bonding electrons in the quartet term, and presumably originates from lattice contributions, but no analysis has been attempted.

Spectra have recently been recorded at room temperature for a total of 28 compounds of this type, namely [R$_2$NCS$_2$]$_2$FeX where X = Cl, Br, I, NCS, and C$_6$F$_5$CO$_2$, and R$_2$ = Me$_2$, Et$_2$, C$_6$H$_{12}$, (C$_6$H$_{11}$)$_2$, C$_8$H$_{14}$, etc. All give very similar spectra [30]. One may therefore assume that all have the $S = \frac{3}{2}$ spin state. The chemical isomer shift is virtually independent of R and X {δ(Fe) = 0·38 $\pm$ 0·02 mm s^{-1}}, but the quadrupole splitting, although independent of R, shows a strong dependence on the nature of X and ranges from 2·4 to 3·0 mm s^{-1}.

8.5 Iron 1,2-Dithiolate Complexes

It is not always possible to make a systematic comparison of the effects on the Mössbauer parameters of variations in the ligand and of the overall charge state for a large series of complexes. One of the few cases where this has proved possible is for the 1,2-dithiolate complexes with ligands such as a substituted *cis*-ethylene-1,2-dithiolate (I), tetrachlorobenzene-1,2-dithiolate

(II), or toluene-3,4-dithiolate (III) [31, 32]. Parameters for 32 such complexes are listed in Table 8.5.

The anion $[Fe\{S_2C_2(CN)_2\}_2]^-$ has been shown by X-ray analysis to be dimeric in the solid state and to have the structure IV.

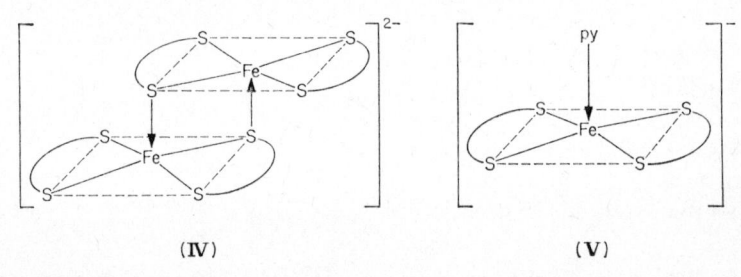

(IV) (V)

Compounds (1–5) are all very similar with one unpaired electron per iron atom, large quadrupole splittings, and a chemical isomer shift compatible with a low-spin Fe(III) $S = \frac{1}{2}$ configuration. All may therefore be presumed to have the same dimeric structure IV. The quadrupole splittings range from 2·37 to 3·02 mm s^{-1} at 77 K and are some of the largest known for this electronic configuration of iron.

Compounds (6–8) contain the basic structural unit $[Fe(py)\{-S_2\}_2]^-$; they have three unpaired electrons and are probably penta-coordinate Fe(III) $S = \frac{3}{2}$ compounds (structure V) similar to the bis-(N,N'-dithiocarbamato)-iron(III) halides discussed in the preceding section. Both series show an approximately systematic variation in Δ with change in the ligand which is not matched by a corresponding variation in the chemical isomer shift, so that it seems unlikely that large changes in delocalisation are occurring. The very small temperature dependence of Δ in the $S = \frac{1}{2}$ complexes makes it difficult to determine the electronic level separations.

Compounds (9) and (10) are believed to be similar to the pyridine adducts, i.e. penta-coordinate with $S = \frac{3}{2}$. Compounds (11–13) contain potentially bidentate ligands, and the smaller quadrupole splittings recorded may well be indicative of 6-coordination as in structure VI for (12) and (13) with a chain-polymeric structure involving both nitrogen atoms of the ligand for (II).

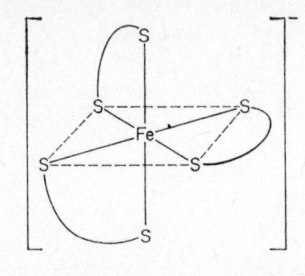

(VI)

Table 8.5 Mössbauer parameters of 1,2-dithiolate complexes [32]

Compound	T/K	Δ /(mm s^{-1})	δ (Fe)† /(mm s^{-1})
(1) [Et$_4$N]$_2$[Fe{S$_2$C$_2$Ph$_2$}$_2$]$_2$	295	2·45	0·27
	77	2·37	0·35
(2) [Et$_4$N]$_2$[Fe{S$_2$C$_2$(CF$_3$)$_2$}$_2$]$_2$	295	2·50	0·23
	77	2·50	0·33
(3) [Et$_4$N]$_2$[Fe{S$_2$C$_2$(CN)$_2$}$_2$]$_2$	295	2·81	0·24
	77	2·76	0·33
(4) [Bu$^n_4$N]$_2$[Fe{S$_2$C$_6$H$_3$Me}$_2$]$_2$	295	2·99	0·28
	77	2·95	0·34
(5) [Bu$^n_4$N]$_2$[Fe{S$_2$C$_6$Cl$_4$}$_2$]$_2$	295	3·03	0·23
	77	3·02	0·32
(6) [Ph$_4$P][Fe(py){S$_2$C$_2$(CN)$_2$}$_2$]	295	2·51	0·27
	77	2·41	0·33
(7) [Et$_4$N][Fe(py){S$_2$C$_2$(CF$_3$)$_2$}$_2$]	295	2·54	0·28
	77	2·61	0·33
(8) [Et$_4$N][Fe(py){S$_2$C$_6$Cl$_4$}$_2$]	295	3·12	0·23
	77	3·02	0·33
(9) [Bu$^n_4$][Fe(γ-pic){S$_2$C$_2$(CN)$_2$}$_2$]	295	2·61	0·26
	77	2·59	0·36
(10) [Bu$^n_4$N][Fe(i-quin){S$_2$C$_2$(CN)$_2$}$_2$]	295	2·44	0·29
	77	2·46	0·36
(11) [Bu$^n_4$N][Fe(4-NH$_2$py){S$_2$C$_2$(CN)$_2$}$_2$]	295	2·20	0·27
	77	2·18	0·37
(12) [Bu$^n_4$N][Fe(phen){S$_2$C$_2$(CN)$_2$}$_2$]	295	1·76	0·25
	77	1·80	0·31
(13) [Bu$^n_4$N][Fe(bipy){S$_2$C$_2$(CN)$_2$}$_2$]	295	1·75	0·24
	77	1·85	0·32*
(14) [Ph$_4$P][Fe(NO){S$_2$C$_2$(CN)$_2$}$_2$]	295	1·69	−0·06
	77	1·68	0·05
(15) [Et$_4$N][Fe(NO){S$_2$C$_2$Ph$_2$}$_2$]	295	1·99	−0·06
	77	1·98	0·01
(16) [Et$_4$N][Fe(NO){S$_2$C$_2$(CF$_3$)$_2$}$_2$]	295	2·01	−0·10
	77	2·11	0·02
(17) [Bu$^n_4$][Fe(NO){S$_2$C$_6$H$_3$Me}$_2$]	295	2·23	−0·04
	77	2·26	0·04

continued

[*Refs.* on p. 219]

Table 8.5 continued

Compound	T/K	Δ /(mm s^{-1})	δ (Fe)† /(mm s^{-1})
(18) [Bu$^n{}_4$][Fe(NO){S$_2$C$_6$Cl$_4$}$_2$]	77	(2·41)	(0·36)
(19) [Ph$_4$P]$_2$[Fe(NO){S$_2$C$_2$(CN)$_2$}$_2$]	77	0·97	0·21*
(20) [Ph$_4$P]$_2$[Fe(NO){S$_2$C$_6$H$_3$Me}$_2$]	77		*
(21) [Bu$^n{}_4$N]$_2$[Fe(NO){S$_2$C$_6$Cl$_4$}$_2$]	77		*
(22) [Bu$^n{}_4$N]$_2$[Fe(CNO){S$_2$C$_2$(CN)$_2$}$_2$]	77		*
(23) [Bu$^n{}_4$N]$_2$[Fe(CN){S$_2$C$_2$(CN)$_2$}$_2$]	77	2·06	0·22*
(24) [Ph$_4$P]$_2$[Fe{S$_2$C$_2$(CN)$_2$}$_3$]	295	1·59	0·16
	77	1·57	0·24
(25) [Ph$_4$P]$_3$[Fe{S$_2$C$_2$(CN)$_2$}$_3$]	190	1·67	0·33*
	77	1·69	0·39*
(26) [Fe{S$_2$C$_2$Ph$_2$}$_2$]$_2$	295	2·05	0·16
	77	2·01	0·25
(27) [Fe{S$_2$C$_2$(CF$_3$)$_2$}$_2$]$_2$	77	2·39	0·25
(28) [Fe{S$_2$C$_6$Cl$_4$}$_2$]$_2$	295	3·15	0·23
	77	3·15	0·32
(29) [Fe$_2$(NO)$_2${S$_2$C$_2$Ph$_2$}$_3$]CHCl$_3$	295	1·58	−0·08
		1·27	0·21
	77	1·57	−0·01
		1·26	0·29
(30) [Et$_4$N][Fe(NO){S$_2$C$_2$Ph$_2$}$_2$]$_n$	295	1·04	0·15
	77	0·98	0·20
(31) [Fe(NO){S$_2$C$_2$Ph$_2$}$_2$]	295	1·65	−0·02
	77	1·65	0·06
(32) [Fe(Ph$_3$P){S$_2$C$_2$Ph$_2$}$_2$]	295	2·75	0·03
	77	2·78	0·12

* Broadened by relaxation processes.
† Converted from sodium nitroprusside by subtraction of 0·26 mm s^{-1}.

The monoanionic nitrosyl complexes (14–18) [Fe(NO){—S$_2$}$_2$]$^-$ are probably diamagnetic iron(II) complexes with structure V, the considerably lower chemical isomer shift being produced by the back donation of electrons to the nitrosyl group. The [Fe(NO){—S$_2$}$_2$]$^{2-}$ complexes (19–23) feature an additional electron which is believed to be in a π^* NO orbital. This decreases the d_π–p_π back donation by the iron atom to these orbitals, effectively increases the 3d-electron density and increases the chemical isomer shift.

The tris-complexes [Fe{—S$_2$}$_3$]$^{2-}$ and [Fe{—S$_2$}$_3$]$^{3-}$ (24) and (25) contain two and one unpaired electrons respectively with nominal oxidation states

[*Refs. on p. 219*]

of +4 and +3. The increase in chemical isomer shift from 0·24 to 0·39 mm s^{-1} at 77 K is consistent with addition of an electron to an orbital of largely metallic character, rather than to a ligand orbital. Suggested configurations for these complexes are

$$a_1'(d_{z^2})^2 e'(d_{xy}, d_{x^2-y^2})^2 \quad \text{and} \quad a_1'(d_{z^2})^2 e'(d_{xy}, d_{x^2-y^2})^3$$

which are not easy to reconcile with the observed quadrupole splittings. Several of the above-mentioned compounds show considerable asymmetry in the spectrum due to spin relaxation.

Complexes (26–28) are probably dimers because they show similar spectra to (1–5). The chemical isomer shifts of compounds (28) and (5), i.e. [Fe{S$_2$C$_6$Cl$_4$}$_2$]$_2$ and [Fe{S$_2$C$_6$Cl$_4$}$_2$]$_2$$^{2-}$, are identical, suggesting that the added electrons are delocalised largely to the ligands. By contrast, the pairs of compounds (26, 1) and (27, 2) show increases in chemical isomer shift on reduction which are similar to that observed for the pair (24, 25) discussed in the preceding paragraph, implying that in these instances electron delocalisation onto ligand-based orbitals is not nearly so extensive as in the pair (28) and (5).

The Mössbauer spectrum of compound (29) shows it to contain two iron environments, and comparison of the parameters with the other data suggests that the dimeric unit contains one 5-coordinated and one 6-coordinated iron atom. The neutral monomers (31) and (32) are equivalent to the penta-coordinated monoanions with an electron removed, possibly from a ligand-based orbital. A ligand-field treatment for the temperature dependence of the quadrupole splitting in some of these complexes has been suggested [33].

8.6 Systems Containing Iron(I), Iron(IV), and Iron(VI)

High-spin Iron(I)

The ionic species Fe$^+$ is not found in stable compounds, but its existence has been shown as a substitutional impurity in ionic lattices. Experiments [34] involving the doping of ^{57}CoCl$_2$ into KCl at 550°C were found to give predominantly a resonance having a room-temperature chemical isomer shift of $\sim$+0·4 mm s^{-1} (relative to iron metal). Quenched samples also showed a weak absorption at +2·09 mm s^{-1}, although no explanation was offered. Subsequent work [35] using NaCl as the source matrix also showed that these two resonances could occur in quenched samples, the chemical isomer shifts being $\sim$+0·4 mm s^{-1} and +1·98 mm s^{-1} at 80 K. The former resonance may be attributed to an iron atom in the Fe^{2+} state, and the small quadrupole splitting found with this resonance is ascribed to the influence of the associated positive ion vacancy. The line at higher velocity is due to an isolated Fe$^+$ ion substitutionally replacing a sodium ion at a site of cubic symmetry. The large chemical isomer shift is compatible with a $3d^7$ configuration.

^{57}Co doped into MgO and CaO produces Fe$^+$ and Fe^{2+} impurities [36].

[Refs. on p. 219]

The Fe$^+$ ground term is $3d^7$ (^4F, $L = 3$, $S = \frac{3}{2}$) which splits in a cubic crystalline field into two orbitally degenerate triplets and an orbital singlet with one of the orbital triplets lying lowest. The spectrum of ^{57}Co/MgO at 4·2 K in external magnetic fields of various magnitudes allows the intrinsic magnetic field at the Fe$^+$ nucleus to be estimated as $+20$ kG. The values of H_L and H_D for cubic symmetry are calculated to be $+231$ and 0 kG respectively, giving the Fermi contact term as $H = -211$ kG. Similar experiments in CaO gave an intrinsic field of $+28$ kG.

Iron(IV)

Very few molecular complexes containing the oxidation state iron(IV) are known, the most familiar being with the ligand *o*-phenylene-bis-(dimethylarsine). Two of these complexes have been examined and show a large temperature-independent quadrupole splitting [37].

	T/K	Δ /(mm s^{-1})	δ (Fe) /(mm s^{-1})
trans-[Fe(das)$_2$Br$_2$][BF$_4$]	298	3·16	0·17
	78	3·19	0·25
	4·2	3·25	0·27
trans-[Fe(das)$_2$Cl$_2$][BF$_4$]	298	3·07	0·12
	78	3·30	0·20
	4·2	3·22	0·20

The sign of e^2qQ was found to be positive in the chloride complex by using the magnetic field perturbation method, and this confirms the ground state of the low-spin $3d^4$ configuration to be a singlet $|d_{xy}\rangle$ state. The comparatively large value for Δ tends to suggest that there is little tendency for back π-donation from iron to the four planar arsenic atoms. The temperature independence of Δ signifies a large crystal field splitting of the lower electronic levels.

Iron(VI)

The iron(VI) compound, K$_2$FeO$_4$, gives a single-line resonance at -0.78 mm s^{-1} at 78 K (-0.88 mm s^{-1} at 298 K) [38]. As might be expected for a tetrahedrally coordinated nominal $3d^2$ configuration, there is no significant quadrupole splitting. Although various electronic configurations have been postulated on the basis of the chemical isomer shift value, they are very dependent on the calibration of the shift scale adopted, and it is therefore difficult to assess the role of covalent bonding with the ligands. Nevertheless, it remains true that compound K$_2$FeO$_4$ has the lowest chemical isomer shift of any iron compound yet studied. The chemical isomer shift increases in the order Fe^{6+} < Fe^{3+} < Fe^{2+} < Fe$^+$ for the high-spin configurations.

Magnetic susceptibility data suggest the onset of the antiferromagnetic ordering below 5 K, and this has been confirmed by observation of a 142-kG field at 1·6 K in K_2FeO_4 [39]. $BaFeO_4$ gives a field of 115 kG at 4 K [40], but $SrFeO_4$ is still paramagnetic at 2 K. The quadrupole splitting is still zero in the magnetically ordered state.

Low-spin Iron(I) with S = ½

Another unusual group of iron compounds are the bis-(N,N-dialkyldithiocarbamato)iron nitrosyl complexes. Several brief investigations have been made [41, 42], but the definitive study is of one compound only, the diethyl derivative [43].

$[Et_2NCS_2]_2FeNO$ may be considered formally as a low-spin iron(I) (d^7) complex, and it has a single unpaired electron. The Mössbauer spectrum in the temperature range 1·3–300 K is a simple quadrupole doublet, the splitting being virtually independent of temperature as shown in Table 8.6. A 15-kG applied field at 4·2 K gives magnetic splitting indicating e^2qQ to be positive, but because of the low symmetry of the molecule, this cannot be used to infer the electronic ground state directly.

Table 8.6 Mössbauer data for $[Et_2NCS_2]_2FeNO$ from Ref. 43

T/K	Δ /(mm s⁻¹)	δ (Fe) /(mm s⁻¹)
300	0·89	0·28
77	0·87	0·34
4·2	0·89	0·35
1·8	1·03	0·36
1·3	1·03	0·38

Application of large external fields H induces strong magnetic splitting (Fig. 8.9), and the effective field H_{eff} may be fitted by the equation

$$H_{eff} = H_n B_s\left(\frac{\mu H}{kT}\right) - H$$

where H_n is the saturation value of the field specified by the Brillouin function $B_s(\mu H/kT)$ for a spin S. Agreement with experiment is only obtained for $S = \frac{1}{2}$ and $H_n = 110$ kG. This establishes that the low-temperature susceptibility is that for a free spin of $S = \frac{1}{2}$, a fact not previously established by magnetic susceptibility measurements. The value of H_n approximates to the $-220\langle S_z \rangle$ rule. Computer calculations of the predicted spectrum in external fields with different assumed electronic ground states were in good agreement with a d_{z^2} configuration but not with $d_{x^2-y^2}$. A d_{z^2} electron would give a negative value for e^2qQ, but one must conclude in this instance that the lattice contributions to the electric field gradient are dominant.

Spectra have also been given for the reduction products of sodium nitro-

[Refs. on p. 219]

prusside, [Fe(CN)$_5$NO]$^{3-}$ and [Fe(CN)$_5$NOH]$^{2-}$ [44]. [Fe(CN)$_5$NO]$^{3-}$ appears to have a configuration close to $(d_{xz}, d_{yz})^4 (d_{xy})^2 d_{\pi^*NO}$. The extra electron resides entirely on the NO as measurements in applied magnetic

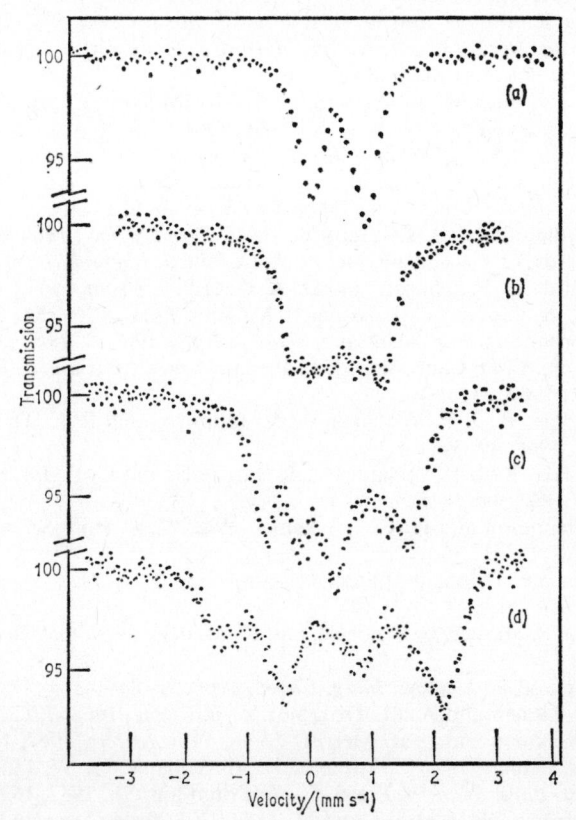

Fig. 8.9 Mössbauer spectra of [Et$_2$NCS$_2$]$_2$FeNO at (a) 4·2 K, (b) 4·2 K in 15-kG field, (c) 4·2 K in 30-kG field, (d) 1·8 K in 30-kG field. [Ref. 43, Fig. 1]

fields show no sign of unpaired spin density on the iron. The gross rearrangement of charge reverses the sign of e^2qQ from positive in nitroprusside to negative in [Fe(CN)$_5$NO]$^{3-}$. [Fe(CN)$_5$NOH]$^{2-}$ gives very complex effects in applied fields which are fully consistent with a $(d_{xz}, d_{yz})^4 (d_{xy})^2 d_{z^2}$ configuration.

REFERENCES

[1] R. L. Martin and A. H. White, 'Transition Metal Chemistry' (Ed. R. L. Carlin), Vol. 4, p. 113, Marcel Dekker Inc., New York.
[2] E. König and K. Madeja, *Chem. Comm.*, 1966, 61.
[3] E. König and K. Madeja, *Inorg. Chem.*, 1967, **6**, 48.

[4] I. Dezsi, B. Molnar, T. Tarnoczi, and K. Tompa, *J. Inorg. Nuclear Chem.*, 1967, **29**, 2486.
[5] G. A. Renovitch and W. A. Baker, Jr, *J. Amer. Chem. Soc.*, 1967, **89**, 6377.
[6] E. König, K. Madeja, and K. J. Watson, *J. Amer. Chem. Soc.*, 1968, **90**, 1146.
[7] E. König, K. Madeja, and W. H. Böhmer, *J. Amer. Chem. Soc.*, 1969, **91**, 4582.
[8] R. J. Dosser, W. J. Eilbeck, A. E. Underhill, P. R. Edwards, and C. E. Johnson, *J. Chem. Soc.* (*A*), 1969, 810.
[9] D. M. L. Goodgame and A. A. S. C. Machado, *Inorg. Chem.*, 1969, **8**, 2031.
[10] D. M. L. Goodgame and A. A. S. C. Machado, *Chem. Comm.*, 1969, 1420.
[11] J. P. Jesson and J. F. Weiher, *J. Chem. Phys.*, 1967, **46**, 1995.
[12] J. P. Jesson, J. F. Weiher, and S. Trofimenko, *J. Chem. Phys.*, 1968, **48**, 2058.
[13] E. Frank and C. R. Abeledo, *Inorg. Chem.*, 1966, **5**, 1453.
[14] R. M. Golding and H. J. Whitfield, *Trans. Faraday Soc.*, 1966, **62**, 1713.
[15] R. Rickards, C. E. Johnson, and H. A. O. Hill, *J. Chem. Phys.*, 1968, **48**, 5231.
[16] R. Rickards, C. E. Johnson, and H. A. O. Hill, *J. Chem. Phys.*, 1969, **51**, 846.
[17] H. H. Wickman and C. F. Wagner, *J. Chem. Phys.*, 1969, **51**, 435.
[18] L. M. Epstein and D. K. Straub, *Inorg. Chem.*, 1969, **8**, 784.
[19] E. Cervone, F. D. Camassei, M. L. Luciani, and C. Furlani, *J. Inorg. Nuclear Chem.*, 1969, **31**, 1101.
[20] B. W. Dale, R. J. P. Williams, C. E. Johnson, and T. L. Thorp, *J. Chem. Phys.*, 1968, **49**, 3441.
[21] B. W. Dale, R. J. P. Williams, P. R. Edwards, and C. E. Johnson, *J. Chem. Phys.*, 1968, **49**, 3445.
[22] A. R. Champion and H. G. Drickamer, *Proc. Natl. Acad. Sci., U.S.A.*, 1967, **58**, 876.
[23] R. Taube, H. Drevs, E. Fluck, P. Kuhn, and F. Brauch, *Z. anorg. Chem.*, 1969, **364**, 297.
[24] E. König, S. Hufner, E. Steichele, and K. Madeja, *Z. Naturforsch.*, 1967, **22A**, 1543.
[25] E. König and K. Madeja, *Inorg. Chem.*, 1968, **7**, 1848.
[26] H. H. Wickman and A. M. Trozzolo, *Inorg. Chem.*, 1968, **7**, 63.
[27] H. H. Wickman and F. R. Merritt, *Chem. Phys. Letters*, 1967, **1**, 117.
[28] H. H. Wickman, A. M. Trozzolo, H. J. Williams, G. W. Hull, and F. R. Merritt, *Phys. Rev.*, 1967, **155**, 563 (erratum loc. cit., 1967, **163**, 526).
[29] R. Rickards, C. E. Johnson, and H. A. O. Hill, *Trans. Faraday Soc.*, 1969, **65**, 2847.
[30] L. M. Epstein and D. K. Straub, *Inorg. Chem.*, 1969, **8**, 560.
[31] T. Birchall, N. N. Greenwood, and J. A. McCleverty, *Nature*, 1967, **215**, 625.
[32] T. Birchall and N. N. Greenwood, *J. Chem. Soc.* (*A*), 1969, 286.
[33] R. M. Golding, F. Jackson, and E. Sinn, *Theor. Chim. Acta*, 1969, **15**, 123.
[34] M. de Coster and S. Amelinckz, *Phys. Letters*, 1962, **1**, 245.
[35] J. G. Mullen, *Phys. Rev.*, 1963, **131**, 1415.
[36] J. Chappert, R. B. Frankel, and N. A. Blum, *Phys. Letters*, 1967, **25A**, 149.
[37] E. A. Paez, W. T. Oosterhuis, and D. L. Weaver, *Chem. Comm.*, 1970, 506.
[38] G. K. Wertheim and R. H. Herber, *J. Chem. Phys.*, 1962, **36**, 2497.
[39] A. Ito and K. Ono, *J. Phys. Soc. Japan*, 1969, **26**, 1548.
[40] T. Shinjo, T. Ichida, and T. Takada, *J. Phys. Soc. Japan*, 1969, **26**, 1547.
[41] J. Danon, *J. Chem. Phys.*, 1964, **41**, 3378.
[42] E. Frank and C. R. Abeledo, *J. Inorg. Nuclear Chem.*, 1969, **31**, 989.
[43] C. E. Johnson, R. Rickards, and H. A. O. Hill, *J. Chem. Phys.*, 1969, **50**, 2594.
[44] W. T. Oosterhuis and G. Lang, *J. Chem. Phys.*, 1969, **50**, 4381.

9 | Covalent Iron Compounds

There is a large class of iron organometallic compounds which can be described as molecular, diamagnetic, and highly covalent. They are derived in the main from the parent compounds iron pentacarbonyl, $Fe(CO)_5$, and the cyclopentadienyl complex ferrocene, $(\pi\text{-}C_5H_5)_2Fe$ or Cp_2Fe. In view of their large numbers, it is not surprising that considerable Mössbauer spectral data are available, but as we shall now see the interpretation of this data introduces considerable problems.

The very fact that there is no effective electronic spin moment ensures that internal magnetic hyperfine fields are not observed. Neither can one expect to find close-lying excited electronic levels which will give a temperature-dependent quadrupole splitting by thermal excitation, as found for example in high-spin iron(II). As a result, the amount of significant information which can be obtained from a given Mössbauer spectrum is proportionately low. Furthermore, the difficulties of estimating electric field gradients and s-electron densities are much greater in environments where there is very marked ligand participation in the bonding. It is, therefore, generally not possible to deduce as much from the spectrum of a single compound as one sometimes can for the paramagnetic iron complexes.

Although the prospects appear discouraging at first sight, it has been convincingly shown that significant information can be deduced from the spectra of large series of closely related compounds, particularly as regards their stereochemistry and the changing trends in σ- and π-bonding from the ligands. It is also possible to use the Mössbauer spectrum as a diagnostic test of sample integrity during the preparation of new complexes. Thus, one can often show the presence of impurity, and perhaps even establish its identity, very quickly by this method. The two primary Mössbauer parameters, the chemical isomer shift and quadrupole splitting, may also give an indication of the type of stereochemistry and bonding occurring in the new complex. The chemical isomer shifts actually recorded range from -0.18 mm s^{-1} to 0.54 mm s^{-1} (relative to iron metal), with quadrupole splittings from zero up to 2.8 mm s^{-1}. 'Although an early attempt was made [1] to establish the concept of a scale of 'partial chemical isomer shifts' for different ligands, it is not sufficiently successful for it to be generally applicable. This is partly a result

[Refs. on p. 237]

222 | COVALENT IRON COMPOUNDS

of the wide range of stereochemistries encountered. The concept is, however, occasionally useful within a limited series of compounds.

9.1 Binary Carbonyls, Carbonyl Anions, and Hydride Anions
Fe(CO)₅, Fe₂(CO)₉, and Fe₃(CO)₁₂

The parent carbonyl Fe(CO)₅ has a trigonal bipyramidal configuration (Fig. 9.1, structure 1). A large electric field gradient and a substantial quadrupole splitting can be anticipated as a result of the non-cubic iron site symmetry, and these are indeed found (see Table 9.1) [2–6]. At liquid nitrogen

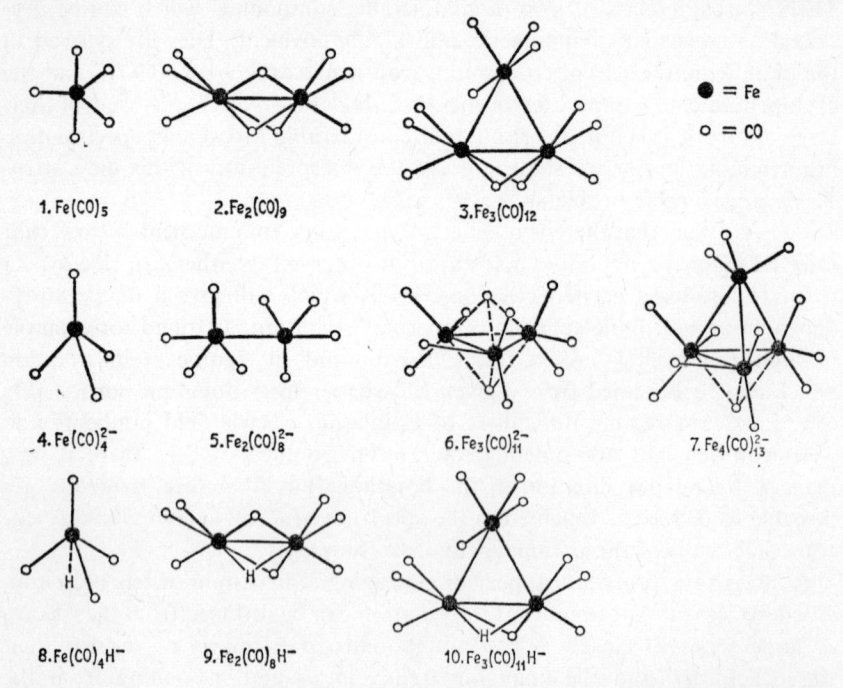

Fig. 9.1 Structures of iron carbonyls, carbonyl anions, and carbonyl hydride anions. [Ref. 5, Fig. 1]

temperature the quadrupole splitting is 2·57 mm s⁻¹, and is virtually independent of temperature. This latter feature is common to all the complexes in this chapter, and the tabulated data are, therefore, given for one temperature only, usually 80 K. Fe(CO)₅ is a liquid at room temperature, and freezing can cause partial orientation in the solid matrix with consequent asymmetry in the quadrupole doublet [2]. The Mössbauer parameters of frozen solutions of Fe(CO)₅ in tetrachlorethane are essentially the same, showing that the principal contributions derive solely from the bonding of the molecular unit rather than from intermolecular or lattice interactions.

[*Refs. on p. 237*]

BINARY CARBONYLS, CARBONYL ANIONS, HYDRIDE ANIONS | 223

Fe$_2$(CO)$_9$ has axial symmetry with three bridging (CO) groups and a presumed Fe–Fe bond (Fig. 9.1, structure 2). Early spectra showed a small quadrupole splitting with considerable asymmetry in intensity [2, 3, 7].

Table 9.1 Mössbauer parameters [5] for iron carbonyls, carbonyl anions, and carbonyl hydride anions

Compound	Δ^* /(mm s^{-1})	δ (Fe)* /(mm s^{-1})
1 Fe(CO)$_5$	2·57	−0·09
2 Fe$_2$(CO)$_9$	0·42	0·16
3 Fe$_3$(CO)$_{12}$	1·13	0·11 (2 atoms)
	0·13	0·05 (1 atom)
4 Na$_2$[Fe(CO)$_4$]	0·00	−0·18
5 [Et$_4$N]$_2$[Fe$_2$(CO)$_8$]	2·22	−0·08
6 [Fe(en)$_3$][Fe$_3$(CO)$_{11}$]	2·11	−0·10 (anion)
	1·19	1·12 (cation)
7 [Et$_4$N]$_2$[Fe$_4$(CO)$_{13}$]	0·27	0·02
8 [Et$_4$N][Fe(CO)$_4$H]	1·36	−0·17
9 [Et$_4$N][Fe$_2$(CO)$_8$H]	0·50	0·07
10 [Et$_4$N][Fe$_3$(CO)$_{11}$H]	1·41	0·04 (2 atoms)
	0·16	0·02 (1 atom)
11 [pyH][Fe$_4$(CO)$_{13}$H]	∼0·67	∼0·07

* All data at 80 K and calibrated from sodium nitroprusside at room temperature. Converted to iron metal by subtracting 0·26 mm s^{-1} from δ.

Careful reinvestigation [8] showed that this was due to a strong tendency towards preferential orientation of the sample. The exact intensity ratio is then a function of absorber inclination as illustrated in Fig. 9.2. Non-oriented absorbers could be prepared to give equal peak intensities. Optical polarisation measurements showed that the crystallites tend to pack with the 'c' axis parallel to the γ-ray direction (assumed to be normal to the absorber plane: i.e. parallel to the Fe–Fe axis of the Fe$_2$(CO)$_9$.) The quadrupole coupling constant e^2qQ was thereby shown to be positive.

One of the more spectacular results of Mössbauer spectroscopy in its formative years was the controversy it reopened over the correct structure of Fe$_3$(CO)$_{12}$. In 1962 the accepted structure, which had been proposed on an incomplete analysis of X-ray data, was an equilateral triangle of three Fe(CO)$_2$ units with two (CO) bridges along each side. On this basis, as there is only one iron environment only one quadrupole doublet would be expected. However, the observed spectrum appears to contain three lines of equal intensity [2, 3, 7] although the central line has recently been shown to be slightly quadrupole split (Fig. 9.3) [5]. Clearly there are two equivalent iron sites, which are distinct from the third. The correct structure (structure 3) was suggested by the close similarity of the Mössbauer spectrum with that

[Refs. on p. 237]

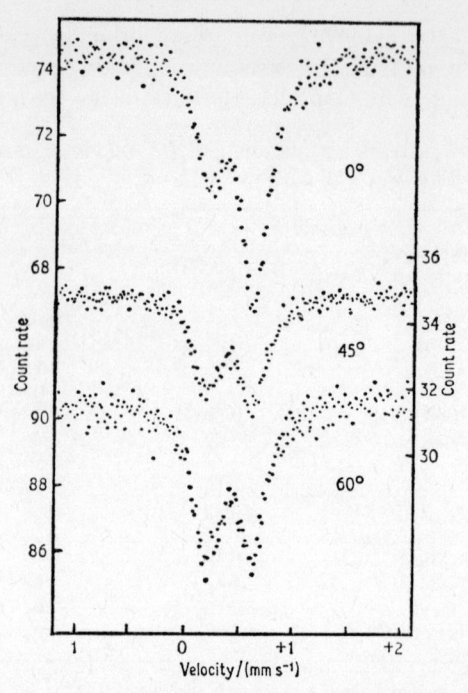

Fig. 9.2 The effect of rotation on the spectrum of an oriented sample of $Fe_2(CO)_9$. [Ref. 8, Fig. 3]

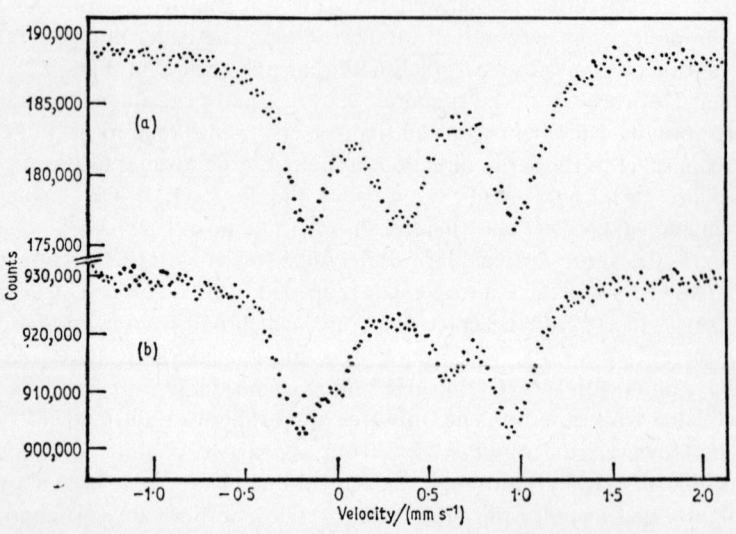

Fig. 9.3 Mössbauer spectra (a) $Fe_3(CO)_{12}$ and (b) $[(OC)_3FePMe_2Ph]_3$ at 80 K. [Ref. 5, Fig. 2]

[*Refs. on p. 237*]

of [Fe$_3$(CO)$_{11}$H]$^-$ [9], and by an X-ray structure determination on the latter [10], and has since been fully confirmed by X-ray methods [11]. It may be pictured as being derived from Fe$_2$(CO)$_9$ by replacing a (CO) bridge with a bridging Fe(CO)$_4$ group.

The Dianions and Hydride Anions

Of the dianions (Fig. 9.1, structures 4–7) only the structure of [Fe$_2$(CO)$_{13}$]$^{2-}$ has been confirmed by full X-ray analysis. The Mössbauer spectra [5, 6, 9] all show only one distinct line or quadrupole doublet, implying a high symmetry. However, as with all other spectroscopic techniques, absence of more than one set of resonance lines does not establish the presence of a single iron environment, but merely that, if differing environments exist, this difference is not resolved spectroscopically. Thus, the Mössbauer resonances from the two confirmed site symmetries in [Fe$_4$(CO)$_{13}$]$^{2-}$ are not resolved. [Fe(CO)$_4$]$^{2-}$ has a zero quadrupole splitting as expected for a tetrahedral geometry, contrasting with the large value for [Fe$_2$(CO)$_8$]$^{2-}$ which was thereby shown to be trigonal bipyramidal like Fe(CO)$_5$. Additional evidence on this point comes from the absence of infrared bands in the bridging carbonyl region for the dianion [6].

The hydride anions are rather unstable at room temperature and are not so well known except for [Fe$_3$(CO)$_{11}$H]$^-$. Structures are shown in Fig. 9.1, that of [Fe$_4$(CO)$_{13}$H]$^-$ being unknown. The structure of [Fe$_2$(CO)$_8$H]$^-$ was suggested from infrared and Mössbauer evidence on the basis of its similarity to Fe$_2$(CO)$_9$ [12]. The Mössbauer data are in Table 9.1.

Viewing the data as a whole, two generalisations can be made: (i) the chemical isomer shift decreases as the anionic charge increases; (ii) the

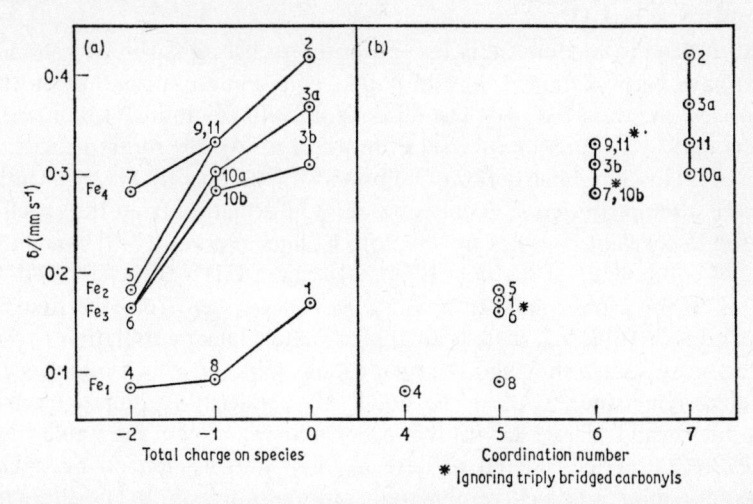

Fig. 9.4 Trends of chemical isomer shift δ with charge and coordination number. [Ref. 5, Fig. 3]

226 | COVALENT IRON COMPOUNDS

chemical isomer shift increases as the coordination number increases. These features are illustrated in Fig. 9.4. Furthermore, the quadrupole splitting is less than 0·6 mm s^{-1} for 4- and 6-coordination; it is in the range 2·1–2·6 mm s^{-1} for 5-coordination, and 0·4–1·4 mm s^{-1} for effective 7-coordination. [Fe(CO)$_4$H]$^-$ shows a smaller quadrupole splitting than expected for 5-coordination and presumably has C_{3v} symmetry as drawn in Fig. 9.1 rather than a planar grouping of the equatorial carbonyls.

Detailed interpretation is prevented by the gross inadequacies in our knowledge of the detailed bonding and electron distribution in this type of complex. However, it can be said that the observed chemical isomer shifts result from the interplay of at least five factors, three of which decrease δ and two of which may act in the opposite direction. Thus, there may be:

(1) an increase in 4s-participation in the bonding by σ-donation from the ligands (decreases δ)
(2) an increase in the effective 3d-population by σ-donation from the ligands (increases δ by increased shielding effects)
(3) an increase in the π-back donation to the ligands (decreases δ by reducing shielding)
(4) radial expansion of any non-bonding electrons as a result of electron correlation (decreases δ)
(5) effective radial expansion of the bonding orbitals upon molecular orbital formation affecting both 3d- and 4s-symmetry types (opposing effect on δ).

9.2 Substituted Iron Carbonyls

Derivatives of Fe(CO)$_5$

In the following sections it is not proposed to list all carbonyl complexes which have been examined, but all papers containing information on these complexes are referred to. The discussion will be limited to important aspects and to a representative selection of data. As already remarked, the study of a closely related series of compounds is frequently more rewarding than the attempt to derive detailed bonding information from the spectrum of a single covalent complex in isolation. Collins and Pettit [13] examined a range of compounds of the type LFe(CO)$_4$ where L is a two-electron donor such as triphenylphosphine or π-allyl. As can be seen from the first nine compounds in Table 9.2, there is an approximately linear correlation between the chemical isomer shift and the quadrupole splitting, which attaches considerable significance to the L–Fe bond. The tendency to donate electrons from the ligand decreases in the order moving down the table. Thus (Ph$_3$P)Fe(CO)$_4$ is considered to have an enhanced 4s-electron occupation from σ-donation, and a correspondingly more negative shift. The alternative process of back donation from the 3d-orbitals to the ligand π-orbitals

Table 9.2 Mössbauer parameters for substituted mono-iron carbonyls

Compound	Δ /(mm s^{-1})	δ /(mm s^{-1})	δ (Fe) /(mm s^{-1})	Reference
Ph$_3$PFe(CO)$_4$	2·54	−0·314 (Cu)	−0·088	13
triethylphosphite Fe(CO)$_4$	2·31	−0·348 (Cu)	−0·122	13
acenaphthalene Fe(CO)$_4$	1·78	−0·236 (Cu)	−0·010	13
trans-cinnamaldehyde:Fe(CO)$_4$	1·75	−0·238 (Cu)	−0·012	13
maleic anhydride:Fe(CO)$_4$	1·41	−0·214 (Cu)	0·012	13
syn-syn-1,3-dimethyl-π-allyl:Fe(CO)$_4^+$BF$_4^-$	1·16	−0·174 (Cu)	0·052	13
syn-1-methyl-π-allyl:Fe(CO)$_4^+$BF$_4^-$	1·01	−0·158 (Cu)	0·068	13
π-allyl:Fe(CO)$_4^+$BF$_4^-$	0·93	−0·165 (Cu)	0·061	13
I$_2$:Fe(CO)$_4$	0·38	−0·084 (Cu)	0·142	13
2-methoxy-3,5-hexadiene-Fe(CO)$_3$	1·71	−0·218 (Cu)	0·008	4
racemic-5,6-dimethyl-1,3,7,9-tetradecaene[Fe(CO)$_3$]$_2$	1·69	−0·216 (Cu)	0·010	4
allo-ocimine-Fe(CO)$_3$	1·75	−0·214 (Cu)	0·012	4
meso-5,6-dimethyl-1,3,7,9-tetradecaene [Fe(CO)$_3$]$_2$	1·58	−0·208 (Cu)	0·018	4
2-hydroxy-3,5-hexadiene-Fe(CO)$_3$	1·56	−0·202 (Cu)	0·024	4
butadiene-Fe(CO)$_3$	1·46	−0·198 (Cu)	0·028	4
7-acetoxy-bicycloheptadiene-Fe(CO)$_3$	2·01	−0·191 (Cu)	0·035	4
1-phenylbutadiene-Fe(CO)$_3$	1·59	−0·178 (Cu)	0·048	4
2,4-hexadienoic acid-Fe(CO)$_3$	1·63	−0·178 (Cu)	0·048	4
1,5-dimethyl-pentadienyl-Fe(CO)$_3^+$BF$_4^-$	1·83	−0·126 (Cu)	0·100	4
1-methyl-pentadienyl-Fe(CO)$_3^+$SbCl$_6^-$	1·69	−0·126 (Cu)	0·100	4
cyclohexadienyl-Fe(CO)$_3^+$ClO$_4^-$	1·66	−0·122 (Cu)	0·104	4
1-methyl-pentadienyl-Fe(CO)$_3^+$ClO$_4^-$	1·70	−0·117 (Cu)	0·109	4
1-methyl-pentadienyl-Fe(CO)$_3^+$BF$_4^-$	1·72	−0·117 (Cu)	0·109	4
cycloheptadienyl-Fe(CO)$_3^+$BF$_4^-$	1·57	−0·111 (Cu)	0·115	4
1-methyl-pentadienyl-Fe(CO)$_3^+$PF$_6^-$	1·67	−0·103 (Cu)	0·123	4
bicyclo-octadienyl-Fe(CO)$_3^+$BF$_4^-$	1·79	−0·098 (Cu)	0·128	4
(Ph$_3$P)$_2$Fe(CO)$_3$	2·76	−0·324 (Cu)	−0·098	4
ffosFe(CO)$_3$	2·336	0·200 (NP)	−0·057	16
f$_6$fosFe(CO)$_3$	2·342	0·197 (NP)	−0·060	16
diphosFe(CO)$_3$	2·124	0·185 (NP)	−0·072	16
diarsFe(CO)$_3$	2·267	0·227 (NP)	−0·030	16
ffosFe(CO)$_4$	2·607	0·188 (NP)	−0·069	16
ffarsFe(CO)$_4$	2·791	0·207 (NP)	−0·050	16
diphosFe$_2$(CO)$_8$	2·463	0·161 (NP)	−0·096	16
diarsFe$_2$(CO)$_8$	2·681	0·188 (NP)	−0·069	16
ffarsFe$_2$(CO)$_8$	2·820	0·207 (NP)	−0·050	16
Fe(CO)$_2$(NO)$_2$	0·328	0·062 (Fe)	0·062	17
	−0·337*	($\eta \sim 0·85$)	0·089*	17
K[Fe(CO)$_3$NO]	0·362	−0·083 (Fe)	−0·083	17
	+0·355*	($\eta \sim 0$)	−0·070*	17
Hg[Fe(CO)$_3$NO]$_2$	1·318	0·023 (Fe)	0·023	17
(Ph$_3$P)Fe(CO)(NO)$_2$	0·553	0·038 (Fe)	0·038	17
(Ph$_3$P)$_2$Fe(NO)$_2$	0·687	0·088 (Fe)	0·088	17
	−0·681*	($\eta \sim 0·76$)	0·110*	17

* Data at 4·2 K. All other data at liquid nitrogen temperature.

[*Refs. on p. 237*]

228 | COVALENT IRON COMPOUNDS

increases in the same relative order. It would increase s-density at the nucleus by a decrease in shielding and give a trend in chemical isomer shift opposite to that found in practice. It was therefore inferred that back donation is not the dominant feature of the bonding. The compatibility of the $I_2Fe(CO)_4$ data with the rest has been taken to imply that this is not a *cis*-octahedral complex but is a true 5-coordinate $LFe(CO)_4$ complex where L is molecular iodine.

Similar measurements have been made for a number of compounds containing the $Fe(CO)_3$ group with diene ligands [4] (see Table 9.2). In this case changes in the chemical isomer shift are quite small in comparison to changes in the quadrupole splitting. Possibly this is a result of stereochemical effects rather than more direct changes in the type of chemical bonding. The corresponding carbonium ion ligands show a considerable increase in chemical isomer shift because the σ-donation to the iron $4s$-orbitals is reduced, causing the s-electron density at the iron nucleus to decrease. Here again the σ-donation is more important than π-back donation. As might be anticipated, the Mössbauer parameters are not very sensitive to the nature of the anion.

Studies have also been made of cyclo-octatetraene derivatives of the type $(COT)Fe(CO)_3$ and $(COT)[Fe(CO)_3]_2$ [14], and solid and frozen solution data show that there is no appreciable change in the conformations of the molecules between the two phases [15].

(11) diphos

(12) diars

(13) ffos

(14) f₆fos

(15) ffars

[*Refs.* on p. 237]

The Fe(CO)$_3$ and Fe(CO)$_4$ moieties are also apparently found in a number of complexes with ligands such as 1,2-bis-(dimethylarsino)benzene (diars). These ligands are shown in structures 11–15. The compound (diars)Fe$_2$(CO)$_8$ for example contains an Fe(CO$_4$) group bonded to each arsenic donor atom. In all cases the Mössbauer parameters are not dissimilar to those of Fe(CO)$_5$ and show little major variation (see Table 9.2) [16]. This prevents reliable assignment of the detailed stereochemistry from the Mössbauer data alone. The arsenic derivatives generally have larger chemical isomer shifts than those containing phosphorus, possibly indicating greater π-back donation from the metal in the latter case.

It is convenient to include here some pseudotetrahedral mixed nitrosyl carbonyl compounds. Data for these are given in Table 9.2. From experience with other diamagnetic carbonyl systems it seems reasonable to assume that any quadrupole splitting will be derived largely from the geometry of the ligand environment. In a molecule with C_{3v} symmetry, FeA$_3$B, the asymmetry parameter η is expected to be zero, and this was confirmed in K[Fe(CO)$_3$NO] by the magnetic perturbation method [17]. The lower symmetry in FeA$_2$B$_2$ should be associated with an η value close to unity, and as seen in the table this is indeed found in Fe(CO)$_2$(NO)$_2$ and Fe(Ph$_3$P)$_2$(NO)$_2$.

Derivatives of Fe$_2$(CO)$_9$

The replacement of the bridging carbonyl groups in Fe$_2$(CO)$_9$ by a range of other ligands containing nitrogen, phosphorus, and sulphur can be used to study the relative effects of different ligands on the carbonyl structure [18, 19]. Di-μ-dimethylphosphido-bis-(tricarbonyliron), (OC)$_3$Fe(PMe$_2$)$_2$Fe(CO)$_3$ has structure 16. Its diamagnetism implies that the spins of the eighteenth

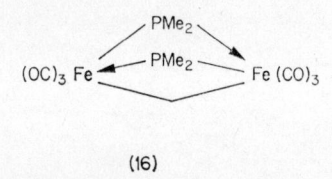

$$(OC)_3 Fe \underset{}{\overset{PMe_2}{\underset{PMe_2}{\rightleftarrows}}} Fe(CO)_3$$

(16)

electron in the valency shell of the two iron atoms are opposed and the small quadrupole splitting establishes the presence of an iron–iron bond. Thus, in this and a number of analogous sulphur-containing compounds, the quadrupole splitting remains relatively small (0·62–1·07 mm s^{-1}), compatible with 6-coordination of the iron (Table 9.3). This is particularly emphasised by the large quadrupole splitting of 2·58 mm s^{-1} for (OC)$_4$Fe(PMe$_2$)$_2$Fe(CO)$_4$ which is 5-coordinate and the small value of 0·99 mm s^{-1} for the complex I(OC)$_3$Fe(PMe$_2$)$_2$Fe(CO)$_3$I which is already 6-coordinate without the iron–iron bond (structures 17–18).

The overall impression is one of insensitivity to substitution in general, but strong dependence on stereochemistry. Substitution of a terminal carbonyl

[Refs. on p. 237]

Table 9.3 Mössbauer parameters for di- and tri-iron carbonyl complexes

Compound	Δ^* /(mm s^{-1})	δ (Fe)* /(mm s^{-1})	Reference
(OC)$_3$Fe(PMe$_2$)$_2$Fe(CO)$_3$	0·685	−0·043	18
(OC)$_3$Fe(PPh$_2$)$_2$Fe(CO)$_3$	0·627	−0·015	18
(OC)$_3$Fe(PMe$_2$)$_2$Fe(CO)$_2$PEt$_3$	0·658	−0·037	18
(OC)$_3$Fe(PPh$_2$)(SPh)Fe(CO)$_3$	0·942	0·036	18
(OC)$_3$Fe(SPh)$_2$Fe(CO)$_3$	1·067	0·061	18
(OC)$_3$Fe(SePh)$_2$Fe(CO)$_3$	1·037	0·042	19
(OC)$_3$Fe(SPh)(SePh)Fe(CO)$_3$	1·032	0·078	19
(OC)$_3$Fe(SPh)$_2$Fe(CO)$_2$PPh$_3$	{1·02 / 0·86}	{0·05 / 0·13}	5
syn-(OC)$_3$Fe(SMe)$_2$Fe(CO)$_3$	0·895	0·028	18
anti-(OC)$_3$Fe(SMe)$_2$Fe(CO)$_3$	1·034	0·022	18
(OC)$_4$Fe(PMe$_2$)$_2$Fe(CO)$_4$	2·580	−0·032	18
I(OC)$_3$Fe(PMe$_2$)$_2$Fe(CO)$_3$I	0·990	−0·001	18
ffarsFe$_2$(CO)$_6$ {A / B}	0·64 / 1·44	0·02 / 0·058	26 / 26
ffosFe$_2$(CO)$_6$ {A / B}	0·66 / 1·30	−0·03 / 0·06	26 / 26
f$_6$fosFe$_2$(CO)$_6$ {A / B}	0·65 / 1·19	−0·04 / 0·06	26 / 26
(ffars)Fe$_3$(CO)$_{10}$	{0 / 1·523}	{0·026 / 0·163}	27
[(OC)$_3$FePMe$_2$Ph]$_3$	{1·15 / 0·57}	{0·09 / 0·02}	5

* All data at 80 K, originally calibrated with NP and converted.

(17)

(18)

group by triethylphosphine has little effect on the Mössbauer parameters, which are also insensitive to substitution of an alkyl or aryl group in the bridging units. Consideration of the possible contributions to the chemical isomer shift suggests that the more positive isomer shifts of the sulphur derivatives are due to this atom being a less effective σ-electron donor than phosphorus [18]. The selenium and mixed sulphur–selenium derivatives are similar [19].

Mössbauer data have provided useful evidence in the assignment of possible structures for new complexes, and have shown both environments to be equivalent in structures 19–21 [20–22] and non-equivalent in 22–23 [23, 24]. A number of triene-Fe$_2$(CO)$_6$ derivatives have also been examined [25].

(19)

(20)

(21)

(22)

(23)

(24)

(25)

The binuclear compounds of which ffarsFe$_2$(CO)$_6$ is an example are believed to have structure 24. The two iron atoms are not identical, although that bonding to the butene ring (atom B) is expected to be less sensitive to

[Refs. on p. 237]

change of the donor atom from arsenic to phosphorus, but to be in the more distorted environment [26]. This assists interpretation of the four-line spectra seen in Fig. 9.5. The outer lines represent the quadrupole splitting from

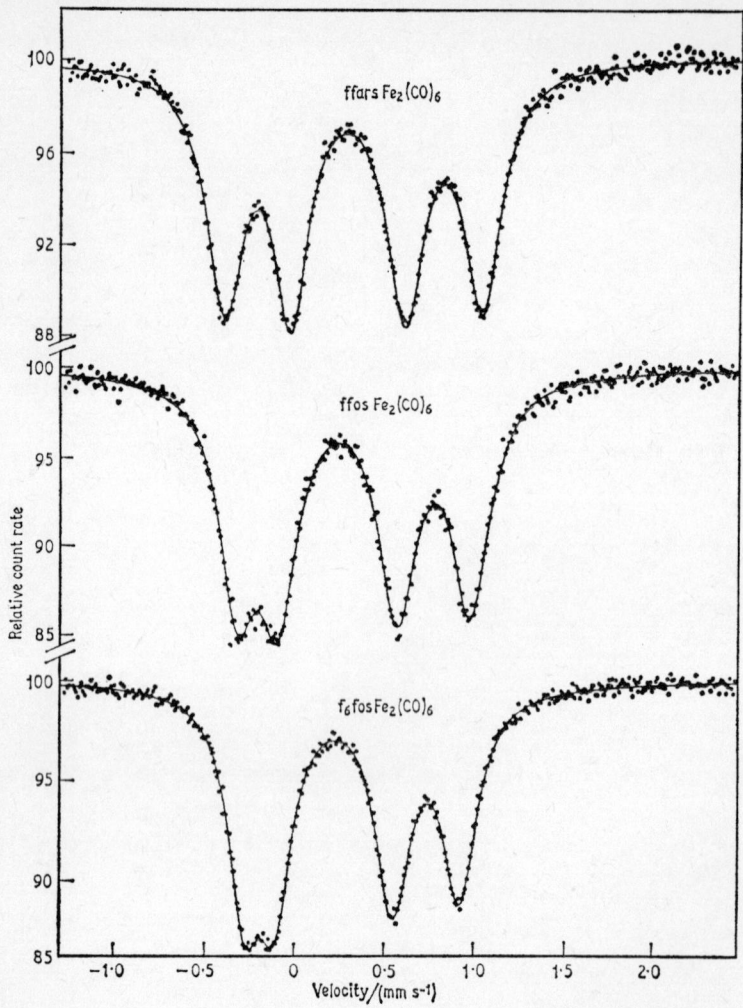

Fig. 9.5 Mössbauer spectra of LFe$_2$(CO)$_6$ compounds at 80 K. [Ref. 26, Fig. 3]

atom B, while the inner lines, which show considerable variation in the chemical isomer shift, are from atom A which is directly bonded to arsenic or phosphorus.

Derivatives of Fe$_3$(CO)$_{12}$

The compound (ffars)Fe$_3$(CO)$_{10}$ shows a spectrum very similar to that of Fe$_3$(CO)$_{12}$ with the exception that there is a small increase in the quadrupole splitting of the two equivalent iron positions [27]. This lends confidence to the assigned structure 25.

The compound now formulated as [(OC)$_3$FePMe$_2$Ph]$_3$ was found to be a trinuclear species rather than di- or tetra-nuclear from the similarity of its Mössbauer spectrum to Fe$_3$(CO)$_{12}$ [5]. All three iron atoms have shown a change in chemical isomer shift, and this leads one to suggest that one terminal carbonyl group on each iron atom has been substituted. This has since been confirmed by an X-ray crystal structure analysis [28].

The polynuclear mixed metal-iron carbonyl complexes Co$_2$FeS(CO)$_9$, Co$_2$FeS(CO)$_8$(PPh$_3$), and Co$_2$FeS(CO)$_7$(PPh$_3$)$_2$ feature a triangle of ML$_3$ units as the base of a prism with the sulphur atom at the apex. The quadrupole splittings of 0·71, 0·78, and 0·86 mm s^{-1} respectively at liquid nitrogen temperature are consistent with the effective 6-coordination of the iron atom [29].

We shall leave discussion of compounds such as (CH$_3$)$_4$Sn$_3$Fe$_4$(CO)$_{16}$ where both the ^{57}Fe and ^{119}Sn resonances are known until Chapter 14.

9.3 Ferrocene and Other π-Cyclopentadienyl Derivatives

The ferrocene type of 'sandwich' complex (structure 26) has always excited interest because of the unusual nature of the bonding involved. The Mössbauer spectrum of the parent ferrocene, (π-C$_5$H$_5$)$_2$Fe, consists of a large

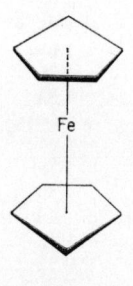

(26)

temperature-independent quadrupole splitting (see Table 9.4). The classic work on ferrocene was by Collins [30], who determined the sign of the electric field gradient tensor by applying a 40-kG field at 4·2 K and noting that the low- and high-energy components split into a triplet and doublet respectively. He used a first-order perturbation theory treatment to show that this indicated e^2qQ to be positive, a fact since verified by more accurate computer calculations (see Chapter 3.6), and by single-crystal Mössbauer data [31].

It had already been observed that ferricinium salts such as (π-C$_5$H$_5$)$_2$FeBr give only very small quadrupole splittings [32, 33]. Although it was suspected

[Refs. on p. 237]

234 | COVALENT IRON COMPOUNDS

Table 9.4 Mössbauer parameters for some cyclopentadienyl derivatives

Compound	Δ /(mm s^{-1})	δ /(mm s^{-1})	δ (Fe) /(mm s^{-1})	Reference
(Cp)$_2$Fe	2·37	0·68 (Cr)	0·53	33
(Cp)$_2$FeBr	~0·2	0·52 (SS)	0·43	33
[CpFe(C$_5$H$_4$)]$_2$ (biferrocenyl)	2·36	0·67 (Cr)	0·52	33
[CpFe(C$_5$H$_4$)]$_2$Hg (diferrocenyl mercury)	2·32	0·69 (Cr)	0·54	33
(C$_5$H$_4$)Fe(C$_5$H$_4$) (1,1'-trimethylene-⎿—(CH$_2$)$_3$—⏌ ferrocene)	2·30	0·66 (Cr)	0·51	33
2-acetyl-1,1'-trimethylene-ferrocene	2·05	0·29 (Cu)	0·52	37
CpFe(C$_5$H$_4$Me) (methylferrocene)	2·39	0·29 (Cu)	0·52	37
CpFe(C$_5$H$_4$CN)	2·30	0·30 (Cu)	0·53	37
CpFe(C$_5$H$_4$OOCCH$_3$)	2·25	0·32 (Cu)	0·55	37
CpFe(C$_5$H$_4$CH$_2$OH)	2·39	0·29 (Cu)	0·52	37
cis-Cp(OC)Fe(PMe$_2$)$_2$Fe(CO)Cp	1·609	0·401 (NP)	0·144	18
trans-Cp(OC)Fe(PMe$_2$)$_2$Fe(CO)Cp	1·638	0·416 (NP)	0·159	18
stable-Cp(OC)Fe(SPh)$_2$Fe(CO)Cp	1·670	0·605 (NP)	0·348	18
unstable-Cp(OC)Fe(SPh)$_2$Fe(CO)Cp	1·720	0·611 (NP)	0·354	18
stable-Cp(OC)Fe(SMe)$_2$Fe(CO)Cp	1·634	0·569 (NP)	0·312	18
CpFe(CO)$_2$I	1·83	0·38 (Cr)	0·23	1
CpFe(CO)$_2$Cl	1·88	0·39 (Cr)	0·24	1
CpFe(CO)$_2$Br	1·87	0·40 (Cr)	0·25	1
[CpFe(CO)$_3$]$^+$PF$_6^-$	1·88	0·23 (Cr)	0·08	1
[CpFe(CO)$_2$]$_2$Hg	1·60	0·28 (Cr)	0·13	1
[CpFe(CO)$_2$Mn(CO)$_5$]	1·68	0·36 (Cr)	0·21	1
[CpFe(CO)$_2$]$_2$	1·91	0·47 (NP)	0·21	39
[CpFe(CO)$_2$]$_2$SnCl$_2$	1·68	0·36 (NP)	0·10	39
[CpFe(CO)$_2$]SnCl$_3$	1·86	0·40 (NP)	0·14	39
[CpFe(CO)$_2$]SnPh$_3$	1·83	0·37 (NP)	0·11	39
[CpFe(CO)$_2$]$_2$GeCl$_2$	1·66	0·36 (NP)	0·10	39
[Cp$_2$Fe$_2$(CO)$_3$]$_2$Ph$_2$PCCPPh$_2$	1·94	0·53 (NP)	0·27	40
[CpFe(CO)]$_4$	1·76	0·66 (NP)	0·40	5
[CpFe(CO)]$_4$Cl	1·38	0·67 (NP)	0·41	5
[CpFe(CO)]$_4$Br$_3$	1·40	0·67 (NP)	0·41	5
[CpFe(CO)]$_4$I$_{\sim 5}$	1·42	0·67 (NP)	0·41	5
(Me$_4$N)Fe(C$_2$B$_9$H$_{11}$)$_2$	—	0·630 (NP)	0·373	41
CpFe(C$_2$B$_9$H$_{11}$)	0·529	0·608 (NP)	0·351	41

that this was the result of an accidental cancellation, confirmation was given by Collins using the molecular-orbital analyses available.

The crystal field model of the bonding in ferrocene predicts the wrong sign for e^2qQ. By contrast, the molecular-orbital model predicts the correct sign. A typical molecular-orbital scheme in Table 9.5 is from the calculations of Dahl and Ballhausen [34]. The molecular-orbital wavefunctions (ψ) are taken as linear combinations of metal (μ) and ligand (ρ) orbitals. The effective occupation of the metal atomic orbitals is then given by the squares of the metal coefficients. The 4s-orbitals give no electric field gradient, and there is

[*Refs. on p.* 237]

good evidence to show that any contribution from the 4p-orbitals will be small in comparison to that from the 3d- because of the greater radial expansion of the former. Each individual orbital generates a contribution to the

Table 9.5 Molecular-orbital calculations for ferrocene

Wavefunction	No. of electrons	$\left(\dfrac{\text{metal}}{\text{coefficient}}\right)^2$	net Δ $\mid 4p_0\rangle$	net Δ^* $\mid 3d_0\rangle$
$\psi(e_{2g}) = 0.898\mu(3d_2) + 0.440\rho(e_{2g}{}^+)$	4	0.8064		3.222
$\psi(e_{1g}) = 0.454\mu(3d_1) + 0.891\rho(e_{1g}{}^+)$	4	0.2061		−0.411
$\psi(a_{1g}) = \mu(3d_0)$	2	1.000		−2.00
$\psi(e_{1u}) = 0.591\mu(4p_1) + 0.807\rho(e_{1u}{}^+)$	4	0.3493	0.693	
$\psi(a_{2u}) = 0.471\mu(4p_0) + 0.882\rho(a_{2u})$	2	0.2218	−0.444	
$\psi(a_{1g}) = 0.633\mu(4s) + 0.774\rho(a_{1g})$	2	0.4007		
			0.249	0.811
$(\pi\text{-}C_5H_5)_2\text{Fe}$	$e^2qQ/2 = 0.249\mid 4p_0\rangle + 0.811\mid 3d_0\rangle$			
$(\pi\text{-}C_5H_5)_2\text{Fe}^+$	$e^2qQ/2 = 0.249\mid 4p_0\rangle + 0.005\mid 3d_0\rangle$			

(increasing energy ↑)

* The quadrupole splitting in units of that from a $\mid 3d_0\rangle$ electron taken to be '−1'.

field gradient and to e^2qQ along the symmetry axis of the molecule, and in the last two columns in the table these are normalised in terms of that from the $\mid 4p_0\rangle$ function and the $\mid 3d_0\rangle$ function (which will be about −3.6 mm s^{-1} per electron). We can now see that in ferrocene one predicts a quadrupole splitting of $-0.811 \mid 3d_0\rangle = \sim 2.9$ mm s^{-1}. Removal of an electron from the $\psi(e_{2g})$ orbitals will effectively remove 0.8064 of a $\mid 3d_0\rangle$ electron and reduce the splitting to $0.005 \mid 3d_0\rangle = \sim 0.2$ mm s^{-1}. The agreement with experiment is excellent, and confirms the validity of the molecular-orbital model for ferrocene. The reduction in the chemical isomer shift from 0.53 mm s^{-1} to 0.43 mm s^{-1} on oxidation of ferrocene to ferricinium ion is also compatible with the removal of 3d-density.

As already mentioned in Chapter 5, calculations using the known ^{59}Co nuclear quadrupole moment and quadrupole coupling constants for the isoelectronic species ferrocene and cobalticinium perchlorate, $(\pi\text{-}C_5H_5)_2\text{CoClO}_4$, have been made to obtain an estimate of +0.175 barn for the nuclear quadrupole moment, Q_e of ^{57}Fe [35].

Application of pressure to ferrocene decreases both the chemical isomer shift and the quadrupole splitting [36]. Molecular-orbital calculations give predictions approximately in agreement with experiment.

A small selection of the large volume of data available for substituted ferrocenes [33, 37, 38] is given in Table 9.4. The iron environment is very insensitive to substitution in the cyclopentadienyl rings which affects primarily the σ-bonding of the ring. A possible exception to this is 2-acetyl-1,1'-trimethylene-ferrocene, where the trimethylene bridge may be inducing

[Refs. on p. 237]

some inclination of the rings away from the mutually parallel disposition in non-linked ferrocenes.

Although the Mössbauer spectra of the sandwich compounds are insensitive to ring substitution, a wide range of parameters is observed for the mixed ferrocene-carbonyls. Values of quadrupole splitting span the range 1·38–1·91 mm s^{-1}, and the chemical isomer shift is between 0·07 and 0·41 mm s^{-1}. Both of these ranges of values fall below the values of Δ and δ which are typical for ferrocene derivatives ($\sim$2·3 and $\sim$0·51 mm s^{-1} respectively) (see Table 9.4 for typical examples). However, interpretation is again difficult. It is interesting to note that the *cis*- and *trans*-isomers of Cp(OC)Fe(PMe$_2$)$_2$Fe(CO)Cp do not show significant differences because the first coordination spheres of the iron atoms are equivalent [18]. Solid and frozen-solution data of the complexes containing Fe–Sn and Fe–Ge bonds do not show conformation changes with the possible exception of [CpFe(CO)$_2$]$_2$SnCl$_2$ [39]. The compound [Cp$_2$Fe$_2$(CO)$_3$]$_2$Ph$_2$PCCPPh$_2$ which contains Cp(OC)Fe(CO)$_2$FeCp— units bonding to the phosphorus ligand is interesting in that no difference is found between the two iron environments [40].

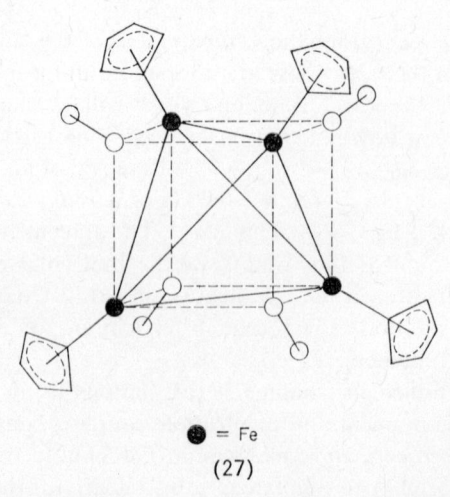

● = Fe
(27)

The compound [CpFe(CO)]$_4$ has the structure 27. The close similarity of its spectrum to those of some otherwise intractable cationic species show that all contain the [CpFe(CO)]$_4$$^+$ cation [5]. Presumably the electron is removed from a molecular orbital involving all four iron atoms equally since there is no indication of more than one iron environment in any of the spectra.

The spectra of the dicarbollide carborane derivatives (Me$_4$N)Fe(C$_2$B$_9$H$_{11}$)$_2$ and CpFe(C$_2$B$_9$H$_{11}$)$_2$ are very similar to those of ferricinium salts [41], and confirm the similarity of the bonding in the dicarbollide to that of ferrocene.

[*Refs. on p. 237*]

REFERENCES

[1] R. H. Herber, R. B. King, and G. K. Wertheim, *Inorg. Chem.*, 1964, **3**, 101.
[2] M. Kalvius, U. Kahn, P. Kienle, and H. Eicher, *Z. Naturforsch.*, 1962, **17A**, 494.
[3] R. H. Herber, W. R. Kingston, and G. K. Wertheim, *Inorg. Chem.*, 1963, **2**, 153.
[4] R. L. Collins and R. Pettit, *J. Amer. Chem. Soc.*, 1963, **85**, 2332.
[5] R. Greatrex and N. N. Greenwood, *Discuss Faraday Soc.*, No. 47, 1969, 126.
[6] K. Farmery, M. Kilner, R. Greatrex, and N. N. Greenwood, *J. Chem. Soc. (A)*, 1969, 2339.
[7] W. Kerler, W. Neuwirth, E. Fluck, P. Kuhn, and B. Zimmermann, *Z. Physik*, 1963, **173**, 321.
[8] T. C. Gibb, R. Greatrex, and N. N. Greenwood, *J. Chem. Soc. (A)*, 1968, 890.
[9] N. E. Erickson and A. W. Fairhall, *Inorg. Chem.*, 1965, **4**, 1320.
[10] L. F. Dahl and J. F. Blount, *Inorg. Chem.*, 1965, **4**, 1373.
[11] C. H. Wei and L. F. Dahl, *J. Amer. Chem. Soc.*, 1969, **91**, 1351.
[12] K. Farmery, M. Kilner, R. Greatrex, and N. N. Greenwood, *Chem. Comm.*, 1968, 593.
[13] R. L. Collins and R. Pettit, *J. Chem. Phys.*, 1963, **39**, 3433.
[14] G. K. Wertheim and R. H. Herber, *J. Amer. Chem. Soc.*, 1962, **84**, 2274.
[15] R. H. Herber, *Symposia Faraday Soc. No. 1*, 1968, 86.
[16] W. R. Cullen, D. A. Harbourne, B. V. Liengme, and J. R. Sams, *Inorg. Chem.*, 1969, **8**, 1464.
[17] R. A. Mazak and R. L. Collins, *J. Chem. Phys.*, 1969, **51**, 3220.
[18] T. C. Gibb, R. Greatrex, N. N. Greenwood, and D. T. Thompson, *J. Chem. Soc. (A)*, 1967, 1663.
[19] E. Kostiner and A. G. Massey, *J. Organometallic Chem.*, 1969, **19**, 233.
[20] M. Dekker and G. R. Knox, *Chem. Comm.*, 1967, 1243.
[21] W. T. Flannigan, G. R. Knox, and P. L. Pauson, *Chem. and Ind.*, 1967, 1094.
[22] R. Greatrex, N. N. Greenwood, and P. L. Pauson, *J. Organometallic Chem.*, 1968, **13**, 533.
[23] R. Benshoshan and R. Pettit, *Chem. Comm.*, 1968, 247.
[24] E. O. Fischer, V. Kiener, D. St P. Bunbury, E. Frank, P. F. Lindley, and O. S. Mills, *Chem. Comm.*, 1968, 1378.
[25] G. F. Emerson, J. E. Mahler, R. Pettit, and R. Collins, *J. Amer. Chem. Soc.*, 1964, **86**, 3590.
[26] W. R. Cullen, D. A. Harbourne, B. V. Liengme, and J. R. Sams, *Inorg. Chem.*, 1969, **8**, 95.
[27] W. R. Cullen, D. A. Harbourne, B. V. Liengme, and J. R. Sams, *J. Amer. Chem. Soc.*, 1968, **90**, 3293.
[28] W. S. McDonald, J. R. Moss, G. Raper, B. L. Shaw, R. Greatrex, and N. N. Greenwood, *Chem. Comm.*, 1969, 1295.
[29] K. Burger, L. Korecz, and G. Bor, *J. Inorg. Nuclear Chem.*, 1969, **31**, 1527.
[30] R. L. Collins, *J. Chem. Phys.*, 1965, **42**, 1072.
[31] J. T. Dehn and L. N. Mulay, *J. Inorg. Nuclear Chem.*, 1969, **31**, 3103.
[32] U. Zahn, P. Kienle, and H. Eicher, *Z. Physik*, 1962, **166**, 220.
[33] G. K. Wertheim and R. H. Herber, *J. Chem. Phys.*, 1963, **38**, 2106.
[34] J. P. Dahl and C. F. Ballhausen, *Mat. Fys. Medd. Dan. Vid. Selsk.*, 1961, **33**, No. 5.
[35] C. B. Harris, *J. Chem. Phys.*, 1968, **49**, 1648.

[36] R. W. Vaughan and H. G. Drickamer, *J. Chem. Phys.*, 1967, **47,** 468.
[37] A. V. Lesikar, *J. Chem. Phys.*, 1964, **40,** 2746.
[38] R. A. Stukan, S. P. Gubin, A. N. Nesmeyanov, V. I. Goldanski, and E. F. Makarov, *Teor. i eksp. Khim.*, 1966, **2,** 805.
[39] R. H. Herber and Y. Goscinny, *Inorg. Chem.*, 1968, **7,** 1293.
[40] A. J. Carty, T. W. Ng, W. Carter, G. J. Palenik, and T. Birchall, *Chem. Comm.*, 1969, 1101.
[41] R. H. Herber, *Inorg. Chem.*, 1969, **8,** 174.

10 | Iron Oxides and Sulphides

The economic significance of iron has resulted in great interest over the years in the products of its oxidation, and more recently in the vast ranges of ternary oxides such as spinel ferrites which have been commercially exploited in their own right.

An oxide (or sulphide) may be considered as a close-packed lattice of oxygen anions containing sufficient metal cations in the largest interstices to achieve electrical neutrality. In this respect the basic structural type is largely determined by the anion. The power of the anion lattice to accommodate widely differing cations results not only in a wide range of different stoichiometric oxides, but also in the very important property of nonstoichiometry. In iron oxides the iron cation is almost invariably in either the high-spin iron(II) or high-spin iron(III) electronic configuration, although a few examples of iron(IV) compounds are known. Disorder phenomena, in which two or more different cations can distribute themselves statistically on a given lattice site, are also frequently found.

Both Fe^{2+} and Fe^{3+} cations possess a large spin moment which frequently results in cooperative magnetic ordering processes. Consequently much of the interpretation of the internal magnetic fields in iron(II) and iron(III) compounds (see Chapter 6) is also applicable to the oxides. However, there are the added complications that there may be more than one distinct type of cation site, and that there may be effects due to order–disorder phenomena in the next-nearest-neighbour coordination sphere of cations. Sometimes it is also found that the site occupancy in a disordered oxide is a function of the conditions of preparation, and this can cause confusion when comparing results from different laboratories. However, the Mössbauer spectrum can provide a great deal of information about individual sites. By way of comparison, bulk magnetisation properties are an averaged function of the whole material. For this reason an antiferromagnetic oxide has a small magnetisation. This averaging process does not occur in the Mössbauer spectrum since each resonant atom is unique, and the result is a statistical summation (*not* an averaging) of all possible spectra. An antiferromagnetic oxide would therefore give a single hyperfine six-line spectrum because each iron atom is

[*Refs. on p. 296*]

identical as regards the resonant process. In consequence, the Mössbauer spectrum provides a means of studying the microstructural properties of these systems in a way which X-ray diffraction and magnetic susceptibility measurements cannot, and it is therefore particularly valuable in the study of order–disorder and other cooperative phenomena. Examples of the powerful capabilities of the technique will be found in this chapter.

The discussion will be limited more to the phenomenological aspects of the Mössbauer spectra, rather than to the broader field of interpreting the spin ordering processes in these oxides in general. However, much has been achieved in the latter direction. No attempt has been made to tabulate data of the selected examples, partly because of the problems introduced by the many phases of variable composition, and also because much of the better-quality original data have been presented graphically rather than in numerical form. The chemical isomer shift is only of limited diagnostic use because of its relative lack of sensitivity to environment in these systems. However, the shift of a tetrahedrally coordinated cation is generally noticeably less than that of an octahedrally coordinated cation because of increased covalency in the former, and this can sometimes be used to differentiate between site symmetries.

The following subdivisions will be considered: (i) binary oxides; (ii) spinel oxides; (iii) other ternary oxides; (iv) iron(IV) oxides; (v) chalcogenides; (vi) silicate minerals; and (vii) lunar samples.

The last two sections listed are included here for convenience, although there are significant differences from the other sections in behaviour of the cations.

10.1 Binary Oxides and Hydroxides

In this section the following compounds will be discussed: $\alpha\text{-}Fe_2O_3$ (haematite), $\gamma\text{-}Fe_2O_3$ (p. 246), $Fe_{1-x}O$ (wüstite) (p. 248), Fe_3O_4 (magnetite) (p. 251), FeOOH (α, β, γ, and δ forms) (p. 254), $Fe(OH)_2$ (p. 256), and FeOF (p. 257).

$\alpha\text{-}Fe_2O_3$ (Haematite)

Historically, $\alpha\text{-}Fe_2O_3$ has attracted the attention of chemists and physicists alike for many years. The crystal structure is that of corundum ($\alpha\text{-}Al_2O_3$) with a close-packed oxygen lattice and Fe^{3+} cations in octahedral sites. Magnetically it is unusually complex, being antiferromagnetic at low temperature, then undergoing a transition above the so-called Morin temperature to a weak ferrimagnetic state as a result of spin canting, before finally becoming paramagnetic at high temperatures. The influence of impurities on bulk magnetic measurements makes them difficult to interpret, but Mössbauer spectroscopy gives a much clearer indication of the ordering processes involved.

[*Refs. on p. 296*]

The first reported Mössbauer spectrum of α-Fe$_2$O$_3$ was by Kistner and Sunyar [1], who thereby recorded the first chemical isomer shift and electric quadrupole hyperfine interactions to be observed by this technique. With a single-line source the room-temperature spectrum comprises six lines from a hyperfine field of 515 kG; the chemical isomer shift (Table 10.1) is

Table 10.1 Mössbauer parameters for binary iron oxides and hydroxides

Compound	T/K	Δ /(mm s^{-1})	H/kG	δ /(mm s^{-1})	δ (Fe) /(mm s^{-1})	Reference
α-Fe$_2$O$_3$	298	0·12*	515	0·47 (SS)	0·38	1
	80	−0·22*	—	—	—	2
	(0)	—	(544)	—	—	3
γ-Fe$_2$O$_3$	RT	—	{488, 499}	0·36 (SS), 0·50 (SS)	{0·27, 0·41}	19
Fe$_{0.93}$O	297	{0·46, 0·78}	0, 0	0·68 (Cu), 0·63 (Cu)	{0·91, 0·86}	22
Fe$_3$O$_4$	300	{—, —}	491, 453	A site, B site	{—, —}	26
	82	{−0·05, 0·50, −0·02, 0·95, −2·62}	511, 533, 516, 473, 374	A site Fe^{3+}, B site Fe^{3+} {0·77, 0·59}, B site Fe^{2+} {0·71, 1·20}	0·37	31
α-FeOOH	405	0·6	0	—	—	44
	291	−0·15*	384	0·35 (Fe)	0·35	44
	4·2	—	504	—	—	44
β-FeOOH	(0)	−0·10*	475	—	—	47
γ-FeOOH	RT	0·55	0	0·30 (Fe)	0·30	48
	4·2	<0·1*	460	—	—	48
δ-FeOOH	80	—	525, 505	—	—	47
Fe(OH)$_2$	90	3·00	0	—	—	49
	4·2	3·06	200	—	—	49
FeOF	(0)	—	(485)	—	—	50

* $\varepsilon = (e^2qQ/8)(1 - 3\cos^2\theta)$

+0.38 mm s^{-1} (rel. to Fe) and there is a small quadrupole interaction of = +0.12 mm s^{-1} [$\varepsilon = (e^2qQ/8)(1 - 3\cos^2\theta)$; see equation 3.48 for definition of ε as a perturbation to H]. At room temperature the spins are directed in the [111] plane of the crystal, but below about 260 K there is a rearrangement which results in the spins pointing along the [111] trigonal direction. Under the assumption that the electric field gradient tensor remains

242 | IRON OXIDES AND SULPHIDES

oriented with the [111] axis, this change in the direction of the magnetic axis causes a change in the value of θ and thence in the sign of ε. At low temperature ε is about -0.22 mm s^{-1} [2]. Typical spectra are shown in Fig. 10.1 and show the reversal in the quadrupole perturbation [3]. The value of ε *apparently* changes sign in a continuous process between 200 and 280 K, reaching zero

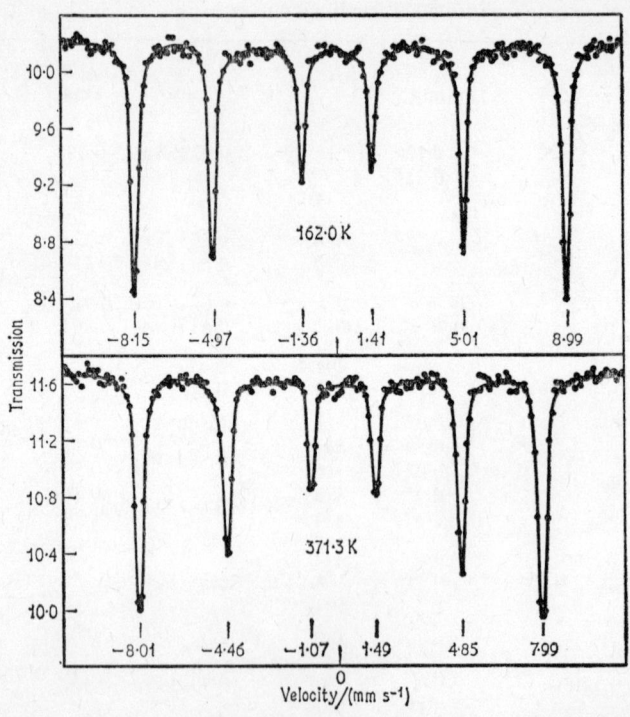

Fig. 10.1 Spectra of α-Fe$_2$O$_3$ at 162·0 K (below Morin temperature) and at 371·3 K (above Morin temperature). Note the reversal in sign of the quadrupole perturbation as shown by the spacings of the outer lines. [Ref. 3, Fig. 1]

at 260 K, the Morin temperature [4]. It had been postulated that the change could be attributed to a continuous rotation of the magnetic axis over the transition region, but the Mössbauer spectra refute this. The line shape of the $|-\frac{1}{2}\rangle \rightarrow |+\frac{1}{2}\rangle$ transition is best described in this region as two components in varying proportions, rather than as an average (Fig. 10.2) [4]. The magnetic axis of each iron atom flips instantaneously from 0° to 90° with increase in temperature, but in the material as a whole the process takes place over a range of temperature because of microscopic inhomogeneities in the crystal. This conclusion was confirmed [5] by use of a single-crystal absorber of natural haematite oriented with the incident γ-rays parallel to

[*Refs.* on p. 296]

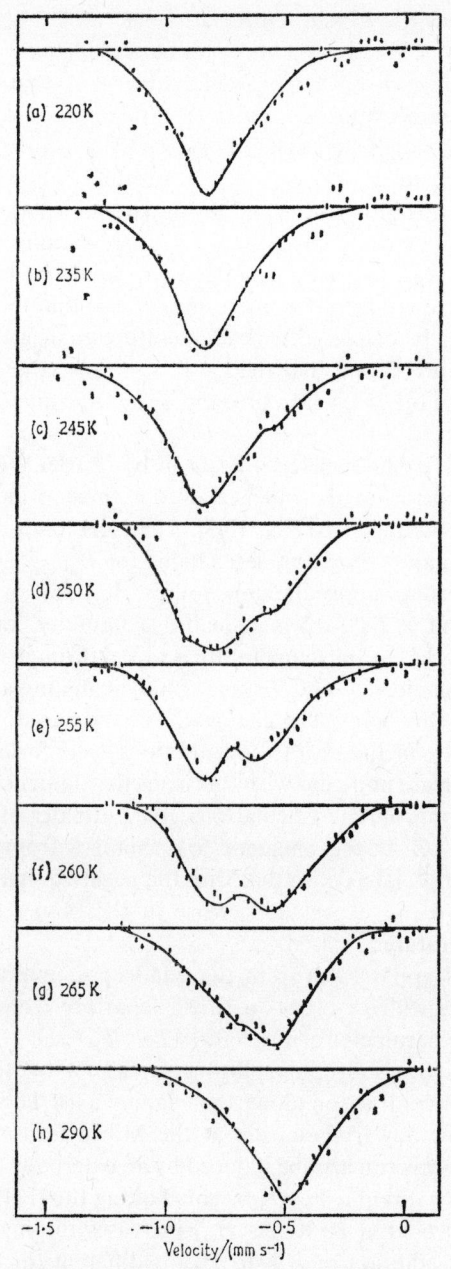

Fig. 10.2 The line shape of the $|-\tfrac{1}{2}\rangle \to |+\tfrac{1}{2}\rangle$ transition in α-Fe$_2$O$_3$ near the Morin temperature. Note the presence of two distinct components, indicating that the spin-flip is not a continuous process. [Ref. 4, Fig. 6]

the [111] axis. Below the Morin temperature the $|\pm\tfrac{1}{2}\rangle \to |\pm\tfrac{1}{2}\rangle$ transitions are then forbidden. Three possibilities can be considered:

(a) θ is a unique angle for all spins at temperature T so that there is only one value of ε at any temperature;

(b) values for θ of 0° and 90° coexist at the same temperature. This means that the $|\pm\tfrac{3}{2}\rangle \to |\pm\tfrac{1}{2}\rangle$ and $|\pm\tfrac{1}{2}\rangle \to |\mp\tfrac{1}{2}\rangle$ transitions will be split (assumed unresolved) in the transition region and the apparent displacement of the lines will vary with T. However, the $|\pm\tfrac{1}{2}\rangle \to |\pm\tfrac{1}{2}\rangle$ transitions only contain the $\theta = 90°$ components so that their displacement should be independent of T;

(c) at each temperature there are many values of θ so that the $|\pm\tfrac{1}{2}\rangle \to |\pm\tfrac{1}{2}\rangle$ transitions will also be displaced with temperature change.

Experimentally, it is found that the $|\pm\tfrac{1}{2}\rangle \to |\pm\tfrac{1}{2}\rangle$ transitions do remain stationary while the other four do not, thereby confirming model (b) as the correct one.

The definitive work on powdered α-Fe$_2$O$_3$ is by van der Woude [3], who has measured the temperature dependence of the three principal Mössbauer parameters very accurately (see Fig. 10.3). The Néel temperature was determined to be 956 K, and the extrapolated value for H_{eff} ($T = 0$) was 544 kG. The magnetic field data approximately follow that for an $S = \tfrac{5}{2}$ Brillouin curve. Just visible at $T/T_N \sim 0.3$ is a small discontinuity corresponding to a decrease in H_{eff} of 8 kG as the temperature is raised through the Morin transition. Both the magnetic field and the quadrupole coupling also show a small hysteresis effect at this point. The change in the measured value of H_{eff} can be accounted for by including the contributions to it from the orbital and dipolar terms which are non-zero when the iron environment is not spherically symmetrical. The appropriate calculations give estimates of $\Delta H_L = -6$ kG and $\Delta H_D = +12$ kG as the magnetic axis changes from parallel to perpendicular to the trigonal axis, so that with the negative sign of the dominant Fermi term H_S, we can expect a decrease in the observed field of about 6 kG as the temperature is raised.

There is a small apparent drop in the quadrupole interaction above the Morin temperature which suggests a small departure from axial symmetry and an asymmetry parameter of $\eta \sim 0.03$ [$\varepsilon(T < T_M)/\varepsilon(T > T_M) = 1.94$, not 2.00 as predicted]. The quadrupole splitting increases above the Néel temperature due to a small contraction along the trigonal axis. The chemical isomer shift does not show any irregularities at the Morin transition (Fig. 10.3).

The spin-flip process can also be induced by an externally applied magnetic field [6]. A single crystal of α-Fe$_2$O$_3$ oriented along the [111] axis was placed in a large external field at 80 K. As H_{ext} increases the magnetic lines split because the vector addition of H_{eff} to H_{ext} is different for the two opposed sublattices. Between 64 kG and 71 kG there is a gradual reversion to a single six-line spectrum as the spins flip to the $\theta = 90°$ position and the sublattices become equivalent in the magnetic field. In a similar type of experiment, the temperature at which the Morin transition takes place was found to be a function of a small external magnetic field perpendicular to the [111] axis [7, 8].

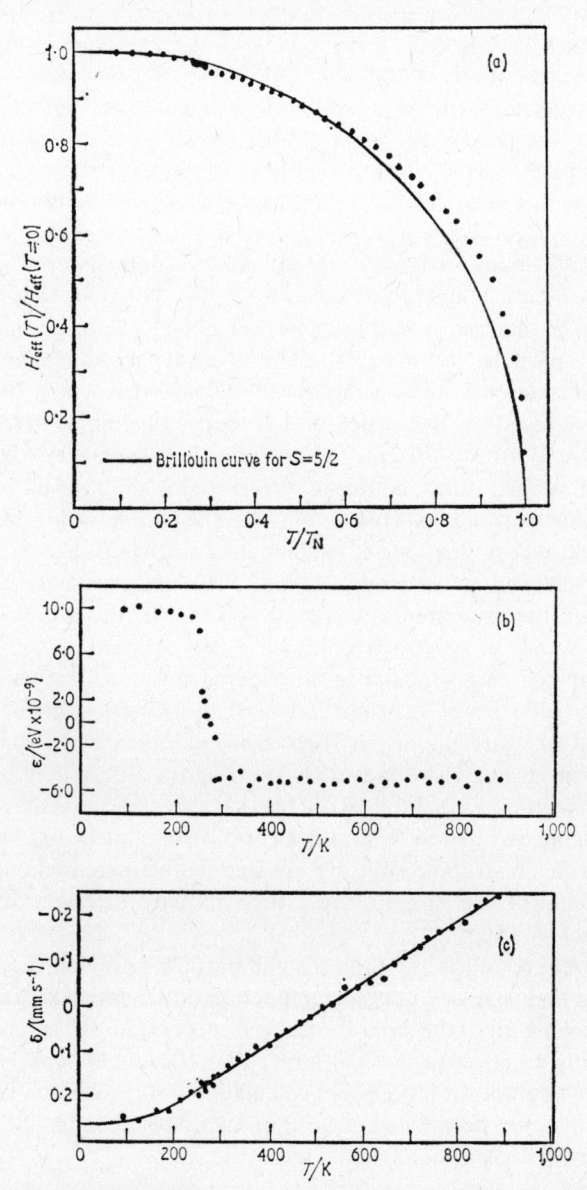

Fig. 10.3 Temperature dependence of the spectrum of α-Fe$_2$O$_3$: (a) the reduced effective hyperfine field and the Brillouin curve for $S = \frac{5}{2}$; (b) the quadrupole interaction showing the reversal in sign at the Morin transition; (c) the chemical isomer shift. [Ref. 3, Fig. 2]

A field of 9 kG for example lowers the transition temperature by 17°, the general pattern of behaviour being consistent with Dzyaloshinsky's thermodynamic theory of weak ferromagnetism.

High pressure applied to α-Fe_2O_3 causes the expected pressure-induced change in the second-order Doppler shift [9]. The electron density at the iron is effectively independent of pressure. However, the sign of the quadrupole splitting reverses at 30 kbar, the data suggesting that, in addition to a spin-flip transition, there is also some alteration in the local site symmetry.

The chemical isomer shift of α-Fe_2O_3 drops discontinuously by about 0·05 mm s^{-1} as the Néel temperature is exceeded [10]. There is no such change in the structurally similar α-Ga_2O_3 doped with iron, which is not magnetically ordered at the same temperature. This clearly demonstrates that the presence of an exchange interaction can influence the chemical isomer shift.

Ultra-fine particles of α-Fe_2O_3 exhibit superparamagnetism due to a decreasing relaxation time with decreasing particle size. The Morin temperature is also lowered as the particle size is reduced below 50–100 nm [11, 12] (1 nanometre = 10 Å). Spectra of apparently paramagnetic α-Fe_2O_3 are observed at temperatures below the Néel point as a result of motional narrowing due to rapid spin relaxation. Detailed studies have been made of small-particle α-Fe_2O_3 supported on high-area silica [13]. The room-temperature spectrum shows varying amounts of superparamagnetic material depending on the estimated average particle size of the sample (Fig. 10.4). A very similar set of spectra is obtained by cooling any given sample. The spins are oriented perpendicular to the trigonal axis at all temperatures above 10 K for particle sizes of less than 18 nm. It is not impossible of course that the preferred spin direction is partly governed in this case by the nature of the supporting substrate. The quadrupole splitting increases slightly as the particle size decreases. A sample with particle size 52·5 nm has the Morin transition depressed to 166 K, and it has been postulated that both this and the increase in quadrupole splitting are due to the increased surface effects causing a *uniform* increase in the lattice spacing throughout the whole microcrystal [14, 15].

The apparent recoilless fraction of small particles is higher than one would predict from their size [16]. It is also a function of the degree of packing of the sample, showing that the recoil energy is dissipated by several particles, thereby giving an effective mass of larger than the single particle mass.

The many attempts at lattice-sum calculations of the electric field gradient tensor have already been listed in Section 5.4 in connection with the evaluation of the iron nuclear quadrupole moment.

γ-Fe_2O_3

The γ-form of Fe_2O_3 has the spinel (AB_2O_4) structure, in which the A cations have regular tetrahedral coordination by oxygen and the B cations are octa-

BINARY OXIDES AND HYDROXIDES | 247

hedrally coordinated by oxygen (with a trigonal local distortion, although the crystal symmetry is cubic). There are insufficient Fe^{3+} cations to fill all the A and B sites so that the stoichiometry corresponds to $Fe_{8/3}\square_{1/3}O$

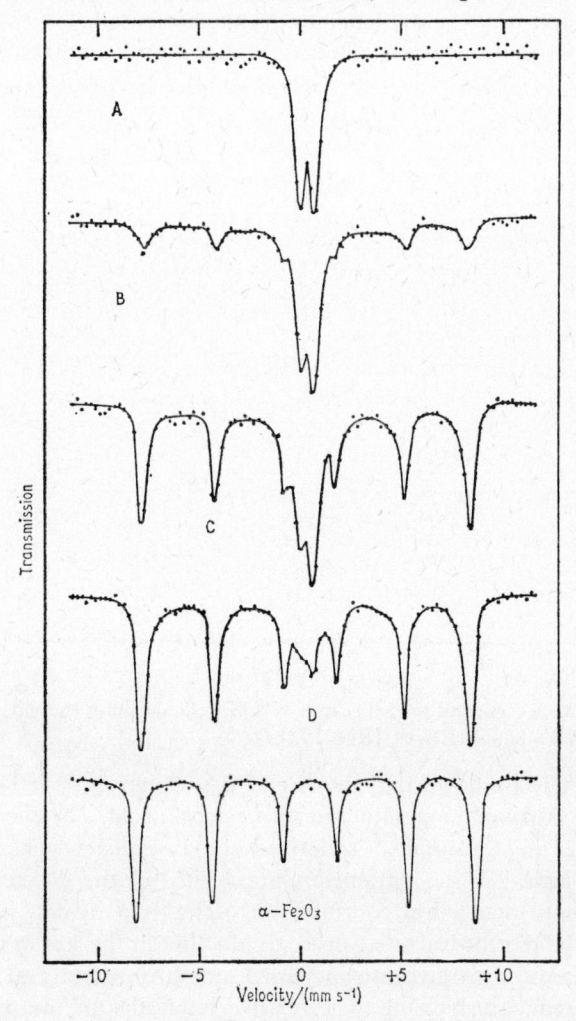

Fig. 10.4 The room-temperature Mössbauer spectra of α-Fe_2O_3 for different particle sizes: (A) <10 nm (1 nanometre = 10 Å), (B) 13·5 nm, (C) 15 nm, (D) 18 nm, and bulk material. The full hyperfine pattern is only seen for large particles. [Ref. 13, Fig. 3]

where $\square$ represents a cation vacancy. Antiparallel alignment of the A and B sublattices makes γ-Fe_2O_3 ferrimagnetic.

The Mössbauer spectrum shows apparently only one hyperfine pattern below the Néel temperature as if the A and B site cations are indistinguishable

[*Refs. on p. 296*]

[17, 18]. This is not untypical of iron(III) oxides because the ^6S state Fe^{3+} ion is not sensitive to environment. However, it is possible to separate the two sublattice contributions slightly by application of an external field [19]. The A and B site fields align parallel and antiparallel to the applied field respectively so that the two resultant fields are now dissimilar (see Fig. 10.5).

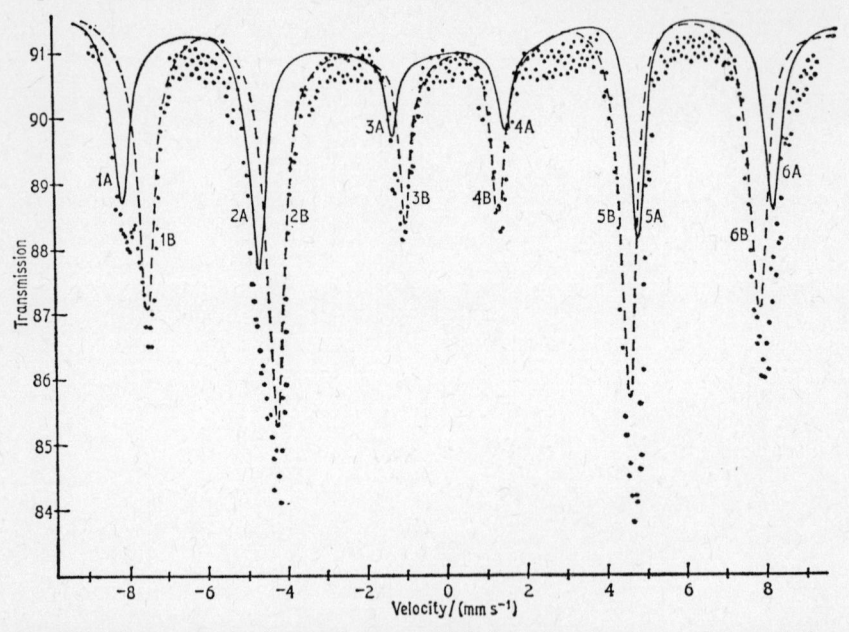

Fig. 10.5 The spectrum of γ-Fe_2O_3 in a 17-kG field showing the partial resolution of the A and B site sublattices. [Ref. 19, Fig. 1]

in this way it was shown that the A and B sites have internal fields of 488 and 499 kG at room temperature in zero external field. The chemical isomer shifts are 0·27 and 0·41 mm s^{-1} (relative to Fe) respectively. The lower values of the tetrahedral A site parameters are typical of the difference between tetrahedral and octahedral coordination of Fe^{3+} in oxides. Increased covalency at the tetrahedral site causes a reduction in the Fermi contact term of the field, and also gives an increased s-electron density at the nucleus because of reduced shielding. The relative intensities of the two hyperfine patterns give a site occupancy ratio of A/B = 0·62 ± 0·05 (assuming that the recoilless fractions are equal). If all the vacancies are located on the B sites, the structure would be $Fe_A(Fe_{5/3}\square_{1/3})_BO_4$ implying a ratio of 0·60, in good agreement with the observed value.

$Fe_{1-x}O$ (Wüstite)

Iron(II) oxide is the most unusual of the binary oxides of iron. The phase corresponding most nearly to FeO is stable only at high temperature, and is

[*Refs. on p. 296*]

always cation deficient with an upper limit of Fe$_{0.94}$O; however, it can be quenched to a metastable state at room temperature which decomposes to Fe and Fe$_3$O$_4$ on annealing. The structure is of the rock-salt type but with cation vacancies and the appropriate proportion of Fe^{3+} ions to achieve electroneutrality: Fe$^{2+}_{1-3x}$Fe$^{3+}_{2x}$□$_x$O. The Mössbauer spectrum shows an apparent Fe^{2+} quadrupole doublet with a slight suggestion of asymmetry due to the Fe^{3+} content [20]. Although the phase is crystallographically cubic, the local site symmetry will be less than cubic because of the vacancies and ferric cations, and one can expect a splitting of the t_{2g} manifold giving rise to the small temperature-dependent quadrupole splitting (∼0·6 mm s^{-1} at 295 K).

More recently two independent investigations have produced more complex interpretations which differ significantly from each other. In both cases the spectra obtained were found to show broad lines with considerable asymmetry at all compositions. In one instance the spectrum envelope was considered to arise from two overlapping quadrupole doublets [21]. The parameters for these were given as:

site I δ(Fe) = 0·76; Δ = 0·73 mm s^{-1}
site II δ(Fe) = 0·88; Δ = 0·50 mm s^{-1}

The site II cations were assumed to occupy octahedral sites and to feature electronic-exchange between Fe^{2+} and Fe^{3+} ions. The site I cations were then assigned to Fe^{3+} ions on tetrahedral sites which exchange with Fe^{2+} ions on the octahedral sites.

The alternative description invokes two Fe^{2+} quadrupole doublets and an Fe^{3+} singlet, and was based on data between 203 and 297 K for 0·905 < x < 0·935 [22]. Parameters at 297 K for Fe^{2+} ions (x = 0·93) were

site A δ(Fe) = 0·91; Δ = 0·46 mm s^{-1}
site B δ(Fe) = 0·86; Δ = 0·78 mm s^{-1}

A typical spectrum is shown in Fig. 10.6. If this interpretation is valid there is no rapid electronic exchange between sites, and the defect structure is such as to generate two distinct types of Fe^{2+} site. However, in a system as complex as this it is difficult to decide as to how many components are actually present under an unresolved envelope.

The results of thermal treatment of FeO to give Fe and Fe$_3$O$_4$ have been followed in the Mössbauer spectrum [23]. In some cases there was an apparent reduction in the quadrupole splitting of a sample which had been heated, and magnetic splitting appeared above the bulk FeO Curie temperature. The Zeeman lines were generally smeared out into a broad envelope. One possible explanation for this is that the microscopic variations in the site symmetry cause large variations in the orbital and dipolar contributions to the internal

[Refs. on p. 296]

magnetic field at the ferrous ion [24]. The result is the existence of a range of internal-field values, rather than a single value as found in a stoichiometric iron(II) compound, with a consequent gross broadening of the spectrum.

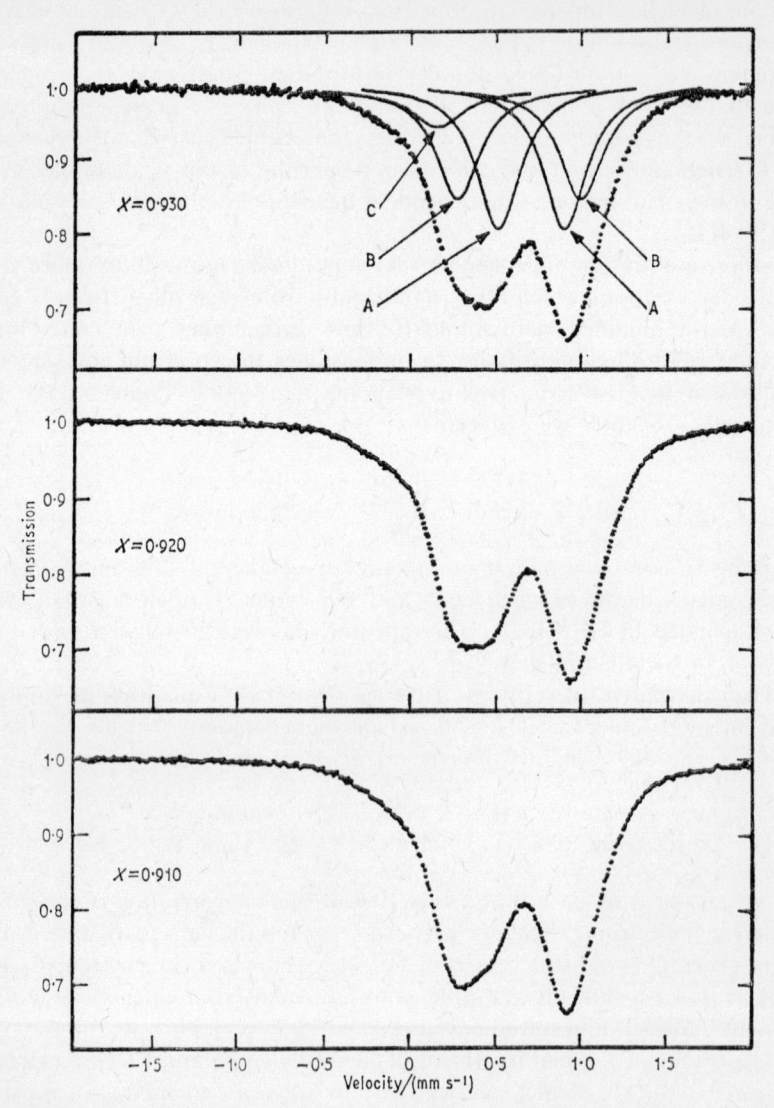

Fig. 10.6 The spectra of three samples in the $Fe_{1-x}O$ phase. The line C represents the Fe^{3+} contribution and the pairs A and B are Fe^{2+} quadrupole doublets. [Ref. 22, Fig. 1]

Application of high pressure at room temperature induces gross broadening of magnetic origins above 50 kbar [9]. There are also signs of a dis-

[*Refs.* on p. 296]

continuity in the chemical isomer shift and quadrupole splitting, but a full explanation has not been formulated.

Fe₃O₄ (Magnetite)

Magnetite is a spinel ferrite which can be written as $Fe^{3+}[Fe^{2+}Fe^{3+}]O_4$. Unlike $\gamma\text{-}Fe_2O_3$ it has no cation vacancies on the octahedral sites, but these sites contain equal numbers of Fe^{2+} and Fe^{3+} ions. A transition in many of its physical properties takes place between 110 and 120 K, and Verwey postulated a fast electron-transfer process (electron hopping) between the Fe^{2+} and Fe^{3+} ions on the octahedral B sites above this temperature. The low-temperature form has discrete valence states and orthorhombic symmetry.

Several Mössbauer investigations have confirmed that this hopping process takes place [17, 25, 26], although the exact interpretations of the spectra differ considerably. At 77 K where the Fe^{2+} and Fe^{3+} states are discrete one apparently sees two partially resolved hyperfine patterns with fields of 503 and 480 kG (see Fig. 10.7). The bar diagrams show a suggested breakdown

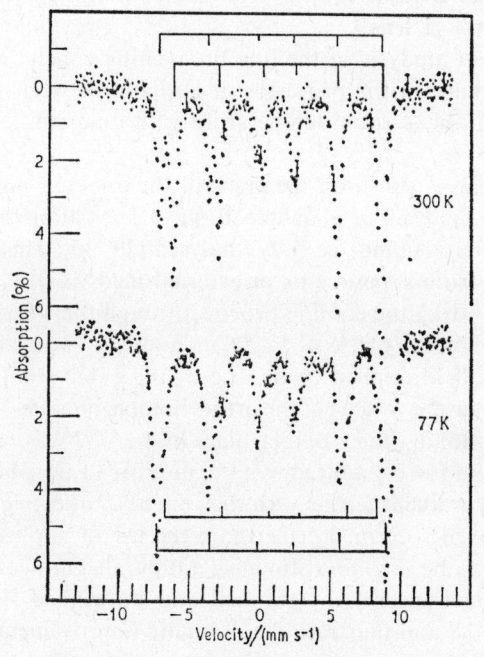

Fig. 10.7 Spectra of Fe_3O_4 above (300 K) and below (77 K) the temperature where electron hopping starts to occur. [Ref. 26, Fig. 1]

into components. The 503-kG field corresponds to the Fe^{3+} ions on both A and B sites which, as already found for $\gamma\text{-}Fe_2O_3$, are very similar. The 480-kG field is due to the B site Fe^{2+} ions, and this is confirmed by the large

[*Refs. on p. 296*]

quadrupole splitting produced by the trigonal site symmetry. At 300 K, above the transition temperature, there are fields of 491 and 453 kG corresponding to the A site Fe^{3+} ions and to the B site (Fe^{2+} + Fe^{3+}) cations respectively. The fast electron hopping produces a completely averaged spectrum from these latter ions which does not show a quadrupole effect. The outer lines reverse their intensity ratio in the transition region [25]. It seems likely that the low-temperature phase has a very regular degree of ordering because of the sharpness of the Fe^{2+} lines.

The temperature dependences of the A and B sites magnetic fields have been recorded from 300 to 800 K and show good agreement with the sublattice curves from neutron diffraction data [27].

Application of an external magnetic field of 12 kG gives better resolution of the two patterns at room temperature and establishes that the tetrahedral A site spins align antiparallel and the B site spins parallel to the applied field [28]. The lines from the B site cations are broader than those from the A site because of the electron-hopping process. Study of the relative line broadening led to a value for the relaxation time of the hopping at room temperature of $\tau = 1 \cdot 1$ ns.

An independent analysis of the line broadening concluded that there is a significant difference in the processes of hopping between 118 K and room temperature and above room temperature with different activation energies [29].

Below the Verwey transition the areas of the lines are not fully consistent with the simple analysis of a unique B site Fe^{2+} cation site, although the powder sample data cannot be fully analysed [29, 30]. The most convincing evidence comes from experiments in external fields [31]. Unless a magnetic field is applied during the cooling process through the Verwey transition the resulting orthorhombic phase of Fe_3O_4 is multiply twinned even in a single crystal. A 9-kG field applied along one of the cubic [100] axes of a single crystal ensures that the 'c' axis of the orthorhombic phase lies along that direction, although twinning can still take place in the 'a'–'b' axes. The removal of 'c' twinning from the crystal reduces the number of possible orientations of the electric field gradient tensor with the spin axis, and one can expect substantially improved spectra. Furthermore the use of applied magnetic fields on this crystal can be used to distinguish antiparallel lattices and to alter line intensities. Typical spectra for twinned and untwinned single crystals are shown in Fig. 10.8 and illustrate the dramatic improvement in resolution so obtained. Detailed computer analysis gives good evidence for four distinct cations on the B site in equal proportion, two being Fe^{2+} and two Fe^{3+}. This indicates a more complicated ordering than originally proposed by Verwey with the unit cell at least doubled in size. Numerical data are given in Table 10.1.

Detailed measurements in the vicinity of the Verwey transition have shown

[Refs. on p. 296]

that the rearrangement of the Mössbauer lines takes place between 95 and 120 K in a sample which showed a maximum in the heat capacity curve at 110 K [32]. The change in electrical conductivity normally takes place over a smaller temperature range, and it may be that the two effects are sensitive to different values of the electron exchange frequency.

The temperature dependence of the areas of the spectrum lines in Fe_3O_4

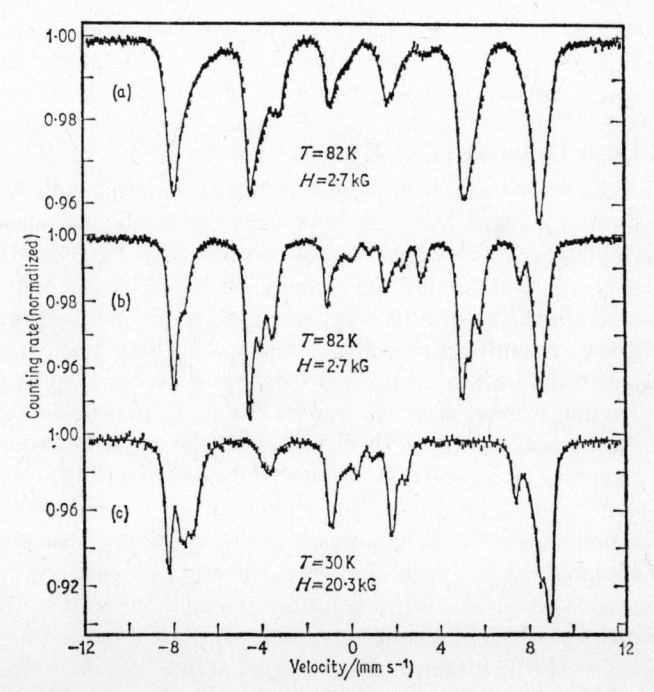

Fig. 10.8 Spectra for single crystal Fe_3O_4 below the Verwey transition: (a) cooled through the transition in zero applied field to 82 K so that 'c' twinning occurs, and polarised by a 2·7-kG field applied transversely; (b) cooled in a 9-kG magnetic field aligned along the [100] axis to 82 K and polarised similarly; (c) cooled as in (b) but with a longitudinal field of 20·3 kG at 30 K. In all cases the [100] direction was parallel to the applied field. Note the dramatic improvement in resolution when 'c' twinning is eliminated. [Ref. 31, Fig. 1]

between 300 and 800 K has been used to measure the effective Debye temperatures of the individual sites [33]. The results were $\theta_A = (334 \pm 10)$ K and $\theta_B = (314 \pm 10)$ K with the ratio of the recoil-free fractions at room temperature $f_B/f_A = 0.94$. These figures emphasise the approximation which may be involved in measuring site occupancy directly from area ratios.

Studies on nonstoichiometric Fe_3O_4 in an external magnetic field at 300 K seem to suggest that the introduction of vacancies at the octahedral B sites prevents electron hopping except where Fe^{2+}–Fe^{3+} octahedral pairs exist [34].

[*Refs. on p. 296*]

Non-paired Fe^{3+} cations do not participate in the exchange process and thereby tend to contribute to the A site (Fe^{3+}) components of the spectrum. An external magnetic field causes broadening of this component because the spin alignment remains antiparallel. The localised nature of the exchange process accounts for the observed drastic decrease in conductivity with increasing nonstoichiometry, although in a perfect stoichiometric lattice the overall effect can become one of delocalisation.

Small-particle Fe_3O_4 shows temperature-dependent motional narrowing to a degree dependent on the mean particle size [35]. A detailed analysis was found to give a satisfactory interpretation of the spectra.

Iron(III) Oxide Hydroxide, FeOOH

The iron oxide hydroxides have been investigated independently by several research groups, and the Mössbauer spectra have highlighted many of the structural differences. Although the various forms of FeOOH have been known for some considerable time, many of the published data are inconsistent, and in particular little was known about their magnetic properties until relatively recently. α-FeOOH, goethite, has the same structure as α-AlOOH with the iron in a distorted octahedral environment of oxygens, and a three-dimensional structure results from the sharing of edges and corners of the octahedra. Lepidocrocite, γ-FeOOH, is similar to α-FeOOH but has a complex layer structure. β-FeOOH differs in that it is clearly non-stoichiometric, containing various quantities of F^- or Cl^- and H_2O depending on the preparation and conditions. It has the α-MnO_2 structure with a three-dimensional oxygen lattice and an octahedral iron environment. Little is known about δ-FeOOH, which is ferromagnetic in contrast to the α-, β-, and γ-forms for which low susceptibilities imply paramagnetism or antiferromagnetism. Little is known about its crystal structure other than that it appears to be based on a hexagonally close-packed oxygen (hydroxyl) lattice, and that high disorder or small crystallite dimensions pertain in the [001] direction. This structure would imply four tetrahedral and two octahedral sites for every iron atom.

Early work on α-FeOOH reported a magnetic hyperfine splitting at room temperature, showing it to be antiferromagnetic [36, 37]. Other data showed evidence for two hyperfine patterns which it was claimed had a different temperature dependence and ordered at 340 K and 370 K respectively [38, 39]. This implies four magnetic sublattices to preserve antiferromagnetism. Further confusion was generated by two reports of superparamagnetism in small-particle α-FeOOH, the incipient broadening of the outer peaks being not unlike the twin-magnetic field data [40, 41]. Better quality data over the range 90–440 K showed only one hyperfine field below the Néel temperature of 393 K with a zero-temperature extrapolated value of 510 kG [42]. Lattice-sum calculations of the small quadrupole effect were used to infer that the spins lie along the 'c' axis, and the sublattice magnetisation was compared with theory.

The inconsistencies were explained by Dezsi and Fodor [43], who found that three out of five deposits of natural α-FeOOH and a synthetic sample all gave a unique hyperfine pattern, while the other two did show clear evidence of two fields. The latter result was found to be true of samples containing excess water; deposits of stoichiometric α-FeOOH do not show evidence for more than two magnetic sublattices. The Néel temperature was found to be only 367 K, and relaxation occurred at up to 30° below this. It would appear that the defect structure and particle size are both important in determining the properties of a specific sample.

A definitive study of the magnetic structure using powdered synthetic α-FeOOH and a natural single crystal has been made by Mössbauer and neutron diffraction methods [44]. The spins are all parallel to the 'c' axis of the crystal (also shown by partial orientation of powder samples [45]) with the major axis of the electric field gradient tensor in the 'ab' plane, making an angle of about 40° with the 'b' axis. The Mössbauer data gave a Néel temperature of 403 K. Neutron diffraction data indicate both ferromagnetic and antiferromagnetic coupling between iron atoms. The Néel temperature will correspond to the point where the weaker of these breaks down, and it is the existence of residual order (which is not three-dimensional) above T_N which causes a reduced susceptibility compared with that predicted for completely disordered $S = \frac{5}{2}$ spin states. The extrapolated magnetic field value of $H(0) = 504$ kG is smaller than that of Fe_2O_3 (544 kG) or FeF_3 (622 kG) and reflects the effect on the Fermi term of increased covalency in this compound. In this sense the magnetic field of ferric ions is a better indication of degree of covalency than the chemical isomer shift.

Deuteration to give α-FeOOD does not affect the electronic structure to any extent, and the Mössbauer spectrum is identical to that of α-FeOOH [46].

β-FeOOH is antiferromagnetic below 295 K with a magnetic hyperfine pattern similar to the stoichiometric α-FeOOH [36, 37, 47]. The extrapolated $H(0)$ value is 475 kG, and relaxation collapse is seen above 200 K in small-grained samples [47]. The paramagnetic quadrupole splitting of 0·70 mm s^{-1} is to be compared with the ε value of -0.10 mm s^{-1} in the magnetic spectrum and shows that the electric field gradient major axis and the magnetic spin axis are not collinear. Oriented powder data show that the spin axis is parallel to the crystal 'c' axis and that the electric field gradient axis is probably in the 'c' plane [45]. Decomposition to α-Fe$_2$O$_3$ at 670 K brings about a reversion to a magnetic spectrum.

γ-FeOOH is paramagnetic down to at least 77 K with a small quadrupole splitting [36, 37, 39, 48]. Below this temperature it becomes antiferromagnetic [48]. The Néel temperature is not clearly defined in either mineral or synthetic samples, and magnetic and paramagnetic spectra coexist over a range

[*Refs. on p. 296*]

256 | IRON OXIDES AND SULPHIDES

of at least 10 K even in large single crystals. This excludes small-particle superparamagnetism as an explanation. Possibly the small magnetic anisotropy in the region of the Néel temperature causes the spin fluctuations to be slow.

At 4·2 K the magnetic field is 460 kG and from single-crystal data the spin axis lies in the '*ac*' plane. Application of a 30-kG field parallel to the '*c*' axis splits the antiparallel sublattices, but has no effect when perpendicular to the '*c*' axis, and confirms that the spins are collinear and aligned in the '*c*' axis. Data with a single crystal at room temperature and a polarised ^{57}Co/iron metal source showed that the major axis of the electric field gradient tensor lies in the '*ac*' plane at 55° to the '*c*' axis, and the sign of e^2qQ is negative.

δ-FeOOH shows a ferromagnetic spectrum with two magnetic fields, probably derived from the tetrahedral and octahedral sites, of 505 and 525 kG at 80 K [47]. The compound begins to decompose above 370 K (i.e. below the extrapolated Curie point) and the product above 510 K can be identified as α-Fe$_2$O$_3$.

Iron(II) Hydroxide, Fe(OH)$_2$

Fe(OH)$_2$ has a hexagonal layer structure of the CdI$_2$ type with iron octahedrally coordinated by OH ions. The '*c*' axis is the major symmetry axis. Susceptibility data suggest antiferromagnetic ordering below 34 K, and this is confirmed by the Mössbauer spectrum [49]. The paramagnetic state shows a large quadrupole splitting of 3·00 mm s^{-1} at 90 K and a chemical isomer shift of 1·16 mm s^{-1} (w.r.t. Fe). At 4·2 K, an apparent four-line magnetic spectrum is seen (Fig. 10.9) although there are in fact eight lines present, and

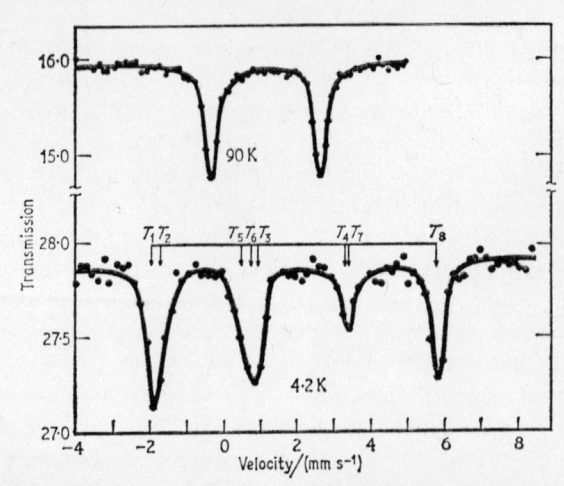

Fig. 10.9 Mössbauer spectra of Fe(OH)$_2$, showing the suggested line positions from Fig. 10.10. [Ref. 49, Fig. 1]

[*Refs.* on p. 296]

an analysis can be made in terms of an axially symmetric field gradient tensor making an angle of θ to the field H (Fig. 10.10). This gives a field of 200 kG

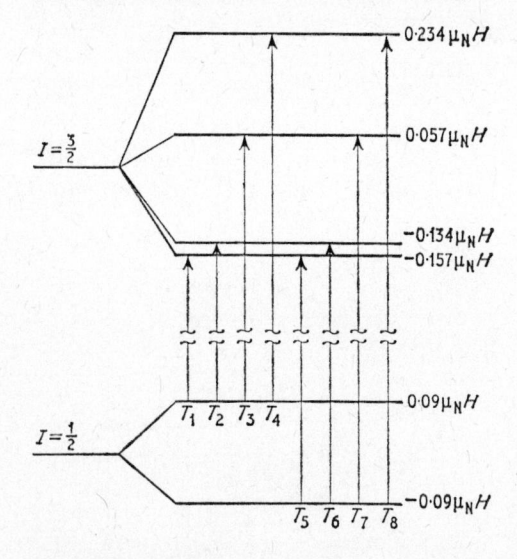

Fig. 10.10 The energy level scheme for $Fe(OH)_2$ with the main axis of the electric field gradient at 90° to the spin axis. [Ref. 49, Fig. 2]

and $e^2qQ/2$ as 3·06 mm s^{-1}. θ is found to be 90°, which suggests that the spin axis lies in the 'c' plane.

Iron(III) Oxide Fluoride, FeOF

The iron(III) oxide fluoride, FeOF adopts the rutile lattice structure, although the X-ray data cannot differentiate between oxygen and fluorine. The Mössbauer spectrum [50] indicates magnetic ordering below about 315 K, and the temperature dependence of the magnetic field follows an $S = \frac{5}{2}$ Brillouin function with a zero-temperature value of 485 kG. The principal axis of the electric field gradient is probably perpendicular to the spin axis, and the intensities of the magnetic patterns suggest that the asymmetry parameter is finite. The observation of a unique quadrupole splitting for the iron (1·18 mm s^{-1} in the paramagnetic state) suggests an ordered distribution of oxygen and fluorine giving a superstructure which is not observed in the X-ray pattern. A neutron diffraction study [51] showed that the 'c' axis is the antiferromagnetic spin axis.

Iron(III) Hydroxide Gels

The gels of $Fe(OH)_3$ precipitated from aqueous solution by reagents such as NH_4OH present a complex system. Recoil-free fraction measurements have been made on such gels before and after drying, and on the organogel obtained by treatment with EtOH [52]. The results for the hydrogel and organogel show a sharp decrease above 270 and 160 K respectively in agreement with the freezing temperature of intermicellar liquids.

[*Refs. on p. 296*]

10.2 Spinel Oxides AB_2O_4

The capacity of the spinel structure type to accommodate a wide variety of metal cations is reflected in the large number of known iron-containing spinels in addition to the binary spinels Fe_3O_4 and γ-Fe_2O_3 which have already been discussed. The iron can occupy either or both of the distinct cation sites, and many systems have wide ranges of solid solution and nonstoichiometry. The idealised normal spinel structure is shown in Fig. 10.11 and consists of

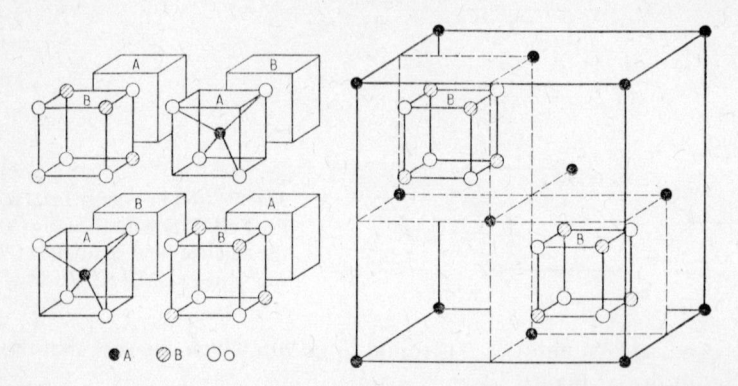

Fig. 10.11 The normal spinel structure, AB_2O_4, contains eight octants of alternating AO_4 and B_4O_4 units as shown on the left; the oxygens build up into a face-centred cubic lattice of 32 ions which coordinate A tetrahedrally and B octahedrally. The unit cell, $A_8B_{16}O_{32}$, is completed by an encompassing face-centred cube of A ions, as shown on the right in relation to two B_4O_4 cubes.

tetrahedral (A) sites and octahedral (B) sites in a face-centred cubic oxide sublattice, i.e. $A[B_2]O_4$. If half the B cations occupy all the tetrahedral sites and the remaining B cations are distributed with the A cations on the octahedral sites, the spinel is said to be inverse, i.e. $B[AB]O_4$. The tetrahedral site in the ideal spinel has cubic (T_d) symmetry and therefore no electric field gradient at the cation, but the octahedral site has trigonal point symmetry and one anticipates a large electric field gradient. The Mössbauer spectrum allows a more intimate study of the effects on individual cations of changes in the species at neighbouring sites than most other experimental methods available. Furthermore, many of these spinel phases are magnetically ordered; the major interaction is of the A–B type, and this may involve Fe–Fe inter-actions as well as Fe–M interactions where M is another paramagnetic cation.

It will be convenient to summarise some of the general results before discussing specific compounds in more detail. Taking iron(II) spinels first in the paramagnetic region: normal 2,3 spinels $Fe^{II}M^{III}_2O_4$ with Fe^{2+} in the

[Refs. on p. 296]

tetrahedral A sites only give a single-line spectrum because of the cubic symmetry. By contrast normal 4,2 spinels $M^{IV}Fe_2^{II}O_4$ with Fe^{2+} on the trigonally distorted octahedral B sites give a well-spaced quadrupole-split spectrum. With inverse or partly inverse 2,3 spinels two pairs of lines are expected, one quadrupole doublet from the B site and one less widely spaced doublet from the A sites which will no longer be strictly cubic due to charge variations on the B sites.

With iron(III) spinels it is not possible to have a normal spinel structure with Fe^{3+} on the A sites because the cationic charge is an odd number but a disordered spinel such as $Fe^{3+}[M_a^{II}M_b^{III}]O_4$ can have the tetrahedral sites occupied by Fe^{3+}; disorder in the B sites generates a small electric field gradient at the A site and a small quadrupole splitting is observed. Normal 2,3 spinels $M^{II}Fe_2^{III}O_4$ with Fe^{3+} on the trigonally distorted B sites show a similar spectrum because the electric field gradient acting on the spherical d^5 ion gives a smaller quadrupole splitting than that observed with the d^6 Fe^{2+} ion on the B sites. Most inverse 2,3 iron(III) spinels are magnetically ordered even above room temperature and the Fe^{3+} occupancy of both A and B sites leads to two superimposed six-line spectra in which only the B site shows a quadrupole hyperfine interaction.

The combined data from chemical isomer shifts, quadrupole splittings and hyperfine magnetic fields, coupled with intensity measurements and the changes which occur in the spectra in applied magnetic fields, provide a powerful means of studying site occupancy, site symmetry, and magnetic exchange interactions. The systematics outlined in earlier chapters lead to the following three diagnostic generalisations which are well illustrated by spinel phases:

(i) chemical isomer shifts increase in the sequence $\delta(\text{tetrahedral Fe}^{3+}) < \delta(\text{octahedral Fe}^{3+}) < \delta(\text{tetrahedral Fe}^{2+}) < \delta(\text{octahedral Fe}^{2+})$
(ii) quadrupole splitting for both Fe^{2+} and Fe^{3+} in sites of accurately cubic symmetry is zero; for non-cubic sites $\Delta(Fe^{3+}) \ll \Delta(Fe^{2+})$
(iii) the magnetic hyperfine field H at Fe^{2+} is less than $H(Fe^{3+})$.

These general remarks will now be illustrated by reference to individual oxide spinel phases. Sulphide spinels are discussed on p. 285.

$Fe^{II}Al_2^{III}O_4$

$FeAl_2O_4$ shows a large iron(II) quadrupole splitting [53], and detailed temperature-dependence data from 0 to 800 K can be explained using a tetrahedral site model, with a splitting of the E level by 300 K (208 cm^{-1}) and a thermal population of the two resultant singlet levels [54]. Data for $Mg_{0.98}Fe_{0.02}Al_2O_4$ are identical. However, it is now known that the compound may be partly inverse [55]. Recent data at 295 K showed the two sites to be resolved giving an estimate for the individual site occupancies as

260 | IRON OXIDES AND SULPHIDES

$Fe^{2+}_{0.77}Al^{3+}_{0.23}[Fe^{2+}_{0.23}Al^{3+}_{1.77}]O_4$, although the degree of inversion is likely to depend on the sample preparation. There are signs of spin ordering at 4·2 K [54].

$Fe^{II}Ga^{III}_2O_4$

FeGa$_2$O$_4$ shows two iron(II) quadrupole doublets, indicating that it is partly inverse [56]. It is unusual in that reheating the product of the original reaction mixture causes an alteration in the site occupancy, apparently involving a change towards a normal spinel structure.

$Ge^{IV}Fe^{II}_2O_4$

GeFe$_2$O$_4$ is a normal spinel with the B site iron(II) ions producing a large quadrupole splitting (2·9 mm s^{-1}) at room temperature [56, 57]. It is antiferromagnetic below 11 K [58–60]. The axially symmetric electric field

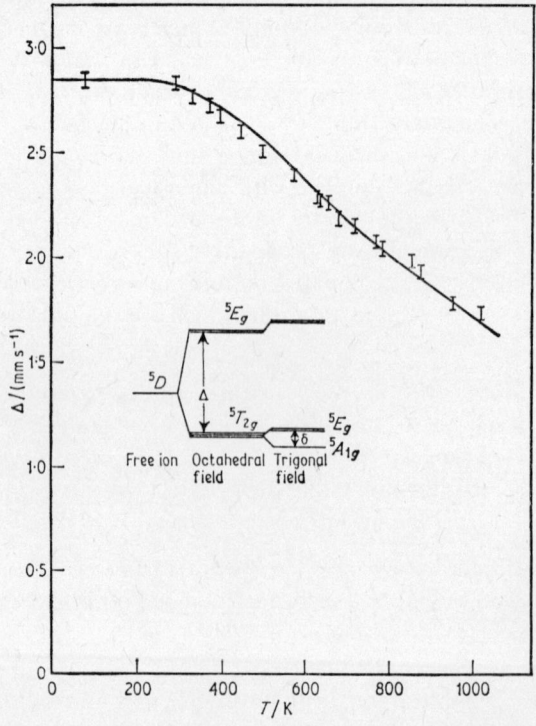

Fig. 10.12 The temperature dependence of the quadrupole splitting in GeFe$_2$O$_4$. The solid line corresponds to a trigonal-field splitting of 1650 K. [Ref. 61, Fig. 1]

gradient produced by the trigonal site symmetry is negative and is perpendicular to the spin axis. The magnetic field has an extrapolated value of 163 kG at 0 K, and the temperature dependence deviates slightly from an

[*Refs.* on p. 296]

$S = 2$ Brillouin function. The observation of a unique environment in the magnetically ordered phase means that the spin direction is perpendicular to the trigonal axis on all three possible orientations of this axis, which refutes a previous suggestion that there were only two sublattices.

The electronic ground state is the singlet $|d_z{}^2\rangle$, and the singlet–doublet separation determined from the temperature dependence of the quadrupole splitting between 77 and 1020 K is 1650 K (1145 cm^{-1}) [61]. Since the spin–orbit coupling (~100 cm^{-1}) is much smaller than the level splitting, satisfactory agreement can be obtained without including it in the calculations (see Fig. 10.12).

Ti^{IV}Fe$_2^{II}$O$_4$ and Related Phases

TiFe$_2$O$_4$ is an inverse spinel, Fe^{2+}[Fe^{2+}Ti^{4+}]O$_4$, and shows a large quadrupole splitting above the Néel temperature of 140 K [24, 26, 62, 63]. The A site spectrum is not resolved from the B site spectrum in the range from 140 to 800 K, and clearly the A site does not have cubic symmetry [63]. Each tetrahedral Fe^{2+} ion has yTi^{4+} and $(12 - y)$Fe^{2+} octahedral sites in the second coordination sphere [$0 < y < 12$], whereas each octahedral Fe^{2+} ion

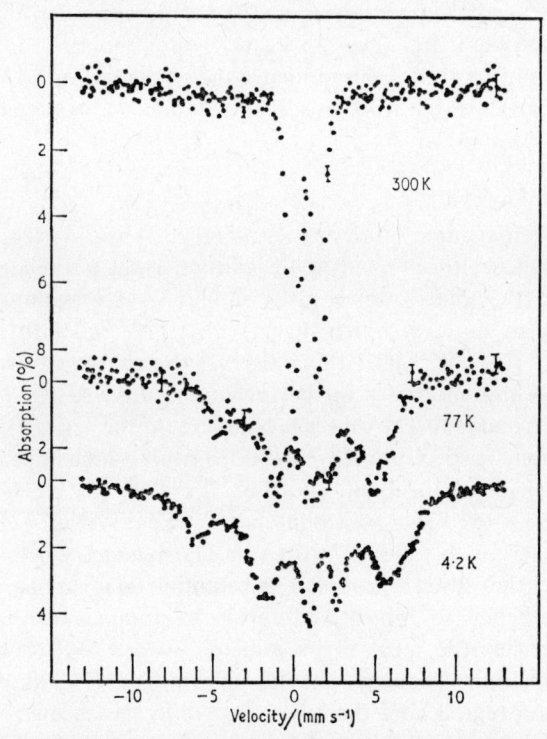

Fig. 10.13 Spectra of TiFe$_2$O$_4$ showing gross broadening of the magnetic hyperfine patterns below the Néel temperature. [Ref. 26, Fig. 2]

[Refs. on p. 296]

has 6 tetrahedral-site Fe^{2+} cations plus $(6-x)Fe^{2+}$ and xTi^{4+} octahedral sites [$0 < x < 6$]. Randomisation of the B site cations renders many of the local A site symmetries non-cubic [24, 64]. The temperature dependence of the quadrupole splitting corresponds to a splitting of the tetrahedral E level by 450 K (333 cm^{-1}) and of the octahedral T_{2g} level by 600 K (417 cm^{-1}) with the singlet lying lowest [63]. Below the Néel temperature, the magnetic spectra are very broad because of the varying effects of the randomisation on the orbital and dipolar contributions to the hyperfine field (see Fig. 10.13).

$TiFe_2O_4$ forms a continuous solid solution with Fe_3O_4, and the spectra of the intermediate compositions all show gross magnetic broadening below the Néel temperature because of disorder and consequent variation of local site symmetry [24, 64]. In $Fe^{3+}_{0.6}Fe^{2+}_{0.4}[Fe^{2+}_{1.2}Fe^{3+}_{0.2}Ti^{4+}_{0.6}]O_4$ there is a considerable improvement in resolution of the spectrum at 77 K, an Fe^{3+} hyperfine pattern being seen, and it has been suggested that the additional broadening of the spectra at higher temperatures is the result of an electron-hopping exchange between Fe^{2+} and Fe^{3+} ions on both A and B sites [26]. Thus local variations in hyperfine fields can arise either from (a) random variations of local symmetry at the iron sites which are characteristic of the local disorder and independent of temperature, or (b) temperature-dependent electron hopping on the octahedral sites. In Fe_3O_4 hopping occurs; in $Ti_{0.6}Fe_{2.4}O_4$ both hopping and local disorder influence the spectrum; in $TiFe_2O_4$ hopping cannot occur since all the iron ions are Fe^{2+} and the broadening therefore arises from local disorder.

$Fe^{II}V^{III}_2O_4$ and Fe_2VO_4

FeV_2O_4 is a normal cubic spinel at room temperature and accordingly shows an Fe^{2+} resonance without quadrupole splitting at the tetrahedral A site [53]. The $Fe_{1+x}V_{2-x}O_4$ solid solution gives similar spectra without quadrupole broadening up to $x = 0.5$, which suggests that Fe^{3+} is substituting for V^{3+} at B sites, $Fe^{2+}[Fe^{3+}_xV^{3+}_{2-x}]O_4$; this preserves the charge equivalence on the octahedral sites and minimises any perturbations at the A site. The linewidths are, however, broad [53]. The other end-member of the series Fe_2VO_4 gives a broadened magnetic spectrum at room temperature which appears to indicate two magnetic fields. A formulation of $Fe^{2+}[Fe^{2+}V^{4+}]O_4$ has been suggested [53], although a more likely possibility is $Fe^{2+}[Fe^{3+}V^{3+}]O_4$ [24].

FeV_2O_4 undergoes a transition to tetragonal symmetry below 140 K, and this results in the appearance of a substantial quadrupole splitting [65]. However, this begins to appear well above the nominal transition temperature, showing that the local site symmetry distorts well before the bulk crystalline transition. The experimental linewidths are significantly broader in the transition region, and there is a change in the chemical isomer shift. Qualitatively, the data can be interpreted in terms of a dynamic reorientation between the three equivalent tetragonal distortions, so that the distortion is

[*Refs. on p. 296*]

'frozen' at lower temperatures and 'time-averages' to an effective cubic symmetry at higher temperatures because of an increased reorientation frequency. FeV_2O_4 becomes ferrimagnetic below 109 K with a slight orthorhombic distortion in the vicinity of 100 K [61, 66, 67]. The electric field gradient tensor has axial symmetry and e^2qQ a positive value, i.e. the ground state is $|d_{x^2-y^2}\rangle$, and together with the spin axis is aligned along the 'c' axis. The extrapolated values $H(0)$ and Δ_0 are 86 kG and 3·15 mm s^{-1} respectively. However, the magnetic structure is not simple, and the anomalous orthorhombic region from about 70 to 100 K seems to involve a small misalignment between the tetragonal and magnetic axis [66, 67].

$Fe^{II}Cr_2^{III}O_4$

$FeCr_2O_4$ is also a normal iron(II) spinel without resolved quadrupole splitting at room temperature, although the resonance line is broad [53, 57]. Like FeV_2O_4, it undergoes a tetragonal distortion below 135 K with a similar quadrupole splitting behaviour [65]. Magnetic ordering is seen below 69 K, although a second phase change occurs at 40 K causing anomalies in the temperature dependence of the magnetic field [60, 68]. At low temperatures the axially symmetric electric field gradient tensor has a negative value for e^2qQ so that the distortion at the A site is in the opposite sense to FeV_2O_4. Above 40 K the spins lie in the 'ab' plane along the 'a' or 'b' axes, or both. Below 40 K there appears to be randomisation of the spin axis within the 'ab' plane.

The spinel $Fe^{3+}_{0.92}Li^{+}_{0.08}[Li^{+}_{0.42}Fe^{3+}_{0.38}Cr^{3+}_{1.20}]O_4$ is magnetically ordered below 503 K, and appears to show broadening from disorder effects [69].

$Li^{I}_{0.5}Fe^{III}_{2.5}O_4$

The spinel $Fe^{3+}[Fe^{3+}_{1.5}Li^{+}_{0.5}]O_4$ contains lithium in the octahedral B sites. Thermal treatment can cause long-range ordering of the B site cations, but the spectra of ordered and disordered samples are identical and show only one apparent magnetic splitting, which is 510 kG at room temperature [18]. The ferrimagnetic ordering will be maintained by A–B exchange coupling between ferric ions. As already seen in connection with γ-Fe_2O_3, the Fe^{3+} ion is not very sensitive to environment, and this accounts for the relatively simple spectrum in this instance despite the large difference in formal cation charges in the second coordination spheres which might have been expected to give a detectable quadrupole effect.

$Mg^{II}Fe_2^{III}O_4$

$MgFe_2O_4$ has been investigated [70] using the Faraday effect to determine the degree of inversion in this ferrimagnetic spinel which can be formulated as $(Mg^{2+}_xFe^{3+}_{1-x})[Mg^{2+}_{1-x}Fe^{3+}_{1+x}]O_4$. The A and B sites both give a magnetic field of about 500 kG at room temperature and they are antiferromagnetically

[*Refs. on p. 296*]

264 | IRON OXIDES AND SULPHIDES

coupled. Application of a 55-kG longitudinal external field separates the two antiparallel lattice hyperfine patterns which in this case are considerably simplified by the absence of the $\Delta m = 0$ resonance lines. The spectrum of a 'thick' absorber of MgFe$_2$O$_4$ isotopically enriched in ^{57}Fe is shown in Fig. 10.14. This absorber was used as a transmitter in a zero-velocity Mössbauer

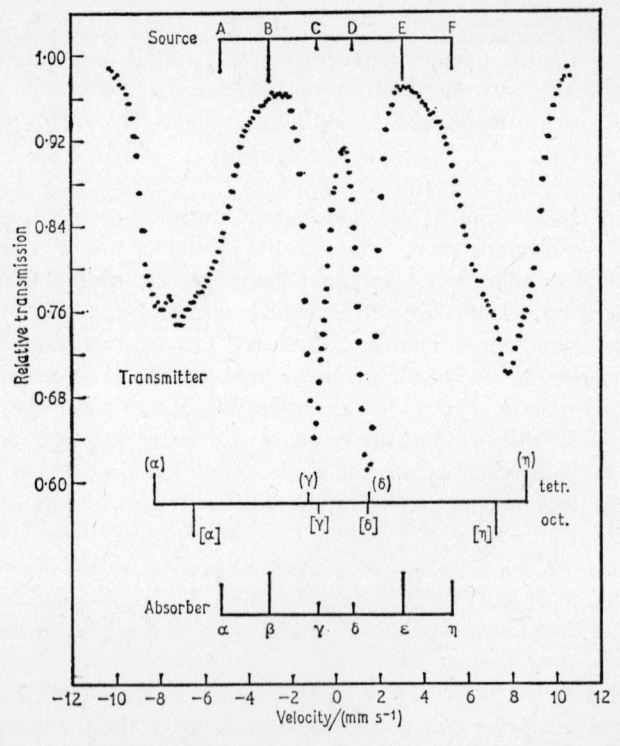

Fig. 10.14 Transmission spectrum of polycrystalline Mg^{57}Fe$_2$O$_4$ in a longitudinal field of 55 kG using a single-line source. The tetrahedral and octahedral sublattices are shown. The top bar diagram is the emission spectrum of the α-iron source, and the bottom diagram is the absorption spectrum for the α-iron absorber. Both are polarised. [Ref. 70, Fig. 2]

polarimeter which employed a ^{57}Co/α-Fe source as polariser and a ^{57}Fe/α-Fe absorber as analyser. The metal foils were both polarised in transverse magnetic fields, the relative orientation between which was ω. The only significant radiations transmitted by this source/transmitter combination are the $|\pm\tfrac{1}{2}\rangle \to |\mp\tfrac{1}{2}\rangle$ components of the source (lines B and E at the top of Fig. 10.14), which are both subject to Faraday rotation. The dependence of the zero-velocity transmission on the angle ω between source and analyser polarisation directions shows a periodic behaviour, which is repeated after insertion of the MgFe$_2$O$_4$ transmitter but with a phase shift of 19° due to the

[*Refs. on p. 296*]

Faraday effect (Fig. 10.15). The relevant theoretical equations show that this rotation is a function of the difference in the oppositely directed polarisations

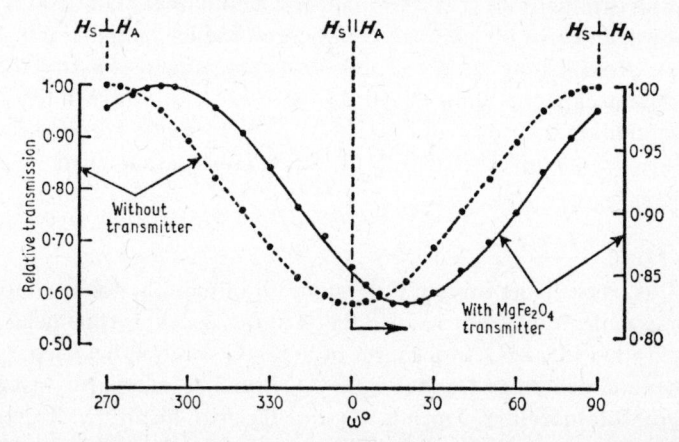

Fig. 10.15 The relative transmission of the polariser (H_S) and absorber (H_A) in transverse magnetic fields both with and without the $MgFe_2O_4$ transmitter as a function of the angle ω between H_S and H_A. [Ref. 70, Fig. 3]

from the two magnetic sublattices which arises because of their unequal occupation, and enables one to derive a value for the occupancy parameter, x, of 0.26.

$Mn^{II}Fe_2^{III}O_4$ and Related Phases

$MnFe_2O_4$ is partly inverse, and application of a 17-kG field separates the A and B site fields which are 483 and 430 kG respectively at room temperature [71]. The site occupancy by Fe^{3+} ions is $n_A/n_B = 0.12$, in good agreement with neutron diffraction data. Inconsistencies in the published data for $MnFe_2O_4$ have been suggested to result from nonstoichiometry with oxygen vacancies and Fe^{2+} cations [72]. However, independent work has shown that at least two forms, formulated as $Mn_{0.8}^{2+}Fe_{0.2}^{3+}[Mn_{0.2}^{3+}Fe_{1.6}^{3+}Fe_{0.2}^{2+}]O_4$ and $Mn_{0.48}^{2+}Fe_{0.52}^{3+}[Mn_{0.06}^{2+}Mn_{0.46}^{3+}Fe_{1.02}^{3+}Fe_{0.46}^{2+}]O_4$, can exist [73].

The $Mn_xFe_{3-x}O_4$ phase has been briefly examined [74]. $Mn_{2.7}Fe_{0.3}O_4$ shows a tetragonal distortion which disappears at ~1370 K and there is a concomitant reduction in the quadrupole splitting of the Fe^{3+} ion as the effect of Jahn–Teller distortion of the Mn^{3+} ion is removed [75]. This effect has been studied at several compositions in the range $2.7 \leqslant x \leqslant 2.95$ [76].

The $Mn_{1-x}^{II}Zn_x^{II}Fe_2^{III}O_4$ phase has been studied to follow the effects of A site randomisation on the iron in the B site [77]. The paramagnetic Mn^{2+} and diamagnetic Zn^{2+} ions will have different effects on the A–B magnetic exchange interactions. A reduction in the number of Mn^{2+} nearest neighbours causes a reduction in the Fe^{3+} hyperfine field. A detailed neutron

[*Refs.* on p. 296]

diffraction and Mössbauer study of the temperature dependence of the sublattice magnetisation in $(Mn_{0.6}Zn_{0.4})[Fe_2]O_4$ has shown good agreement between the two methods [78]. The magnetic field of 490 kG at 100 K suffers a considerable degree of relaxation collapse at higher temperatures because of the local weakening of the Fe–Fe exchange interactions by the Zn^{2+} cations. Site occupancy studies have also been made on Mg/Mn ferrites [79], and with additional Cr doping [80].

Small-particle samples of $MnFe_2O_4$ show superparamagnetic relaxation effects [81].

$Co^{II}Fe_2^{III}O_4$

$CoFe_2O_4$ is basically an inverse Fe^{3+} spinel, although the room-temperature magnetic spectrum does not resolve the A and B site hyperfine fields, which average 516 kG [81, 82]. Application of a 55-kG external field separates the sublattices and demonstrates that $CoFe_2O_4$ is in fact only partly inverse [83]. The degree of inversion depends on the thermal history of the sample; samples which have been slowly cooled show two octahedral site symmetries, but at least four probably exist in quenched specimens. Small-particle $CoFe_2O_4$ shows superparamagnetic behaviour with partial or complete collapse of the hyperfine spectrum at temperatures below the Néel temperature [81, 84].

$Ni^{II}Fe_2^{III}O_4$ and Related Phases

$NiFe_2O_4$ is also an inverse Fe^{3+} spinel which is magnetically ordered at room temperature. The A and B site hyperfine fields are very similar so that the two six-line patterns are but poorly resolved ($H_A = 506$ kG, $H_B = 548$ kG) [85, 86]. It was proposed [85] that in $NiFe_2O_4$, $NiFe_{1.9}Cr_{0.1}O_4$, and $NiFeCrO_4$ the magnetic ordering follows the Yafet–Kittel model; that is, the A and B sublattices are each further divided into sub-lattices such that their vector resultants are aligned in opposite directions to each other, giving ferrimagnetism. This contrasts with the Néel model of antiferromagnetically coupled collinear A and B sublattices. However, application of a 70-kG longitudinal field at 4·2 K fully resolves the A and B sublattices, and the $\Delta m = 0$ lines are absent [86]. This can only be consistent with the Néel ordering type. $NiFe_{0.3}Cr_{1.7}O_4$, which was formerly believed to have Fe^{3+} on the A sites, was found, from experiments in external fields, to have $\frac{1}{3}$ of the Fe^{3+} on B sites [86]. In this instance the $\Delta m = 0$ lines do not disappear, and one can conclude that in this case the ordering is of the Yafet–Kittel type. Typical spectra and the spin-vector additions are illustrated in Fig. 10.16.

The sublattice magnetisation has been studied as a function of temperature up to the Néel temperature, and good agreement can be obtained with a Weiss-field model if a strong positive B–B interaction between the Ni^{2+} and

SPINEL OXIDES AB_2O_4 | 267

Fe^{3+} ions on octahedral sites is assumed [87]. Superparamagnetic effects are seen in ultra-fine powders [84]. The hyperfine spectrum can be partially

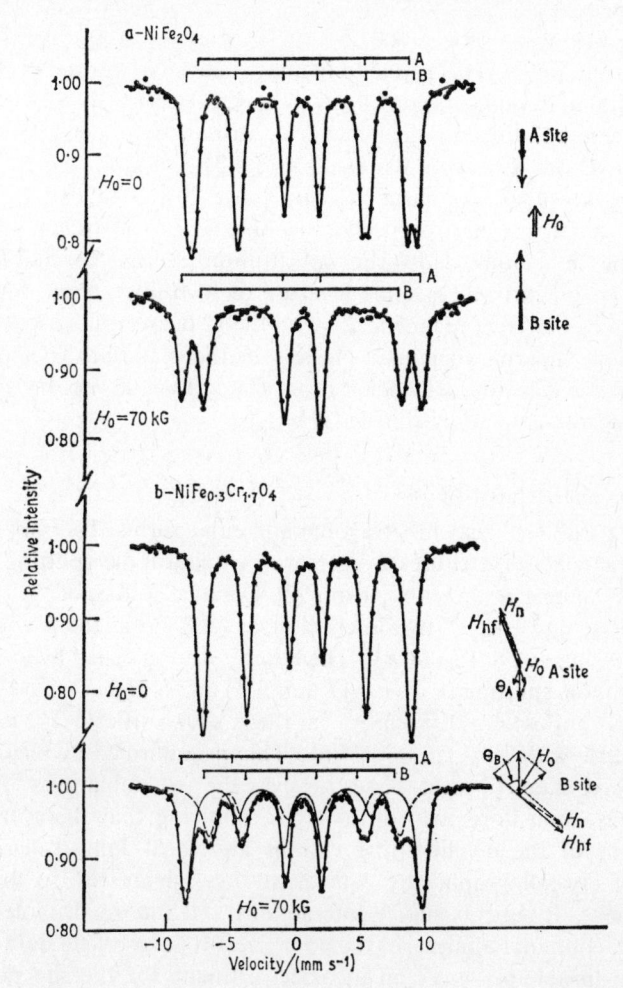

Fig. 10.16 Mössbauer spectra at 4·2 K of (a) $NiFe_2O_4$ and (b) $NiFe_{0.3}Cr_{1.7}O_4$. Note the absence of $\Delta m = 0$ lines in a 70-kG field in the former (Néel ordering) but not in the latter (Yafet–Kittel ordering). [Ref. 86, Fig. 1]

restored by application of an external magnetic field, apparently by polarisation of the particles in the field [88].

$FeNiCrO_4$ in the paramagnetic state at 721 K shows a quadrupole splitting of 0·40 mm s^{-1} [89]. The Fe^{3+} cations are predominantly on A sites which should have tetrahedral symmetry, but the randomisation of Ni^{2+} and Cr^{3+} cations on the B sites causes local distortions and a reduction in symmetry at

[*Refs. on p. 296*]

the A site. Similar behaviour is shown in FeNiAlO$_4$ with a splitting of 0·53 mm s^{-1} at 534 K [89].

The spinels Ni$^{2+}_{1+x}$Ge$^{4+}_x$Fe$^{3+}_{2-3x}$O$_4$ ($0 \leqslant x \leqslant 0.41$) show no quadrupole splitting at the A site Fe^{3+} ions, and the field at the A site is significantly less than at the B [90]. This is due to increased covalency at the A site, and it is suggested that 3d-polarisation effects on any 4s-electron density originating from covalent bonding will generate a contribution to the hyperfine field opposite in sign to that from the 1s-, 2s-, and 3s-electrons.

Ni$_{0.1}$Zn$_{0.9}$Fe$_2$O$_4$ is superparamagnetic above 20 K. The diamagnetic zinc ions on the A sites in ZnFe$_2$O$_4$ itself allow only weak B–B exchange coupling, between the Fe^{3+} ions. However, substitution of zinc by nickel to give Ni$_{0.1}$Zn$_{0.9}$Fe$_2$O$_4$ introduces some measure of A–B interaction. Each A site Ni^{2+} cation can interact magnetically with the 12 nearest B site Fe^{3+} cations, causing small superparamagnetic clusters in the material. The Mössbauer spectrum of the substituted spinel at room temperature is very broad because of the effects of this randomisation [91].

Cu^{II}Fe$^{III}_2$O$_4$ and Related Phases

The spinel CuFe$_2$O$_4$ exists in tetragonal and cubic forms. Thermal treatment of the tetragonal form causes a decrease in quadrupole splitting at about 363 K as it changes to the cubic form [92]. Detailed studies of the tetragonal form show the presence of two internal fields (508 kG at the A site and 538 kG at the B site, at 298 K) [93, 94]. The latter is distinguished by the presence of a quadrupole splitting ($\varepsilon = -0.17$ mm s^{-1}) and a higher chemical isomer shift (0·49 as opposed to 0·37 mm s^{-1} at the A site). CuFe$_2$O$_4$ is known to be largely inverse with 10% Cu on A sites. This is confirmed by measurements in an applied magnetic field which resolves the two sublattices. The B site patterns have non-Lorentzian line-shapes, implying that there are various distributions of the neighbouring cations and local Jahn–Teller-type distortions of crystallographically equivalent sites, giving rise to fluctuations in both the A–B superexchange interactions and the quadrupole splitting. As a result, computer analysis of the poorly resolved zero-field data assuming Lorentzian line-shapes gives an incorrect estimate for the site population.

Cu$_{0.5}$Zn$_{0.5}$Fe$_2$O$_4$ and Cu$_{0.5}$Cd$_{0.5}$Fe$_2$O$_4$ have all the zinc and cadmium ions on A sites [94], and are partly inverse with respect to iron. All three spinels show no $\Delta m = 0$ transitions in longitudinal fields, and are therefore ordered in the Néel arrangement with only two sublattices.

Zn^{II}Fe$^{III}_2$O$_4$ and Cd^{II}Fe$^{III}_2$O$_4$

ZnFe$_2$O$_4$ is a normal spinel and shows a small quadrupole splitting (0·36 mm s^{-1}) at room temperature from the trigonal distortion [89, 95]. CdFe$_2$O$_4$ is similar with a splitting of 0·83 mm s^{-1}. Estimations of the electric field gradient by lattice-sum calculations are complicated by the sensitivity of the result to the distortion parameter of the anion lattice [95].

The chromite $Zn^{II}[Cr^{III}_{1\cdot96}Fe^{III}_{0\cdot04}]O_4$ shows an abrupt disappearance of magnetic ordering above about 10 K due to a first-order phase transition and a change in crystallographic symmetry [96]. Both paramagnetic and antiferromagnetic contributions occur in the spectrum from 9 to 13 K, and similar effects have been found in $Mg^{II}[Cr^{III}_{1\cdot98}Fe^{III}_{0\cdot02}]O_4$.

10.3 Other Ternary Oxides

In this section the results on high-spin iron(II) systems are presented before those on iron(III). The latter dominate, and are ordered approximately as follows: the major structural classes of Fe_2O_3/M_2O_3 solid solutions, $MFeO_3$ perovskites, $MFeO_3$ orthoferrites, $M_3Fe_5O_{12}$ garnets; and other iron(III) oxides approximately in the periodic table classification of the second metal. Any quaternary oxides are included with the most appropriate ternary system.

$FeSb_2O_4$

The quadrupole splitting of the Fe^{2+} ion in the ternary oxide $FeSb_2O_4$ decreases from 2·88 mm s^{-1} at 50 K to 1·02 mm s^{-1} at 817 K as a result of thermal population of the t_{2g} levels [97]. The iron is in a site of 6-coordination with three mutually perpendicular twofold axes, and it is possible to describe the three lowest states as $|yz\rangle$, $|xz\rangle$, and primarily $|x^2 - y^2\rangle$, although the theoretical fit shown in Fig. 10.17 does not determine their relative order uniquely.

$FeNb_2O_6$

$FeNb_2O_6$ is a ternary oxide with a nearly regular octahedral coordination about the Fe^{2+} ion and a relatively small crystal-field splitting of the t_{2g} levels. Consequently the quadrupole splitting is strongly temperature dependent (2·35 mm s^{-1} at 80 K; 0·64 mm s^{-1} at 1000 K) [98]. Full theoretical analysis indicates an $|xz\rangle$ ground state and a doubly degenerate level at 520 K, the spin–orbit coupling parameter being -90 K.

$FeTi_2O_5$

$FeTi_2O_5$ is derived from Fe_2TiO_5 and is a ferrous oxide. The cation distribution is not known, but the broad Mössbauer spectrum suggests that it may be partly 'inverse' [20].

The System Al_2O_3–Fe_2O_3

Early data on a sample of $Al_{0\cdot46}Fe_{1\cdot54}O_3$ showed it to be magnetically ordered and to give a room-temperature Mössbauer spectrum comparable to that of α-Fe$_2$O$_3$ [20]. An orthorhombic phase of Al$_{2-x}$Fe$_x$O$_3$ exists for $0.6 < x < 1.0$. AlFeO$_3$ itself gives a magnetically ordered spin system at 4·2 K which shows at least two distinct antiparallel sublattices in an external

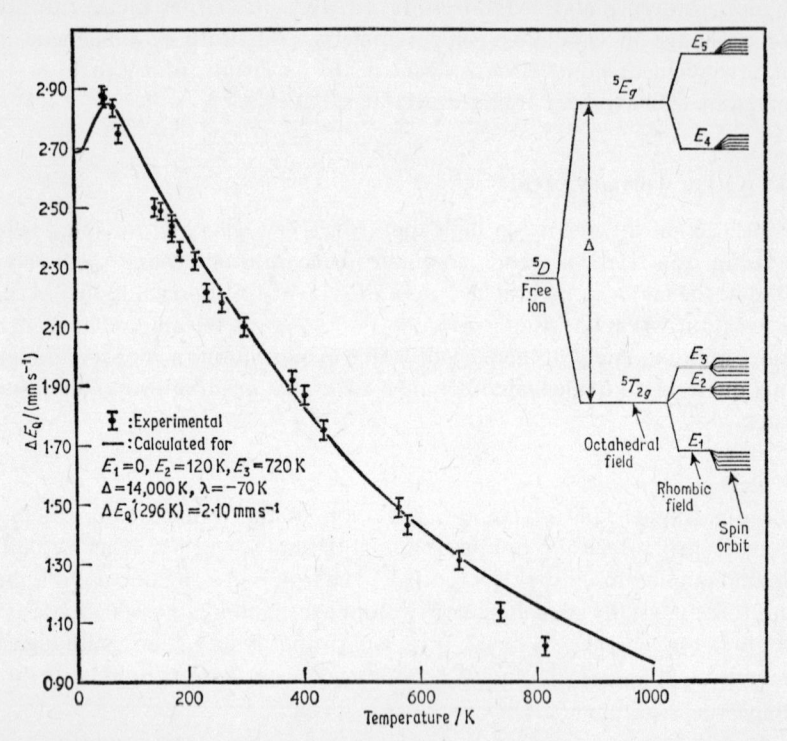

Fig. 10.17 The temperature dependence of the quadrupole splitting of FeSb$_2$O$_4$ showing a theoretical fit, using the energy level scheme inset. [Ref. 97, Fig. 1]

field [99, 100]. Rhombohedral Al$_{1.6}$Fe$_{0.4}$O$_3$ is paramagnetic at 80 K with a quadrupole doublet spectrum, but at 4·2 K is magnetically ordered. The spectrum is, however, very broad and presumably represents the effect of disorder of the Fe^{3+} and Al^{3+} cations which are not of comparable size.

The System Ga$_2$O$_3$–Fe$_2$O$_3$

The spectrum of β-Ga$_2$O$_3$ doped with ^{57}Fe shows two quadrupole doublets [100]. The oxygen atoms form a distorted cubic close-packed array with one tetrahedral and one octahedral cation site. Fe^{3+} substitutes more easily in the octahedral sites ($\Delta = 0.525$ mm s^{-1}, $\delta = 0.35$ mm s^{-1}) than in the tetrahedral sites ($\Delta = 0.98$ mm s^{-1}, $\delta = 0.17$ mm s^{-1}).

The phase Ga$_{2-x}$Fe$_x$O$_3$ ($0.7 < x < 1.4$) is both ferrimagnetic and piezoelectric. The double hexagonal close-packed array of oxygen ions has four sites: Ga(i) which has nearly regular tetrahedral symmetry, and three dis-

[*Refs. on p. 296*]

torted octahedral coordinations, Ga(ii), Fe(i), and Fe(ii). The paramagnetic spectra of samples in this phase show two detectable quadrupole splittings [101]. It is believed that the Fe^{3+} and Ga^{3+} cations are randomised on the octahedral sites, with gallium showing a preference for the Ga(ii) site. Magnetic ordering at low temperature allows site populations to be estimated approximately [102, 103]. The spins are oriented close to the 'c' axis [104], and neutron diffraction data show antiparallel sublattices, Fe(i) + Ga(i) and Fe(ii) + Ga(ii) [103].

The System $FeTiO_3$–Fe_2O_3

The solid solution $(1-x)FeTiO_3.xFe_2O_3$ has been extensively studied. Ilmenite, $FeTiO_3$, is derived from α-Fe_2O_3 by alternation of Fe and Ti cation layers in the [111] plane. When magnetically ordered the Fe spins are all parallel in one plane but antiparallel in adjacent planes. $FeTiO_3$ shows a hyperfine field of only 70 kG at 20 K, superimposed on the quadrupole interaction of $1 \cdot 14$ mm s^{-1} [105]. This follows from the orbital properties of the Fe^{2+} ion, and the contributing terms have been estimated as H_S (-455 kG), H_L ($+338$ kG), H_D ($+47$ kG), giving H_{eff} (-70 kG), although the sign of H_{eff} has not been confirmed experimentally [106].

Since the end-members of the phase are $Fe_2^{3+}O_3$ and $Fe^{2+}Ti^{4+}O_3$, the intermediate compositions must contain mixed valence states of iron. As x decreases from $1 \cdot 0$ to about $0 \cdot 5$, the basic six-line spectrum of α-Fe_2O_3 is retained, although it is noticeably broadened by the proliferation of site symmetries. At $x = 0 \cdot 33$ only a very broad spectrum envelope is seen, and this has been claimed to indicate the presence of ferrimagnetic clusters and general superparamagnetic behaviour [105].

The chemical isomer shift and quadrupole splitting of $FeTiO_3$ show substantial pressure dependence [9]. The temperature dependence of the quadrupole splitting has also been used to study the orbital state of the trigonally distorted Fe^{2+} ion [107].

Spectra for 22 natural ilmenite samples from different deposits show that ilmenite itself is normally deposited with a composition close to the stoichiometry $FeTiO_3$ [108]. The weathering products which occur particularly in beach sands contain Fe^{3+} ions, but are ill defined and of uncertain composition. The degree of weathering can be measured in any given sample from the Mössbauer spectrum. The spectra of lunar ilmenite at various temperatures is discussed on p. 294.

The Systems V_2O_3–Fe_2O_3 and Cr_2O_3–Fe_2O_3

Brief details have been given for two compositions in the phase $(Fe_xV_{1-x})_2O_3$ ($x = 0 \cdot 3, 0 \cdot 5$) which is related to α-Fe_2O_3 [20].

The system $(Fe_xCr_{1-x})_2O_3$ is also related to α-Fe_2O_3 and brief details have been given for $x = 0 \cdot 2$ and $0 \cdot 5$ [20]. α-Fe_2O_3 and Cr_2O_3 both have the same

[*Refs. on p. 296*]

crystal structure but differ in the type of antiferromagnetic ordering. In solid solution these ordered structures are connected through a cone spiral arrangement to preserve continuity. In $(Fe_{0.035}Cr_{0.965})_2O_3$ ($T_N = 289$ K) the apparent magnitude of the quadrupole interaction in the magnetic phase decreases as the temperature rises due to an alteration in the cone half-angle [109]. The magnetic field approximately follows a Brillouin function. Spin relaxation causes an increasing degree of motional narrowing as the temperature increases from 100 K to T_N [110].

The System Mn_2O_3–Fe_2O_3

The Mn_2O_3–Fe_2O_3 system is very complex [111–113]. α-Mn_2O_3 is orthorhombic at room temperature but is cubic above 308 K. Room-temperature spectra of the intermediate compositions show two quadrupole doublets from the two sites. The manganese tends to enter the more distorted site. The lines are broad due to disorder, and the low-temperature spectra show magnetic ordering (but with site symmetry disorder effects) on both sites. Redistribution of the site populations was observed on thermal treatment of $(Mn_{0.0075}Fe_{0.9925})_2O_3$ [113]. $MnFeO_3$ has the β-Mn_2O_3 structure, and the Mössbauer spectrum shows it to contain only Fe^{3+} iron. The latter does not distribute statistically over the two cation sites, but shows a distinct preference for one of them. The distribution is $\{Mn_{2.7}Fe_{5.3}\}\{Mn_{13.3}Fe_{10.7}\}O_{48}$ to a first approximation.

The System Rh_2O_3–Fe_2O_3

The system $(Rh_xFe_{1-x})_2O_3$ has been examined for $x = 0.11, 0.22, 0.25, 0.41$, and 0.81 [114]. It has the homogeneous haematite structure with randomisation of the cations. The Néel temperature falls with increasing x while the Morin transition rises. For α-Fe_2O_3 the Morin transition is 257 K, but at $x = 0.11$ it is 507 K and at $x = 0.22$ it is 544 K. Significant collapse of the hyperfine splitting is seen below the Néel temperatures.

$BiFeO_3$ and Related Perovskites

The ferroelectric perovskite $BiFeO_3$ shows an Fe^{3+} ion quadrupole split spectrum above the Néel temperature of 645 K [115]. Both the chemical isomer shift and the quadrupole splitting show a discontinuity at the ferroelectric phase transition in $(PbTiO_3)_{0.95}(BiFeO_3)_{0.05}$ [116].

The substituted perovskite phases $Sr(TaFe)_{\frac{1}{2}}O_3$, $Pb(NbFe)_{\frac{1}{2}}O_3$ and $SrTiO_3 + 1.15Sr(TaFe)_{\frac{1}{2}}O_3$ give quadrupole spectra which can be simulated theoretically by assuming a random distribution of cations on the iron site and an electric field gradient which arises solely from the cation randomisation [117].

Preliminary data at room temperature for the ternary perovskite phase

LaFe$_x$Al$_{1-x}$O$_3$ ($0 < x < 1$) have shown that magnetic ordering is destroyed as the aluminium content increases [118].

The solid solution Sr$_2$(FeMo$_x$W$_{1-x}$)O$_6$ also has the perovskite structure and the end-member Sr$_2$Fe^{3+}Mo^{5+}O$_6$ is ferrimagnetically ordered at room temperature with a field of 315 kG [119]. Appreciable Fe^{2+} content is not found until above 70% Sr$_2$Fe^{2+}W^{6+}O$_6$ in composition because of appreciable reduction to W^{5+}.

Orthoferrites, RFeO$_3$

The orthoferrites, RFeO$_3$, where R is yttrium or a rare-earth element have a structure based on a distorted perovskite lattice with only one iron environment. Most of them possess weak ferrimagnetic character as a result of spin canting from the ideal two-sublattice antiferromagnetic ordering.

Several studies have been made including detailed comparisons of the full series, RFeO$_3$ (R = Y, La, Pr, Nd, Sm, Eu, Gd, Tb, Dy, Ho, Er, Tm, Yb, Lu) [120–122] with spectra recorded between 85 and 770 K. All show a single magnetic hyperfine splitting, and the magnetic field (extrapolated to 0 K) decreases regularly with the atomic number of R (see Table 10.2) from 564

Table 10.2 Mössbauer and other parameters of orthoferrites [120, 121]

Compound	H_{eff}*/kG	T_N/K	Lattice parameters in Å		
			a	b	c
LaFeO$_3$	564	740	5·556	5·565	7·862
PrFeO$_3$	559	707	5·495	5·578	7·810
NdFeO$_3$	557	687	5·441	5·573	7·753
SmFeO$_3$	552	674	5·394	5·592	7·711
EuFeO$_3$	552	662	5·371	5·611	7·686
GdFeO$_3$	551	657	5·346	5·616	7·668
TbFeO$_3$	550	647	5·326	5·602	7·635
DyFeO$_3$	548	645	5·302	5·598	7·623
YFeO$_3$	549	640	5·283	5·592	7·603
HoFeO$_3$	548	639	5·278	5·591	7·602
ErFeO$_3$	546	636	5·263	5·582	7·591
TmFeO$_3$	545	632	5·251	5·576	7·584
YbFeO$_3$	546·5	627	5·233	5·557	7·570
LuFeO$_3$	545·5	623	5·213	5·547	7·565

* Extrapolated to 0 K with an estimated error of ±2 kG.

kG (LaFeO$_3$) to 545·5 kG (LuFeO$_3$), a result of the steadily changing ionic radius of the rare earth which is also reflected in the crystal lattice parameters [120]. The enforced change in the iron–oxygen distance presumably affects the degree of covalency and thence the magnetic field. The Néel temperature also decreases from 740 K (LaFeO$_3$) to 623 K (LuFeO$_3$), but the chemical isomer shift proves insensitive to the nature of R. Detailed

[*Refs. on p. 296*]

analysis of the sublattice magnetisation was given using several possible theories of interpretation. Relation of the sublattice magnetisation to the observed weak ferromagnetism enables the spin canting angle to be derived, and in all cases it is independent of temperature. Lattice-sum calculations of the electric field gradient in $GdFeO_3$ and $YFeO_3$ agree with the observed quadrupole splitting.

$YFeO_3$ shows considerable collapse of the hyperfine spectrum in the temperature range of 6° below the Néel point [123], the data being characteristic of electronic spin relaxation.

The oxide $CeFeO_3$ has been prepared and shown to be analogous to the other rare-earth orthoferrites [124]. The Curie temperature is 719 K, and the quadrupole perturbation on the hyperfine spectrum (ε) changes sign at 230 K as the result of electronic spin reorientation with respect to the crystal axes and hence the electric field gradient tensor. The spins are perpendicular to the 'c' axis above 230 K, but parallel to it below that temperature.

Measurements of the relative line intensities in oriented $SmFeO_3$ seem to suggest that the spin reorientation at ~160°C occurs by a rotational process rather than by a discontinuous jump [125]. However, it was not possible to decide whether the rotation occurs coherently. Similar work on $ErFeO_3$ using a polarised iron metal source gave unequivocal evidence for a continuous and coherent spin rotation in this compound [126].

A reversal in the sign of the small quadrupole splitting in $HoFeO_3$ between 77 K and 290 K as a result of a change in the magnetic axis has been claimed [127], but the three differently ordered structures of $TbFeO_3$ (Néel temperatures of 3·1 K, 8·4 K, and 681 K) are virtually indistinguishable in the Mössbauer spectrum [128]. The chemical isomer shift in $HoFeO_3$ has been measured from 99 to 875 K and shows no discontinuity at the Néel temperature [129]. However, the temperature dependence is different in the magnetic and paramagnetic phases, presumably as a result of the effects of magnetic ordering on the s-electron density at the nucleus.

The derived system $Fe_xMn_{1-x}YO_3$ has a hexagonal structure for $0 \leqslant x \leqslant 0·15$ and a very deformed perovskite structure for $0·2 \leqslant x \leqslant 1$, and the effect of composition on the Mössbauer parameters has been followed [130, 131].

Iron Garnets, $R_3Fe_5O_{12}$

Considerable interest has been shown in the rare-earth iron garnets, $R_3Fe_5O_{12}$, because of their interesting ferrimagnetic properties. The unit cell is large and contains 16 octahedral (a) site and 24 tetrahedral (d) site Fe^{3+} cations. These sites can be differentiated in the magnetically ordered compounds $Y_3Fe_5O_{12}$ and $Dy_3Fe_5O_{12}$ at room temperature by the lower magnetic field and chemical isomer shift at the tetrahedral site. Although they have only axial symmetry, there are three possible orientations of the tetra-

[Refs. on p. 296]

hedral site and four for the octahedral site in the crystal, which effectively averages out the small quadrupole interaction in polycrystalline samples [132]. The fact that Y^{3+} is diamagnetic and Dy^{3+} is paramagnetic has little effect.

A single crystal of $Y_3Fe_5O_{12}$ cut normal to the [110] direction and magnetised in the [100] direction has all the trigonal axes of the octahedral sites oriented at 55° to the magnetic field, while the tetrahedral sites are aligned parallel or perpendicular to the field. Magnetisation in the [111] direction gives two effective octahedral but only one tetrahedral orientation although one of the octahedral orientations contributes only weakly to the spectrum. In this way the small quadrupole effects can be resolved in the magnetic spectra [133].

Detailed measurements of the temperature dependence of the hyperfine field in single-crystal $Dy_3Fe_5O_{12}$ have been made in a study of the sublattice magnetisation [134]. Similar results have also been obtained for $Y_3Fe_5O_{12}$ and $Sm_3Fe_5O_{12}$. In the former the magnetisation lies along the [111] direction so that two of the octahedral site orientations become non-equivalent. In the latter, it is along the [110] direction so that two of the tetrahedral orientations become non-equivalent. The resultant line splittings enable this difference in ordering to be verified [135].

High-temperature measurements of the paramagnetic quadrupole splittings have been made in Y, Sm, Gd, Dy, Yb, and Lu garnets, but comparison with estimates from point-charge lattice-sum calculations gives only poor agreement [136].

In the system $Eu_3Ga_xFe_{5-x}O_{12}$ ($0 \leqslant x \leqslant 3.03$) about 80% of the Ga substitutes in tetrahedral sites as shown by the change in line intensities. At $x \sim 1.4$ the net magnetisation of the iron sublattices vanishes due to compensation of the spins [137].

In $Y_3Al_xFe_{5-x}O_{12}$ small amounts of aluminium (which enters tetrahedral sites) show no appreciable effect on the magnetic order, but for $x > 0.8$ there is a drastic collapse of the hyperfine splitting as a result of a decrease in the number of (a)–(d) antiferromagnetic interactions, caused by dilution of the iron with a diamagnetic cation [138]. Similar effects are seen in $Y_{3-x}Ca_xFe_{5-x}Sn_xO_{12}$ ($0 \leqslant x \leqslant 2$), where Sn^{4+} replaces iron in octahedral sites. Collapse of the hyperfine structure occurs above $x = 1.0$ because of the randomisation of magnetic and non-magnetic cations [139–141].

In the phase $Y_3Ga_xFe_{5-x}O_{12}$, which has been studied above the ordering temperatures, the gallium substitutes initially with a preference for the tetrahedral sites and this decreases as the value of x increases [142].

Non-equivalent octahedral site Fe^{3+} ions are observed in the garnets $Y_{3-2x}Ca_{2-x}Fe_{5-x}V_xO_{12}$ because of variations in the distribution of Fe^{3+} and V^{5+} ions on the tetrahedral sites surrounding the Fe^{3+} on octahedral sites [143, 144]. Other multi-cation garnets which have been studied include $V_{3-x}Ca_xFe_{5-x}Si_xO_{12}$ [145] and $Y_3Fe_{4.7}Ga_{0.15}In_{0.15}O_{12}$ [146].

[*Refs. on p. 296*]

The garnets $Ca_3Fe_2Si_3O_{12}$, $Ca_3Fe_2Ge_3O_{12}$, and $Gd_3Fe_2Ge_3O_{12}$ are paramagnetic at room temperature; they contain Fe^{3+} ions on octahedral sites and show small quadrupole splittings [147].

LiFeO$_2$

There are three forms of $LiFeO_2$: the α-form has a sodium chloride lattice with complete cation disorder; the tetragonal γ (Q_I)-form is ordered; and the β (Q_{II})-form is also tetragonal but with alternating layers of Li^+ and Fe^{3+}. The γ-form is antiferromagnetic below 315 K, but motional narrowing is seen below the Néel temperature [148]. The disordered α-form seems to show short-range ordering effects, and the small quadrupole splitting at room temperature emphasises that the local asymmetry can still be non-cubic although the overall crystal symmetry is cubic. The β-form becomes antiferromagnetically ordered below 42 K [149]. The internal magnetic fields in the three forms are: α-$LiFeO_2$ 495 kG at 4 K, β-$LiFeO_2$ 480 kG at 6·5 K, γ-$LiFeO_2$ 515 kG at 77 K.

CuFeO$_2$

$CuFeO_2$ has a rhombohedral structure with iron in octahedral coordination. The Mössbauer spectrum confirms the oxidation states as $Cu^+Fe^{3+}O_2$ [150]. The paramagnetic state shows a small quadrupole splitting which has been compared with lattice-sum calculations. Below 19 K the compound becomes antiferromagnetically ordered. The spins align along the crystallographic 'c' axis (from single-crystal Mössbauer data), and the field at 4·4 K is 520 kG. It is therefore suggested that the magnetic structure consists of ferromagnetic layers normal to the 'c' axis with the spin direction alternating from layer to layer.

Ca$_2$Fe$_2$O$_5$

The oxide $Ca_2Fe_2O_5$ contains one octahedral and one tetrahedral iron site. It is antiferromagnetic and, despite an earlier report of different ordering temperatures for the two sites (i.e. a four-sublattice ordering) [151], it is now known that all sites order together at 730 K [152]. The internal magnetic fields have been measured from 80 K up to this temperature [153]. Substitution of the type $Ca_2Fe_{2-x}M_xO_5$ has been studied for M = Al^{3+} ($0 < x < 1·4$), [151, 154, 155], Ga^{3+} ($0 \leqslant x \leqslant 1·5$) [152, 156, 157], and Sc^{3+} ($0 \leqslant x \leqslant 0·5$) [152, 156, 157]. In the former case there is a change in the crystal structure near $x = 0·6$. As a result the magnetic structure of Ca_2FeAlO_5 differs from that of $Ca_2Fe_2O_5$ by a change in the spin axis from being parallel to the crystallographic 'c' axis to being parallel to the 'a' axis [152]. Scandium substitution takes place almost entirely at the octahedral sites, whereas gallium shows some preference for tetrahedral sites.

$Ca_2Fe_2O_5$ itself gives well-resolved internal magnetic splittings from both

OTHER TERNARY OXIDES | 277

sites, and it is possible to study site occupancy in substituted derivatives from the relative intensities of the lines (see Fig. 10.18). In the scandium derivative

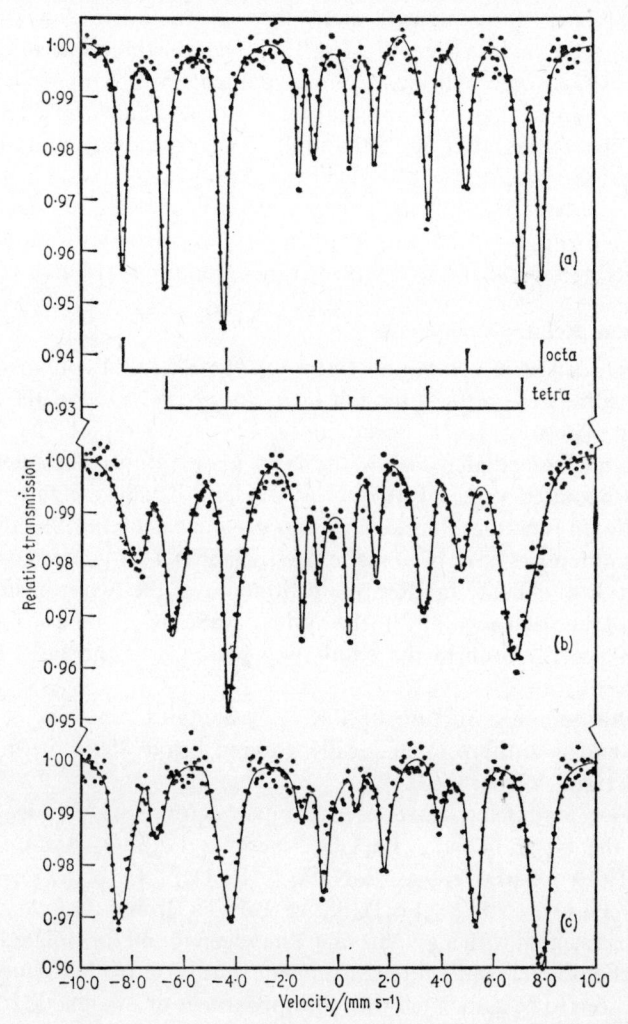

Fig. 10.18 Mössbauer spectra at room temperature of (a) $Ca_2Fe_2O_5$ and (b) $Ca_2Sc_{0.5}Fe_{1.5}O_5$; and at 5 K of (c) Ca_2FeGaO_5. Note how scandium replaces iron in octahedral sites, whereas gallium substitutes in tetrahedral sites. [Ref. 156, Fig. 1]

the outer (octahedral site) lines are weaker than are the neighbouring tetrahedral site lines, and conversely for the gallium derivative. The retention of a low spontaneous magnetisation indicates that the octahedral and tetrahedral sublattices must each be antiferromagnetic [156], but the unique Néel temperatures also show a strong *inter*-sublattice exchange interaction [157].

Ca_2FeAlO_5 also shows preferential substitution of the tetrahedral Fe^{3+} cations [154]. Lattice summation calculations for the electric field gradients observed in the paramagnetic state have been made for $Ca_2Fe_2O_5$, $Ca_2Fe_{1.5}Al_{0.5}O_5$, and Ca_2FeAlO_5 [158, 159]. The spin directions in $Ca_2Fe_2O_5$ and Ca_2FeAlO_5 were confirmed by measurements at 5 K using a polarised iron metal source [155]. The internal fields at the octahedral [ ] and tetrahedral ( ) sites in the latter are 502 and 454 kG respectively, and the relative site population is $Ca_2[Fe_{0.68}Al_{0.32}]$ $(Fe_{0.32}Al_{0.68})O_5$.

Detailed analysis of the electric field gradient tensors has shown that both sites in both oxides have the principal value V_{zz} directed along the crystal 'b' axis, correcting several earlier studies on these materials [160].

$CaFe_2O_4$ and Related Compounds

The antiferromagnetic compound $CaFe_2O_4$ ($T_N \sim 200$ K) contains two distinct octahedral Fe^{3+} sites, although these are not resolved in the magnetic Mössbauer spectrum [161]. Some motional narrowing of the hyperfine pattern occurs immediately below the Néel temperature. The internal field has an extrapolated value of 485 kG at 0 K and deviates slightly from the $S = \frac{5}{2}$ Brillouin function. Point-charge lattice-sum calculations of the electric field gradient tensors have been made [159]. $NaScTiO_4$ is related to $CaFe_2O_4$ except that Sc and Ti are randomly substituted over the two iron sites. There is a continuous solid solution in the system $NaScTiO_4$–$NaFeTiO_4$, and the Mössbauer spectra confirm the continuing presence of the Fe^{3+} oxidation state [162].

The orthorhombic structure of $BaFe_2O_4$ contains only one tetrahedral Fe^{3+} site, and is antiferromagnetically ordered below 880 K with the spins along the crystal 'c' axis [163].

Mössbauer parameters have been compared for a wide range of Fe^{3+} oxides of the types $NaSc_{1-x}Fe_xTiO_4$, $NaSc_{1-x}Fe_xSnO_4$, $CaSc_{2-x}Fe_xO_4$, $Sc_{2-x}Fe_xTiO_5$, $NaFeTi_3O_8$, α-$NaSc_{1-x}Fe_xO_2$, $Cs_xSc_{x-y}Fe_yTi_{2-x}O_4$, β-$NaAl_{1-x}Fe_xO_2$, $Cs_xSc_{x-y}Fe_yTi_{4-x}O_8$, $LiFeTiO_4$, and $CsFeSi_2O_6$ [164]. A strong e.s.r. signal with a g-value of 4·3 has been found for isolated $Fe^{3+}3d^5$ ions in a rhombic crystal field environment in several of the dilute systems, and is also seen to be associated with the appearance of a magnetic Mössbauer spectrum due to long spin-relaxation times. Increase in Fe^{3+} content causes g to revert to a value of 2·00 and the Mössbauer spectrum to a more common paramagnetic form. Such behaviour was found in $NaSc_{0.99}Fe_{0.01}TiO_4$, α-$NaSc_{0.995}Fe_{0.005}O_2$, β-$NaAl_{0.99}Fe_{0.01}O_2$, $Cs_{0.67}Sc_{0.66}Fe_{0.01}Ti_{1.33}O_4$, and $Cs_{0.63}Sc_{0.62}Fe_{0.01}Ti_{3.37}O_8$. Magnetic ordering was found in $NaFeTiO_4$, $CaFe_2O_4$, Fe_2TiO_5, and the β-$NaAl_{1-x}Fe_xO_2$ and $LiFeTiO_4$ phases at low temperature.

[*Refs.* on p. 296]

CaFe$_4$O$_7$

CaFe$_4$O$_7$ is not well characterised but has been shown to be magnetically ordered at 77 K with at least two distinct iron sites [165].

MFe$_{12}$O$_{19}$ (M = Ba, Sr, Pb) and Related Compounds

The ferrimagnetic oxide BaFe$_{12}$O$_{19}$ contains four distinct cation sites. An attempted analysis of the complicated hyperfine spectra shows that the site with the unusual trigonal bipyramidal configuration has a saturation magnetic field of about 620 kG [166]. This 5-coordinated site is the only one to feature substantial anisotropy of the recoil-free fraction as shown by single crystal studies [167]. The direction with the lowest f-factor corresponds to the trigonal axis of the bipyramid, and it was proposed that there may be an oscillation of the iron atom along this axis through the equatorial plane of oxygen neighbours. The lead analogue PbFe$_{12}$O$_{19}$ shows considerable paramagnetic character below the Curie temperature [168].

Comparison of the Mössbauer spectra of SrFe$_{12}$O$_{19}$ and the substituted oxide SrFe$_{11}$GaO$_{18}$ shows that gallium substitution in the $4f_{VI}$ sites has a detectable influence on the hyperfine field of Fe^{3+} in the $12k$ sublattice, presumably due to a change in the superexchange interactions [169]. The substituted oxide SrCo$_{0.42}$Ti$_{0.42}$Fe$_{11.16}$O$_{19}$ is a similar system [170].

Studies of the Ba$_2$Zn$_2$Fe$_{12}$O$_{22}$ ferrites are complicated by the presence of at least six iron sublattices, resulting in very complex magnetic spectra [171]. Another complex system is BaCo$_{1.75}$Fe$_{16.25}$O$_{27}$ [172].

Fe$_2$TiO$_5$

Fe$_2$TiO$_5$, pseudo-brookite, has orthorhombic symmetry and contains Fe^{3+} in a distorted octahedral environment [20]. The quadrupole splitting at room temperature is 0·75 mm s^{-1} [108].

Fe$_2$TeO$_6$

Fe$_2$TeO$_6$ has the trirutile structure and is a collinear antiferromagnet. Mössbauer spectra indicate a Néel temperature of 218 K, the saturation magnetic field at the Fe^{3+} site being 520 kG. The principal axis of the electric field gradient is perpendicular to the spin axis [173]. An independent measurement of the Néel temperature gives 201 K [174].

Fe$_2$(MoO$_4$)$_3$

The ferric molybdate, Fe$_2$(MoO$_4$)$_3$, has a structure related to the garnet type but without cations in the 8-coordinated sites. The quadrupole splitting of 0·25 mm s^{-1} at room temperature is only half that found in conventional garnets, implying that the octahedral Fe^{3+} site has a lower distortion in the molybdate structure [175].

UFeO$_4$

The oxide UFeO$_4$ orders magnetically below 54·5 K to give a typical Fe^{3+} pattern without quadrupole splitting [176]. There is a phase change at 42 K which causes a discontinuity in the temperature dependence of the hyperfine field and the appearance of a small quadrupole splitting. The magnetic ordering is complex.

Fe$_3$BO$_6$

The iron borate Fe$_3$BO$_6$ is a weak ferromagnet with a Curie temperature of 508 K [177]. The Mössbauer spectra show that below 415 K the spins of both types of iron site are perpendicular to the [100] axis, but above 415 K they are parallel to this axis. The ordering is basically antiferromagnetic but with spin canting.

10.4 Iron(IV) Oxides

A small but interesting group of oxides contain appreciable quantities of iron in oxidation state IV. First of these to be investigated was the phase SrFe^{IV}O$_3$–SrFe^{III}O$_{2·50}$, spectra being recorded for SrFeO$_3$, SrFeO$_{2·86}$, SrFeO$_{2·60}$, and SrFeO$_{2·50}$ at several temperatures [178]. Strontium ferrate(IV), SrFeO$_3$, has the perovskite structure and gives a single-line spectrum at 298 K (Fig. 10.19). The chemical isomer shift of 0·147 mm s^{-1} at 4 K and 0·055 mm s^{-1} at 298 K (relative to iron metal) is lower than that for Fe^{3+} in oxide systems and falls into the general scheme for oxygen compounds of iron Fe(VI) < Fe(IV) < Fe(III) < Fe(II) < Fe(I). Detailed analysis of this chemical isomer shift value is not possible; the Fe(IV) is nominally in a low-spin configuration, but the 3d-electrons are collective rather than localized and no attempt has been made to estimate the effects of this on the shift.

SrFeO$_3$ is antiferromagnetic below 134 K (Fig. 10.19) and gives a field of 331 kG at 4 K. The iron site symmetry is cubic so that there is no quadrupole splitting from the nominal 3d^4 configuration.

SrFeO$_{2·86}$ is still a single-phase perovskite but is deficient in oxygen, thereby causing the formation of some Fe^{3+} and inducing a consequent tetragonal distortion of the lattice. Consistent with this the Mössbauer spectra show the appearance of extra lines appropriate to the Fe^{3+} cations. The strontium ferrate(III), SrFeO$_{2·50}$, contains both tetrahedral and octahedral Fe^{3+} ions; it is isostructural with brownmillerite, and gives magnetic spectra similar to those of the iron garnets (see p. 274). A sample of composition SrFeO$_{2·60}$ proved to be a mixture of SrFeO$_{2·50}$ and an iron(IV)-rich phase. The Mössbauer spectra show signs of short-range ordering in the otherwise complex spectra. Application of pressure to SrFeO$_{2·86}$ causes reversible reduction of Fe^{4+} to Fe^{3+} [179].

The barium ferrate(IV) phase is only partly related to the strontium compound as it has a hexagonal structure instead of the cubic perovskite structure of the latter [180]. The chemical isomer shift of BaFeO$_{2 \cdot 95}$ is $-0 \cdot 13$ mm s^{-1} at 90 K and $-0 \cdot 07$ mm s^{-1} at 298 K. This is contrary to the expected thermal red shift, but the structure of the compound is not well established, and magnetic ordering probably induces phase changes. The hyperfine field at 4 K is only 255 kG in this instance. By contrast, barium ferrate(III), BaFeO$_{2 \cdot 50}$, is completely analogous to the strontium compound, and is

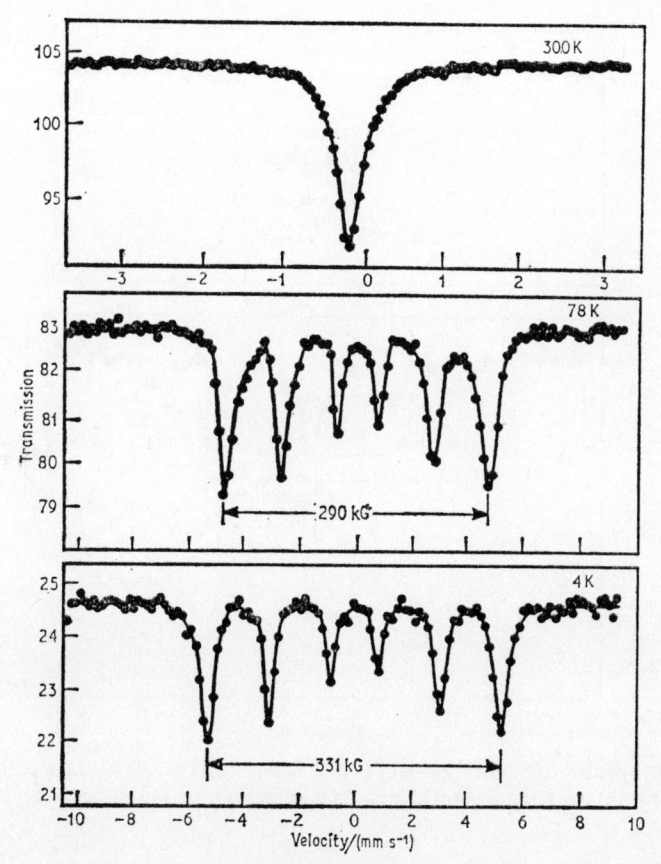

Fig. 10.19 Mössbauer spectra of SrFeO$_3$ at 300 K, 78 K, and 4 K. [Ref. 178, Fig. 1]

magnetically ordered. The samples of intermediate composition are rather complex because of the effects of increased anion vacancy concentration and the resultant quadrupole splitting, and in some cases contain at least two phases.

The system Sr$_3$Fe$_2$O$_{6-7}$ is closely related to SrFeO$_3$ [181]. In this instance in a sample of Sr$_3$Fe$_2$O$_{6 \cdot 2}$ it is possible to detect an appreciable quadrupole splitting at the Fe^{3+} ion as a result of oxygen ion vacancies (Fig. 10.20). The fact that the iron(IV) spectrum is not quadrupole split implies that vacancies

[Refs. on p. 296]

are confined to the vicinity of the trivalent ions. There is also some suggestion of two slightly different types of iron(III) environment. The chemical isomer shift in $Sr_3Fe_2O_{6\cdot 9}$ is 0·00 mm s^{-1} at 298 K. The whole phase including the trivalent iron compound $SrFe_2O_6$ shows antiferromagnetic ordering. The compound $Sr_2FeO_{3\cdot 7}$ from the $Sr_2FeO_{3\cdot 5-4\cdot 0}$ phase is basically similar with an iron(IV) chemical isomer shift of $-0\cdot 03$ mm s^{-1} at 298 K and a magnetic

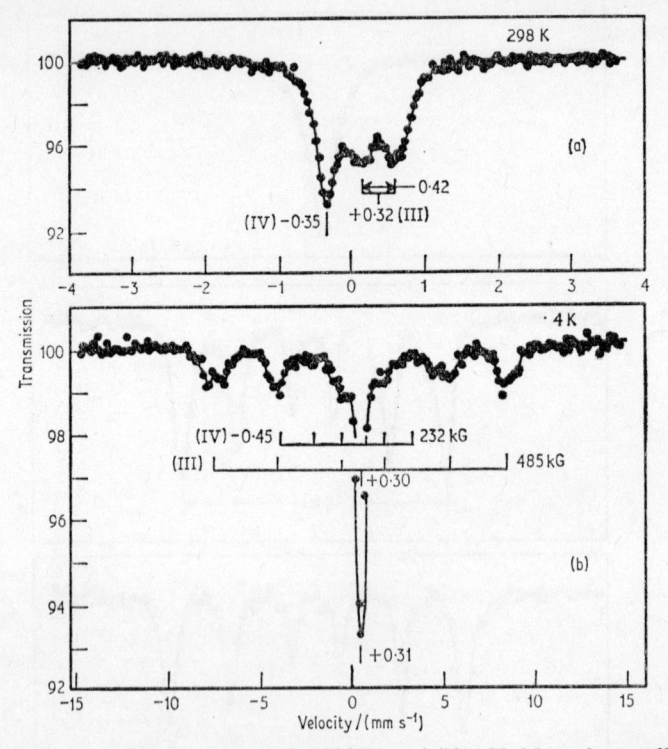

Fig. 10.20 Spectra of $Sr_3Fe_2O_{6\cdot 2}$ at (a) 298 K and (b) 4 K. Note the small quadrupole splitting at the ferric ion produced by a neighbouring anion vacancy. [Ref. 181, Fig. 4]

field of 276 kG at 4 K. In $Sr_3Fe_2O_{6\cdot 9}$ the field is 279 kG at 4 K. All of these compounds appear to show strong covalency effects.

In the phase $La_{1-x}Sr_xFeO_3$ ($0 \leqslant x \leqslant 1$) substitution for trivalent La by divalent Sr produces iron(IV) and considerable effects on the magnetic ordering [182]. In contrast with the previous systems, charge compensation is achieved without introduction of anion vacancies. Increase in x up to 0·3 causes considerable broadening of the antiferromagnetic spectrum and lowers the Néel temperature. In the range where the crystal symmetry is rhombohedral ($x \sim 0\cdot 5$) it appears that Fe(III) and Fe(IV) do not have a separate distinct existence, and the single resonance line observed has an intermediate chemical isomer shift value. Cubic phase samples can be prepared up to $x = 0.6$. A close examination [183] of the temperature dependence of the spectra for $x = 0.5$ and $x = 0.6$ shows that at low temperatures these compositions show magnetic hyperfine patterns from both iron(III) and iron(IV) oxidation states. In the cubic phase sample ($x = 0.6$), the magnetic ordering process coincides with a marked reduction in the rate of electronic exchange; conversely, on warming through the Néel temperature the Mössbauer spectrum indicates there is a loss of separate identity of the two oxidation states and simultaneously there is a large increase in the electrical conductivity of the compound. The rhombohedral phase sample ($x = 0.5$) shows evidence for both oxidation states above the Néel temperature, but the fast electronic relaxation begins at about 270 K with consequent narrowing of the spectrum.

Preliminary results on the $SrFe_{1-x}Cr_xO_{3-y}$ phase suggest that compositions with $x = 0.20$ and 0.30 are cubic perovskites with Fe^{3+} and Fe^{4+} cations magnetically ordered [184]. The spectrum lines are very broad in the magnetic region. When $x = 0.5$ only Fe^{3+} ions are present.

10.5 Iron Chalcogenides

There are comparatively few data available for iron sulphides and other chalcogenides, though interesting comparisons can be made in those cases where there are equivalent oxide systems. The same broad trends in the Mössbauer systematics are observed. Thus for chemical isomer shifts

$$\delta(Fe^{3+})_{tet} < \delta(Fe^{3+})_{oct} < \delta(Fe^{2+})_{tet} < \delta(Fe^{2+})_{oct};$$

for quadrupole interactions at non-cubic sites

$$\Delta(Fe^{3+}) \ll \Delta(Fe^{2+});$$

and for magnetic hyperfine fields

$$H(Fe^{2+}) < H(Fe^{3+}).$$

Further, when direct comparison between oxides and chalcogenides can be made the chemical isomer shifts and magnetic hyperfine fields observed for the chalcogenides are less than those in the corresponding oxide phases because of covalency effects.

Iron(II) Sulphide, FeS

Stoichiometric FeS has the NiAs structure. It is antiferromagnetic at room temperature [185–188], with a magnetic field of 309 kG. The quadrupole interaction e^2qQ is probably about -0.85 mm s^{-1} with the spin and eq axes angled at 48° to each other [187]. It was believed to undergo a change in the direction of the magnetic axis from along the 'c' axis at low temperature to perpendicular to this axis at high temperatures, analogous to the Morin transition in α-Fe_2O_3. This has been confirmed in samples of composition $Fe_{0.95}S$ and $Fe_{0.935}S$ by observation of a change in the sign of the second-order quadrupole interaction parameter ε as the temperature is raised. Troilite, FeS, has also been detected in lunar samples as mentioned on p. 295.

The composition Fe_7S_8 (i.e. $Fe_{0.875}S$) is derived from FeS by subtraction of one iron atom per 8 (FeS) units. The cation vacancies are ordered to give a pseudo-hexagonal unit cell with a slight monoclinic distortion and there are three distinct iron environments. The room-temperature Mössbauer spectrum shows three well-resolved hyperfine fields of 298, 252, and 221 kG, all with a chemical isomer shift of $\delta(Fe) \sim 0.65$ mm s^{-1} indicating predominantly Fe^{2+} iron [189]. There is no evidence at all for the separate existence of trivalent iron, as implied by the ionic formulation $(Fe^{3+})_2(Fe^{2+})_5(S^{2-})_8$. This suggests a time-averaging of the oxidation states over times of the order of the excited-state lifetime ($\sim 10^{-7}$ s) and this factor, together with the increased covalency of the sulphide phase when compared to the oxide, results in the low value for the chemical isomer shift. The 298-kG field is similar to the value for FeS and aligns antiparallel to an external field. It is therefore taken to originate from the A_1 and A_2 sites which will be least affected by the vacancies. The other two fields come from the B and C sites in the crystal and align parallel to an external field. The sharpness of the three patterns is good evidence for a high degree of ordering.

Tetragonal FeS contains tetrahedrally coordinated Fe^{2+} and no evidence was found, using Mössbauer and neutron diffraction data on synthetic samples, for magnetic order even at 1·7 K [190]. The chemical isomer shift of $\delta(Fe) = 0.35$ mm s^{-1} at room temperature is lower than for the octahedrally coordinated form, and indicates a large increase in covalent character, possibly accounting for the lack of magnetic interaction. However, independent data have indicated magnetic ordering at room temperature [188].

FeS_2 and Related Phases

FeS_2 occurs in two forms, pyrite and marcasite, and both give a simple quadrupole split spectrum from the octahedrally coordinated Fe(II) (pyrite: $\delta(Fe) = 0.314$ mm s^{-1}, $\Delta = 0.614$ mm s^{-1} at 300 K; marcasite $\delta(Fe) = 0.277$ mm s^{-1}, $\Delta = 0.506$ mm s^{-1}) [188, 191]. Basically similar results are obtained for the marcasites $FeSe_2$ ($\delta = 0.395$ mm s^{-1}, $\Delta = 0.584$ mm s^{-1}), $FeTe_2$ ($\delta = 0.471$ mm s^{-1}, $\Delta = 0.502$ mm s^{-1}), and the löllingites $FeAs_2$ ($\delta = 0.314$ mm s^{-1}, $\Delta = 1.68$ mm s^{-1}) and $FeSb_2$ ($\delta = 0.455$ mm s^{-1}, $\Delta = 1.281$ mm s^{-1}) [191, 192]. There is some correlation between chemical isomer shift and unit-cell volume, but although the electronic configurations are low-spin, there is little evidence to decide whether they are formally iron(II) or iron(IV) compounds. Such difficulty in interpretation is often found in highly covalent systems.

[*Refs. on p. 296*]

Both the chemical isomer shift and the quadrupole splitting in FeS$_2$ show appreciable pressure dependence [193].

The Co$_{1-x}$Fe$_x$S$_2$ phase is ferromagnetic and contains low-spin cobalt(II) [194]. The ^{57}Fe resonance shows no magnetic hyperfine splitting below the Curie temperature for $x = 0.01, 0.25, 0.50$, and 0.75. One can therefore conclude that the iron atoms do not participate in the magnetic order, which is consistent with a proposed model of localised $3d$-electrons on the cobalt ions.

Spinel Sulphides

Several iron sulphides adopt the spinel structure. Little information is available for Fe$_3$S$_4$ which has a distorted spinel structure and is ferromagnetic. The room-temperature spectrum shows a broadened magnetic hyperfine pattern with a superimposed doublet which may indicate spin relaxation [188, 195]. FeIn$_2$S$_4$ is an inverse spinel with Fe^{2+} iron on octahedral B sites [196, 197]. The room-temperature shift of $\delta(\text{Fe}) = 0.80$ mm s^{-1} and a quadrupole splitting of $\Delta = 3.23$ mm s^{-1} are typical of the cation. The temperature dependence of the quadrupole splitting between 80 and 640 K assuming a trigonal site distortion gives a singlet–doublet splitting of the t_{2g} level of the order of 1700 K (1180 cm^{-1}) with $\Delta(0\text{ K}) = 3.30$ mm s^{-1} [196]. FeSb$_2$S$_4$ and FeNb$_2$S$_4$ have been briefly studied [198].

FeCr$_2$S$_4$ is a normal spinel with Fe^{2+} in tetrahedral coordination and a chemical isomer shift of only 0.50 mm s^{-1} because of increased covalency to iron [198]. Magnetic ordering occurs at 180 K. Above this temperature the resonance shows no quadrupole splitting because of the cubic site symmetry. Magnetic ordering appears to effectively lower the site symmetry as in RbFeF$_3$ (see p. 119), and a quadrupole splitting appears although the crystallographic symmetry is still basically cubic. Detailed attempts to explain the experimental data have been made [199–201], but the observed effects are apparently more complicated than a simple theory would lead one to predict. The magnetic field is 203 kG at 61 K.

The phase Fe$_{1-x}$Cu$_x$Cr$_2$S$_4$ has been found to contain both Fe^{2+} and Fe^{3+} iron for $0 < x < 0.5$, but only Fe^{3+} iron for $0.5 \leqslant x \leqslant 1$ [202]. In the former case there is a fast electron exchange at higher temperatures which gives an averaged spectrum. Fe$_{0.5}$Cu$_{0.5}$Cr$_2$S$_4$ is magnetically ordered below $T_\text{C} = 360$ K [203]. The Fe^{3+} and Cu$^+$ cations on the A site are ordered so that local cubic symmetry is retained.

Other Sulphides

Chalcopyrite, CuFeS$_2$, is antiferromagnetic below 550°C and gives a magnetic spectrum appropriate to Fe^{3+} iron [204–208]. The field of ∼370 kG at 77 K is smaller than usually found for tetrahedral Fe^{3+} because of covalency factors. Data are also available for the sulphides Cu$^\text{I}_5$Fe$^\text{III}$S$_4$ [206], and Cu$^\text{I}_2$Fe$^\text{II}$Sn$^\text{IV}$S$_4$ [198, 205, 206], and the Mössbauer parameters have been used to establish the oxidation states shown. In the case of Cu_2FeSnS_4, the presence of Sn(IV) was also established by Mössbauer spectroscopy. The temperature dependence of the ^{57}Fe quadrupole splitting gives a singlet–doublet crystal field splitting of 1700 K (1180 cm^{-1}) and a value of $\Delta(0\ K)$ of 2·92 mm s^{-1} [198].

$CuFe_2S_3$, cubanite, is nominally a mixed Fe^{2+}/Fe^{3+} sulphide, but its orthorhombic phase shows fast electronic exchange and a single hyperfine pattern which is an average [205]. Data on the disordered cubic form are conflicting. One source quotes an essentially paramagnetic spectrum at room temperature with some magnetic broadening at lower temperatures [205], while the other gives a magnetic six-line spectrum with evidence for antiferromagnetic ordering [209].

A study of the tetrahedrally coordinated iron(III) compounds $NaFeS_2$, $KFeS_2$, $RbFeS_2$, and $CsFeS_2$ has shown that the quadrupole splitting is dependent on the cationic radius of the second metal [210]. The previously reported magnetic transition in $KFeS_2$ [211] is not confirmed by susceptibility measurements, and spin relaxation is suggested as an alternative explanation.

10.6 Silicate Minerals

Silicate minerals frequently contain iron. The presence of chains, layers, or networks of silicon bonding to oxygen is one of the most significant structural features. Electric neutrality is maintained by including the appropriate number of metal cations, usually in sites with fourfold or sixfold coordination.

The rigidity of the basic silicate structures exerts a greater controlled influence on cation site symmetries than do the individual cation charges, an important difference from the oxides where the anions have more freedom of movement. The ease with which Fe^{2+} and Fe^{3+} cations can be distinguished and site occupancies determined in oxides by Mössbauer spectroscopy suggests a logical extension to similar studies in silicates, and considerable progress has been made in this direction.

The application of Mössbauer spectroscopy to silicate mineralogy has been well described in two papers by Bancroft *et al.*, in which the influences of electronic configuration, oxidation state, and coordination symmetry of the iron cations were correlated firstly with silicates of known structures [212], and latterly with silicates of unknown and complex structures [213]. Most of the data discussed here are taken from these works, but references are given where appropriate to other data available. Silicate minerals of lunar origin are discussed on p. 294.

Silicate minerals provide a wide range of site symmetries, some of which are not usually found in conventional oxide systems, and frequently the phases have wide ranges of composition enabling studies of cation substitution effects. The principal types of silicate are listed in Table 10.3.

[*Refs.* on p. 296]

Table 10.3 Mössbauer parameters at room temperature for cations in silicates of known structure [212]

Name of mineral or series	Formula	Oxidation state of Fe	δ (Fe)* /(mm s^{-1})	Δ/(mm s^{-1})	Coordination number of Fe
olivine	(Mg,Fe,Mn)$_2$SiO$_4$	2	1·16–1·18	2·80–3·02 (M$_1$M$_2$)	6
garnet group (pyrope-almandine)	(Mg,Fe)$_3$Al$_2$(SiO$_4$)$_3$	2	1·31	3·53–3·56	8
(andradite)	Ca$_3$(Fe,Al)$_2$(SiO$_4$)$_3$	3	0·41	0·58	6
orthopyroxene	(Mg,Fe,Mn)$_2$Si$_2$O$_6$	2	1·15–1·18 1·12–1·16	2·35–2·69 (M$_1$) 1·91–2·13 (M$_2$)	6
diopside–hedenbergite	Ca(Fe,Mg)Si$_2$O$_6$	2	1·16	2·15 (M$_1$)	6
cummingtonite–grunerite	(Fe,Mg,Mn)$_7$Si$_8$O$_{22}$(OH)$_2$	2	1·14–1·18 1·05–1·11	2·76–2·90 (M$_1$M$_2$M$_3$) 1·50–1·68 (M$_4$)	6
anthophyllite	(Mg,Fe)$_7$Si$_8$O$_{22}$(OH)$_2$	2	1·12–1·13 1·09–1·11	2·58–2·61 (M$_1$M$_3$) 1·80–1·81 (M$_4$)	6
actinolite	Ca$_2$(Mg,Fe)$_5$Si$_8$O$_{22}$(OH)$_2$	2	1·15–1·16 1·13–1·16	2·81–2·82 (M$_1$M$_3$) 1·89–2·03 (M$_2$)	6
epidote	Ca$_2$(Al,Fe,Mn)AlOH.AlO.Si$_2$O$_7$.SiO$_4$	3	0·34–0·36	2·01–2·02	6
staurolite	(Fe,Mg)(Al,Fe)$_9$O$_6$(SiO$_4$)$_8$(O,OH)$_2$	2	0·97	2·30	4 tetr.
gillespite	BaFeSi$_4$O$_{10}$	2	0·76	0·51	4 sq. planar

* Converted from stainless steel by subtracting 0·09 mm s^{-1}.

Olivines, (SiO_4^{2-})

Several measurements have been made in the forsterite–fayalite (Mg_2SiO_4–Fe_2SiO_4) and fayalite–tephroite (Fe_2SiO_4–Mn_2SiO_4) series [212, 214–217]. All show a simple quadrupole doublet with a splitting in the range 2·80–3·02 mm s^{-1} at room temperature. The fayalite structure involves independent SiO_4 tetrahedra surrounded by 6-coordinated metal cations in two distinct sites, M_1 and M_2. These are not always distinguished in the Mössbauer spectrum. Fayalite itself in the paramagnetic state shows two large quadrupole splittings (both 3·05 mm s^{-1} at 80 K; 1·56 and 1·40 mm s^{-1} at 1000 K) [214]. It is not possible to distinguish which site is which from the Mössbauer spectrum, and analysis of the temperature dependence of the quadrupole splittings in terms of crystal-field parameters was not attempted. The related compound $CaFeSiO_4$ gives a quadrupole splitting smaller than either of the two distorted octahedral sites in Fe_2SiO_4, and cannot therefore be used for site identification in the latter.

Magnetic ordering occurs in Fe_2SiO_4 below 66 K, and at 9 K there are two magnetic patterns with fields of 120 kG and 323 kG which have different orientations with respect to their electric field gradient tensor [215].

The Garnet Group

Spectra for the pyrope–almandine series $(Mg,Fe^{2+})_3Al_2(SiO_4)_3$ and an andradite $Ca_3(Al,Fe^{3+})_2(SiO_4)_3$ show several interesting features. The structures contain independent SiO_4 tetrahedra with the trivalent cations in 6-coordination and the divalent cations in 8-coordination. The almandine series shows one of the largest quadrupole splittings recorded for Fe^{2+} ions at room temperature (3·55 mm s^{-1}). The chemical isomer shift ($\delta(Fe) = 1·29$ mm s^{-1} at room temperature) is higher than the value of $\sim$1·15 mm s^{-1} normally recorded for Fe^{2+} on octahedral sites in silicates.

Orthopyroxenes, $(SiO_3^{2-})_n$

Several measurements have been made on orthopyroxenes [212, 216–218], which are a group of silicates containing single-stranded chains of SiO_4 groups of overall composition $(SiO_3^{2-})_n$. In an orthopyroxene containing 23·1% of Fe^{2+} only a single quadrupole doublet is seen ($\Delta = 2·11$ mm s^{-1}) with a chemical isomer shift $\delta(Fe) = 1·14$ mm s^{-1}, a value which is typical of 6-coordination in silicates. However, for 85·9% Fe^{2+} two such doublets are visually resolved ($\Delta = 1·91, 2·46$ mm s^{-1}). In pyroxenes the infinite silicate chains are linked by bands of cations; there are two sites of 6-coordination designated M_1 and M_2 and the latter are more distorted, the distortion increasing as the Fe^{2+} content increases. The spectrum of the orthopyroxene containing 85·9% Fe^{2+} is shown in Fig. 10.21a. Comparison of line intensities with the known cation populations shows that the smaller quadrupole

[Refs. on p. 296]

splitting is associated with the M_2 site. Presumably there is a larger lattice term in the electric field gradient which partly cancels the contribution from

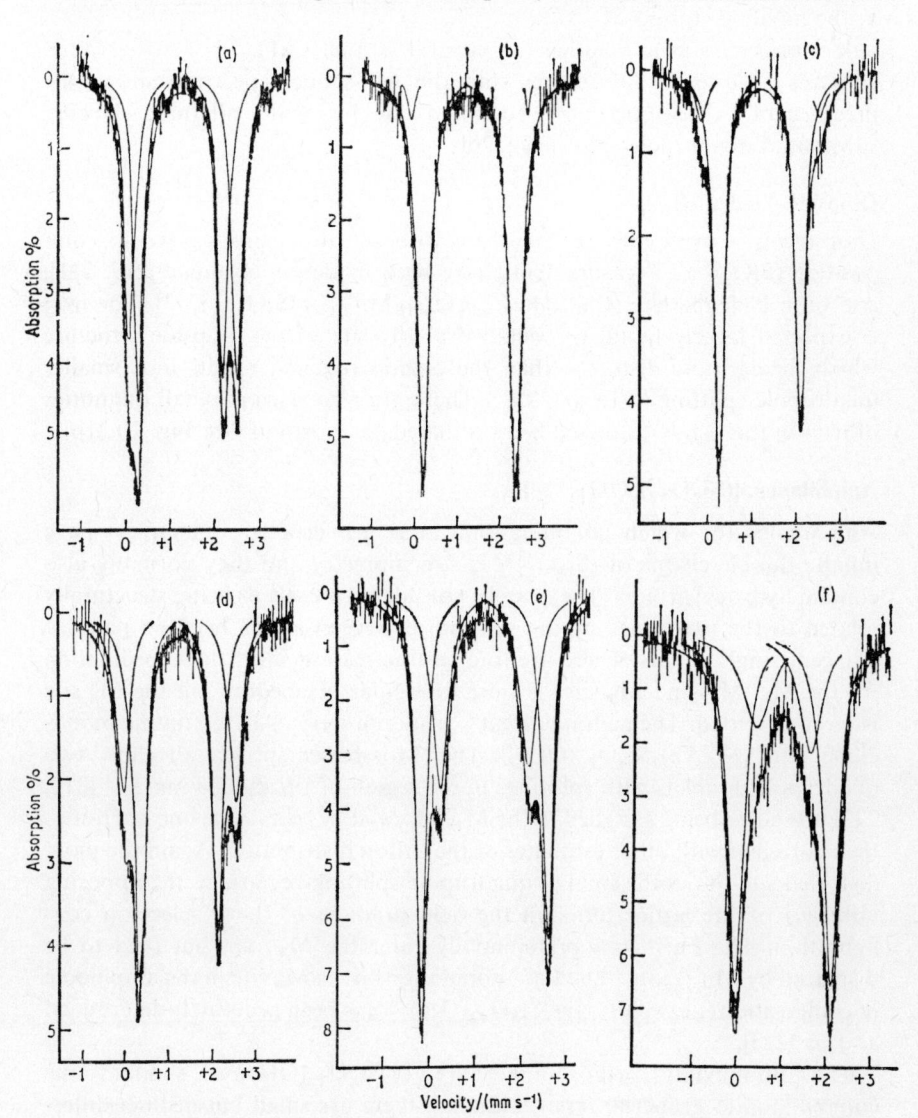

Fig. 10.21 Spectra of iron silicates. (a) 85·9% Fe^{2+} orthopyroxene, (b) hedenbergite, (c) 23·0% Fe^{2+} anthophyllite, (d) 31·6% Fe^{2+} anthophyllite, (e) 47·9% Fe^{2+} actinolite, (f) 12·22 wt % FeO staurolite. [Ref. 212, Fig. 1]

the valence electron. Although the iron prefers the M_2 site, it can be selectively displaced from it by introduction of manganese [212, 218].

The orthopyroxene $Fe_{0·269}Mg_{0·731}SiO_3$ shows no magnetic ordering

[*Refs. on p. 296*]

down to 1·7 K [219]. However, the quadrupole splitting shows substantial broadening due to an unusually slow spin relaxation rate for the Fe^{2+} ions at the highly distorted M_2 sites.

It is uncertain as to whether the mineral ilvaite, $CaFe_2^{2+}Fe^{3+}(SiO_4)_2OH$, contains SiO_4 or Si_2O_7 groups, but the Mössbauer spectrum shows the presence of two distinct Fe^{2+} sites and one Fe^{3+} site, all three showing substantial quadrupole splittings [220].

Diopside–Hedenbergite

Diopside is a pyroxene-type structure with silicate chains of overall composition $(SiO_3^{2-})_n$. Measurements have been made on an augite [217, 221], and on a hedenbergite $(Ca_{0·95}Fe_{0·85}Mg_{0·18}Mn_{0·02})(SiO_3)_2$ [213]. The iron is expected largely to fill the octahedral M_1 site of the diopside structure which, being more distorted than the orthopyroxenes, results in a smaller quadrupole splitting (2·15 mm s^{-1}). There are also signs of small quantities of iron in the highly distorted 8-coordinated M_2 position (see Fig. 10.21b).

Amphiboles, $[(Si_8O_{22})(OH)_2^{14-}]_n$

Amphiboles (of which cummingtonite and grunerite are examples) have infinite double chains of $(Si_4O_{11}^{6-})_n$ stoichiometry and they normally also contain hydroxyl groups. They can be considered loosely as being structurally related to the pyroxenes by the sharing of oxygen atoms between pairs of adjacent single chains. There are four distinct cation sites: the coordination for the M_1, M_2, and M_3 sites is close to regular octahedral, but the M_4 site is more distorted. The cummingtonite–grunerite series is based on the monoclinic unit $(Mg,Fe)_7Si_8O_{22}(OH)_2$. The Mössbauer spectra all show two quadrupole doublets with splittings in the region of 1·6 and 2·8 mm s^{-1} [212, 222], the latter being assigned to the M_1, M_2, and M_3 sites. The line intensities then correlate with other estimates of the cation distribution. Again the more distorted site shows the smaller quadrupole splitting because of the opposing influence of the lattice term on the field gradient of the d^6-electron configuration. The Fe^{2+} ions preferentially enter the M_4 sites but tend to be displaced by Mn^{2+} ions. The Fe^{2+} population on the M_4 site in the amphibole of composition $(Fe_{0·35}Mg_{0·65})_7Si_8O_{22}(OH)_{22}$ has been accurately determined as 1·65 [223].

The anthophyllites, orthorhombic $(Mg,Fe)_7Si_8O_{22}(OH)_2$, are similar to the cummingtonite–grunerite series, although there are small but distinct differences in the Mössbauer parameters [212, 224]. The crystal structure is basically identical, and once again Fe^{2+} cations favour the M_4 site. Spectra for anthophyllites with 23·0% and 31·6% Fe^{2+} are given in Figs. 10.21c and 10.21d.

The actinolite minerals, $Ca_2(Mg,Fe)_5Si_8O_{22}(OH)_2$, also show two distinct iron quadrupole doublets [212, 217]. Actinolite is structurally similar to cummingtonite, but with iron on the near-octahedral M_1 and M_3 sites and on the distorted octahedral M_2 site. Known site populations allow the inner doublet to be assigned to the latter (Fig. 10.21e).

Aluminosilicates

The epidote group [$Ca_2(Al,Fe,Mn)AlOH.AlO.Si_2O_7.SiO_4$] have a structure consisting of chains of AlO_6 and $AlO_4(OH)_2$ octahedra sharing edges and linked by SiO_4 and Si_2O_7 groups. The Fe^{3+} ions appear to occupy an irregular $(Al,Fe)O_6$ polyhedron, and the Mössbauer spectrum [212, 217] shows an unusually large quadrupole splitting from the Fe^{3+} ions (2·02 mm s^{-1}).

The staurolites $(Mg,Fe^{2+})_4(Al,Fe^{3+})_9O_6(SiO_4)_8(O,OH)_2$ consist of independent SiO_4 tetrahedra, and chains of AlO_6 octahedra and FeO_4 tetrahedra sharing edges. The spectrum of tetrahedral Fe^{2+} has a chemical isomer shift of only $\delta(Fe) = 0.94$ mm s^{-1} which is less than that for octahedral Fe^{2+} in silicates, and confirms the tetrahedral coordination. There is also some suggestion of a previously unsuspected small quantity of iron in the octahedrally coordinated sites (Fig. 10.21f).

Gillespite $(Si_8O_{20}{}^{8-})_n$

Gillespite, $BaFeSi_4O_{10}$, is unusual in that it contains Fe^{2+} in a square-planar environment. The structure is based on $(Si_8O_{20}{}^{8-})_n$ sheets. The Fe–O bond distance is 197 pm (1 picometre = 10^{-2} Å), the next-nearest oxygen atoms being at 398 pm. The chemical isomer shift of $\delta(Fe) = 0.75$ mm s^{-1} is the lowest value recorded for Fe^{2+} in any silicate [212, 225, 226]. The quadrupole splitting (0·51 mm s^{-1}) is also unusually small. Full ligand field analysis shows that the lattice term is opposite in sign to the valence-electron contribution and is also larger in magnitude. The small observed splitting is thus a result of partial cancellation. The temperature dependence of the spectrum has been followed from 80 to 650 K, and lattice-dynamical studies made [225]. The asymmetric line broadening found is attributed to partial quenching of the orbit-lattice interaction and the consequent increase in the spin-lattice relaxation time [226].

Silicates with Complex Structures

From the data presented in the preceding sections on silicates of known structure the following generalisations can be made:

(i) the Fe^{2+} and Fe^{3+} oxidation states are easily distinguished;
(ii) the chemical isomer shift for Fe^{2+} ions in silicates depends both on coordination number and symmetry:
δ(square-planar) < δ(tetrahedral) < δ(octahedral) < δ(8-coordinate);
(iii) the quadrupole splitting of 6-coordinate Fe^{2+} ions is very sensitive to site symmetry. This last point is illustrated in Fig. 10.22. The quadrupole splitting also varies slightly in a regular manner with composition in any

given series. It is not uncommon to find that a more distorted site symmetry gives a smaller splitting, presumably because distortions from cubic symmetry are large enough for the lattice terms to become important. The rigidity of the silicate structures probably allows greater distortions of the iron cation sites than is normally found in oxides. Some correlation can be drawn between the quadrupole splitting at a given site in related silicates and the cation occupancy of neighbouring sites.

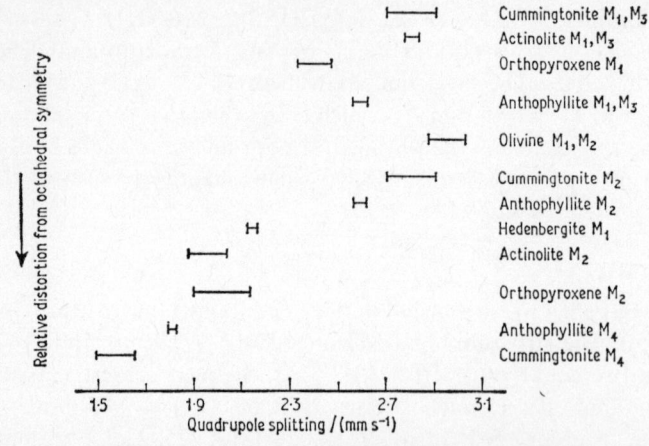

Fig. 10.22 The dependence of six-coordinate Fe^{2+} quadrupole splitting on the degree of distortion from octahedral symmetry. [Ref. 212, Fig. 6]

These correlations have enabled diagnostic analysis of several less well-known complex silicates [213]. Zussmanite, howieite, deerite, sapphirine, and crocidolite have been shown to contain only Fe^{2+} and Fe^{3+} iron cations. The first two named contain at least one distinct type of 6-coordinated Fe^{2+} and the sapphirines at least two, whilst deerite contains 6-coordinated and 4-coordinated Fe^{2+} ions. Howieite and deerite also contain smaller proportions of 6-coordinated Fe^{3+} ions, and the sapphirines contain two types of 4-coordinated Fe^{3+} ions.

The spectrum of a crocidolite (Fig. 10.23) having the composition $(Na_{1.85}Ca_{0.14}K_{0.01})(Mg_{0.30}Fe^{2+}_{2.76}Mn_{0.04})(Fe^{3+}_{1.93}Al_{0.03})Si_{7.94}O_{22}(OH)_{2.34}$ can be analysed in terms of two Fe^{2+} quadrupole doublets (AA', CC') and one from Fe^{3+}(BB'). The fraction of Fe^{3+} from the spectrum area is 0·41 which agrees with the chemical analysis. Crocidolite is an amphibole, and the spectrum assignment of AA' to M_1 sites, CC' to $M_3 + (M_2)$ sites, and BB' to Fe^{3+} in M_2 sites is consistent with data from earlier studies on crocidolite and amosite, which also examined effects of oxidation and reduction [227, 228].

[Refs. on p. 296]

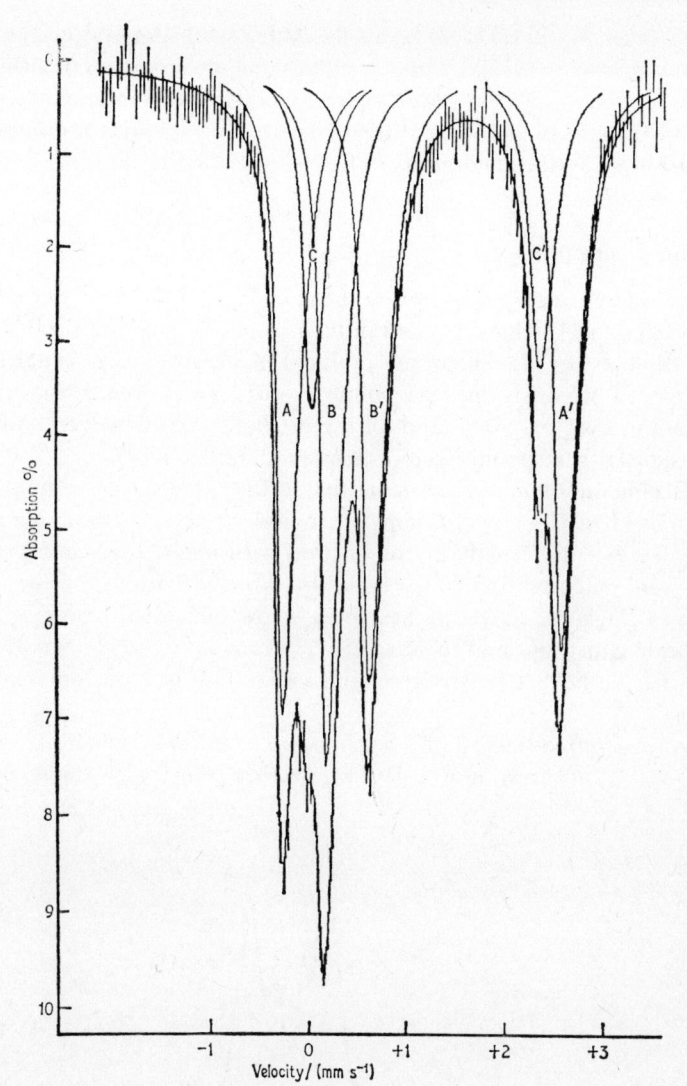

Fig. 10.23 Mössbauer spectrum of a crocidolite mineral. [Ref. 213, Fig. 3]

The silicate neptunite LiNa$_2$K(Fe,Mn,Mg)$_2$Ti$_2$O$_2$(Si$_8$O$_{22}$) has been shown to contain at least 95% of the iron in the Fe^{2+} state, confirming that the titanium is present as Ti^{4+} [229].

Clay Minerals

Preliminary data for 14 clay minerals with sheet and chain silicate structures suggest that oxidation states and site symmetries can also be investigated in these complex materials [230]. Similar measurements have been made on micas such as biotite [231–234], zinnwaldite KLiFeAl(AlSi$_3$O$_{10}$)(OH,F)$_2$ [232], and muscovite [235]. Chemical studies include ferric cation adsorption on the three-layer type iron-bearing clay minerals, illite and montmorillonite [236], and the effects of chemical treatment to remove potassium ions from dioctahedral and trioctahedral micas [237].

10.7 Lunar Samples

It is hardly necessary to comment on the epic flight of three American astronauts in 'Apollo 11' which resulted in their return to earth on the 24th July, 1969, with the first 22 kilograms of lunar material. These samples were distributed to scientists in nine countries for one of the most intensive investigations ever prepared, and all the preliminary results were published simultaneously in a special issue of *Science* in January 1970.

The likelihood of finding iron in the surface rocks had prompted the adoption of Mössbauer spectroscopy as a non-destructive method of analysis. Such measurements, usually in conjunction with microscopic mineralogical studies, were made on different samples by three laboratories in the U.S.A. and two in England. These are described below individually because of the variations in constitution of the samples. The reference numbers quoted were assigned by the NASA Lunar Receiving Laboratory at Houston, Texas.

(A) A typical sample 10087,4 of lunar fines (i.e. fine particle material or dust) gave the spectrum shown in Fig. 10.24 [238]. The inner quadrupole doublet

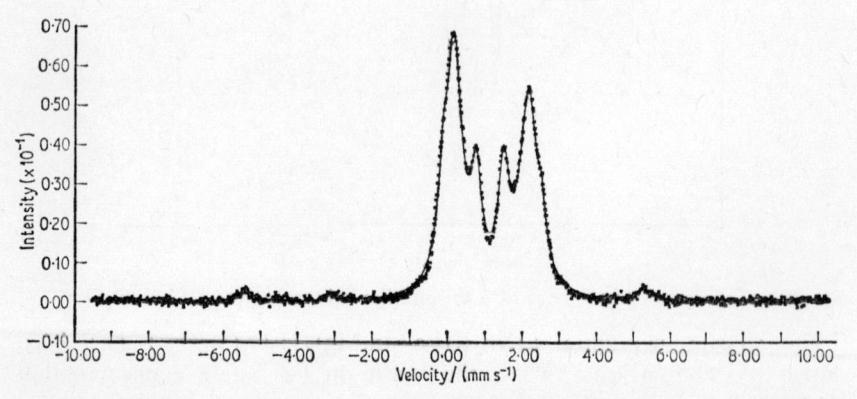

Fig. 10.24 Mössbauer transmission spectra from a sample of lunar fines (10087,4). Note the three weak lines from iron metal. [Ref. 238, Fig. 1]

is from the high-spin Fe^{2+} cations in ilmenite which mineralogical studies had shown to be a major component of the lunar material. The outer broadened doublet is from high-spin Fe^{2+} in silicate phases (including iron-bearing

[*Refs.* on p. 296]

glasses and monoclinic pyroxenes). The three weak resonances well separated from the main group are three of the lines of metallic iron (about 5 wt % of the total iron content of the sample). In some samples additional weak lines from troilite (FeS) were found. The lines from the silicates are broad because of several sites and mineral components.

A number of the documented rock samples were examined by Mössbauer scattering methods. Specimens 10020,20 and 10003,22 both contained ilmenite and silicates, although the dominance of pyroxene in the latter sample resulted in narrower resonance lines. Breccia 10065 proved essentially similar to the fines, as indeed expected if they originate as shock lithified fines.

An important point is the failure to identify any high-spin Fe^{3+} cations to within a few per cent.

(B) Data on the bulk dust sample 10084,85 showed it to contain high-spin Fe^{2+} in ilmenite, iron-bearing glass, and pyroxene (M1 and M2 sites), as well as small quantities of iron metal [239]. Fractionation of the dust according to density established that the metal was associated primarily with the glass. The rock samples 10017,17, 10058,24, 10057,59, and 10046,17 were all basically mixtures of ilmenite and pyroxene, although the spectra did show clear differences in chemical composition. Metallic iron was only found in 10046,17, with troilite being seen in the other three. The distribution of these magnetic materials in the rocks was ascertained by microscopic examination.

The Néel temperature of the ilmenite was measured in rocks 10017 and 10058 and the heavy fraction of bulk sample 10084 for comparison with that of a reference terrestrial ilmente. In all cases T_N was 57 ± 2 K, indicating no significant departure from stoichiometry.

A magnetically separated sample is illustrated in Fig. 10.25 showing the Fe and FeS components. Exposure to air for two months produced no detectable oxidation, and this investigation also failed to find any trace of Fe^{3+} iron.

(C) Specimens of clinopyroxenes separated from 10003 and 10044 showed clearly resolved Fe^{2+} quadrupole doublets from the M1 and M2 sites [240]. The 10044 sample appeared to be exclusively augite, while 10003 also contained some pigeonite. Relative site occupancies were estimated in several samples, and again no evidence for high-spin Fe^{3+} was found.

(D) Spectra of sample 10084,14 were basically similar to those mentioned under (A) and (B) [241]. Ilmenite and pyroxene (mainly augite) provided the main components, but with smaller contributions from olivine (unresolved from the pyroxene), iron metal, troilite, and magnetite. A rock sample 10045,24 gave the same basic details. Of two pyroxene separates from 10044,43 one proved to have ordered Fe^{2+} cations, while the other was disordered.

[Refs. on p. 296]

296 | IRON OXIDES AND SULPHIDES

This was clearly shown by a change in the M1/M2 ratio obtained from the line intensities.

(E) Spectra of a sample of fines 10084,13 and a magnetic separate from this also showed the presence of ilmenite, pyroxenes, olivine, and iron metal [242]. Concentrations of FeS and Fe_3O_4 were less than 1% of the total iron.

Viewing the results as a whole they present a closely unified picture, the main differences being the inevitable variations in the composition of samples. The iron was present as either Fe(II) or Fe(0) and no significant concentration of Fe(III) was detected. Separation of individual components was only partly feasible because several minerals were often embedded in a single grain, and, for example, iron metal was seen lining cracks in ilmenite and silicates.

Full details of these investigations have since been published [243-247].

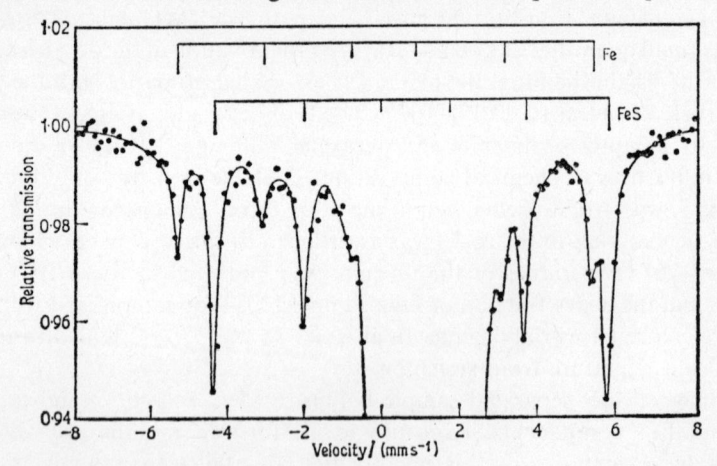

Fig. 10.25 Mössbauer spectrum at room temperature of a magnetically separated material from lunar rock 10057. [Ref. 239, Fig. 3]

REFERENCES

[1] O. C. Kistner and A. W. Sunyar, *Phys. Rev. Letters*, 1960, **4**, 412.
[2] J. Gastebois and J. Quidort, *Compt. rend.*, 1961, **253**, 1257.
[3] F. van der Woude, *Phys. Stat. Sol.*, 1966, **17**, 417.
[4] K. Ono and A. Ito, *J. Phys. Soc. Japan*, 1962, **17**, 1012.
[5] P. Imbert and A. Gerard, *Compt. rend.*, 1963, **257**, 1054.
[6] N. Blum, A. J. Freeman, J. W. Shaner, and L. Grodzins, *J. Appl. Phys.*, 1965, **36**, 1169.
[7] D. J. Simkin and R. A. Bernheim. *Phys. Rev.*, 1967, **153**, 621.
[8] G. Cinader, P. J. Flanders, and S. Shtrikman, *Phys. Rev.*, 1967, **162**, 419.
[9] R. W. Vaughan and H. G. Drickamer, *J. Chem. Phys.*, 1967, **47**, 1530.
[10] Sh. Sh. Bashkirov, G. J. Selyutin, and V. A. Christyakov, *Doklady Akad. Nauk S.S.S.R.*, 1968, **180**, 567.

[11] T. Nakamura, T. Shinjo, Y. Endoh, N. Yamamoto, M. Shiga, and Y. Nakamura, *Phys. Letters*, 1964, **12**, 178.
[12] Y. Bando, M. Kiyama, N. Yamamoto, T. Takada, T. Shinjo, and H. Takaki, *J. Phys. Soc. Japan*, 1965, **20**, 2086.
[13] W. Kundig, H. Bömmel, G. Constabaris, and R. H. Lindquist, *Phys. Rev.*, 1966, **142**, 327.
[14] D. Schroeer and R. C. Nininger, Jr, *Phys. Rev. Letters*, 1967, **19**, 632.
[15] D. Schroeer, *Phys. Letters*, 1968, **27A**, 507.
[16] J. S. van Wieringen, *Phys. Letters*, 1968, **26A**, 370.
[17] R. Bauminger, S. G. Cohen, A. Marinov, S. Ofer, and E. Segal, *Phys. Rev.*, 1961, **122**, 1447.
[18] W. H. Kelly, V. J. Folen, M. Hass, W. N. Schreiner, and G. B. Beard, *Phys. Rev.*, 1961, **124**, 80.
[19] R. J. Armstrong, A. H. Morrish, and G. A. Sawatzky, *Phys. Letters*, 1966, **23**, 414.
[20] G. Shirane, D. E. Cox, and S. L. Ruby, *Phys. Rev.*, 1962, **125**, 1158.
[21] D. J. Elias and J. W. Linnett, *Trans. Faraday Soc.*, 1969, **65**, 2673.
[22] D. P. Johnson, *Solid State Commun.*, 1969, **7**, 1785.
[23] H. Schechter, P. Hillman, and M. Ron, *J. Appl. Phys.*, 1966, **37**, 3043.
[24] S. K. Banerjee, W. O'Reilly, T. C. Gibb, and N. N. Greenwood, *J. Phys. and Chem. Solids*, 1967, **28**, 1323.
[25] A. Ito, K. Ono, and Y. Ishikawa, *J. Phys. Soc. Japan*, 1963, **18**, 1465.
[26] S. K. Banerjee, W. O'Reilly, and C. E. Johnson, *J. Appl. Phys.*, 1967, **38**, 1289.
[27] F. van der Woude, G. A. Sawatzky, and A. H. Morrish, *Phys. Rev.*, 1968, **167**, 533.
[28] W. Kundig and R. S. Hargrove, *Solid State Commun.*, 1969, **7**, 223.
[29] G. A. Sawatzky, J. M. D. Coey, and A. H. Morrish, *J. Appl. Phys.*, 1969, **40**, 1402.
[30] B. J. Evans and S. S. Hafner, *J. Appl. Phys.*, 1969, **40**, 1411.
[31] R. S. Hargrove and W. Kundig, *Solid State Commun.*, 1970, **8**, 303.
[32] V. P. Romanov, V. D. Checherskii, and V. V. Eremenko, *Phys. Stat. Sol.*, 1969, **31**, K153.
[33] G. A. Sawatzky, F. van der Woude, and A. H. Morrish, *Phys. Rev.*, 1969, **183**, 383.
[34] J. M. Daniels and A. Rosencwaig, *J. Phys. and Chem. Solids*, 1969, **30**, 1561.
[35] T. K. McNab, R. A. Fox, and A. J. F. Boyle, *J. Appl. Phys.*, 1968, **39**, 5703.
[36] M. J. Rossiter and A. E. M. Hodgson, *J. Inorg. Nuclear Chem.*, 1965, **27**, 63.
[37] T. Takada, M. Kiyama, Y. Bando, T. Nakamura, M. Shija, T. Shinjo, N. Yamamoto, Y. Endoh, and H. Takaki, *J. Phys. Soc. Japan*, 1964, **19**, 1744.
[38] A. Z. Hrynkiewicz, D. S. Kulgawczuk, and K. Tomala, *Phys. Letters*, 1965, **17**, 93.
[39] A. Z. Hrynkiewicz and D. S. Kulgawczuk, *Acta Phys. Polon.*, 1963, **24**, 689.
[40] T. Shinjo, *J. Phys. Soc. Japan*, 1966, **21**, 917.
[41] A. M. van der Kraan and J. J. van Loef, *Phys. Letters*, 1966, **20**, 614.
[42] F. van der Woude and A. J. Dekker, *Phys. Stat. Sol.*, 1966, **13**, 181.
[43] I. Dezsi and M. Fodor, *Phys. Stat. Sol.*, 1966, **15**, 247.
[44] J. B. Forsyth, I. G. Hedley, and C. E. Johnson, *J. Phys. C*, 1968, Ser. 2, **1**, 179.

[45] N. Yamamoto, T. Shinjo, M. Kiyama, Y. Bando, and T. Takada, *J. Phys. Soc. Japan*, 1968, **25**, 1267.
[46] A. Burewicz, D. Kulgawczuk, and A. Szytuza, *Phys. Stat. Sol.*, 1967, **21**, K85.
[47] I. Dezsi, L. Keszthelyi, D. Kulgawczuk, B. Moinar, and N. A. Eissa, *Phys. Stat. Sol.*, 1967, **22**, 617.
[48] C. E. Johnson, *J. Phys. C*, 1969, Ser. 2, **2**, 1996.
[49] H. Miyamoto, T. Shinjo, Y. Bando, and T. Takada, *J. Phys. Soc. Japan*, 1967, **23**, 1421.
[50] J. Chappert and J. Portier, *Solid State Commun.*, 1966, **4**, 185.
[51] J. Chappert and J. Portier, *Solid State Commun.*, 1966, **4**, 395.
[52] S. L. Kordyuk, I. P. Suzdalev, and V. I. Lisichenko, *Ukr. Fiz. Zhur.*, 1969, **16**, 692.
[53] M. J. Rossiter, *J. Phys. and Chem. Solids*, 1965, **26**, 775.
[54] K. Ono, A. Ito, and Y. Syono, *Phys. Letters*, 1966, **19**, 620.
[55] C. M. Yagnik and H. B. Mathur, *J. Phys. C*, 1968, Ser. 2, **1**, 469.
[56] M. J. Rossiter, *Phys. Letters*, 1966, **21**, 128.
[57] H. B. Mathur, A. P. B. Sinha, and C. M. Yagnik, *Solid State Commun.*, 1965, **3**, 401.
[58] F. Hartmann-Boutron, *Compt. rend.*, 1966, **263**, B188.
[59] P. Imbert, *Compt. rend.*, 1966, **263**, B184.
[60] F. Hartmann-Boutron and P. Imbert, *J. Appl. Phys.*, 1968, **39**, 775.
[61] M. Eibschutz, U. Ganiel, and S. Shtrikman, *Phys. Rev.*, 1966, **151**, 245 (erratum loc. cit., 1967, **158**, 566).
[62] M. J. Rossiter and P. T. Clarke, *Nature*, 1965, **207**, 402.
[63] K. Ono, L. Chandler, and A. Ito, *J. Phys. Soc. Japan*, 1968, **25**, 174.
[64] S. K. Banerjee, W. O'Reilly, T. C. Gibb, and N. N. Greenwood, *Phys. Letters*, 1966, **20**, 455.
[65] M. Tanaka, T. Tokoro, and Y. Aiyama, *J. Phys. Soc. Japan*, 1966, **21**, 262.
[66] P. Imbert, *Compt. rend.*, 1966, **263**, B767.
[67] F. Hartmann-Boutron, *Compt. rend.*, 1966, **263**, B 1131.
[68] P. Imbert and E. Martel, *Compt. rend.*, 1965, **261**, 5404.
[69] G. Ritter, *Z. Physik*, 1966, **189**, 23.
[70] R. M. Housley and U. Gonser, *Phys. Rev.*, 1968, **171**, 480.
[71] G. A. Sawatzky, F. van der Woude, and A. H. Morrish, *Phys. Letters*, 1967, **25A**, 147.
[72] I. Bunget, *Phys. Stat. Sol.*, 1968, **28**, K39.
[73] H. Yasuoka, A. Hirai, T. Shinjo, M. Kiyama, Y. Bando, and T. Takada, *J. Phys. Soc. Japan*, 1967, **22**, 174.
[74] M. Tanaka, T. Mizoguchi, and Y. Aiyama, *J. Phys. Soc. Japan*, 1963, **18**, 1091.
[75] M. Bornaz, G. Filoti, A. Gelberg, and M. Rosenberg, *Phys. Letters*, 1967, **24A**, 449.
[76] M. Bornaz, G. Filoti, A. Gelberg, and M. Rosenberg, *J. Phys. C*, 1969, Ser. 2, **2**, 1008.
[77] L. Cser, I. Dezsi, I. Gladkih, L. Keszthelyi, D. Kulgawczuk, N. A. Eissa, and E. Sterk, *Phys. Stat. Sol.*, 1968, **27**, 131.
[78] U. Konig, Y. Gros, and G. Chol, *Phys. Stat. Sol.*, 1969, **33**, 811.
[79] E. Wieser, V. A. Povitskii, E. F. Makarov, and K. Kleinstuck, *Phys. Stat. Sol.*, 1968, **25**, 607.
[80] V. F. Belov, P. P. Kurichok, G. S. Podvalnykh, T. A. Khimich, E. V. Korneyev, and D. A. Bondarev, *Fiz. Tverd. Tela*, 1969, **11**, 2675.

[81] T. K. McNab and A. J. F. Boyle, p. 957 of 'Hyperfine Structure and Nuclear Radiations', Ed. by E. Matthias and D. A. Shirley, North-Holland, 1968.
[82] I. Dezsi, A. Z. Hrynkiewicz, and D. S. Kulgawczuk, *Acta Phys. Polon.*, 1963, **24**, 283.
[83] G. A. Sawatzky, F. van der Woude, and A. H. Morrish, *J. Appl. Phys.*, 1968, **39**, 1204.
[84] W. J. Schuele, S. Shtrikman, and D. Treves, *J. Appl. Phys.*, 1965, **36**, 1010.
[85] D. Kedem and T. Rothem, *Phys. Rev. Letters*, 1967, **18**, 165.
[86] J. Chappert and R. B. Frankel, *Phys. Rev. Letters*, 1967, **19**, 570.
[87] J-P. Morel, *J. Phys. and Chem. Solids*, 1967, **28**, 629.
[88] M. Eibschutz and S. Shtrikman, *J. Appl. Phys.*, 1968, **39**, 997.
[89] T. Mizoguchi and M. Tanaka, *J. Phys. Soc. Japan*, 1963, **18**, 1301.
[90] G. S. Fatseas and R. Krishnan, *J. Appl. Phys.*, 1968, **39**, 1256.
[91] Y. Ishikawa, *J. Phys. Soc. Japan*, 1962, **17**, 1877.
[92] T. Yamadaya, T. Mitui, T. Okada, N. Shikazono, and Y. Hamaguchi, *J. Phys. Soc. Japan*, 1962, **17**, 1897.
[93] B. J. Evans, S. S. Hafner, and G. M. Kalvius, *Phys. Letters*, 1966, **23**, 24.
[94] B. J. Evans and S. S. Hafner, *J. Phys. and Chem. Solids*, 1968, **29**, 1573.
[95] A. Hudson and H. J. Whitfield, *Mol. Phys.*, 1967, **12**, 165.
[96] F. Hartmann-Boutron, A. Gerard, P. Imbert, R. Kleinberger, and F. Varret, *Compt. rend.*, 1969, **268**, B906.
[97] M. Eibschutz and U. Ganiel, *Solid State Commun.*, 1968, **6**, 775.
[98] M. Eibschutz, U. Ganiel, and S. Shtrikman, *Phys. Rev.*, 1967, **156**, 259.
[99] M. Schieber, R. B. Frankel, N. A. Blum, and S. Foner, *J. Appl. Phys.*, 1967, **38**, 1282.
[100] J. M. Trooster and A. Dymanus, *Phys. Stat. Sol.*, 1967, **24**, 487.
[101] J. M. Trooster, *Phys. Letters*, 1965, **16**, 21.
[102] F. Bertaut, G. Buisson, J. Chappert, and G. Bassi, *Compt. rend.*, 1965, **260**, 3355.
[103] E. F. Bertaut, G. Bassi, G. Buisson, J. Chappert, A. Delapalme, R. Pauthenet, H. P. Rebouillat, and R. Aleonard, *J. Phys. Radium*, 1966, **27**, 433.
[104] R. B. Frankel, N. A. Blum, S. Foner, A. J. Freeman, and M. Schieber, *Phys. Rev. Letters*, 1965, **15**, 958.
[105] G. Shirane, D. E. Cox, W. J. Takei, and S. L. Ruby, *J. Phys. Soc. Japan*, 1962, **17**, 1598.
[106] A. Okiji and J. Kanamori, *J. Phys. Soc. Japan*, 1964, **19**, 908.
[107] Sh. Sh. Bashkirov, G. D. Kurbatov, R. A. Manapov, I. N. Penkov, E. K. Sadykov, and V. A. Chistyakov, *Doklady Akad. Nauk S.S.S.R.*, 1967, **173**, 407.
[108] T. C. Gibb, N. N. Greenwood, and W. Twist, *J. Inorg. Nuclear Chem.*, 1969, **31**, 947.
[109] J. K. Srivastava, G. K. Shenoy, and R. P. Sharma, *Solid State Commun.*, 1968, **6**, 73.
[110] J. K. Srivastava and R. P. Sharma, *Phys. Stat. Sol.*, 1969, **35**, 491.
[111] R. R. Chevalier, G. Roult, and E. F. Bertaut, *Solid State Commun.*, 1967, **5**, 7.
[112] S. Geller, R. W. Grant, J. A. Cape, and G. P. Espinosa, *J. Appl. Phys.*, 1967, **38**, 1457.
[113] R. W. Grant, S. Geller, J. A. Cape, and G. P. Espinosa, *Phys. Rev.*, 1968, **175**, 686.
[114] I. Dezsi, G. Erlaki, and L. Keszthelyi, *Phys. Stat. Sol.*, 1967, **21**, K121.
[115] V. G. Bhide and S. Multani, *Solid State Commun.*, 1965, **3**, 271.

[116] C. M. Yagnik, J. P. Canner, R. Gerson, and W. J. James, *J. Appl. Phys.*, 1969, **40**, 4713.
[117] R. O. Bell, *J. Phys. and Chem. Solids*, 1968, **29**, 1.
[118] J. Traff, *Phys. Stat. Sol.*, 1969, **34**, K139.
[119] T. Nakagwa, K. Yoshikawa, and S. Nomura, *J. Phys. Soc. Japan*, 1969, **27**, 880.
[120] M. Eibschutz, S. Shtrikman, and D. Treves, *Phys. Rev.*, 1967, **156**, 562.
[121] D. Treves, *J. Appl. Phys.*, 1965, **36**, 1033.
[122] M. Eibschutz, G. Gorodetsky, S. Shtrikman, and D. Treves, *J. Appl. Phys.*, 1964, **35**, 1071.
[123] L. M. Levinson, M. Lubau, and S. Shtrikman, *Phys. Rev.*, 1969, **177**, 864.
[124] M. Robbins, G. K. Wertheim, A. Menth, and R. C. Sherwood, *J. Phys. and Chem. Solids*, 1969, **30**, 1823.
[125] G. Gorodetsky and L. M. Levinson, *Solid State Commun.*, 1969, **7**, 67.
[126] R. W. Grant and S. Geller, *Solid State Commun.*, 1969, **7**, 1291.
[127] J. Chappert and P. Imbert, *J. Phys. Radium*, 1963, **24**, 412.
[128] E. F. Bertaut, J. Chappert, J. Mareschal, J. P. Rebouillat, and J. Sivardiere, *Solid State Commun.*, 1967, **5**, 293.
[129] J. M. D. Coey, G. A. Sawatzky, and A. H. Morrish, *Phys. Rev.*, 1969, **184**, 334.
[130] J. Chappert, *J. Phys. Radium*, 1967, **28**, 81.
[131] J. Chappert, *Phys. Letters*, 1965, **18**, 229.
[132] R. Bauminger, S. G. Cohen, A. Marinov, and S. Ofer, *Phys. Rev.*, 1961, **122**, 743.
[133] C. Alff and G. K. Wertheim, *J. Appl. Phys.*, 1961, **122**, 1414.
[134] G. Crecelius, D. Quitmann, and S. Hufner, *Solid State Commun.*, 1967, **5**, 817.
[135] J. J. van Loef, *J. Appl. Phys.*, 1968, **39**, 1258.
[136] W. J. Nicholson and G. Burns, *Phys. Rev.*, 1964, **133**, A1568.
[137] I. Nowik and S. Ofer, *Phys. Rev.*, 1967, **153**, 409.
[138] V. F. Belov and L. A. Aliev, *Fiz. Tverd. Tela*, 1966, **8**, 2791.
[139] I. S. Lyubutin, E. F. Makarov, and V. A. Povitskii, *Zhur. eksp. teor. Fiz.*, 1967, **53**, 65.
[140] V. I. Goldanskii, V. A. Trukhtanov, M. N. Devisheva, and V. F. Belov, *Phys. Letters*, 1965, **15**, 317.
[141] I. S. Lyubutin, E. F. Makarov, and V. A. Povitskii, *Symposia Faraday Soc. No. 1*, 1968, 31.
[142] E. R. Czerlinsky, *Phys. Stat. Sol.*, 1969, **34**, 483.
[143] V. A. Bokov, G. V. Popov, and S. I. Yushchuk, *Fiz. Tverd. Tela*, 1969, **11**, 593.
[144] V. A. Bokov, S. I. Yushchuk, and G. V. Popov, *Solid State Commun.*, 1969, **7**, 373.
[145] V. A. Bokov, G. V. Popov, and S. I. Yushchuk, *Fiz. Tverd. Tela*, 1969, **11**, 1994.
[146] G. N. Belozerskii, V. N. Gittsovich, and A. N. Murin, *ZETF Letters*, 1969, **9**, 352.
[147] L. M. Belyaev, I. S. Lyubutin, B. V. Mill, and V. A. Povitskii, *Fiz. Tverd. Tela*, 1969, **11**, 795.
[148] D. E. Cox, G. Shirane, P. A. Flinn, S. L. Ruby, and W. J. Takei, *Phys. Rev.*, 1963, **132**, 1547.
[149] G. A. Fatseas and S. Lefevre, *Compt. rend.*, 1968, **266B**, 374.
[150] A. H. Muir, Jr, and H. Wiedersich, *J. Phys. and Chem. Solids*, 1967, **28**, 65.
[151] H. J. Whitfield, *Austral. J. Chem.*, 1967, **20**, 859.

[152] S. Geller, R. W. Grant, U. Gonser, H. Wiedersich, and G. P. Espinosa, *Phys. Letters*, 1967, **25A**, 722.
[153] M. Eibschutz, U. Ganiel, and S. Shtrikman, *J. Mater. Sci.*, 1969, **4**, 574.
[154] F. Pobell and F. Wittmann, *Phys. Letters*, 1965, **19**, 175.
[155] R. W. Grant, S. Geller, H. Wiedersich, U. Gonser, and L. D. Fullmer, *J. Appl. Phys.*, 1968, **39**, 1122.
[156] S. Geller, R. W. Grant, U. Gonser, H. Wiedersich, and G. P. Espinosa, *Phys. Letters*, 1966, **20**, 115.
[157] R. W. Grant, H. Wiedersich, S. Geller, U. Gonser, and G. P. Espinosa, *J. Appl. Phys.*, 1967, **38**, 1455.
[158] F. Wittmann, *Phys. Letters*, 1967, **24A**, 252.
[159] A. Hudson and H. J. Whitfield, *J. Chem. Soc.* (A), 1967, 376.
[160] R. W. Grant, *J. Chem. Phys.*, 1969, **51**, 1156.
[161] H. Yamamoto, T. Okada, H. Watanabe, and M. Fukase, *J. Phys. Soc. Japan*, 1968, **24**, 175.
[162] A. F. Reid, H. K. Perkins, and M. J. Sienko, *Inorg. Chem.*, 1968, **7**, 119.
[163] C. Do-Dinh, E. F. Bertaut, and J. Chappert, *Journal de Physique*, 1969, **30**, 566.
[164] T. Birchall, N. N. Greenwood, and A. F. Reid, *J. Chem. Soc.* (A), 1969, 2382.
[165] N. Ichinose and T. Yoshioka, *J. Phys. Soc. Japan*, 1966, **21**, 1471.
[166] J. J. van Loef and P. J. M. Franssen, *Phys. Letters*, 1963, **7**, 225.
[167] J. G. Rensen and J. S. van Wieringen, *Solid State Commun.*, 1969, **7**, 1139.
[168] V. W. Zinn, S. Hufner, M. Kalvius, P. Kienle, and W. Wiedemann, *Z. angew. Physik*, 1964, **17**, S147.
[169] G. Albanese, G. Asti, and P. Batti, *Nuovo Cimento*, 1968, **54B**, 339.
[170] T. A. Khimich, V. F. Belov, M. N. Shipko, and E. V. Korneyev, *Zhur. eksp. teor. Fiz.*, 1969, **57**, 395.
[171] G. Albanese, G. Asti, and C. Lamborizio, *J. Appl. Phys.*, 1968, **39**, 1198.
[172] T. A. Khimich, V. F. Belov, M. N. Shipko, and E. V. Korneyev, *Fiz. Tverd. Tela*, 1969, **11**, 2093.
[173] M. C. Montmory, M. Belakhovsky, R. Chevalier, and R. Newnham, *Solid State Commun.*, 1968, **6**, 317.
[174] J. T. Dehn, R. E. Newnham, and L. N. Mulay, *J. Chem. Phys.*, 1968, **49**, 3201.
[175] C. L. Herzenberg and D. L. Riley, *J. Phys. and Chem. Solids*, 1969, **30**, 2108.
[176] M. Bacmann, E. F. Bertaut, A. Blaise, R. Chevalier, and G. Roult, *J. Appl. Phys.*, 1969, **40**, 1131.
[177] R. Wolfe, R. D. Pierce, M. Eibschutz, and J. W. Nielsen, *Solid State Commun.*, 1969, **7**, 949.
[178] P. K. Gallagher, J. B. MacChesney, and D. N. E. Buchanan, *J. Chem. Phys.*, 1964, **41**, 2429.
[179] V. N. Panyushkin, G. de Pasquali, and H. G. Drickamer, *J. Chem. Phys.*, 1969, **51**, 3305.
[180] P. K. Gallagher, J. B. MacChesney, and D. N. E. Buchanan, *J. Chem. Phys.*, 1965, **43**, 516.
[181] P. K. Gallagher, J. B. MacChesney, and D. N. E. Buchanan, *J. Chem. Phys.*, 1966, **45**, 2466.
[182] U. Shimony and J. M. Knudsen, *Phys. Rev.*, 1966, **144**, 361.
[183] P. K. Gallagher and J. B. MacChesney, *Symposia Faraday Soc. No. 1*, 1968, 40.
[184] E. Banks and M. Mizushima, *J. Appl. Phys.*, 1969, **40**, 1408.
[185] K. Ono, A. Ito, and E. Hirahara, *J. Phys. Soc. Japan*, 1962, **17**, 1615.

[186] K. Ono, Y. Ishikawa, A. Ito, and E. Hirahara, *J. Phys. Soc. Japan*, 1962, **17**, Suppl. B-I, 125.
[187] S. S. Hafner, B. J. Evans, and G. M. Kalvius, *Solid State Commun.*, 1967, **5**, 17.
[188] J. A. Morice, L. V. C. Rees, and D. T. Rickard, *J. Inorg. Nuclear Chem.*, 1969, **31**, 3797.
[189] L. M. Levinson, and D. Treves, *J. Phys. and Chem. Solids*, 1968, **29**, 2227.
[190] E. F. Bertaut, P. Burlet, and J. Chappert, *Solid State Commun.*, 1965, **3**, 335.
[191] A. A. Temperly and H. W. Lefevre, *J. Phys. and Chem. Solids*, 1966, **27**, 85.
[192] P. Imbert, A. Gerard, and M. Winterberger, *Compt. rend.*, 1963, **256**, 4391.
[193] R. W. Vaughan and H. G. Drickamer, *J. Chem. Phys.*, 1967, **47**, 468.
[194] P. K. Gallagher, J. B. MacChesney, and R. C. Sherwood, *J. Chem. Phys.*, 1969, **50**, 4417.
[195] E. F. Makarov, A. S. Marfunin, A. R. Mkrtchyan, G. N. Nadzharyan, V. A. Povitskii, and R. A. Stukan, *Fiz. Tverd. Tela*, 1969, **11**, 495.
[196] M. Eibschutz, E. Hermon, and S. Shtrikman, *Solid State Commun.*, 1967, **5**, 529.
[197] C. M. Yagnik and H. B. Mathur, *Solid State Commun.*, 1967, **5**, 841.
[198] M. Eibschutz, E. Hermon, and S. Shtrikman, *J. Phys. and Chem. Solids*, 1967, **28**, 1633.
[199] M. Eibschutz, S. Shtrikman, and Y. Tenenbaum. *Phys. Letters*, 1967, **24A**, 563.
[200] G. R. Hoy and S. Chandra, *J. Chem. Phys.*, 1967, **47**, 961.
[201] G. R. Hoy and K. P. Singh, *Phys. Rev.*, 1968, **172**, 514.
[202] G. Haacke and A. J. Nozik, *Solid State Commun.*, 1968, **6**, 363.
[203] F. K. Lotgering, R. P. Van Stapele, G. H. A. M. van der Steen, and J. S. van Wieringen, *J. Phys. and Chem. Solids*, 1969, **30**, 799.
[204] D. Raj, K. Chandra, and S. P. Puri, *J. Phys. Soc. Japan*, 1968, **24**, 39.
[205] N. N. Greenwood and H. J. Whitfield, *J. Chem. Soc. (A)*, 1968, 1697.
[206] B. V. Borshagovskii, A. S. Marfunin, A. R. Mkrtchyan, R. A. Stukan, and G. N. Nadzharyan, *Izvest. Akad. Nauk S.S.S.R.*, 1968, 1267.
[207] C. L. Herzenberg, *Nuovo Cimento*, 1968, **53B**, 516.
[208] R. Deo and S. P. Puri, *Nuovo Cimento*, 1969, **60B**, 261.
[209] E. F. Makarov, A. S. Marfunin, A. R. Mkrtchyan, V. A. Povitskii, and R. A. Stukan, *Fiz. Tverd. Tela*, 1968, **10**, 913.
[210] D. Daj and S. P. Puri, *J. Chem. Phys.*, 1969, **50**, 3184.
[211] W. Kerler, W. Neuwirth, E. Fluck, P. Kuhn, and B. Zimmermann, *Z. Physik*, 1963, **173**, 321.
[212] G. M. Bancroft, A. G. Maddock, and R. G. Burns, *Geochim. Cosmochim. Acta*, 1967, **31**, 2219.
[213] G. M. Bancroft, R. G. Burns, and A. J. Stone, *Geochim. Cosmochim. Acta*, 1968, **32**, 547.
[214] M. Eibschutz and U. Ganiel, *Solid State Commun.*, 1967, **5**, 267.
[215] W. Kundig, J. A. Cape, R. H. Lindquist, and G. Constabaris, *J. Appl. Phys.*, 1967, **38**, 947.
[216] E. L. Sprenkel-Segal and S. S. Hanna, *Geochim. Cosmochim. Acta*, 1964, **27**, 1913.
[217] M. de Coster, H. Pollak, and S. Amelinckx, *Phys. Stat. Sol.*, 1963, **3**, 283.
[218] G. M. Bancroft, R. G. Burns, and R. A. Howie, *Nature*, 1967, **213**, 1221.
[219] G. K. Shenoy, G. M. Kalvius, and S. S. Hafner, *J. Appl. Phys.*, 1969, **40**, 1314.
[220] C. L. Herzenberg and D. L. Riley, *Acta Cryst.*, 1969, **25A**, 389.
[221] Y. Takashima and S. Ohashi, *Bull. Chem. Soc. Japan*, 1968, **41**, 88.

[222] G. M. Bancroft, R. G Burns, and A. G. Maddock, *Amer. Min.*, 1967, **52**, 1009.
[223] G. M. Bancroft, *Phys. Letters*, 1967, **26A**, 17.
[224] G. M. Bancroft, R. G. Burns, A. G. Maddock, and R. F. J. Strens, *Nature*, 1966, **212**, 913.
[225] M. G. Clark, G. M. Bancroft, and A. J. Stone, *J. Chem. Phys.*, 1967, **47**, 4250.
[226] M. G. Clark, *J. Chem. Phys.*, 1968, **48**, 3246.
[227] H. J. Whitfield and A. G. Freeman, *J. Inorg. Nuclear Chem.*, 1967, **29**, 903.
[228] T. C. Gibb and N. N. Greenwood, *Trans. Faraday Soc.*, 1965, **61**, 1317.
[229] G. M. Bancroft, R. W. Burns, and A. G. Maddock, *Acta Cryst.*, 1967, **22**, 934.
[230] C. E. Weaver, J. M. Wampler, and T. E. Pecuil, *Science*, 1967, **156**, 204.
[231] H. Pollak, M. de Coster, and S. Amelinckx, *Phys. Stat. Sol.*, 1962, **2**, 1653.
[232] C. L. Herzenberg, D. L. Riley, and R. Lamoreaux, *Nature*, 1968, **219**, 364.
[233] L. Häggström, R. Wäppling, and H. Annersten, *Chem. Phys. Letters*, 1969, **4**, 107.
[234] L. Häggström, R. Wäppling, and H. Annersten, *Phys. Stat. Sol.*, 1969, **33**, 741.
[235] P. J. Malden and R. E. Meads, *Nature*, 1967, **215**, 844.
[236] N. Malathi, S. P. Puri, and I. P. Saraswat, *J. Phys. Soc. Japan*, 1969, **26**, 680.
[237] H. L. Bowen, S. B. Weed, and J. G. Stevens, *Amer. Min.*, 1969, **54**, 72.
[238] C. L. Herzenberg and D. L. Riley, *Science*, 1970, **167**, 683.
[239] A. H. Muir, R. M. Housley, R. W. Grant, M. Abdel-Gawad, and M. Blander, *Science*, 1970, **167**, 688.
[240] H. Fernandez-Moran, S. S. Hafner, M. Ohtsuki, and D. Virgo, *Science*, 1970, **167**, 686.
[241] P. Gay, G. M. Bancroft, and M. G. Brown, *Science*, 1970, **167**, 626.
[242] S. K. Runcorn, D. W. Collinson, W. O'Reilly, A. Stephenson, N. N. Greenwood, and M. H. Battey, *Science*, 1970, **167**, 697.
[243] C. L. Hertzenberg and D. L. Riley, Proc. Apollo 11 Lunar Science Conference, Vol. 3, *Geochim. Cosmochim. Acta, Suppl. I*, 1970, 2221.
[244] R. M. Housley, M. Blander, M. Abdel-Gawad, R. W. Grant, and A. H. Muir, Proc. Apollo 11 Lunar Science Conference, Vol. 3, *Geochim. Cosmochim. Acta, Suppl. I*, 1970, 2251.
[245] S. S. Hafner and D. Virgo, Proc. Apollo 11 Lunar Science Conference, Vol. 3, *Geochim. Cosmochim. Acta, Suppl. I*, 1970, 2183.
[246] P. Gay, G. M. Bancroft, and M. G. Brown, Proc. Apollo 11 Lunar Science Conference, Vol. 1, *Geochim. Cosmochim. Acta, Suppl. I*, 1970, 481.
[247] N. N. Greenwood and A. T. Howe, Proc. Apollo 11 Lunar Science Conference, Vol. 3, *Geochim. Cosmochim. Acta, Suppl. I*, 1970, 2163.

11 | Alloys and Intermetallic Compounds

One of the more active areas of Mössbauer spectroscopy has been in the study of metallic systems. Primarily this has been a result of the interest in magnetic ordering properties. The hyperfine field at an iron nucleus in a metallic matrix is produced by the mechanisms already described in Chapter 3.5. The unpaired $3d$-band electron spin density induces an imbalance in the s-electron spin density at the nucleus by exchange polarisation. This is referred to as the core-polarisation term and corresponds to the Fermi contact term, H_S. In addition, there may also be a small s-electron spin-density imbalance which arises directly from the conduction band of electrons which usually has considerable s-character. This is referred to as conduction-electron polarisation. Systematic study of the ^{57}Fe hyperfine field with change in alloy composition provides a means of determining the relative importance of these two contributions. The chemical isomer shift can also be related in principle to the electronic band structure of the alloy, but little success has been achieved in this direction.

The effective range and type of the magnetic exchange interactions has an important bearing on the spectrum. In disordered iron alloys the number of iron atoms in the nearest-neighbour coordination sphere to the resonant nucleus may not be constant. If the exchange interactions involve essentially only the nearest neighbours, a number of hyperfine fields are seen corresponding to the different symmetry combinations. In the event that the exchange interactions are long-range, the magnetic field is not a function of the immediate environment to the same extent, and only a single averaged field is found. Magnetic dilution of the iron provides an interesting method of studying the distance over which exchange interactions occur.

In view of the considerable volume of data available for ^{57}Fe in metallic systems and the rather tenuous connection of much of it to chemistry, we propose to present in this chapter a selective synopsis. This will illustrate the application of Mössbauer spectroscopy to metallic phases in general from a phenomenological aspect, but a detailed treatment of metal phase-relationships, magnetic ordering, and lattice dynamics is not attempted. The sections considered are:

(i) metallic iron
(ii) iron alloys
(iii) intermetallic compounds of iron

Discussion of ^{57}Co doping of non-ferrous alloys is deferred to Chapter 12.

11.1 Metallic Iron

The spectrum of iron metal featured in many of the early papers in which the magnetic hyperfine interactions of ^{57}Fe were first analysed, but these are now of historical interest only. A definitive work [1] on the temperature dependence of the spectrum showed that the hyperfine field at 293·9 K was $H_{\text{eff}} = 330$ kG. The magnetic moment ratio μ_1/μ_0 was found to be $-1\cdot715 \pm 0\cdot004$, giving the excited-state magnetic moment as $\mu_1 = -0\cdot1549$

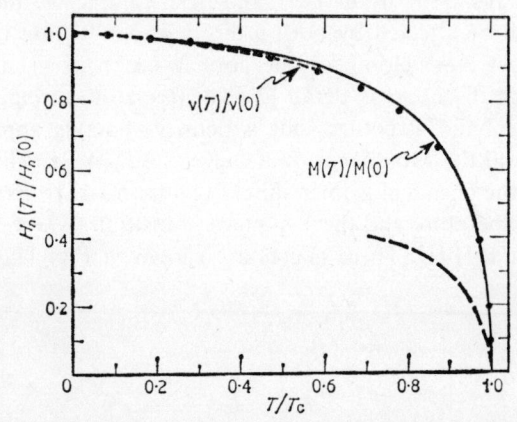

Fig. 11.1 The reduced internal magnetic field $H_n(T)/H_n(0)$ plotted as a function of the reduced temperature T/T_C. $H_n(0)$ is the value of the field at absolute zero and T_C is the Curie temperature. The dots represent Mössbauer data, the solid line saturation magnetisation data, and the upper dashed line the n.m.r. measurements. The lower dashed line is drawn with the temperature scale expanded tenfold. [Ref. 1, Fig. 8]

$\pm\ 0\cdot0013\mu_N$. The temperature dependence of H_{eff} below the Curie point of $T_C = 773°\text{C}$ agrees well with saturation–magnetisation data and n.m.r. data (Fig. 11.1). Independent measurement of the spectrum at room temperature in terms of frequency calibration gave $\mu_1/\mu_0 = -1\cdot716$ and a value for g_0 ($= 2\mu_0$) of $45\cdot47 \pm 0\cdot04$ MHz [2]. This value refers of course to the bulk material, and it is very encouraging to note the excellent agreement with the value derived from n.m.r. measurements [3] of 45·46 MHz which by contrast refers to iron nuclei in the domain walls. If the magnetic domains are randomly oriented, the expected 3 : 2 : 1 : 1 : 2 : 3 intensity ratios are found, but deviations from these are often found because of the ease with which iron metal foils can be magnetised and thence polarised, and because of saturation effects.

The sign of the magnetic field was found to be negative by some very early applied magnetic field measurements [4], i.e. the hyperfine field aligns antiparallel to the magnetisation; this contradicted prediction and indicated the dominance of core-polarisation terms. *Ab initio* calculations of field values in metals using Hartree–Fock calculations are not very successful, however, because the resultant field is determined by the difference between opposing terms which are greater by orders of magnitude [5]. Measurements in external fields of up to 135 kG at 4·2 K have been used in a study of band ferromagnetism in iron metal [6].

Detailed interpretation of the chemical isomer shift value (which is used in this book as the calibration zero at room temperature) is difficult. There is no defined orbital state as in the Fe^{2+} and Fe^{3+} complexes, and the chemical isomer shift will be affected by both a direct 4s-conduction band contribution and indirect 3d-shielding of the core 3s-electrons. The problems involved have been discussed in detail [7]. The free-atom configuration $3d^74s^1$ agrees with the band structure, but a density-of-states approach favours $3d^74s^{0.5}4p^{0.5}$, and the Mössbauer data suggests $3d^{6.5}4s^{1.5}$. The temperature dependence of the chemical isomer shift is regular apart from discontinuities at the Curie temperature and the α–γ phase transition (where it decreases by about 0·05 mm s^{-1}) [1]. These effects are shown in Fig. 11.2. The discon-

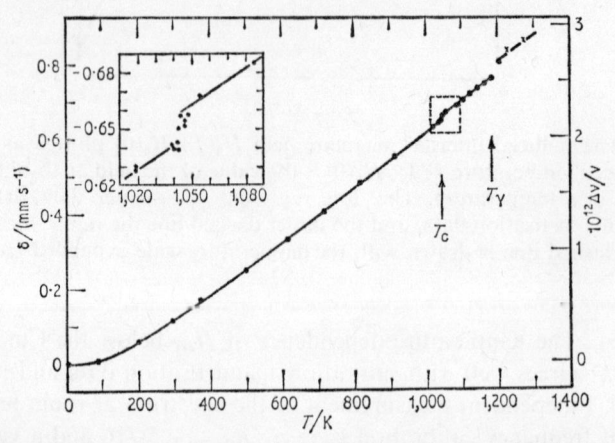

Fig. 11.2 Temperature dependence of the chemical isomer shift in metallic ^{57}Fe. The shift is arbitrarily set at zero at 0 K. The Curie temperature and the temperature of the α–γ phase transition are designated T_C and T_γ and the insert shows a magnified view in the vicinity of T_C. [Ref. 1, Fig. 9]

[*Refs.* on p. 325]

tinuity at the Curie temperature has been interpreted as due to the disappearance of magnetic splitting of the 3d-band, causing a shift of the absolute position of the Fermi level and a consequent change in the number of 4s-electrons [8]. More accurate measurements show that the change takes place within a temperature range of less than 0·3 K, suggesting the possibility of a first-order transition [9]. Detailed analysis of the temperature dependence of the chemical isomer shift shows that in addition to the expected second-order Doppler shift and decrease in s-density due to thermal expansion, there is a third contribution related to the magnetisation, consistent with interpretation of the Curie point discontinuity [10].

Discontinuities in both the chemical isomer shift and temperature dependence of the recoil-free fraction have been found at the α–γ (~1200 K) and γ–δ (~1660 K) phase transitions [11]. The changes in recoil-free fraction of $\delta f_{\alpha\gamma} = 0·030 \pm 0·008$ and $\delta f_{\gamma\delta} = -0·06 \pm 0·01$ correspond to a difference in the Debye temperatures of the different phases of $\delta\theta_{\alpha\gamma} = +(8 \pm 3 \text{ K})$ and $\delta\theta_{\gamma\delta} = -(70 \pm 15 \text{ K})$.

The internal magnetic field and chemical isomer shift both decrease linearly with increasing pressure [12–14].

$$\partial(H/H_0)/\partial(V/V_0) = 0·34 \pm 0·01$$
$$\partial(\delta)/\partial(V/V_0) = 1·50 \pm 0·05 \text{ mm s}^{-1}$$

Above 130 kbar there is a phase change to a non-magnetic hexagonal phase which has a chemical isomer shift 0·17 mm s^{-1} more negative than for the body-centred cubic α-iron [13, 15]. The relative line intensities are also affected by pressure due to a pressure-induced polarisation of the iron foil [16].

The Néel temperature of precipitated γ-iron (face-centred cubic) was determined to be of the order of 60 K using samples precipitated in copper matrices by quenching [17]. Measurements have also been made of the chemical isomer shift near the $\alpha\varepsilon\gamma$ triple-point of iron under conditions of high temperature and pressure [18]. The properties of thin films of iron have been studied by evaporation of Fe onto the surface of SiO$_2$, the process being repeated several times with intermediate layers of SiO$_2$ to give an adequate absorption cross-section [19]. The spectrum was that of bulk iron metal at room temperature for a layer thickness of greater than 6 Å. A 4·6-Å sample gave only a two-line spectrum at room temperature but showed magnetic broadening in the 77–20 K region, and reverted to the full six-line spectrum at 4 K.

Another particularly interesting type of experiment gives information about the [57]Co parent atom. At temperatures of < 1K the Zeeman levels of [57]Co ($I = \frac{7}{2}$) atoms in iron metal are not equally occupied as their separation is ~kT. Assuming that the spin-lattice relaxation times are longer than the total nuclear decay time, the preferential orientation of the nucleus

[Refs. on p. 325]

is preserved through the decay, and the hyperfine spectrum becomes asymmetric because of preferential population via the lower ^{57}Co Zeeman levels. This was first shown at temperatures between 0·2 K and 1·4 K [20], and more recently at 0·128 K (see Fig. 11.3) [21]. The field at the ^{57}Co nucleus is found

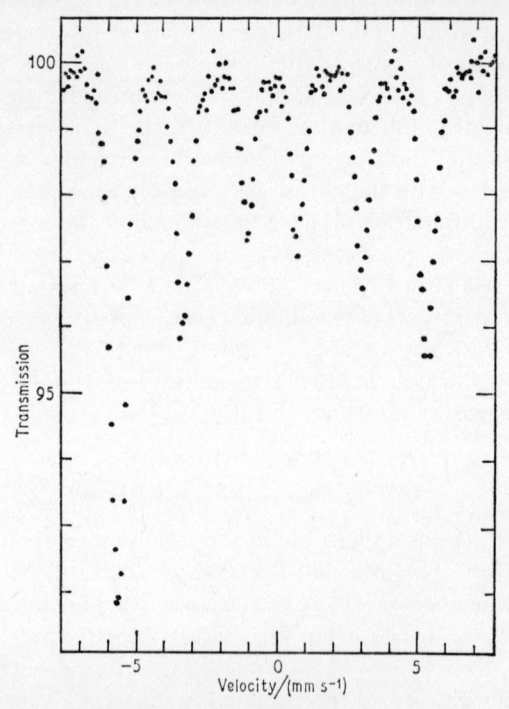

Fig. 11.3 The observed spectrum from a ^{57}Co/iron source at 0·128 K with a stainless steel absorber at room temperature. All three pairs of $| m_z |$ lines show asymmetric intensities as a result of unequal population of the ^{57}Co levels. [Ref. 21, Fig. 1]

to be parallel to that of the ^{57}Fe daughter. It is interesting to note that temperatures of <0·01 K would be needed to observe this effect in an ^{57}Fe atom in an iron absorber because the magnetic moment $\mu_g(^{57}\text{Fe})$ is 50 times smaller than $\mu_g(^{57}\text{Co})$. Selective level population in stable ^{57}Fe can therefore be achieved only at extremely low temperatures. Similar experiments using ^{57}Co in palladium metal at 0·15 K were able to show that the sign of the field at the ^{57}Co nucleus is antiparallel to that of the ^{57}Fe daughter [22].

11.2 Iron Alloys

Aluminium

The Fe-Al alloys are a good example of magnetically ordered alloys in

[*Refs. on p. 325*]

which the solute makes no appreciable magnetic contribution of its own and in which the exchange forces are largely dominated by core polarisation.

Particular interest has centred about the order–disorder properties of Fe$_3$Al [23]. Its structure consists of two interpenetrating simple-cubic sublattices, one containing $\frac{2}{3}$ of the Fe atoms and the other containing the remaining $\frac{1}{3}$ of the Fe atoms and all the Al atoms. The A site, containing only Fe atoms, has an environment of 4 Fe and 4 Al nearest neighbours. The D site has 8 Fe nearest neighbours in cubic array and, in the disordered state, has a random distribution of the iron and aluminium atoms. Ordering can be achieved with up to 94% of the Al atoms on the correct (alternate) D sites. The Mössbauer spectrum of the ordered alloy (below the Curie temperature of 713 K) consists of two overlapping six-line spectra with intensities in the ratio of 2 : 1 [24]. H_{eff} is 294 kG for the D sites and 211 kG for the A sites at room temperature compared with 330 kG for pure iron. The reduced internal field H_T/H_0 as a function of reduced temperature T/T_C is identical for the two sites. Theoretical analysis of the magnetic field data is consistent with a long-range coupling of the atomic spins by itinerant d-electrons, the exchange interaction being dominated by core polarisation to the effective exclusion of conduction-electron polarisation.

In the disordered Fe$_3$Al alloy, paramagnetic and ferromagnetic peaks coexist in the spectrum for a considerable range below the Curie temperature (which is higher than in ordered Fe$_3$Al) as a result of the dependence of the spin relaxation time on the (variable) number of iron atom nearest neighbours [25, 26]. More than 3 Al nearest neighbours are considered to result in paramagnetic spectra. Consequently when ordering takes place between 780 and 800 K to give a uniform 4 Al nearest neighbours, the magnetic spectrum starts to disappear. The chemical isomer shift shows an appreciable temperature hysteresis in the phase transformation region [26, 27].

The internal magnetic field at a given iron atom in the disordered Fe–Al alloys is strongly dependent on the number of aluminium nearest neighbours [28]. In the concentration range 19–28 at. % Al as many as three distinct internal magnetic fields are observed corresponding to 0, 3, and 4 Al nearest neighbours [29]. The disordered alloys with 0–21 at. % Al show separate internal magnetic fields from iron atoms with 0–4 Al nearest neighbours (see Fig. 11.4) [30]. The field is strongly dependent on the number of aluminium neighbours, but almost independent of the aluminium concentration (the small variation is due to second nearest neighbours etc.). This indicates that the core-polarisation contribution at an iron nucleus is remaining constant for a given immediate environment. Similar behaviour is shown by ordered alloys in the concentration range 21–30 at. % Al.

The chemical isomer shift in Fe–Al alloys increases regularly with increase in Al nearest neighbours; probably indicating transfer of electrons to the Fe 3d-band which would also cause the reduction in core polarisation observed through H_{eff} [30]. The second-order Doppler shift has been studied in a quenched Al–0·01 % Fe alloy [33].

The change in magnetic field relative to pure α-Fe for any given site, ΔH,

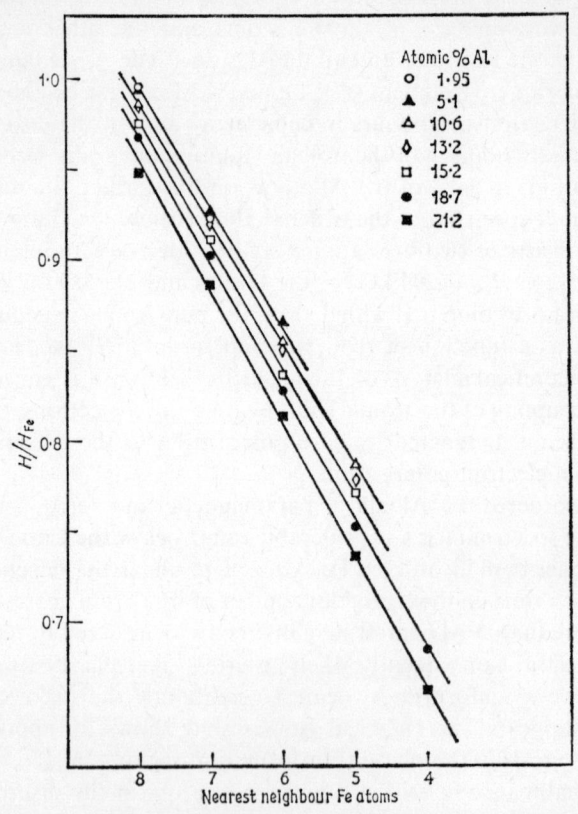

Fig. 11.4 Variation in the internal magnetic field at an iron nucleus in a disordered Fe–Al alloy with the number of nearest-neighbour iron atoms. 8 and 4 nearest-neighbour iron correspond to 0 and 4 nearest-neighbour aluminiums respectively. [Ref. 30, Fig. 2]

can be expressed (using terms up to and including the 6th nearest neighbours) as

$$\Delta H = \sum_{n=1}^{6} m_n \Delta H_n$$

where m_n is the number of the nth nearest-neighbour sites occupied by solute (Al) atoms and ΔH_n is the contribution to ΔH produced by a single aluminium atom at the nth nearest-neighbour site [31, 32]. The probability for all possible environments in a randomised 5·1 at. % Al alloy have been cal-

[*Refs. on p. 325*]

culated and summed to derive the ΔH_n values. Experimental and simulated line profiles for the outer pair of peaks in the ^{57}Fe spectrum are shown in Fig. 11.5 and the dependence of ΔH_n on distance is seen in Fig. 11.6. Basically

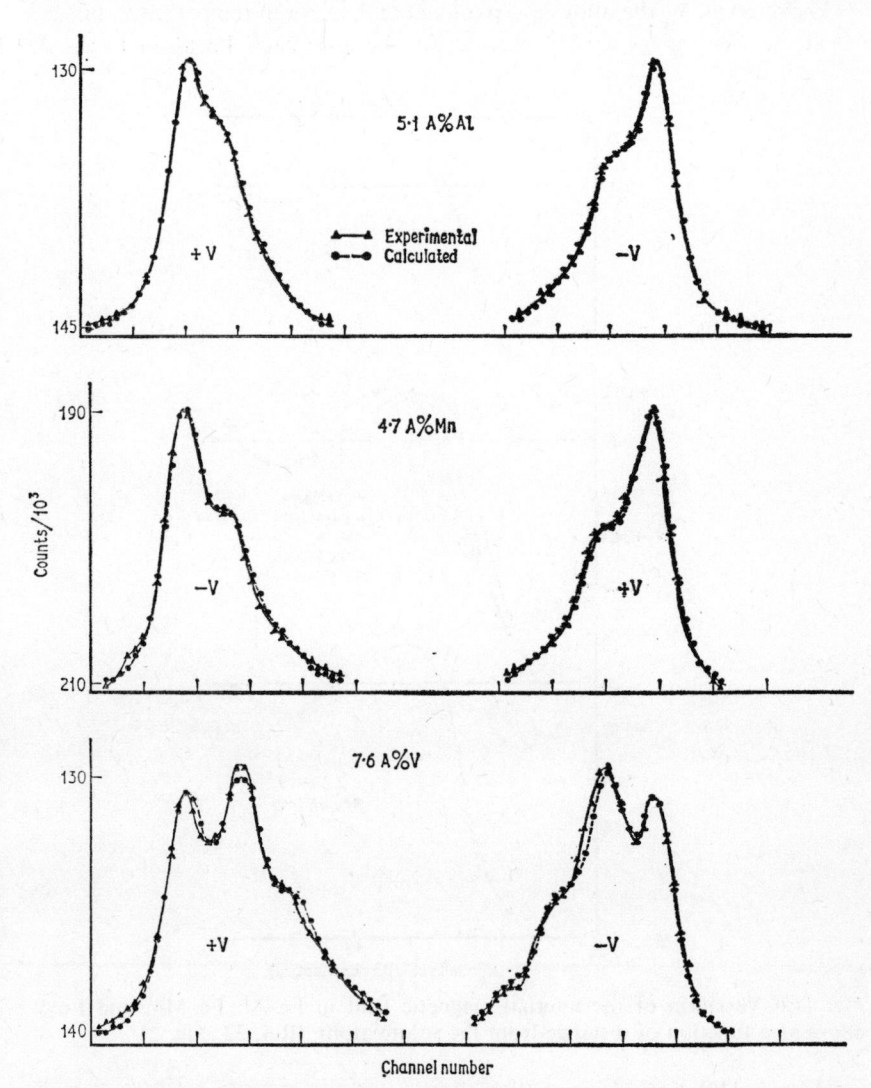

Fig. 11.5 Experimental and simulated curves for the outer peaks of Fe–Al, Fe–Mn, and Fe–V alloys. [Ref. 32, Fig. 1]

identical figures are obtained for a 10.6 at. % Al alloy, showing the ΔH_n values to be independent of concentration, i.e. the core polarisation is not a function of concentration. The conduction-electron spin density oscillates

[*Refs. on p. 325*]

about the value for pure α-Fe with increasing distance, and it appears that the Al atom is not contributing to the magnetic or band structure but acts very much like a hole in the lattice.

Above 30 at. % the alloys are paramagnetic at room temperature, but the 50 at. % alloy gives a very narrow line because each Fe atom has 8 Al

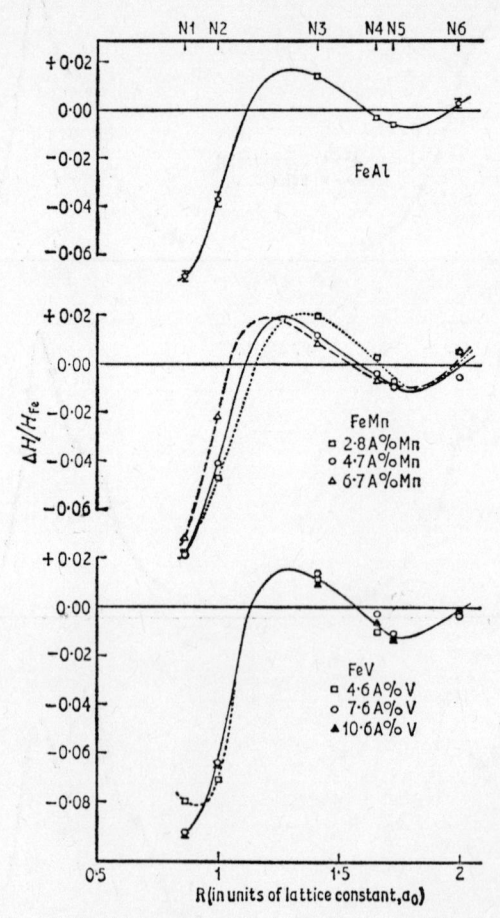

Fig. 11.6 Variation of the internal magnetic field in Fe–Al, Fe–Mn, and Fe–V alloys as a function of distance from the solute atom. [Ref. 32, Fig. 2]

neighbours [30]. Crushing the alloys in this region causes ferromagnetic lines to appear [34]. Plastic deformation induces ferromagnetism by creating large numbers of antiphase boundaries across which the number of Fe–Fe nearest neighbours is significantly greater than in the ordered alloy.

Preliminary data on iron–gallium alloys (0–10 at. % Ga) show multiple hyperfine fields analagously to the aluminium alloys [35].

[*Refs. on p. 325*]

First-row Transition Metals

The transition-metal alloys present several new features. Many of them have a spin moment on both the iron and solute atoms, and the simple non-magnetic impurity hole model as applied to aluminium will no longer be necessarily valid. The magnetic moment on the solute atom will affect the spin density at the iron nuclei and hence may also change the magnetic moment on the iron. The saturation magnetic moment is an average over the two components. It is dependent on the total electron concentration in

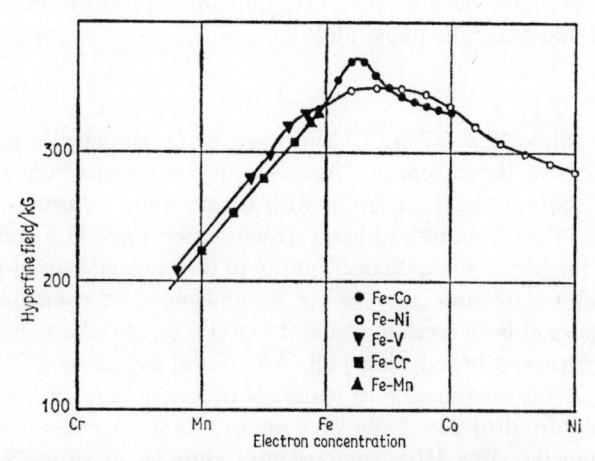

Fig. 11.7 The variation of hyperfine field at the ^{57}Fe nucleus in transition-metal alloys as a function of electron concentration. [Ref. 38, Fig. 6]

the alloy, and is approximately related to the saturation magnetic field at the iron. However, this hyperfine field is generally intermediate between that predicted from the local moment on the iron atom and that from the saturation moment, although it is difficult to decide whether the anomaly is due to changes in core polarisation or in conduction polarisation. The variation in the hyperfine field with electron concentration is shown in Fig. 11.7. The values plotted often represent an average field because of fluctuations in the nearest-neighbour environment.

Titanium

Titanium alloys have been referred to but briefly [36]. TiFe$_2$ is discussed at the end of the chapter.

Vanadium

The Fe–V alloys (4·6–10·6 at. % V) are similar to Fe–Al, and the oscillation of conduction-electron spin density has been illustrated in Figs. 11.5 and 11.6 [32]. Once again the hyperfine field is reduced by a fraction proportional to the number of impurity nearest neighbours [36, 37], so that alloys in the 0–16 at. % V range show complicated structure in the outer lines of the hyperfine spectrum which can be simulated by computer analysis assuming a statistical model of randomisation.

Chromium

Dilute Fe–Cr alloys show similar nearest-neighbour effects to the corresponding vanadium alloys [36–38]. Quenched samples of the Fe$_3$Cr alloy containing ~0·01% of N$_2$ have been found to contain small proportions of an as yet unidentified non-magnetic phase [39].

Manganese

The Fe–Mn alloys (2·8–6·7 at. % Mn) have many similarities to the Fe–Al system [36], with the exception that the conduction-electron spin-density oscillations show a phase variation with concentration corresponding to a change in the Fermi radius and band structure (see Figs. 11.5 and 11.6) [32]. In this case the Mn atoms *are* contributing to the magnetic structure.

Solid solutions of iron in β-Mn are paramagnetic at room temperature. Iron substitutes at both types of site in the structure, and shows no more than a slight quadrupole broadening [40]. Additional broadening at 4·2 K for 10–30 at. % Fe is attributable to magnetic ordering with fields of <20 kG.

The γ-Fe–Mn disordered alloys with face-centred cubic structure are antiferromagnetic. The Mössbauer spectra show a hyperfine field (extrapolated to 0 K) of 40 kG throughout the range 30–50 at. % Mn. [41–43]. The sublattice magnetisation on the other hand is dependent on concentration, indicating that the magnetic moments on Fe and Mn atoms behave rather differently with temperature, and complicating any interpretation of the magnetic properties.

Similar studies of the α-Mn alloys (70–95 at. % Mn) show fields of 5 and 16 kG at the two site symmetries occupied by iron with little or no dependence on manganese concentrations. The aligned 3d-moment on the iron atoms must therefore be small [44].

Cobalt

The 0–73 at. % Co body-centred cubic alloys are unusual in that there is an initial increase in the magnetic field at room temperature with increasing Co concentration, reaching a maximum at 25 at. % [36, 38, 45]. This signifies an *increase* in the local exchange potential and the spin density. Much less line broadening is seen in this case.

The TiFe$_x$Co$_{1-x}$ (0·3 ⩽ x ⩽ 0·7) alloys are known to be ferromagnetic below 50 K, but the Mössbauer spectra below this temperature show no magnetic field in the ^{57}Fe resonance [46, 47]. There is therefore no localised moment at the iron sites. Studies have also been made of the $(ZrCo_2)_x(ZrFe_2)_{1-x}$ alloys [48].

Nickel

The body-centred cubic alloys of nickel (0–24 at. % Ni) also show a slight rise in the hyperfine field with increasing nickel content, but this is less marked than for the cobalt alloys [38].

Several investigations of γ-FeNi 'Invar' type alloys (20–34 at. % Ni) have reported the coexistence of paramagnetic and ferromagnetic contributions in the spectra [49–54].

A considerable difference is found between ordered and disordered alloys containing 50 at. % Ni [55]. The ordered alloy shows much narrower resonance lines because of the more uniform site environments.

Copper

Iron is soluble in copper at concentrations of less than 4·5 at. % Fe. A typical 2 at. % Fe foil has been found to contain precipitated ferromagnetic α-Fe and precipitated superparamagnetic α-iron and γ-iron as well as single iron atoms embedded in the Cu matrix [56]. Unresolved magnetic interactions are found at low temperatures [57]. The inhomogeneities in these phases cause difficulties, and an interpretation of the 'quadrupole doublet' observed above the Curie temperature (7 K for 0·6 at. % Fe) of the Cu–Fe and Cu–Ni–Fe alloys as being of magnetic origin rather than from quadrupole interactions [58] has been countered by an explanation involving clustering effects [59].

A particularly interesting series of experiments has recorded the direct observation of diffusion broadening of the Mössbauer line [60]. A ^{57}Co/Cu source *or* a ^{57}Fe/Cu absorber shows the onset of considerable line broadening between 1000° and 1060°C due to solid-state diffusion, although the data show considerable discrepancies from prediction by a simple diffusion-jump model.

Second-row Transition Metals

Molybdenum

Iron alloys containing 0–6·0 at. % Mo show complex hyperfine spectra which can be analysed in terms of fields from iron atoms with 0, 1, and 2 Mo nearest neighbours [61, 62]. The decrease in field per Mo atom is unusually large at about 40 kG. Fe_3Mo_2 gives a simple quadrupole doublet at room temperature, and Fe_2Mo has been identified as a precipitate in an alloy containing 20 at. % Mo [62].

Ruthenium

Dilute iron–ruthenium alloys show a decrease in the internal magnetic field with increasing Ru content [36]. Above 13 at. % Ru they adopt a hexagonal

[Refs. on p. 325]

close-packed structure, and are magnetically ordered below 100 K. The internal field at 6 K is about 11 kG from 15–30 at. % Ru [63].

Rhodium

The iron–rhodium alloys are not well characterised. The disordered range (0–25 at. % Rh) shows only one hyperfine field slightly greater than that in α-Fe but with broader lines [64, 65]. The ordered phase of the CsCl type (20–50 at. % Rh) shows two distinct hyperfine fields from Fe on Fe sites and on Rh sites. A paramagnetic γ-phase can be present in the 25 at. % disordered alloy. At 50 at. % Rh there is a transition from an antiferromagnetic phase to a ferromagnetic phase at 338 K with a concomitant diminution in H of 17 kG.

Palladium

The Fe–Pd system shows two ordered superlattices, face-centred cubic $FePd_3$ (25–30 at. % Fe) and face-centred tetragonal FePd (43–50 at. % Fe). The Mössbauer spectra of these phases are somewhat sensitive to the degree of ordering, but the hyperfine fields are 334 and 281 kG respectively at 4·2 K in ordered samples [66]. The temperature dependence of the fields close to their Curie points was analysed.

In a range of alloys from 57 to 99·6% Pd there is no sign of variation or substructure in the hyperfine field (at 4.2) K within a given sample though the field does vary sharply with overall change in Pd concentration [67]. The conduction-electron polarisation presumably changes only very slowly over extremely long ranges and the exchange interaction does not oscillate in sign spatially. More detailed measurements have been made of the magnetic properties of a 97·35 at. % Fe–Pd alloy [68]. In the disordered Fe–Pd alloys up to 40 at. % Pd there is no variation in the internal field with concentration [67].

Mixed palladium–gold–iron alloys have also been briefly examined [69].

Third-row Elements

Dilute Fe–Re and Fe–Os alloys show clear evidence of satellite lines due to impurities in the nearest-neighbour sites [70]. Ir, Pt, and Au alloys show no satellites, although the lines are broadened.

The 85–99 at. % Pt alloys show a comparatively sharp magnetic spectrum with no appreciable broadening of the lines at temperatures well below the magnetic ordering temperatures [71]. General collapse of the spectrum occurs with temperature increase due to either relaxation effects or a wide distribution in hyperfine field values. The complex magnetic behaviour of the platinum alloys near Pt_3Fe has been studied [72].

An alloy containing 95 at. % Au has been shown to have antiferromagnetic ordering with an almost random distribution of spin orientations [73]. More extensive data have covered the range 75·3–96·7 at. % Au [57]. In alloys with more than 89·5 at. % Au, the Fe atoms are distributed randomly [74]. The room-temperature spectra show clear indication of iron atoms with no Fe nearest neighbours and with one or more Fe neighbours, the latter causing a larger electric field gradient. Magnetic ordering in the region 92–99·2 at. % Au has also been followed [75]. The most recent study of 87·2–99·5 at. % alloys in the range 0·4–300 K has confirmed the basic details and extended the study of the disorder in considerable depth [76].

11.3 Intermetallic Compounds

There are several elements which do not readily form alloy phases with iron, but which give one or more compounds which are metallic in character, and have a closely defined composition corresponding to a simple stoichiometric formula. Such intermetallic compounds are important in enabling the magnetic environment of metallic iron to be modified without introducing the statistical complications of alloys. They also give some insight into the way in which chemical bonding modifies the magnetic properties of the sublattices.

Beryllium

Solutions of up to 0·8 at. % Fe in beryllium show a small quadrupole splitting ($\sim$0·58 mm s^{-1}) which is a function of concentration [77].

The intermetallic compound FeBe$_2$ has a lattice of the MgZn$_2$ type. Each iron is surrounded by 12 Be atoms, the nearest iron neighbours being at the corners of a tetrahedron. FeBe$_2$ is ferromagnetic below 521°C and shows a room-temperature hyperfine field of 192 kG [78]. Single-crystal data indicate that the spins are parallel to the 'c' axis. Increase in the Be content above 66 at. % causes substitution at the iron sites. The introduction of iron sites with only 3 iron next-nearest neighbours results in the presence of additional satellite lines; e.g. for Fe$_{0.3}$Be$_{0.7}$ the fields are 175 and 137 kG for 4 and 3 Fe neighbours respectively. From 78 to 85 at. % Be the specimens are all paramagnetic above 80 K.

(Fe$_{0.75}$Mn$_{0.25}$)Be$_2$ also shows signs of structure in the ^{57}Fe spectrum, but the effects are not as marked as in the beryllium-rich samples, implying that, unlike Be, the Mn atoms interact magnetically with the iron [78].

Boron

The ferromagnetic iron borides are often referred to as interstitial compounds because the boron atoms occupy interstices in an otherwise close-packed iron lattice. The internal fields at room temperature of Fe$_2$B and FeB are 242 and 118 kG respectively compared with a value of 330 kG for iron itself [79, 80]. Data available concerning the detailed band structure of these compounds are consistent with 3d-populations on the iron atoms of 3$d^{8.2}$ and

$3d^{8\cdot 9}$ electrons respectively compared with $3d^{\sim 7}$ in iron metal. The augmented 3d-population in the borides should give a lower spin moment, a smaller magnetic field, and a small positive chemical isomer shift with increasing boron content, as are actually observed. This type of interpretation involving donation of electrons by the interstitial atom to the iron 3d-band is appropriate to most of the compounds in this section.

A reinvestigation of the magnetic structure of Fe_2B has shown the existence of two superimposed magnetic splittings of 244 and 252 kG at 4·2 K [81]. The mixed boride $Fe_{1\cdot 2}Co_{0\cdot 8}B$ has a field at room temperature of 232 kG [82]. The iron atom is apparently insensitive to cobalt substitution, although magnetic susceptibility data suggest that the cobalt is strongly affected by the total iron content. Temperature-dependence data for FeB give a saturation hyperfine field of 131 kG [83].

Manganese doping of Fe_2B produces little broadening of the outer peaks [84]. This implies that the magnetic field at the iron is produced by the short-range core-polarisation effects, and that the conduction-electron polarisation is negligible because of there being little 4s density at the Fermi surface. This is consistent with the successful interpretations of the hyperfine field values using a 3d-model only.

Carbon

Cementite, Fe_3C, is an interstitial solid solution of carbon in iron which is frequently found in steels and cast iron containing carbon. It is ferromagnetic below $T_C \approx 210°C$, the magnetic field being 208 kG at room temperature; the chemical isomer shift is $+0\cdot 19$ mm s^{-1} above that of α-Fe [79, 85]. This corresponds to a 3d-band population for iron of $3d^{\sim 8}$. The presence of a precipitate of Fe_3C in cast iron is readily detected by Mössbauer spectroscopy as shown in Fig. 11.8. The separate cementite (Fig. 11.8c) obtained by dissolving the steel in hydrochloric acid shows a central paramagnetic doublet in addition to the six-line magnetic spectrum, and it has proved possible to prepare from carbon steel a paramagnetic form of Fe_3C which appears from X-ray measurements to have a slightly modified structure in the lattice [85].

The borocarbides $Fe_3B_xC_{1-x}$ also have the cementite structure. As x increases to 0·54 the average magnetic field increases to 240 kG (partly due to an increase in T_C) and the lines broaden and indicate some structure [86]. The effect of boron or carbon neighbours on the iron atoms are predominantly short-range, and the effective 3d-population remains constant.

Fe_5C_2 gives a complex spectrum because of the presence of three different iron sites, and shows fields of 222, 184, and 110 kG at room temperature [86].

Martensite is a solid solution of carbon in iron with a body-centred tetragonal lattice. A freshly prepared martensite with 1·87 wt % carbon gave a

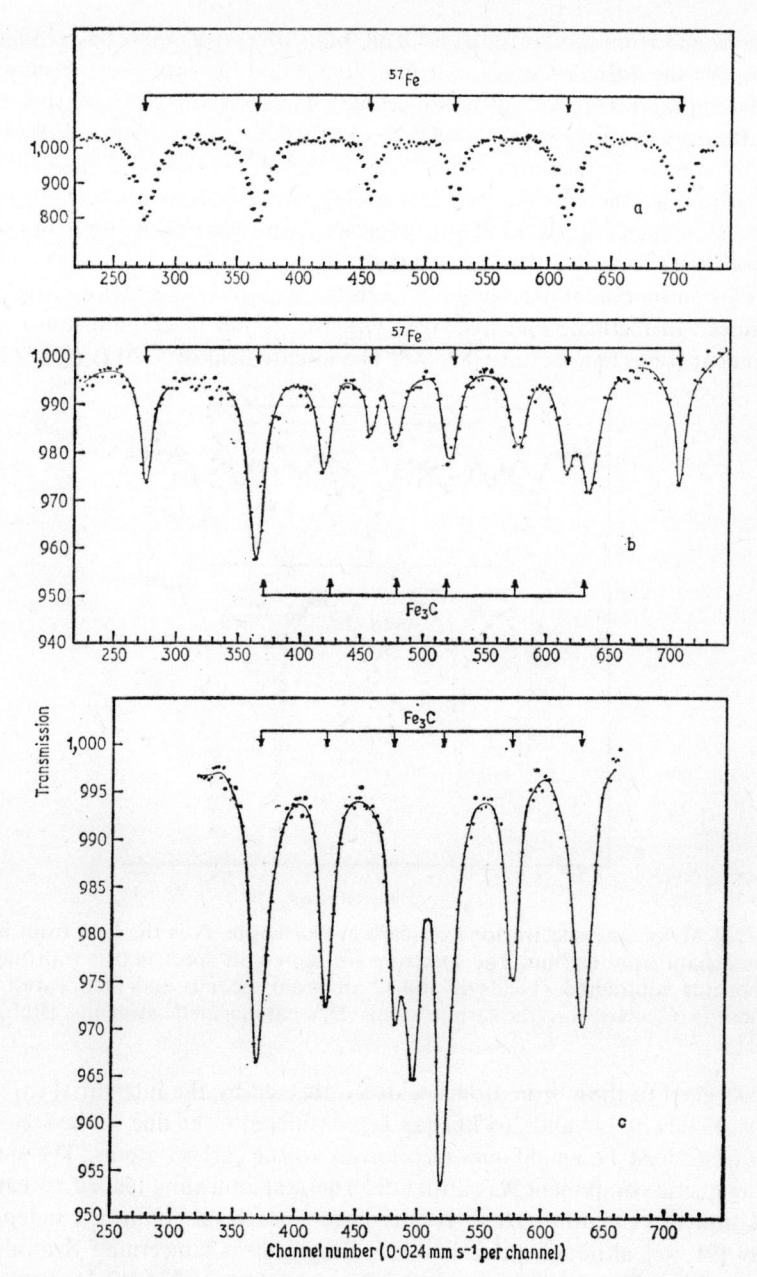

Fig. 11.8 Mössbauer spectra of (a) α-Fe, (b) cast iron, (c) cementite separated from eutectoid carbon steel. [Ref. 85, Fig. 1]

magnetic spectrum similar to that of iron, but with considerable line broadening due to the different local environments around the iron [87]. Ageing at room temperature produced a noticeable sharpening of the spectrum but with the formation of weak satellites corresponding to a smaller hyperfine field. This possibly indicates a clustering of the carbon atoms in the martensite on ageing. Increase in the carbon content from zero up to 1·7 wt % causes a steady increase in the average magnetic field above that of α-Fe of about 5·0 kG per wt % of carbon [88].

A martensite containing 0·9 wt % carbon, prepared from high-purity iron carburised in methane and hydrogen gas, was found to contain four components at room temperature [89, 90]. The intense field of 332 kG (Fig. 11.9)

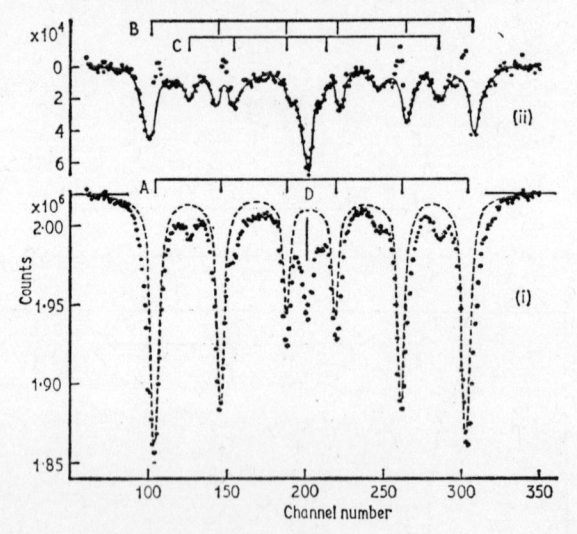

Fig. 11.9 Mössbauer spectra for iron–carbon martensite. A is the field from iron atoms distant from carbon. The spectrum (ii) represents spectrum (i) with the A components subtracted. Fields B and C are from second- and first-nearest Fe neighbours respectively to the carbon atoms. D is paramagnetic austenite. [Ref. 89, Fig. 1]

was assigned to those iron atoms scarcely affected by the interstitial carbon atoms. Fields of 342 and 265 kG can be considered to be due to the second- and first-nearest Fe neighbours respectively to the carbon atoms. The small paramagnetic component was austenite. Thermal annealing tended to cause formation of cementite. These results have since been confirmed independently [91, 92], although there is some disagreement concerning the values registered for the minority hyperfine fields, e.g. 304 and 274 kG for martensites prepared from commercial carbon steels [92]. The possibility of impurities other than carbon atoms having an important role has also been suggested [90]. The presence of three distinct iron site symmetries in the marten-

[*Refs. on p. 325*]

site phase might well suggest that the overall structure is not tetragonal and that it is probably inhomogeneous. The effects of tempering on the Mössbauer spectrum of martensite are extremely difficult to analyse with certainty [93].

As already briefly mentioned many specimens of quenched carbon steels retain a proportion of the paramagnetic austenitic phase [94]. At room temperature the resonance consists of a single line due to atoms without carbon nearest neighbours, and a quadrupole doublet from those with one carbon neighbour which removes the local cubic symmetry at the iron site [95]. The doublet shows a shift of 0·06 mm s^{-1} relative to the singlet. The same sample (1·6 wt % C) at 895°C showed a much narrower single line because the jump-diffusion time of the carbon atoms becomes less than the excited-state lifetime as the temperature rises.

The alloy $Fe_{80}C_{7.5}P_{12.5}$ may be considered to be amorphous because the lack of strict periodicity in the atomic arrangement gives the diffuse X-ray diffraction pattern of a liquid. The short range of exchange interactions means that strict periodicity is not essential to the existence of ferromagnetism, and this alloy was found to be ferromagnetic below 586 K [96]. The magnetic spectra are very broad, implying a non-unique value of the internal magnetic field, but the temperature dependence of the average magnetic field follows a $J = 1$ Brillouin function quite closely.

Nitrogen

The nitride Fe_4N has a face-centred cubic lattice of iron atoms with a nitrogen atom at the body centre. Each corner Fe_A atom has 12 Fe_B neighbours at 2·96 Å, and each face-centre Fe_B has 2 nitrogens at 1·90 Å. The room-temperature spectrum shows two hyperfine fields of 345 and 215 kG with intensities in the ratio 1 : 3 allowing assignment to Fe_A and Fe_B respectively [97]. The chemical isomer shifts are +0·13 and +0·28 mm s^{-1} relative to iron metal respectively. Clearly the Fe_A site is similar to that of iron metal itself, while the B site shows the typical effects of nearest-neighbour interstitial atoms. $Fe_{3.6}Ni_{0.4}N$ is similar, the Ni substitution taking place at the Fe_A sites but with little effect on the other iron atoms.

Recent data have shown that the face-centre Fe_B atoms are not all equivalent because of the presence of the magnetic axis. Field values of 340·6, 215·5, and 219·2 kG at room temperature were given [98]. However, the electric field gradient axis at the Fe_B atom is normal to the face-centre, so that with the magnetic axis parallel to one of the cube edges there are $\frac{2}{3}$ of the Fe_B sites with the electric field gradient perpendicular to the magnetic field and $\frac{1}{3}$ with the two parallel. This ensures three distinct iron site symmetries without having to invoke significantly differing electronic configurations or fields at the two types of face-centre site [99].

The ε and ζ phases of Fe_2N both show a small quadrupole splitting at room temperature with a positive chemical isomer shift from α-Fe [100]. Ferromagnetic ordering takes place below about 90 K.

Iron–nitrogen austenite and martensite are basically similar to the corresponding carbon systems [91].

Silicon

The Fe–Si system appears to be on the dividing line between the metallic-alloy and intermetallic compound classifications.

The ordered Fe_3Si alloy behaves in an identical manner to ordered Fe_3Al with two fields of 310 kG (D site; 8 Fe nearest neighbours) and 198 kG (A site; 4 Fe 4 Si) at room temperature [24, 101]. The discussion for Fe_3Al given on p. 309 is thus also appropriate here.

In the disordered iron-rich alloys (0–10 at. % Si) nearest-neighbour effects from Fe atoms with 8, 7, and 6 Fe neighbours are found, analogously to the dilute Fe–Al disordered alloys [31, 102, 103]. The internal magnetic field decreases stepwise as the number of Si neighbours increases. The alloys with 14–27 at. % Si are ordered in the Fe_3Al-type lattice. Once again the decrease in field and small increase in chemical isomer shift are compatible with a transfer of electrons to the iron 3d-band from the silicon.

Fe_5Si_3 has been only cursorily examined but is magnetically ordered at low temperatures [101].

FeSi is paramagnetic but shows an unusual decrease in the quadrupole splitting (0·74 mm s^{-1} at 4 K to about 0·1 mm s^{-1} at 1000 K) [104]. This implies a thermal excitation to low-lying levels as in, for example, Fe^{2+} complexes, but this behaviour is most unexpected in metallic phases. Magnetic susceptibility and ^{29}Si n.m.r. data lend weight to the idea of an activation energy analysis in which electrons are excited from filled states across an energy gap of about 0·05 eV to unfilled states of a different symmetry. Data have also been given for the system $Co_{1-x}Fe_xSi$ and for $Fe_{0.9}Ni_{0.1}Si$ and $Fe_{0.9}Rh_{0.1}Si$ [105]. The iron environment is not very sensitive to substitution, possibly because the nearest neighbours are seven Si atoms, although the quadrupole splitting does decrease gradually.

β-$FeSi_2$ appears to contain only one distinct iron site and this shows a small quadrupole splitting [106].

Phosphorus

The phosphides Fe_3P, Fe_2P, and FeP are similar in behaviour to the borides and carbides. The increasing phosphorus content causes an increase in the number of electrons donated by the phosphorus into the 3d-band, resulting in a progressive decrease in the magnetic moment and internal magnetic field [107]. Fe_3P has three iron environments giving fields of 295, 265, and 185 kG at 90 K. Fe_2P has a complex structure with two iron sites having tetrahedral and square-pyramidal coordination by phosphorus respectively;

[Refs. on p. 325]

this is reflected in the quadrupole interactions and the spectrum at 90 K is also magnetic with fields of 140 and 115 kG. FeP is paramagnetic even at 30 K with a quadrupole split spectrum from the iron which is in a distorted octahedral environment of phosphorus atoms.

More detailed study of the magnetic moments and internal magnetic fields of the two iron sites in Fe_2P suggests that it is ferromagnetically ordered, rather than ferrimagnetically with spin canting [108].

In the ternary phosphides $(Fe_{1-x}M_x)_2P$, when M = Mn or Cr it replaces iron at square-pyramidal sites, but when M = Co or Ni it replaces at tetrahedral sites. The magnetically ordered phases give complex spectra which have not been fully interpreted as yet [109].

Germanium

Germanium forms a large number of ordered intermetallic phases, and the data regarding these are rather confusing. The hexagonal and cubic forms of Fe_3Ge both show an internal field of about 240 kG at room temperature, presumably because both phases have 8 Fe and 4 Ge nearest neighbours at 2·59 Å, although with different symmetry [110, 111].

Fe_5Ge_3 (i.e. $Fe_{1.67}Ge$) gives a complicated magnetic spectrum (derived from several site symmetries) which has been only partially analysed [111].

The $Fe_{1.5}Ge$–$Fe_{1.68}Ge$ phase shows at least two internal fields of approximately 240 and 180 kG both of which are strongly dependent on concentration [112].

The hexagonal phase of FeGe has been shown to be antiferromagnetic below 400 K with a saturation magnetic field of 155 kG [113, 114]. The cubic FeGe phase is antiferromagnetic below about 275 K with a field of 110 kG at 80 K [115].

The magnetic structure of $FeGe_2$ was established by combined neutron diffraction and Mössbauer measurements as being antiferromagnetic below 287 K with two equivalent but independent sublattices inclined at 71° to each other [116]. The hyperfine field collapses very slowly below the Néel point in an unusual manner. Once again the parameters are consistent with increased 3d occupation at the iron in the intermetallic phases.

Tin

The tin intermetallic compounds are very closely related to those of germanium. Fe_3Sn is similar to Fe_3Ge and shows two hyperfine fields [110].

Fe_5Sn_3 is magnetically ordered below 310°C and, like Fe_5Ge_3, shows a complex spectrum with probably three distinct iron sites [117].

FeSn proves to be antiferromagnetic below 370 K with two fields of 146 and 158 kG at 4 K. There is only one crystallographic iron site, but the spin axis and the electric field gradient tensor can take two distinct relative orientations [118].

[*Refs. on p. 325*]

FeSn$_2$ is antiferromagnetic below 377 K with a field of 115 kG at 295 K. In all cases (except Fe$_3$Sn$_2$) a field of between 30 and 110 kG is also recorded at the tin nucleus in the ^{119}Sn Mössbauer resonance (see Chapter 14). This results from exchange polarisation of the diamagnetic tin atoms by the iron [118–120].

Dilute Fe–Sn alloys (0–8 at. % Sn) show nearest-neighbour effects giving a reduction in the internal field [36, 118].

Rare Earths

The cubic Laves phases, MFe$_2$ (M = Ti, Zr, Y, Ce, Sm, Gd, Dy, Ho, Er, Tm, Lu), have been examined by several groups [121–124]. The iron electronic configuration is basically the same in all these compounds with a saturation field of about 210 kG; this implies that the hyperfine field does not depend significantly on the magnetic properties of the rare-earth metal used. The hyperfine field is appreciably smaller than that observed for the 3d-metal alloys (see Fig. 11.6) but the chemical isomer shifts of both groups of compound are very similar, implying that the electron configuration of the iron is the same in the two cases. The lower hyperfine field for the rare-earth Laves phases therefore suggests that in these compounds there is an appreciable contribution to the field from conduction-electron polarisation. The iron atoms occupy corner-sharing tetrahedral networks with a threefold axis lying along one of the [111] directions. If the crystal magnetises along one of the [111] directions then the local magnetic field and electric field gradient tensor axes will not be oriented identically on all sites and two six-line patterns in the ratio 1 : 3 are expected. It is therefore possible to say that Tb, Er, Y, and Zr, which give two distinct patterns, magnetise along the [111] directions, while Dy and Ho, which give only one, magnetise along the [001] direction. Sm, Gd, and Ce are found to be more complicated [124]. TiFe$_2$ contains two distinct types of iron, and also shows complex behaviour [121]. The transition from ferromagnetism in iron-rich to antiferromagnetism in titanium-rich material appears to be due to the excess iron entering titanium sites [125].

Actinides

The binary actinide compound UFe$_2$ seems to show small unresolved fields of less than 60 kG [126]. Both U$_6$Fe and Pu$_6$Fe show a single quadrupole splitting at all temperatures above 16 K [127]. Brief details have also been given for PuFe$_2$ [128].

REFERENCES

[1] R. S. Preston, S. S. Hanna, and J. Heberle, *Phys. Rev.*, 1962, **128,** 2207.
[2] T. E. Cranshaw and P. Reivari, *Proc. Phys. Soc.*, 1967, **90,** 1059.
[3] J. O. Listner and G. B. Benedek, *J. Appl. Phys.*, 1963, **34,** 688.
[4] S. S. Hanna, J. Heberle, G. J. Perlow, R. S. Preston, and D. H. Vincent, *Phys. Rev. Letters*, 1960, **4,** 513.
[5] D. A. Goodings and V. Heine, *Phys. Rev. Letters*, 1960, **5,** 370.
[6] S. Foner, A. J. Freeman, N. A. Blum, R. B. Frankel, E. J. McNiff, Jr, and H. C. Praddaude, *Phys. Rev.*, 1969, **181,** 863.
[7] R. Ingalls, *Phys. Rev.*, 1967, **155,** 157 (erratum loc. cit., 1967, **162,** 518).
[8] S. Alexander and D. Treves, *Phys. Letters*, 1966, **20,** 134.
[9] R. S. Preston, *Phys. Rev. Letters*, 1967, **19,** 75.
[10] R. M. Housley and F. Hess, *Phys. Rev.*, 1967, **164,** 340.
[11] T. A. Kovats and J. C. Walker, *Phys. Rev.*, 1969, **181,** 610.
[12] R. V. Pound, G. B. Benedek, and R. Drever, *Phys. Rev. Letters*, 1961, **7,** 405.
[13] D. N. Pipkorn, C. K. Edge, P. Debrunner, G. de Pasquali, H. G. Drickamer, and H. Frauenfelder, *Phys. Rev.*, 1964, **135,** A1604.
[14] J. A. Moyzis, Jr, and H. G. Drickamer, *Phys. Rev.*, 1968, **171,** 389.
[15] M. Nicol and G. Jura, *Science*, 1963, **141,** 1035.
[16] W. H. Southwall, D. L. Decker, and H. B. Vanfleet, *Phys. Rev.*, 1968, **171,** 354.
[17] U. Gonser, C. J. Meechan, A. H. Muir, and H. Wiedersich, *J. Appl. Phys.*, 1963, **34,** 2373.
[18] L. E. Millet and D. L. Decker, *Phys. Letters*, 1969, **29A,** 7.
[19] E. L. Lee, P. E. Bolduc, and C. E. Violet, *Phys. Rev. Letters*, 1964, **13,** 800.
[20] J. G. Dash, R. D. Taylor, P. P. Craig, D. E. Nagle, D. R. F. Cochran, and W. E. Keller, *Phys. Rev. Letters*, 1960, **5,** 152.
[21] G. J. Ehnholm, T. E. Katila, O. V. Lounasmaa, and P. Reivari, *Phys. Letters*, 1967, **25A,** 758.
[22] P. Reivari, *Phys. Rev. Letters*, 1969, **22,** 167.
[23] K. Ono, Y. Ishikawa, and A. Ito, *J. Phys. Soc. Japan*, 1962, **17,** 1747.
[24] M. B. Stearns, *Phys. Rev.*, 1968, **168,** 588.
[25] L. Cser, I. Dezsi, L. Keszthelyi, J. Ostanevich, and L. Pal, *Phys. Letters*, 1965, **19,** 99.
[26] L. Cser, J. Ostanevich, and L. Pal, *Phys. Stat. Sol.*, 1967, **20,** 581.
[27] L. Cser, J. Ostanevich, and L. Pal, *Phys. Stat. Sol.*, 1967, **20,** 591.
[28] P. A. Flinn and S. L. Ruby, *Phys. Rev.*, 1961, **124,** 34.
[29] E. A. Fridman and W. J. Nicholson, *J. Appl. Phys.*, 1963, **34,** 1048.
[30] M. B. Stearns, *J. Appl. Phys.*, 1964, **35,** 1095.
[31] M. B. Stearns, *J. Appl. Phys.*, 1965, **36,** 913.
[32] M. B. Stearns and S. S. Wilson, *Phys. Rev. Letters*, 1964, **13,** 313.
[33] S. Nasu, M. Nishio, Y. Tsuchida, Y. Murakami, and T. Shinjo, *J. Phys. Soc. Japan*, 1969, **27,** 1363.
[34] G. P. Huffman and R. M. Fisher, *J. Appl. Phys.*, 1967, **38,** 735.
[35] I. Vincze and L. Cser, *Phys. Stat. Sol.*, 1969, **35,** K25.
[36] G. K. Wertheim, V. Jaccarino, J. H. Wernick, and D. N. E. Buchanan, *Phys. Rev. Letters*, 1964, **12,** 24.
[37] M. Rubinstein, G. H. Stauss, and M. B. Stearns, *J. Appl. Phys.*, 1966, **37,** 1334.
[38] C. E. Johnson, M. S. Ridout, and T. E. Cranshaw, *Proc. Phys. Soc.*, 1963, **81,** 1079.

[39] R. B. Roy, B. Solly, and R. Wäppling, *Phys. Letters*, 1967, **24A**, 583.
[40] C. W. Kimball, J. K. Tison, and M. V. Nevitt, *J. Appl. Phys.*, 1967, **38**, 1153.
[41] Y. Ishikawa and Y. Endoh, *J. Phys. Soc. Japan*, 1967, **23**, 205.
[42] Y. Ishikawa and Y. Endoh, *J. Appl. Phys.*, 1968, **39**, 1318.
[43] C. Kimball, W. D. Gerber, and A. Arrott, *J. Appl. Phys.*, 1963, **34**, 1046.
[44] C. W. Kimball, W. C. Phillips, M. V. Nevitt, and R. S. Preston, *Phys. Rev.*, 1966, **146**, 375.
[45] C. E. Johnson, M. S. Ridout, T. E. Cranshaw, and P. E. Madsen, *Phys. Rev. Letters*, 1961, **6**, 450.
[46] L. H. Bennett and L. J. Swartzendruber, *Phys. Letters*, 1967, **24A**, 359.
[47] L. H. Bennett, L. J. Swartzendruber, and R. E. Watson, *Phys. Rev.*, 1968, **165**, 500.
[48] L. J. Swartzendruber and L. H. Bennett, *J. Appl. Phys.*, 1968, **39**, 1323.
[49] H. Asano, *J. Phys. Soc. Japan*, 1968, **25**, 286.
[50] Y. Nakamura, Y. Takeda, and M. Shiga, *J. Phys. Soc. Japan*, 1968, **25**, 287.
[51] Y. Tino and T. Maeda, *J. Phys. Soc. Japan*, 1968, **24**, 729.
[52] I. I. Dekhtiar, B. G. Egiazarov, L. M. Isakov, V. S. Mikhalenkov, and V. P. Romasho, *Doklady Akad. Nauk S.S.S.R.*, 1967, **175**, 556.
[53] Y. Nakamura, M. Shiga, and N. Shikazono, *J. Phys. Soc. Japan*, 1964, **19**, 1177.
[54] H. Asano, *J. Phys. Soc. Japan*, 1969, **27**, 542.
[55] Y. Gros and J. C. Pebay-Peyroula, *Phys. Letters*, 1964, **13**, 5.
[56] M. Ron, A. Rosencuraig, H. Shechter, and A. Kidron, *Phys. Letters*, 1966, **22**, 44.
[57] U. Gonser, R. W. Grant, C. J. Meechan, A. H. Muir, and H. Wiedersich, *J. Appl. Phys.*, 1965, **36**, 2124.
[58] L. J. Swartzendruber and L. H. Bennett, *Phys. Letters*, 1968, **27A**, 141.
[59] B. Window and C. E. Johnson, *Phys. Letters*, 1969, **29A**, 703.
[60] R. C. Knauer and J. G. Mullen, *Phys. Rev.*, 1968, **174**, 711.
[61] H. L. Marcus and L. H. Schwartz, *Phys. Rev.*, 1967, **162**, 259.
[62] H. L. Marcus, M. E. Fine, and L. H. Schwartz, *J. Appl. Phys.*, 1967, **38**, 4750.
[63] H. Ohno, M. Mekata, and H. Takaki, *J. Phys. Soc. Japan*, 1968, **25**, 283.
[64] G. Shirane, C. W. Chen, P. A. Flinn, and R. Nathans, *J. Appl. Phys.*, 1963, **34**, 1044.
[65] G. Shirane, C. W. Chen, P. A. Flinn, and R. Nathans, *Phys. Rev.*, 1963, **131**, 183.
[66] G. Longworth, *Phys. Rev.*, 1968, **172**, 572.
[67] P. P. Craig, B. Mozer, and R. Segnan, *Phys. Rev. Letters*, 1965, **14**, 895.
[68] P. P. Craig, R. C. Perisho, R. Segnan, and W. A. Steyert, *Phys. Rev.*, 1965, **138**, A1460.
[69] G. Longworth, *Phys. Letters*, 1969, **30A**, 180.
[70] H. Bernas and I. A. Campbell, *Solid State Commun.*, 1966, **4**, 577.
[71] R. Segnan, *Phys. Rev.*, 1967, **160**, 404.
[72] D. Palaith, C. W. Kimball, R. S. Preston, and J. Crangle, *Phys. Rev.*, 1969, **178**, 795.
[73] P. P. Craig and W. A. Steyert, *Phys. Rev. Letters*, 1964, **13**, 802.
[74] C. E. Violet and R. J. Borg, *Phys. Rev.*, 1967, **162**, 608.
[75] C. E. Violet and R. J. Borg, *Phys. Rev.*, 1966, **149**, 540.
[76] M. S. Ridout, *J. Phys. C*, 1969, Ser. 2, **2**, 1258.
[77] C. Janot and G. le Caer, *Compt. rend.*, 1968, **267B**, 954.
[78] K. Ohta, *J. Appl. Phys.*, 1968, **39**, 2123.

[79] T. Shinjo, F. Itoh, H. Takaki, Y. Nakamura, and N. Shikazono, *J. Phys. Soc. Japan*, 1964, **19**, 1252.
[80] J. D. Cooper, T. C. Gibb, N. N. Greenwood, and R. V. Parish, *Trans. Faraday Soc.*, 1964, **60**, 2097.
[81] I. D. Weisman, L. J. Swartzendruber, and L. H. Bennett, *Phys. Rev.*, 1969, **177**, 465.
[82] M. C. Cadeville, R. Wendling, E. Fluck, P. Kuhn, and W. Neuwirth, *Phys. Letters*, 1965, **19**, 182.
[83] J. B. Jeffries and N. Hershkowitz, *Phys. Letters*, 1969, **30A**, 187.
[84] H. Bernas and I. A. Campbell, *Phys. Letters*, 1967, **24A**, 74.
[85] M. Ron, H. Shechter, A. A. Hirsch, and S. Niedzwiedz, *Phys. Letters*, 1966, **20**, 481.
[86] H. Bernas, I. A. Campbell, and R. Fruchart, *J. Phys. and Chem. Solids*, 1967, **28**, 17.
[87] J. M. Genin and P. A. Flinn, *Phys. Letters*, 1966, **22**, 392.
[88] T. Zemcik, *Phys. Letters*, 1967, **24A**, 148.
[89] H. Ino, T. Moriya, F. E. Fujita, and Y. Maeda, *J. Phys. Soc. Japan*, 1967, **22**, 346.
[90] T. Moriya, H. Ino, F. E. Fujita, and Y. Maeda, *J. Phys. Soc. Japan*, 1968, **24**, 60.
[91] P. M. Gielen and R. Kaplow, *Acta Met.*, 1967, **15**, 49.
[92] M. Ron, A. Kidron, H. Shechter, and S. Niedzwiedz, *J. Appl. Phys.*, 1967, **38**, 590.
[93] H. Ino, T. Moriya, F. E. Fujita, Y. Maeda, Y. Ono, and Y. Inokuti, *J. Phys. Soc. Japan*, 1968, **25**, 88.
[94] C. W. Kocher, *Phys. Letters*, 1965, **14**, 287.
[95] S. J. Lewis and P. A. Flinn, *Phys. Stat. Sol.*, 1968, **26**, K51.
[96] C. C. Tsuei, G. Longworth, and S. C. H. Lin, *Phys. Rev.*, 1968, **170**, 603.
[97] G. Shirane, W. J. Takei, and S. L. Ruby, *Phys. Rev.*, 1962, **126**, 49.
[98] A. J. Nozik, J. C. Wood, and G. Haacke, *Solid State Commun.*, 1969, **7**, 1677.
[99] M. J. Clauser, *Solid State Commun.*, 1970, **8**, 781.
[100] M. Chabanel, C. Janot, and J. P. Motte, *Compt. rend.*, 1968, **266B**, 419.
[101] T. Shinjo, Y. Nakamura, and N. Shikazono, *J. Phys. Soc. Japan*, 1963, **18** 797.
[102] T. E. Cranshaw, C. E. Johnson, M. S. Ridout, and G. A. Murray, *Phys. Letters*, 1966, **21**, 481.
[103] M. B. Stearns, *Phys. Rev.*, 1963, **129**, 1136.
[104] G. K. Wertheim, V. Jaccarino, J. H. Wernick, J. A. Seitchik, H. J. Williams, and R. C. Sherwood, *Phys. Letters*, 1965, **18**, 89.
[105] G. K. Wertheim, J. H. Wernick, and D. N. E. Buchanan, *J. Appl. Phys.*, 1966, **37**, 3333.
[106] R. Wäppling, L. Häggström, and S. Rundqvist, *Chem. Phys. Letters*, 1968, **2**, 160.
[107] R. E. Bailey and J. F. Duncan, *Inorg. Chem.*, 1967, **6**, 1444.
[108] K. Sato, K. Adachi, and E. Ando, *J. Phys. Soc. Japan*, 1969, **26**, 855.
[109] R. Fruchart, A. Roger, and J. P. Senateur, *J. Appl. Phys.*, 1969, **40**, 1250.
[110] G. A. Fatseas and P. Lecocq, *Compt. rend.*, 1966, **262B**, 107.
[111] H. Yamamoto, *J. Phys. Soc. Japan*, 1965, **20**, 2166.
[112] E. Germagnoli, C. Lamborizio, S. Mora, and I. Ortalli, *Nuovo Cimento*, 1966, **42B**, 314.

[113] V. I. Nikolaev, S. S. Yakimov, I. A. Dubovtsev, and Z. G. Gavrilova, *ZETF Letters*, 1965, **2**, 373.

[114] S. Tomiyoshi, H. Yamamoto, and H. Watanabe, *J. Phys. Soc. Japan*, 1966, **21**, 709.

[115] R. Wäppling and L. Häggström, *Phys. Letters*, 1968, **28A**, 173.

[116] J. B. Forsyth, C. E. Johnson, and P. J. Brown, *Phil. Mag.*, 1964, **10**, 713.

[117] H. Yamamoto, *J. Phys. Soc. Japan*, 1966, **21**, 1058.

[118] E. Both, G. Trumpy, J. Traff, P. Ostergaard, C. Djega-Mariadasson, and P. Lecocq, p. 487 of 'Hyperfine Structure and Nuclear Radiations', Ed. E. Matthias and D. A. Shirley, North-Holland, Amsterdam, 1968.

[119] V. I. Nikolaev, Yu. I. Shcherbina, and A. I. Karchevskii, *Soviet Physics – JETP*, 1963, **17**, 524.

[120] V. I. Nikolaev, Yu. I. Shcherbina, and S. S. Yakimov, *Soviet Physics – JETP*, 1964, **18**, 878.

[121] E. W. Wallace, *J. Chem. Phys.*, 1964, **41**, 3857.

[122] E. W. Wallace and L. M. Epstein, *J. Chem. Phys.*, 1961, **35**, 2238.

[123] G. K. Wertheim and J. H. Wernick, *Phys. Rev.*, 1962, **125**, 1937.

[124] G. J. Bowden, D. St P. Bunbury, A. P. Guimares, and R. E. Snyder, *J. Phys. C*, 1968, Ser. 2, **1**, 1376.

[125] G. K. Wertheim, J. H. Wernick, and R. C. Sherwood, *Solid State Commun.*, 1969, **7**, 1399.

[126] S. Komura, N. Kunitomi, P. K. Tseng, N. Shikazono, and H. Takekoshi, *J. Phys. Soc. Japan*, 1961, **16**, 1479.

[127] S. Blow, *J. Phys. and Chem. Solids*, 1969, **30**, 1549.

[128] S. Blow, *Phys. Letters*, 1969, **29A**, 676.

12 | ^{57}Fe – Impurity Studies

Measurement of the energy of a γ-ray emitted (or absorbed) in a recoilless event registers information only about the immediate chemical environment of the emitting (or absorbing) nucleus. Mössbauer experiments are performed by comparison of a source and an absorber, and it is usual to utilise a source without hyperfine effects as a reference standard for a series of absorber experiments. This is not the only option available, however. If we use a reference absorber, it is possible to study hyperfine effects in the source. These are produced by the chemical environment in which the excited-state ^{57}Fe nucleus finds itself when it is generated in the source matrix. The total concentration of the ^{57}Co parent atoms is very small indeed, and each atom can therefore be considered as an isolated impurity in the source matrix.

^{57}Co doping techniques have been mainly used in the study of metallic systems. The theory of an impurity atom vibrating in a crystal lattice has been developed to describe in detail the temperature dependence of the recoilless fraction and the chemical isomer shift. The chemical isomer shift is a function of the band structure of the metal, and in magnetic hosts the conduction-electron polarisation can also be studied via the magnetic hyperfine splitting. This type of work is of borderline interest within the scope of this book, but we shall endeavour to indicate some of the more important points later in this chapter without attempting to be comprehensive in coverage.

Some interest has also centred on the ^{57}Co doping of iron and cobalt compounds, although this technique is probably not so useful because of the unpredictable chemical effects of the ^{57}Co decay. The doping of non-iron compounds with enriched ^{57}Fe is also discussed where appropriate.

Also included for convenience in this chapter is a small section on miscellaneous applications of Mössbauer spectroscopy to subjects such as surface states, chemical reactions, and diffusion in liquids, which are of chemical interest but do not as yet represent major areas of study.

[*Refs. on p. 348*]

12.1 Chemical Compounds

General Considerations

In a number of instances, the Mössbauer spectrum recorded using a source of ^{57}Co doped into a cobalt or an iron chemical compound shows contributions from both Fe^{2+} and Fe^{3+} charge states. The interpretation formulated initially was a logical deduction based on the known effects of the electron-capture decay process in ^{57}Co. The Auger cascade which follows the capture of a K-electron causes ionisation of the emitting atom with resultant momentary charge states of up to $+7$. In metallic matrices, the high mobility of the conduction electrons results in an extremely rapid return to equilibrium conditions in which the iron nucleus again has a 'normal' charge state, but one might anticipate that equilibration would be slower by orders of magnitude in insulating materials. Thus it came to be assumed that the presence of Fe^{2+} and Fe^{3+} charge states in the Mössbauer spectrum of ^{57}Co sources was a direct observation of this dynamic process, the higher charge state still being in process of reverting to the more stable Fe^{2+} state. With this hypothesis the relative proportions of the two states gives information regarding the rate of electron transfer.

This assumption was tested in a classic series of experiments [1, 2] in which the Mössbauer spectrum was recorded using 14·4-keV γ-rays counted in delayed coincidence with the preceding 123-keV γ-ray. If the Fe^{3+} oxidation state was decaying, its proportional contribution to the spectrum would decrease as the delay-time was increased. Data for $Fe(NH_4)_2(SO_4)_2 \cdot 6H_2O$, CoO, NiO, $CoSO_4 \cdot 7H_2O$, and $CoCl_4 \cdot 4H_2O$ prove conclusively that the Fe^{2+}/Fe^{3+} ratio is not significantly time dependent after a lapse of 10^{-7} s following the ^{57}Co decay. The immediate implication, therefore, is that the Fe^{2+} and Fe^{3+} charge states can be described as being at least in metastable equilibrium on their respective lattice sites. This may in part be due to the fact that the ionic radius of Co^{2+} (0·74 Å) is intermediate between those of Fe^{2+} (0·76 Å) and Fe^{3+} (0·64 Å) so that both can be accommodated fairly easily in the original lattice. The ultimate balance is probably controlled by the influence of localised lattice defects and nonstoichiometry on the lattice energies of individual cations.

Both Fe^{2+} and Fe^{3+} daughter atoms are also sometimes observed in the decay of cobalt(III) compounds. This has resulted in the suggestion that the generally smaller atomic dimensions of cobalt complexes can result in an effective pressure at the impurity site [3]. A parallel is thus drawn between the observation of pressure-induced $Fe^{3+} \rightarrow Fe^{2+}$ reduction and the Fe^{2+} generated by the decay of Co^{3+}.

Inconsistencies between results from different laboratories have highlighted the necessity for careful purification of the compounds concerned. The final balance between the daughter oxidation states can be sensitive to impurities

[*Refs. on p. 348*]

and lattice inhomogeneities, and the ^{57}Co impurity may show clustering effects when the host matrix is not a cobalt compound.

^{57}Co Doped into Cobalt Compounds

Although ^{57}Co/CoCl$_2$ gives a spectrum appropriate to Fe^{2+} cations only, the hydrates ^{57}Co/CoCl$_2$.2H$_2$O and ^{57}Co/CoCl$_2$.4H$_2$O both show significant Fe^{3+} content (see Fig. 12.1) [4]. ^{57}Co/CoSO$_4$, ^{57}Co/CoSO$_4$.H$_2$O, and ^{57}Co/CoSO$_4$.4H$_2$O all show both species, and in each series the Fe^{3+} concentration is seen to increase with increasing degree of hydration. It is suggested that this results from the oxidising action of OH· radicals formed during the decay processes [4]. This is supported by the observation of Fe^{3+} ions following the radiolysis of Fe(NH$_4$)$_2$(SO$_4$)$_2$.6H$_2$O and FeSO$_4$.7H$_2$O [5]. The relative proportion of Fe^{3+} formed by ^{57}Co/CoCl$_2$.6H$_2$O and ^{57}Co/CoSO$_4$.7H$_2$O sources decreases at higher temperatures because dehydration alters the relative probability for achieving stability of the two species [6]. ^{57}Co/CoF$_2$ and ^{57}Co/Co$_2$(SO$_4$)$_3$.18H$_2$O also show both Fe^{2+} and Fe^{3+} cations [4], as does ^{57}Co/CoF$_2$.4H$_2$O [7].

Initial experiments on antiferromagnetic ^{57}Co/CoF$_3$ were thought to show evidence for production of some paramagnetic Fe^{2+} ions [4]. Later work found that this was the result of partial decomposition of the fluoride, and that pure CoF$_3$ gives Fe^{3+} ions only which give a large internal magnetic field of 613 kG at 94 K [8]. The temperature dependence of the latter showed the Néel temperature to be 460 K.

Independent data confirm the production of only Fe^{2+} ions in ^{57}Co/CoCl$_2$ [9]. At 10 K, which is below the Néel temperature of CoCl$_2$, a field of 49 kG is found at the Fe^{2+} ions, compared with 245 kG for Fe^{2+} ions in ^{57}Co/CoF$_2$ at 11 K (T_N = 38 K). In the latter case the Fe^{3+} field was 545 kG. It is also interesting to note that whereas in CoCl$_2$ (doped with 2% Fe^{2+}) the Fe^{2+} spins lie in the 'c' plane, in isomorphous NiCl$_2$ they lie along the 'c' axis [10]. This is a result of the difference in anisotropy energy of the host spins and shows how impurity doping can be used to study magnetic ordering in the host material.

The creation of a small proportion of Fe^{4+} iron has been claimed from computer-fitting of data for the sources ^{57}Co/Co(NH$_4$SO$_4$)$_2$.6H$_2$O and ^{57}Co/CoSiF$_6$.6H$_2$O [6], but this was not reported in an independent investigation [11]. As already mentioned the Fe^{3+} fraction is not time dependent in the spectra of ^{57}Co/Fe(NH$_4$SO$_4$)$_2$.6H$_2$O, ^{57}Co/CoSO$_4$.7H$_2$O, and ^{57}Co/CoCl$_2$.4H$_2$O [2]. The stability of Fe^{3+} in ^{57}Co/CoSO$_4$.7H$_2$O and ^{57}Co/CoSO$_4$ decreases with increasing temperature, and in the latter compound with increasing pressure [12]. The latter effect is perhaps unexpected since the Fe^{3+} ionic radius is much smaller than for Fe^{2+} ion, but presumably the increased overlap with ligand ions facilitates electron transfer from nearby cations.

[*Refs. on p. 348*]

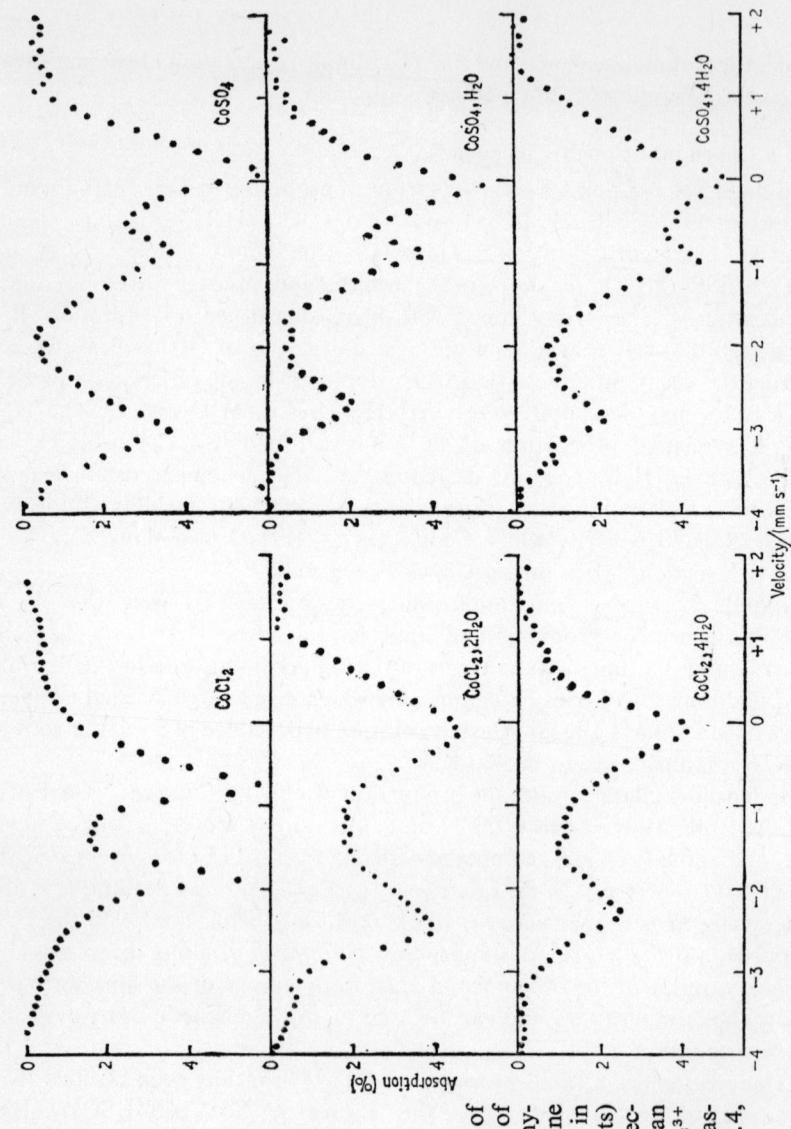

Fig. 12.1 Mössbauer spectra of ^{57}Co sources in matrices of CoCl$_2$, CoSO$_4$, and their hydrates. Note that the hyperfine spectrum (which originates in the source in these experiments) is the mirror image of the spectrum normally observed in an absorber experiment. The Fe^{3+} content increases with increasing degree of hydration. [Ref. 4, Figs. 1 and 2]

[*Refs. on p. 348*]

^{57}Co/CoCO$_3$ shows an asymmetric Fe^{2+} quadrupole spectrum which is attributable to a Karyagin effect, rather than to Fe^{3+} formation or to partial orientation of the source [13]. Studies on CoCO$_3$ doped with ^{57}Fe have investigated its magnetic structure below the Néel temperature of 17·6 K [14]. The magnetic field at the Fe^{2+} nucleus (72 kG at 4·9 K) is perpendicular to the principal axis of the electric field gradient tensor. Partial collapse of the magnetic spectrum occurs immediately below T_N, and can only be interpreted in terms of a *spatial* distribution of the average magnetic hyperfine field over the nuclear observation time, rather than by a field fluctuating *linearly* between $+H$ and $-H$. A ^{57}Co/CoC$_2$O$_4$.2H$_2$O source showed no signs of the Fe^{3+} charge state, presumably due to the high reducing power of the oxalate anion preventing stabilisation of anything other than Fe^{2+} cations [15]. Surface adsorption on oxalates has also been studied [16].

^{57}Co/cobalt(III) acetylacetonate shows a complex spectrum with lines attributable to iron(III) acetylacetonate and an iron(II) species [17]. The ^{57}Co/cobalticinium tetraphenylborate shows evidence for both iron(II) and iron(III) nominal oxidation states, although the iron(II) quadrupole splitting is only $\frac{2}{3}$ that of ferrocene, indicating that the environment of the daughter atom is not identical to that of ferrocene [18]. Electron capture in ^{57}Co/Co^{III}(phen)$_3$(ClO$_4$)$_3$.2H$_2$O appears to cause considerable disruption in the electronic configuration, the chief product being an Fe^{2+} species, but with evidence for other low-spin products [19]. The violet form of ^{57}Co/Co(py)$_2$Cl$_2$ has a bridged polymer structure, and as a result shows a higher recoil-free fraction than the blue ^{57}Co/Co(py)$_2$Cl$_2$ which has a tetrahedral monomeric structure [20]. The spectrum of the cobalt(III) complex, ^{57}Co/Co(bipy)$_3$(ClO$_4$)$_3$, is similar to that of the equivalent iron(III) complex, there being no evidence for other oxidation states [21], whereas ^{57}Co/Co(bipy)$_3$(ClO$_4$)$_3$.3H$_2$O gives some Fe^{2+} daughter atoms [22].

Cobalt Oxides

Early experiments on ^{57}Co/CoO showed the production of Fe^{2+} and Fe^{3+} ions [23]. The lattice symmetry is cubic above the Néel temperature of 291 K, and the Fe^{2+} and Fe^{3+} contributions were not quadrupole split. Below T_N there were two magnetic hyperfine splittings, the Fe^{2+} pattern showing a quadrupole splitting in keeping with the known tetragonal distortion. The fraction of Fe^{2+} ions produced is strongly dependent on temperature, and is vanishingly small by 800 K (Fig. 12.2) [24]. As already detailed, the Fe^{2+}/Fe^{3+} ratio is not time dependent [1, 2] and Fe^{3+} ion can be detected in absorbers of CoO doped with 2% ^{57}Fe, showing that Fe^{3+} ion can be chemically stable in the CoO lattice [1].

Further investigation unexpectedly established that two entirely different forms of CoO exist [25]. CoO(I) has the NaCl structure, and a density of 6·4 g cm^{-3}; when doped with ^{57}Co it shows only Fe^{2+} daughter atoms. The

Néel temperature is 288 K, and it is the stable form under normal laboratory conditions. CoO(II) also has the NaCl structure, but a density of only 3·2 g cm^{-3} showing that half of the cation and half of the anion sites are vacant; this form gives only Fe^{3+} ions after decay. The Néel temperature is 270 K and it reacts with oxygen at room temperature. Many chemical preparations contain simple mixtures of CoO(I) and CoO(II) at room temperature, but as the temperature is raised the anion vacancies of CoO(II) diffuse into the CoO(I) lattice, causing the apparent change to an all Fe^{3+} spectrum. Detailed Mössbauer studies of CoO(I) and CoO(II) both above and below the Néel

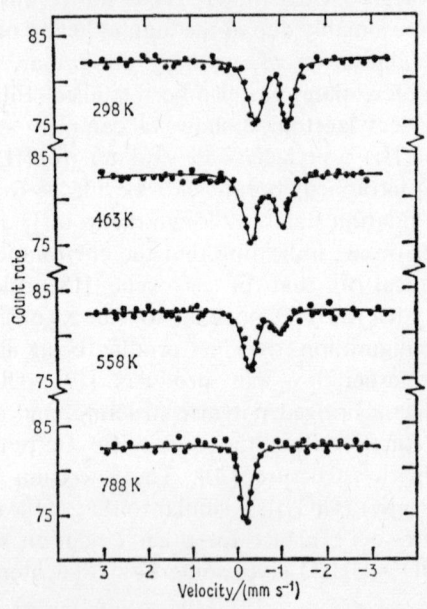

Fig. 12.2 Mössbauer spectra of ^{57}Co/CoO at various temperatures showing the decrease in Fe^{2+} produced with increase in temperature. [Ref. 24, Fig. 1]

temperatures have been made, and the crystal-field parameters for Fe^{2+} in CoO(I) determined from the temperature-dependence data below T_N [25, 26]. A counter proposal that Co(II) consists of microcrystals of CoO(I) [27] was shown to be inconsistent with the experimental data [28]. The Fe^{2+}/Fe^{3+} ratio in CoO(I, II) is reversibly temperature dependent, and a direct relaxation experiment has confirmed that the stabilisation of ionic states follows localised lattice heating after electron capture in ^{57}Co [29]. Atoms in the vicinity of vacancies stabilise as Fe^{3+}, whereas atoms not in the vicinity of vacancies become Fe^{2+} ions. Above 400 K the defect mobility increases, leading to dominance of the Fe^{3+} state. Application of pressure to CoO induces magnetic splitting as a result of an increase in the Néel temperature [30].

[*Refs. on p. 348*]

Brief investigation of $^{57}Co/Co_2O_3$ and $^{57}Co/Co_3O_4$ has shown both Fe^{2+} and Fe^{3+} to be present with only an approximate relationship to the Co^{3+} content [31].

^{57}Co Doped into Other Compounds

It is also feasible to dope ^{57}Co into compounds containing neither iron nor cobalt for impurity studies. Precipitates of $^{57}CoCl_2$ in single crystals of NaCl (produced by thermal diffusion of $^{57}CoCl_2$ into NaCl followed by quenching) show purely an Fe^{2+} spectrum with a quadrupole splitting of only 1·14 mm s^{-1} at 80 K. This has been interpreted in terms of a doublet ground state in the iron orbital configuration [32]. By contrast, very dilute solutions of ^{57}Co in NaCl crystals produce two types of impurity iron site [33]; an intense quadrupole doublet considered to arise from Fe^{2+} or Fe^{3+} ions associated with a positive ion vacancy which causes splitting by lowering the local symmetry, and an additional weak broad resonance at $\sim -1·9$ mm s^{-1} attributed to isolated Fe^+ ions substituted on sodium sites (N.B. Fe^+ in the source will give a δ value of $-1·9$ mm s^{-1} relative to iron metal, whereas if it were in an absorber, δ would be $+1·9$ mm s^{-1}). A similar investigation of $^{57}CoCl_2$ in KCl had previously reported the line at 1·9 mm s^{-1} but had not recognised its origin [34].

^{57}Co diffused into single-crystal NaF in an atmosphere of HF gives a spectrum almost entirely attributable to Fe^{2+} ions with several distinct quadrupole splittings [11]. These presumably correspond to the different

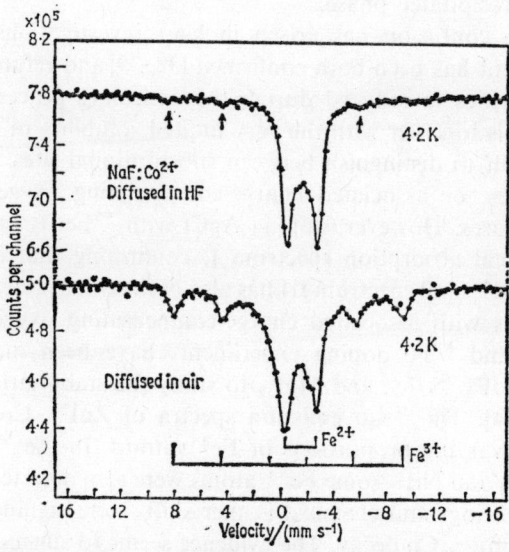

Fig. 12.3 Diffusion of ^{57}Co into NaF produces little Fe^{3+} contribution to the spectrum if the atmosphere is HF, but gives a significant concentration of Fe^{3+} iron if air is admitted. [Ref. 11, Fig. 3]

[*Refs.* on p. 348]

positions for the charge-compensating sodium vacancy near each divalent cobalt atom. Diffusion in air instead of HF greatly enhances the amount of trivalent iron (see Fig. 12.3), possibly due to oxygen-anion charge compensation near the cobalt sites. The extreme magnetic dilution contributes to a long spin-relaxation time, and the Fe^{3+} ion is seen as a paramagnetic hyperfine spectrum at low temperatures. $^{57}Co/ZnF_2$ produces mainly Fe^{2+} ion [11].

^{57}Co diffused into single crystals of AgCl in a chlorine/argon atmosphere decays to Fe^{2+} daughter atoms [35]. The freshly prepared samples showed a substantial quadrupole splitting at 80 K which has been attributed to Fe^{2+} ions in substitutional sites with positive ion vacancies at unique (100) positions [35–37]. The temperature dependence of the quadrupole splitting appears to imply considerable covalency in the iron environment. $^{57}Co/AgBr$ behaves similarly [35].

As the temperature of the $^{57}Co/AgCl$ crystal is raised above 230 K, the quadrupole splitting diminishes and a broad symmetrical line is produced which narrows continuously up to 320 K (Fig. 12.4). This may be the result of thermal motion of the vacancy about the impurity. Ageing at room temperature introduces two new components into the spectrum [35]. Component II with a larger quadrupole splitting is thought to be due to the introduction of oxygen anions rather than cation vacancies for charge compensation. The spectra are not very reproducible from sample to sample because of the undetermined effects of other impurities. Component III may be a precipitated phase.

Considerable confusion has arisen in later investigations in which the above assignment has been both confirmed [38, 39] and refuted [40–43]. The effect of impurities introduced during the annealing process is difficult to control, and this together with the very limited solubility of $CoCl_2$ in AgCl makes it difficult to distinguish between substitutional sites with associated cation vacancies, or associated charge-compensating oxygen anions, and precipitated phases. However, doping AgCl with $^{57}FeCl_2$ (<0·05 mol. %) gives an identical absorption spectrum I, confirming that it is due to an Fe^{2+} impurity [37], and spectrum III has also been reinterpreted as originating from Fe^{3+} ions with associated charge-compensating oxygen anions [38].

Both ^{57}Co and ^{57}Fe doping experiments have been made on MnF_2, FeF_2, ZnF_2, CoF_2, NiF_2, and MgF_2 to study the stabilisation of the Fe^{3+} charge state [44]. The ^{57}Co emission spectra of ZnF_2, CoF_2, NiF_2, and MgF_2 showed varying proportions of Fe^{3+} atoms. In the ^{57}Fe absorption spectra of CoF_2 and NiF_2 some Fe^{3+} atoms were also detected, although the quadrupole splittings and chemical isomer shifts were significantly different to those found for ^{57}Co decay. The evidence seems to suggest the close presence of a charge-compensating defect in the ^{57}Fe doped materials.

$^{57}Co/NiO$ has also been observed to produce both Fe^{2+} and Fe^{3+} oxidation

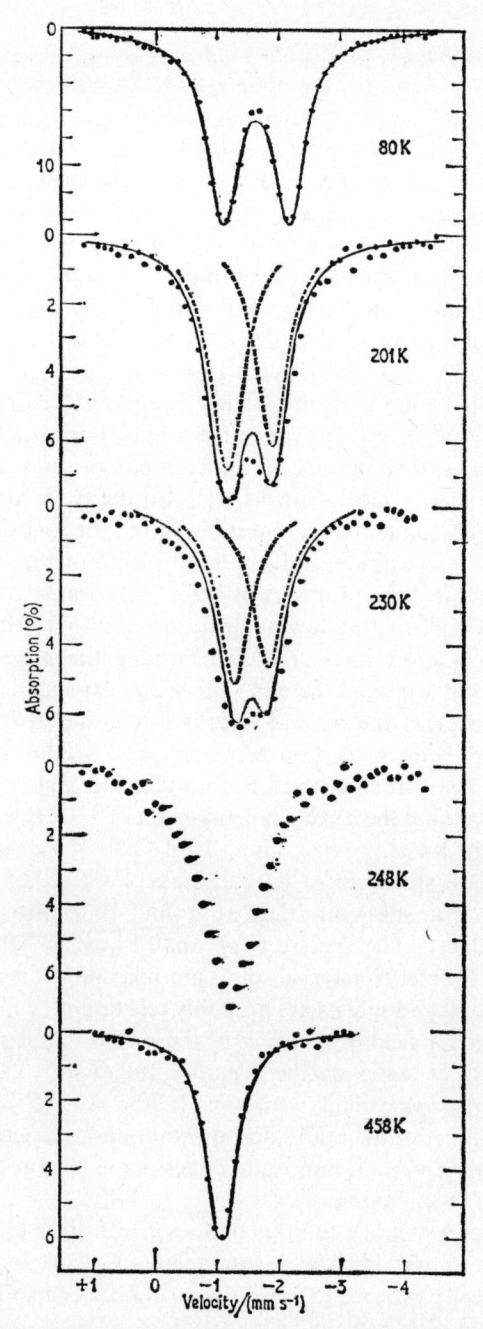

Fig. 12.4 Mössbauer spectra at 5 temperatures for a freshly prepared sample of ^{57}Co/AgCl. The electric field gradient disappears with the onset of thermal motion of the associated cation vacancy about the impurity. [Ref. 35, Fig. 1]

[*Refs. on p. 348*]

states as stable entities [1, 2, 45], although samples giving only the divalent spectrum can be obtained by doping in a CO_2 atmosphere [46]. In the latter case, a Néel temperature of 525 K was derived. Doped MnO similarly shows two charge states at room temperature, but the Fe^{3+} contribution disappears as the temperature falls to the Néel temperature which is 117 K. Detailed analysis of the magnetic spectra shows that the Ni^{2+} and Mn^{2+} spins (to which the Fe^{2+} impurity spins must be parallel) lie along the [112] directions [46]. Whether NiO and MnO also exist in two forms has not yet been investigated. However, ultra-fine NiO particles prepared on high-area silica show Fe^+, Fe^{3+} lines, and superparamagnetism is seen below the Néel temperature [47].

^{57}Co doped MgO and CaO show both Fe^+ and Fe^{2+} charge states. Application of an external magnetic field induces hyperfine structure, analysis of which confirms that both cations have a core-polarisation of $-254\langle S_z\rangle$ kG [48]. Continuing studies have shown that ^{57}Co decay in MgO can produce Fe^+, Fe^{2+}, and Fe^{3+} daughter charge states in proportions dependent on the particular sample preparation and the temperature of the Mössbauer measurement [49]. Dilute concentrations of ^{57}Fe ($\sim$0·03 at. %) in MgO show a small quadrupole splitting at low temperature [50, 51]. This can be interpreted in terms of crystal-field theory assuming the presence of random strains in the crystal although the Fe^{2+} site symmetry may be perfectly cubic [50]. The degeneracy of the ground state is lifted, and at low temperatures there is a slow electronic relaxation between levels. A Jahn–Teller effect need not be invoked. The theory evolved to interpret the MgO data has led to a method for determining the quadrupole moment of ^{57}Fe (see Chapter 5) [52].

Doping of up to 1·4 at. % Fe^{3+} in an α-Al_2O_3 absorber has been used to study the hyperfine spectrum of paramagnetic Fe^{3+} [53, 54]. The trigonal symmetry of the iron site splits the ^6S ground state into the $S_z = |\pm\frac{5}{2}\rangle$, $|\pm\frac{3}{2}\rangle$, $|\pm\frac{1}{2}\rangle$ doublets. The spectrum of an 0·14 at. % sample (Fig. 12.5) shows the $|\pm\frac{5}{2}\rangle$ doublet to have a spin–spin relaxation time of 7×10^{-8} s, but the $|\pm\frac{3}{2}\rangle$ and $|\pm\frac{1}{2}\rangle$ doublets are probably relaxing at a higher frequency. The external magnetic field dependence of the spectrum is extremely complex because of mixing of states by the hyperfine interaction, extensive changes being produced by very small variations in the field [55]. The relaxation properties are a function of both concentration and temperature. The effect of heat treatment in various atmospheres results in the production of some Fe^{2+} content [56] in some cases.

V_2O_3 undergoes a metallic-to-semiconductor transition at 168 K. Doping with 1% Fe_2O_3 shows that the low-temperature semiconductor phase is antiferromagnetically ordered [57]. Spin relaxation has also been found in a 0·22 wt % solution of $^{57}Fe^{3+}$ in TiO_2 [58].

CrO_2 doped with ^{57}Fe produces only Fe^{3+} substitutional atoms which magnetise collinear and antiparallel to the magnetisation of the host matrix below the Curie temperature of 397 K [59]. SnO_2 doped with ^{57}Fe shows paramagnetic relaxation effects from Fe^{3+} ions which are strongly dependent on the concentration of iron present [60].

^{57}Co doped into single crystals of UO_2 tends to show pronounced ageing effects including the appearance of a hyperfine field. Reordering of interstitial oxygen around the ion [61], or precipitation of iron-rich phases [62] have been suggested as explanations.

The ferroelectric compounds $BaTiO_3$ and $SrTiO_3$ have both been studied

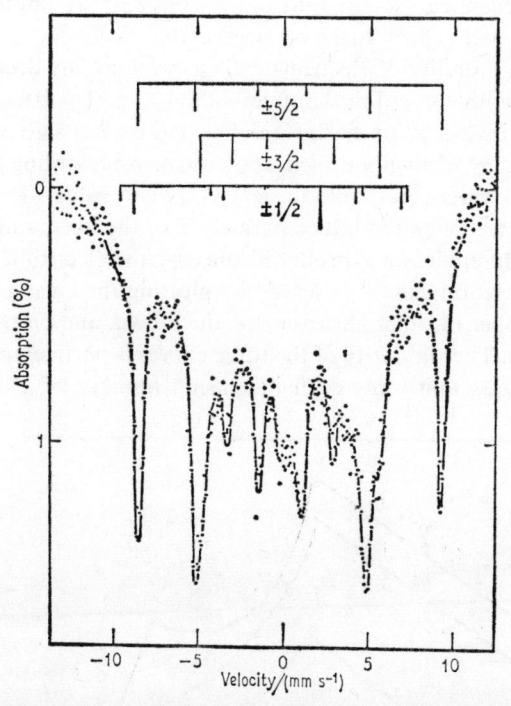

Fig. 12.5 Spectrum at 77 K of 0·14 at. % Fe^{3+} in Al_2O_3. The predicted lines are shown as bar diagrams, and the $|\pm\frac{5}{2}\rangle$ and $|\pm\frac{3}{2}\rangle$ states can be detected in the spectrum. [Ref. 54, Fig. 2]

by ^{57}Co and ^{57}Fe doping techniques. The effects of thermal treatment in oxidising and reducing atmospheres on the charge states produced and on site occupancy and clustering have been considered in detail [63-67].

Although several attempts have been made to study semiconducting materials by ^{57}Co doping, these have not generally been very successful because the iron is almost invariably incorporated identically as iron(III) in both *n*- and *p*-type semiconductors, and the major proportion is apparently electrically inactive. This was found for instance with ^{57}Co in *n*- and *p*-type silicon and germanium [68, 69], the lines being extremely broad. ZnS and

ZnSe differ in that they contain iron(II) [69]. Other systems studied include InAs [70], GaAs [71, 72], InSb [73, 74], and InP [72], and several tellurides [75, 76].

12.2 Metals

As already indicated at the beginning of this chapter, it is not intended to treat ^{57}Co doping of metals in detail, but to discuss typical illustrative examples. The chemical isomer shift of an isolated ^{57}Fe impurity atom in a metallic lattice will reflect to some degree the electronic structure of the metal. The high mobility of electrons will ensure that any decay after-effects are completely nullified within the time-scale of the Mössbauer event.

An extensive investigation has given chemical isomer shift values at room temperature for 32 elements doped with ^{57}Co, representing the most self-consistent set of values available to date [77]. The values do not show any systematic variation with the lattice parameter of the host, and it is therefore likely that the differences are produced only by the electronic structure. An interesting correlation can be observed by plotting the chemical isomer shift against the number of outer electrons for the 3d, 4d, and 5d transition-metal series (Fig. 12.6). The similarity of the three curves is particularly noteworthy. The shift increases uniformly with increasing number of electrons until a

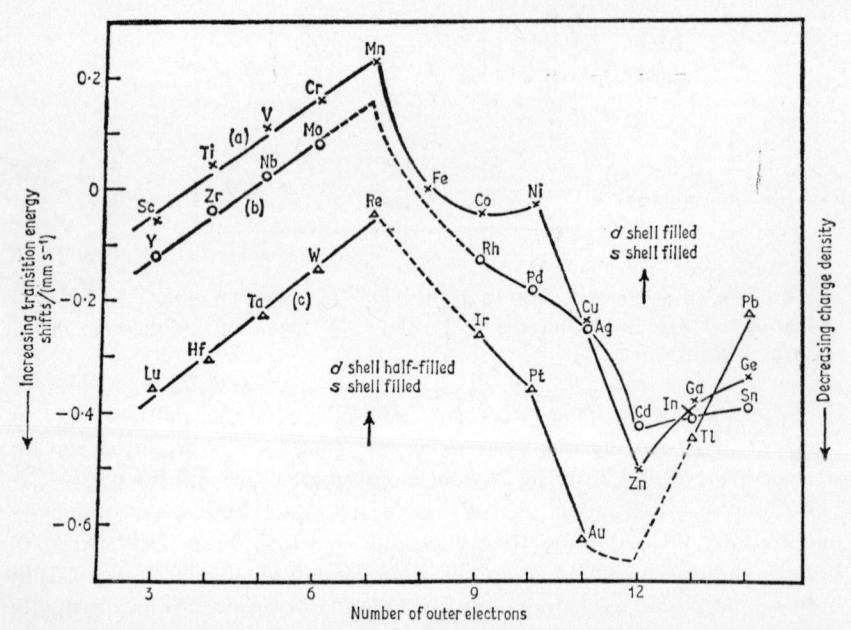

Fig. 12.6 The chemical isomer shift of ^{57}Co impurities in metals of (a) the 3d (b) the 4d, and (c) the 5d transition-metal series. [Ref. 77, Fig. 2]

[*Refs. on p. 348*]

configuration corresponding to d^5s^2 with a half-filled d-shell is achieved. It then decreases again until the $3d^{10}s^2$ configuration, whereupon a further increase is observed. Empirically one can say that the variation in shift is associated with the filling of the d-bands of the host. However, detailed interpretation would require knowledge of the metal band structures and the mechanisms for s-electron shielding.

The ^{57}Co-doped close-packed hexagonal metals Y, Hf, Lu, Ti, Zr, Tl, Re, Zn, and Cd all show a quadrupole splitting in the Mössbauer resonance which correlates to the axial ratio c/a of the host metal [78].

The chemical isomer shift of ^{57}Co impurities in metals generally decreases with increasing pressure, although quantitative interpretation is not possible [79].

^{57}Co doped into single crystals of cobalt metal shows a magnetic hyperfine pattern similar to that of α-iron with a field of -316 kG at room temperature and a value of e^2qQ of -0.064 mm s^{-1} [80]. The latter value is non-zero because of the hexagonal symmetry of the cobalt. Application of an external magnetic field parallel and perpendicular to the 'c' axis showed that the external field was strictly additive in both directions. This implies that the hyperfine field is isotropic, an unexpected result when the local environment is non-cubic as seen from the finite quadrupole splitting.

Although the temperature dependence of the magnetic field of ^{57}Fe in iron metal is exactly proportional to the magnetisation, this is not found to be the case for ^{57}Co (^{57}Fe) impurities in other ferromagnetic materials, e.g. nickel metal [81, 82]. The deviations are less than 15% but are significant. One can visualise a locally large density of states at the impurity, due to the exchange field between the impurity and the host being weaker or stronger than between host atoms. Such an effect is referred to as a quasi-independent localised moment [82].

The discontinuity found in the chemical isomer shift of ^{57}Fe in iron metal at the Curie point is paralleled by ^{57}Co in nickel metal, and although the change is of opposite sign it is probably also due to a change in the absolute position of the Fermi level as the magnetic interaction splits the d-band into spin-up and spin-down states [83]. Measurement of the relative line intensities of the hyperfine spectrum of ^{57}Co in single crystals of nickel with change in temperature shows clearly that the direction of easy magnetisation changes from the [111] to the [100] direction at 490 K [84].

In ^{57}Co/aluminium the ^{57}Fe resonance gives magnetic fields which are independent of temperature over wide ranges, implying a paramagnetic hyperfine coupling with a long relaxation time [85].

In many metallic hosts it is possible for the impurity atom to have a localised electronic spin moment which interacts magnetically neither with the host nor with other distant impurities. While such an impurity will not in general show a magnetic spectrum while unperturbed, the localised moment will tend to align in an external magnetic field. In the event that the moment is non-zero, the internal field recorded at the impurity nucleus will not equal the applied field, due to induced alignment, and the discrepancy will reach a saturation value as the external field is raised. ^{57}Co impurities in V, Nb, and Ta show no localised moment and the applied and observed magnetic fields correspond [86]. Mo, W, Rh, Pd, Pt, Cu, Ag, and Au all show a significant localised moment which aligns antiparallel to the applied field so that the observed internal field is smaller than expected. The magnetic moment of the impurity can be quite large, e.g. $12 \cdot 6\mu_B$ in Pd [87]. The size of the moment is closely related to the number of electrons in the valence shell so that, for example, Mo and W show similar behaviour. The mechanisms of spin polarisation of the ^{57}Fe by the spin-density waves of the host have been described in detail [86].

Particular interest has been expressed in the ^{57}Co/copper system because of the existence of a bound state formed between the localised magnetic impurity moments and the conduction-electron spins below a critical temperature [88]. The impurity moment is significantly quenched by spin compensation from the conduction electrons below the critical temperature, but this bound state can be perturbed by application of large external magnetic fields.

Major examples of lattice dynamical studies involving measurements of the recoil-free fraction and second-order Doppler shift over wide ranges of temperature, and followed by theoretical analysis of the vibrational properties of the impurity atoms, include the systems ^{57}Co/V [89]; ^{57}Co/Au, Cu, Ir, Pd, Pt, Rh, and Ti [90]; and ^{57}Co/Pt, Pd, and Cu [91]. Anisotropy of the mean-squared displacement has been shown in, for example, ^{57}Co/zinc [92].

A rather unusual experiment was devised to study the penetration of conduction-electron spin polarisation into non-magnetic metallic films [93]. The surface of a bulk copper matrix was coated with palladium to a required thickness by an evaporative method. ^{57}Co was then evaporated onto the palladium and covered with a second layer of the latter. The absence of magnetic effects in the spectrum verified that the ^{57}Co was not clustering magnetically. Similar preparations using an iron matrix gave magnetic hyperfine spectra, the temperature dependence of which feature relaxation narrowing in proportion to the thickness of the palladium interlayer. The magnetic interactions were taking place at distances of 20–120 lattice spacings of Pd, indicating that the interaction is not oscillatory but is more likely to be a conduction spin polarisation through the palladium. Copper and silver interlayers gave very much smaller interactions.

A nickel hydride of composition close to $NiH_{0.7}$ can be prepared by an electrolytic process. It has been suggested that the electron associated with the hydrogen enters the 3d-band of the metal which has ~ 0.6 holes/atom.

[Refs. on p. 348]

This would then explain the lack of magnetic ordering. The hydrogen occupies the octahedral sites in the face-centred cubic lattice. The Mössbauer spectrum (Fig. 12.7) of a sample of nickel doped with ^{57}Co which has been

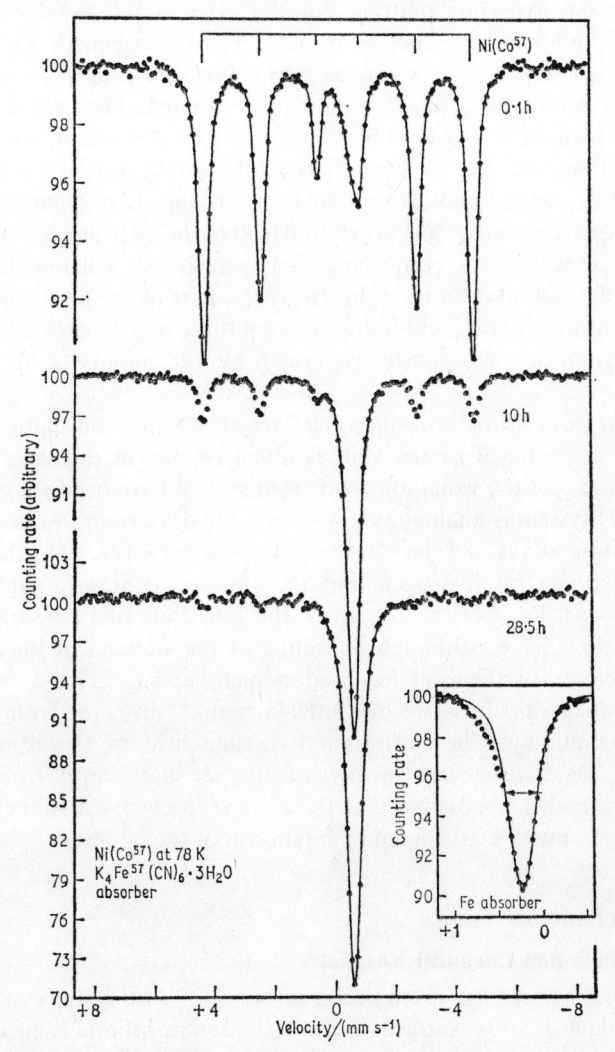

Fig. 12.7 The spectrum of ^{57}Fe in the nickel/nickel hydride system as a function of hydrogen content. The insert shows one line of the source with an iron absorber, showing the intrinsic broadening. [Ref. 94, Fig. 1]

partially converted to the hydride shows the original ferromagnetic ^{57}Co/nickel spectrum, plus a new paramagnetic component [94, 96]. The relative intensity of the latter increases with increasing electrolysis time and thus increasing hydrogen content. These results show dramatically that the partially reduced sample is a two-phase material. Introduction of hydrogen throughout the lattice on a variable concentration basis would be expected to broaden the hyperfine pattern, but the ^{57}Fe–nickel metal spectrum is retained throughout the reduction process. The paramagnetic line tends to show some asymmetry which can be attributed to hydrogen nearest-neighbour effects as not all octahedral sites are occupied. The chemical isomer shift of the hydride is +0·44 mm s^{-1} relative to nickel metal, which, under the assumption that any changes in the ^{57}Fe configuration affect the 3d-electrons only, would indicate an increase in the 3d-population by 0·35 electrons. Application of a 50-kG external field to the hydride gives a magnetic splitting of about 90 kG, confirming that there is still a magnetic moment (probably of ∼140 kG) on the iron. However, part of the paramagnetic line does not split in this way, and it appears that there may be at least two basic ^{57}Fe configurations which are governed by the numbers of hydrogen neighbours.

Hydrogen can also be introduced electrolytically into austenitic steel, but the change in chemical isomer shift is much smaller in this case [94]. The lattice dynamics of the vanadium/hydrogen system have been studied [95].

The Pd–H system is analogous to the nickel hydride, and has been studied by using dilute alloys of ^{57}Fe (2 or 5 at. %) with Pd [97, 98]. The α-phase contains virtually no hydrogen and the β-phase contains ∼0·65 atoms of hydrogen per Pd. For the 2% alloy, the β-hydride phase becomes ferromagnetic below 3·7 K, although the filling of the 4d-band of the alloy produces a decrease in the total localised moment and a decrease in T_C. The chemical isomer shift increases by +0·045 mm s^{-1} in going from the α- to the β-form, although the saturation hyperfine field of the iron remains unchanged. The hydrogen atoms do not interact significantly with the iron 3d-electrons, and it is possible that the change in chemical isomer shift on hydrogenation may be attributed to expansion of the lattice.

12.3 Miscellaneous Topics

Surface Studies and Chemical Reactions

The Mössbauer effect has given useful information in the study of chemical reactions which involve surface states and chemisorption. Reproducibility of data and their interpretation are difficult in this type of work because the adsorbing material may be affected by its pre-history. Comparisons have to be made with iron cations in compounds of known site symmetry.

A sample of high-area silica gel impregnated to 3 wt % Fe with ^{57}Fe-enriched iron(III) nitrate, and then calcined to produce small-particle Fe$_2$O$_3$, and subsequently reduced in hydrogen can be used as a supported iron catalyst material. The Mössbauer spectrum shows the presence of both Fe^{2+}

and Fe^{3+} iron in distorted site symmetries [99]. Chemisorption of ammonia reduces the Fe^{3+} to Fe^{2+}, the original spectrum being regenerated by outgassing at elevated temperature. If one assumes that ammine radicals are formed by chemisorption of ammonia, then these will have a strong tendency to transfer electrons to the adsorption site, thus causing the reduction. The ease of reduction shows that the Fe^{3+} ions must be very near the surface of the catalyst. The Fe^{3+}/Fe^{2+} ratio in the catalyst depends strongly on the conditions of the hydrogen reduction [100].

Although hydrated magnetite can be precipitated from aqueous solutions containing Fe^{2+} and Fe^{3+} ions at pH >5, this reaction is inhibited by the presence of magnesium ions. Examination of typical precipitates by Mössbauer spectroscopy shows that replacement of only 20% of the Fe^{2+} by magnesium virtually destroys the ferromagnetic character of the Fe_3O_4, presumably by weakening the number of A–B site interactions [101]. In chemical preparations like this where the spectrum of the final product is strongly dependent on the exact conditions employed, it is feasible to use the Mössbauer spectrum as a quality-control check and to help establish the optimum conditions for production of samples with given characteristics.

An attempt has been made to study ^{57}Co atoms on the surfaces of single crystals of tungsten and silver by measuring the anisotropy of the recoil-free fraction, but lack of detailed knowledge of the type of site occupied by the impurity nucleus at the surface precludes an unambiguous interpretation [102].

Zeolites also have considerable application as catalysts. Iron in the +3 oxidation state is introduced by ion-exchange methods [103]. However, attempted adsorption of Fe^{2+} ions causes complete breakdown of the structure with any retained iron being in the +3 state. Dehydration of the zeolite causes non-reversible reduction of the iron. Adsorption of Fe^{3+} salts on ion-exchange resins of the sulphonated styrene–divinyl benzene and quaternary ammonium types has little effect on the iron resonances and indicates very weak binding of the ions to the resin [104]. Spin-relaxation effects and temperature-dependent paramagnetic hyperfine structure have been recorded and interpreted in detail for Fe^{3+} ions adsorbed on exchange resins [105, 106], and a number of other recent papers have shown interest in this new field [107].

Solid-state Systems

Studies with portland cement have shown it to contain iron in the form of the oxide $4CaO.Al_2O_3.Fe_2O_3$ [108, 109]. The oxide particles are generally small enough to show superparamagnetic behaviour, but lowering the temperature or increasing the ferrite concentration can induce magnetic splitting.

The luminescent properties of zinc sulphide phosphors change considerably with the introduction of iron into the crystal. In particular, if the iron is introduced at 600°C it gives rise to a weak red luminescence band, whereas the luminescence is quenched if the crystal is fired at 900°C. If the Fe^{3+} ions are in a local environment of cubic symmetry, red luminescence results, whereas the substitution of Fe^{3+} ions on Zn^{2+} sites with a charge-compensating cation vacancy on the next Zn site acts as a quencher to the luminescence centres [110, 111].

Mössbauer spectra of $FeCl_3$/graphite intercalation compounds are not consistent with previous postulates of $FeCl_2$ and $FeCl_4^-$ species [112]. The single broad resonance line gives a more positive chemical isomer shift than $FeCl_3$, suggesting donation of graphite π-electrons into the iron $3d$-shell, although nothing can be derived concerning the possible role of the chlorine atoms in determining the stability of these compounds. Intercalation compounds of boron nitride with $FeCl_3$ have also been examined [113].

The presence of an infinite repeating lattice is not a prerequisite for the observation of a Mössbauer resonance, and it is possible to obtain good spectra from glassy materials. An example of this is a study of reduced and oxidised sodium trisilicate glass containing iron [114]. The iron can be incorporated as Fe^{2+} or Fe^{3+} ions, and the site symmetry appears to be distorted octahedral. At low temperatures glasses containing Fe^{3+} ions show paramagnetic hyperfine structure due to long relaxation times [115]. In samples with a high iron content it is possible to identify magnetic precipitates of Fe_2O_3 and Fe_3O_4 [116].

An interesting application to the study of 'foreign bodies' is the detection of iron in the Orgueil and cold Bokkeveld meteorites [117]. The former was shown to contain magnetite together with paramagnetic Fe^{3+} iron, possibly as a silicate. The latter did not contain the magnetic phase.

Diffusion in Liquids and Solids

Although the Mössbauer effect is normally regarded as being restricted to the solid state, it is possible to observe a resonance in viscous liquids. Hydrated iron(III) sulphate when dissolved in glycerol and then dehydrated with P_4O_{10} shows a mixture of Fe^{2+} and Fe^{3+} ions at $-100°C$ at which temperature the matrix is frozen [118]. The resonance becomes steadily broader with rise in temperature, but at 0°C when the sample is clearly liquid it is still detectable albeit broadened by a factor of more than 5 (Fig. 12.8). Later experiments have used glycerol doped with $^{57}Co/HCl$ [119]. ^{57}Co-doped haemin (MW = 620) and ^{57}Co-doped haemoglobin (MW = 64,450) in glycerol were also used to study the effects of particle size in the recoil and diffusion processes. The relationship between the viscosity of the liquid and the broadening of the Mössbauer line can be related to the relative importance of 'jump' and 'continuous' types of diffusion process [120, 121].

Mössbauer spectroscopy can also be used to study diffusion and related processes in solid systems [122] since the appearance of the spectrum is

[Refs. on p. 348]

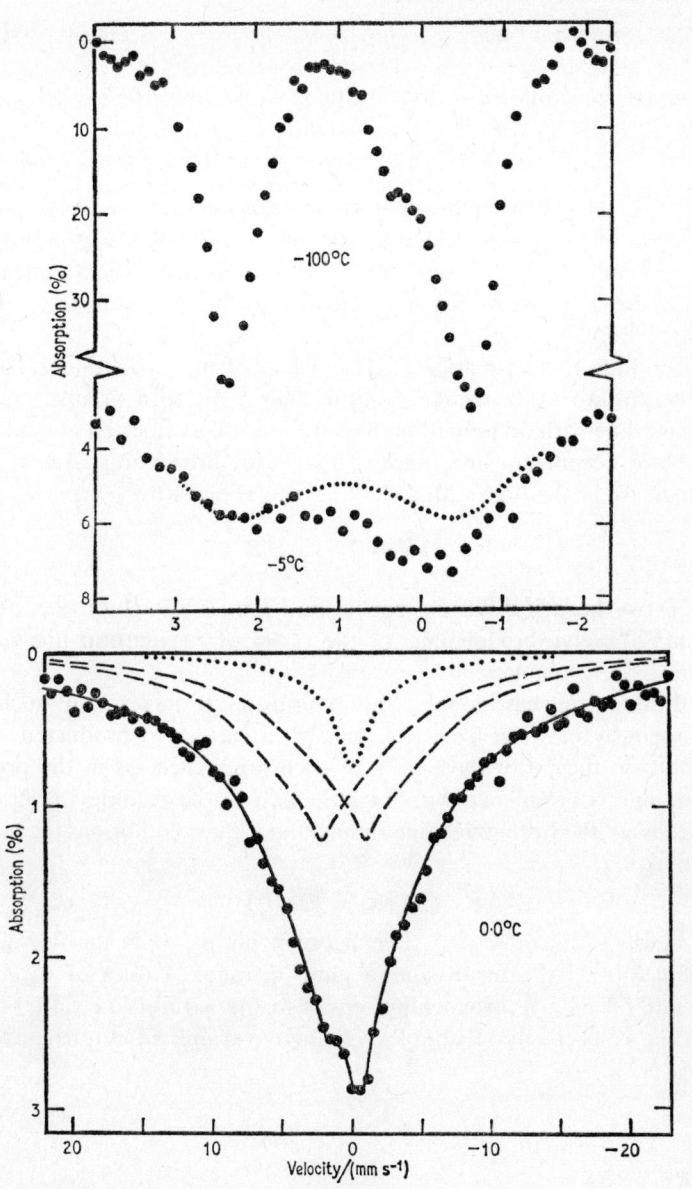

Fig. 12.8 The resonance of iron(III) sulphate in glycerol at $-100°C$, $-5°C$, and $0°C$. Note the very broad lines (Fe^{2+} doublet + Fe^{3+} singlet) of the $0°C$ spectrum when the sample is liquid. [Ref. 118, Fig. 1]

profoundly influenced by the time scale of the motion executed by the resonant atom. If the nucleus of interest is undergoing harmonic vibration about a fixed site at a frequency which is high compared with the inverse

[*Refs. on p. 348*]

excited-state lifetime then, as seen in Section 1.4, this simply produces a reduction in intensity of the resonance without any line broadening, the decrease in the recoilless fraction f being given by the Debye–Waller formula (p. 11)

$$f = e^{-4\pi^2 x^2/\lambda^2}$$

where x^2 is the mean square amplitude of oscillation in the direction of observation of the γ-ray and λ is the wavelength of the γ-photon. This means that values of x of the order of 10 pm are conveniently measurable with Mössbauer isotopes having excited-state lifetimes similar to those of ^{57}Fe or ^{119}Sn.

A second limiting case arises if the nucleus of interest is undergoing continuous diffusion or Brownian motion. This leads to a broadening of the Mössbauer line with no reduction in overall recoilless fraction and no change in the basic Lorentzian line shape. This energy broadening, $\Delta\Gamma$, expressed as an increase in the linewidth at half height, is given by

$$\Delta\Gamma = 2h\left(\frac{2\pi}{\lambda}\right)^2 D$$

where D is the self-diffusion coefficient. Values of D of the order of 10^{-15} m^2 s^{-1} give broadenings of the order of the natural linewidth for ^{57}Fe or ^{119}Sn resonances.

The third limiting case concerns jump diffusion. If the resonant nuclei jump instantaneously between fixed sites in such a way as to produce a random phase shift in the γ-ray wave train at each jump then, as in the preceding case, the line is broadened without reduction in the recoilless fraction and without loss of the Lorentzian line shape; under these conditions the broadening is given by

$$\Delta\Gamma = 2h/\tau_0 = 12hD/l^2$$

where τ_0 is the mean resting time between jumps, D is the self-diffusion coefficient and l^2 the mean square jump distance. Values of τ_0 equal to 280 ns and 53 ns give broadenings equal to the natural linewidth for ^{57}Fe and ^{119}Sn respectively. Examples of these various situations have been reviewed [122].

REFERENCES

[1] W. Triftshauser and P. P. Craig, *Phys. Rev. Letters*, 1966, **16**, 1161.
[2] W. Triftshauser and P. P. Craig, *Phys. Rev.*, 1967, **162**, 274.
[3] Y. Hazony and R. H. Herber, *J. Inorg. Nuclear Chem.*, 1969, **31**, 321.
[4] J. M. Friedt and J. P. Adloff, *Compt. rend.*, 1967, **264C**, 1356.
[5] G. K. Wertheim and D. N. E. Buchanan, *Chem. Phys. Letters*, 1969, **3**, 87.
[6] R. Ingalls and G. de Pasquali, *Phys. Letters*, 1965, **15**, 262.

REFERENCES | 349

[7] J. M. Friedt and J. P. Adloff, *Compt. rend.*, 1969, **268C,** 1342.
[8] J. M. Friedt and J. P. Adloff, *Inorg. Nuclear Chem. Letters*, 1969, **5,** 163.
[9] J. F. Cavanagh, *Phys. Stat. Sol.*, 1969, **36,** 657.
[10] T. Fujita, A. Ito, and K. Ono, *J. Phys Soc. Japan*, 1969, **27,** 1143.
[11] G. K. Wertheim and H. J. Guggenheim, *J. Chem. Phys.*, 1965, **42,** 3873.
[12] R. Ingalls, C. J. Coston, G. de Pasquali, H. G. Drickamer, and J. J. Pinajian, *J. Chem. Phys.*, 1966, **45,** 1057.
[13] J. M. Friedt and J. P. Adloff, *Compt. rend.*, 1968, **266C,** 1733.
[14] H. N. Ok, *Phys. Rev.*, 1969, **181,** 563.
[15] H. Sano and F. Hashimoto, *Bull. Chem. Soc. Japan*, 1965, **38,** 1565.
[16] P. R. Brady and J. F. Duncan, *J. Chem. Soc.*, 1964, 653.
[17] G. K. Wertheim, W. R. Kingston, and R. H. Herber, *J. Chem. Phys.*, 1962, **37,** 687.
[18] G. K. Wertheim and R. H. Herber, *J. Chem. Phys.*, 1963, **38,** 2106.
[19] R. Jagannathan and H. B. Mathur, *Inorg. Nuclear Chem. Letters*, 1969, **5,** 89.
[20] H. Sano, M. Aratini, and H. A. Stöckler, *Phys. Letters*, 1968, **26A,** 559.
[21] A. Nath, R. D. Agarwal, and P. K. Mathur, *Inorg. Nuclear Chem. Letters*, 1968, **4,** 161.
[22] P. K. Mathur, *Indian J. Chem.*, 1969, **7,** 183.
[23] G. K. Wertheim, *Phys. Rev.*, 1961, **124,** 764.
[24] V. G. Bhide and G. K. Shenoy, *Phys. Rev.*, 1966, **147,** 306.
[25] J. G. Mullen and H. N. Ok, *Phys. Rev. Letters*, 1966, **17,** 287; H. N. Ok and J. G. Mullen, *Phys. Rev.*, 1968, **168,** 550.
[26] H. N. Ok and J. G. Mullen, *Phys. Rev.*, 1968, **168,** 563.
[27] D. Schroeer and W. Triftshauser, *Phys. Rev. Letters*, 1968, **20,** 1242.
[28] H. N. Ok and J. G. Mullen, *Phys. Rev. Letters*, 1968, **21,** 823.
[29] W. Trousdale and P. P. Craig, *Phys. Letters*, 1968, **27A,** 552.
[30] C. J. Coston, R. Ingalls, and H. G. Drickamer, *Phys. Rev.*, 1966, **145,** 409.
[31] A. N. Murin, B. G. Lure, and P. P. Seregin, *Fiz. Tverd. Tela*, 1968, **10,** 1254.
[32] J. G. Mullen, *Phys. Rev.*, 1963, **131,** 1410.
[33] J. G. Mullen, *Phys. Rev.*, 1963, **131,** 1415.
[34] M. de Coster and S. Amelinckx, *Phys. Letters*, 1962, **1,** 245.
[35] D. H. Lindley and P. Debrunner, *Phys. Rev.*, 1966, **146,** 199.
[36] K. Hennig, W. Meisel, and H. Schnorr, *Phys. Stat. Sol.*, 1966, **13,** K9.
[37] K. Hennig, W. Meisel, and H. Schnorr, *Phys. Stat. Sol.*, 1966, **15,** 199.
[38] K. Hennig, *Phys. Stat. Sol.*, 1968, **27,** K115.
[39] K. Hennig, K. Yung, and S. Skorchev, *Phys. Stat. Sol.*, 1968, **27,** K161.
[40] A. N. Murin, B. G. Lure, P. P. Seregin, and N. K. Cherezov, *Fiz. Tverd. Tela*, 1966, **8,** 3291.
[41] A. N. Murin, B. G. Lure, and P. P. Seregin, *Fiz. Tverd. Tela*, 1967, **9,** 1424.
[42] A. N. Murin, B. G. Lure, and P. P. Seregin, *Fiz. Tverd. Tela*, 1967, **9,** 2428.
[43] A. N. Murin, B. G. Lure, and P. P. Seregin, *Fiz. Tverd. Tela*, 1968, **10,** 923.
[44] G. K. Wertheim, H. J. Guggenheim, and D. N. E. Buchanan, *J. Chem. Phys.*, 1969, **51,** 1931.
[45] V. G. Bhide and G. K. Shenoy, *Phys. Rev.*, 1966, **143,** 309.
[46] J. D. Siegwarth, *Phys. Rev.*, 1967, **155,** 285.
[47] K. J. Ando, W. Kundig, G. Constabaris, and R. H. Lindquist, *J. Phys. and Chem. Solids*, 1967, **28,** 2291.
[48] J. Chappert, R. B. Frankel, and N. A. Blum, *Phys. Letters*, 1967, **25A,** 149.
[49] J. Chappert, R. B. Frankel, A. Misetich, and N. A. Blum, *Phys. Rev.*, 1969, **179,** 578.

350 | ^{57}Fe – IMPURITY STUDIES

[50] F. S. Ham, *Phys. Rev.*, 1967, **160**, 328.
[51] H. R. Leider and D. N. Pipkorn, *Phys. Rev.*, 1968, **165**, 494.
[52] J. Chappert, R. B. Frankel, A. Misetich, and N. A. Blum, *Phys. Letters*, 1969, **28B**, 406.
[53] G. K. Wertheim and J. P. Remeika, *Phys. Letters*, 1964, **10**, 14.
[54] C. E. Johnson, T. E. Cranshaw, and M. S. Ridout, Proc. Int. Conf. on Magnetism, Nottingham, 1964 (Inst. Phys. Physical Soc., London, 1965), p. 459.
[55] H. H. Wickman and G. K. Wertheim, *Phys. Rev.*, 1966, **148**, 211.
[56] V. G. Bhide and S. K. Date, *Phys. Rev.*, 1968, **172**, 345.
[57] T. Shinjo, K. Kosuge, M. Shiga, Y. Nakamura, S. Kachi, and H. Takaki, *Phys. Letters*, 1965, **19**, 91.
[58] M. Alam, S. Chandra, and G. R. Hoy, *Phys. Letters*, 1966, **22**, 26.
[59] T. Shinjo, T. Takada, and N. Tamagawa, *J. Phys. Soc. Japan*, 1969, **26**, 1404.
[60] V. G. Bhide and S. K. Date, *J. Inorg. Nuclear Chem.*, 1969, **31**, 2397.
[61] M. de Coster, L. Mewissen, and M. Verschueren, *Phys. Letters*, 1964, **8**, 293.
[62] S. Nasu, N. Shikazono, and H. Takekoshi, *J. Phys. Soc. Japan*, 1964, **19**, 2351.
[63] V. G. Bhide and M. S. Multani, *Phys. Rev.*, 1966, **149**, 289.
[64] V. G. Bhide and M. S. Multani, *Phys. Rev.*, 1965, **139**, A1983.
[65] V. G. Bhide and H. C. Bhasin, *Phys. Rev.*, 1967, **159**, 586.
[66] V. G. Bhide, H. C. Bhasin, and G. K. Shenoy, *Phys., Letters*, 1967 **24A**, 109.
[67] V. G. Bhide and H. C. Bhasin, *Phys. Rev.*, 1968, **172**, 290.
[68] P. C. Norem and G. K. Wertheim, *J. Phys. and Chem. Solids*, 1962, **23**, 1111.
[69] G. N. Belozerskii, Yu. A. Nemilov, S. B. Tomilov, and A. V. Shvedchikov, *Soviet Physics – Solid State*, 1966, **8**, 485.
[70] G. Bemski and J. C. Fernandes, *Phys. Letters*, 1963, **6**, 10.
[71] G. Albanese, G. Fabri, C. Lamborizio, M. Musci, and I. Ortalli, *Nuovo Cimento*, 1967, **50B**, 149.
[72] G. N. Belozerskii, Yu. A. Nemilov, S. B. Tomilov, and A. V. Shvedchikov, *Fiz. Tverd. Tela*, 1965, **7**, 3607.
[73] G. N. Belozerskii, I. A. Gusev, A. N. Murin, and Yu. A. Nemilov, *Fiz. Tverd. Tela*, 1965, **7**, 1254.
[74] G. N. Belozerskii, Yu. A. Nemilov, and S. S. Tolkachev, *Fiz. Tverd. Tela*, 1966, **8**, 451.
[75] C. W. Franck, G. de Pasquali, and H. G. Drickamer, *J. Phys. and Chem. Solids*, 1969, **30**, 2321.
[76] V. Fano and I. Ortalli, *Phys. Stat. Sol.*, 1969, **33**, K109.
[77] S. M. Quaim, *Proc. Phys. Soc.*, 1967, **90**, 1065.
[78] S. M. Quaim, *J. Phys. C*, 1969, Ser. 2, **2**, 1434.
[79] R Ingalls, H. Drickamer, and G. de Pasquali, *Phys. Rev.*, 1967, **155**, 165.
[80] G. J. Perlow, C. E. Johnson, and W. Marshall, *Phys. Rev.*, 1965, **140**, A875.
[81] D. G. Howard, B. D. Dunlap, and J. G. Dash, *Phys. Rev. Letters*, 1965, **15**, 628.
[82] J. G. Dash, B. D. Dunlap, and D. G. Howard, *Phys. Rev.*, 1966, **141**, 376.
[83] D. G. Howard and J. G. Dash, *J. Appl. Phys.*, 1967, **38**, 991.
[84] G. Chandra and T. S. Radhakrishnan, *Phys. Letters*, 1968, **28A**, 323.
[85] J. Bara, H. U. Hrynkiewicz, A. Z. Hrynkiewicz, M. Karapandzic, and T. Matlak, *Phys. Stat. Sol.*, 1966, **17**, K53.
[86] T. A. Kitchens, W. A. Steyert, and R. D. Taylor, *Phys. Rev.*, 1965, **138**, A467.
[87] P. P. Craig, D. E. Nagle, W. A. Steyert, and R. D. Taylor, *Phys. Rev. Letters*, 1962, **9**, 12.
[88] R. B. Frankel, N. A. Blum, B. B. Schwartz, and D. J. Kim, *Phys. Rev. Letters*, 1967, **18**, 1051.

[89] P. D. Mannheim and A. Simopoulos, *Phys. Rev.*, 1968, **165**, 845.
[90] W. A. Steyert and R. D. Taylor, *Phys. Rev.*, 1964, **134**, A716.
[91] R. H. Nussbaum, D. G. Howard, W. L. Nees, and C. F. Steen, *Phys. Rev.* 1968, **173**, 653.
[92] R. M. Housley and R. H. Nussbaum, *Phys. Rev.*, 1965, **138**, A753.
[93] W. L. Trousdale and R. A. Lindgren, *J. Appl. Phys.*, 1965, **36**, 968.
[94] G. K. Wertheim and D. N. E. Buchanan, *J. Phys. and Chem. Solids*, 1967, **28**, 225.
[95] A. Simopoulos and I. Pelah, *J. Chem. Phys.*, 1969, **51**, 5691.
[96] G. K. Wertheim and D. N. E. Buchanan, *Phys. Letters*, 1966, **21**, 255.
[97] G. Bemski, J. Danon, A. M. de Graaf, and Z. A. da Silva, *Phys. Letters*, 1965, **18**, 213.
[98] A. E. Jech and C. R. Abeledo, *J. Phys. and Chem. Solids*, 1967, **28**, 1371.
[99] M. C. Hobson, *Nature*, 1967, **214**, 79.
[100] M. C. Hobson and A. D. Campbell, *J. Catal.*, 1967, **8**, 294.
[101] G. J. Kakabadse, J. Riddoch, and D. St P. Bunbury, *J. Chem. Soc. (A)*, 1967, 576.
[102] J. W. Burton and R. P. Godwin, *Phys. Rev.*, 1967, **158**, 218.
[103] J. A. Morice and L. V. C. Rees, *Trans. Faraday Soc.*, 1968, **64**, 1388.
[104] J. L. Mackey and R. L. Collins, *J. Inorg. Nuclear Chem.*, 1967, **29**, 655.
[105] I. P. Suzdalev, A. M. Afanasev, A. S. Plachinda, V. I. Goldanskii, and E. F. Makarov, *Zhur. eksp. teor. Fiz.*, 1968, **55**, 1752.
[106] V. I. Goldanskii, A. P. Suzdalev, A. S. Plachinda, and V. P. Korneyev, *Doklady Akad. Nauk S.S.S.R.*, 1969, **185**, 629.
[107] W. N. Delgass, R. L. Garten, and M. Boudart, *J. Phys. Chem.*, 1969, **73**, 2970;
W. N. Delgass, R. L. Garten, and M. Boudart, *J. Chem. Phys.*, 1969, **50**, 4603;
A. Johansson, *J. Inorg. Nuclear Chem.*, 1969, **31**, 3273;
Y. Takashima, Y. Maeda, and S. Umemoto, *Bull. Chem. Soc. Japan*, 1969, **42**, 1760;
R. W. J. Wedd, B. V. Liengme, J. C. Scott, and J. R. Sams, *Solid State Commun.*, 1969, **7**, 1091.
[108] F. Pobell and F. Wittmann, *Z. angew. Physik*, 1966, **20**, 5488.
[109] F. Wittmann, F. Pobell, and W. Wiedemann, *VDI - Zeitschrift*, 1966, **108**, 5676.
[110] K. Luchner and J. Dietl, *Z. Physik*, 1963, **176**, 261.
[111] K. Luchner and J. Dietl, *Acta Phys. Polon.*, 1964, **210**, 697.
[112] A. G. Freeman, *Chem. Comm.*, 1968, 193.
[113] A. G. Freeman and J. P. Larkindale, *J. Chem. Soc. (A)*, 1969, 1307.
[114] J. P. Gosselin, U. Shimony, L. Grodzins, and A. R. Cooper, *Phys. and Chem. Glasses*, 1967, **8**, 56.
[115] C. R. Kirkjian and D. N. E. Buchanan, *Phys. and Chem. Glasses*, 1964, **5**, 63.
[116] A. A. Belyustin, Yu. M. Ostanevich, A. M. Pisarevskii, S. B. Tomilov, U. Bai-shi, and L. Cher, *Fiz. Tverd. Tela*, 1965, **7**, 1447.
[117] A. Gerard and M. Delmelle, *Compt. rend.*, 1964, **259**, 1756.
[118] D. St P. Bunbury, J. A. Elliott, H. E. Hall, and J. M. Williams, *Phys. Letters*, 1963, **6**, 34.
[119] P. P. Craig and N. Sutin, *Phys. Rev. Letters*, 1963, **11**, 460.
[120] K. S. Singwi and J. E. Robinson, *Phys. Letters*, 1964, **8**, 248.
[121] J. A. Elliott, H. E. Hall, and D. St P. Bunbury, *Proc. Phys. Soc.*, 1966, **89**, 595.
[122] D. C. Champeney, *Phys. Bulletin*, 1970, **21**, 248.

13 | Biological Compounds

Although one tends to associate living organisms with organic carbon compounds containing largely the elements H, C, N, O, P, and S, it is now recognised that many essential biological functions are controlled by large protein molecules containing at least one metal atom. An example of this is the haemoglobin type of protein which acts as an oxygen carrier in the blood of many forms of animal life. The iron atom is involved in direct bonding to the molecule being transported. Likewise, catalytic enzymes may use metal oxidation–reduction properties to facilitate chemical reaction. The large protein body usually controls among other things the specificity of the reaction. The metal atom in such a protein is therefore of vital importance, and a study of its immediate environment should yield considerable information about the types of chemical reactions involved in biological processes.

The Mössbauer effect can play a unique role here, because each resonance is specific to a single isotope. Accordingly it has been used to examine several of the iron-containing proteins; the information obtained being appropriate to the immediate environment of the active site. In this respect a parallel can be drawn with the use of electron-spin resonance spectroscopy. A large protein residue can lower the effective cross-section of the iron-containing complex considerably, but isotopic enrichment can be achieved by 'feeding' the organism producing the iron-protein with a ^{57}Fe-enriched diet, and in this way high-quality spectra can be obtained. The large distance between individual iron atoms in biological systems has an important bearing on the phenomena observed, particularly in the case of high-spin iron(III) compounds where the spin-lattice relaxation time can be long enough to produce paramagnetic hyperfine effects. Interpretation can sometimes be complicated for this reason, but enables the spin Hamiltonian for the iron to be studied in greater detail.

In this account we shall restrict attention to the phenomenological aspects of the Mössbauer spectra, and neglect the deeper physiological viewpoint. Many of the earlier references have now been superseded by later data, and are accordingly omitted.

[*Refs. on p. 369*]

13.1 Haemeproteins

The iron in haemoglobin, the protein which carries oxygen in the blood, is bound in the centre of a planar porphyrin structure, referred to as protoporphyrin IX (Fig. 13.1). The compound of this porphyrin with iron is called haeme (ferroprotoporphyrin IX), and in the haemoglobin this unit is attached to the protein by the nitrogen of a histidine unit below the plane of

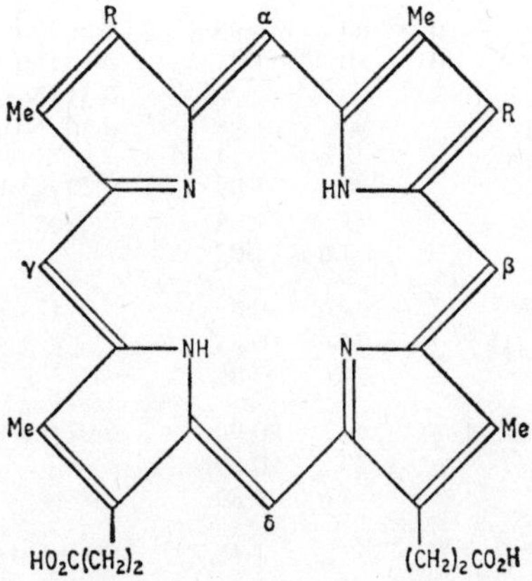

Fig. 13.1 Structure of some common porphyrins: in protoporphyrin the two R groups are vinyl (—CH=CH$_2$); in deuteroporphyrin R = H; and in mesoporphyrin R = Et.

the four nitrogen ligands. The sixth coordination position may be left vacant and provides a site for compound formation with small molecules. The iron–iron distance is about 25 Å in haemoglobin. The oxidation- and spin-states of the iron atom are very sensitive to the bonding at the sixth coordination site; e.g. oxygenated and reduced haemoglobins both contain iron(II) but in the $S = 0$ and $S = 2$ spin-states respectively.

Haeme and Haemin Derivatives

It might at first be thought that a study of the isolated porphyrins would be instructive, but in fact it transpires that their properties are somewhat different from those of the proteins. Accepted nomenclature is to use haeme for ferroprotoporphyrin, haemin to refer to ferriprotoporphyrin chloride, and haematin for ferriprotoporphyrin hydroxide.

The bispyridine haeme iron(II) compounds are low-spin $S = 0$ complexes and give only a single quadrupole split spectrum [1]. The bis-adducts of iron(II) and iron(III) protoporphyrin, α, β, γ, δ-tetraphenylporphyrin, and protoporphyrin dimethyl ester with bases such as imidazole and pyridine are also low-spin [2]. The quadrupole splittings at 80 K were found to be

[*Refs. on p. 369*]

354 | BIOLOGICAL COMPOUNDS

generally larger in the Fe(III) compounds than in the corresponding Fe(II), but interpretation of the small differences within a given oxidation state can only be semi-empirical. The results are summarised in Table 13.1.

Table 13.1 Spectra of haeme derivatives

Compound	S	T/K	Δ /(mm s^{-1})	δ /(mm s^{-1})	δ (Fe) /(mm s^{-1})	Reference
Fe(II) haeme (imidazole)$_2$	0	80	0·95	0·69 (NP)	0·43	2
Fe(II) haeme (pyridine)$_2$	0	80	1·21	0·72 (NP)	0·46	2
Fe(III) haemin (imidazole)$_2$	$\frac{1}{2}$	80	2·30	0·51 (NP)	0·25	2
Fe(III) haemin (pyridine)$_2$	$\frac{1}{2}$	80	1·88	0·50 (NP)	0·24	2
Haemin	$\frac{5}{2}$	298	1·06	0·20 (Cu)	0·43	1
		77	1·04	—	—	
		4·6	1·02	—	—	
2,4-diacetyldeuterohaemin chloride dimethyl ester (IV)	$\frac{5}{2}$	298	0·89	0·09 (Cu)	0·32	1
		77	0·81	—	—	
		4·6	0·85	—	—	
mesohaemin chloride dimethyl ester (V)	$\frac{5}{2}$	298	0·94	0·07 (Cu)	0·30	1
		77	0·89	—	—	
		4·6	0·89	—	—	
bispyridyl derivative of (IV)	0	298	1·18	0·11 (Cu)	0·34	1
		77	1·11	—	—	
		4·6	1·08	—	—	
bispyridyl derivative of (V)	0	77	0·63	—	—	1
		4·6	0·65	—	—	

Several haemin type compounds have been examined [1, 3, 4], and selected Mössbauer parameters are given in Table 13.1. The haemin chlorides of protoporphyrin IX, 2,4-diacetyldeuteroporphyrin IX dimethyl ester, and mesoporphyrin IX, and haematin all show a characteristic broadening of one of the resonance lines with rising temperature. This is dramatically illustrated in Fig. 13.2 for the deuterohaemin compound. In all cases they are Fe(III) $S = \frac{5}{2}$ compounds.

Application of an external magnetic field of 30 kG to haemin at 1·6 K produces an observed magnetic splitting of 345 kG and establishes e^2qQ to be positive [5]. This is consistent with the $|\pm\frac{1}{2}\rangle$ Kramers' doublet lying lowest. The field aligns the spins in the direction perpendicular to the axis of the haeme plane, resulting in a comparatively simple spectrum. This in turn leads to an explanation of the increase in asymmetry with temperature rise [6]. The spin–spin relaxation process involves simultaneously an $S_z = +\frac{1}{2}$ to $-\frac{1}{2}$ transition on one ion and the reverse process on a neighbouring ion. Since the $|\pm\frac{3}{2}\rangle$ and $|\pm\frac{5}{2}\rangle$ levels are only slightly above the $|\pm\frac{1}{2}\rangle$ levels, the population of the latter will decrease with thermal excitation so that the

[*Refs. on p. 369*]

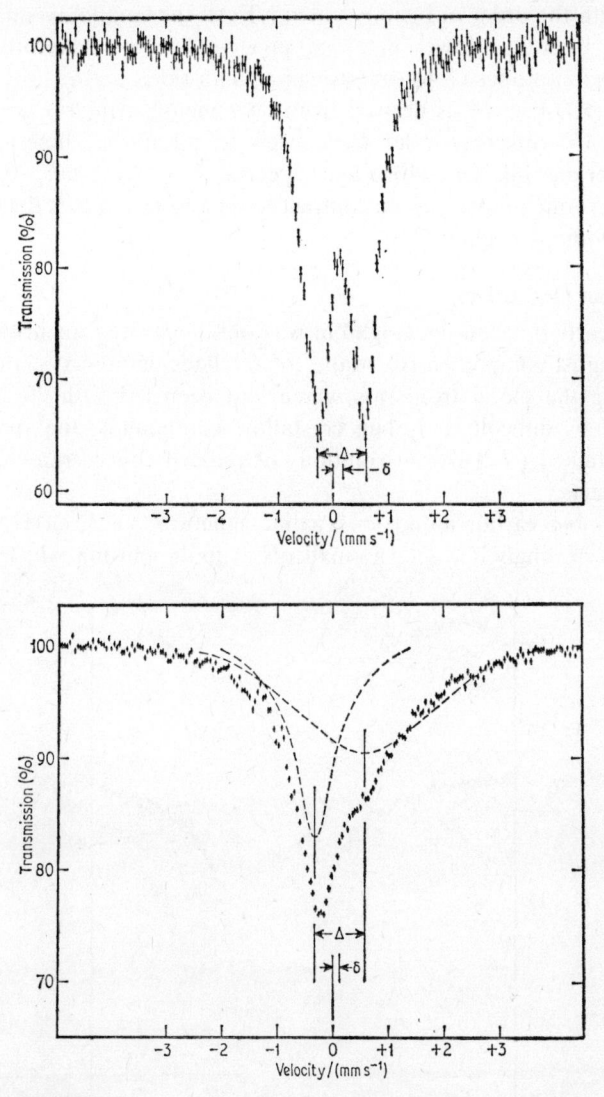

Fig. 13.2 Mössbauer spectra of 2,4-diacetyldeuterohaemin chloride dimethyl ester at (a) 4·6 K, (b) 77 K, showing the characteristic broadening of one of the lines with temperature rise. [Ref. 1, Figs. 2 and 3]

probability of spin–spin relaxation via this level is reduced. This factor when taken in conjunction with the longer relaxation times of the excited states can be used to simulate the observed spectra [6].

The penta-coordinated fluoro-, acetato-, azido-, chloro-, and bromo-derivatives of haemin-type compounds show an increase in the quadrupole

[*Refs. on p. 369*]

splitting with the order of ligands stated [7]. All the compounds have $S = \frac{5}{2}$ with the $|\pm\frac{1}{2}\rangle$ doublet lowest and e^2qQ positive, and are therefore similar in their relaxation properties to deuterohaemin chloride.

The iron(III) haematins derived from tetraphenylporphyrin and deuteroporphyrin IX dimethyl ester have been found to be dimeric with an Fe–O–Fe bridge [8]. The Mössbauer spectra show very sharp quadrupole doublets at room temperature in contrast to the relaxation-broadened spectra of the monomeric haemins.

Haemoglobin Derivatives

Although early data on haemoglobin was obtained using unenriched blood [3, 9], the most comprehensive studies of the haemoglobin system were initiated using the blood from rats which had been fed with 80% enriched ^{57}Fe. As it is difficult to isolate crystalline compounds, the spectra were usually obtained from frozen solutions of red-cell concentrates or haemoglobin extracts.

Haemoglobin–carbon monoxide is a diamagnetic $S = 0$ iron(II) compound [10–12]. Accordingly one sees a small quadrupole splitting which has little

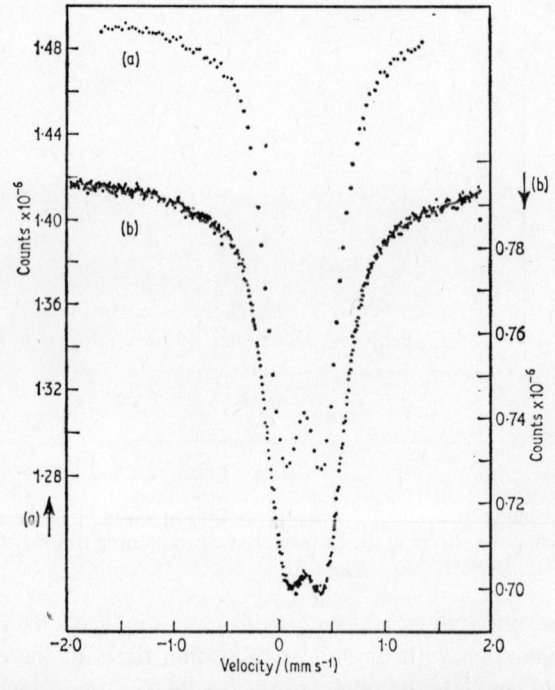

Fig. 13.3 Mössbauer spectra of haemoglobin carbon monoxide at (a) 77 K and (b) 4 K, showing the absence of magnetic interactions in an $S = 0$ complex. [Ref. 10, Fig. 8]

[*Refs. on p. 369*]

temperature dependence, and there is a complete absence of magnetic interactions. Typical spectra are shown in Fig. 13.3 for comparison with some of the more complex spectra, and parameters are tabulated in Table 13.2.

Table 13.2 Mössbauer spectra for haemoglobin derivatives [10]

Compound*	S	T/K	Δ /(mm s^{-1})	δ (Fe) /(mm s^{-1})
HbCO	0	195	0·36	0·18
		4	0·36	0·26
Hb reduced	2	195	2·40	0·90
		4	2·40	0·91
HbNO	?	195–1·2	magnetically broadened	
HbO$_2$	0	195	1·89	0·20
		77	2·19	0·26
		1·2	2·24	0·24
HiF	$\frac{5}{2}$	195–1·2	magnetically broadened	
HiH$_2$O	$\frac{5}{2}$	195	2·00	0·20
HiOH	$\frac{1}{2}$?	195	1·57	0·18
		77	1·9	0·2
HiN$_3$	$\frac{1}{2}$	195	2·30	0·15
HiCN	$\frac{1}{2}$	195	1·39	0·17

* The abbreviation Hb is used for a Fe(II) haemoglobin compound and Hi for a Fe(III) haemoglobin compound.

Reduced haemoglobin is an $S = 2$ iron(II) compound with a singlet ground state and no suggestion of paramagnetic hyperfine splitting at 4 K. Application of an external magnetic field of 30 kG generates gross magnetic broadening as the result of mixing of the levels of the spin-orbit multiplet (see Fig. 13.4). The spectrum breadth corresponds to an effective magnetic field of about 150 kG which can be crudely correlated with the predicted spin-Hamiltonian.

Haemoglobin–nitric oxide shows considerable magnetic broadening at all temperatures between 195 and 1·2 K (Fig. 13.5). It cannot therefore be considered simply as a low-spin iron(II) compound, as there is obviously strong covalent bonding and a large spin transfer from the paramagnetic NO molecule. Detailed analysis has not been attempted.

Oxyhaemoglobin is a low-spin $S = 0$ iron(II) complex and shows a large quadrupole splitting of 2·24 mm s^{-1} at 1·2 K. The interaction between the metal d_{yz}-orbital and the π^*-antibonding orbitals of the oxygen is responsible for the loss of the oxygen molecule's paramagnetism, and the covalent bonding effectively removes an electron from the d_{yz}-orbital. The geometry

[Refs. on p. 369]

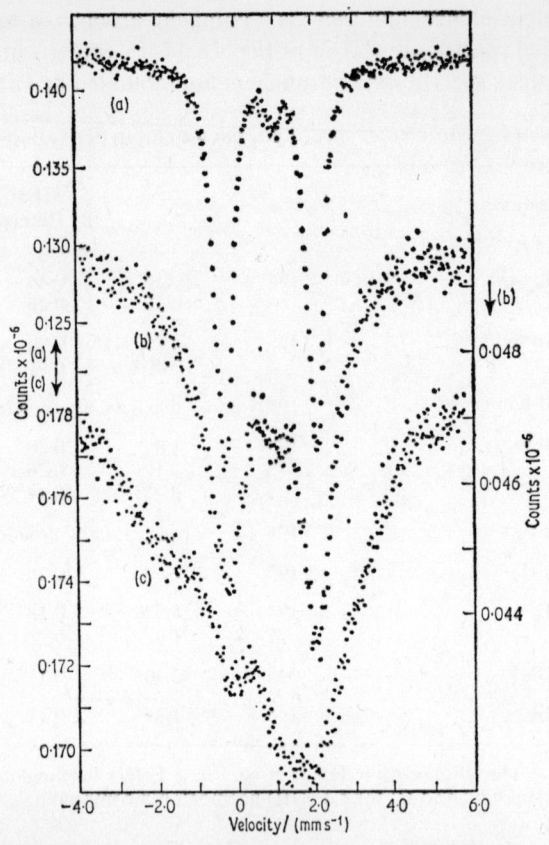

Fig. 13.4 Mössbauer spectra of reduced red cells at 4 K in (a) zero magnetic field, (b) a field of 7·5 kG perpendicular to the direction of observation, (c) a field of 30 kG. [Ref. 10, Fig. 5]

and schematic molecular-orbital scheme are shown in Fig. 13.6. The molecular-orbital $(d_{yz} + \pi_z^*)/\sqrt{2}$ has effectively only half the character of a d_{yz}-orbital. The sign of e^2qQ measured by the applied magnetic field method is negative, consistent with the implied planar deficit of electronic charge (note that eq corresponds to V_{xx} in the above axis definition and results from an electron 'hole' in the d_{yz}-orbital). The quadrupole splitting shows more temperature dependence than expected, and no explanation has yet been advanced for this.

Haemoglobin fluoride is an $S = \frac{5}{2}$ iron(III) compound known from electron-spin resonance data to have the $|S_z = \pm\frac{1}{2}\rangle$ Kramers' doublet lying lowest. Spectra between 195 K and 1·2 K show appreciable spin–spin and spin-lattice relaxation effects. The examples in Fig. 13.7 can be compared with the predicted line spectrum. Surprisingly, the spectrum sharpens in

[Refs. on p. 369]

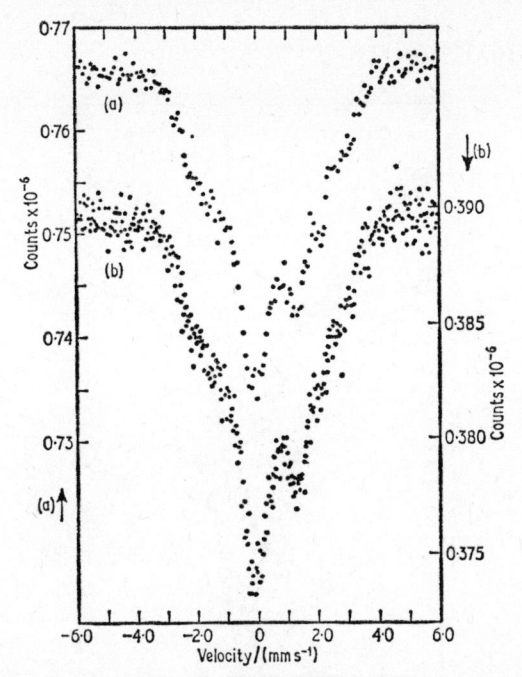

Fig. 13.5 Mössbauer spectra of haemoglobin–nitric oxide at (a) 4 K and (b) 1·2 K. [Ref. 10, Fig. 10]

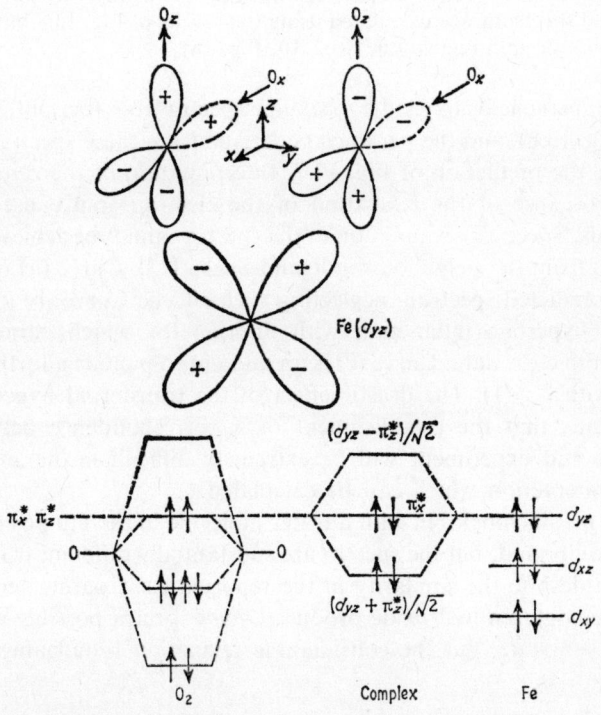

Fig. 13.6 Schematic representation of the bonding in oxyhaemoglobin showing how the interaction of the oxygen π_z^* molecular orbital with the iron d_{yz} orbital removes the degeneracy and causes spin pairing on the oxygen molecule.

360 | BIOLOGICAL COMPOUNDS

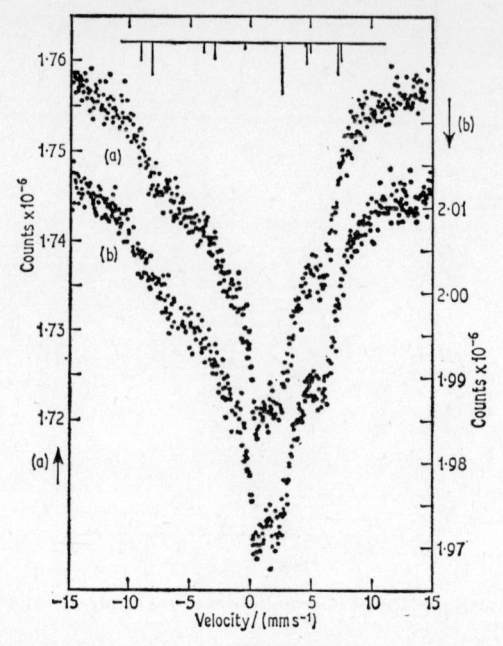

Fig. 13.7 Mössbauer spectra of haemoglobin fluoride at (a) 4 K and (b) 1·2 K. The predicted spectrum was calculated using the results of Fig. 13.8 but is not seen clearly because of spin relaxation. [Ref. 10, Fig. 16]

external magnetic fields to yield two six-line spectra from the split $|S_z = \pm\frac{1}{2}\rangle$ doublet (Fig. 13.8), and the parameters derived from these spectra were used to calculate the prediction of Fig. 13.7. Interpretation of the zero-field data is difficult because of the relaxation of the electron spin caused by local nuclear fields. Successful simulation of the spectra cannot be achieved without interactions from the neighbouring ligand nuclei [13]. Curve (a) in Fig. 13.9 shows the predicted spectrum neglecting such effects. Curve (b) includes the transferred hyperfine interaction with fluorine for which parameters are available from e.s.r. data. Curve (c) is an attempt to include a further nucleus (nitrogen with $I = 1$). The drastic effect of the transferred hyperfine interactions means that the establishment of a correspondence between such calculations and experiment will be extremely difficult in the absence of a dominant interaction which can be calculated.

The acid methaemoglobin with a water molecule in the 6th position is also an $S = \frac{5}{2}$ compound, but the spectra are substantially different from those of the fluoride despite the similarity in the reported e.s.r. parameters. At high pH, the haemoglobin hydroxide produced gives spectra possibly compatible with an $S = \frac{1}{2}$ state, but the considerable relaxation broadening precludes accurate analysis.

[*Refs.* on p. 369]

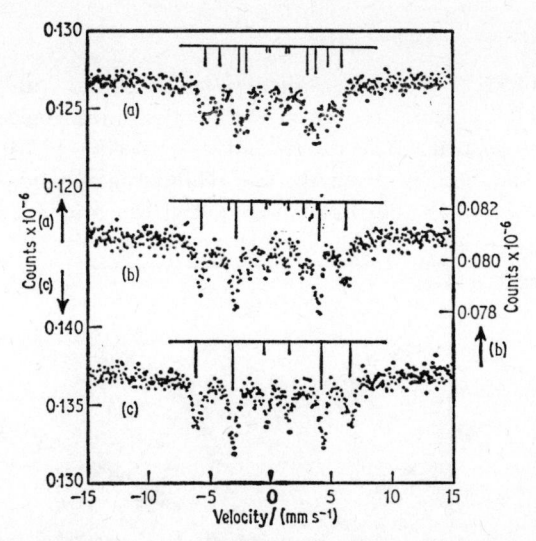

Fig. 13.8 Mössbauer spectra of haemoglobin fluoride in magnetic fields of (a) 7·5 kG, (b) 15 kG, (c) 30 kG, all applied perpendicular to the direction of observation. Note the sharpness of the lines in comparison with the zero-field spectra of Fig. 13.7. [Ref. 10, Fig. 17]

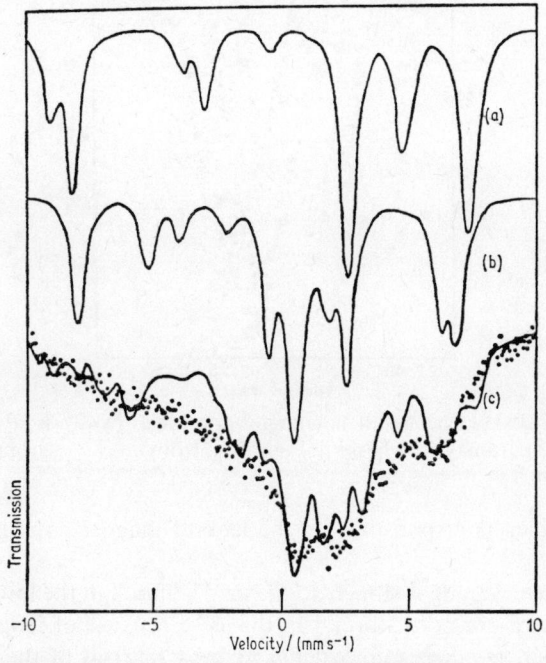

Fig. 13.9 Curve (a) shows the predicted zero-field Mössbauer spectrum of haemoglobin fluoride neglecting ligand–nucleus interactions. In curve (b) the transferred hyperfine interaction with fluorine is included, and in curve (c) a second nucleus with $I = 1$ is introduced. The experimental data shown for comparison were recorded at 1·2 K. [Ref. 13, Fig. 1]

362 | BIOLOGICAL COMPOUNDS

Haemoglobin azide is a low-spin iron(III) compound with $S = \frac{1}{2}$. Complicated magnetic structure is seen at low temperatures, replacing the comparatively sharp quadrupole doublet found at 195 K (Fig. 13.10). Good agreement can be obtained by using the spin-Hamiltonian predicted from the e.s.r. g-values, although the broad lines reflect the presence of some spin

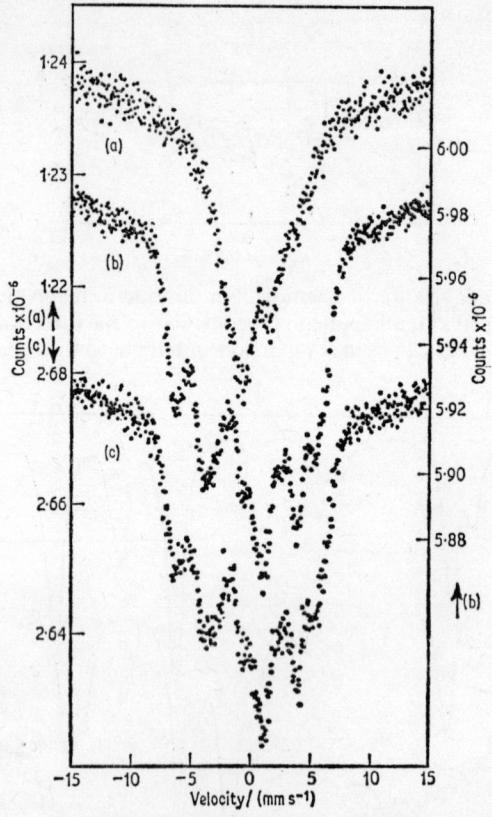

Fig. 13.10 Mössbauer spectra of haemoglobin azide at (a) 77 K, (b) 4 K, (c) 1·2 K, showing the paramagnetic hyperfine splitting from an $S = \frac{1}{2}$ iron(III) ion. [Ref. 10, Fig. 14]

relaxation, which is responsible for the lack of magnetic splitting at higher temperatures.

Haemoglobin cyanide is similar to the azide, although the low-temperature magnetic spectra are not as broad in this case because of longer relaxation times. The e.s.r. g-values were calculated by a reversal of the computation process for the azide, and simulation of the asymmetric spectrum at 77 K has been attempted [14].

Detailed molecular-orbital calculations have been performed for some of

[*Refs.* on p. 369]

these haemoglobin derivatives in an attempt to correlate the observed quadrupole splittings with the electronic structures [15, 16].

Metmyoglobin

Metmyoglobin has been prepared enriched in ^{57}Fe by combination of enriched protoporphyrin IX iron(III) chloride with apomyoglobin. The Mössbauer spectrum of lyophilised metmyoglobin shows two types of iron at 298 K, with quadrupole splittings of 1·96 and 0·36 mm s^{-1} and chemical isomer shifts of 0·04 and 0·06 (Fe) mm s^{-1} respectively [17]. Cooling causes these doublets to broaden and sharpen respectively so that at 4·6 K the spectrum comprises a sharp doublet with a splitting of 0·44 mm s^{-1} superimposed on a very diffuse background. It is suggested that the low shift and large splitting of the dominant component is from Fe(III) in an $S = \frac{1}{2}$ spin configuration, an unexpected conclusion considering that metmyoglobin is in an $S = \frac{5}{2}$ state in solution. The weaker sharp lines were probably an impurity such as free haemin, or metmyoglobin still retaining attached water.

Peroxidase

The peroxidase enzyme, Japanese-radish peroxidase a (JRP-a), has been prepared with enriched ^{57}Fe by combination of the apo-peroxidase with

Table 13.3 Mössbauer parameters for peroxidase derivatives [18–20]

Compound	S	T/K	Δ /(mm s^{-1})	δ (Fe) /(mm s^{-1})
peroxidase (JRP-a)	$\frac{1}{2}$	243	1·67	0·31
		195	1·67	0·31
		120	2·12	0·33
		77	2·20	0·37
		4·2	magnetically broadened	
peroxidase hydroxide	$\frac{1}{2}$	243	1·84	0·27
		120	2·36	0·22
peroxidase azide	$\frac{1}{2}$	120	2·22	0·31
peroxidase cyanide	$\frac{1}{2}$	195	1·61	0·11
		77	1·81	0·15
reduced peroxidase	2	243	1·79	0·79
		77	2·32	0·81
peroxidase carbon monoxide	0	77	0·21	0·32
peroxidase fluoride	$\frac{5}{2}$	195–4·2	magnetically broadened	
peroxidase – H$_2$O$_2$ (I)	1	77	1·33	0·10
peroxidase – H$_2$O$_2$ (II)	1	77	1·46	0·11
peroxidase – H$_2$O$_2$ (III)	0	77	2·37	0·29

[Refs. on p. 369]

protohaemin chloride [18, 19]. The Mössbauer spectrum shows a large quadrupole splitting which is strongly temperature dependent (Table 13.3). It is symmetrical at 243 K, but develops pronounced asymmetry below 120 K and is grossly broadened by magnetic interactions at 4·2 K. The similarity of the results to those for dehydrated metmyoglobin and methaemoglobin confirms the $S = \frac{1}{2}$ iron(III) configuration. The peroxidase hydroxide, azide, and cyanide are basically similar in behaviour, although there is more magnetic broadening in the azide and hydroxide than in the cyanide because of longer relaxation times. Peroxidase fluoride shows substantial magnetic broadening between 4·2 K and 195 K, and appears to be analogous to haemoglobin fluoride with an $S = \frac{5}{2}$ spin state. Reduced peroxidase gives a quadrupole split spectrum characteristic of an $S = 2$ iron(II) compound and is analogous to reduced haemoglobin. The peroxidase carbon monoxide compound is similarly in a low-spin $S = 0$ iron(II) state.

The reaction of peroxidase with H_2O_2 produces at least three intermediates (signified by I, II, and III). All three give a sharp quadrupole splitting with only a small temperature dependence (Table 13.3) [20]. Compounds I and II give an unusually low chemical isomer shift and are indistinguishable. It has been proposed that II is an iron(IV) compound with two unpaired electrons, and this would be compatible with the Mössbauer results. Compound III appears to be similar to oxyhaemoglobin and may be structurally analogous.

Cytochrome

The enzyme cytochrome c also contains a haeme unit attached to a protein and is associated with the oxidation of nutrients in cells. Samples of horse-heart cytochrome c have been studied without using ^{57}Fe enrichment [21]. Frozen solutions of oxidised cytochrome c show a quadrupole splitting of about 2·3 mm s^{-1} between 225 and 100 K. Asymmetric broadening of the spectrum occurs at lower temperatures, and a broad magnetic spectrum is found at 4·2 K. Dry cytochrome c gives a smaller splitting of $\sim$1·9 mm s^{-1}. The chemical isomer shift (relative to Fe) of 0·18 mm s^{-1} at 100 K and the magnetic broadening are compatible with an $S = \frac{1}{2}$ iron(III) configuration with an electron hole in a d_{xz}-orbital. The difference between solid and solution spectra is thought to be a result of hydration of the various protein chains in the vicinity of the haeme. Reduced cytochrome c gives a quadrupole splitting of about 1·2 mm s^{-1} and a shift of 0·38 mm s^{-1}, and is an $S = 0$ iron(II) compound.

^{57}Fe-enriched cytochrome c from *Torula utilis* yeast strain 321 has been similarly examined at low temperatures, and the $S = \frac{1}{2}$ oxidised form is basically similar to haemoglobin cyanide [22]. Spectrum simulation using the e.s.r. g-values for horse-heart cytochrome c was partially successful. The applied-field technique gave a positive value for e^2qQ in the $S = 0$ iron(II) cytochrome c, with an asymmetry parameter of $\eta \sim$0·5.

[Refs. on p. 369]

Photosynthetic bacteria can also be used to prepare ^{57}Fe-enriched cytochromes, and cytochrome cc' and cytochrome c_{552} have been isolated from *Chromatium* strain D, and cytochrome cc' and cytochrome c_2 from *Rhodospirillum rubrum* [23]. The oxidised and reduced forms were studied in frozen solutions. Although all the oxidised cytochromes discussed in this section have an $S = \frac{1}{2}$ configuration, the two cc' cytochromes are unusual in that the low-velocity line in the quadrupole splitting broadens first on cooling, whereas in all the others it is the high-velocity line which broadens. The CO complexes of the two *Chromatium* cytochromes are both $S = 0$ compounds with quadrupole splittings of <0.4 mm s^{-1}. The reduced cc' cytochromes are high-spin $S = 2$ iron(II) compounds, whereas the c_{552} and c_2 cytochromes are low-spin $S = 0$ compounds, an interesting distinction between the cc' and c classes of cytochrome.

Detailed measurements have also been made on a cytochrome c peroxidase containing protohaemin or mesohaemin and obtained from baker's yeast [24]. The fluoride derivatives of both forms of the enzyme are similar to those of haemoglobin fluoride, i.e. are $S = \frac{5}{2}$ ferric compounds, and appear to have the $|\pm\frac{1}{2}\rangle$ Kramers' doublet lowest. The very broad hyperfine spectra seen at 4·2 K in zero field become sharp in external fields of less than 1 kG because of a change in relaxation. Interpretation of the zero-field spectra is extremely complex because of ligand interactions of the type already mentioned under haemoglobin fluoride. Broad unresolved spectra involving relaxation are seen at higher temperatures. The two parent enzymes in acid conditions show a mixture of high- and low-spin species, and are even more complicated to understand than the fluoride derivatives. The ratio of the spin-states is a function of conditions.

13.2 Metalloproteins

The readiness with which haemeprotein samples can be obtained and the considerable data available on them from other types of investigation has resulted in a comparative neglect of other types of iron-bearing protein. However, several useful studies have been made, and the interesting properties of the haemeproteins are found to extend to other systems.

Ferrichrome A

The well-characterised metalloprotein ferrichrome A has been studied between 0·98 and 300 K using a sample prepared with ^{57}Fe-enriched material [25]. The iron is in a distorted octahedral environment of oxygens, and is in the $S = \frac{5}{2}$ iron(III) oxidation state. A single broad line is seen at 300 K and this gradually widens into a fully resolved six-line pattern at 0·98 K. The spectrum can be sharpened considerably by using an external magnetic field. The method of freezing solutions of the protein in ethyl alcohol was used as a

[*Refs. on p. 369*]

means of increasing the Fe–Fe separation, and this also sharpens the spectrum by slowing the dipole–dipole relaxation process. Detailed calculations of relaxation effects on iron(III) ions were also made.

Ferredoxin

The ferredoxins are enzymes known to catalyse photochemical reactions in plants and photosynthetic bacteria. The iron is of non-haeminic character and is believed to be close to the active centre of the enzyme, but its chemical form is not known with certainty. Spinach ferredoxin has a molecular weight of about 12,000 with two iron atoms per molecule. The oxidised ferredoxin shows a two-line spectrum corresponding to a quadrupole splitting of 0·6 mm s^{-1} and a chemical isomer shift of $\sim$0·2 mm s^{-1} [26, 27]. Application of a 30-kG field at 1·5 K produces little magnetic broadening, confirming the electronic state to be non-magnetic, i.e. an $S = 0$ iron(II) state [26]. Reduction of spinach ferredoxin produces a more complex spectrum from one $S = 0$ iron(II) ion and one $S = 2$ iron(II) ion (Fig. 13.11) [26, 27]. The latter component is broadened by a 30-kG field, whereas the former remains comparatively sharp and is thus non-magnetic. Neither of these species is consistent with the observed e.s.r. spectrum, and it is possible that a third configuration can occur in small amounts.

The ferredoxin from the green alga *Euglena* has been isolated containing enriched ^{57}Fe [28]. The two iron atoms appear to be identical, and in the oxidised *Euglena* ferredoxin give a quadrupole splitting of 0·65 mm s^{-1} at 4·2 K. The chemical isomer shift of 0·22 mm s^{-1} (rel. to Fe) and the lack of magnetic interaction in an external field confirm an $S = 0$ configuration. The sign of e^2qQ appears to be positive. The reduced form gives an unusual complex magnetic spectrum at low temperature, which may be interpreted approximately as the result of a single unpaired electron being shared by both iron atoms.

A ferredoxin with a molecular weight of 6000 and seven atoms of iron per molecule was extracted from *Clostridium pasteurianum* and oxidised samples show similar spectra [29], although with some evidence for at least two iron environments. An $S = \frac{1}{2}$ configuration was proposed, but the confirmatory experiment of applying a magnetic field and looking for magnetic effects has not yet been carried out.

Chromatium high-potential iron protein contains four non-haeme irons per formula weight of 10,074. Its oxidised form gives a quadrupole splitting of 0·82 mm s^{-1} at 77 K, shows extensive electronic relaxation broadening at 4·6 K, and is possibly in an $S = \frac{1}{2}$ state [27]. The reduced form contains the low-spin $S = 0$ iron(II) ion. *Chromatium* ferredoxin has between three and five iron atoms per formula weight of about 6000. The oxidised form contains at least two distinct iron sites with slightly different quadrupole splittings, but the oxidation state has not been distinguished as between $S = 0$ or $S = \frac{1}{2}$; by analogy with spinach ferredoxin it is probably the former. The reduced ferredoxin contains about one-sixth of

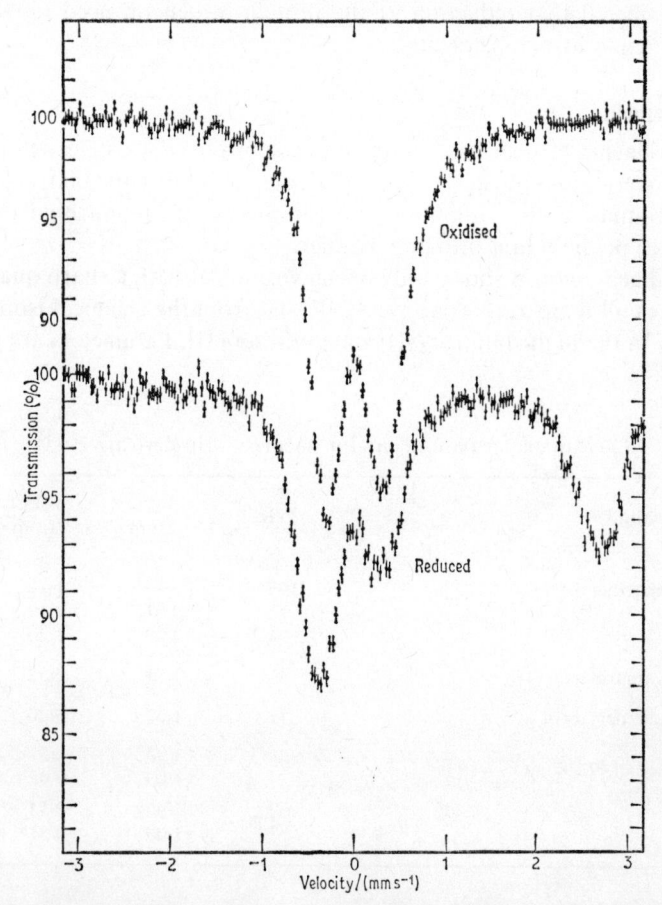

Fig. 13.11 The Mössbauer spectra of oxidised and reduced spinach ferrodoxin at 4·6 K, showing the change in electronic configuration of only one of the iron atoms. [Ref. 27, Fig. 5]

the iron in the $S = 2$ state. Clearly high-potential iron protein is quite distinct from the ferredoxin series.

Xanthine Oxidase

The iron in xanthine oxidase is related to the ferredoxins and there are eight iron atoms per molecule [30]. The oxidised xanthine oxidase appears to contain all eight atoms in the low-spin ($S = 0$) Fe(II) state as the small unique quadrupole split spectrum is not affected by magnetic fields. Reduction with sodium dithionite ($Na_2S_2O_4$) results in severe magnetic broadening

[*Refs. on p. 369*]

which appears to arise from long relaxation times in a low-spin ($S = \frac{1}{2}$) iron(III) ion. Although there is an apparent paradox in these results, it must be remembered that reduction of the protein in general need not coincide with reduction at the metal site.

Haemerythrin

The non-haeme protein haemerythrin is responsible for oxygen transport in certain invertebrate phyla. It has a molecular weight of 108,000, comprising eight sub-units, each containing two iron atoms which can bind one mole of oxygen, possibly in a bridging manner.

Deoxyhaemerythrin shows only one environment with a sharp quadrupole splitting at all temperatures above 4·2 K and from the chemical isomer shift is clearly in the high-spin iron(II) configuration [31]. Parameters are given in Table 13.4.

Table 13.4 Mössbauer parameters for haemerythrin derivatives [31]

Compound	S	T/K	Δ /(mm s^{-1})	δ (Fe) /(mm s^{-1})
deoxyhaemerythrin	2	195	2·75	1·11
		77	2·81	1·19
		4·2	2·89	1·20
thiocyanatohaemerythrin	$\frac{5}{2}$	77	1·81	0·52
aquohaemerythrin	$\frac{5}{2}$	77	1·57	0·46
oxyhaemerythrin	$\frac{5}{2}$	77	$\begin{cases}1·93\\1·03\end{cases}$	0·51 / 0·48
		4·2	$\begin{cases}1·92\\1·09\end{cases}$	0·54 / 0·51

The iron(III) thiocyanato- and aquo-haemerythrin derivatives both contain only a single iron environment, despite the addition of only one thiocyanate per two iron atoms in the former. The simple quadrupole split spectrum is typical of a pair of coupled $S = \frac{5}{2}$ spins as found for example in the [Fe(salen)Cl]$_2$ and [Fe(salen)$_2$]O systems already discussed in Chapter 6. The diamagnetism of the coupled system was confirmed by the lack of magnetic splitting in a 5-kG field.

Oxyhaemerythrin cannot be prepared pure, and two high-spin ferric quadrupole splittings were found whose relative intensity varied with the treatment of the sample. Again, however, there was considerable evidence in favour of two iron atoms coupled by an oxygen bridging unit to cause overall diamagnetism.

[*Refs. on p. 369*]

Nucleotides

Metals such as iron appear to play an essential part in the action of such materials as ribonucleic acid (RNA), but as yet their precise role is not clear. Some preliminary studies of the iron complexes with nucleotides have been made [32, 33]. The iron(II) and iron(III) complexes of adenosine mono-, di-, and tri-phosphate for example have been studied in solutions at various pH values by freezing the solutions [33]. The iron(III) spectra show no quadrupole splitting at low pH, but a detectable splitting of $\sim$0·6 mm s^{-1} at high pH. It is presumed that at low pH the structure is a chelate involving only the phosphate moieties of the adenosine nucleotide. At higher pH the ring nitrogen atoms can become effective ligands. The iron(II)-adenosine tri-phosphate complex appears to suffer partial oxidation as the pH is raised.

Nitrogen Fixation

Studies of the iron-bearing non-haeminic protein found in the cells of the nitrogen-fixing bacteria *Azotobacter vinelandii* have shown that the spectrum changes considerably if the bacteria are incubated under nitrogen instead of argon [34, 35]. It thus appears that the iron plays a major role in the fixation process.

In summary, it can be said that Mössbauer spectroscopy has already proved of considerable use in the study of active centres in proteins, complementing the data obtained from e.s.r. and other measurements, and that it shows every promise for major application in the near future.

REFERENCES

[1] A. J. Bearden, T. H. Moss, W. S. Caughey, and C. A. Beaudreau, *Proc. Natl. Acad. Sci., U.S.A.*, 1965, **53**, 1246.
[2] L. M. Epstein, D. K. Straub, and C. Maricondi, *Inorg. Chem.*, 1967, **6**, 1720.
[3] U. Gonser and R. W. Grant, *Biophys. J.*, 1965, **5**, 823.
[4] C. I. Wynter, P. Hembright, C. H. Cheek, and J. J. Spijkerman, *Nature*, 1967, **216**, 1105.
[5] C. E. Johnson, *Phys. Letters*, 1966, **21**, 491.
[6] M. Blume, *Phys. Rev. Letters*, 1967, **18**, 305.
[7] T. H. Moss, A. J. Bearden, and W. S. Caughey, *J. Chem. Phys.*, 1969, **51**, 2624.
[8] I. A. Cohen, *J. Amer. Chem. Soc.*, 1969, **91**, 1980.
[9] U. Gonser, R. W. Grant, and J. Kregzde, *Science*, 1964, **143**, 680.
[10] G. Lang and W. Marshall, *Proc. Phys. Soc.*, 1966, **87**, 3.
[11] G. Lang and W. Marshall, *J. Mol. Biol.*, 1966, **18**, 385.
[12] G. Lang, *J. Appl. Phys.*, 1967, **38**, 915.
[13] G. Lang, *Phys. Letters*, 1968, **26A**, 223.
[14] E. Bradford and W. Marshall, *Proc. Phys. Soc.*, 1966, **87**, 731.
[15] M. Weissbluth and J. E. Maling, *J. Chem. Phys.*, 1967, **47**, 4166.
[16] H. Eicher and A. Trautwein, *J. Chem. Phys.*, 1969, **50**, 2540.

[17] W. S. Caughey, W. Y. Fujimoto, A. J. Bearden, and T. H. Moss, *Biochemistry*, 1966, **5**, 1255.
[18] Y. Maeda, T. Higashimura, and Y. Morita, *Biochem. Biophys. Res. Comm.*, 1967, **29**, 362.
[19] Y. Maeda, *J. Phys. Soc. Japan*, 1968, **24**, 151.
[20] Y. Maeda and Y. Morita, *Biochem. Biophys. Res. Comm.*, 1967, **29**, 680.
[21] R. Cooke and P. Debrunner, *J. Chem. Phys.*, 1968, **48**, 4532.
[22] G. Lang, D. Herbert, and T. Yonetani, *J. Chem. Phys.*, 1968, **49**, 944.
[23] T. H. Moss, A. J. Bearden, R. G. Bartsch, and M. A. Cusanovich, *Biochemistry*, 1968, **7**, 1583.
[24] G. Lang, T. Asakura, and T. Yonetani, *J. Phys. C*, 1969, Ser. 2, **2**, 2246.
[25] H. H. Wickman, M. P. Klein, and D. A. Shirley, *Phys. Rev.*, 1966, **152**, 345.
[26] C. E. Johnson and D. O. Hall, *Nature*, 1968, **217**, 446.
[27] T. H. Moss, A. J. Bearden, R. G. Bartsch, M. A. Cusanovich, and A. S. Pietro, *Biochemistry*, 1968, **7**, 1591.
[28] C. E. Johnson, E. Elstner, J. F. Gibson, B. Banfield, M. C. W. Evans, and D. O. Hall, *Nature*, 1968, **220**, 1291.
[29] D. C. Blomstrom, E. Knight, W. D. Phillips, and J. F. Weiher, *Proc. Natl. Acad. Sci., U.S.A.*, 1964, **51**, 1085.
[30] C. E. Johnson, R. C. Bray, and P. F. Knowles, *Biochem. J.*, 1967, **103**, 10C.
[31] M. Y. Okamura, I. M. Klotz, C. E. Johnson, M. R. C. Winter, and R. J. P. Williams, *Biochemistry*, 1969, **8**, 1951.
[32] R. A. Stukan, A. N. Ilina, Yu. Sh. Moshkovskii, and V. I. Goldanskii, *Biofizika*, 1965, **10**, 343.
[33] I. N. Rabinowitz, F. F. Davis, and R. H. Herber, *J. Amer. Chem. Soc.*, 1968, **88**, 4346.
[34] Yu. Sh. Moshkovskii, Y. D. Ivanov, R. A. Stukan, G. T. Makhanov, S. S. Mardanyan, Yu. M. Belov, and V. I. Goldanskii, *Doklady Akad. Nauk S.S.S.R.*, 1967, **174**, 215.
[35] G. V. Novikov, L. A. Syrtsova, G. I. Likhtenstein, V. A. Trukhtanov, V. F. Pachek, and V. I. Goldanskii, *Doklady Akad. Nauk S.S.S.R.*, 1968, **181**, 1170.

14 | Tin-119

The total quantity of data available on the ^{119}Sn Mössbauer resonance is now quite considerable, but the application of this resonance has not had quite the same impact on chemistry as has that of ^{57}Fe. In the latter isotope many of the effects observed result from the sensitivity of the $3d$-orbitals to the chemical environment. However, these remain essentially non-bonding or antibonding in all but the most covalent compounds, and thus retain a degree of identity which allows full and successful use of ligand field theory. We have already seen in Chapter 9 that the results on diamagnetic covalent compounds of iron are more difficult to interpret in detail, and this is also true for most of the compounds of tin since our knowledge of the chemical binding involved is equally rudimentary. The tin atom does not carry an intrinsic spin moment in any of its compounds, and consequently magnetic hyperfine splitting is only found in rather rare circumstances. In the absence of large applied magnetic fields, therefore, this considerably limits the information obtainable.

It is also unfortunate that some of the early ^{119}Sn data proved to be of dubious reliability when compared with later, more accurate measurements, and this has caused considerable difficulties in the present review and correlation of the data. Nevertheless, the problems which have arisen in the interpretation have stimulated considerable interest in many of the aspects of tin chemistry, and have undoubtedly inspired more intensive study of chemical binding in heavy elements of the Main Groups of the Periodic Table.

14.1 γ-Decay Scheme and Sources

As shown in Fig. 14.1 the γ-ray transition used for Mössbauer work is the 23·875-keV decay [1, 2] from the first excited state. The precursor is metastable ^{119m}Sn which has a half-life of 250 days and can be prepared in adequate activity by neutron capture in isotopically enriched ^{118}Sn. Because the capture reaction also results in considerable production of ^{119}Sn in the ground state the preparation is not 'carrier-free' in terms of the resonant isotope. The 23·88-keV transition is a $\frac{3}{2} \rightarrow \frac{1}{2}$ magnetic dipole transition and as such may

[Refs. on p. 424]

be expected to show similar hyperfine properties to ^{57}Fe. The excited-state lifetime of $18\cdot3 \pm 0\cdot5$ ns [3] corresponds to a Heisenberg width of $\Gamma_r = 0\cdot626$ mm s^{-1}. In practice it is found that the velocity range needed for observation of hyperfine effects is about the same as that for ^{57}Fe, and therefore the larger half-width for the resonance line causes a proportionate reduction in the resolution which can be achieved in complex spectra.

The 23·88-keV γ-ray is rather highly internally converted ($\alpha_T = 5\cdot12$ [2], 5·13 [4]) but the basic decay scheme is uncomplicated and efficient counting can be achieved. The 65·66-keV γ-ray is even more strongly converted and is consequently weak in intensity; the resulting 25·04- and 25·27-keV X-rays can

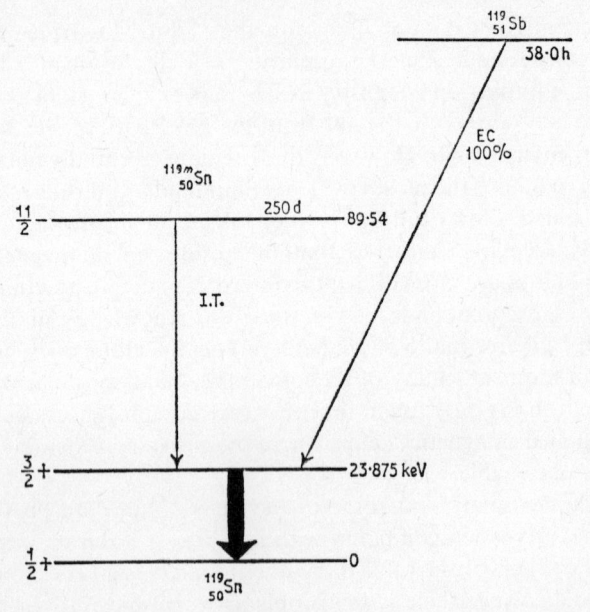

Fig. 14.1 The decay scheme for ^{119}Sn.

be preferentially absorbed by using a palladium filter. The latter has a K-edge for photoelectric absorption of 24·35 keV, intermediate between the unwanted X rays and the required 23·88-keV γ-ray.

The nuclear parameters determining the probability of a recoilless event are closely matched to those of ^{57}Fe, although the preponderance of organometallic compounds in tin chemistry has meant that most measurements are made at liquid nitrogen temperature because of the low effective Debye temperature of these materials. It is not, however, difficult to obtain a source with an adequate recoilless fraction for use at room temperature.

As already discussed in Chapter 12, it is not unusual for anomalous charge states to be produced following nuclear transitions involving electron

[Refs. on p. 424]

capture. Fortunately, this complication is not important in ^{119m}Sn. In principle the internal conversion of the preceding 65·66-keV isomeric transition might be expected to produce such charge states in the subsequent Auger cascade, but in only one case has this effect been observed: the oxalate K$_6$^{119m}Sn^{IV}$_2$(C$_2$O$_4$)$_7$.4H$_2$O shows substantial tin(II) content in the spectrum obtained using this compound as a source and barium stannate as the resonant absorber. This is illustrated in Fig. 14.2, which includes spectra for

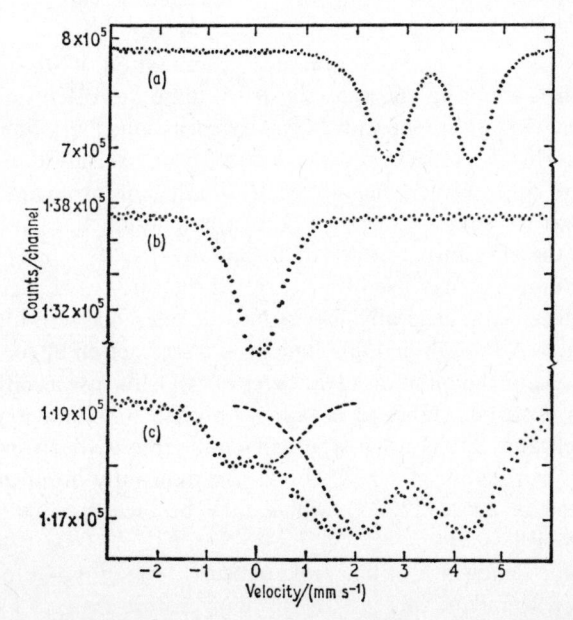

Fig. 14.2 Mössbauer spectra of (a) SnC$_2$O$_4$ and (b) K$_6$Sn$_2$(C$_2$O$_4$)$_7$.4H$_2$O absorbers with Ba^{119m}SnO$_3$ source; (c) K$_6$^{119m}Sn$_2$(C$_2$O$_4$)$_7$.4H$_2$O source and BaSnO$_3$ absorber. Note the formation of both tin(II) and tin(IV) in the oxalate source. [Ref. 5]

absorbers of K$_6$Sn^{IV}$_2$(C$_2$O$_4$)$_7$.4H$_2$O and Sn^{II}C$_2$O$_4$ obtained with a barium stannate source for comparison [5]. Electronic changes in the oxalate ligands result from photochemical processes associated with the internal conversion which precedes the Mössbauer event; this results in the decomposition of the oxalate ions to carbon dioxide and the reduction of Sn(IV) to Sn(II).

An alternative source preparation which has not been widely adopted is the ^{118}Sn(d, n)^{119}Sb reaction using 10-MeV deuterons in a cyclotron [6]. The ^{119}Sb decays by electron capture with a 38-hour half-life to the 23·88-keV level as shown in Fig. 14.1. 25-keV K-X-rays are also produced in the electron-capture process, so that in most respects the decay is similar to the ^{119m}Sn transition.

Historically, the ^{119}Sn resonance was first reported in 1960 by Barloutaud et al. [7] for a source and absorber of β-tin at liquid nitrogen temperature. They used the ^{119m}Sn precursor, and their early studies included the detection of the recoilless Rayleigh scattering of the Mössbauer γ-rays by non-resonant materials such as aluminium [8, 9].

Several source matrices have subsequently found widespread use, and their chemical isomer shifts with respect to SnO_2 are:

SnO_2	0·00 mm s^{-1}	Mg_2Sn	1·82 mm s^{-1}
$BaSnO_3$	0·00	α-Sn (grey)	2·00
Pd/Sn (3%)	1·46	β-Sn (white)	2·56

These figures are a personal selection after examination of all the data available, and values differing substantially from these are often quoted. As an example, the SnO_2–β-Sn separation has been assigned a value as high as 2·70 mm s^{-1}. This is a reflection of inadequate standardisation in the velocity calibration of equipment which is still a significant problem. The recent tendency towards using ^{57}Co/iron foil as a magnitude/linearity calibration is improving the absolute accuracy of ^{119}Sn data.

The two source matrices used for nearly all the early work were SnO_2 and β-Sn, the former being generally adopted as the reference zero for the chemical isomer shift. Although giving a high recoilless fraction at room temperature ($\sim$0·5), one of the main disadvantages of SnO_2 for use as either a source matrix or an absorber standard is that its precise stoichiometry is difficult to define accurately. It is an n-type semiconductor with an excess of tin. SnO_2 roasted in vacuum at 600°C shows a significantly broader line than a sample roasted in air at 1200°C, which may be presumed to be closer to stoichiometry [10].

The resonance linewidth of a sample of SnO_2 (extrapolated to zero thickness) has been found to increase in the temperature range 78–645 K [11], although there was no evidence for a unique quadrupole splitting which in any case would be unlikely to decrease with rise in temperature. This phenomenon is also likely to derive from changes in stoichiometry. The second-order Doppler shift approaches a limiting value of $3·5 \times 10^{-4}$ mm s^{-1} per degree K in agreement with prediction.

Claims for the conclusive detection of a unique quadrupole splitting in SnO_2 have been made several times [12, 13], and one can certainly be generated under high pressure [14], but if such a splitting does exist it is on the limit of experimental resolution as determined by the Heisenberg width. The discrepancies reported in the literature must be due in part to differences in the stoichiometry of the samples used.

Both α-tin [15] and β-tin [16] have been used as sources, the latter giving a broader line because of a small unresolved quadrupole interaction, but both suffer from a very low recoil-free fraction at room temperature ($f = 0·039$ for β-tin).

Mg_2Sn has been occasionally used as a source but also suffers from a rather

[Refs. on p. 424]

low recoil-free fraction (0·28 at 295 K, rising to 0·77 at 77 K) [17], which necessitates either cooling of the source or long counting times. However, the linewidth for an $Mg_2{}^{119m}Sn/Mg_2Sn$ experiment extrapolated to zero absorber thickness is 0·68 mm s^{-1}, compared with a natural linewidth of 0·63 mm s^{-1}.

A source matrix which shows no line broadening due to unresolved quadrupole splitting and which gives a large recoil-free fraction at room temperature is barium stannate, $BaSnO_3$ [18], and this is rapidly becoming the most popular source for tin Mössbauer spectroscopy. The f-fraction is 0·6 at 293 K and 0·46 at 690 K. The source linewidth is close to the natural width [19]. A method of preparation has been detailed [19], and the material can also be used with high efficiency in a resonant counter [18].

Another useful source can be made using a 3% Sn–97% Pd alloy [20]. The latter has a face-centred cubic structure which places the tin atoms in a cubic environment. The linewidth is consequently close to natural, and the recoil-free fraction is ~0·42 at 297 K.

Although the conventional experimental technique uses a transmission geometry and a scintillation or proportional counter, spectra have also been recorded using counting of the internal conversion electrons in a double-lens β-spectrometer [21]; the source and absorber matrices were β-tin.

A resonance scintillation counter in which the absorber is incorporated into a plastic scintillator (thereby enabling the conversion electrons produced following Mössbauer absorption to be counted), was developed for ^{119}Sn work (see Chapter 3). It has not, however, been widely used, and the difficulties in application have been outlined [22]. The method does have the significant advantage of giving a linewidth of somewhat less than the Heisenberg width, thereby conferring high resolution in cases where it can be applied.

14.2 Hyperfine Interactions

The Sign and Magnitude of $\delta R/R$

The observed range of chemical isomer shifts for the ^{119}Sn resonance in compounds and alloys of tin falls within the limits of about $-0·5$ to $+4·5$ mm s^{-1} with respect to SnO_2. Shifts of greater than 2·9 mm s^{-1} are diagnostic for tin in oxidation state II, while tin(IV) compounds give shifts below 2·0 mm s^{-1}. (The term 'tin(IV)' as used here also embraces 4-coordinate tin compounds which formally have lower oxidation states because of the presence of tin–tin bonds, e.g. $Ph_3SnSnPh_3$, which is a 4-coordinate tin(III) compound, and the cyclic compounds $(Ph_2Sn)_x$, which are again 4-coordinate but contain tin in the formal oxidation state II.) Metals and alloys fall in the region 1·3–3·0 mm s^{-1}. The characteristic difference between tin(II) and tin(IV) has already been illustrated in Fig. 4.2.

[Refs. on p. 424]

As already discussed in some detail in Chapter 3, the magnitude of the chemical isomer shift depends not only on the values of the electron density at the tin nucleus for the compounds being compared, but also on the value of the fractional change in nuclear radius on excitation to the 23·88-keV level, $\delta R/R$. Determination of the sign and magnitude of the nuclear constant $\delta R/R$ has proved more difficult for ^{119}Sn than for ^{57}Fe because of the initial lack of accurate electronic wavefunctions for tin compounds. The various values which have been proposed are given in Table 14.1 in approximate chronological order [1, 23–29].

Table 14.1 Estimates for $\delta R/R$ in ^{119}Sn

$10^4 \, \delta R/R$	Method	Reference
+1·1	assumed ionic configurations	23 (1962)
+1·9	electronegativity of X in SnX$_4$	24 (1962)
−1·6	molecular-orbital calculations	25 (1965)
+3·3	experimental:	
	internal-conversion electrons	1 (1966)
+1·2	Hartree–Fock SCF calculations	26 (1967)
+0·92	estimation of $4d^{10}$ and $4d^{10}5s^2$ shifts	27 (1968)
+3·5	molecular-orbital calculations	28 (1968)
+3·6	molecular-orbital calculations	29 (1968)

The first estimate was made by approximating to the 5s-electron density with the Fermi–Segré equation and adopting configurations of $5s^0 5p^0$ and $5s^2 5p^0$ for SnF$_4$ and SnCl$_2$ respectively [23]. The value derived was $+1·1 \times 10^{-4}$, which is smaller than the currently accepted value by a factor of three. Similar consideration of the tin(IV) compounds SnI$_4$, SnBr$_4$, SnCl$_4$, SnO$_2$, and SnF$_4$ was used to obtain $\delta R/R = +1·9 \times 10^{-4}$ [24]. In this series there is an approximately linear correlation between the chemical isomer shift and the electronegativity differences between tin and the other element in each compound, but this is not as regular as was previously supposed. Deviations no doubt reflect the substantial variation in bond type throughout the series.

Although one might naively conclude that an increase in 5s-occupation would lead to an increase in $|\psi_s(0)|^2$, it has been pointed out that this need not necessarily be the case [30]. If there is also a concomitant radial expansion of the 5s-orbital due to an increase in covalency in the bonding, it is also feasible that the net result might be the reverse effect. This postulate was put forward for ^{119}Sn using semi-quantitative reasoning on molecular-orbital principles, and would require that $\delta R/R < 0$ in order to interpret the chemical isomer shift of SnCl$_4$, SnBr$_4$, and SnI$_4$ [25]. However, approximations are necessarily involved in the calculations, and the result is determined by a partial cancellation of opposed terms, so that the sign of $\delta R/R$ derived

[Refs. on p. 424]

is controlled by the numerical values adopted for the various disposable parameters.

Controversies of this nature are best resolved by an independent determination which avoids using the same assumptions. Such a method was devised using data on the internal conversion properties of ^{119m}Sn. During γ-emission the probability that internal conversion will take place, causing expulsion of an extra-nuclear electron, is proportional to the probability of finding that electron at the nucleus, i.e. on the value of $|\psi_{ns}(0)|^2$. Therefore, only s-electrons are normally involved in internal conversion, and their kinetic energy signifies in which shell they originated. The intensities of the various possibilities decrease sharply in the order $K > L > M > N > O$... In general the total probability for conversion will be almost independent of chemical environment in the heavy elements because of the dominance of the inner closed shells, but in ^{119}Sn, for example, one might expect that the O conversion which involves the 5s-electrons might be dependent on environment.

Such an effect has been found by comparing the intensities of the N and O conversion lines from ^{119}Sn in β-Sn and SnO$_2$ [1]. Conversion in the O shell is about 30% smaller in SnO$_2$ than in white tin, thereby implying a similar decrease in $|\psi_{5s}(0)|^2$. The total value of $\sum_{n=1}^{5}|\psi_{ns}(0)|^2$ must be calculated using non-relativistic Hartree–Fock wave functions to allow for shielding of the inner shells by the 5s-electrons, and this reduces the value to 16%. Taking this in conjunction with the smaller chemical isomer shift of SnO$_2$ as compared to β-Sn enables $\delta R/R$ to be calculated as $+3\cdot 3 \times 10^{-4}$. The possible error is of the order of 30%, but the sign is established unequivocally as positive.

The earlier molecular-orbital work was then revised to conform with this value, although the possibility of shielding effects giving the opposite chemical shift to that anticipated in a *particular* series of compounds still cannot be ruled out [31].

A value of $+0\cdot 92 \times 10^{-4}$ was derived by Lees and Flinn who compared the difference in $|\psi_s(0)|^2$ for the $4d^{10}$(Sn^{4+}) and $4d^{10}5s^2$(Sn^{2+}) atomic configurations with the chemical isomer shift values of K$_2$SnF$_6$ (assumed to be wholly ionic), α-tin (assumed to be precisely $5s^15p^3$) and an estimated value for the chemical isomer shift of the completely ionic Sn^{2+} configuration, which was taken to be 5·22 mm s^{-1} higher than that for K$_2$SnF$_6$ [27]. However, the assumed configurations cannot be adequately justified on chemical grounds; e.g. α-tin is not restricted to an s-orbital occupation of unity by its tetrahedral coordination and more detailed calculations [28] suggest a configuration $5s^{1\cdot 2}5p^{2\cdot 8}$.

The first of a series of more extensive molecular-orbital calculations used Hartree–Fock self-consistent field calculations, thereby deriving a value of

$\delta R/R = +1\cdot 2 \times 10^{-4}$ [26], although the values of the chemical isomer shift established as reference points were basically those of Lees and Flinn and thus subject to the same criticisms.

Probably the most significant orbital calculations at the present time are those which were made by the Pople–Segal–Santry self-consistent molecular-orbital method using the complete set of 5s-, 5p-, and 5d-orbitals of tin [28]. The occupancies of the 5s-, 5p-, and 5d-orbitals were obtained, and the 5s-electron density at the nucleus calculated from the Fermi–Segré equation (using a modified form of Burn's screening rules) for the assumed tetrahedral

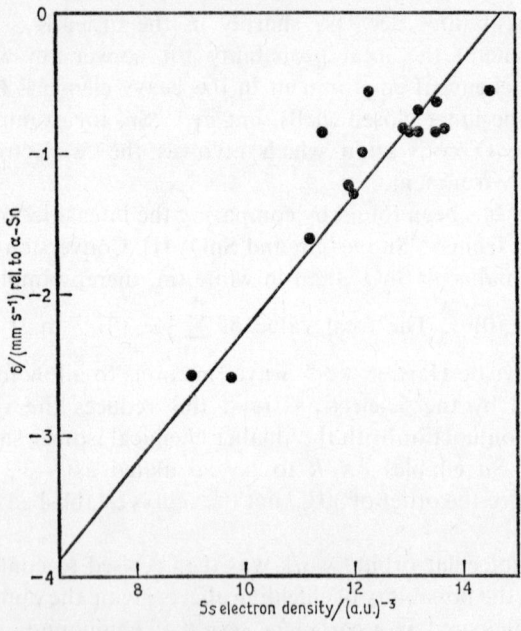

Fig. 14.3 The chemical isomer shift (relative to α-tin) plotted against the estimated 5s-electron density at the nucleus for a series of tin(IV) compounds. The slope of the line signifies that $\delta R/R$ is positive. [Ref. 28, Fig. 2]

species $SnCl_4$, $SnBr_4$, SnI_4, SnH_4, $SnMe_4$, $SnClMe_3$, $SnBrMe_3$, $SnIMe_3$, $SnCl_2Me_2$, SnH_2Me_2, SnH_3Me, and the octahedral species SnF_6^{2-}, $SnCl_6^{2-}$, $SnBr_6^{2-}$, and SnI_6^{2-}. The experimental chemical isomer shift shows an approximately linear relationship with both the 5s-occupation number and the derived 5s-electron density at the nucleus as illustrated in Fig. 14.3. It is interesting to note that these results imply a direct proportionality of the 5s-electron density at the nucleus and the occupation number in tin compounds. As already noted α-Sn has the configuration $5s^{1\cdot 25}5p^{2\cdot 8}$, and SnF_6^{2-} has the configuration $5s^{0\cdot 75}5p^{1\cdot 3}5d^{0\cdot 1}$ in contrast with the configurations $5s^1 5p^3$ and $5s^0 5p^0$ used in earlier, more qualitative arguments. The value of

[Refs. on p. 424]

$\delta R/R$ derived was $+3\cdot 5 \times 10^{-4}$, which agrees well with the 'experimental' value derived from internal conversion data. The doubtful validity of using the Fermi–Segré equation in this context has since been circumvented by direct conversion of the orbital occupation numbers to electron densities at the nucleus using Hartree–Fock wavefunctions, although the value obtained for $\delta R/R$ remains virtually unaffected at $+3\cdot 6 \times 10^{-4}$ [29].

An alternative approach using a special type of LCAO molecular-orbital wavefunction proved to be less successful [32]. In both cases calculations are hampered by lack of adequate structural and wavefunction data. Analysis of the chemical isomer shifts in tin(II) compounds has not yet been successfully achieved by such calculations.

It is interesting to note an apparent anomaly in the data for ferroelectric barium titanate doped with either iron or tin. The lattice distortion on going to the ferroelectric phase affects the ^{57}Fe resonance, the increase in chemical isomer shift being interpreted as a decrease in the s-electron density at the iron nucleus [33]. By contrast the corresponding increase in the chemical isomer shift which is seen in the ^{119}Sn spectrum, implies an increase in the s-electron density at the tin nucleus [34]. Although one might naively be tempted to assume that the ferroelectric transition would influence the s-electron density at the iron and tin nuclei in the same sense, thereby implying that $\delta R/R$ also had the same sign for the two nuclei, there is in fact no reason why the changes in chemical binding of tin and iron in the two phases should be even remotely similar, particularly in their influence on the s-electron density at the nucleus. The result cannot therefore throw serious doubt on the assignment of a positive value of $\delta R/R$ for ^{119}Sn.

Quadrupole Moment

Few attempts have been made to determine the excited-state quadrupole moment of ^{119}Sn, in marked contrast to the numerous experiments on ^{57}Fe discussed in Chapter 5. The only numerical estimate comes from the early work of Boyle *et al.* [23]. Tetragonal SnO can be partially oriented, and the resultant asymmetry in the quadrupole splitting shows that e^2qQ is positive and lies along the crystal 'c' axis. The structure is known, and the tin atom has four equidistant oxygen neighbours which form the corners of a square, the tin itself being at the apex of a square pyramid. These observations were used to argue that any unequal occupancy of the p-orbitals would involve only the p_z-orbital participating in the 'lone-pair' of electrons so that Q is negative in sign with a value of $\sim -0\cdot 08$ barn. However, our present knowledge of the bonding in stannous compounds is still too empirical to confirm this result analytically, and confirmatory evidence would be welcomed, though the sign of Q may well be correct. The quadrupole splitting in tin compounds is expected to be largely a function of the unequal occupancy of the $5p$-orbitals resulting from covalent bonding in environments of low

[*Refs. on p. 424*]

symmetry. Comparatively few data are available for inorganic compounds and these have not been interpreted in detail. Several unusual features are found in organotin compounds, and it is convenient to defer a full discussion of the quadrupole splitting in these compounds until Section 14.5.

Magnetic Hyperfine Interactions

Magnetic hyperfine splitting of the ^{119}Sn resonance is only found in matrices where there is a polarising field external to the tin atom. This can be either an external applied magnetic field or a large internal magnetic field at a nearby magnetic ion. In the latter case the field at the tin atom is produced via induced spin polarisation of the otherwise spin-paired molecular orbitals involved in covalent bonding.

Early attempts to observe magnetic splitting involved the use of external fields on non-magnetic metals, but gave poor results because of lack of resolution [35, 36]. The various estimates [36–44] of the excited-state magnetic moment made since then are given in Table 14.2, and were calculated assum-

Table 14.2 Estimates of the excited state magnetic moment, μ_e ($\mu_g = -1.046$ n.m.)

μ_e/(n.m.)	Method	Reference
$+0.78 \pm 0.08$	Mn$_2$Sn alloy	37
$+0.83 \pm 0.031$	Mn$_2$Sn alloy; α-Sn in 16-kG field	36
$+0.672 \pm 0.025$	^{119m}Sn/iron foil	38
$+0.75 \pm 0.04$	1·7% tin/iron alloy	39
$+0.67 \pm 0.01$	Ca$_{0.25}$Y$_{2.75}$Sn$_{0.25}$Fe$_{4.75}$O$_{12}$	40
$+0.68$	Co$_2$MnSn	42
$+0.70 \pm 0.02$	^{119m}Sn/Co metal	43
$+0.67 \pm 0.01$	Co$_2$MnSn	44

ing the value of -1.046 n.m. for the ground-state moment. The most probable value for μ_e appears to be $+0.67$ n.m., obtained from a particularly well-resolved spectrum (see Fig. 14.9) [40, 41]. A typical determination used ^{119m}Sn doped into an iron foil [38]. Magnetisation of the foil both parallel and perpendicular to the direction of the γ-ray allows identification of the various components of the spectrum (Fig. 14.4). The large value of the ground-state moment causes significantly larger line splitting for the observed 78·5-kG internal field at 283 K than would be found in ^{57}Fe.

Combined Quadrupole/Magnetic Interactions

Although intrinsic magnetic splitting is rarely found in the ^{119}Sn resonance, the comparatively large magnetic moments of this isotope result in a large splitting from a field of the order of 50 kG. The applied magnetic field method can therefore be used to determine the sign of e^2qQ in compounds with a resolved quadrupole splitting.

[Refs. on p. 424]

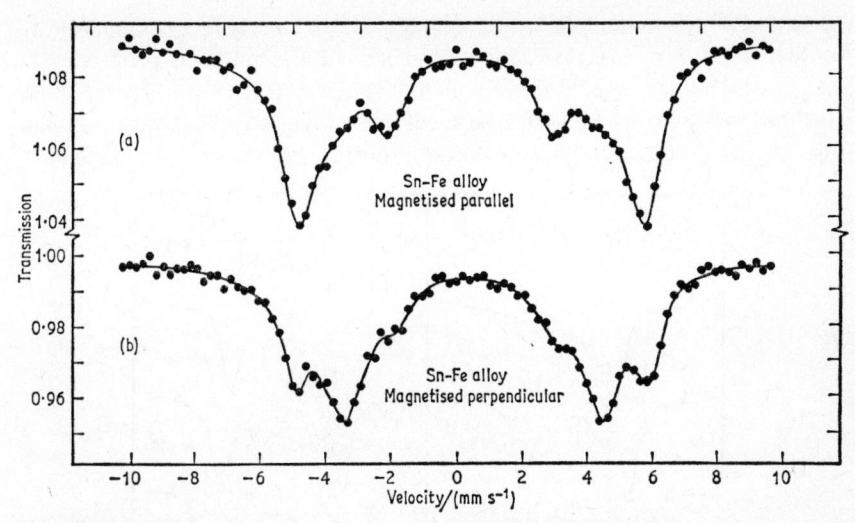

Fig. 14.4 Spectra of an ^{119m}Sn/iron alloy magnetised parallel and perpendicular to the γ-ray. [Ref. 38, Fig. 3]

In the case of ^{57}Fe the external field acts as a pertubation on the quadrupole splitting, but this is not necessarily true for ^{119}Sn so that the observed spectra are more complicated than the 'doublet–triplet' spectra of the former. In Figs. 14.5 and 14.6 are shown some typical computed curves for a random polycrystalline sample with e^2qQ positive in sign, $\eta = 0$, and H either parallel or perpendicular to the axis of observation [45]. Appreciable asymmetry is not seen for quadrupole splittings of less than 1 mm s^{-1} and fields of less than 30 kG, so that the sign of e^2qQ is not determined under these conditions. The asymmetry parameter has almost no effect for values of η less than 0·6 and can only be determined in rare circumstances.

Under the assumption that Q is negative in sign, one finds that e^2qQ is positive for p_z, d_{z^2}, d_{xz}, d_{yz} and negative for p_x, p_y, $d_{x^2-y^2}$, and d_{xy}. This interpretation has been used for several tin(II) and organotin(IV) compounds (see appropriate sections) and gives a satisfactory qualitative description of the bonding involved.

14.3 Tin(II) Compounds

The tin(II) oxidation state is comparatively unstable with respect to tin(IV), and relatively few compounds have been isolated. They are easily characterised in the Mössbauer spectrum by their high positive chemical isomer shifts with respect to SnO$_2$ ($\sim$2·3–4·1 mm s^{-1}). However, the detailed interpretation of stannous chemical isomer shifts has so far not been achieved.

Several interpretations have adopted the naive assumption that tin(II)

382 | TIN-119

chloride has an outer electronic configuration closely approximating to $5s^2$ [23, 46]. This is inconsistent with many of the physical properties of $SnCl_2$, which undoubtedly shows much covalent character, and in particular with the crystal structure [47]. The local environment of the tin atom comprises two chlorine atoms at 2·78 Å and a third at 2·66 Å, so that the effective

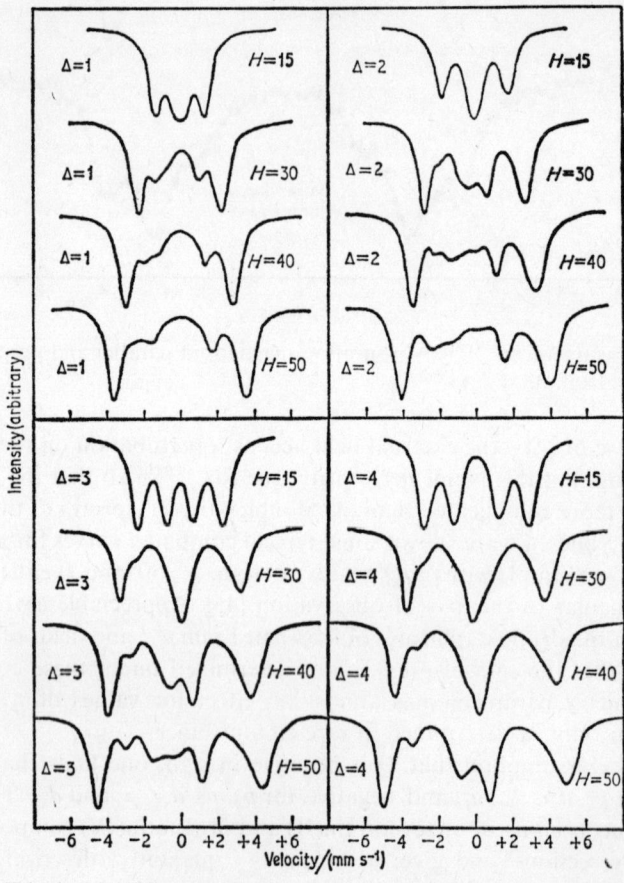

Fig. 14.5 Calculated ^{119}Sn spectra for selected values of the quadrupole splitting [Δ/(mm s^{-1}] and magnetic field [H/kG]. In all cases H is parallel to the direction of observation, $\eta = 0$, e^2qQ is positive, and the assumed linewidth is 0·8 mm s^{-1}. [Ref. 45, Fig. 1]

geometry is a chlorine-bridged pyramidal structure, although there are other chlorine neighbours at distances >3 Å.

On chemical grounds alone, a decrease in positive shift towards α-tin in a series of tin(II) compounds is compatible with an effective loss of 5s-electron density from the hypothetical $5s^2$ ion [48, 49]. The use of valence-bond theory and *s–p* hybridisation concepts give an approximate correlation

[*Refs.* on p. 424]

between the chemical isomer shift and the % sp^3 character in the bonding, but the anomalous positions of SnO and SnF_2 suggest that, in cases where the Sn–ligand bond distance is very short, additional factors such as s–p mixing caused by crystal-field effects are involved [48]. In this way it is possible to derive [48] a hypothetical shift for the $Sn^{2+}(5s^2)$ ion of 7·6 mm s^{-1} (rel. to

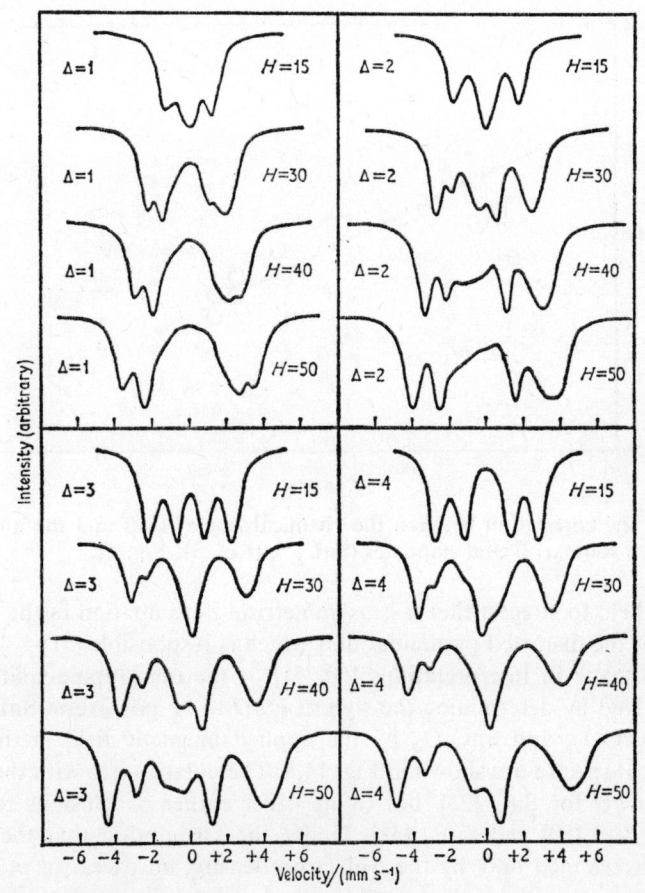

Fig. 14.6 The calculations of Fig. 14.5 repeated with H perpendicular to the direction of observation. [Ref. 45, Fig. 2]

SnO_2), which is probably more realistic than other lower values given earlier [50]. In this connection it may be noted that the suggested correlation between chemical isomer shift and quadrupole splitting [50] makes erroneous assumptions about the stereochemistry of tin(II) compounds. There is also no reason why one should assume a general relationship between s-electron density at the nucleus and the imbalance in the p-electron density in structurally unrelated types. However, if one considers the tri-ligand stannates

384 | TIN-119

(SnL$_3^-$) alone, such a relationship does appear to hold [51] and is illustrated in Fig. 14.7. Although it has been suggested that the observed quadrupole splitting can be related to the amount of *p*-character in the lone-pair orbital [23, 46] comparison of the known bond-lengths in SnF$_2$, Na$_2$Sn$_2$F$_5$, SnS, K$_2$SnCl$_4$.H$_2$O, SnSO$_4$, and SnCl$_2$ with the observed quadrupole splitting

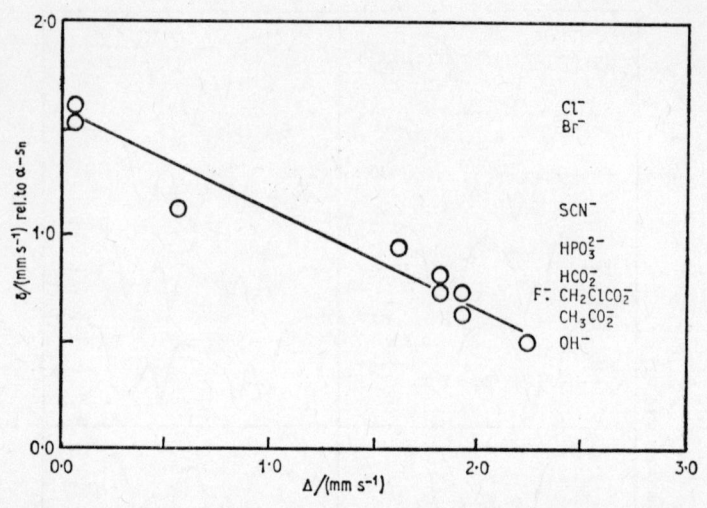

Fig. 14.7 The correlation between the chemical isomer shift and the quadrupole splitting for some tri-ligand stannates (SnL$_3^-$). [Ref. 51, Fig. 1]

has been held to suggest that it is asymmetrical *p*-occupation in the bonding orbitals of the distorted pyramidal unit which is responsible [51].

Inconsistency in interpretations [50, 51] of the quadrupole splitting has been resolved by determining the sign of e^2qQ to be positive in SnF$_2$, SnO, SnS, Sn$_3$(PO$_4$)$_2$, and SnC$_2$O$_4$ by the applied magnetic field method [52]. Two typical spectra are shown in Fig. 14.8. These data agree with the assignment of e^2qQ for SnO [23], but refute other earlier conclusions regarding SnO and SnS [50] and SnF$_2$ [51]. Under the assumption that the sign of e^2qQ is determined only by the *p*-electron density, an excess of p_z electron density over p_x, p_y is indicated in all five compounds. This leads to the further conclusion that the dominant contribution to the electric field gradient arises from appreciable p_z character in the non-bonding orbital of the tin(II).

Tin(II) Halides

Two crystalline modifications of tin(II) fluoride are known [53]. The orthorhombic form of SnF$_2$ shows significantly different Mössbauer parameters from the better-known monoclinic form. In particular the smaller chemical isomer shift in the former is indicative of increased covalent character, while the larger quadrupole splitting suggests a more distorted environment about

[Refs. on p. 424]

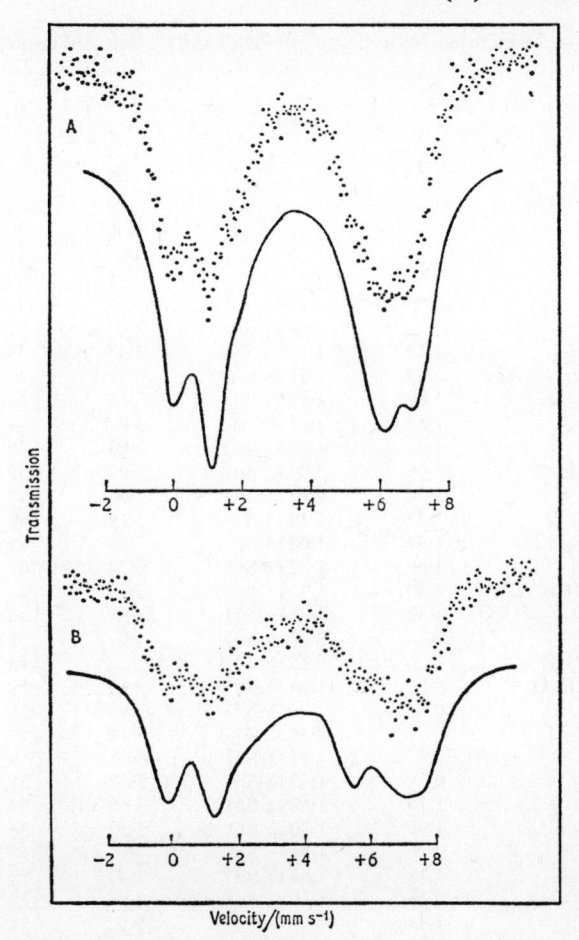

Fig. 14.8 ^{119}Sn spectra and simulated curves for (A) SnS at 4·2 K with a field of 50 kG applied perpendicular to the axis of observation, and (B) SnC$_2$O$_4$ under the same conditions. [Ref. 52]

the tin. The X-ray structure shows it to have the pyramidal coordination to three fluorine neighbours [54]. The Mössbauer data are in Table 14.3. Since the majority of the tin(II) data has come from a single laboratory, values from this source are quoted where possible to maintain self-consistency. The chemical isomer shifts are converted to SnO$_2$ using the values on p. 374.

The tin atoms in the pentafluorodistannate ion of the complex NaSn$_2$F$_5$ also adopt the distorted pyramidal geometry commonly found for tin(II) by having a bridging fluorine atom. This is at 2·22 Å from the tin compared to 2·07 Å for the two non-bridging fluorines [55]. This phenomenon of two short and one long bond also appears to be a characteristic feature of tin(II).

[*Refs. on p. 424*]

Table 14.3 Mössbauer parameters (80 K) of tin(II) halogen compounds

Compound	Δ /(mm s^{-1})	δ/(mm s^{-1})	δ (SnO$_2$) /(mm s^{-1})	Reference
SnF$_2$ (orthorhombic)	2·20	1·20 (α-Sn)	3·20	53
SnF$_2$ (monoclinic)	1·80	1·60 (α-Sn)	3·60	53
NH$_4$Sn$_2$F$_5$	1·94	1·28 (α-Sn)	3·28	49
NaSn$_2$F$_5$	1·86	1·27 (α-Sn)	3·27	49
KSn$_2$F$_5$	1·96	1·21 (α-Sn)	3·21	49
RbSn$_2$F$_5$	2·03	1·13 (α-Sn)	3·13	49
CsSn$_2$F$_5$	2·06	1·07 (α-Sn)	3·07	49
Sr(Sn$_2$F$_5$)$_2$	1·69	1·34 (α-Sn)	3·34	49
Ba(Sn$_2$F$_5$)$_2$	1·69	1·31 (α-Sn)	3·31	49
NH$_4$SnF$_3$	1·88	1·18 (α-Sn)	3·18	49
NaSnF$_3$	1·84	1·07 (α-Sn)	3·07	49
KSnF$_3$	1·92	1·02 (α-Sn)	3·02	49
RbSnF$_3$	1·96	0·97 (α-Sn)	2·97	49
CsSnF$_3$	2·00	0·93 (α-Sn)	2·93	49
Sr(SnF$_3$)$_2$	1·75	1·19 (α-Sn)	3·19	49
Ba(SnF$_3$)$_2$	1·87	1·08 (α-Sn)	3·08	49
PbF.SnF$_3$	1·66	1·22 (α-Sn)	3·22	49
Sr$_2$Sn$_2$NO$_3$F$_7$.2H$_2$O	1·30	1·35 (α-Sn)	3·35	56
Pb$_2$SnNO$_3$F$_5$.2H$_2$O	1·32	1·38 (α-Sn)	3·38	56
SnCl$_2$	0	2·07 (α-Sn)	4·07	58
SnCl$_2$.2H$_2$O	?	3·55 (SnO$_2$)	3·55	15
K$_2$SnCl$_4$.H$_2$O	<0·6	1·64 (α-Sn)	3·64	51
SnBr$_2$	0	1·93 (α-Sn)	3·93	58
SnBr$_2$.2H$_2$O	0	3·74 (SnO$_2$)	3·74	59
SnI$_2$	0	1·85 (α-Sn)	3·85	58
Sn(NCS)$_2$	0	1·42 (α-Sn)	3·42	58
SnClF	1·10	1·68 (α-Sn)	3·68	58
Sn$_2$ClF$_3$	1·29	1·30 (α-Sn)	3·30	58
SnBrF	0·96	1·54 (α-Sn)	3·54	58
Sn$_3$BrF$_5$	1·18	1·64 (α-Sn)	3·64	58
SnIF	0·76	1·56 (α-Sn)	3·56	58
Sn$_2$IF$_3$	1·34	1·52 (α-Sn)	3·52	58
Sn$_2$BrCl$_3$	0	1·94 (α-Sn)	3·94	58
SnICl	0·55	1·66 (α-Sn)	3·66	58
SnIBr	0	1·66 (α-Sn)	3·66	58
Sn$_2$(NCS)F$_3$	1·02	1·53 (α-Sn)	3·53	58
Sn$_2$(NCS)Cl$_3$	0	1·88 (α-Sn)	3·88	58
Sn$_2$(NCS)Br$_3$	0	1·67 (α-Sn)	3·67	58
Sn$_2$(NCS)I$_3$	0	1·69 (α-Sn)	3·69	58
SnF$_2$.py	1·80	1·14 (α-Sn)	3·14	62
SnCl$_2$.py	0·98	1·19 (α-Sn)	3·19	62
SnCl$_2$.2py	1·49	1·02 (α-Sn)	3·02	62
pyH.SnCl$_3$	1·07	1·02 (α-Sn)	3·02	62
SnBr$_2$.py	0·70	1·35 (α-Sn)	3·35	62
SnBr$_2$.2py	1·00	1·26 (α-Sn)	3·26	62
pyH.SnBr$_3$	<0·6	1·55 (α-Sn)	3·55	62
Sn(NCS)$_2$.2py	1·39	1·19 (α-Sn)	3·29	62
pyH.Sn(NCS)$_3$	0·90	1·18 (α-Sn)	3·38	62

[*Refs.* on p. 424]

The order of the chemical isomer shifts $SnF_3^- < Sn_2F_5^- < SnF_2$ implies an increasing 5s-density and decreasing sp^3 character in the order stated [49]. For both SnF_3^- and $Sn_2F_5^-$ complexes there is a small variation in the chemical isomer shift with the polarising power of the co-cation. As the polarising power of the cation rises the covalent bond between tin and fluorine is weakened and the chemical isomer shift rises. Data are summarised in Table 14.3.

The similarity of the spectra of the compounds $Sr_2Sn_2NO_3F_7.2H_2O$ and $Pb_2SnNO_3F_5.2H_2O$ with those of the SnF_3^- complexes suggests that the tin environment is similar and that both contain the SnF_3^- anion [56].

The anhydrous halides $SnCl_2$, $SnBr_2$, and SnI_2 probably have the same halide-bridged structure [57]. They show no significant quadrupole splitting and larger chemical isomer shifts than the fluoride [58]. Of the hydrated halides, $SnCl_2.2H_2O$ contains pyramidal molecules of $SnCl_2.H_2O$ as the structural unit [60], whereas in $K_2SnCl_4.H_2O$ one finds the $SnCl_3^-$ anion [61].

A considerable number of ternary (mixed) tin(II) halides have been identified [58]. The Mössbauer parameters (Table 14.3) and X-ray data suggest that they are structurally derived from the parent compounds with a similar type of bridging structure. In the case of the ternary fluorides, the bridging atom is probably fluorine.

The tin(II) halides form a number of stable 1:1 and 1:2 adducts with pyridine, as well as pyridinium tri-ligand stannates [62]. The $SnX_2.2py$ complexes (X = Cl, Br, NCS) lose one molecule of pyridine quite readily, and since there is only one empty p-acceptor orbital in molecular tin(II) complexes, this implies that it is either weakly bonded to the tin, or is present as lattice pyridine. The pyridine complexes and tri-ligand stannates show lower chemical isomer shifts (Table 14.3) than the parent halides because the replacement of the bridging anion by a unidentate ligand or anion causes an increase in the use of 5s-electrons in bonding. There is a large change in shift with formation of the 1:1 compound, but only a small decrease with addition of the second pyridine, consistent with the lability of the second group.

Tin(II) Compounds Containing Oxygen

Rather less information is available for the oxide derivatives of tin(II). The crystal structure of black, tetragonal SnO is known [63], and was referred to in Chapter 14.1 in the discussion of the nuclear quadrupole moment. The Mössbauer parameters are given in Table 14.4 together with those for SnS, which has a considerably distorted NaCl lattice [64], SnSe (isostructural with SnS) [65], and SnTe, which has a cubic NaCl lattice [66]. Application of high pressure to SnO causes the formation of some SnO_2 and tin metal [67]. A detailed lattice dynamical study of SnS between 60 and 320 K has shown evidence for a Karyagin effect [68].

[Refs. on p. 424]

Table 14.4 Tin(II) compounds bonding to oxygen

Compound	Δ /(mm s^{-1})	δ/(mm s^{-1})	δ (SnO$_2$) /(mm s^{-1})	Reference
SnO (black)	1·45	0·71 (α-Sn)	2·71	69
SnO (red)	2·20	0·60 (α-Sn)	2·60	69
SnS	0·8	1·16 (α-Sn)	3·16	51
SnSe	—	1·30 (α-Sn)	3·30	48
SnTe	—	1·21 (α-Sn)	3·21	48
NaSn(OH)$_3$	2·29	0·60 (α-Sn)	2·60	69
Ba[Sn(OH)$_3$]$_2$	2·22	0·49 (α-Sn)	2·49	69
Ba[Sn(OH)$_3$]$_2$.2H$_2$O	2·00	0·34 (α-Sn)	2·34	69
Sr[Sn(OH)$_3$]$_2$.2H$_2$O	2·00	0·56 (α-Sn)	2·56	69
Sr[Sn(OH)$_2$OSn(OH)$_2$]	1·85	0·45 (α-Sn)	2·45	69
Ba[Sn(OH)$_2$OSn(OH)$_2$]	1·86	0·41 (α-Sn)	2·41	69
5SnO.2H$_2$O	2·04	0·85 (α-Sn)	2·85	69
Sn$_3$O(OH)$_2$SO$_4$	2·00	0·57 (α-Sn)	2·57	69
Sn$_3$O$_4$SO$_4$	1·94	0·91 (α-Sn)	2·91	69
Sn$_3$(OH)$_4$(NO$_3$)$_2$	1·85	1·43 (α-Sn)	3·43	69
Sn$_4$(OH)$_6$Cl$_2$	1·59	0·81 (α-Sn)	2·81	69
Sn$_4$O$_3$Cl$_2$	1·46	1·09 (α-Sn)	3·09	69
SnSO$_4$	1·00	1·90 (α-Sn)	3·90	62
SnSO$_4$.py	1·07	1·42 (α-Sn)	3·42	62
(HCO$_2$)$_2$Sn	1·56	1·05 (α-Sn)	3·05	62
(HCO$_2$)$_2$Snpy	1·56	0·95 (α-Sn)	2·95	62
(HCO$_2$)$_2$Snpy$_2$	1·80	0·94 (α-Sn)	2·94	62
(MeCO$_2$)$_2$Sn	1·77	1·21 (α-Sn)	3·21	71
(MeCH$_2$CO$_2$)$_2$Sn	1·89	1·21 (α-Sn)	3·21	71
(MeCH$_2$CH$_2$CO$_2$)$_2$Sn*	1·86	1·20 (α-Sn)	3·20	71
(Me$_2$CHCO$_2$)$_2$Sn*	1·84	1·23 (α-Sn)	3·23	71
(Me$_3$CCO$_2$)$_2$Sn*	1·88	1·25 (α-Sn)	3·25	71
(Et$_2$CHCO$_2$)$_2$Sn*	1·86	1·24 (α-Sn)	3·24	71
(CH$_2$ClCO$_2$)$_2$Sn	1·66	1·10 (α-Sn)	3·10	71
(CHCl$_2$CO$_2$)$_2$Sn	1·64	1·38 (α-Sn)	3·38	71
(CCl$_3$CO$_2$)$_2$Sn*	1·78	1·19 (α-Sn)	3·19	71
(CH$_2$FCO$_2$)$_2$Sn	1·76	1·00 (α-Sn)	3·00	71
(CHF$_2$CO$_2$)$_2$Sn	1·75	1·15 (α-Sn)	3·15	71
(CF$_3$CO$_2$)$_2$Sn*	1·76	1·06 (α-Sn)	3·06	71
(MeCHClCO$_2$)$_2$Sn*	1·70	1·12 (α-Sn)	3·12	71
(CH$_2$ClCH$_2$CO$_2$)$_2$Sn*	1·83	1·17 (α-Sn)	3·17	71
KSn(HCO$_2$)$_3$	1·95	1·03 (α-Sn)	3·03	71
RbSn(HCO$_2$)$_3$	1·95	0·95 (α-Sn)	2·95	71
CsSn(HCO$_2$)$_3$	1·86	0·80 (α-Sn)	2·80	71
NH$_4$Sn(HCO$_2$)$_3$	1·95	0·92 (α-Sn)	2·92	71
KSn(MeCO$_2$)$_3$	2·02	0·73 (α-Sn)	2·73	71
RbSn(MeCO$_2$)$_3$	1·83	0·90 (α-Sn)	2·90	71
CsSn(MeCO$_2$)$_3$	1·75	0·97 (α-Sn)	2·97	71
NH$_4$Sn(MeCO$_2$)$_3$	1·75	0·74 (α-Sn)	2·74	71
Mg[Sn(MeCO$_2$)$_3$]$_2$	1·94	0·80 (α-Sn)	2·80	71
Ca[Sn(MeCO$_2$)$_3$]$_2$	2·03	0·80 (α-Sn)	2·80	71
Sr[Sn(MeCO$_2$)$_3$]$_2$	1·94	0·64 (α-Sn)	2·64	71
Ba[Sn(MeCO$_2$)$_3$]$_2$	1·87	0·77 (α-Sn)	2·77	71
KSn(CH$_2$ClCO$_2$)$_3$	1·97	0·84 (α-Sn)	2·84	71

[*Refs.* on p. 424]

Table 14.4 (continued)

Compound	Δ /(mm s⁻¹)	δ/(mm s⁻¹)	δ (SnO₂) /(mm s⁻¹)	Reference
RbSn(CH₂ClCO₂)₃	1·90	0·76 (α-Sn)	2·76	71
CsSn(CH₂ClCO₂)₃	1·93	0·75 (α-Sn)	2·75	71
NH₄Sn(CH₂ClCO₂)₃	1·88	0·79 (α-Sn)	2·79	71
Ca[Sn(CH₂ClCO₂)₃]₂	1·99	0·86 (α-Sn)	2·86	71
Sr[Sn(CH₂ClCO₂)₃]₂	1·96	0·73 (α-Sn)	2·73	71
Ba[Sn(CH₂ClCO₂)₃]₂	1·92	0·75 (α-Sn)	2·75	71
KSn(CH₂FCO₂)₃	1·97	0·81 (α-Sn)	2·81	71
NH₄Sn(CH₂FCO₂)₃	1·93	0·74 (α-Sn)	2·74	71
SnHPO₃	1·60	1·10 (α-Sn)	3·10	72
Na₄Sn(HPO₃)₃	1·54	0·67 (α-Sn)	2·67	72
K₄Sn(HPO₃)₃	1·74	1·05 (α-Sn)	3·05	72
Rb₄Sn(HPO₃)₃	1·80	0·83 (α-Sn)	2·83	72
Cs₄Sn(HPO₃)₃	1·70	0·70 (α-Sn)	2·70	72
(NH₄)₄Sn(HPO₃)₃	1·70	0·89 (α-Sn)	2·89	72
o-phenylenedioxytin(II)	1·76	2·95 (SnO₂)	2·95	73
3-Me-1,2-phenylenedioxytin(II)	1·89	3·13 (SnO₂)	3·13	73
2,2′-diphenylenedioxytin(II)	1·98	3·13 (SnO₂)	3·13	73
2,3-naphthalenedioxytin(II)	1·82	3·08 (SnO₂)	3·08	73

* Frozen solutions.

Addition of hydroxide ion to solutions of tin(II) compounds precipitates a basic salt in which extra hydroxide replaces the retained anion to give hydrous tin oxide, itself soluble in excess of alkali to yield trihydroxostannates (II) [69]. The predominant species in solutions at low pH is known to be $Sn_3(OH)_4^{2+}$. The chemical isomer shifts for the trihydroxostannates(II) are the lowest values observed for tin(II) compounds, and at the same time are associated with some of the largest quadrupole splittings. The compounds are similar in some respects to the trifluorostannates(II). The compounds $Sn_3(OH)_4(NO_3)_2$, $Sn_3O(OH)_2SO_4$, and $Sn_4(OH)_6Cl_2$ give very different spectra and obviously do not contain the same structural unit about the tin atoms.

Tin(II) sulphate, $SnSO_4$, has a complex structure in which the tin has twelve oxygen nearest neighbours, although the three shortest interatomic distances are in a pyramidal arrangement [70]. It has a very large chemical isomer shift, comparable with those of the dihalides, and a moderate quadrupole interaction (1·00 mm s⁻¹).

The tin(II) carboxylates $(RCO_2)_2Sn$ all show a quadrupole splitting in the range 1·6–1·9 mm s⁻¹ with a chemical isomer shift of 3·0–3·3 mm s⁻¹ [71]. Other chemical evidence favours identical structures for the series with a 3-coordinated distorted pyramidal tin environment, probably involving intermolecular carboxylate bridging with one bond longer than the other two. This would also be compatible with the large quadrupole splitting.

[Refs. on p. 424]

The tricarboxylatostannates(II) show smaller chemical isomer shifts than the normal salts as indeed do the fluorides, chlorides, and phosphites [71]. The parameters show a small dependence on the cation. Once again a distorted pyramidal structure is thought probable.

The phosphite complexes $M_4Sn(HPO_3)_3$ (M = Na, K, Rb, Cs, and NH_4) also give values compatible with a distorted pyramidal geometry [72].

Spectra have been given for four tin(II) complexes with oxygen heterocyclic ligands, and the parameters (summarised in Table 14.4) are typical of the oxidation state [73], although definite structural data are lacking.

14.4 Inorganic Tin(IV) Compounds

The chemistry of tin(IV) is dominated by the organometallic derivatives discussed in the next section. The inorganic compounds fall into two main classes: (a) the tetrahalides and their derivatives, and (b) oxide phases including nonstoichiometric compounds.

Tin(IV) Halides

Of the simple tetrahalides, SnF_4 is exceptional in that the tin atom is 6-coordinated, being surrounded by two *trans*-non-bridging and four bridging fluorines [74]. The non-cubic symmetry results in a large quadrupole splitting (see Table 14.5). The other halides, $SnCl_4$, $SnBr_4$, and SnI_4 are 4-coordinate in the solid state and, although the structure of $SnBr_4$ shows the four bromine atoms to be inequivalent [75], all three compounds show an essentially unsplit resonance line at 80 K. As can be seen from the figures in Table 14.5, the chemical isomer shift increases as the electronegativity of the halogen decreases, and as detailed in Section 14.2, the nearly linear relationship found was used in the early derivation of values for $\delta R/R$.

SnI_4 is one of the few compounds for which the Mössbauer spectra of all the elements comprising the compound have been recorded [77]. The intensity of the ^{119}Sn resonance between 85 and 220 K was used to derive an effective Debye temperature of 166 K which contrasts with the equivalent value for the iodine of 85 K. The structural implications will be discussed more fully on p. 478 in connection with the latter isotope. The results were later used to predict the specific heat C_P of SnI_4 [78].

Salts of the hexa-halogeno anions, SnF_6^{2-}, $SnCl_6^{2-}$, $SnBr_6^{2-}$, and SnI_6^{2-} show an increase in the chemical isomer shift in the order given because of decreasing electronegativity of the ligand, although the values are lower than in the corresponding SnX_4 compound. No quadrupole splitting is found, which is consistent with octahedral symmetry, and the cations do not show a significant influence [79–82]. The $[Me_4N]^+$ salts of the octahedral anions $SnCl_6^{2-}$, $SnBr_6^{2-}$, SnI_6^{2-}, $SnCl_4Br_2^{2-}$, $SnCl_2Br_4^{2-}$, $SnCl_2I_4^{2-}$, and $SnBr_2I_4^{2-}$ all show single lines (although splitting would be anticipated in

Table 14.5 Inorganic tin(IV) halide derivatives

Compound	Δ /(mm s^{-1})	δ/(mm s^{-1})	δ (SnO$_2$) /(mm s^{-1})	Reference
SnF$_4$	1·66	−0·47 (SnO$_2$)	−0·47	76
SnCl$_4$	0	0·85 (SnO$_2$)	+0·85	15
SnCl$_4$.5H$_2$O	0	0·25 (SnO$_2$)	+0·25	15
SnBr$_4$	0	1·15 (SnO$_2$)	+1·15	15
SnI$_4$	0	1·55 (SnO$_2$)	+1·55	77
Li$_2$SnF$_6$	0	−3·15 (β-Sn)	−0·59	79
Na$_2$SnF$_6$	0	−0·48 (SnO$_2$)	−0·48	80
K$_2$SnF$_6$	0	−3·15 (β-Sn)	−0·59	79
Rb$_2$SnF$_6$	0	−3·01 (β-Sn)	−0·45	79
CuSnF$_6$	0	−3·05 (β-Sn)	−0·49	79
FeSnF$_6$	0	−3·09 (β-Sn)	−0·53	79
SrSnF$_6$	0	−3·02 (β-Sn)	−0·46	79
BeSnF$_6$	0	−2·96 (β-Sn)	−0·40	79
(NO$_2$)$_2$SnF$_6$	0	−2·9 (β-Sn)	−0·3	79
(NO)$_2$SnF$_6$	0	−3·0 (β-Sn)	−0·4	79
(H$_3$O)$_2$SnF$_6$	0	−3·0 (β-Sn)	−0·4	79
H$_2$SnCl$_6$	0	+0·50 (SnO$_2$)	+0·50	80
K$_2$SnCl$_6$	0	+0·45 (SnO$_2$)	+0·45	80
Rb$_2$SnCl$_6$	0	+0·43 (SnO$_2$)	+0·43	80
Cs$_2$SnCl$_6$	0	+0·45 (SnO$_2$)	+0·45	80
(NH$_4$)$_2$SnCl$_6$	0	+0·48 (SnO$_2$)	+0·48	80
MgSnCl$_6$.6H$_2$O	0	+0·49 (SnO$_2$)	+0·49	81
CaSnCl$_6$.6H$_2$O	0	+0·46 (SnO$_2$)	+0·46	81
[MeNH$_3$]$_2$SnCl$_6$	0	+0·50 (SnO$_2$)	+0·50	82
[Me$_4$N]$_2$SnCl$_6$	0	+0·50 (SnO$_2$)	+0·50	82
(NH$_4$)$_2$SnBr$_6$	0	+0·80 (SnO$_2$)	+0·80	80
K$_2$SnBr$_6$	0	+0·75 (SnO$_2$)	+0·75	80
[Me$_4$N]$_2$SnBr$_6$	0	+0·87 (SnO$_2$)	+0·87	82
[Et$_4$N]$_2$SnBr$_6$	0	+0·86 (SnO$_2$)	+0·86	83
Rb$_2$SnI$_6$	0	+1·35 (SnO$_2$)	+1·35	80
[Me$_4$N]$_2$SnI$_6$	0	+1·25 (SnO$_2$)	+1·25	82
[Me$_4$N]$_2$Sn(N$_3$)$_6$	0	+0·48 (SnO$_2$)	+0·48	83
[Me$_4$N]$_2$SnCl$_4$Br$_2$	0	+0·66 (SnO$_2$)	+0·66	83
[Me$_4$N]$_2$SnCl$_4$I$_2$	0	+0·53 (SnO$_2$)	+0·53	83
[Me$_4$N]$_2$SnCl$_2$Br$_4$	0	+0·74 (SnO$_2$)	+0·74	83
[Me$_4$N]$_2$SnBr$_4$I$_2$	0	+0·89 (SnO$_2$)	+0·89	83
[Me$_4$N]$_2$SnCl$_2$I$_4$	0	+1·17 (SnO$_2$)	+1·17	83
[Me$_4$N]$_2$SnBr$_2$I$_4$	0	+1·35 (SnO$_2$)	+1·35	83
(Ph$_3$C)SnCl$_5$	0·460	+0·63 (SnO$_2$)	+0·63	84
(Ph$_3$C)SnBr$_5$	0	+0·99 (SnO$_2$)	+0·99	84
[(*p*-MeC$_6$H$_4$)Ph$_2$C]SnCl$_5$	0·671	+0·47 (SnO$_2$)	+0·47	84
SnF$_4$.2ClF$_3$	1·5	−3·05 (β-Sn)	−0·49	79
SnF$_4$.2BrF$_5$	1·15	−3·02 (β-Sn)	−0·46	79
SnF$_4$.2BrF$_3$	0·87	−3·00 (β-Sn)	−0·44	79
SnF$_4$.2IF$_5$	1·08	−3·36 (β-Sn)	−0·80	79
SnF$_4$.BrF$_3$	1·3	−3·0 (β-Sn)	−0·4	79
SnF$_4$(oxinH)$_2$	0	−0·08 (SnO$_2$)	−0·08	91
SnCl$_4$.2(Me$_2$N)$_3$PO	0·70	−1·21 (Pd/Sn)	+0·31	89
SnCl$_4$.2Ph$_3$PO	0·50	−1·17 (Pd/Sn)	+0·35	89

[*Refs. on p.* 424]

Table 14.5 (continued)

Compound	Δ /(mm s⁻¹)	δ/(mm s⁻¹)	δ (SnO₂) /(mm s⁻¹)	Reference
SnCl₄.2Me₂SO	0	−1·15 (Pd/Sn)	+0·37	89
SnCl₄.2(Me₂N)₂CS	0	−0·82 (Pd/Sn)	+0·70	89
SnCl₄.2Bu^{n}₃P	1·0	−0·65 (Pd/Sn)	+0·87	89
SnCl₄.2py	0	−1·01 (Pd/Sn)	+0·51	89
SnCl₄.(C₄H₈N)₃PS	0	−1·09 (Pd/Sn)	+0·43	89
SnCl₄.bipy	0	+0·42 (SnO₂)	+0·42	82
SnCl₄.en₂	0	+0·50 (SnO₂)	+0·50	76
SnCl₄.thf₂	0	+0·70 (SnO₂)	+0·70	76
SnCl₄.ths₂	0	+0·81 (SnO₂)	+0·81	76
SnCl₄.(oxinH)₂	0	+0·30 (SnO₂)	+0·30	82
SnCl₄(salH)₂	1·10	+0·43 (SnO₂)	+0·43	91
SnCl₃(oxin)	0	+0·34 (SnO₂)	+0·34	91
SnCl₂(sal)₂	0	+0·23 (SnO₂)	+0·23	91
SnCl₄.2MeCN	0·70	+0·43 (SnO₂)	+0·43	90
SnCl₄.2MeOH	0·70	+0·43 (SnO₂)	+0·43	90
SnCl₄.2Me₂CO	0	+0·40 (SnO₂)	+0·40	90
SnCl₄.2Et₂O	1·10	+0·45 (SnO₂)	+0·45	90
SnBr₄.2(Me₂N)₃PO	0·70	−0·96 (Pd/Sn)	+0·56	89
SnBr₄.2Ph₃PO	0·61	−0·89 (Pd/Sn)	+0·63	89
SnBr₄.2Me₂SO	0	−0·86 (Pd/Sn)	+0·66	89
SnBr₄.2(Me₂N)CS	0	−0·58 (Pd/Sn)	+0·94	89
SnBr₄.2Ph₃P	0·66	−0·89 (Pd/Sn)	+0·63	89
SnBr₄.2py	0	−0·78 (Pd/Sn)	+0·74	89
SnBr₄.(C₄H₈N)₃PS	0	−0·90 (Pd/Sn)	+0·62	89
SnBr₄.bipy	0	+0·66 (SnO₂)	+0·66	82
SnBr₄.en₂	0	+0·43 (SnO₂)	+0·43	76
SnBr₄(oxinH)₂	0	+0·65 (SnO₂)	+0·65	91
SnBr₄(salH)₂	1·22	+0·73 (SnO₂)	+0·73	91
SnBr₂(oxin)₂	0	+0·44 (SnO₂)	+0·44	91
SnBr₂(sal)₂	0	+0·28 (SnO₂)	+0·28	91
SnI₄.en₂	0	+0·43 (SnO₂)	+0·43	76
SnI₂(oxin)₂	0	+0·61 (SnO₂)	+0·61	91
SnI₂(sal)₂	0	+0·41 (SnO₂)	+0·41	91

en = ethylenediamine ths = tetrahydrothiophene salH = salicylaldehyde
thf = tetrahydrofuran oxinH = 8-hydroxyquinoline

the mixed species) with chemical isomer shifts which are a linear function of the sum of the Pauling electronegativities [83]. SnF_6^{2-}, $SnBr_4I_2^{2-}$, and $SnCl_4I_2^{2-}$ deviate from this relationship, however.

A number of complexes of the $SnCl_5^-$ and $SnBr_5^-$ anions with carbonium cations have been prepared and show a quadrupole splitting which is usually below resolvable limits [84], the highest recorded value being 0·67 mm s⁻¹ in the *p*-tolyldiphenylcarbonium salt of $SnCl_5^-$.

The bis-adducts of the tetrahalides with various organic and inorganic ligands, SnX_4L_2, may all be presumed to have an octahedral geometry.

[*Refs. on p. 424*]

SnCl$_4$.2py is known from its X-ray structure [85] to have a *trans*-configuration, and is isomorphous with SnBr$_4$.2py. By contrast the compounds SnCl$_4$.2MeCN [86], SnCl$_4$.2POCl$_3$, [87], and SnCl$_4$.2SeOCl$_2$ [88] all have a *cis*-configuration. In all cases one would expect to find a sizeable quadrupole splitting, but in practice a splitting is resolved in only a small fraction of the large range of complexes studied. A selection of the available data is given in Table 14.5. Quadrupole splittings have been recorded for SnF$_4$ with halogen fluorides, and for SnCl$_4$ and SnBr$_4$ with oxygen and phosphorus donor atoms but not with nitrogen and sulphur. In no case is the splitting large. This rather unexpected result will be considered more fully in Chapter 14.5 in the discussion of organometallic compounds.

A considerable number of derivatives of 8-hydroxyquinoline (oxinH) and salicylaldehyde (salH) have been examined [91]. The basic types are SnX$_2$(oxin)$_2$, SnX$_2$(sal)$_2$, SnCl$_{4-n}$(oxin)$_n$ ($n = 0$–4), SnX$_4$.2oxinH, and SnX$_4$.2salH, the majority of them showing no detectable quadrupole splitting. Possible stereochemistries have been suggested.

From the chemical isomer shift data it can be seen that the coordination of ligands reduces the *s*-electron density at the tin nucleus; the effect depends on the donor atom and decreases in the approximate sequence O > N > S > P for the tetrachloride. The effect is less regular though still pronounced in the bromide. The suggestion has been made that the lack of a quadrupole splitting probably indicates a *cis*-configuration, but this does not explain the lack of splitting in SnCl$_4$.2py, which is known to be *trans*-, or the comparatively large splitting in SnCl$_4$.2MeCN, which is known to have a *cis*-configuration.

The adducts of SnF$_4$ with solvents such as diethyl ether, pyridine, and dimethylsulphoxide have been shown to be of the type SnF$_4$L$_2$ in frozen solutions [92], and these also give no quadrupole splitting. Complexes which are nominally SnF$_4$L are probably mixtures of SnF$_4$ and SnF$_4$L$_2$, with the exception of SnF$_4$.thf, which is characteristically different and may be 5-coordinate.

The reaction of LiAlH$_4$ with SnCl$_4$ in diethyl ether at low temperature can be followed in the Mössbauer spectrum [93]. The initial SnCl$_4$-etherate complex has a chemical isomer shift intermediate between the final product SnH$_4$ and an intermediate compound assumed to be Sn(AlH$_4$)$_4$. At $-100°$C the reaction is slow, but at $-80°$C the Sn(AlH$_4$)$_4$ is gradually formed, and this decomposes slowly to SnH$_4$ at $-40°$C. β-tin is the end-product at $-35°$C.

The spectrum of a frozen aqueous solution of 0·2M SnCl$_4$ shows similar anomalous changes in the line intensity with temperature to those previously observed in solutions of iron salts [94].

Silver chloride doped with ^{119}Sn gives an unsplit resonance line [95]. A model has been proposed involving Sn^{4+} at an interstitial site surrounded by four cation vacancies.

[*Refs. on p. 424*]

Tin(IV) Oxides

The use of SnO_2 and $BaSnO_3$ as source matrices has already been mentioned in Section 14.1 and their Mössbauer characteristics are discussed there. The perovskite stannates $BaSnO_3$, $SrSnO_3$, and $CaSnO_3$ show a single resonance line unshifted from SnO_2 [96]. Spectra have been recorded over the temperature range 78 K to 1020 K, and the recoil-free fraction remains substantial throughout this range. Some anomalous line broadening occurs above 650 K, possibly as a result of a change in structure or partial chemical reduction. At high temperatures the observed second-order Doppler shift is 3.2×10^{-4} mm s^{-1} K^{-1}.

The stannates $M_2Sn_2O_7$ (R = La, Y, Pr, Nd, Sm, Eu, Gd, Tb, Dy, Ho, Er, Tm, Yb, Bi) show varying degrees of line broadening which are probably due to small unresolved quadrupole interactions [97, 98].

The spinel oxides $SnCo_2O_4$, $SnMg_2O_4$, $SnZn_2O_4$, and $SnMn_2O_4$ are all inverse with Sn(IV) on the trigonally distorted B sites [99]. Accordingly the ^{119}Sn spectra show quadrupole splittings of 1·00, 1·10, 0·75, and 1·10 mm s^{-1} respectively, the chemical isomer shifts being about 0·1 mm s^{-1} higher than in SnO_2.

In the phase xMg_2SnO_4–$(1 - x)MgFe_2O_4$, magnetic interaction is seen in both the ^{57}Fe and ^{119}Sn resonances for $x \geqslant 0.3$ (in the case of tin as a result of an induced polarisation by the magnetic cations) [100]. However, in both instances randomisation effects prevent the appearance of a unique magnetic field, and the resonance lines are broadened. Similar effects are found in the system $MgFe_2O_4$–$MnFe_2O_4$ when doped with ^{119}Sn [101].

In the ferrite $Ni_{1.125}Fe_{1.75}Sn_{0.125}O_4$ the ^{119}Sn shows fields of >250 kG at 80 K [102]. These have been ascribed to a combination of both indirect A–B and direct B–B exchange interactions. The field strength is dependent on the number of Ni^{2+} ions on B sites in the neighbourhood of the tin atom.

The lattice dynamical properties of the $MnFe_2O_4$–$ZnFe_2O_4$ phase have been studied above the Curie temperatures by measuring the recoil-free fraction in samples doped with 5% $^{119}SnO_2$ impurity [103]. It was found that f is higher in those samples with high Curie points, and that it decreases considerably with increasing zinc content.

Substitution of Sn^{4+} for Fe^{3+} in yttrium iron garnet gives the oxide $Ca_{0.3}Y_{2.7}Sn_{0.3}Fe_{4.7}O_{12}$. At room temperature and liquid nitrogen temperature the tin is found to be in an internal magnetic field of 152 and 210 kG respectively [104]. The fields are generated by the polarisation of the electrons of the tin atom by the exchange fields of the unpaired 3d-electrons on the iron atoms, thereby causing an imbalance in the spin density at the tin nucleus and a Fermi contact interaction. An independent investigation of the composition $Ca_{0.25}Y_{2.75}Sn_{0.25}Fe_{4.75}O_{12}$ using both the ^{57}Fe and ^{119}Sn resonances showed that the hyperfine fields at the two iron sublattices and at the tin all decrease with increasing temperature and reach zero simultaneously at the Curie temperature [40, 41]. The published spectrum at 77 K for this sample is the best example of a resolved magnetic hyperfine spectrum in ^{119}Sn available, and is shown in Fig. 14.9. The value of the magnetic field is apparently unique despite the nonstoichiometry of the oxide.

Continued substitution of tin in the octahedral a sublattice in the phase $Ca_xY_{3-x}Sn_xFe_{5-x}O_{12}$ might be expected to weaken the a–d exchange interactions, and in Fig. 14.10 this effect is clearly seen in a series of spectra at

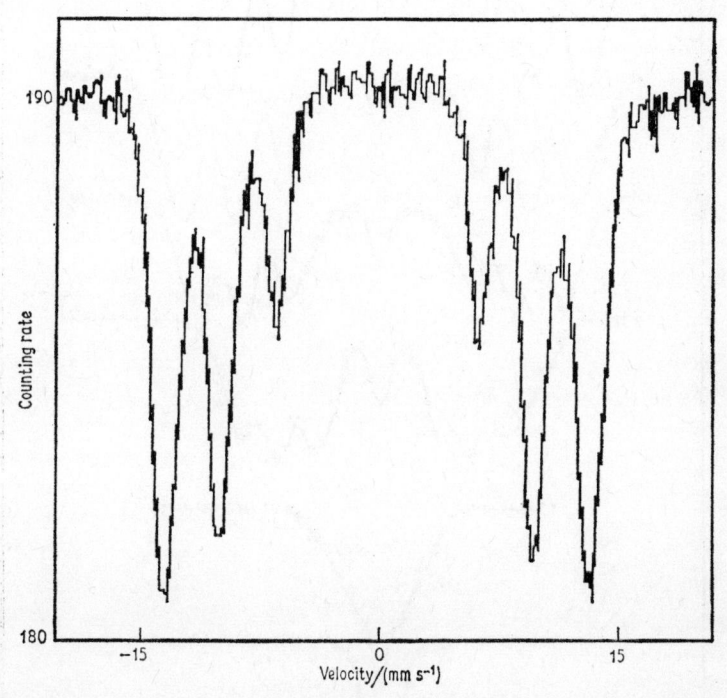

Fig. 14.9 Spectrum of $Ca_{0.25}Y_{2.75}Sn_{0.25}Fe_{4.75}O_{12}$ at 77 K, showing induced magnetic splitting of the ^{119}Sn resonance. [Ref. 41, Fig. 1]

77 K by a collapse of the magnetic hyperfine splitting with increasing tin substitution [105]. The variation of the magnetic field with composition is summarised in Table 14.6. At high tin content the hyperfine field arises mainly from polarisation at the tin by an a–d exchange interaction, but at low tin content there also appears to be an a–a exchange which is of opposite sign to and smaller than the a–d interaction. The observed field at the tin nuclei in a sample with $x = 0.25$ increases in an applied field of 15 kG [106]. The ferrite is completely polarised in such a field with the larger moment (d site) parallel to the applied field and the a site moment antiparallel to it. The inference which can be drawn is that the field at the ^{119}Sn nuclei is positive in sign.

[*Refs. on p. 424*]

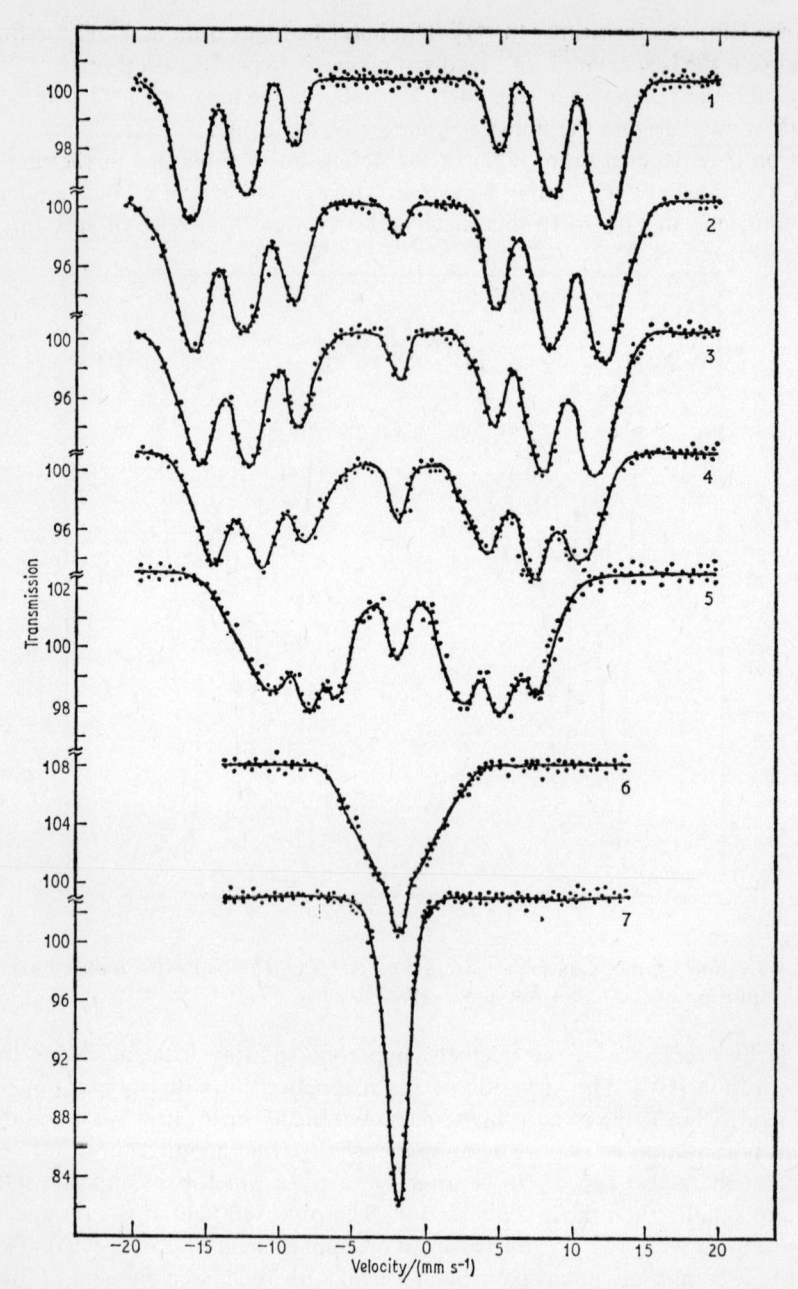

Fig. 14.10 Spectra for $Ca_xY_{3-x}Sn_xFe_{5-x}O_{12}$ showing the collapse in the hyperfine field as x increases: (1) $x = 0.1$, (2) $x = 0.3$, (3) $x = 0.5$, (4) $x = 0.7$, (5) $x = 0.9$, (6) $x = 1.2$, (7) $x = 1.5$. [Ref. 105, Fig. 1]

The gadolinium iron garnets $Ca_xGd_{3-x}Sn_xFe_{5-x}O_{12}$ are similar to the yttrium system [107], the main difference being that both the gadolinium (c) and iron (a and d) sublattices carry a magnetic moment. That on the c-lattice is opposite in direction to that of the a-sublattice and any c–a exchange would reinforce the a–a exchange and thereby reduce the observed field. The experimental values for the yttrium and gadolinium systems are identical (Table 14.6), clearly demonstrating that the c–a exchange interaction in these garnets is very weak.

The temperature dependence of the magnetic properties of these systems is

Table 14.6 Variation of the magnetic field with composition in garnets

$Ca_xY_{3-x}Sn_xFe_{5-x}O_{12}$

x	H (295 K) /kG	H (77 K) /kG
0·1	159	210
0·3	152	210
0·5	128	201
0·7	94	188
0·9	35	135
1·2	0	26
1·5	0	0

$Ca_xGd_{3-x}Sn_xFe_{5-x}O_{12}$

x	H (290 K) /kG	H (77 K) /kG
0·1	163	208
0·3	155	204
0·9	30	137

also of considerable interest. At low temperatures the total magnetisation is dominated by the rare-earth element but at high temperatures it is dominated by the contribution from the iron, there being a so-called compensation point at which temperature the net magnetisation is zero. The value of the field at the tin nucleus in a sample with $x = 0·3$ proves to be -184 kG at 95 K and $+148$ kG at 300 K [108]. This means that the field at the tin nucleus is strongly linked to the iron sublattice so that reversal in direction of the total magnetic moment of the sample produces an effective reversal in sign of the field at the tin site.

The ^{57}Fe and ^{119}Sn resonances have been recorded in the sulphide Cu_2FeSnS_4 and establish the oxidation states as $Cu^+_2Fe^{2+}Sn^{4+}S_4$ [109] (see also p. 285). The tin chemical isomer shift is $+1\cdot48$ mm s^{-1}, at room temperature and the quadrupole splitting zero because of the tetrahedral site symmetry.

The broad linewidth of the ^{119}Sn resonance has discouraged the use of this resonance in impurity studies, but some work has been done. Dilute solid solutions of ^{119}Sn in NiO, Cr$_2$O$_3$, V$_2$O$_5$, Sb$_2$O$_5$, and MoO$_3$ give single lines unshifted from SnO$_2$ [110]. In the case of NiO and Cr$_2$O$_3$ magnetic broadening is found below the Néel temperatures.

The occurrence of spontaneous polarisation in ferroelectric crystals of the type BaTiO$_3$ is usually attributed to an anomalous decrease in the frequency of long-wave oscillations in the crystal lattice on approaching the phase-transition point from the paraelectric region. This change in the optical branch of the phonon spectrum might be expected to cause a significant change in the Mössbauer recoil-free fraction and the effect has been demonstrated in BaTiO$_3$ doped with 1–2% Sn [111, 112]. The recoil-free fraction decreases sharply by about 10% on approaching the Curie point (390 K) from the paraelectric region and passes through a minimum. Similar studies on the BiFeO$_3$–SrSnO$_3$ perovskite phase [113] and the Mn-substituted series BiFeO$_3$–Sr(Sn$_{0.33}$Mn$_{0.67}$)O$_3$ [114] appeared to show a sharp drop in f at the Curie point, but more careful examination revealed the presence of fields of up to 200 kG at low temperatures which, because of inhomogeneities, cause drastic smearing of the spectrum [114]. Such gross broadening was not found in BaTiO$_3$, but a small 'quadrupole broadening' was reported which decreased to zero in the region immediately below the Curie temperature.

A source of Mg$_2$SnO$_4$ produced by neutron irradiation shows auxiliary lines at chemical isomer shifts of $-5\cdot0$, $+3\cdot6$, and $+6\cdot2$ mm s^{-1} rel. to SnO$_2$ (signs are those for conventional absorber definition) due to radiation damage [115]. The line at $+3\cdot6$ mm s^{-1} (also reported in independent investigations [116, 117]) is the strongest impurity and derives from a tin(II) oxidation state, but the other two lines have not yet been identified. Similar radiation damage has been seen in SnO$_2$ [118].

The spectra of ^{119}Sn in alkali–tin–silicate and borate glasses show small chemical isomer shifts from the SnO$_2$ resonance as well as line broadening which can sometimes be resolved as a quadrupole splitting [119]. This provides some evidence for the presence of non-equivalent Sn–O–Si and Sn–O–B bonds. Several other studies of glasses and glass-fibres have been made [120, 121].

The adsorption of Sn^{2+} atoms on zeolite and silica-gel surfaces has been studied [122]. The bonding appears to become stronger as the pore size of the material decreases towards molecular dimensions. The asymmetry of the tin(II) quadrupole splitting was held to indicate a Karyagin effect because of the anisotropy of surface atoms.

A number of tin-bearing minerals have been examined, but the broad line of the ^{119}Sn resonance decreases the extent to which structural information can be derived in comparison with, for example, the iron silicate minerals. Arandisite (ideally $3SnSiO_4.2SnO_2.4H_2O$), cassiterite (mineral SnO_2), hulsite ($[Fe^{2+}, Mg^{2+}]_2[Fe^{3+}, Al^{3+}]BO_3O_2$), and nordenskioldine($CaSnB_2O_6$) give spectra [123] very similar to that of SnO_2 in common with most tin-bearing oxides. The sulphide canfieldite (Ag_8SnS_6) gives two unsplit resonances at 1·07 and 1·74 mm s^{-1} (rel. to SnO_2) at 80 K. The former which is more intense is similar to SnS_2. The major components of cylindrite ($Pb_3Sn_3Sb_2S_{14}$) and franckeite ($Pb_5Sn_3Sb_2S_{14}$) have chemical isomer shifts of +1·27 and +1·07 mm s^{-1} at 80 K and are typical of tin(IV) sulphides. Herzenbergite (nominally [SnPb]SnS$_2$) contains both tin(II) and tin(IV). Tealite (PbSnS$_2$) contains tin(II).

Very little data are available for molecular compounds with tin–oxygen bonds except for the adducts SnX_4L_2 given in Table 14.5. A number of compounds with ligands chelating through oxygen (4-bonds) and nitrogen (1–2-bonds) to give a 5- or 6-coordinate structure, e.g. [MeN(CH$_2$CH$_2$O)$_2$]$_2$Sn and [N(CH$_2$CH$_2$O)$_3$]SnOMe, give single broadened lines with chemical isomer shifts in the range 0·18–0·45 mm s^{-1} at 78 K [124]. Of particular interest is the compound [(SnCl$_3$POCl$_3$)$^+$(PO$_2$Cl$_2$)$^-$]$_2$. The crystal structure indicates a 6-coordinated tin environment with the (PO$_2$Cl$_2$)$^-$ ions forming bidentate bridges between the two cations [125]. The Mössbauer parameters at 77 K for this dimer ($\delta = 0·21$; $\Delta = 0·75$ mm s^{-1}) can be compared with those of monomeric SnCl$_4$.2POCl$_3$ ($\delta = 0·44$, $\Delta = 1·13$ mm s^{-1}). Only the dimer shows a significant absorption at room temperature.

14.5 Organotin(IV) Compounds

A substantial part of the chemistry of tin is concerned with organometallic compounds in which there is at least one tin–carbon bond. The availability of hundreds of these compounds has resulted in the recording of large numbers of Mössbauer spectra. The two principal parameters determined are the chemical isomer shift and the quadrupole splitting, and unfortunately these show no significant temperature-dependent properties and have proved singularly difficult to interpret in detail.

The factors influencing the chemical isomer shift in organo-tin(IV) compounds have already been discussed in detail in Chapter 14.2. The values observed in practice range from 0·55 to 2·20 mm s^{-1} (relative to SnO$_2$). The small differences observed within related series of compounds cannot in general be interpreted with certainty in chemically meaningful terms.

The magnitude of the electric field gradient tensor is determined in principle by several factors:

(1) the stereochemistry of the compound. If the geometry is regular tetrahedral or octahedral there will be no electric field gradient;

(2) differences in the σ-bonding between various ligands. This includes the effect of unequal occupancy in both the 5p- and 5d-orbitals. The $\langle r^{-3} \rangle$ term for the 5d-orbitals will be smaller than for the 5p-orbitals, so that the observed electric field gradient is more likely to reflect the unequal occupancy of the latter. In this respect, for a 5-coordinate compound which can naively be represented as $5s5p^3 5d$, any small asymmetry in the 5p-orbitals may outweigh the contribution from the 5d-orbitals;

(3) differences in the π-bonding between various ligands. This can alter the relative occupancy of both the 5p- and the 5d-orbitals and will have the same effects as in (2).

Complete analysis of the various factors listed under (2) and (3) can only be carried out if the stereochemistry has been accurately defined and if accurate atomic wavefunctions are available for the appropriate molecular-orbital calculations to be made. This latter condition has not as yet been fully met, and structural data are available for only a few compounds. Many authors have discussed qualitatively the relative importance of σ- and π-bonding in the production of the electric field gradient [82, 126–129]. However, concepts such as hybridisation of atomic orbitals were originally devised for the lighter elements where the distinction between different types of bonding is often easy to make, and it is probably unwise to pursue qualitative arguments too far when interpreting data for heavy elements such as tin. The lack of structural data is particularly serious in view of the fact that nearly all organotin compounds with nominally four ligands which have been subjected to X-ray analysis have proved to possess a higher coordination number in the solid state. Although attempts have been made to link the magnitude of the quadrupole splitting with the stereochemistry [127, 130], distinctions are probably insufficiently clear to justify the use of the method with any confidence.

Effective Debye temperatures have been calculated for a number of alkyl and aryl tin halides [131], and although for instance the Debye temperature for polymeric Me$_3$SnF (102·8 K at 70 K) is larger than that for monomer Me$_4$Sn (71 K at 70 K), it remains true in general that lattice characteristics do not provide a means of distinguishing polymeric structures [131, 132].

One disturbing feature of the Mössbauer spectra of organotin compounds which was noticed at an early stage was the lack of a significant (i.e. >0·3 mm s^{-1}) quadrupole splitting in a number of compounds such as Ph$_3$SnI and (Ph$_3$Sn)$_4$Ge in which the point symmetry of the tin environment is lower than T_d [126]. There is no correlation between the quadrupole splitting and the electronegativity difference in the atoms bonding to tin. If at least one of the tin–ligand bonds is to carbon, one can make the generalisation [82, 126] that (a) if one or more of the other atoms bonding to tin possesses lone-pair non-bonding electrons (i.e. F, Cl, Br, I, O, S, N, and P) then a large quadrupole splitting is observed, (b) if none of the other atoms possesses a lone-pair of non-bonding electrons (i.e. H, Na, C, Ge, Sn, and Pb) then the quadrupole splitting will be either less than the experimental resolution or else no greater than about 1·5 mm s^{-1}. The distinction is probably caused by a large difference in the σ-bonding to tin between carbon and similar elements on the one hand and those elements to the right of Group IV in the Periodic Table on the other. The exact explanation remains obscure.

Although this rule is entirely empirical, examination of the tabulated data in this section shows that it is valid for the great majority of the compounds studied. It can therefore be used as an effective safeguard against incorrect assumptions regarding the tin environments in new compounds [82]. One can also include those 6-coordinate inorganic compounds which have a zero or unresolved quadrupole splitting (as mentioned without detailed comment in the preceding section) by adding the qualification: (c) if all the atoms bonding to tin possess lone-pair non-bonding electrons then the quadrupole splitting will once again be small.

Having made this categorisation it is tempting to devise a theoretical interpretation of the observation. The original explanation for the large electric field gradients observed in compounds of type (a) invoked either π-overlap between the ligand lone-pair orbitals and the $5p$- and $5d$-orbitals on the tin, or alternatively an increase in the coordination to five [126]. With regard to the latter concept, bridging structures are more likely to be formed by those elements with pairs of donor electrons and it is now known that several apparently 4-coordinate tin compounds are, in fact, 5-coordinate, e.g. Me$_3$SnF etc. Neither explanation seems entirely satisfactory, and σ-bonding is clearly a significant factor. However, more information about the electronic structure is needed.

Because the value of the excited-state quadrupole moment of the ^{119}Sn nucleus is not well established, the actual values of the electric field gradients cannot be derived from the observed quadrupole splitting. We cannot say therefore what degree of asymmetry of $5p$-orbital occupancy is indicated, or what the upper limit of quadrupole splitting for an asymmetric environment is likely to be (the maximum value known to date is 5·54 mm s^{-1} in Me$_2$Sn(SO$_3$F)$_2$ in which the tin atom is probably 6-coordinated with bridging O–SO(F)–O groups [133].

One of the helpful approaches is to establish the sign of e^2qQ in a large number of compounds of known stereochemistry so that the asymmetry at the tin can be related to particular orbitals. The applied magnetic field method is suitable for this, and has been used recently to establish the sign of e^2qQ in 6-coordinate Me$_2$SnMoO$_4$ and Me$_2$SnCl$_2$ as positive [134]. Typical spectra are shown in Fig. 14.8. An earlier attempt [135] to establish the sign of e^2qQ in Ph$_2$SnCl$_2$ by this method was unsuccessful because failure to allow for free rotation of the electric field gradient about the magnetic direction prevented full mathematical analysis of the experimental data, and because the method as shown on p. 381 is largely ineffective for fields below about 30 kG.

More significantly, the sign of e^2qQ has been determined in three compounds for which the X-ray structure is known, and is positive in Me_2SnF_2 (*trans*-octahedral) and negative in Me_3SnNCS and Me_3SnOH (trigonal bipyramidal) [136]. In each case an explanation can be formulated in terms of an excess of electron density in the Sn–C directions over the Sn–X directions. Furthermore, in the case of the two D_{3h} symmetry structures, the results show that asymmetric σ-bonding, rather than π-donation from X to tin or lattice effects, is primarily responsible for the quadrupole interaction. Similar conclusions had previously been drawn from the systematics of chemical isomer shift and quadrupole splitting [89, 127, 128, 149] and have subsequently been confirmed by application of magnetic fields to determine the sign of e^2qQ in numerous other compounds [259–263].

The discussion which follows tabulates data for a wide selection of compounds and includes data for all types which have been studied. Many compounds have been examined in several laboratories but only one set of values is listed, and references are not necessarily given unless a full series of compounds of a given type were studied. An exhaustive tabulation of data published up to January 1966 is now available [137].

Alkyl and Aryl Derivatives of Tin: R_4Sn and R_3SnR'

The tetraalkyl and tetraaryl tin derivatives R_4Sn can all be presumed to have a tetrahedral geometry, and therefore show no significant quadrupole splitting (Table 14.7). Differences in chemical isomer shift are small and of the order of the reproducibility between laboratories. In less symmetric compounds of the type R_3SnR', quadrupole splitting is only found when R' is a strongly electron-withdrawing group such as C_6F_5, C_6Cl_5, or $MeCOCH_2$, and even in such cases is rarely much greater than the experimental linewidth. The comparative insensitivity of the spectrum to the nature of X in compounds of the type R_3SnCH_2X shows that these groups influence the tin environment only by their relative electron-withdrawing power. The possibility of higher coordination using donor groups in the ligand must be considered. Thus R_3SnCH_2OMe may be involved in inter- or intra-molecular bonding to tin via oxygen, a factor which could increase the electric field gradient substantially. The carborane derivatives (structures 1–3) show strong similarities to the C_6F_5 derivatives, and they also have the largest quadrupole splittings of this class of compound in general [147].

Hydrides and Distannanes

Detailed recoil-free fraction and linewidth measurements have been made on SnH_4 [150]. The recoil-free fraction is only $f = 0.025$ at 78 K, and the resonance is a single line as expected for a tetrahedral molecule.

R₃Sn—C—CR with B₁₀H₁₀ below (triangular cage)

(1)

R₃Sn—C—C—SnR₃ with B₁₀H₁₀ below

(2)

R₂Sn and SnR₂ bridged by two C—C units each capped with B₁₀H₁₀

(3)

The alkyl and aryl derivatives of tin with at least one bond to hydrogen, sodium, tin, germanium, or lead also show no detectable quadrupole splitting (Table 14.8). The derivatives Me_nSnH_{4-n} ($n = 1$, 2, or 3) all show a chemical isomer shift of 1·24 mm s⁻¹, while the phenyl and n-butyl derivatives fall in the narrow range 1·38–1·44 mm s⁻¹. One might have expected a difference between the alkyltin derivatives on the one hand and the aryltin compounds on the other, but this is not found. It is also unexpected that the chemical isomer shift is insensitive to the degree of hydrogen substitution on the tin. However, molecular-orbital calculations [28] on the methyl derivatives give very similar values in all cases for both the s-electron density at the tin nucleus and for the p-orbital imbalance; the latter is very small, in agreement with the observed absence of quadrupole splitting. Attempted correlation of the nuclear magnetic resonance $J(^{119}Sn-^{1}H)$ and $J(^{119}Sn-C-^{1}H)$ coupling constants with the chemical isomer shift for the methyl series is frustrated by the fact that differences in the Mössbauer parameters are well within the estimated experimental error [152].

The chlorohydride Bu^n_2SnHCl has a large quadrupole splitting (note that the chlorine is directly bonded to the tin) [151]. Also noteworthy in Table 14.8 is the lack of significant splitting in derivatives of the distannane type with a tin–tin bond, e.g. $(Ph_3Sn)_2$. The diphenyltin compound, Ph_2Sn, is also 4-coordinate with an oligomeric structure involving Sn–Sn bonds [153]. The crystal structure establishes it to be a hexamer [154]. Dibutyltin is similar, and oxidises quite easily to Bu_2SnO. The reaction is comparatively slow, and the kinetics of the change can be investigated using the Mössbauer spectrum [153].

[Refs. on p. 424]

Table 14.7 Organotin compounds: R$_4$Sn, R$_3$SnR′, and R$_2$SnR′$_2$ (data at 80 K)

Compound	Δ /(mm s^{-1})	δ/(mm s^{-1})	δ (SnO$_2$) /(mm s^{-1})	Reference
Me$_4$Sn	—	1·21 (SnO$_2$)	1·21	138
Et$_4$Sn	—	1·33 (SnO$_2$)	1·33	140
Pr$^n{}_4$Sn	—	1·30 (SnO$_2$)	1·30	140
Bu$^n{}_4$Sn	—	1·35 (SnO$_2$)	1·35	140
[PhC(Me$_2$)CH$_2$]$_4$Sn	—	1·34 (SnO$_2$)	1·34	142
Ph$_4$Sn	—	1·22 (SnO$_2$)	1·22	138
(C$_6$F$_5$)$_4$Sn	—	1·04 (SnO$_2$)	1·04	138
(C$_6$Cl$_5$)$_4$Sn	—	−0·88 (α-Sn)	1·12	139
(*m*-CF$_3$.C$_6$H$_4$)$_4$Sn	—	1·28 (SnO$_2$)	1·28	141
(*p*-CF$_3$.C$_6$H$_4$)$_4$Sn	—	1·29 (SnO$_2$)	1·29	141
Me$_3$SnPh	—	1·25 (SnO$_2$)	1·25	144
Me$_3$SnCH=CH$_2$	—	1·30 (SnO$_2$)	1·30	144
Me$_3$SnC$_6$H$_4$.CH=CH$_2$	—	1·30 (SnO$_2$)	1·30	144
Et$_3$SnMe	—	1·35 (SnO$_2$)	1·35	145
(C$_6$F$_5$)$_3$SnPh	0·92	1·16 (SnO$_2$)	1·16	138
(C$_6$Cl$_5$)$_3$SnPh	0·8	−0·99 (α-Sn)	1·01	139
(C$_6$F$_5$)$_2$SnPh$_2$	1·11	1·22 (SnO$_2$)	1·22	138
(C$_6$Cl$_5$)$_2$SnPh$_2$	1·05	−0·60 (α-Sn)	1·40	139
(C$_6$F$_5$)SnPh$_3$	0·98	1·25 (SnO$_2$)	1·25	138
(C$_6$F$_5$)$_3$SnMe	1·14	1·19 (SnO$_2$)	1·19	138
(C$_6$F$_5$)$_2$SnMe$_2$	1·48	1·25 (SnO$_2$)	1·25	138
(C$_6$F$_5$)$_3$Sn(C$_6$H$_4$.*p*-Me)	1·02	1·18 (SnO$_2$)	1·18	138
(C$_6$F$_5$)$_2$Sn(C$_6$H$_4$.*p*-Me)	1·18	1·22 (SnO$_2$)	1·22	138
(*o*-Me.C$_6$H$_4$)SnPh$_3$	—	1·30 (SnO$_2$)	1·30	145
(*p*-Me.C$_6$H$_4$)SnPh$_3$	—	1·30 (SnO$_2$)	1·30	145
(*p*-Me.C$_6$H$_4$)SnMe$_3$	—	−0·92 (α-Sn)	1·18	146
(*p*-MeO.C$_6$H$_4$)SnMe$_3$	—	−0·90 (α-Sn)	1·10	146
(*p*-Cl.C$_6$H$_4$)SnMe$_3$	—	−0·86 (α-Sn)	1·14	146
(*p*-F.C$_6$H$_4$)SnMe$_3$	—	1·23 (SnO$_2$)	1·23	148
(*p*-H.C$_6$F$_4$)SnMe$_3$	1·08	1·24 (SnO$_2$)	1·24	148
C$_6$Cl$_5$.SnMe$_3$	1·09	1·32 (SnO$_2$)	1·32	149
C$_6$Cl$_5$.SnPh$_3$	0·84	1·27 (SnO$_2$)	1·27	149
(*o*-CF$_3$.C$_6$H$_4$)SnMe$_3$	0·66	1·21 (SnO$_2$)	1·21	148
Me$_3$SnCF$_3$	1·38	1·31 (SnO$_2$)	1·31	149
1,4-(Me$_3$Sn)$_2$.C$_6$F$_4$	1·20	1·20 (SnO$_2$)	1·20	148
1,4-(Me$_3$Sn)$_2$C$_6$Cl$_4$	1·10	1·26 (SnO$_2$)	1·26	148
(*o*-Br.C$_6$F$_4$)$_2$SnMe$_2$	1·41	1·25 (SnO$_2$)	1·25	148
Me$_3$SnCH$_2$F	—	1·38 (SnO$_2$)	1·38	145
Me$_3$SnCH$_2$Cl	—	1·32 (SnO$_2$)	1·32	145
Et$_3$SnCH$_2$Cl	—	1·43 (SnO$_2$)	1·43	145
Et$_3$SnCH$_2$CN	—	1·29 (SnO$_2$)	1·29	145
Et$_3$SnCH$_2$NMe$_2$	—	1·35 (SnO$_2$)	1·35	145
Me$_3$SnCH$_2$OMe	—	1·38 (SnO$_2$)	1·38	145
Me$_3$SnCH$_2$COMe	1·13	1·38 (SnO$_2$)	1·38	145
Et$_3$SnCH$_2$COMe	1·00	1·35 (SnO$_2$)	1·35	145
Et$_3$SnC≡CSnEt$_3$	1·00	−0·70 (α-Sn)	1·30	137
Pr$^n{}_3$SnC(B$_{10}$H$_{10}$)CH	1·65	−0·60 (α-Sn)	1·40	147
[Pr$^n{}_3$SnC]$_2$(B$_{10}$H$_{10}$)	1·50	−0·65 (α-Sn)	1·35	147
[Pr$^n{}_3$SnC$_2$(B$_{10}$H$_{10}$)]$_2$	1·58	−0·90 (α-Sn)	1·10	147
Ph$_3$SnC(B$_{10}$H$_{10}$)CPh	1·20	−0·80 (α-Sn)	1·20	147

Additional data which duplicate the above are given in refs. 12, 126, 127, and 143.

Table 14.8 Tin hydrides, distannane derivatives, and related compounds (data at 80 K)

Compound	Δ /(mm s^{-1})	δ (SnO$_2$) /(mm s^{-1})	Reference
SnH$_4$	—	1·27	151
MeSnH$_3$	—	1·24	151
Me$_2$SnH$_2$	—	1·23	151
Me$_3$SnH	—	1·24	151
Bu^{n}SnH$_3$	—	1·44	151
Bu$^n_2$SnH$_2$	—	1·42	151
Bu$^n_3$SnH	—	1·41	151
PhSnH$_3$	—	1·40	151
Ph$_2$SnH$_2$	—	1·38	151
Ph$_3$SnH	—	1·39	151
Bu$^i_3$SnH	—	1·45	144
Pr$^n_3$SnH	—	1·45	144
Pr$^i_3$SnH	—	1·40	151
Me$_3$SnNa*	—	1·28	146
Bu$^n_2$SnHCl	3·34	1·56	151
(Ph$_3$Sn)$_2$	—	1·35	12
[(p-Cl.C$_6$H$_4$)$_3$Sn]$_2$	—	1·44	141
[(m-F.C$_6$H$_4$)$_3$Sn]$_2$	—	1·40	141
(Et$_3$Sn)$_2$	—	1·55	144
(Ph$_3$Sn)$_4$Sn	—	1·33	126
(Ph$_3$Sn)$_4$Ge	—	1·13	126
(Ph$_3$Sn)$_4$Pb	—	1·39	126
Ph$_2$Sn	—	1·56	12

*Converted from a-Sn by adding 2·00 mm s^{-1}.

Organotin Halides

Although considerable Mössbauer data are available for the organotin halides, lack of structural data frequently makes parameter interpretation difficult and often purely speculative. Me$_3$SnF is 5-coordinate in the solid state with planar Me$_3$Sn units and an axial non-linear unsymmetrical Sn–F . . . Sn bridge [155]. Me$_2$SnF$_2$ is closely related to SnF$_4$ except that the *trans*-non-bridging fluorine atoms are replaced by methyl groups, the overall coordination being 6-coordinate [156]. However, distinct differences are found in the corresponding chloride derivatives [157]. Me$_3$SnCl shows a linear symmetrical Sn–Cl–Sn bridge, while Me$_2$SnCl$_2$ has a considerably distorted *trans*-octahedral geometry. The extent to which higher coordination persists as the organic groups increase in size has not been established. Most solid-state spectroscopic techniques have proved rather unreliable in this respect. The derivatives of the trineophyltin group, [PhC(Me$_2$)CH$_2$]$_3$SnX, are likely to involve considerable steric hindrance from the large enveloping alkyl groups, and chemical and spectroscopic evidence favours 4-coordination [141, 142]. However, the Mössbauer parameters are not sufficiently different from the known polymeric alkyltin halides to make any distinction with confidence. Typical values for the organotin halides are given in Table 14.9.

Table 14.9 Organotin halides (data at 80 K)

Compound	Δ /(mm s^{-1})	δ/(mm s^{-1})	δ (SnO$_2$) /(mm s^{-1})	Reference
Me$_3$SnF	3·47	1·18 (SnO$_2$)	1·18	130
Et$_3$SnF	3·50	−0·75 (α-Sn)	1·25	137
(i-C$_5$H$_{11}$)$_3$SnF	3·77	−0·76 (α-Sn)	1·24	137
Ph$_3$SnF	3·34	1·17 (SnO$_2$)	1·17	130
[PhC(Me$_2$)CH$_2$]$_3$SnF	2·79	1·33 (SnO$_2$)	1·33	142
Me$_3$SnCl	3·32	1·43 (SnO$_2$)	1·43	128
(ClCH$_2$)$_3$SnCl	2·55	−0·58 (α-Sn)	1·42	137
Et$_3$SnCl	3·24	−0·80 (α-Sn)	1·20	137
Ph$_3$SnCl	2·45	1·37 (SnO$_2$)	1·37	130
(p-Cl.C$_6$H$_4$)$_3$SnCl	2·49	1·37 (SnO$_2$)	1·37	130
[PhC(Me$_2$)CH$_2$]$_3$SnCl	2·65	1·39 (SnO$_2$)	1·39	142
[PhC(B$_{10}$H$_{10}$)C]$_3$SnCl	0·40	−0·93 (α-Sn)	1·07	147
(C$_6$F$_5$)$_3$SnCl	1·55	−0·99 (α-Sn)	1·01	143
Me$_3$SnBr	3·28	1·38 (SnO$_2$)	1·38	138
Et$_3$SnBr	3·45	−0·48 (α-Sn)	1·52	160
Pr$^n{}_3$SnBr	2·92	−0·64 (α-Sn)	1·36	137
Bu$^n{}_3$SnBr	3·30	−0·40 (α-Sn)	1·60	137
Ph$_3$SnBr	2·51	−0·80 (α-Sn)	1·20	143
(C$_6$F$_5$)$_3$SnBr	1·60	−0·94 (α-Sn)	1·06	143
(m-CF$_3$.C$_6$H$_4$)$_3$SnBr	1·94	1·22 (SnO$_2$)	1·22	130
[PhC(Me$_2$)CH$_2$]$_3$SnBr	2·65	1·42 (SnO$_2$)	1·42	142
Me$_3$SnI	3·05	−0·62 (α-Sn)	1·38	146
Pr$^n{}_3$SnI	2·70	−0·64 (α-Sn)	1·36	137
Ph$_3$SnI	2·05	1·41 (SnO$_2$)	1·41	145
(p-F.C$_6$H$_4$)$_3$SnI	1·92	1·23 (SnO$_2$)	1·23	130
[PhC(Me$_2$)CH$_2$]$_3$SnI	2·40	1·41 (SnO$_2$)	1·41	142
Me$_2$SnF$_2$	4·65	1·34 (BaSnO$_3$)	1·34	164
Bu$^n{}_2$SnF$_2$	3·90	−0·60 (α-Sn)	1·40	137
Me$_2$SnCl$_2$	3·62	1·52 (SnO$_2$)	1·52	138
Et$_2$SnCl$_2$	3·4	1·6 (SnO$_2$)	1·6	158
Pr$^n{}_2$SnCl$_2$	3·6	1·7 (SnO$_2$)	1·7	158
Bu$^n{}_2$SnCl$_2$	3·25	1·6 (SnO$_2$)	1·6	159
Ph$_2$SnCl$_2$	2·89	1·34 (SnO$_2$)	1·34	138
Me$_2$SnBr$_2$	3·41	1·59 (SnO$_2$)	1·59	138
Bu$^n{}_2$SnBr$_2$	3·15	1·7 (SnO$_2$)	1·7	159
Bu$^n{}_2$SnI$_2$	2·9	1·8 (SnO$_2$)	1·8	159
Bu^{n}SnCl$_3$	3·40	−0·40 (α-Sn)	1·60	137
PhSnCl$_3$	1·80	1·27 (SnO$_2$)	1·27	138
MeSnBr$_3$	1·91	1·41 (SnO$_2$)	1·41	138

[*Refs. on p. 424*]

It is interesting to note that in any series R_xSnX_{4-x} ($x = 0, \ldots, 4$, R = alkyl or aryl, X = halogen) the chemical isomer shift and quadrupole splitting are both considerably greater when $x = 2$ or 3. In particular, there is no suggestion of a monotonic relationship between the chemical isomer shift and the degree of substitution as one might predict if the only factor acting were the electron-withdrawing power of the group bonding to tin. Thus the parameters for Me_4Sn, Me_3SnBr, Me_2SnBr_2, $MeSnBr_3$, and $SnBr_4$ are $\delta = 1\cdot21, 1\cdot38, 1\cdot59, 1\cdot41, 1\cdot15$; $\Delta = 0, 3\cdot28, 3\cdot41, 1\cdot91, 0$ mm s^{-1} respectively. The lack of information concerning coordination numbers and stereochemistry makes any interpretation purely speculative, particularly when so many mutually opposing factors can be enumerated using the semi-empiricism of conventional valence-bond treatments.

Following the postulation of the existence of the Karyagin effect (see Chapter 3.9) in polycrystalline materials as a result of anisotropy of the recoilless fraction, it was claimed that this effect existed in Ph_3SnCl [161]. However, the subsequent demonstration of a similar effect in Ph_2SnCl_2 caused by small amounts of impurity raised some degree of doubt [162]. Furthermore many tin compounds tend to pack with preferential orientation even when finely ground. The resultant intensity asymmetry in the quadrupole spectrum can easily be mistaken for a Karyagin effect. Reorientation of the absorber to check for anisotropy can guard against this particular difficulty, but the Karyagin effect is unfortunately a phenomenon which can only be verified when all other possible explanations have been eliminated. Nevertheless, a Karyagin effect has been claimed in the polymeric compounds Me_3SnF, Ph_3SnF, and $Me_3SnCOOH$, and for $Me_3SnCl.py$ [163]. The effect in Me_2SnF_2 [164] was described in Chapter 3.9.

Organotin Halide Adducts

The 1:1 adduct of Me_3SnCl with pyridine has a trigonal bipyramidal structure with the three methyl groups in the equatorial plane [165]. The quadrupole splitting in this and similar adducts with organic bases (Table 14.10) is essentially constant at about 3·5 mm s^{-1}, and is slightly greater than the value of 3·2 mm s^{-1} for the parent compound, Me_3SnCl [128].

The parameters for the complexes of Me_3SnBr, Bu^n_3SnCl, and Et_3SnBr with seven aromatic bases (examples given in Table 14.10) show no obvious correlation with the structural or electronic parameters of the base [166].

Known structures of octahedral complexes are $Me_2SnCl_2(Me_2SO)_2$ [167], which is *cis*-dichloro-*cis*-bis(dimethylsulphoxide)-*trans*-dimethyltin(IV), and $[Me_2SnClterpy]^+[Me_2SnCl_3]^-$ [168], which features an octahedral cation with *trans*-methyl groups and a trigonal bipyramidal anion with equatorial methyl groups. Mössbauer spectra are not yet available, however, for these particular compounds.

The different effect on the quadrupole splitting of carbon ligands (R) on the

Table 14.10 Organotin halide adducts (data at 80 K)

Compound	Δ /(mm s^{-1})	δ (SnO$_2$) /(mm s^{-1})	Reference
Me$_3$SnCl.py	3·52	1·53	128
Me$_3$SnCl.hexamethylphosphoramide	3·52	1·44	128
Me$_3$SnCl.N,N-dimethylacetamide	3·69	1·50	128
Me$_3$SnCl.p-methylpyridine-N-oxide	3·45	1·44	128
Me$_3$SnCl.Ph$_3$PO	3·49	1·45	128
Me$_3$SnBr.quinoline	3·20	1·34	166
Me$_3$SnBr.pyridine	3·18	1·30	166
Bu$^n{_3}$SnCl.quinoline	2·76	1·25	166
Bu$^n{_3}$SnCl.pyridine	2·84	1·36	166
Et$_3$SnBr.quinoline	2·60	1·11	166
Et$_3$SnBr.pyridine	3·06	1·43	166

one hand and ligands with non-bonding pairs of electrons on donor atoms (X) such as Cl, Br, I, N, and O finds practical application in the determination of stereochemistry [169–173]. For example, the compound Ph$_2$SnCl$_2$.phen can be considered to be of the type R$_2$SnX$_4$ and one can distinguish two stereochemistries depending on whether the R groups are cis- or trans- to each other. The fact that all the X groups are not chemically identical seems not to affect the issue. In Table 14.11 several compounds of this type are listed. The compounds known to be trans-isomers from other evidence have quadrupole splittings in the range 3·75–4·32 mm s^{-1} with the exception of Ph$_2$Sn(acac)$_2$, which is thought to have been incorrectly assigned. The known cis-isomers have values of 1·78–1·98 mm s^{-1}. The 2:1 ratio in quadrupole splittings is that predicted, and verified for low-spin iron(II) complexes in Chapter 7. Since the relationship appears to hold here also, it is possible to assign the previously unknown stereochemistries given in the final column of the table on the basis of the Mössbauer evidence.

However, the relationship will only be valid for 'similar' groups of compounds. Thus Bu$^n{_2}$Sn[O$_2$P(OEt)$_2$]$_2$ and Me$_2$Sn(OCHO)$_2$ are probably octahedral but have quadrupole splittings of 4·79 mm s^{-1} and 4·72 mm s^{-1}, which are abnormally high; it may be relevant that both contain donor atoms involved in localised π-bonding elsewhere in the ligand molecule [170].

It is comparatively easy to obtain spectra of frozen solutions of organotin compounds. The spectrum of e.g. Ph$_2$SnCl$_2$ dissolved in non-polar solvents such as benzene and hexane is essentially that of the parent solid [174]. However, for the same compound in polar solvents such as ethyl alcohol, pyridine, and acetone, there is a substantial increase in the quadrupole splitting. The latter is concentration independent in dilute solution, but concentration dependent at low solvent/solute molar ratio. The solute/solvent interaction may be assumed to effectively cause an increase in the coordination number of the tin [174]. Similar effects were found in $Bu^n{}_2SnCl_2$

Table 14.11 Octahedral complexes: R_2SnX_4 (data at 80 K)

Compound	Δ /(mm s^{-1})	δ (SnO$_2$) /(mm s^{-1})	Stereochemistry from other techniques	Stereochemistry from Mössbauer	Reference
$Bu^n{}_2SnCl_2$.phen	4·07	1·69	*trans*	*trans*	171
$Bu^n{}_2SnCl_2$.bipy	3·83	1·56	*trans*	*trans*	171
$Bu^n{}_2SnBr_2$.phen	3·94	1·63	*trans*	*trans*	171
$Bu^n{}_2SnI_2$.phen	3·75	1·69	*trans*	*trans*	171
$Bu^n{}_2SnBr_2$.bipy	3·95	1·62	*trans*	*trans*	171
Ph_2SnCl_2.phen	3·70	1·28	—	*trans*	170
Ph_2SnCl_2.bipy	3·90	1·35	—	*trans*	170
Ph_2SnCl_2.py$_2$	3·39	1·32	—	*trans*	172
Ph_2SnBr_2.py$_2$	3·49	1·34	—	*trans*	172
Ph_2SnBr_2.bipy	3·52	1·33	—	*trans*	172
Ph_2SnI_2.bipy	3·35	1·41	—	*trans*	172
Ph_2SnCl_2.dipyam	3·58	1·23	—	*trans*	172
Ph_2SnBr_2.dipyam	3·45	1·34	—	*trans*	172
Ph_2SnCl_2.tripyam	3·59	1·29	—	*trans*	172
Ph_2SnBr_2.tripyam	3·62	1·38	—	*trans*	172
Et_2SnCl_2.dipyam	3·78	1·68	—	*trans*	172
Et_2SnBr_2.dipyam	3·64	1·72	—	*trans*	172
Me_2SnCl_2.bipy	4·02	1·35	*trans*	*trans*	146
Me_2SnCl_2.py$_2$	3·83	1·27	—	*trans*	146
$MeSnCl_2$.phen	4·03	1·32	—	*trans*	170
$[Me_2SnCl_4][pyH]_2$	4·32	1·59	*trans*	*trans*	170
$[Ph_2SnCl_4][pyH]_2$	3·80	1·44	—	*trans*	170
Me_2Sn.oxin$_2$	1·98	0·88	*cis*	*cis*	173
$Pr^n{}_2Sn$.oxin$_2$	2·20	1·02	—	*cis*	170
Ph_2Sn.oxin$_2$	1·78	0·83	*cis*	*cis*	170
$Bu^n{}_2Sn$.oxin$_2$	2·21	1·10	—	*cis*	170
Et_2Sn.oxin$_2$	2·02	0·99	—	*cis*	173
n-octyl$_2$Sn.oxin$_2$	1·86	1·13	—	*cis*	173
Ph_2Sn.acac$_2$	2·14	0·74	*trans*	*cis*	170
Me_2Sn.acac$_2$	3·93	1·18	*trans*	*trans*	170

py = pyridine
phen = 1,10-phenanthroline
bipy = 2,2'-bipyridyl
dipyam = di-2-pyridylamine
tripyam = tri-2-pyridylamine
oxin = 8-hydroxyquinoline

[175] and are illustrated in Fig. 14.11. On this basis the sequence of solvent donor strengths would be

$$Et_2O < EtOCH_2CH_2OEt < C_4H_8O < MeOCH_2CH_2OMe$$
$$< (Me_2N)_3PO < HC(O)NMe_2 < Me_2SO_2.$$

[*Refs. on p. 424*]

However, the detailed interaction in the solution cannot be established from these data above, and the prime assumption must always be made that no significant change occurs during the freezing process.

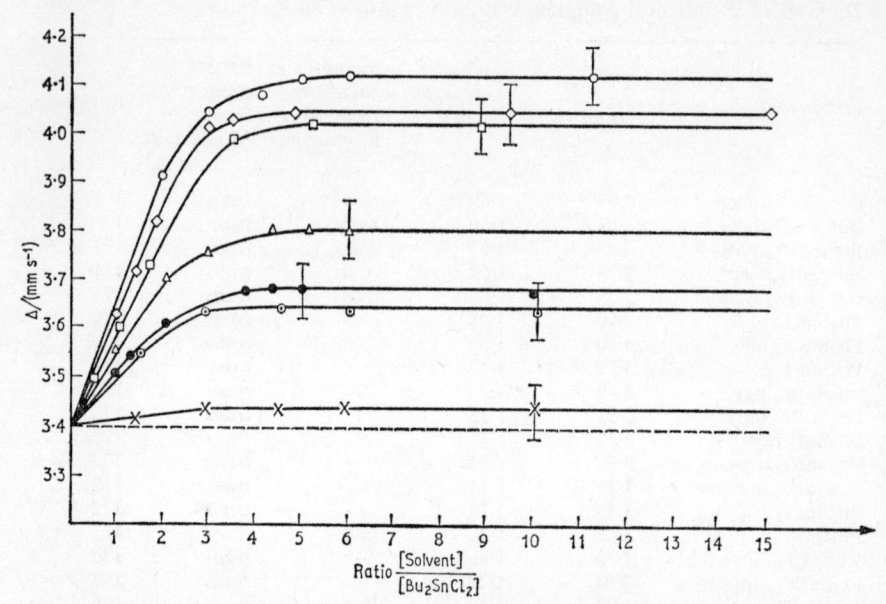

Fig. 14.11 Concentration dependence of quadrupole splitting of Bu$^n{}_2$SnCl$_2$ in aprotic donor solvents. ○ Dimethyl sulphoxide, ◇ dimethyl-formamide, ☐ hexamethyltriamidophosphate, △ dimethoxyethane, ● tetrahydrofuran, ▨ diethoxyethane, × diethyl ether. [Ref. 175, Fig. 1]

Cyanides and Thiocyanates

The crystal structure of Me$_3$SnNCS shows the tin to be 5-coordinate with a nearly linear S–Sn–N–C–S skeleton and a planar Me$_3$Sn group [176]. The Sn–N bond distance is 2·15 Å, and the Sn–S distance is 3·13 Å. Me$_3$SnCN has a similar structure with Sn–CN and Sn–NC bond-lengths of 2·48 Å [177]. Mössbauer parameters for several CN and NCS compounds have been given [178, 179], and these are shown in Table 14.12. The R$_3$SnNCS and R$_3$SnCN derivatives all show similar parameters and probably have the same basic structure, whereas other chemical evidence suggests that the R$_2$Sn(NCS)$_2$ compounds are 6-coordinate with four bridging NCS groups. The absence of quadrupole interactions and the low chemical isomer shift value in Sn(NCS)$_4$ itself is noteworthy.

Also included in Table 14.12 are some adducts with organic ligands [178]. By analogy with the halide complexes in Table 14.11, the quadrupole splittings in Bu$^n{}_2$Sn(NCS)$_2$.bipy and Bu$^n{}_2$Sn(NCS)$_2$.phen imply a *trans*-octahedral

[Refs. on p. 424]

geometry, while those in the corresponding phenyl derivatives suggest a *cis*-octahedral configuration.

Table 14.12 Organotin cyanides and thiocyanates (data at 80 K)

Compound	Δ /(mm s^{-1})	δ/(mm s^{-1})	δ (SnO$_2$) /(mm s^{-1})	Reference
Me$_3$SnNCS	3·77	1·40 (SnO$_2$)	1·40	179
Et$_3$SnNCS	3·80	1·57 (SnO$_2$)	1·57	179
Bu^n$_3$SnNCS	3·69	1·60 (SnO$_2$)	1·60	179
Ph$_3$SnNCS	3·50	1·35 (SnO$_2$)	1·35	179
Me$_2$Sn(NCS)$_2$	3·87	1·48 (SnO$_2$)	1·48	179
Et$_2$Sn(NCS)$_2$	3·96	1·56 (SnO$_2$)	1·56	179
Bu^n$_2$Sn(NCS)$_2$	3·88	1·56 (SnO$_2$)	1·56	179
Bu^{n}Sn(NCS)$_3$	1·46	1·43 (SnO$_2$)	1·43	179
Sn(NCS)$_4$	0	0·56 (SnO$_2$)	0·56	179
Me$_3$SnCN	3·12	1·39 (SnO$_2$)	1·39	179
Et$_3$SnCN	3·19	1·41 (SnO$_2$)	1·41	179
Bu^n$_3$SnCN	3·27	1·37 (SnO$_2$)	1·37	179
Bu^n$_2$Sn(NCS)$_2$.bipy	4·04	−0·09 (Pd/Sn)	1·35	178
Bu^n$_2$Sn(NCS)$_2$.phen	4·18	−0·10 (Pd/Sn)	1·36	178
Ph$_2$Sn(NCS)$_2$.bipy	2·13	−0·70 (Pd/Sn)	0·76	178
Ph$_2$Sn(NCS)$_2$.phen	2·34	−0·71 (Pd/Sn)	0·75	178
Bu$_2$Sn(NCS).oxin	3·25	−0·19 (Pd/Sn)	1·27	178
Ph$_2$Sn(NCS).oxin	2·48	−0·54 (Pd/Sn)	0·92	178

phen = 1,10-phenanthroline bipy = 2,2′-bipyridyl oxin = 8-hydroxyquinoline

Oxygen and Sulphur Derivatives

The organotin hydroxides are probably nearly all polymeric, and indeed Me$_3$SnOH is known to have the oxygen atoms midway between planar Me$_3$Sn groups [180]. A possible exception is the trisneophyl derivative [PhC(Me$_2$)CH$_2$]$_3$SnOH in which steric hindrance may preserve fourfold coordination [142], consistent with this the quadrupole splitting is much lower than in other alkyl and aryl derivatives (see Table 14.13) [130, 137, 146]. A Karyagin effect has been claimed in Me$_3$SnOH [181]. The alkoxy derivatives, R$_3$SnOR′, and trialkyltin oxides, [R$_3$Sn]$_2$O, have also been studied and sample parameters are given in Table 14.13.

The dialkyltin oxides, R$_2$SnO, are believed to be network coordination polymers with 5-coordinate tin atoms [137, 184, 185], and the effect of the structure on the recoil-free fraction has been discussed.

Organotin Carboxylates

Considerable data are available for the organotin carboxylates (see Table 14.14) [130, 144]. Both polymeric and monomeric compounds occur, e.g. the X-ray structure of (PhCH$_2$)$_3$SnOC(O)Me shows a polymeric trigonal bipyramidal stereochemistry [188], whereas (C$_6$H$_{11}$)$_3$SnOC(O)Me has sufficient

[*Refs. on p. 424*]

Table 14.13 Mössbauer parameters for organotin oxygen and sulphur derivatives (data at 80 K)

Compound	Δ /(mm s^{-1})	δ/(mm s^{-1})	δ (SnO$_2$) /(mm s^{-1})	Reference
Me$_3$SnOH	2·71	1·07 (SnO$_2$)	1·07	130
Et$_3$SnOH	3·24	−0·75 (α-Sn)	1·25	137
Bu$^n_3$SnOH	3·24	−0·64 (α-Sn)	1·36	137
Ph$_3$SnOH	2·68	1·18 (SnO$_2$)	1·18	130
(neophyl)$_3$SnOH	1·08	1·13 (SnO$_2$)	1·13	142
Et$_3$Sn(OMe)	2·86	1·41 (SnO$_2$)	1·41	182
Et$_3$Sn(OCMe$_3$)	2·59	1·40 (SnO$_2$)	1·40	182
Me$_2$Sn(OMe)$_2$	2·31	0·99 (SnO$_2$)	0·99	130
Bu$^n_2$Sn(OEt)$_2$	2·00	1·30 (SnO$_2$)	1·30	130
(Bu$^n_3$Sn)$_2$O	2·40	1·10 (SnO$_2$)	1·10	158
(Ph$_3$Sn)$_2$O	2·15	1·08 (SnO$_2$)	1·08	130
[(C$_6$F$_5$)$_3$Sn]$_2$O	2·13	−0·99 (α-Sn)	1·01	143
[Bu$^n_2$SnCl]$_2$O	3·24	−0·70 (α-Sn)	1·30	137
[Bu$^n_2$SnOCO(CH$_2$)$_3$Me]$_2$O	3·28	1·53 (SnO$_2$)	1·53	183
(Ph$_3$Sn)$_2$S	1·17	1·22 (SnO$_2$)	1·22	130
Me$_2$SnO	1·82	0·92 (SnO$_2$)	0·92	184
Et$_2$SnO	2·10	1·05 (SnO$_2$)	1·05	185
Bu$^n_2$SnO	2·06	1·08 (SnO$_2$)	1·08	184
Ph$_2$SnO	1·73	0·88 (SnO$_2$)	0·88	184
(*p*-I.C$_6$H$_4$)$_2$SnO	1·73	0·84 (SnO$_2$)	0·84	184
Bu$^n_2$SnS	1·9	0·9 (SnO$_2$)	0·9	159

steric hindrance from the cyclohexyl groups to remain 4-coordinate [189]. The triphenyltin carboxylates Ph$_3$SnOC(O)R' have been shown to retain the bridged structure as the chain length of R' increases from —Me to —(CH$_2$)$_{14}$Me and there is almost no change in the Mössbauer spectrum [186]. However, when the R' alkyl group is branched at the α-position there is sufficient steric hindrance to prevent polymer formation and there is a significant reduction in both the quadrupole splitting and the chemical isomer shift [186].

The compounds Me$_3$SnOC(O)R (R = Me, CH$_2$I, CH$_2$Br, CH$_2$Cl, CHCl$_2$, CBr$_3$, CCl$_3$, and CF$_3$) show a linear correspondence between the quadrupole splitting and both the pK of the haloacetic acid and the Taft inductive factor σ* [187]. Mössbauer and infrared data are consistent with a *penta*-coordinate polymer structure. The products RSn(O)OC(O)R' have unusually low chemical isomer shifts, presumably because of coordination through several oxygens [186].

The organotin derivatives of inorganic acids are basically similar, but give very large quadrupole splittings (Table 14.14). The neophyl derivatives [PhC(Me$_2$)CH$_2$]$_3$SnNO$_3$ and [PhC(Me$_2$)CH$_2$]$_3$SnClO$_4$ have been suggested

Table 14.14 Mössbauer parameters for organotin carboxylates (data at 80 K)

Compound	Δ /(mm s^{-1})	δ/(mm s^{-1})	δ (SnO$_2$) /(mm s^{-1})	Reference
Me$_3$SnOC(O)Me	3·68	1·35 (SnO$_2$)	1·35	186
Bu$^n{}_3$SnOC(O)Me	3·64	1·46 (SnO$_2$)	1·46	186
Ph$_3$SnOC(O)H	3·58	1·37 (SnO$_2$)	1·37	186
Ph$_3$SnOC(O)Et	3·42	1·33 (SnO$_2$)	1·33	186
Ph$_3$SnOC(O)(CH$_2$)$_6$Me	3·35	1·29 (SnO$_2$)	1·29	186
Ph$_3$SnOC(O)(CH$_2$)$_{14}$Me	3·44	1·25 (SnO$_2$)	1·25	186
Ph$_3$SnOC(O)CMe:CH$_2$	2·26	1·21 (SnO$_2$)	1·21	186
Ph$_3$SnOC(O)CH(Et)Bu	2·26	1·21 (SnO$_2$)	1·21	186
Ph$_3$SnOC(O)CMe$_3$	2·40	1·21 (SnO$_2$)	1·21	186
[PhC(Me$_2$)CH$_2$]$_3$SnOC(O)Me	2·45	1·35 (SnO$_2$)	1·35	142
Me$_3$SnOC(O)CH$_2$Cl	3·89	1·41 (SnO$_2$)	1·41	187
Me$_3$SnOC(O)CHCl$_2$	4·08	1·37 (SnO$_2$)	1·37	187
Me$_3$SnOC(O)CF$_3$	4·22	1·38 (SnO$_2$)	1·38	187
Bu$^n{}_2$Sn[OC(O)Me]$_2$	3·50	1·34 (SnO$_2$)	1·34	184
Bu$^n{}_2$Sn[OC(O)CCl$_3$]$_2$	4·00	−0·50 (α-Sn)	1·50	184
Bu$^n{}_2$Sn[OC(O)Et]$_2$	3·70	1·49 (SnO$_2$)	1·49	183
Bu$^n{}_2$Sn[OC(O)(CH$_2$)$_{16}$Me]$_2$	3·56	1·36 (SnO$_2$)	1·36	184
Bu$^n{}_2$Sn(OMe)[OC(O)Me]	3·24	−0·80 (α-Sn)	1·20	137
PhSn(O)OC(O)(CH$_2$)$_8$CH:CH$_2$	2·31	0·57 (SnO$_2$)	0·57	186
PhSn(O)OC(O)(CH$_2$)$_{16}$Me	2·32	0·56 (SnO$_2$)	0·56	186
PhSn(O)OC(O)Me	2·26	0·70 (SnO$_2$)	0·70	186
PhSn(O)OH	1·71	0·65 (SnO$_2$)	0·65	186
Me$_2$SnSO$_4$	5·00	1·61 (SnO$_2$)	1·61	133
Bu$^n{}_2$SnO$_4$	4·8	1·8 (SnO$_2$)	1·8	159
Bu$^n{}_2$SnSO$_3$	4·0	1·3 (SnO$_2$)	1·3	159
[PhC(Me$_2$)CH$_2$]$_3$SnNO$_3$	3·18	1·40 (SnO$_2$)	1·40	142
[PhC(Me$_2$)CH$_2$]$_3$SnClO$_4$	3·83	1·57 (SnO$_2$)	1·57	142
Me$_2$Sn(SO$_3$F)$_2$	5·54	1·82 (SnO$_2$)	1·82	133
Me$_2$Sn(SO$_3$CF$_3$)$_2$	5·51	1·79 (SnO$_2$)	1·79	133
Me$_2$Sn(SO$_3$Cl)$_2$	5·20	1·75 (SnO$_2$)	1·75	133
Me$_2$Sn(SO$_3$Me)$_2$	5·05	1·52 (SnO$_2$)	1·52	133
Me$_2$Sn(SO$_3$Et)$_2$	4·91	1·52 (SnO$_2$)	1·52	133

Additional data contained in refs. 130, 144.

to be polymeric [142]. The dimethyltin derivatives of substituted sulphonic acids (Table 14.14) give the largest known quadrupole splittings in ^{119}Sn, e.g. Δ for Me$_2$Sn(SO$_3$CF$_3$)$_2$ is 5·51 mm s^{-1}. The compounds are believed to adopt a polymeric *trans*-octahedral structure [133].

Derivatives of 8-Hydroxyquinoline

Data for several 8-hydroxyquinoline (oxin) derivatives were given in Table 14.11 for convenient comparison with other octahedral compounds. Me$_2$Sn.oxin$_2$ is known from X-ray measurements to have a highly distorted

[*Refs. on p. 424*]

octahedral coordination with *cis*-methyl groups [190]. All the $R_2Sn.oxin_2$ compounds appear to have this stereochemistry [173]. The Mössbauer parameters of all the $R_2SnX.oxin$ compounds where X is a halogen (Table 14.15) are similar with quadrupole splittings in the range 2·40–3·36 mm s^{-1};

Table 14.15 Organotin compounds with 8-hydroxyquinoline (data at 80 K)

Compound	Δ /(mm s^{-1})	δ (SnO$_2$) δ/(/(mm s^{-1})	Reference
Me$_2$SnCl.oxin	3·12	1·26	173
Et$_2$SnCl.oxin	3·13	1·34	173
Et$_2$SnBr.oxin	3·08	1·39	173
Et$_2$SnI.oxin	2·85	1·43	173
Et$_2$Sn(NCS).oxin	3·07	1·31	173
Pr$^n{}_2$SnCl.oxin	2·78	1·31	173
Bu$^n{}_2$SnCl.oxin	3·21	1·40	173
Oct$^n{}_2$SnCl.oxin	3·36	1·56	173
Ph$_2$SnCl.oxin	2·40	1·12	173
Me$_2$Sn.oxin$_2$	1·93	0·85	82
BuSnCl.oxin$_2$	1·67	0·84	173
PhSnCl.oxin$_2$	1·48	0·66	173
Ph$_3$Sn.oxin	1·75	1·07	173
BuSn.oxin$_3$	1·82	0·69	173

they probably contain a penta-coordinate structure with a bidentate oxin-group. The $RSnX.oxin_2$ and $RSn.oxin_3$ compounds are also believed to feature high coordination [173].

Organotin–Transition-metal Compounds

A number of organotin compounds with a tin–transition-metal bond have been examined; this bond has usually been to iron to facilitate study

Table 14.16 Iron and tin Mössbauer parameters for iron–organotin complexes (data at 80 K and values for δ(Fe) converted from nitroprusside)

Compound	^{57}Fe Δ /(mm s^{-1})	δ (Fe) /(mm s^{-1})	^{119}Sn Δ /(mm s^{-1})	δ (SnO$_2$) /(mm s^{-1})	Reference
Me$_2$Sn[Fe(CO)$_4$]$_2$SnMe$_2$	0·15	−0·11	1·22	1·47	191
{Me$_2$Sn[Fe(CO)$_4$]$_2$}$_2$Sn	0·30	−0·10	{1·24 —	1·45 2·20}	191
[(π-C$_5$H$_5$)Fe(CO)$_2$]$_2$SnCl$_2$	1·68	0·10	2·38	1·95	194
[(π-C$_5$H$_5$)Fe(CO)$_2$]$_2$GeCl$_2$	1·66	0·10	—	—	194
[(π-C$_5$H$_5$)Fe(CO)$_2$]SnCl$_3$	1·86	0·14	1·77	1·74	194
[(π-C$_5$H$_5$)Fe(CO)$_2$]SnPh$_3$	1·83	0·11	0	1·50	194

[*Refs.* on p. 424]

with both the ^{57}Fe and ^{119}Sn resonances. Parameters are given in Table 14.16.

Me$_2$Sn[Fe(CO)$_4$]$_2$SnMe$_2$ and Me$_2$Sn[Fe(CO)$_4$]$_2$Sn[Fe(CO)$_4$]$_2$SnMe$_2$ are closely related [191], and the crystal structure of the latter is known (struc-

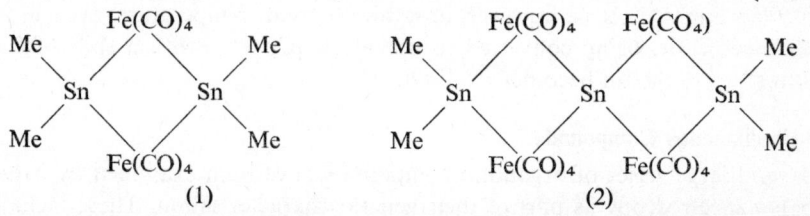

(1)   (2)

ture 1) [192]. The iron atoms feature 6-coordination and only a small quadrupole splitting. The central tin atom in the second compound has four bonds to iron and an approximately tetrahedral environment, consequently showing no quadrupole splitting and a chemical isomer shift close to FeSn$_2$ and β-Sn.

The crystal structure of [(π-C$_5$H$_5$)Fe(CO)$_2$]$_2$SnCl$_2$ is also known [193] and shows unusually short Sn–Fe bonds and long Sn–Cl bonds in the 4-coordinate structure (see structure 2). The Mössbauer spectra of this and other compounds in the solid state and in frozen solution tend to confirm that the spectra reflect the internal bonding of the molecules and are largely uninfluenced by intermolecular or crystal packing effects [194, 195]. The compound shown in structure 2 also gives evidence for the presence of two rotational isomers in the frozen solutions [194]. The compound triphenyltin derivative [(π-C$_5$H$_5$)Fe(CO)$_2$]SnPh$_3$ is unusual in that the ^{119}Sn resonance shows zero quadrupole splitting; the Fe–Sn bond is thus similar in effect to the Sn–Sn and Pb–Sn bonds discussed on p. 403.

Brief details have also been given of other transition-metal–tin compounds [196–198] (see Table 14.17), and although several contain nominal SnCl$_3^-$

Table 14.17 Mössbauer parameters for other transition-metal complexes (data at 80 K)

Compound	Δ /(mm s^{-1})	δ (SnO$_2$) /(mm s^{-1})	Reference
Sn[Co(CO)$_4$]$_4$	0	1·96	196
[Et$_4$N]$_3$Pt(SnCl$_3$)$_5$	1·53	1·64	196
[Et$_4$N]$_2$PtCl$_2$(SnCl$_3$)$_2$	1·61	1·56	196
[Me$_4$N]$_2$PtCl$_2$(SnCl$_3$)$_2$	1·61	1·80	196
[Me$_4$N]$_4$Rh$_2$Cl$_2$(SnCl$_3$)$_4$	1·62	1·90	196
(PPh$_3$)$_3$Rh(SnCl$_3$)	1·73	1·78	196
(C$_8$H$_{12}$)$_2$Ir(SnCl$_3$)	1·64	1·80	196
[Me$_4$N]$_2$RuCl$_2$(SnCl$_3$)$_2$	1·64	1·93	196
Mn(CO)$_5$SnCl$_3$	1·56	1·73	197
Mn(CO)$_5$SnBr$_3$	1·44	1·84	197
Mn(CO)$_5$SnPh$_3$	0	1·45	197
[PdCl$_2$(SnCl$_3$)$_2$]$^{2-}$	2·21	1·42	198

units and are prepared from a tin(II) starting material, they are invariably tin(IV) in character.

The only real exception to this is dicyclopentadienyltin(II) [199], Cp_2Sn, which has a chemical isomer shift of 3·73 mm s^{-1} and a quadrupole splitting of 0·65 mm s^{-1}. It is, however, unstable at room temperature, even in the absence of air, being converted to polymeric $[Cp_2Sn]_n$ with a shift of 0·72 mm s^{-1}, i.e. the tin becomes oxidised.

Miscellaneous Compounds

Several large series of organotin compounds have been examined by Mössbauer spectroscopy as part of their general characterisation. These include:

(i) substituted mercapto- and hydroxo-compounds of the types (3)–(9), in which R = Et or Ph.

The ranges of parameters observed were [200]:

		$\delta(SnO_2)/(mm\ s^{-1})$	$\Delta/(mm\ s^{-1})$
sulphur derivatives	R = Et	1·55–1·62	2·07–3·05
	R = Ph	1·36–1·40	1·16–2·61
oxygen derivatives	R = Et	1·34–1·56	2·86–3·77
	R = Ph	1·31–1·36	2·35–3·20

The quadrupole splitting is clearly more sensitive to substitution than is the chemical isomer shift. It was concluded that the mercaptopyridines are probably 5-coordinated with both sulphur *and* nitrogen bonding.

(ii) *p*-substituted phenolate and thiophenolate derivatives [201] of the types $Et_3SnO(p\text{-}C_6H_4X)$ where X = H, MeO, F, Br, CHO, SCN, NO_2, Me_2N, Me, Cl, I, $MeCO_2$, CN [$\delta(SnO_2)$ in the range 1·31–1·41 mm s^{-1}, Δ 2·96–3·82 mm s^{-1}]; $Et_3SnS(p\text{-}C_6H_4X)$, X = H, MeO, F, NO_2, Me_2N, Me, Cl [$\delta(SnO_2)$ in the range 1·35–1·52 mm s^{-1}, Δ 2·07–2·41 mm s^{-1}].

(iii) dithiolate complexes with 1,2-ethanedithiolate (EDT) and 3,4-toluenedithiolate (TDT), examples being $Sn(EDT)_2$, $Sn(EDT)_2phen$, $Sn(TDT)_2$,

[*Refs. on p. 424*]

Sn(TDT)$_2$py, Me$_2$SnEDT, Me$_2$SnEDTpy, and Ph$_2$SnTDTpy [202]. The Mössbauer parameters vary considerably [δ(SnO$_2$) in the range 0·50–1·36 mm s^{-1}; Δ 0·84–2·62 mm s^{-1}], and most of the complexes are believed to have polymeric structures.

(iv) some *bis*-dithiocarbamate derivatives of the types Ph$_2$Sn[S$_2$CNR$_2$]$_2$ where R = Ph, Et, CH$_2$Ph, (CH$_2$)$_4$; [δ(SnO$_2$) = 1·08–1·19 mm s^{-1}; Δ = 1·66–1·76 mm s^{-1}] and R$_2$'Sn[S$_2$CNR$_2$]$_2$ where R' = Bu^n, Me [δ(SnO$_2$) = 1·54–1·69 mm s^{-1}; Δ = 2·85–3·38 mm s^{-1}]. 4- and 6-coordination have been proposed for the respective groups [203].

(v) parameters from various sources have also been listed without comment for a number of compounds [137]. Examples include Me$_3$Sn(1,2,3-triazole) and Ph$_3$Sn(1,2,4-triazole).

14.6 Metals and Alloys
Tin Metal

Properties of α-(grey) and β-(white) tin relevant to their use as source matrices and the interpretation of the chemical isomer shift were discussed in Sections 14.1 and 14.2.

β-tin, which is the form stable at room temperature, has a tetragonal unit cell in which each tin atom is surrounded by a distorted tetrahedron of neighbours at 3·02 Å and by two more only slightly further away at 3·17 Å. The linewidth of the β-Sn resonance is broader than expected for an unsplit resonance, and quadrupole splitting of $\sim$0·3 mm s^{-1} has been suggested [13]. This results in a small change in the apparent position of the line in single crystals oriented successively in the [100] and [001] directions [204–206]. An estimate of 0·2 mm s^{-1} for the quadrupole splitting has also been obtained from the anisotropy of the angular dependence of resonantly scattered γ-radiation [207].

A thermal red-shift in β-Sn was first reported in 1960 [16]. A more accurate study in the range 3·6–90 K has taken into account the contributions due to relativistic time-dilation, temperature dependence of the chemical isomer shift from volume changes, and the effect of unresolved quadrupole asymmetry on the effective line position [208].

The recoil-free fraction of β-tin has been measured in many laboratories, but with little consistency of results. Probably the most accurate results are by Hohenemser [209], who tabulates previous measurements, and his own data for the temperature range 1·3–370 K: $f = 0·72 \pm 0·01$ at 4·2 K, 0·455 $\pm$ 0·010 at 77·3 K and 0·039 $\pm$ 0·010 at 300 K. The effective Debye temperature of about 135 K is almost independent of temperature in this region.

The recoil-free fraction is anisotropic, and the temperature dependence of f

for different orientations of single crystals is not the same [206]. Application of pressures of up to 110 kbar at room temperature causes a significant increase in f and a change in the chemical isomer shift of 0.25×10^{-2} mm s^{-1} per kbar [210]. The recoil-free fraction is smaller in samples of β-tin with small particle size (1550–250 Å), there being some correlation with the total surface area which increases as the particle diameter decreases [211]. Other earlier discussions of the lattice-dynamical properties of β-tin are available [6, 212–214].

A premelting anomaly observed as a reduction in the intensity of the Mössbauer line within 3° of the melting point was originally attributed to solid-state diffusion [214]. However, this has now been shown to be due to the presence of impurity, there being no such effect in 99·999% purity samples [215].

Monolayers of tin can be formed on platinum electrodes in an acid solution of the metal cation if the potential is more anodic than required to produce a multilayer deposit. The spectrum of such a monolayer was observed using ^{119m}Sn activity in solution, and showed the following features [216]:
(a) the chemical isomer shift was 2·25 mm s^{-1} (SnO$_2$) compared with 2·56 mm s^{-1} for β-Sn and is closer to the value for ^{119}Sn in platinum;
(b) the recoilless fraction of the adsorbed tin was greater than for β-tin because of firmer binding to the platinum. This is also implied by the use of an anodic potential;
(c) there is a quadrupole splitting of 1·4 mm s^{-1} as a result of an appreciable electric field gradient perpendicular to the surface of the platinum. However, no Karyagin effect due to anisotropy was seen.

Fractional tin layers produced at higher potential showed a larger quadrupole splitting and smaller chemical isomer shift, the former a result of lower symmetry in the plane of the surface and the latter suggesting that isolated adsorbed atoms are more firmly bound than the monolayer.

Another unusual surface layer experiment has studied the interaction of tin foil with bromine vapour [217]. The energy of conversion electrons emitted following resonant absorption is a function of the depth of the emitting nucleus below the material surface. A focussing β-spectrometer was used to characterise the different layers near the surface of the Mössbauer absorber. An outer coating of SnBr$_4$ was found, together with an intermediate layer of SnBr$_2$.

It is commonly assumed in the study of impurity atoms in metals that the properties of the host are effectively unchanged. β-Sn containing $\sim$1% of Ge, In, Sb, Pb, or Bi impurity atoms shows no change in the chemical isomer shift, but 0·5% of Zn, Cd, or Na can cause an increase in shift of as much as 0·06 mm s^{-1} [218]. This implies the existence of a long-range interaction between the impurity and host atoms which perturbs the electron density in the alloy.

[Refs. on p. 424]

The recoil-free fraction, effective Debye temperature, and quadrupole interaction show no change within experimental error between the normal and superconducting states of tin and tin–indium alloys [6, 219].

Binary and Ternary Alloys

The alloy Mn_4Sn is magnetically ordered below about 150 K, and at low temperatures shows an induced magnetic field (at saturation) of about −45 kG at the tin nucleus [37, 220]. Mn_2Sn is also magnetically ordered, but with a larger field of about +200 kG. The signs were determined by application of an external magnetic field.

The temperature dependence of the internal field at ^{119}Sn in $MnSn_2$ has been followed up to the Néel temperature of 324 K. It follows a Brillouin function, the value at 83 K being 65 kG [221].

$FeSn_2$ shows a ^{57}Fe field of 115 kG at 295 K, the temperature dependence of which gives a Néel temperature of 377 K. The ^{119}Sn spectrum shows a field of approximately 25 kG at 295 K with signs of a small quadrupole interaction [222–223]. The results have been confirmed independently, and the ^{119}Sn chemical isomer shift found to be $\sim 2 \cdot 1$ mm s^{-1} [224]. Brief details have also been given for the alloys Fe_3Sn, Fe_5Sn_3, Fe_3Sn_2, and $FeSn$ [225].

The copper–tin system shows a non-linear dependence of the chemical isomer shift on the tin concentration [226]. The copper-rich α-phase (<10 at. % Sn) has a face-centred cubic structure, and the shift remains nearly constant at about 1·61 mm s^{-1} until the β-phase is reached, whereupon there is an almost linear concentration dependence encompassing the β (~ 15%), γ (~ 16–20%), ε (Cu_3Sn), and η (Cu_6Sn_5) phases. This is illustrated in Fig. 14.12, which also incorporates the value for β-tin itself. The conduction

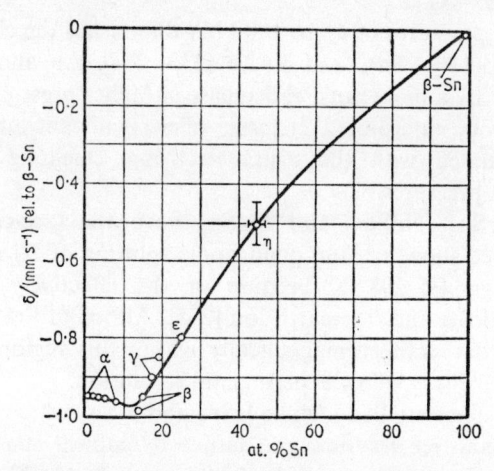

Fig. 14.12 The chemical isomer shift of ^{119}Sn in the Sn–Cu phases. [Ref. 226]

[*Refs. on p. 424*]

electrons of copper may be described as a half-filled collectivised s-band, and the tin atoms initially introduced give up their valence electrons to the conduction band as shown by the low s-electron density with respect to β-Sn. The electron density at the tin remains essentially constant until this band is filled, and it is only then that the tin plays a significant part in the band structure of the metals.

By contrast, the chemical isomer shift of the alloys Pd_3Sn, Pd_2Sn, Pd_3Sn_2, PdSn, $PdSn_2$, $PdSn_4$, and a solid solution of tin in Pd (>9·5 at. % Sn) show a strictly linear relationship with the atomic percentage of tin [227, 228]. Palladium metal has 9·4 electrons per atom in the 4d-band with 0·6 electrons in the 5s-band. An increase in the tin content causes a gradual filling of the palladium d-band by the tin valence electrons, and again the negative chemical isomer shifts with respect to β-tin indicate a lowering of the s-electron density at the tin nucleus. As this d-band fills, the proportion of electrons localised in tin 5s-orbitals increases and causes the observed increase in chemical isomer shift. A similar situation is found in the platinum alloys Pt_3Sn, PtSn, Pt_2Sn_3, $PtSn_2$, and $PtSn_4$ [229].

The chemical isomer shift of ^{119}Sn has been measured over the entire range of composition of tin with the six metals Cd, In, Sb, Tl, Pb, and Bi [230]. Considerable correlation with the known phase diagrams of the binary alloys was found and in the cases of In, Tl, and Pb, large changes in chemical isomer shift were recorded in the vicinity of phase boundaries.

The Heusler alloys Cu_2FeSn, Cu_2CoSn, and Cu_2NiSn give only a single resonance line [42], but Co_2MnSn, Ni_2MnSn, and Cu_2MnSn show substantial magnetic splitting [42, 44, 231]. Published data are very inconsistent because disordering causes inhomogeneities in the magnetic ordering, but fields of up to 235 kG are reported. The field in Co_2MnSn at 4·2 K has been shown to be +107 kG [232].

Application of pressures of up to 100 kbar diminishes the chemical isomer shift of β-Sn, Pd(12% Sn), and SnAu [233, 234]. An anomalous initial increase followed by a more normal decrease at higher pressures is found for Mg_2Sn and may be interpreted in terms of a significant amount of band repopulation coupled with the usual electronic shielding effects which dominate at high pressures.

The alloys $PrSn_3$, $NdSn_3$, and $SmSn_3$ have the face-centred $AuCu_3$ structure. All three show a ^{119}Sn quadrupole splitting of 1·1 mm s^{-1} in the temperature range 1·5–298 K because of the effectively square-planar coordination of tin to the rare-earth ion [235]. Although $PrSn_3$ and $NdSn_3$ are known to order antiferromagnetically within this region, any induced field at the tin nucleus is below experimental resolution.

Other metal systems studied include $IrSn_2$ and $PtSn_2$ which have the CaF_2 lattice [236]. Recoil-free fraction and lattice dynamical studies have been made on SnAs, SnSb, SnTe, and SnPt [237] and on Nb_3Sn [238, 239].

[Refs. on p. 424]

Tin Impurity Atoms

The use of ^{119m}Sn doping in metals for lattice-dynamical studies is considered to be beyond the scope of this book, but notable contributions include studies of the metals Au, Pt, Th [240], and V [241], and the alloys Sn/Bi, Sn/Cd, Sn/In, Sn/Sb [242], and Pd/Ag, Au/Ag, In/Ag [243], and Pd/H, Pd/Ag [244].

It is interesting to note, however, that there is a relation between the chemical isomer shift of a tin impurity atom and the effective force constant

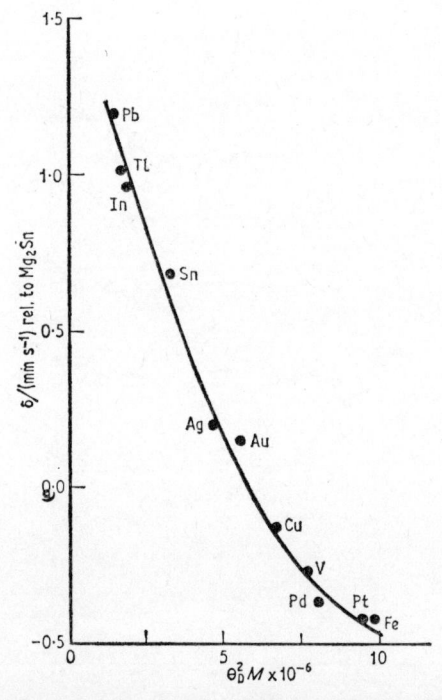

Fig. 14.13 Relation between the chemical isomer shift of ^{119}Sn impurity atoms in various metals and the effective force constant of the host, $\theta_D^2 M$. [Ref. 245, Fig. 1]

of the lattice defined as $\theta_D^2 M$, where M is the mass of the host atom and θ_D is the Debye temperature of the lattice [245]. This is illustrated in Fig. 14.13. Similarly there is a linear relation between the chemical isomer shift and the compressibility of the host matrix [246].

The behaviour of ^{119}Sn in chromium is rather unusual. At temperatures below the Néel temperature one does not see a single unique magnetic field as expected, but rather a broad distribution of fields which results in a diffuse spectrum [247]. Examples are shown in Fig. 14.14. The narrow central line is due to precipitated metallic tin and is not part of the ^{119}Sn/Cr spectrum. The solid curves were calculated assuming that the field was directly proportional to the linearly polarised spin-density waves of the conduction electrons, which in this case are incommensurate with the lattice. This was verified by repeating the measurements for an alloy of 0·8 wt % Mn in Cr which is known from neutron diffraction measurements

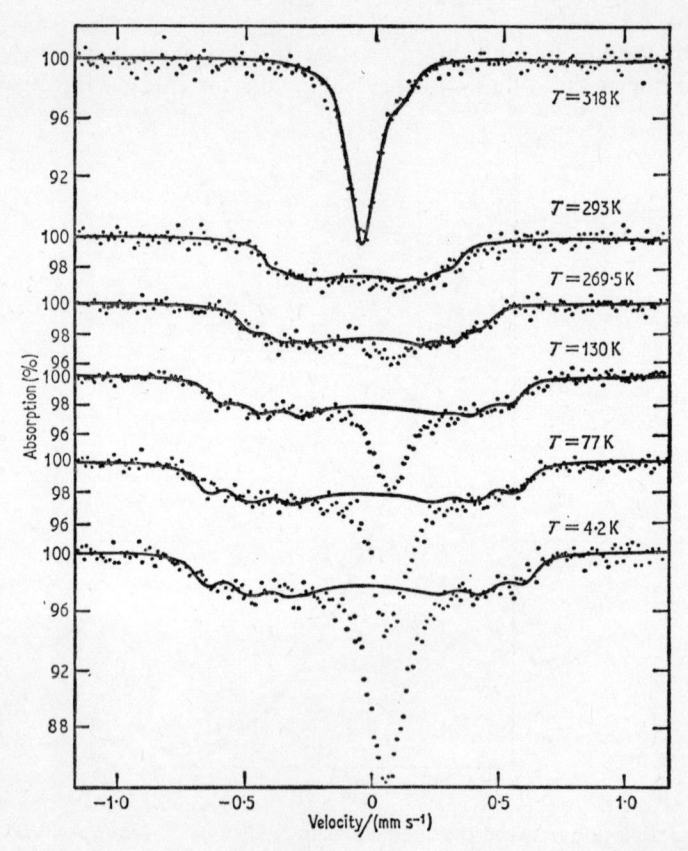

Fig. 14.14 Mössbauer spectra of ^{119}Sn in chromium. Note the very diffuse magnetic spectrum. [Ref. 247, Fig. 1]

to have a commensurate structure; the anticipated single-valued magnetic field (103 kG at 200 K) was found in this sample.

The internal field at ^{119}Sn nuclei in face-centred cubic cobalt metal falls from −22 kG at 4·2 K to −10 kG at 500 K [248]. Above 800 K the field increases in value again, and it is believed that the smooth temperature-dependence curve (Fig. 14.15), which was drawn assuming that the value of the field at first decreases and then reverses in sign at about 700 K, to approach a new positive maximum at 1100 K, is genuine [249].

On the assumption that the field at the tin is proportional to the $3d \mid 5s$-

[*Refs. on p. 424*]

electron overlap, it is possible to calculate an average value for $|\langle 3d | 5s \rangle|^2$ ($= \alpha$) in the presence of lattice vibrations. A simple Einstein-model treatment gives remarkably good agreement with experiment. If α is taken to be a function of internuclear distance such that $\alpha \propto r^{-n}$ where n is large, and if $r = d - x$ where d is the mean distance between Sn and Co such that

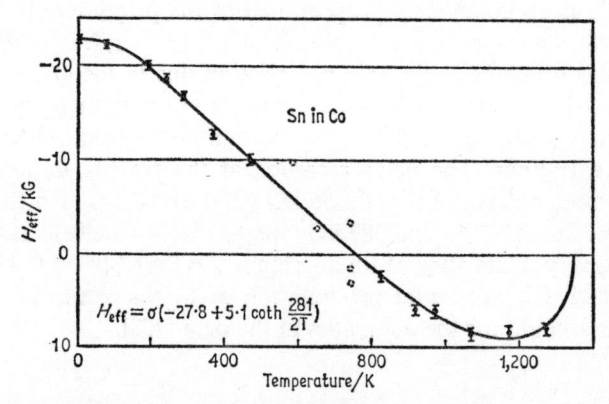

Fig. 14.15 The effective field of ^{119}Sn in cobalt as a function of temperature with a theoretical curve for H_{eff} superimposed. [Ref. 249, Fig. 1]

x is the change in d, then α can be expanded as a power series to give

$$\langle \alpha \rangle \propto A + B\langle x \rangle + C\langle x^2 \rangle$$

which with an Einstein model for $\langle x^2 \rangle$ gives [249] the temperature-dependence function as

$$H_T = \sigma\left[H_0 + h \coth\left(\frac{T_E}{2T}\right)\right]$$

where H_0, h, and T_E are constants. The fit to the experimental data is indicated by the solid line in Fig. 14.15.

By comparison, the behaviour of the field at ^{119}Sn in nickel and iron is more regular and approximates quite closely to the reduced magnetisation curve [250]. Cobalt is thus anomalous and this has not been fully explained. ^{119m}Sn in iron metal as already seen in Chapter 14.2 gives a field of 78·5 kG at 283 K[38]. Co/Pd and Fe/Pd alloys have also been studied [251]. Magnetic hyperfine broadening has been found at low temperatures for ^{119m}Sn nuclei in Au/Cr, Au/Mn, and Au/Fe alloys, but not in Au/Co or Au/V [252, 253]. Similar studies have been made for Cu/Mn and Ag/Mn alloys [254]. Detailed work on the Cu/Mn system showed long-range ordering with the ^{119}Sn field determined by a probability function rather than a unique value [255].

The magnetic field at ^{119}Sn nuclei in a 0·2 at. % alloy with gadolinium ranges from -329 kG at 4·2 K to -206 kG at 200 K and appears to follow the $J = \frac{7}{2}$ Brillouin function appropriate to the gadolinium [256]. Earlier work on 1% alloys of tin with rare earths had given the following values for $|H|$ at 4·2 K: Gd 238 kG, Tb 169 kG, Dy 125 kG, Ho 62 kG, Er 45 kG, and Tm 54 kG. The sign of the field at the tin nucleus in gadolinium was positive and in erbium negative, but the other signs were not determined [257]. The internal field is approximately proportional to the projection of the spin of the rare-earth ion on its angular momentum, i.e. $(g - 1)J$, implying that the induced fields originate in the spin polarisation due to the 4f-electrons. The Tm and Er fields would then be opposite in sign to the others, but there is a clear conflict with the later work as to the sign of the field in gadolinium at lower concentrations. The fields recorded at the tin site in several intermetallic compounds at 4·2 K were 55 kG (Er_2Sn), 68 kG (Ho_2Sn), 142 kG (Dy_2Sn), 184 kG (Tb_2Sn), and 289 kG (Gd_2Sn) [257]. The field at the ^{119m}Sn nucleus in a very dilute alloy of tin in erbium was found to be 124 kG, and it was suggested that accidental precipitation of M_2Sn phases in the earlier work might account for the anomalies in the data [258].

REFERENCES

[1] J. P. Bocquet, Y. Y. Chu, O. C. Kistner, M. L. Perlman, and G. T. Emery, *Phys. Rev. Letters*, 1966, **17**, 809.
[2] J. P. Bocquet, Y. Y. Chu, G. T. Emery, and M. L. Perlman, *Phys. Rev.*, 1968, **167**, 1117; 1968, **174**, 1538.
[3] C. Hohenemser, *Phys. Rev.*, 1965, **139**, A185.
[4] V. A. Kostroun and B. Crasemann, *Phys. Rev.*, 1968, **174**, 1535.
[5] H. Sano and M. Kanno, *Chem. Comm.*, 1969, 601.
[6] M. Yaqub and C. Hohenemser, *Phys. Rev.*, 1962, **127**, 2028.
[7] R. Barloutaud, E. Cotton, J. L. Picou, and J. Quidort, *Compt. rend.*, 1960, **250**, 319.
[8] C. Tzara and R. Barloutaud, *Phys. Rev. Letters*, 1960, **4**, 405.
[9] R. Barloutaud, J. L. Picou, and C. Tzara, *Compt. rend.*, 1960, **250**, 2705.
[10] V. S. Zykov, E. V. Petrovich, and Yu. P. Smirnov, *Zhur. eksp. teor. Fiz.*, 1965, **49**, 1019 (*Soviet Physics – JETP*, 1966, **22**, 708).
[11] P. Z. Hien and V. S. Shpinel, *Zhur. eksp. teor. Fiz.*, 1963, **44**, 393 (*Soviet Physics – JETP*, 1963, **17**, 268).
[12] H. A. Stöckler, H. Sano, and R. H. Herber, *J. Chem. Phys.*, 1966, **45**, 1182.
[13] K. P. Mitrofanov, M. V. Plotnikova, and V. S. Shpinel, *Zhur. eksp. teor. Fiz.*, 1965, **48**, 791 (*Soviet Physics – JETP*, 1965, **21**, 524).
[14] R. H. Herber and J. Spijkerman, *J. Chem. Phys.*, 1965, **42**, 4312.
[15] V. A. Bukarev, *Zhur. eksp. teor. Fiz.*, 1963, **44**, 249 (*Soviet Physics – JETP*, 1963, **17**, 579).
[16] A. J. F. Boyle, D. St P. Bunbury, C. Edwards, and H. E. Hall, *Proc. Phys. Soc.*, 1960, **76**, 165.
[17] V. A. Bryukhanov, N. N. Delyagin, and R. N. Kuzmin, *Zhur. eksp. teor. Fiz.*, 1964, **46**, 137 (*Soviet Physics – JETP*, 1964, **19**, 98).
[18] M. V. Plotnikova, K. P. Mitrofanov, and V. S. Shpinel, *ZETF Letters*, 1966, **3**, 323 (*JETP Letters*, 1966, **3**, 209).

[19] H. Sano and R. H. Herber, *J. Inorg. Nuclear Chem.*, 1968, **30**, 409.
[20] R. H. Herber and J. J. Spijkerman, *J. Chem. Phys.*, 1965, **43**, 4057.
[21] K. P. Mitrofanov and V. S. Shpinel, *Zhur. eksp. teor. Fiz.*, 1961, **40**, 983 (*Soviet Physics – JETP*, 1961, **13**, 686).
[22] L. Levy, L. Mitrani, and S. Ormandjiev, *Nuclear Instr. Methods*, 1964, **31**, 233.
[23] A. J. F. Boyle, D. St P. Bunbury, and C. Edwards, *Proc. Phys. Soc.*, 1962, **79**, 416.
[24] V. I. Goldanskii, G. M. Gorodinskii, S. V. Karyagin, L. A. Korytko, L. M. Krizhanskii, E. F. Makarov, I. P. Suzdalev, and V. V. Khrapov, *Doklady Akad. Nauk S.S.S.R.*, 1962, **147**, 127.
[25] I. B. Bersuker, V. I. Goldanskii, and E. F. Makarov, *Zhur. eksp. teor. Fiz.*, 1965, **49**, 699 (*Soviet Physics – JETP*, 1966, **22**, 485).
[26] S. L. Ruby, G. M. Kalvius, G. E. Beard, and R. E. Snyder, *Phys. Rev.*, 1967, **159**, 239.
[27] J. K. Lees and P. A. Flinn, *J. Chem. Phys.*, 1968, **48**, 882.
[28] N. N. Greenwood, P. G. Perkins, and D. H. Wall, *Symposia Faraday Soc.* No. 1, 1968, 51.
[29] N. N. Greenwood, P. G. Perkins, and D. H. Wall, *Phys. Letters*, 1968, **28A**, 339.
[30] V. I. Goldanskii and E. F. Makarov, *Phys. Letters*, 1965, **14**, 111.
[31] V. I. Goldanskii, E. F. Makarov, and R. A. Stukan, *J. Chem. Phys.*, 1967, **47**, 4048.
[32] M. L. Unland and J. H. Letcher, *J. Chem. Phys.*, 1968, **49**, 2706.
[33] V. G. Bhide and M. S. Multani, *Phys. Rev.*, 1965, **139A**, 1983.
[34] L. M. Krizhanskii, B. I. Rogozev, and G. V. Popov, *ZETF Letters*, 1966, **3**, 382 (*JETP Letters*, 1966, **3**, 248).
[35] N. N. Delyagin, V. S. Shpinel, V. A. Bryukhanov, and B. Zvenglinskii, *Zhur. eksp. teor. Fiz.*, 1960, **39**, 894 (*Soviet Physics – JETP*, 1961, **12**, 619).
[36] A. J. F. Boyle, D. St P. Bunbury, and C. Edwards, *Proc. Phys. Soc.*, 1961, **77**, 1062.
[37] S. S. Hanna, L. Meyer-Schutzmeister, R. S. Preston, and D. H. Vincent, *Phys. Rev.*, 1960, **120**, 2211.
[38] O. C. Kistner, A. W. Sunyar, and J. B. Swan, *Phys. Rev.*, 1961, **123**, 179.
[39] V. A. Bryukhanov, N. N. Delyagin, and V. S. Shpinel, *Zhur. eksp. teor. Fiz.*, 1962, **42**, 1183 (*Soviet Physics – JETP*, 1962, **15**, 818).
[40] V. I. Goldanskii, V. A. Trukhtanov, M. N. Devisheva, and V. F. Belov, *Phys. Letters*, 1965, **15**, 317.
[41] V. I. Goldanskii, V. A. Trukhtanov, M. N. Devisheva, and V. F. Belov, *ZETF Letters*, 1965, **1**, 19 (*JETP Letters*, 1965, **1**, 19).
[42] R. N. Kuzmin, N. S. Ibraimov, and G. S. Zhdanov, *Zhur. eksp. teor. Fiz.*, 1966, **50**, 330 (*Soviet Physics – JETP*, 1966, **23**, 219).
[43] A. P. Jain and T. E. Cranshaw, *Phys. Letters*, 1967, **25A**, 425.
[44] J. M. Williams, *J. Phys. C.*, 1968, Ser. 2, **1**, 473.
[45] T. C. Gibb, *J. Chem. Soc.* (*A*), 1970, 2503.
[46] M. Cordey-Hayes, *J. Inorg. Nuclear Chem.*, 1964, **26**, 915.
[47] J. M. van den Berg, *Acta Cryst.*, 1961, **14**, 1002.
[48] J. D. Donaldson and B. J. Senior, *J. Chem. Soc.* (*A*), 1966, 1796.
[49] J. D. Donaldson and B. J. Senior, *J. Chem. Soc.* (*A*), 1966, 1798.
[50] J. Lees and P. A. Flinn, *Phys. Letters*, 1965, **19**, 186.
[51] J. D. Donaldson and B. J. Senior, *J. Inorg. Nuclear Chem.*, 1969, **31**, 881.

[52] T. C. Gibb, B. A. Goodman, and N. N. Greenwood, *Chem. Comm.*, 1970, 774.
[53] J. D. Donaldson, R. Oteng, and B. J. Senior, *Chem. Comm.*, 1965, 618.
[54] J. D. Donaldson and R. Oteng, *Inorg. Nuclear Chem. Letters*, 1967, **3**, 163.
[55] R. R. McDonald, A. C. Larson, and D. T. Cromar, *Acta Cryst.*, 1964, **17**, 1104.
[56] J. D. Donaldson and B. J. Senior, *J. Chem. Soc.* (*A*), 1967, 1821.
[57] J. M. Van den Berg, *Acta Cryst.*, 1961, **14**, 1002.
[58] J. D. Donaldson and B. J. Senior, *J. Chem. Soc.* (*A*), 1969, 2358.
[59] J. J. Zuckerman, *J. Inorg. Nuclear Chem.*, 1967, **29**, 2191.
[60] B. Kamenar and B. Grdenic, *J. Chem. Soc.*, 1961, 3954.
[61] B. Kamenar and B. Grdenic, *J. Inorg. Nuclear Chem.*, 1962, **24**, 1039.
[62] J. D. Donaldson, D. G. Nicholson, and B. J. Senior, *J. Chem. Soc.* (*A*), 1968, 2928.
[63] W. J. Moore and L. Pauling, *J. Amer. Chem. Soc.*, 1941, **63**, 1392.
[64] W. Hofmann, *Z. Krist.*, 1935, **92A**, 161.
[65] A. Oakzaki and I. Ueda, *J. Phys. Soc. Japan*, 1956, **11**, 470.
[66] A. J. Panson, *Inorg. Chem.*, 1964, **3**, 940.
[67] A. A. Bekker, V. N. Panyushkin, V. I. Gotlib, A. M. Babeshkin, and A. N. Nesmeyanov, *Vestnik Moskov. Univ., Khim.*, 1969, **24**, 127.
[68] H. A. Stöckler and H. Sano, *J. Chem. Phys.*, 1969, **50**, 3813.
[69] C. G. Davies and J. D. Donaldson, *J. Chem. Soc.* (*A*), 1968, 946.
[70] R. J. Rentzeperis, *Z. Krist.*, 1962, **117**, 431.
[71] J. D. Donaldson and A. Jelen, *J. Chem. Soc.* (*A*), 1968, 1448.
[72] C. G. Davies, J. D. Donaldson, and W. B. Simpson, *J. Chem. Soc.* (*A*), 1969, 417.
[73] A. J. Bearden, H. S. Marsh, and J. J. Zuckerman, *Inorg. Chem.*, 1966, **5**, 1260.
[74] R. Hoppe and W. Dähne, *Naturwiss.*, 1962, **49**, 254.
[75] P. Brand and H. Sackmann, *Acta Cryst.*, 1963, **16**, 446.
[76] V. I. Goldanskii, E. F. Makarov, R. A. Stukan, T. N. Sumarokova, V. A. Trukhtanov, and V. V. Khrapov, *Doklady Akad. Nauk S.S.S.R.*, 1964, **156**, 400.
[77] S. Bukshpan and R. H. Herber, *J. Chem. Phys.*, 1967, **46**, 3375.
[78] Y. Hazony, *J. Chem. Phys.*, 1968, **49**, 159.
[79] V. F. Sukhovenkhov and B. E. Dzevitskii, *Doklady Akad. Nauk S.S.S.R.*, 1967, **177**, 611.
[80] V. S. Shpinel, V. A. Bryukhanov, V. Kotkhekar, and B. Z. Iofa, *Zhur. eksp. teor. Fiz.*, 1967, **53**, 23.
[81] G. Bliznakov and K. Petrov, *Z. anorg. Chem.*, 1967, **354**, 307.
[82] N. N. Greenwood and J. N. R. Ruddick, *J. Chem. Soc.* (*A*), 1967, 1679.
[83] R. H. Herber and H-S. Cheng, *Inorg. Chem.*, 1969, **8**, 2145.
[84] K. M. Harmon, L. Hesse, L. P. Klemann, C. W. Kocher, S. V. McKinley, and A. E. Young, *Inorg. Chem.*, 1969, **8**, 1054.
[85] I. R. Beattie, M. Milne, M. Webster, H. E. Blayden, P. Jones, R. C. G. Killean, and J. L. Lawrence, *J. Chem. Soc.* (*A*), 1969, 482.
[86] M. Webster and H. E. Blayden, *J. Chem. Soc.* (*A*), 1969, 2443.
[87] I. Branden, *Acta Chem. Scand.*, 1963, **17**, 759.
[88] Y. Hermondsson, *Acta Cryst.*, 1960, **13**, 656.
[89] J. Philip, M. A. Mullins, and C. Curran, *Inorg. Chem.*, 1968, **7**, 1895.
[90] S. Ichiba, M. Mishima, H. Sakai, and H. Negita, *Bull. Chem. Soc. Japan*, 1968, **41**, 49.

[91] K. M. Ali, D. Cunningham, M. J. Fraser, J. D. Donaldson, and B. J. Senior, *J. Chem. Soc.* (*A*), 1969, 2836.
[92] V. I. Goldanskii, V. Ya. Rochev, V. V. Khrapov, B. E. Dzevitskii, and V. F. Sukhovenkhov, *Izvest. Sibirsk, Otdel, Akad. Nauk, Ser. Khim. Nauk*, 1968, **4**, 22.
[93] Z. Bontschev, D. Christov, K. L. Burin, and Iv. Mandzukov, *Z. anorg. Chem.*, 1966, **347**, 199.
[94] I. Pelah and S. L. Ruby, *J. Chem. Phys.*, 1969, **51**, 383.
[95] W. Meisel, K. Hennig, and H. Schnorr, *Phys. Stat. Sol.*, 1969, **34**, 577.
[96] Pham Zuy Hien, V. S. Shpinel, A. S. Viskov, and Yu. N. Venevtsev, *Zhur. eksp. teor. Fiz.*, 1963, **44**, 1889 (*Soviet Physics – JETP*, 1963, **17**, 1271).
[97] L. M. Belyaev, I. S. Lyubutin, L. N. Demyanets, T. V. Dmitrieva, and L. P. Mitina, *Fiz. Tverd. Tela*, 1969, **11**, 528.
[98] A. M. Babeshkin, E. N. Efremov, A. A. Bekker, and A. N. Nesmeyanov, *Vestnik Moskov. Univ., Khim.*, 1969, **24**, 78.
[99] M. P. Gupta and H. B. Mathur, *J. Phys. and Chem. Solids*, 1968, **29**, 1479.
[100] V. A. Bokov, G. V. Novikov, O. B. Proskuryakov, Yu. G. Saksonov, V. A. Trukhtanov, and S. I. Yuschuk, *Fiz. Tverd. Tela*, 1968, **10**, 1080.
[101] G. N. Shlokov, P. L. Grusin, and E. J. Kuprianova, *Doklady Akad. Nauk S.S.S.R.*, 1967, **174**, 1144.
[102] G. V. Novikov, B. A. Trukhtanov, L. Cher, S. I. Yuschuk, and V. I. Goldanskii, *Zhur. eksp. teor. Fiz.*, 1969, **56**, 743.
[103] P. L. Grusin, G. N. Shlokov, and L. A. Alekseev, *Doklady Akad. Nauk S.S.S.R.*, 1967, **176**, 362.
[104] K. P. Belov and I. S. Lyubutin, *ZETF Letters*, 1965, **1**, 26 (*JETP Letters*, 1965, **1**, 16).
[105] K. P. Belov and I. S. Lyubutin, *Zhur. eksp. teor. Fiz.*, 1965, **49**, 747 (*Soviet Physics – JETP*, 1966, **22**, 518).
[106] V. I. Goldanskii, M. N. Levisheva, E. F. Makarov, G. V. Novikov, and V. A. Trukhtanov, *ZETF Letters*, 1966, **4**, 63 (*JETP Letters*, 1966, **4**, 42).
[107] I. S. Lyubutin, *Fiz. Tverd. Tela*, 1966, **8**, 643 (*Soviet Physics – Solid State*, 1966, **8**, 519).
[108] I. S. Lyubutin, V. A. Makarov, E. F. Makarov, and V. A. Povitskii, *ZETF Letters*, 1968, **7**, 370 (*JETP Letters*, 1968, **7**, 291).
[109] M. Eibschutz, E. Hermon, and S. Shtrikman, *J. Phys. and Chem. Solids*, 1967, **28**, 1633.
[110] A. N. Karasev, L. Ya. Margolis, and L. S. Polak, *Fiz. Tverd. Tela*, 1966, **8**, 287 (*Soviet Physics – Solid State*, 1966, **8**, 238).
[111] V. V. Chekin, V. P. Romanov, B. I. Verkin, and V. A. Bokov, *ZETF Letters*, 1965, **2**, 186 (*JETP Letters*, 1965, **2**, 117).
[112] V. A. Bokov, V. P. Romanov, and V. V. Chekin, *Fiz. Tverd. Tela*, 1965, **7**, 1886 (*Soviet Physics – Solid State*, 1965, **7**, 1521).
[113] P. Z. Hien, A. S. Viskov, V. S. Shpinel, and Yu. N. Venevtsev, *Zhur. eksp. teor. Fiz.*, 1963, **44**, 2182 (*Soviet Physics – JETP*, 1963, **17**, 1465).
[114] K. P. Mitrofanov, A. S. Viskov, G. Ya. Driker, M. V. Plotnikova, P. Z. Hien, Yu. N. Venevtsev, and V. S. Shpinel, *Zhur. eksp. teor. Fiz.*, 1964, **46**, 383 (*Soviet Physics – JETP*, 1964, **19**, 260).
[115] T. Anderson and P. O. Ostergaard, *Trans. Faraday Soc.*, 1968, **64**, 3014.
[116] P. Hannaford, C. J. Howard, and J. W. G. Wignall, *Phys. Letters*, 1965, **19**, 257.
[117] P. Hannaford and J. W. G. Wignall, *Phys. Stat. Sol.*, 1969, **35**, 809.

[118] A. N. Nesmeyanov, A. M. Babeshkin, A. A. Bekker, and V. Fano, *Radiokhimiya*, 1966, **8**, 264.
[119] K. P. Mitrofanov and T. A. Sidorov, *Fiz. Tverd. Tela*, 1967, **9**, 890 (*Soviet Physics – Solid State*, 1967, **9**, 693).
[120] G. M. Bartenev and A. D. Tsyganov, *Doklady Akad. Nauk S.S.S.R.*, 1968, **181**, 627.
[121] I. P. Polozova and P. P. Seregin, *Fiz. Tverd. Tela*, 1968, **10**, 2536.
[122] I. P. Suzdalev, A. S. Plachinda, and E. F. Makarov, *Zhur. eksp. teor. Fiz.*, 1967, **53**, 1556.
[123] D. L. Smith and J. J. Zuckerman, *J. Inorg. Nuclear Chem.*, 1967, **29**, 1203.
[124] R. G. Kostyanovsky, A. K. Prokofev, V. I. Goldanskii, V. V. Khrapov, and V. Ya. Rochev, *Izvest. Akad. Nauk S.S.S.R., Ser. Khim.*, 1968, 270.
[125] D. Moras and R. Weiss, *Acta Cryst.*, 1969, **25B**, 1726.
[126] T. C. Gibb and N. N. Greenwood, *J. Chem. Soc. (A)*, 1966, 43.
[127] R. V. Parish and R. H. Platt, *J. Chem. Soc. (A)*, 1969, 2145.
[128] J. C. Hill, R. S. Drago, and R. H. Herber, *J. Amer. Chem. Soc.*, 1969, **91**, 1644.
[129] V. Kotkhekar and V. S. Shpinel, *Zhur. strukt. Khim.*, 1969, **10**, 37.
[130] R. H. Herber, H. A. Stöckler, and W. T. Reichle, *J. Chem. Phys.*, 1965, **42**, 2447.
[131] H. A. Stöckler and H. Sano, *Chem. Comm.*, 1969, 954.
[132] H. A. Stöckler, H. Sano, and R. H. Herber, *J. Chem. Phys.*, 1967, **47**, 1567.
[133] P. A. Yeats, B. F. E. Ford, J. R. Sams, and F. Aubke, *Chem. Comm.*, 1969, 791.
[134] B. A. Goodman and N. N. Greenwood, *Chem. Comm.*, 1969, 1105.
[135] G. A. Bykov, G. K. Ryasnyin, and V. S. Shpinel, *Fiz. Tverd. Tela*, 1965, **7**, 1657 (*Soviet Physics – Solid State*, 1965, **7**, 1343).
[136] B. A. Goodman, and N. N. Greenwood, *J. Chem. Soc. (A)*, 1971, 1862.
[137] V. I. Goldanskii, V. V. Khrapov, O. Yu. Okhlobystin, and V. Ya. Rochev, Chap. 6 in 'Chemical Applications of Mössbauer Spectroscopy', Ed. V. I. Goldanskii and R. H. Herber, Academic Press, N.Y., 1968.
[138] H. A. Stöckler and H. Sano, *Trans. Faraday Soc.*, 1968, **64**, 577.
[139] M. Cordey-Hayes, R. D. W. Kammith, R. D. Peacock, and G. D. Rimmer, *J. Inorg. Nuclear Chem.*, 1969, **31**, 1515.
[140] A. Yu. Aleksandrov, K. P. Mitrofanov, L. S. Polak, and V. S. Shpinel, *Doklady Akad. Nauk S.S.S.R.*, 1963, **148**, 126.
[141] R. H. Herber and H. A. Stöckler, *Trans. N.Y. Acad. Sciences*, Ser. 2, 1964, **26**, 929.
[142] W. T. Reichle, *Inorg. Chem.*, 1966, **5**, 87.
[143] M. Cordey-Hayes, *J. Inorg. Nuclear Chem.*, 1964, **26**, 2306.
[144] A. Yu. Aleksandrov, O. Yu. Okhlobystin, L. S. Polak, and V. S. Shpinel, *Doklady Akad. Nauk S.S.S.R.*, 1964, **157**, 934.
[145] V. V. Khrapov, V. I. Goldanskii, R. G. Kostyanovskii, and A. K. Prokofev, *Zhur. Obshch. Khim.*, 1967, **37**, 3.
[146] M. Cordey-Hayes, R. D. Peacock, and M. Vucelic, *J. Inorg. Nuclear Chem.*, 1967, **29**, 1177.
[147] A. Yu. Aleksandrov, V. I. Bregadse, V. I. Goldanskii, L. I. Zakharkin, O. Yu. Okhlobystin, and V. V. Khrapov, *Doklady Akad. Nauk S.S.S.R.*, 1965, **165**, 593.
[148] T. Chivers and J. R. Sams, *Chem. Comm.*, 1969, 249.
[149] R. V. Parish and R. H. Platt, *Chem. Comm.*, 1968, 1118.
[150] Zv. Bontschev, D. Christov, K. Burin, and Iv. Mandzukov, *Z. Physik*, 1966, **190**, 278.

[151] R. H. Herber and G. I. Parisi, *Inorg. Chem.*, 1966, **5**, 769.
[152] L. May, and J. J. Spijkerman, *J. Chem. Phys.*, 1967, **46**, 3272.
[153] V. I. Goldanskii, V. Ya. Rochev, and V. V. Khrapov. *Doklady Akad. Nauk S.S.S.R.*, 1964, **156**, 909.
[154] D. H. Olsen and R. E. Rundle, *Inorg. Chem.*, 1963, **2**, 1310.
[155] H. C. Clark, R. J. O'Brien, and J. Trotter, *J. Chem. Soc.*, 1964, 2332.
[156] E. O. Schlemper and W. C. Hamilton, *Inorg. Chem.*, 1966, **5**, 995.
[157] A. G. Davies, H. J. Milledge, D. C. Puxley, and P. J. Smith, *J. Chem. Soc.* (A), 1970, 2862.
[158] V. A. Bryukhanov, V. I. Goldanskii, N. N. Delyagin, L. A. Korytko, E. F. Makarov, I. P. Suzdalev, and V. S. Shpinel, *Zhur. eksp. teor. Fiz.*, 1962, **43**, 448 (*Soviet Physics – JETP*, 1963, **16**, 321).
[159] Yu. A. Aleksandrov, N. N. Delyagin, K. P. Mitrofanov, L. S. Polak, and V. S. Shpinel, *Zhur. eksp. teor. Fiz.*, 1962, **43**, 1242 (*Soviet Physics – JETP*, 1963, **16**, 879).
[160] E. V. Bryuchova, G. K. Semin, V. I. Goldanskii, and V. V. Khrapov, *Chem. Comm.*, 1969, 491.
[161] V. I. Goldanskii, E. F. Makarov, and V. V. Khrapov, *Zhur. eksp. teor. Fiz.*, 1963, **44**, 752 (*Soviet Physics – JETP*, 1963, **17**, 508).
[162] V. S. Shpinel, A. Yu. Aleksandrov, G. K. Ryasnyin, and O. Yu. Okhlobystin, *Zhur. eksp. teor. Fiz.*, 1965, **48**, 69 (*Soviet Physics – JETP*, 1965, **21**, 47).
[163] H. A. Stöckler and H. Sano, *Chem. Phys. Letters*, 1968, **2**, 448; H. A. Stöckler and H. Sano, *Phys. Rev.*, 1968, **165**, 406.
[164] R. H. Herber and S. Chandra, *J. Chem. Phys.*, 1970, **52**, 6045.
[165] R. Hulme, *J. Chem. Soc.*, 1963, 1524.
[166] J. Nasielski, N. Sprecher, J. de Vooght, and S. Lejeune, *J. Organometallic Chem.*, 1967, **8**, 97.
[167] N. W. Isaacs, C. H. L. Kennard, and W. Kitching, *Chem. Comm.*, 1968, 820.
[168] F. W. B. Einstein and B. R. Penfold, *J. Chem. Soc.* (A), 1968, 3019.
[169] B. W. Fitzsimmons, N. J. Seeley, and A. W. Smith, *Chem. Comm.*, 1968, 390.
[170] B. W. Fitzsimmons, N. J. Seeley, and A. W. Smith, *J. Chem. Soc.* (A), 1969, 143.
[171] M. A. Mullins and C. Curran, *Inorg. Chem.*, 1967, **6**, 2017.
[172] R. C. Poller, J. N. R. Ruddick, M. Thevarasa, and W. R. McWhinnie, *J. Chem. Soc.* (A), 1969, 2327.
[173] R. C. Poller and J. N. R. Ruddick, *J. Chem. Soc.* (A), 1969, 2273.
[174] Yu. Aleksandrov, Ya. G. Dorfman, O. L. Lependina, K. P. Mitrofanov, M. V. Plotnikov, L. S. Polak, A. Ya. Temkin, and V. S. Shpinel, *Zhur. Fiz. Khim.*, 1964, **38**, 2190 (*Russ. J. Phys. Chem.*, 1964, **38**, 1185).
[175] V. I. Goldanskii, O. Yu. Okhlobystin, V. Ya. Rochev, and V. V. Khrapov, *J. Organometallic Chem.*, 1965, **4**, 160.
[176] R. A. Forder and G. M. Sheldrick, *Chem. Comm.*, 1969, 1125.
[177] E. O. Schlemper and D. Britton, *Inorg. Chem.*, 1966, **5**, 507, 2244.
[178] M. A. Mullins and C. Curran, *Inorg. Chem.*, 1968, **7**, 2584.
[179] B. Gassenheimer and R. H. Herber, *Inorg. Chem.*, 1969, **8**, 1120.
[180] N. Kasai, K. Yasuda, and R. Okawara, *J. Organometallic Chem.*, 1965, **3**, 172.
[181] H. A. Stöckler and H. Sano, *Phys. Letters*, 1967, **25A**, 550.
[182] V. V. Khrapov, V. I. Goldanskii, A. K. Prokofev, V. Ya. Rochev, and R. G. Kostyanovsky, *Izvest. Akad. Nauk S.S.S.R., Ser. Khim.*, 1968, 1261.

[183] L. M. Krizhanskii, O. Yu. Okhlobystin, A. V. Popov, and B. I. Rogozev, *Doklady Akad. Nauk S.S.S.R.*, 1965, **160**, 1121.
[184] V. I. Goldanskii, E. F. Makarov, R. A. Stukan, V. A. Trukhtanov, and V. V. Khrapov, *Doklady Akad. Nauk S.S.S.R.*, 1963, **151**, 357.
[185] A. Yu. Aleksandrov, K. P. Mitrofanov, O. Yu. Okhlobystin, L. S. Polak, and V. S. Shpinel, *Doklady Akad. Nauk S.S.S.R.*, 1963, **153**, 370.
[186] B. F. E. Ford, B. V. Liengme, and J. R. Sams, *J. Organometallic Chem.*, 1969, **19**, 53.
[187] C. Poder and J. R. Sams, *J. Organometallic Chem.*, 1969, **19**, 67.
[188] N. W. Alcock and R. E. Timms, *J. Chem. Soc. (A)*, 1968, 1873.
[189] N. W. Alcock and R. E. Timms, *J. Chem. Soc. (A)*, 1968, 1876.
[190] E. O. Schlemper, *Inorg. Chem.*, 1967, **6**, 2012.
[191] M. T. Jones, *Inorg. Chem.*, 1967, **6**, 1249.
[192] R. M. Sweet, C. J. Fritchie, J. Schunn, and R. A. Schunn, *Inorg. Chem.*, 1967, **6**, 749.
[193] J. E. O'Connor and E. R. Corey, *Inorg. Chem.*, 1967, **6**, 968.
[194] R. H. Herber and Y. Goscinney, *Inorg. Chem.*, 1968, **7**, 1293.
[195] R. H. Herber, *Symposia Faraday Soc. No. 1*, 1968, 1886.
[196] D. E. Fenton and J. J. Zuckerman, *J. Amer. Chem. Soc.*, 1968, **90**, 6227; *Inorg. Chem.*, 1969, **8**, 1771.
[197] A. N. Karasev, N. E. Kolobova, L. S. Polak, V. S. Shpinel, and K. N. Anisimov, *Teor. i eksp. Khim.*, 1966, **2**, 126.
[198] V. I. Baranovskii, V. P. Sergeev, and B. E. Dzevitskii, *Doklady Akad. Nauk S.S.S.R.*, 1969, **184**, 632.
[199] P. G. Harrison and J. J. Zuckerman, *J. Amer. Chem. Soc.*, 1969, **91**, 6885.
[200] A. N. Nesmeyanov, V. I. Goldanskii, V. V. Khrapov, V. Ya. Rochev, D. N. Kravtsov, and E. M. Rokhlina, *Izvest. Akad. Nauk S.S.S.R., Ser. Khim.*, 1968, 793.
[201] A. N. Nesmeyanov, V. I. Goldanskii, V. V. Khrapov, V. A. Rochev, D. N. Kravtsov, V. M. Pachevskaya, and E. M. Rokhlina, *Doklady Akad. Nauk S.S.S.R.*, 1968, **181**, 321.
[202] L. M. Epstein and D. K. Straub, *Inorg. Chem.*, 1965, **4**, 1551.
[203] B. W. Fitzsimmons, *Chem. Comm.*, 1968, 1485.
[204] N. E. Alekseevskii, P. Z. Hien, V. G. Shapiro, and V. S. Shpinel, *Zhur. eksp. teor. Fiz.*, 1962, **43**, 790 (*Soviet Physics – JETP*, 1963, **16**, 559).
[205] V. G. Shapiro and V. S. Shpinel, *Zhur. eksp. teor. Fiz.*, 1964, **46**, 1960 (*Soviet Physics – JETP*, 1964, **19**, 1321).
[206] N. E. Alekseevskii, A. P. Kiryanov, V. I. Nizhankovskii, and Yu. A. Samarskii, *ZETF Letters*, 1965, **2**, 269 (*JETP Letters*, 1965, **2**, 171).
[207] B. A. Komissarova, A. A. Sorokin, and V. S. Shpinel, *Zhur. eksp. teor. Fiz.*, 1966, **50**, 1205 (*Soviet Physics – JETP*, 1966, **23**, 800).
[208] N. S. Snyder, *Phys. Rev.*, 1969, **178**, 537.
[209] C. Hohenemser, *Phys. Rev.*, 1965, **139**, A185.
[210] V. N. Panyushkin and F. F. Voronov, *ZETF Letters*, 1965, **2**, 153 (*JETP Letters*, 1965, **2**, 97).
[211] I. P. Suzdalev, M. Ya. Gen, V. I. Goldanskii, and E. F. Makarov, *Zhur. eksp. teor. Fiz.*, 1966, **51**, 118 (*Soviet Physics – JETP*, 1967, **24**, 79).
[212] J. L. Feldman and G. K. Horton, *Phys. Rev.*, 1963, **132**, 644.
[213] R. E. Wames, T. Wolfram, and G. W. Lehman, *Phys. Rev.*, 1963, **131**, 529.
[214] A. J. F. Boyle, D. St P. Bunbury, C. Edwards, and H. E. Hall, *Proc. Phys. Soc.*, 1961, **77**, 129.

[215] G. Longworth and R. H. Packwood, *Phys. Letters*, 1965, **14**, 75.
[216] B. J. Bowles and T. E. Cranshaw, *Phys. Letters*, 1965, **17**, 258.
[217] Zw. Bonchev, A. Jordanov, and A. Minovka, *Nuclear Instr. Methods*, 1969, **70**, 36.
[218] B. I. Verkin, V. V. Chekin, and A. P. Vinnikov, *Zhur. eksp. teor. Fiz.*, 1966, **51**, 25 (*Soviet Physics – JETP*, 1967, **24**, 16).
[219] F. Pobell, *Z. Physik*, 1965, **188**, 57.
[220] L. Meyer-Schutzmeister, R. S. Preston, and S. S. Hanna, *Phys. Rev.*, 1961, **122**, 1717.
[221] M. A. Abidov, R. N. Kuzmin, and S. V. Nikitina, *Zhur. eksp. teor. Fiz.*, 1969, **56**, 1785.
[222] V. I. Nikolaev, Yu. I. Shcherbina, and A. I. Karchevskii, *Zhur. eksp. teor. Fiz.*, 1963, **44**, 775 (*Soviet Physics – JETP*, 1963, **17**, 524).
[223] V. I. Nikolaev, Yu. I. Shcherbina, and S. S. Yakimov, *Zhur. eksp. teor. Fiz.*, 1963, **45**, 1277 (*Soviet Physics – JETP*, 1964, **18**, 878).
[224] G. Fabri, E. Germagnoli, M. Musci, and G. C. Locati, *Nuovo Cimento*, 1965, **B40**, 178.
[225] E. Both, G. Trumpy, J. Traff, P. Ostergaard, C. Djega-Mariadasson, and P. Lecocq, p. 487 of 'Hyperfine Structure and Nuclear Radiations', Ed. E. Matthias and D. A. Shirley, North-Holland, Amsterdam, 1968.
[226] V. V. Chekin and V. G. Naumov, *Zhur. eksp. teor. Fiz.*, 1966, **50**, 534 (*Soviet Physics – JETP*, 1966, **23**, 355).
[227] N. S. Ibraimov and R. N. Kuzmin, *Zhur. eksp. teor. Fiz.*, 1965, **48**, 103 (*Soviet Physics – JETP*, 1965, **21**, 70).
[228] M. Cordey-Hayes and T. R. Harris, *Phys. Letters*, 1967, **24A**, 80.
[229] C. R. Kanekar, K. R. P. Rao, and V. J. S. Rao, *Phys. Letters*, 1965, **19**, 95.
[230] F. Pobell, *Phys. Stat. Sol.*, 1966, **13**, 509.
[231] V. V. Chekin, L. E. Danilenko, and A. I. Kaplienko, *Zhur. eksp. teor. Fiz.*, 1966, **51**, 711 (*Soviet Physics – JETP*, 1967, **24**, 472).
[232] J. M. Williams, *J. Phys. C*, 1969, Ser. 2, **2**, 2037.
[233] H. S. Möller and R. L. Mössbauer, *Phys. Letters*, 1967, **24A**, 416.
[234] H. S. Möller, *Z. Physik*, 1968, **212**, 107.
[235] F. Borsa, R. G. Barnes, and R. A. Reese, *Phys. Stat. Sol.*, 1967, **19**, 359.
[236] N. S. Ibraimov, R. N. Kuzmin, and G. S. Zhadanov, *Zhur. eksp. teor. Fiz.*, 1965, **49**, 1389 (*Soviet Physics – JETP*, 1966, **22**, 956).
[237] V. A. Bryukhanov, N. N. Delyagin, R. N. Kuzmin, and V. S. Shpinel, *Zhur. eksp. teor. Fiz.*, 1964, **46**, 1996 (*Soviet Physics – JETP*, 1964, **19**, 1344).
[238] J. S. Shier and R. D. Taylor, *Solid State Commun.*, 1967, **5**, 147.
[239] J. S. Shier and R. D. Taylor, *Phys. Rev.*, 1968, **174**, 346.
[240] V. A. Bryukhanov, N. N. Delyagin, and Yu. Kagan, *Zhur. eksp. teor. Fiz.*, 1964, **46**, 825 (*Soviet Physics – JETP*, 1964, **19**, 563).
[241] V. A. Bryukhanov, N. N. Delyagin, and Yu. Kagan, *Zhur. eksp. teor. Fiz.*, 1963, **45**, 1372 (*Soviet Physics – JETP*, 1964, **18**, 945).
[242] L. A. Alekseev and P. L. Grusin, *Doklady Akad. Nauk S.S.S.R.*, 1965, **160**, 376.
[243] V. A. Bryukhanov, N. N. Delyagin, and V. S. Shpinel, *Zhur. eksp. teor. Fiz.*, 1964, **47**, 2085 (*Soviet Physics – JETP*, 1965, **20**, 1400).
[244] V. V. Chekin and V. G. Naumov, *Zhur. eksp. teor. Fiz.*, 1966, **51**, 1048 (*Soviet Physics – JETP*, 1967, **24**, 699).
[245] V. A. Bryukhanov, N. N. Delyagin, and V. S. Shpinel, *Zhur. eksp. teor. Fiz.*, 1964, **47**, 80 (*Soviet Physics – JETP*, 1965, **20**, 55).

[246] N. N. Delyagin, *Fiz. Tverd. Tela*, 1966, **8**, 3426 (*Soviet Physics – Solid State*, 1967, **8**, 2748).
[247] R. Street and B. Window, *Proc. Phys. Soc.*, 1966, **89**, 587.
[248] A. P. Jain and T. E. Cranshaw, *Phys. Letters*, 1967, **25A**, 421.
[249] T. E. Cranshaw, *J. Appl. Phys.*, 1969, **40**, 1481.
[250] G. P. Huffman, F. C. Schwerer, and G. R. Dunmyre, *J. Appl. Phys.*, 1969, **40**, 1487.
[251] A. E. Balabanov, N. N. Delyagin, A. L. Yerzinkian, V. P. Parfenova, and V. S. Shpinel, *Zhur. eksp. teor. Fiz.*, 1968, **55**, 2136.
[252] B. Window, *Phys. Letters*, 1967, **24A**, 659.
[253] I. R. Williams, G. V. H. Wilson, and B. Window, *Phys. Letters*, 1967, **25A**, 144.
[254] A. P. Jain and T. E. Cranshaw, *Phys. Letters*, 1967, **25A**, 425.
[255] B. Window, *J. Phys. C*, 1969, Ser. 2, **2**, 2380.
[256] V. Gotthardt, H. S. Möller, and R. L. Mössbauer, *Phys. Letters*, 1969, **28A**, 480.
[257] D. Bosch, F. Pobell, and P. Kienle, *Phys. Letters*, 1966, **22**, 262.
[258] D. C. Price and R. Street, *J. Phys C*, 1968, Ser. 2, **1**, 1258.
[259] R. V. Parish and C. E. Johnson, *Chem. Phys. Letters*, 1970, **6**, 239.
[260] N. E. Erickson, *Chem. Comm.*, 1970, 1349.
[261] B. W. Fitzsimmons, *J. Chem. Soc. (A)*, 1970, 3235.
[262] B. A. Goodman, N. N. Greenwood, K. L. Jaura, and K. K. Sharma, *J. Chem. Soc. (A)*, 1971, 1865.
[263] B. A. Goodman, R. Greatrex, and N. N. Greenwood, *J. Chem. Soc. (A)*, 1971, 1868.

15 Other Main Group Elements

Having discussed at length the data for ^{57}Fe and ^{119}Sn which have undoubtedly dominated the first decade of chemical Mössbauer spectroscopy, we now turn to other elements. In this chapter we discuss the remaining Main Group elements in which the Mössbauer effect has been observed in order of increasing atomic number. The elements are potassium, germanium, krypton, antimony, tellurium, iodine, xenon, caesium, and barium, although not all of these have been used for chemical studies. Particular attention will be paid to the experimental difficulties involved, and the potential of each will be assessed with regard to further study in the near future. Where estimates have been given for the errors in the original data we quote these in the form 3·90(35), indicating 3·90 ± 0·35. This is to allow the significance of small differences in values to be easily assessed.

A full tabulation of relevant nuclear properties for all Mössbauer nuclides is given in Appendix 1.

15.1 Potassium (^{40}K)

The nuclide with the lowest mass for which a Mössbauer resonance has been recorded is ^{40}K. This isotope has a natural abundance of 0·012% and is itself radioactive with a half-life $t_{\frac{1}{2}} = 1·26 \times 10^9$ y. The 29·4-keV first excited state is not populated by any radioactive parent, but can be reached during the course of nuclear reactions on the predominant potassium isotope ^{39}K. Two basic methods have been used: a (d, p) reaction and an (n, γ) reaction.

A 0·2-μA beam of 3·5-MeV deuterons incident on a potassium metal target induces the ^{39}K$(d, p)^{40}$K reaction [1]. The resonance obtained at liquid nitrogen temperature with both the source and a KCl absorber enriched in ^{40}K was ~1% and had a linewidth of 4·10(35) mm s^{-1} compared to the natural width of 2·4 mm s^{-1} derived from the excited-state lifetime of 3·90(35) ns [2]. The broadening can be attributed mainly to absorber-thickness effects. *In situ* experiments of this type are usually conducted with incident radiation at 90° to the direction of observation through the absorber and, provided that adequate shielding is used, the efficiency of the detection system will be high. However, the requirement of a primary radiation source, in this case a van der Graaff accelerator, limits the number of laboratories equipped for these experiments.

In contrast to the resonance in KCl, no resonance was detected with KBr or KOH targets. However, the metal itself gave a greater effect than expected. The kinetic energy of the nuclear reaction is about four times that required to eject an atom from its lattice site in the metal, and it would appear from the experiments that there is an extremely rapid recovery from the local thermal effects due to this radiation damage.

The alternative method uses the ^{39}K$(n, \gamma)^{40}$K reaction in a thermal neutron beam [3]. In this case the recoil energy of the reaction is up to 30 times the displacement energy, but once again radiation damage has no visible effects on the spectra. The method is undoubtedly superior for Mössbauer experiments and was adopted in subsequent work [4]. A series of experiments in which the targets KF, KCl, KC$_8$, KNiF$_3$, KCoF$_3$, KO$_2$, KAlSi$_3$O$_8$, KN$_3$, KOH, KCN, KH, and KHF$_2$ were used successively as sources and enriched ^{40}KCl was the absorber gave single lines with no chemical isomer shifts at liquid nitrogen temperature within experimental error [4]. Likewise there was no chemical shift when a KF target was the source and KI the absorber. Theoretical calculations of the nuclear excited states of ^{40}K predict that $\delta\langle R^2\rangle/\langle R^2\rangle$ ($= 2\delta R/R$) will be approximately zero, which is unfortunate from a chemist's point of view since this inevitably dictates a universally minute or zero chemical isomer shift. Measurements of KF and potassium metal targets against KCl at temperatures between 10 K and 78 K showed a second-order Doppler shift in agreement with prediction. An upper limit for $\delta\langle R^2\rangle/\langle R^2\rangle$ was derived as $\delta\langle R^2\rangle/\langle R^2\rangle < 5 \times 10^{-4}$. Calculation of the recoil-free fraction for potassium metal suggests that there is some reduction in the experimental recoil-free fraction as a result of radiation damage [5].

No quadrupole or magnetic hyperfine interactions have been detected, and the only general application would appear to be the use of the recoil-free fraction in lattice-dynamical studies.

15.2 Germanium (^{73}Ge)

Germanium-73 has several low-lying excited levels, of which the 13·5-keV first excited state suffers from excessive internal conversion ($\alpha_T > 1000$) and a long lifetime (4 μs), as does the 66·8-keV second excited state (0·53 s). The 67·03-keV third excited state has convenient properties for Mössbauer spectroscopy but is only weakly populated by the decay of ^{73}Ga. All Mössbauer experiments have therefore used a direct population of this level by Coulomb excitation in an ion beam. The ground state has $I = \frac{9}{2}+$, while the 67·03-keV level is probably $I = \frac{7}{2}+$ and decays directly to the ground state.

[Refs. on p. 489]

The resonance was first demonstrated by Czjzek et al. in 1966 using 25-MeV oxygen ions from a van der Graaff accelerator to excite a germanium target which, together with a germanium absorber, was at liquid nitrogen temperature [6]. However, the observed intensity was very weak, the nominal recoil-free fraction of the target being only 0·009 compared to the predicted value of 0·125 for a germanium Debye temperature of $\theta_D = 360$ K. The low recoil-free fraction obtained from a germanium target can be attributed directly to radiation damage, and in some cases an initially crystalline target may become completely converted to the amorphous state [7].

More satisfactory results have been achieved using the Coulomb recoil-implantation technique [7, 8]. The Coulomb-excited nuclei have sufficient kinetic energy to leave their lattice sites and travel a distance of the order of 10^{-4} cm in a solid matrix before coming to rest. This provides a mechanism for transferring the excited ^{73}Ge nuclei to a stronger crystalline lattice. A very thin layer of germanium (~0·3 mg cm^{-2}) is deposited on a matrix such as chromium so that the recoiling excited atoms are ejected from the germanium layer and come to rest in the chromium matrix as impurity atoms. This takes place on a time-scale (10^{-12} s) shorter than the excited-state lifetime (10^{-9} s), and the recoil-free fraction of a ^{73}Ge atom in chromium is then 0·06 at 78 K compared to 0·008 for a thicker germanium target [7]. A detailed mathematical description of the recoil-implantation processes has been given [8].

The first measurements using this technique [7, 8] utilised a 3-5-μA beam of oxygen ions, producing a thermal effect of up to 10 K in the host matrices of Ge, Cu, Fe, or Cr, which were cooled with liquid nitrogen. Typical spectra obtained are shown in Fig. 15.1. No line splitting was observed, and the linewidth was close to the natural width (see below). Chromium was found to be the most effective host.

Fig. 15.1 shows clear evidence for a chemical isomer shift between the metal and GeO$_2$ of $-0·98(7)$ mm s^{-1}. Assuming electronic configurations of $4s^14p^3$ for germanium and $4s^04p^0$ for Ge^{4+} ions in GeO$_2$ it is possible to calculate [8] $\delta\langle R^2 \rangle / \langle R^2 \rangle = +1·8 \times 10^{-3}$. However, because of similar arguments to those used for ^{119}Sn, this is unlikely to be more than an 'order of magnitude' result. The chemical isomer shift between a target of ^{73}Ge in chromium and a germanium absorber is $+0·17(7)$ mm s^{-1}. Extrapolation of the experimental linewidths to zero absorber thickness gave a width of 2·2(1) mm s^{-1}, and thence an excited-state lifetime of 1·86(10) ns compared to the value of 1·62(14) ns from other methods [9]. An absorber of Fe$_5$Ge$_3$ showed a chemical isomer shift of $+0·06(6)$ mm s^{-1} relative to the chromium host, and although slight broadening was found, magnetic hyperfine splitting was not resolved.

Independent work [10], also using recoil-implantation into chromium, recorded a shift in the Ge(II) compound, GeSe, and in the two modifications of GeO_2. The observed shifts relative to germanium metal were: GeSe +1·34(39) mm s^{-1}, hexagonal GeO_2 −0·99(13) mm s^{-1}, tetragonal GeO_2 −0·86(9) mm s^{-1}. On the assumption that GeO_2 is 62% ionic (i.e. a configuration 38% $4s^14p^3$ and 62% $4s^04p^0$), that GeSe has the configuration

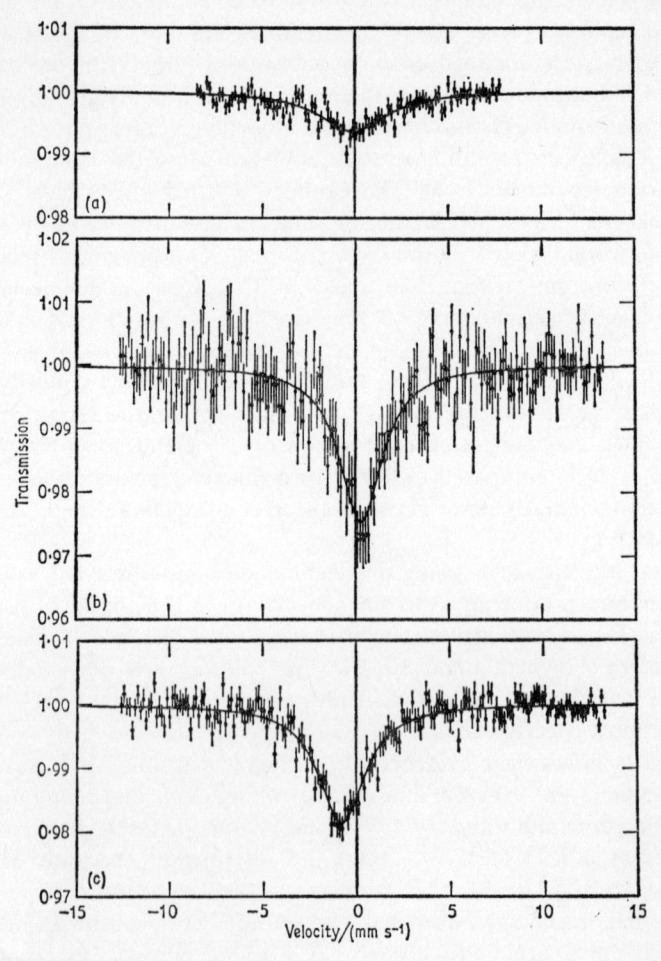

Fig. 15.1 Mössbauer spectra for (a) a ^{73}Ge target and a natural Ge absorber; (b) a ^{73}Ge/chromium target and Ge absorber; (c) a ^{73}Ge/chromium target and GeO_2 absorber. Note the significantly greater effect with the recoil-implantation technique and also the chemical isomer shift between Ge and GeO_2. [Ref. 7, Fig. 1]

15% $4s^24p^0$ + 85% $4s^24p^2$, and that Ge metal is $4s^14p^3$, a value for $\delta\langle R^2\rangle/\langle R^2\rangle$ of +2·24 × 10^{-3} was derived.

Detailed measurement of the recoil-free fraction in germanium metal [8] has since been reinterpreted [11] using a different lattice-dynamical model.

[*Refs. on p. 489*]

15.3 Krypton (^{83}Kr)

Krypton-83 has several nuclear properties which are ideal for a Mössbauer nucleus. It has a low γ-ray energy and a natural linewidth and isotopic abundance similar to those of ^{57}Fe. Unfortunately from the chemical point of view, the degree to which these can be exploited is limited by the lack of compounds. Nevertheless, some interesting work has been done.

The ^{83}Kr (9·3-keV) resonance was first reported in 1962 by Hazony et al. [12] for both krypton gas (frozen) and a hydroquinone clathrate. The decay scheme of ^{83}Kr has already been given in Fig. 2.7 in the general section on source preparation. There are three possible routes to the 9·3-keV level, and the first work utilised the ^{82}Kr$(n, \gamma)^{83m}$Kr reaction. Krypton sources can be made by irradiating the frozen gas or the clathrate. The metastable 41·8-keV state decays with a half-life of 1·86 h via the 9·3-keV level ($I = \frac{7}{2}$) to the ground state ($I = \frac{9}{2}$). The lifetime of the excited level is 147 ns [13], giving a resonance natural linewidth of 0·20 mm s^{-1}. Detection of the γ-rays to the exclusion of the 12·6-keV X-ray can be achieved using gas-counters and a zinc selective filter. In this early work it was shown that the linewidths observed are 8–10 times greater than expected, and that the temperature dependence of the resonance intensity in the clathrate is anomalous, remaining nearly constant above 125 K.

The low activity of ^{83m}Kr generated by neutron irradiation, as well as handling difficulties, have led to the development of other sources. The ^{81}Br$(\alpha, 2n)^{83}$Rb reaction gives the 83-d ^{83}Br isotope, and this has been incorporated in a CaBr$_2$ matrix [13]. Unfortunately the electron-capture decay is mainly via the ^{83m}Kr level, and it was found that there is a considerable chance of the excited krypton daughter atoms escaping from a CaBr$_2$ matrix before they can populate the 9·3-keV level. In addition, the linewidth is about 7 times greater than natural.

The third method (Fig. 2.7) is via the β-decay of ^{83}Br ($t_{\frac{1}{2}} = 2·41$ h). In the first instance this was produced by neutron irradiation of SeO$_2$, followed by chemical separation [14]. The activity was then incorporated in matrices of LiBr, NaBr, KBr, CsBr, NH$_4$Br, and KBrO$_3$. The decay is efficient, high specific activities can be obtained, and such sources are not self-resonant. The chemical isomer shift of these sources at 90 K was zero within experimental error relative to a solid krypton absorber at 22 K. The linewidths were of the order of ten times the natural width, but with no signs of a static electric field gradient, and the broadening was attributed mainly to the sources. The decay of KBrO$_3$ showed no evidence for the formation of a 'KrO$_3$' daughter molecule (reference to Chapter 15.7 and the production of XeO$_3$ by decay of IO$_3^-$ will explain this point). The electric field gradient predicted by assuming that 'KrO$_3$' is analogous to BrO$_3^-$ should give a clear line splitting, but none was found. Any krypton molecular species is thus unstable within the ^{83m}Kr lifetime of 1·86 h. The ^{83}Rb decay in RbF$_2$, RbCl$_2$, and RbBr$_2$ gives a krypton atom in all cases, but small chemical isomer shifts can be detected which are the result of a change in the s-electron density due to orbital overlap [15].

The characteristic symmetric but very broad line found for all three types of ^{83}Kr source was used to argue against the production of different valence-state configurations [14]. This would be unlikely to give symmetry. Instead it was suggested that the isomeric transition decay of ^{83m}Kr which involves the ejection of an L or K electron penultimately produces an electron hole somewhere in the Kr atomic shell which is finally filled from neighbouring ligands. The ligand environment will probably then distort, and in the event that it has high initial symmetry, e.g. octahedral, there will be several degenerate distortion modes with electric field gradients of opposite sign. The situation parallels to some extent the Jahn–Teller effect.

This problem was overcome [16] by incorporating the ^{83m}Kr into a semiconducting lattice of Zn^{82}Se or Pb^{82}Se which might be expected to provide electrons more readily than insulating materials, and which may also be conveniently irradiated to give ^{83}Br. As shown in Fig. 15.2, the line broadening was immediately reduced to only 25% with a ZnSe source and a krypton absorber. Substantial resonance effects can be obtained at room temperature. Particularly interesting is the resolution for the first time of the predicted small quadrupole splitting in a clathrate absorber. An ^{119}Sn^{82}Se source matrix was less successful because of additional broadening which was probably a result of nonstoichiometry in the source. Particular interest attaches to the lattice-dynamical properties of krypton [17, 18]. The van der Waals 'particle-in-a-box' model for krypton clathrate assumes that the dynamics of the occluded atom are entirely decoupled from the dynamics of the host lattice. The 'rattling' frequency of the Kr atom has been studied using infrared techniques, but gives about half the value of the mean-square displacement, $\langle x^2 \rangle$, calculated from the Mössbauer recoil-free fraction data [18]. A large contribution from the low-frequency lattice vibrations thus plays an important part, and although the simpler model is valid when the mass of the occluded atom is small in relation to the unit cell of the host, it is not valid in the case of krypton where the mass difference is smaller.

Recoil-free fraction measurements for solid krypton between 5 K and 85 K are also available [19, 20], but in this case the f values are lower than expected. Some of the first calculations [20] omitted the effects of anharmonicity upon the phonon frequency spectrum. Inclusion of an harmonic effect [21] gives better agreement between experiment and theory, but the zero-temperature limit of f is anomalously low and has still not been explained.

A 77·8-kG external field has been applied to solid krypton along the direction of observation to determine the excited-state magnetic moment, μ_e [22]. In these circumstances there are 16 allowed transitions with $\Delta m = \pm 1$

[Refs. on p. 489]

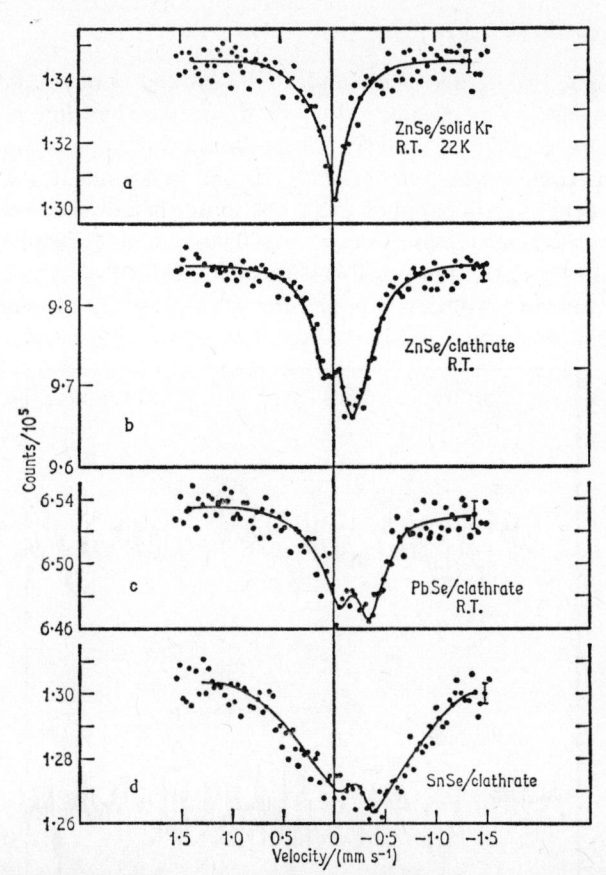

Fig. 15.2 The ^{83}Kr resonance using semiconducting sources: (a) A ZnSe source/solid krypton absorber gives close to the natural linewidth, (b) and (c) the hydroquinone clathrate is quadrupole split, (d) an SnSe source is broadened because of interactions in the source. [Ref. 16, Fig. 1]

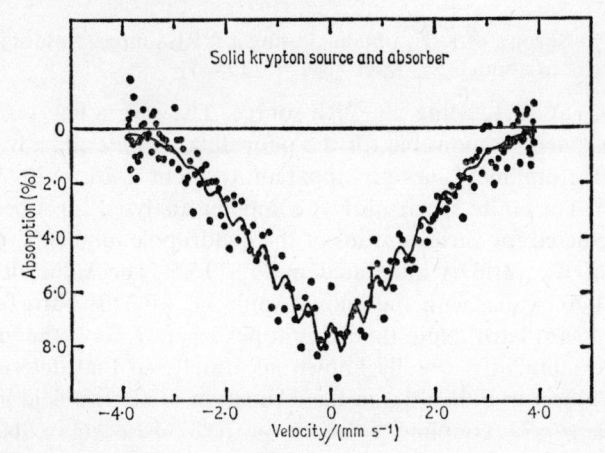

Fig. 15.3 Mössbauer spectrum of ^{83}Kr in solid krypton in a field of 77·8 kG (absorber only) at 4·2 K. [Ref. 22, Fig. 3]

between the $I = \frac{7}{2}$ excited state and $I = \frac{9}{2}$ ground state. A very broad-spectrum envelope was obtained (Fig. 15.3) and the solid line is the calculated result for $g^{\frac{7}{2}}/g^{\frac{9}{2}} = +1\cdot 249(2)$. Using the known value of $\mu_g = -0\cdot 967$ n.m. gives a value of $\mu_e = -0\cdot 939(2)$ n.m. An independent determination [23] using a field of 19 kG applied to a ZnSe source in a direction perpendicular to the observation axis gave $\mu_e = -0\cdot 99(8)$ n.m., which agrees well despite the lower resolution possible with such a field.

The only known krypton compound for which data are available is KrF_2 [13]. Fig. 15.4 shows two spectra (plotted with an X-Y recorder) which were

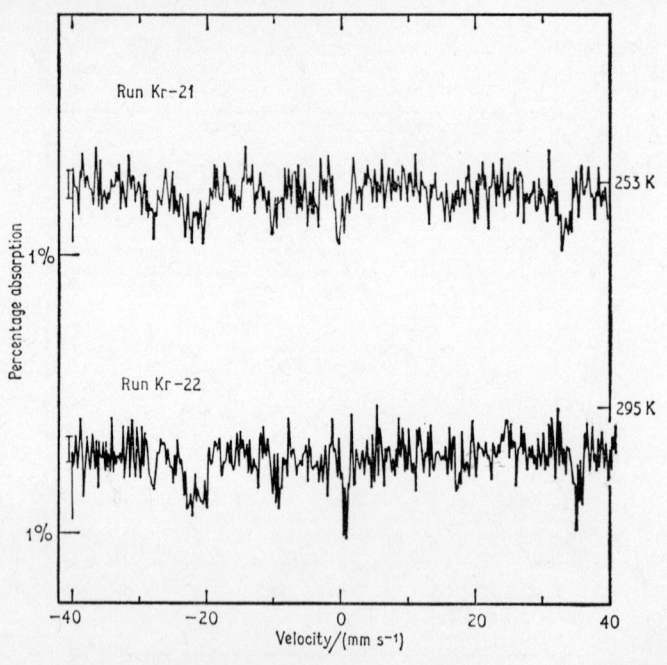

Fig. 15.4 Two spectra of KrF_2 obtained using a ^{83}Rb source. Note signals presen in both spectra at about 0·5% level. [Ref. 13, Fig. 3]

accumulated at 78 K using an ^{83}Rb source. The severe line broadening of the latter is partly responsible for the poor data, despite the advantage of a longer source lifetime. Lines are apparent at about -28, $-20/22$, -10, 0, 34 mm s^{-1}. The entire spectrum was computer analysed for a best fit to the spectra predicted for various ratios of the quadrupole moments, Q_e/Q_g. The influence of the ratio is illustrated in Fig. 15.5. The value deduced was $Q_e/Q_g = 1\cdot 70$, which with the known value of $+0\cdot 2701$ barn for Q_g gives $Q_e = +0\cdot 459(6)$ barn. Note that in isotopes where $I_g > \frac{1}{2}$, the ground-state quadrupole moment is usually known accurately, so that determination of Q_e is not subject to difficulties in the estimation of electric field gradients.

The quadrupole coupling constant in KrF_2 is $e^2qQ = 960(30)$ MHz

[*Refs.* on p. *489*]

(converted from velocity using $v = 0.806vE_\gamma$, where E_γ is in keV and v is in mm s^{-1}). Atomic beam measurements have given a value of 904·4 MHz for

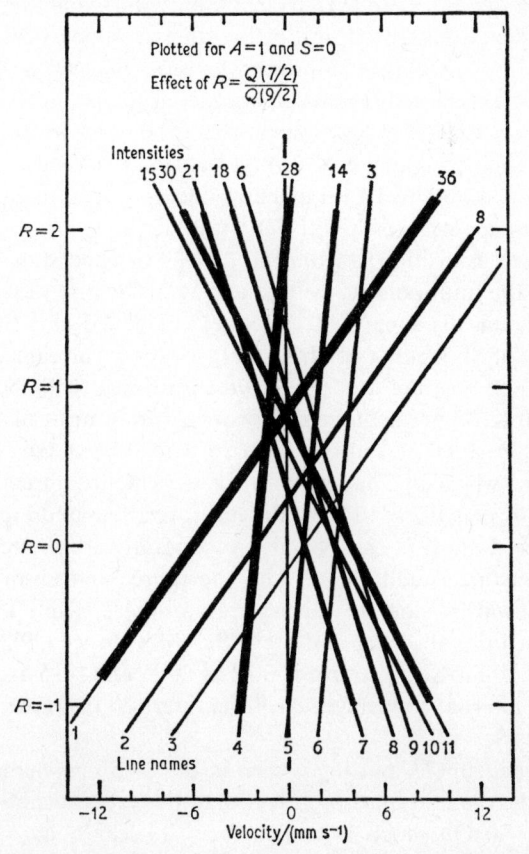

Fig. 15.5 Schematic illustration of how the ratio Q_e/Q_g affects the ^{83}Kr spectrum. [Ref. 13, Fig. 2]

the electric field gradient due to a single 4p-electron in krypton, so that the number of unbalanced p-electrons, U_p, is 1·06. This implies the removal of 0·53 electrons from krypton to each fluorine atom.

The chemical isomer shift was 1·50(5) mm s^{-1} with respect to the source. Decreased shielding of the 4s-electrons by the effective $4p^5$(Kr$^+$) configuration in KrF$_2$ is responsible, allowing an approximate calculation of $\delta\langle R^2\rangle/\langle R^2\rangle = +8 \times 10^{-4}$.

15.4 Antimony (^{121}Sb)

The 37·15-keV resonance of ^{121}Sb was first recorded in antimony metal by Snyder and Beard in 1965 [24]. This level is conveniently populated directly with 100% efficiency by the β^-decay of ^{121m}Sn which has a half-life of 76 y. The latter can be produced by the reaction ^{120}Sn$(n, \gamma)^{121m}$Sn, but the shorter-lived contaminants of ^{113}Sn (115 d), ^{119m}Sn (250 d), and ^{123m}Sn (125 d) are also produced from small quantities of other isotopes in the ^{120}Sn. Consequently, the ^{121m}Sn source is most effective after a lapse of some months when these have decayed. Chemical separation is also advisable to remove the ^{125}Sb activity. The major unwanted radiations in the γ-ray energy spectrum are then the intense K X-rays centred at 26·4 keV, but adequate resolution may be obtained with a xenon-filled proportional counter [24], or by using the escape peak of a NaI(Tl) crystal [25].

The 37·15-keV transition is from the $I = \frac{7}{2}+$ excited state to the $I = \frac{5}{2}+$ ground state, the multipolarity being pure M1. The first measurements used a source of chemically separated ^{121m}Sn electroplated as β-tin onto copper, and absorbers of Sb metal and Sb_2O_3 [24]. Isotopic enrichment of absorbers was not required to obtain a significant absorption with both source and absorber at 80 K. The tin source matrix was used in most of the early work, although ^{121m}Sn/SnO$_2$ has also been used with success [26]. A third source is Ca^{121m}SnO$_3$, which by analogy with the same sources used in ^{119}Sn work (see Chapter 14.1) is likely to have the narrowest line and highest recoil-free fraction of the three [27, 28]. Detailed comparative data are not available, but experiments are usually made with the source and absorber at 80 K in all cases. Linewidths close to the natural width (2·1 mm s^{-1}) can be obtained with an InSb absorber and a ^{121m}Sn/β-Sn source [29]. The recoil-free fraction of a ^{121m}Sn/SnO$_2$ source is 0·32 at 80 K and 0·16 at room temperature [26]. The internal conversion coefficient derived from the same measurements is $\alpha_T \sim 10$.

Although quadrupole splitting is seen in the ^{121}Sb resonance in appropriate compounds, the line splitting is less than the experimental linewidth [29], even in cubic Sb_2O_3 which is known from nuclear quadrupole resonance (n.q.r.) data to have one of the largest electric field gradient values. Eight quadrupole transitions are predicted for a $\frac{7}{2} \to \frac{5}{2}$ M1 decay (see Appendix 2). The known value of e^2qQ from n.q.r. data is 555·5 MHz (18·64 mm s^{-1}), and a series of calculated spectra for different values of $R = Q_e(\frac{7}{2})/Q_g(\frac{5}{2})$ is given in Fig. 15.6, together with an actual spectrum at 4·2 K in Fig. 15.7. Particularly disturbing is the ambiguity as to the value of R. Equally good fits (judged by χ^2 statistical criteria) can be obtained in this instance for $R = 1·42$ and $R = 0·62$, and provide a salutory example of how a statistically acceptable analysis need not be unique. However, e^2qQ_g was left as a free parameter, and the low value of 16·68 mm s^{-1} derived when $R = 0·62$ compares unfavourably with 18·76 mm s^{-1} when $R = 1·42$. Analysis of several spectra led to the adoption of $Q_e(\frac{7}{2})/Q_g(\frac{5}{2}) = 1·38(2)$. Several quadrupole splittings have been subsequently deduced from unresolved spectra, and are given in Table 15.1. Estimations of the electric field gradients in Sb_2O_3 and SbF_3

[Refs. on p. 489]

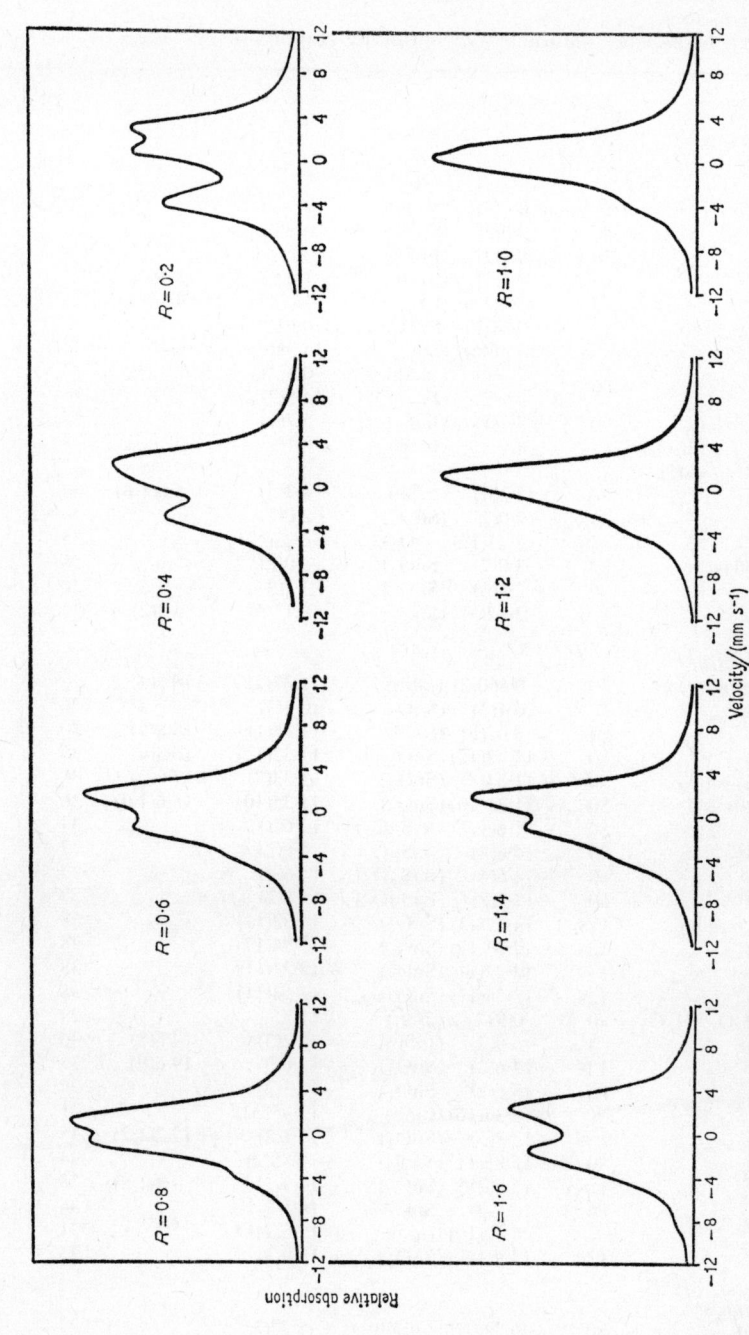

Fig. 15.6 Calculated Mössbauer spectra for ^{121}Sb in an electric field gradient appropriate to cubic-Sb_2O_3 for various ratios of the excited-state quadrupole moment to the ground-state moment, $R = Q_e(\tfrac{7}{2})/Q_g(\tfrac{5}{2})$. [Ref. 29, Fig. 1]

[Refs. on p. 489]

Table 15.1 Mössbauer parameters of antimony compounds (^{121}Sb)

Compound	T/K	δ*/(mm s^{-1})		δ (SnO$_2$)*/(mm s^{-1})	e^2qQ_g/(mm s^{-1})	Reference
Antimony (V)						
NaSb(OH)$_6$	80	+0·5(2)	(SnO$_2$)	+0·5(2)		26
Sb$_2$O$_5$	80	0·0(2)	(SnO$_2$)	0·0(2)		26
	80	0·1(3)	(BaSnO$_3$)	0·1(2)		37
	4·2	+9·9(2)	(InSb)	+1·6(3)	−8·4(8)	30
	80	+1·06(2)	(SnO$_2$)	+1·06(2)	−4·3(1·1)	36
α-Sb$_2$O$_4$	80	$\begin{cases} +0·61(3) \\ -14·36(6) \end{cases}$	(SnO$_2$)(SnO$_2$)	+0·61(3)−14·36(6)	−6·1(1·0)+16·4(6)	3636
	80	$\begin{cases} +0·3(2) \\ -14·5(2) \end{cases}$	(BaSnO$_3$)(BaSnO$_3$)	+0·3(2)−14·5(2)		3737
Na$_3$SbS$_4$.9H$_2$O	80	−5·7(1)	(BaSnO$_3$)	−5·7(1)		37
Ni$_{1+2x}$Fe$_{2-3x}$Sb$_x$O$_4$ (0·33 > x > 0·05)	80	+8·0	(InSb)	−0·15		28
KSbF$_6$	4·2	+12·3(4)	(InSb)	+4·2(5)	+8·0(1·6)	30
NaSbF$_6$	80	+2·0(2)	(SnO$_2$)	+2·0(2)		26
SbF$_5$	LN	+2·23(12)	(SnO$_2$)	+2·23(12)		35
HSbCl$_6$.xH$_2$O	80	−3·0(2)	(SnO$_2$)	−3·0(2)		26
SbCl$_5$	80	−3·5(3)	(SnO$_2$)	−3·5(3)		26
	LN	−3·12(3)	(SnO$_2$)	−3·12(3)	−4·4(2)	35
Antimony (III)						
Sb$_2$O$_3$ (cubic)	80	−0·560(21)	(β-Sn)	−11·37(2)	+18·8(4)	29
	80	−10·4(3)	(SnO$_2$)	−10·4(3)		26
	4	−3·02(5)	(InSb)	−11·17(1)	+18·8(4)	30
(cubic)	80	−11·32(12)	(SnO$_2$)	−11·32(12)	+18·3(4)	36
(orthorhombic)	80	−11·33(7)	(SnO$_2$)	−11·33(7)	+17·0(2)	36
(amorphous)	80	−11·35(10)	(SnO$_2$)	−11·35(10)	+18·6(1·0)	36
	80	−11·6(1)	(BaSnO$_3$)	−11·6(1)		37
Sb$_2$S$_3$	80	−14·6(2)	(BaSnO$_3$)	−14·6(2)		37
Sb$_2$Se$_3$	80	−14·6(4)	(BaSnO$_3$)	−14·6(4)		37
Sb$_2$Te$_3$	80	−15·3(2)	(BaSnO$_3$)	−15·3(2)		37
Sb$_2$S$_{2·97}$ (stibnite)	LN	−14·32(11)	(SnO$_2$)	−14·32(11)		38
Sb$_2$S$_{3·55}$	LN	−12·74(17)	(SnO$_2$)	−12·74(17)		38
Sb$_2$S$_{4·4}$	LN	−12·27(11)	(SnO$_2$)	−12·27(11)		38
Sb$_2$S$_{24·3}$	LN	−12·54(11)	(SnO$_2$)	−12·54(11)		38
KSbC$_4$H$_4$O$_7$.½H$_2$O	80	−0·91	(β-Sn)	−11·72		29
SbF$_3$	4	−6·0(2)	(InSb)	−14·2(3)	+19·6(8)	30
	LN	−14·6(2)	(SnO$_2$)	−14·6(2)	+19·6(9)	35
SbCl$_3$	80	−15·5(2)	(SnO$_2$)	−15·5(2)		34
	?	−5·30(16)	(InSb)	−13·45(22)		31
	LN	−13·8(2)	(SnO$_2$)	−13·8(2)	+12·2(1·7)	35
SbBr$_3$	80	−15·85(2)	(SnO$_2$)	−15·85(2)		34
	LN	−13·9(2)	(SnO$_2$)	−13·9(2)	+9·4(1·8)	35
SbI$_3$	80	−16·5(3)	(SnO$_2$)	−16·5(3)		34
	?	−7·40(10)	(InSb)	−15·55(16)		31
	LN	−15·9(3)	(SnO$_2$)	−15·9(3)		35
Metallic phases						
Sb/nickel	80	−6·29(3)	(CaSnO$_3$)	−6·29(3)		27
Sb/iron	80	−6·39(3)	(CaSnO$_3$)	−6·39(3)		27
InSb	80	−8·15(6)	(SnO$_2$)	−8·15(6)		36
	80	−8·4(3)	(SnO$_2$)	−8·4(3)		26

[*Refs. on p. 489*]

Table 15.1 (continued)

Compound	T/K	δ*/(mm s⁻¹)	δ (SnO₂)*/(mm s⁻¹)	e^2qQ_g/(mm s⁻¹)	Reference
InSb	80	−8·65(8) (CaSnO₃)	−8·65(8)		27
	LN	−8·56(8) (SnO₂)	−8·56(8)		35
Sb/β-Sn	4	−2·66(1) (InSb)	−10·81(2)		26
Sb metal	80	−11·2(2) (SnO₂)	−11·2(2)		26
Organoantimony					
Ph₃SbCl₂	80	−6·9(2) (SnO₂)	−6·9(2)		25
Ph₃SbF₂	80	−5·5(3) (SnO₂)	−5·5(3)		25
(o-MeO.C₆H₄)₃SbCl₂	80	−6·2(2) (SnO₂)	−6·2(2)		25
Ph₄SbBr	80	−5·6(2) (SnO₂)	−5·6(2)		25
Ph₄SnBF₄	80	−6·0(1) (SnO₂)	−6·0(1)		25
(p-Cl.C₆H₄)₃Sb	80	−9·3(2) (SnO₂)	−9·3(2)		25
(p-MeO.C₆H₄)₃Sb	80	−9·0(2) (SnO₂)	−9·0(2)		25

* The chemical isomer shifts are quoted with respect to their original standard, and on a unified scale relative to ^{121}Sb/SnO₂. Conversions used are InSb to ^{121}Sb/SnO₂, −8·15 (6) mm s⁻¹; ^{121}Sb/β-Sn to ^{121}Sb/SnO₂, −10·81 (20) mm s⁻¹; ^{121}Sb/CaSnO₃ to ^{121}Sb/SnO₂, 0·0 mm s⁻¹; ^{121}Sb/BaSnO₃ to ^{121}Sb/SnO₂, assumed 0·0 mm s⁻¹.
No temperature corrections have been made to data at different temperatures.

have led to an estimate of −0·26 barn for $Q_g(\frac{5}{2})$, giving $Q_e(\frac{7}{2}) = -0\cdot36$ barn, although from the aspect of spectrum analysis, R is the important parameter [30].

Magnetic hyperfine splitting was first recorded in ferromagnetic MnSb at 4·2 K [32]. The magnetic field was already known from n.m.r. data to be $H = 352\cdot6$ kG. There is also a small quadrupole interaction, estimated to be $e^2qQ_g = 73$ MHz (2·43 mm s⁻¹), probably directed perpendicular to the spin axis. The nuclear spin-states generate 18 magnetic transitions, although as already seen from ^{83}Kr, the interrelation of their positions and intensities is described by fewer parameters and it is possible to predict the spectrum envelope. In Fig. 15.8 is shown a schematic representation of the line positions for different ratios of the nuclear g factors, $R = g(\frac{7}{2})/g(\frac{5}{2})$. The experimental data for MnSb together with three computed spectra for different values of $R = g(\frac{7}{2})/g(\frac{5}{2})$ are given in Fig. 15.9. The known values of H and e^2qQ, and $\mu(\frac{5}{2}) = +3\cdot359(1)$ n.m. can then be used to derive $\mu(\frac{7}{2}) = +2\cdot35(3)$ n.m. Not all 18 magnetic transitions are resolved, and by chance the ratio R is such that the magnetic spectrum approximates to eight equispaced lines with the relative intensities 1, 9, 37, 37, 37, 37, 9, 1, of which the inner quartet are the prominent feature, and the outer pair are not always detected [28].

Subsequent experiments have used these values to determine the hyperfine fields at Sb nuclei (1%) in iron and nickel, which were 245 and 75 kG at 80 K respectively [27]. The chemical isomer shifts recorded are given in Table 15.1. The spectrum of ^{121}Sb in cobalt did not give a unique magnetic field, and was not analysed.

[Refs. on p. 489]

446 | OTHER MAIN GROUP ELEMENTS

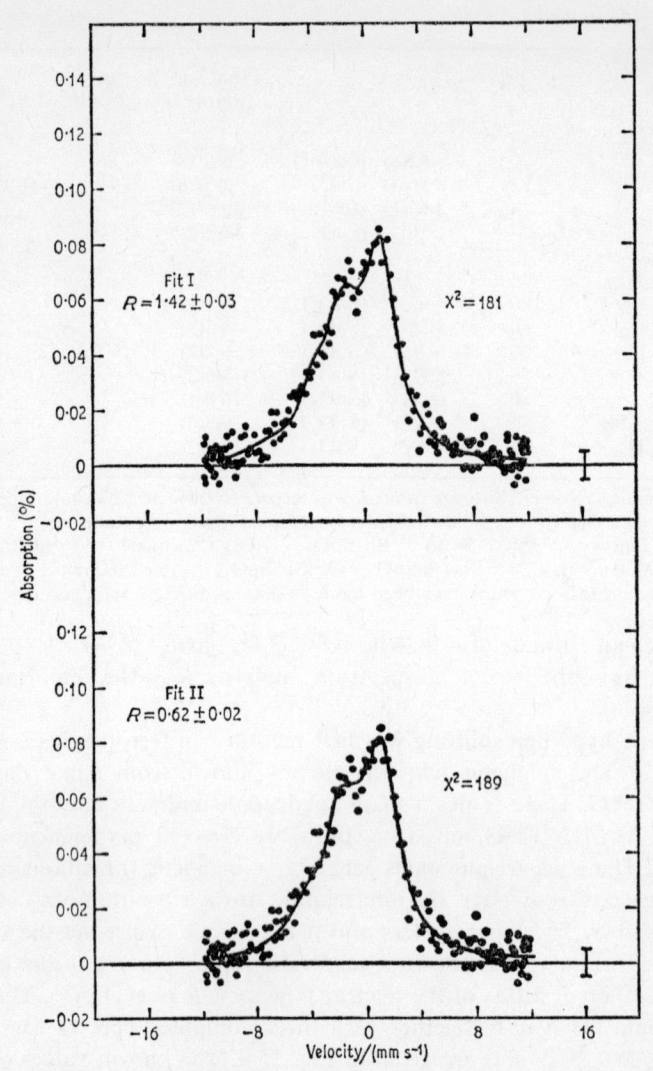

Fig. 15.7 An experimental spectrum for cubic–Sb_2O_3 with two attempted computer fits which are statistically valid but yield different values of $R = Q_c(\frac{7}{2})/Q_g(\frac{5}{2})$. [Ref. 29, Fig. 2]

The detection of transferred hyperfine interaction at diamagnetic Sn^{4+} cations in oxides of the garnet type (see Chapter 14) prompted a study of isoelectronic Sb^{5+} in magnetic oxides of the phase $Ni_{1+2x}Fe_{2-3x}Sb_xO_4$ [28]. These are spinel oxides with antimony at the octahedral B sites and Fe^{3+} and Ni^{2+} distributed on both the tetrahedral A sites and the B sites. The preliminary data show considerable magnetic interactions at the ^{121}Sb sites, but

[Refs. on p. 489]

as seen in Fig. 15.10, the spectra do not correspond to a unique hyperfine field as seen by the deviations from the expected intensity ratios. The symmetry of the spectra precludes a large quadrupole interaction, and indeed none was found in diamagnetic $Zn_{2.33}Sb_{0.67}O_4$ and $Co_{2.33}Sb_{0.67}O_4$. However, the spectra can be analysed successfully assuming that the magnetic field at a given antimony nucleus is directly proportional to the number of Fe^{3+} ions in the six A site nearest-neighbours, and that there is a random probability distribution of cations. The solid curves in Fig. 15.10 were calculated on these assumptions, confirming that the field is due primarily to the A site neighbours and that each Fe^{3+} cation contributes independently of the

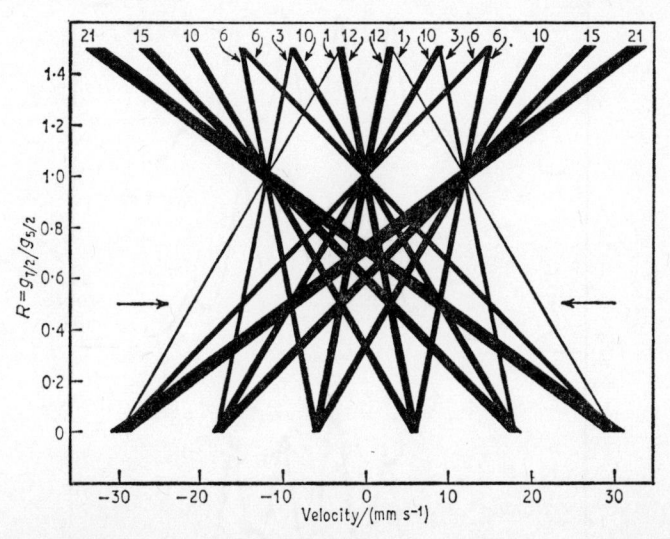

Fig. 15.8 Calculated line positions and relative intensities for an ^{121}Sb nucleus in an internal magnetic field of 353 kG for various values of $R = g(\frac{7}{2})/g(\frac{5}{2})$. The intensities are given at the top of the diagram, and the horizontal arrows indicate the experimentally derived value of R. [Ref. 32, Fig. 2]

nickel. The maximum field corresponding to six Fe^{3+} neighbours is 315(10) kG. Although at first glance the interactions in these spinels are not comparable with the Sn^{4+} garnet data, it should be noted that in the latter the Sn^{4+} ions at d sites are all surrounded by six Fe^{3+} ions at a sites, so that a unique exchange field can be expected.

Chemical isomer shifts in ^{121}Sb are quite large, as may be seen in Table 15.1, the range being 20·7 mm s^{-1} between SbI_3 (Sb^{3+}) and $KSbF_6$ (Sb^{5+}) [30]. The first serious study compared the values of the shifts in the pairs of antimony and tin compounds $KSbF_5/K_2SnF_6$, Sb_2O_5/SnO_2, $InSb/\alpha$-tin, Sb in β-tin/β-tin, Sb_2O_3/SnO, and SbF_3/SnF_2. It was proposed [30] that there should be considerable similarity in the electron-density behaviour in these compounds because the Sb and Sn oxidation states are isoelectronic, although it should be noted that the compounds are not isostructural and will therefore differ markedly in their covalent bonding. This was felt to be

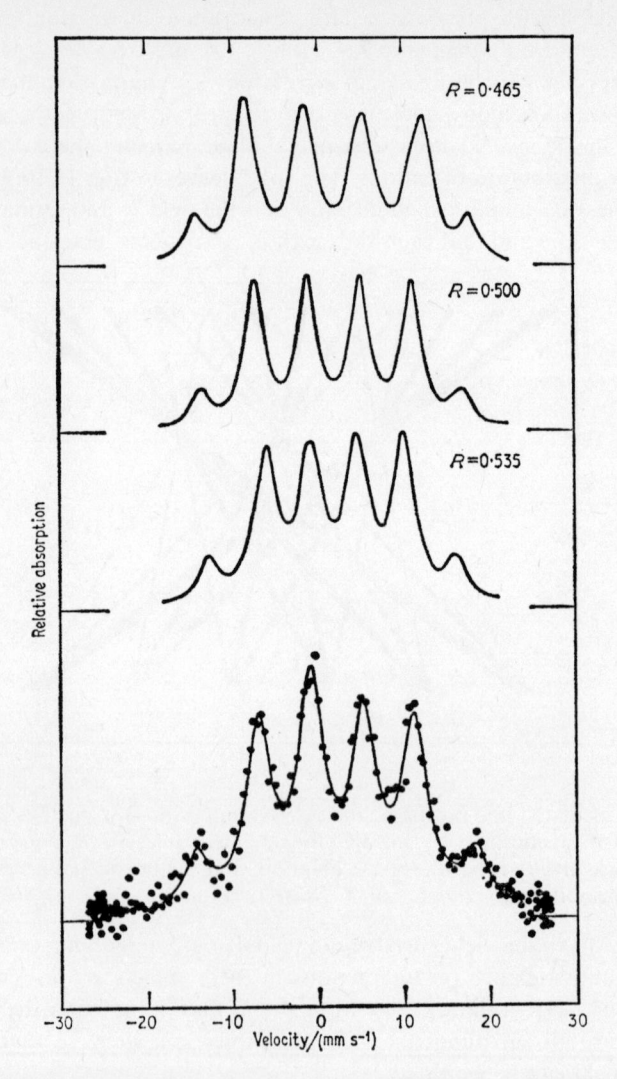

Fig. 15.9 The ^{121}Sb spectrum of MnSb at 4·2 K with three computed envelopes and the final fit which was calculated using the n.m.r. data for H and e^2qQ. [Ref. 32, Fig. 1]

valid after comparing Hartree–Fock self-consistent field atomic wavefunction calculations for the two elements. An approximately linear correlation is found between the shifts for the two isotopes in these pairs of compounds which lends support to the concept. However, an exact interpretation

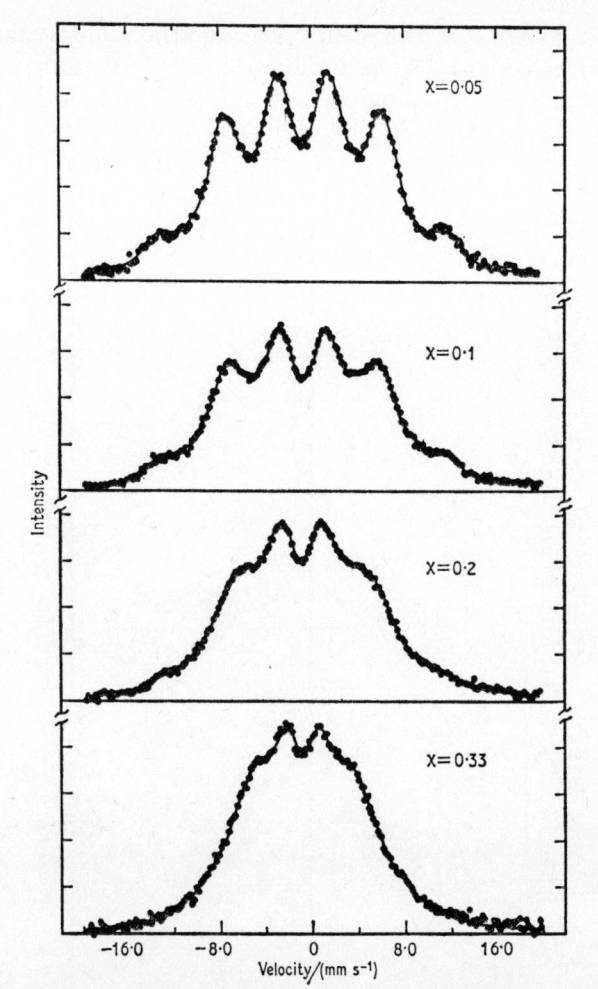

Fig. 15.10 Spectra with calculated curves for the ^{121}Sb resonance in spinels of the type $Ni_{1+2x}Fe_{2-3x}Sb_xO_4$. [Ref. 28, Fig. 1]

of the ^{121}Sb chemical isomer shifts is subject to the same difficulties already outlined for ^{119}Sn in Chapter 14. Nevertheless, it is clear that Sb(III) has a lower shift than Sb(V), as illustrated in the correlation diagram in Fig. 15.11. Since Sn(II) has a *higher* shift than Sn(IV) it is intuitively obvious and this is confirmed by the detailed calculations [30] that the sign of $\delta \langle R^2 \rangle / \langle R^2 \rangle$ in ^{121}Sb is opposite to that in ^{119}Sn. An order of magnitude estimate of $\delta \langle R^2 \rangle / \langle R^2 \rangle$ (^{121}Sb) $= -17 \times 10^{-4}$ was derived. A more recent revision of the calculations embracing data for ^{119}Sn, ^{121}Sb, ^{125}Te, ^{127,129}I, and ^{129}Xe has given $\delta \langle R^2 \rangle / \langle R^2 \rangle$ (^{121}Sb) $= -14.6 \times 10^{-4}$ [31].

A similar comparison of ^{119}Sn, ^{121}Sb, and ^{125}Te data gave a value of

[*Refs. on p.489*]

$\delta\langle R^2\rangle/\langle R^2\rangle$ (^{121}Sb) $= -36\cdot8 \times 10^{-4}$ [26, 33]. The values of the shifts for the antimony compounds are included in Table 15.1. A further estimate of -19×10^{-4} was later given by the same workers [34] from data for SbCl$_3$ and HSbCl$_6$.xH$_2$O. The chemical isomer shifts for SbF$_3$, SbCl$_3$, SbBr$_3$, and

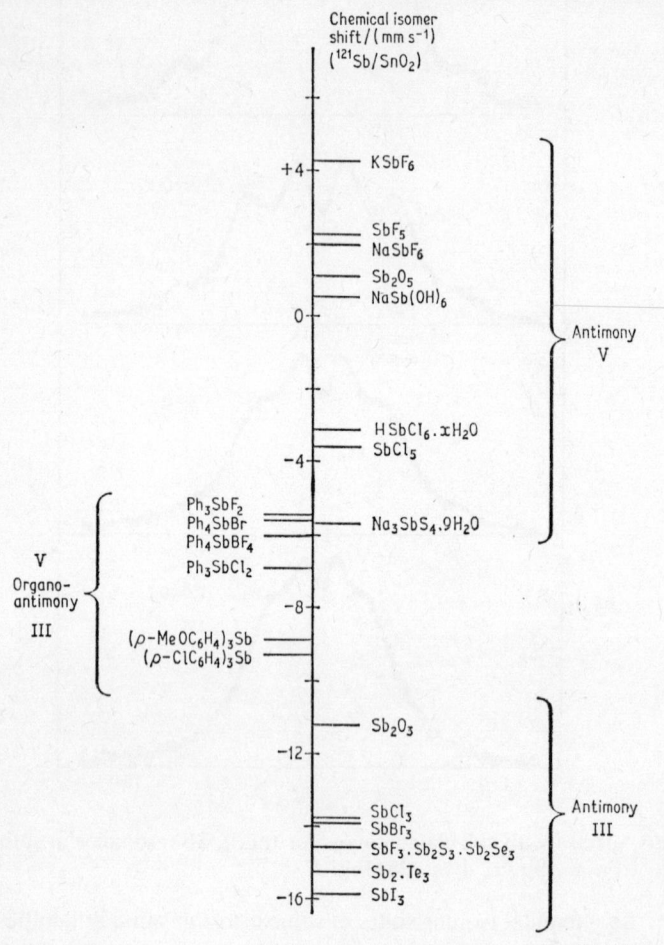

Fig. 15.11 The chemical isomer shifts of antimony compounds (^{121}Sb).

SbI$_3$ show a similar dependence on bond ionicity to that found in the tin halides [34], although more recent data has shown significant differences in the detailed values [35]. SbF$_3$ does not conform to the general pattern of a linear relationship between chemical isomer shift and the ligand electronegativity difference, presumably because of a radically different geometry. The results were considered to indicate a constant *s*-character in the bonding in SbCl$_3$, SbBr$_3$, SbI$_3$, and Sb$_2$O$_3$.

[*Refs. on p. 489*]

Chemical isomer shifts for several aryl organo-antimony compounds [25] are particularly interesting in that there is only a small shift between Sb(III) and Sb(V) compounds, and all are intermediate between typical inorganic compounds of these states (Fig. 15.11). Many of the compounds for which no quadrupole splitting is quoted in Table 15.1 show an asymmetric resonance, but no analysis has been attempted.

One of the first direct chemical applications of the ^{121}Sb resonance was the verification of the oxidation states in α-Sb_2O_4 [36]. This oxide is isostructural with $Sb^{III}Nb^VO_4$ and $Sb^{III}Ta^VO_4$, and the ^{121}Sb spectrum (Fig. 15.12) shows

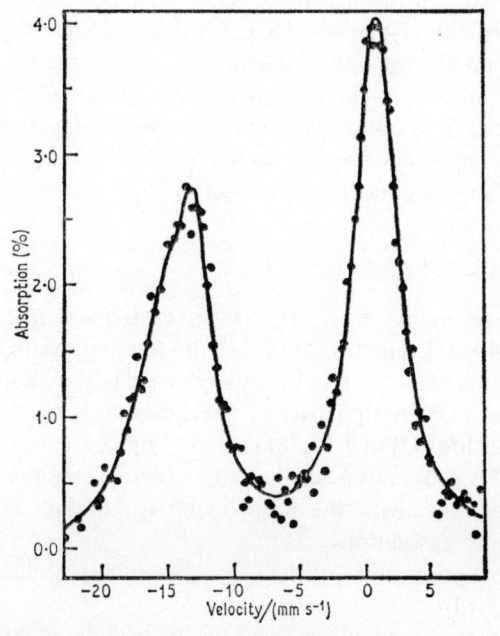

Fig. 15.12 ^{121}Sb spectrum of α-Sb_2O_4. [Ref. 36, Fig. 1]

two distinct resonances clearly identifiable as Sb^{III} and Sb^V. The alternative structure of $Sb^{IV}_2O_4$ is thereby eliminated.

Rather surprisingly the three forms of Sb_2O_3 give very similar spectra (Table 15.1). This is presumably a result of the close similarity in the immediate environment of the antimony. The considerable change to more negative velocities in the sequence Sb_2O_3, Sb_2S_3, Sb_2Se_3, and Sb_2Te_3 indicates an increase in s-electron density at the antimony nucleus, there being apparently a closer approximation to a bare Sb^{3+} in the telluride than the oxide, as may also be inferred from their structures [37].

Mössbauer spectroscopy has also demonstrated that the compound previously claimed to be Sb_2S_5 has only one absorption, clearly in the Sb(III) region of the spectrum [37, 38]. No Sb(IV) resonance was detected and this

[*Refs. on p. 489*]

was confirmed by subsequent phase studies. Preparations containing more sulphur than required by the formula Sb_2S_3 presumably feature polysulphide linkages [38]. Thioantimonate(V) salts can, however, be prepared: the symmetrical spectrum obtained for this compound was consistent with the known tetrahedral coordination of the SbS_4^{3-} ion and the chemical isomer shift of $-5 \cdot 7$ mm s^{-1}, which is the most negative yet reported for a compound of Sb(V), was taken to imply considerable covalent character in the Sb–S bonds.

Frozen solutions of antimony(V) in aqueous HCl contain $SbCl_6^-$ only at very high HCl concentrations above 11M [39]. $SbCl_5(OH)^-$ is formed in 9M HCl and $SbCl_2(OH)_2$ in 6M HCl. All three species show a narrow line without significant electric field gradient and are assumed to be octahedral.

Paucity of adequate data prevents more detailed comment on the significance of the Mössbauer parameters at the present time, but it appears certain that this resonance can give as much information as its better-known counterpart [119]Sn has done in tin chemistry.

15.5 Tellurium ([125]Te)

The 35·48-keV resonance of [125]Te was first observed in two laboratories independently [40, 41]. The transition is from the first excited state ($I = \frac{3}{2}+$) to the ground state ($I = \frac{1}{2}+$) and in many respects shows similarities to the [119]Sn resonance. The multipolarity of the radiation is almost pure M1. The excited-state half-life [42] of $1 \cdot 535(81) \times 10^{-9}$ s gives a natural linewidth of 5·02 mm s^{-1}. This proves to be of the same order of magnitude as the largest hyperfine interactions, with the unfortunate result that [125]Te spectra inevitably show poor resolution.

There are three convenient nuclear decays which populate the 35·48-keV level (see Fig. 15.13):
(a) [125]Sb decays by β-emission with a half-life of 2·7 y in a complex scheme which includes the metastable state [125m]Te. Although the complexity of this decay is the prime disadvantage, it has been successfully used by incorporating [125]Sb in copper matrices [41].
(b) The [125m]Te state can be populated by a neutron irradiation using the [124]Te(n, γ)[125m]Te reaction. It decays with a half-life of 58 days and is equivalent to the [119m]Sn isotope. Single-line sources have been obtained from matrices of PbTe [43], ZnTe [44], electrodeposited Te/Pt [45], and TeO_3 [46], but, as described shortly, there are recorded instances of radiation damage affecting the emission line in some cases. The β-TeO_3 matrix gives a narrow line ($\Gamma_r \leqslant 5 \cdot 5$ mm s^{-1}) with a substantial recoil-free fraction ($>0 \cdot 54$ at 80 K and $>0 \cdot 30$ at room temperature) [47]. A scattering technique has also been used [40].
(c) The commercially available isotope ^{125}I decays with a 60-day half-life

[Refs. on p. 489]

by electron capture directly to the 35·48-keV level, and is therefore the most efficient precursor. It is however difficult to incorporate into a source matrix without decay after-effects. A satisfactory source can be made by diffusing the ^{125}I into a copper foil [48], the recoilless fraction being ~0·2 at 82 K [49].

The bulk of the available data has been obtained with both source and

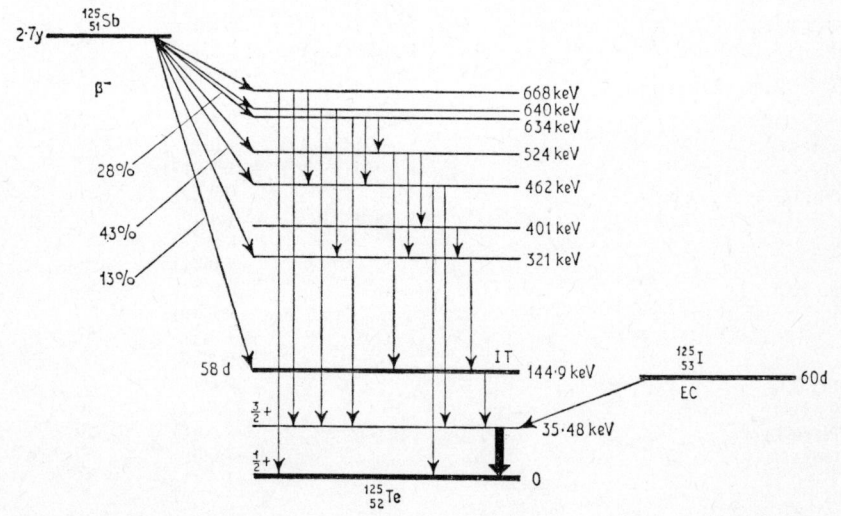

Fig. 15.13 The decay schemes for ^{125}Sb, ^{125m}Te and ^{125}I which all populate the 35·48-keV level of ^{125}Te.

absorber at liquid nitrogen temperature, often with absorbers enriched in ^{125}Te. Under these conditions a resonance of several per cent can usually be obtained. The 35·48-keV γ-ray is highly converted ($\alpha_T = 12·7$) and is difficult to resolve from the intense K_α (27·4-keV) and K_β (31·0-keV) X-rays, but a copper selective filter can be used to advantage [48].

Inorganic Tellurium Compounds

A number of laboratories have observed the ^{125}Te resonance in a variety of inorganic tellurium compounds. Chemical isomer shift and quadrupole splitting values are summarised in Table 15.2. Considerable duplication of measurements has taken place, but the self-consistency of results within the quoted experimental errors is encouraging. The chemical isomer shift has been converted where necessary to a ^{125}I/Cu source, the latter being chosen solely because most values available were already referred to this. The overall range of shifts is only 3 mm s^{-1}, which is less than the natural linewidth of 5·02 mm s^{-1}. However, if spectra of high quality can be obtained, differences of only 0·1 mm s^{-1} may be shown to be significant [50]. The low resolution precludes useful analysis of data where multiple tellurium environments are

[Refs. on p. 489]

Table 15.2 Mössbauer parameters of tellurium compounds (^{125}Te)

Compound	T/K	Δ/(mm s^{-1})	δ†/(mm s^{-1})	δ (^{125}I/Cu)† /(mm s^{-1})	Reference
Tellurium (VI)					
Te(OH)$_6$	78	—		−1·15(4)	51
	80	—		−0·98(6)	50
	LN	—	−1·44(5) (Te/Pt)	−1·19(15)	45
H$_6$TeO$_6$	LN	—	−1·6(2) (Te)	−1·1(3)	52
Na$_2$H$_4$TeO$_6$	LN	—	−1·37(2) (Te/Pt)	−1·12(12)	45
BaH$_4$TeO$_6$	80	—		−0·87(9)	50
'TeO$_3$'	78	2·6(4)		−1·07(5)	51
	LN	—	−0·9(2) (Te)	−0·4(3)	52
α-TeO$_3$	LN	—	−1·33(2) (Te/Pt)	−1·08(17)	45
β-TeO$_3$	LN	—	−1·44(2) (Te/Pt)	−1·19(17)	45
Na$_2$TeO$_4$	78	—		−0·99(3)	51
	LN	—	−1·26(2) (Te/Pt)	−1·41(4)	45
	80	—		−0·95(7)	50
	(?)	—		−1·00(1)	31
K$_2$TeO$_4$	78	—		−0·98(5)	51
	80	—		−0·87(8)	50
(NH$_4$)$_2$TeO$_4$	LN	—	−1·7(2)	−1·2(3)	52
CaTeO$_4$	80	—		−0·97(7)	50
SrTeO$_4$	80	—		−0·89(10)	50
CoTeO$_4$	80	—		−0·74(9)	50
NiTeO$_4$	80	—		−0·78(8)	50
CuTeO$_4$	80	—		−1·09(9)	50
TeF$_6$	(?)	—		−1·40(13)	31
Tellurium (IV)					
TeO$_2$	82	7·76(18)		+0·69(9)	49
	78	6·54(12)		+0·72(7)	51
	LN	6·8(4)	+0·5(3) (Te)	+1·0(4)	52
	LN	?	+0·8(3) (Sb/Cu)	+0·6(4)	55
	80	6·25(3)		+0·91(12)	50
	(?)			+0·7(1)	31
	LN	6·63(6)	+0·48(2) (Te/Pt)	+0·73(12)	45
Te$_2$O$_4$.HNO$_3$	LN	6·0(5)	+0·9(3) (Te)	+1·4(4)	52
	80	6·65(6)		+0·89(19)	50
H$_2$TeO$_3$	LN	7·7(1·2)	+0·5(3) (Te)	+1·0(4)	52
(NH$_4$)$_2$TeO$_3$	LN	6·8(7)	+0·1(4) (Te)	+0·6(5)	52
Na$_2$TeO$_3$	82	6·6(2)		+0·4(1)	49
	78	5·78(8)		+0·22(5)	51
	80	6·65(16)		+0·42(29)	50
	LN	5·94(7)	−0·08(7) (Te/Pt)	+0·17(22)	45
Na$_2$TeO$_3$.5H$_2$O	78	6·72(20)		+0·13(10)	51
K$_2$TeO$_3$	(?)			+0·14(7)	31
BaTeO$_3$	80	6·28(20)		−0·22(38)	50
SrTeO$_3$	80	5·97(28)		+0·66(28)	50
CuTeO$_3$	80	6·45(15)		+0·37(28)	50
TeF$_4$	82	6·8(1·8)		+0·4(9)	49
TeCl$_4$	78	4·0(1·6)		+1·2(1)	51
	LN	5·4(8)	+1·9(4) (Te)	+2·4(5)	52
	(?)			+1·1(1)	31
TeBr$_4$	78	3·8(4·0)		+1·1(1)	51

[*Refs.* on p. 489]

Compound	T/K	Δ/(mm s⁻¹)	δ†/(mm s⁻¹)	δ (^{125}I/Cu)† /(mm s⁻¹)	Reference
TeBr₄	LN	5·0(1·4)	+0·8(3) (Te)	+1·3(4)	52
TeI₄	82	6·0(1·8)		+1·0(9)	49
	78	4·0		+1·8(9)	51
	(?)			+1·0(2)	31
(NH₄)₂TeCl₆	80	—		+1·95(5)	50
	LN	—	+1·0(3) (Te)	+1·5(4)	52
Rb₂TeCl₆	80	—		+1·95(4)	50
Cs₂TeCl₆	80	—		+1·94(9)	50
(Ph₄As)₂TeCl₆	LN	—	+1·5(4) (Te)	+2·0(5)	52
TeCl₆²⁻	LN	—	+1·4(3) (Sb/Cu)	+1·2(4)	53
(NH₄)₂TeBr₆	80	—		+1·73(4)	50
(NMe₄)₂TeBr₆	80	—		+1·75(17)	50
K₂TeBr₆	80	—		+1·72(5)	50
	LN	—	+1·4(4) (Te)	+1·9(5)	52
Rb₂TeBr₆	80	—		+1·72(4)	50
Cs₂TeBr₆	80	—		+1·80(3)	50
TeBr₆²⁻	LN	—	+1·7(3) (Sb/Cu)	+1·5(4)	53
(NH₄)₂TeI₆	80	—		+1·54(9)	50
K₂TeI₆	80	—		+1·65(6)	50
Rb₂TeI₆	80	—		+1·53(8)	50
Cs₂TeI₆	80	—		+1·65(4)	50
TeI₆²⁻	LN	—	+2·0(3) (Sb/Cu)	+1·8(4)	53
NH₄TeF₅	80	6·25(16)		+1·09(33)	50
CsTeF₅	80	5·85(9)		+0·93(21)	50
CsTeF₅*	80	5·58(42)		+0·75(65)	50
[TeF₆²⁻]	LN	—	0·0(3) (Sb/Cu)	0·2(4)	53

Tellurium (II)
TeCl₂	78	6·5(4)		+0·48(20)	51
	LN	6·5(6)	+0·5(3) (Te)	+1·0(4)	52
	(?)			+0·5(2)	31
TeBr₂	LN	6·3(2·7)	−0·2(4) (Te)	+0·3(5)	52

* Frozen solution.

Alloys and tellurides
Te metal	80	7·3(2)			62
	80	7·5(1)	+1·7(1) (TeO₃)	+0·6(2)	46
	80	7·6(2)		+0·7(1)	48
	4·2	7·6(2)		+0·5(1)	48
	86	7·4(1)		+0·50(5)	61
	4·8	7·68(6)		+0·51(4)	49
	77·9	7·36(12)		+0·53(7)	49
	78	7·10(16)		+0·78(10)	51
	LN	7·1(6)	0·0(3) (Te)	+0·5(4)	52
ZnTe	86	—		−0·50(5)	61
	82	—		+0·08(7)	49
	LN	—	+0·2(3) (Sb/Cu)	0·0(4)	54
	80	—		0·00(7)	50
	(?)	—		+0·10(5)	31
CdTe	86	—		−0·50(5)	61
HgTe	86	—		−0·50(5)	61
PbTe	78	—		−0·15(10)	51

[*Refs. on p. 489*]

Table 15.2 (continued)

Compound	T/K	Δ/(mm s⁻¹)	δ†/(mm s⁻¹)	δ (^{125}I/Cu)† /(mm s⁻¹)	Reference
SnTe	(?)			+0·28(4)	31
CaTe	78	—		−0·14(7)	51
MnTe	82	3·2(1·4)		+0·2(7)	49
AuTe$_2$ (cubic)	LN	—		+0·3(1)	62
AuTe$_2$ (monoclinic)	LN	—		+0·4(1)	62
FeTe	86	6·5(1)		+1·75(5)	61
β-FeTe	82	3·0(1·4)		+0·5(7)	49
FeTe$_2$	86	4·5(1)		+0·25(5)	61
Te$_{70}$Cu$_{25}$Au$_5$	LN	7·4(2)			62
Al$_2$Te$_3$	78	4·2(6)		+0·46(20)	51
Cu$_2$Te	78	?		+0·01(8)	51
MnTe$_2$	90	−7·7(1)	0·0(4) (ZnTe)	0·0(5)	44
	77·3	−7·7(1)	+0·2(4) (ZnTe)	0·2(5) (H=55(3) kG)	
	4·2	−7·7(1)	+0·0(4) (ZnTe)	0·0(5) (H=114(7) kG	
CuCr$_2$Te$_4$	LN	−0·08(8) (H=148 kG)		—	66

† Where shifts are originally quoted with respect to a standard other than ^{125}I/Cu, the following conversions have been adopted:

Te metal	+0·52(10) mm s⁻¹
ZnTe	+0·0(1) mm s⁻¹
^{125}Sb/Cu	−0·2(1) mm s⁻¹
TeO$_3$	−1·1(1) mm s⁻¹
^{125m}Te/Pt	+0·25(10) mm s⁻¹

expected, and the presence of impurity in an absorber is unlikely to be detected with any certainty.

Several interpretations of the chemical isomer shift have been put forward [51–53]. The diversity of opinion is considerable, but several points have emerged. Firstly, the Te(IV) complexes with an octahedral coordination (i.e. TeCl$_6^{2-}$, TeBr$_6^{2-}$, TeI$_6^{2-}$) show the largest chemical isomer shifts (1·53–1·95 mm s⁻¹) [50]. Their unique feature is a stereochemically inactive pair of non-bonding electrons which may be presumed to have a considerable s-character and to be highly contracted towards the Te nucleus. Of the other Te(IV) compounds, the square-pyramidal TeF$_5^-$ anion, pyramidal TeX$_3^+$ cations in TeCl$_4$, TeBr$_4$ and TeI$_4$, TeO$_3^{2-}$ anions in the tellurites, and the bridged square-pyramidal geometry of TeF$_4$ all feature a stereochemically active pair of non-bonding electrons. It is proposed that the lower chemical isomer shift found in these compounds is compatible with a reduction in the contribution of the lone-pair to the s-electron density at the nucleus, thus implying that the sign of $\delta\langle R^2\rangle/\langle R^2\rangle$ is positive.

The tellurium(VI) compounds (such as Te(OH)$_6$, H$_6$TeO$_6$, TeO$_4^{2-}$, and TeO$_3$) show negative shifts, which by analogy with tin and antimony is also

[Refs. on p. 489]

consistent with a positive sign for $\delta\langle R^2\rangle/\langle R^2\rangle$. Intuitively one might expect that the 5s-orbital which contributes to the lone-pair in tellurium(IV) would be more extensively involved in covalent bonding in tellurium(VI), thereby reducing both the *s*-electron density at the nucleus and the chemical isomer shift [50].

Extensive data are now available for the TeX_6^{2-} complexes [50]. The shifts observed are characteristic of the anion and independent of the cation

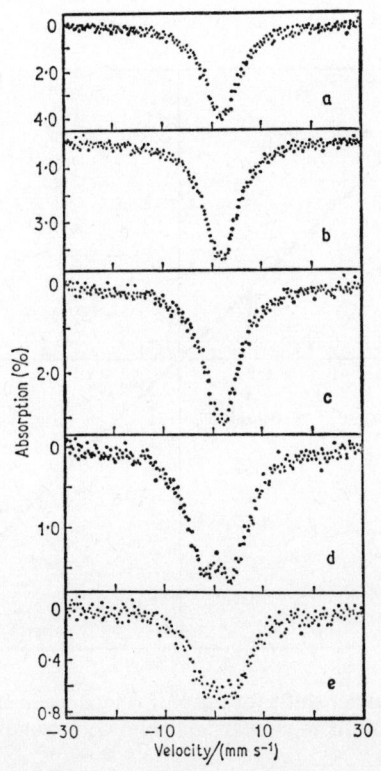

Fig. 15.14 ^{125}Te spectra for a series of halogen complexes: (a) Cs_2TeCl_6, (b) Cs_2TeBr_6, (c) Cs_2TeI_6, (d) $CsTeF_5$, (e) $CsTeF_5$ in frozen solution. [Ref. 50]

($TeCl_6^{2-}$ $\delta = +1.95(1)$ mm s^{-1}; $TeBr_6^{2-}$ $\delta = +1.74^{+0.06}_{-0.02}$ mm s^{-1}; TeI_6^{2-} $\delta = +1.59(6)$ mm s^{-1}) [50]. Typical spectra are shown in Fig. 15.14, and are representative of ^{125}Te spectra in general. The shift values show the opposite trend to that deduced from earlier data [53, 54], and the earlier derived argument concerning $\delta\langle R^2\rangle/\langle R^2\rangle$ is therefore invalid. The report of a spectrum from [TeF_6^{2-}] in solution is also erroneous, as all experimental evidence favours the TeF_5^- anion. The square-pyramidal geometry of TeF_5^- confers a quadrupole splitting (Fig. 15.14), which is still present in a frozen solution

[Refs. on p. 489]

of TeO$_2$ and CsF in aqueous HF in the proportions appropriate to give the hypothetical Cs$_2$TeF$_6$ [50].

It is difficult on present data to estimate a numerical value for $\delta\langle R^2\rangle/\langle R^2\rangle$. Jung and Triftshäuser [51] compared shifts for a series of 'isoelectronic' tellurium (^{125}Te) and iodine (^{127}I) compounds. As seen in Fig. 15.15, a linear plot is obtained, implying a proportionality between the s-electron density at the nucleus in the two series of compounds. They then derived a value for $\delta\langle R^2\rangle/\langle R^2\rangle$ (^{125}Te) of $+4\cdot8 \times 10^{-5}$. Whilst it is true that TeO$_3^{2-}$/IO$_3^-$, Te^{2-}/I$^-$, and TeO$_4^{2-}$/IO$_4^-$ are isoelectronic, it is not at all clear that

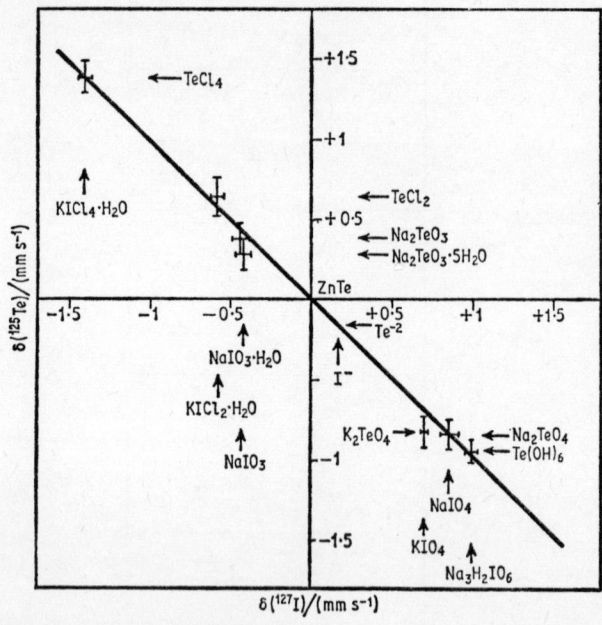

Fig. 15.15 Chemical isomer shifts for the ^{127}I resonance in some iodine compounds plotted against ^{125}Te shifts in similar tellurium compounds. [Ref. 51, Fig. 5]

tellurites are isostructural with iodates or tellurates with periodates. Furthermore, TeCl$_4$/KICl$_4$.H$_2$O, TeCl$_2$/KICl$_2$.H$_2$O, and Te(OH)$_6$/Na$_3$H$_2$IO$_6$, are neither isoelectronic nor isostructural, so that the linear relationship may be largely fortuitous. Attempts have been made to calculate s-electron densities [51, 52], but the problems already discussed at length for ^{119}Sn are also found with tellurium.

Comparison of data for ^{119}Sn, ^{121}Sb, ^{125}Te, ^{127,129}I, and ^{129}Xe has led [31] to a value of $\delta\langle R^2\rangle/\langle R^2\rangle$ (^{125}Te) $= +1\cdot9 \times 10^{-4}$, but the same basic criticisms apply.

In a very recent paper various oxides and oxyanions of tellurium were reinvestigated with a source of ^{125m}Te electrodeposited on to platinum [45].

Absorptions (uncorrected) were in the range 5–15% and the source seems superior to those previously used. Tetragonal TeO_2 was suggested as a standard for chemical isomer shift data since it gives an excellent quadrupole split absorption and is perhaps the most readily obtainable pure, well-characterised compound of tellurium. Results on the orange amorphous α-form of TeO_3 were especially interesting since they suggested that this substance contained up to 17% of Te(IV); it was pointed out that Jung and Triftshäuser's specimen of orange TeO_3 gave an obvious quadrupole splitting and must have contained even more Te(IV). The pure $β$-TeO_3 is apparently unsplit.

Little analysis of quadrupole splitting data has been attempted with the exception of tellurium metal, which is discussed later. ^{129}I data have been obtained from ^{129m}Te sources in Te metal, TeO_2, and $Te(NO_3)_4$ [55]. Under the assumption that the electric field gradient at ^{129}I in TeO_2 is produced by the same imbalance in the *number* of p-electrons as the electric field gradient at ^{125}Te in TeO_2, it is possible to derive a value for the excited-state quadrupole moment of $Q = -0·19(2)$ barn. However, as discussed in the next few paragraphs, the processes which occur at the site which undergoes a radioactive transmutation are likely to be complex.

Decay After-effects

We have already mentioned that the ^{125}I electron-capture decay is prone to decay after-effects. Comparative experiments were first made with ^{125}I in copper, $NaIO_3$, NaI, and I_2 matrices [49]. The recoil-free fractions of the last two named proved too small for serious use, whereas the copper matrix is the narrow-line source commonly used for this isotope. The $NaIO_3$ source gave an emission profile which was complex. An analysis was suggested involving at least two distinct charge states of ^{125}Te formed by the Auger cascade which follows the electron capture.

Subsequent work showed similar complex spectra from $Na^{125}IO_3$, $Na^{125}IO_3.H_2O$, $Na_2Mn(^{125}IO_3)_6$, and $Na^{125}IO_4$ [51]. The resonances are characteristically broader than expected with a shape which must comprise more than two Lorentzian lines. $Na^{125}I$ and $Na_3H_2{}^{125}IO_6$ also gave broader lines than expected, although $K^{125}I$ was least affected. Rather than assume multiple Te charge states, it was proposed that the excited Te atom produced by EC has sufficient energy to cause a reorganisation of the local bonding to the neighbouring atoms during the extensive electronic redistribution which takes place. The complex spectrum observed is then the sum of the different 'compounds' finally produced.

Comparative measurements using the $β$-decay of ^{125}Sb in Sb_2O_3, $NaSbO_3$, $SbCl_3$, $KSbCl_6$, and $HSbCl_6$ showed less evidence for multiple products, $SbCl_3$ being the only source to show substantial quadrupole splitting [51]. However, although $β$-decay in ^{125}Sb is likely to produce less electronic disturbance than EC-decay in ^{125}I, the manner in which the [125]Te daughter atom is incorporated in the different matrices is by no means established, particularly because of the lack of isoelectronic *and* isostructural compounds from the different elements.

Less data are available on decay after-effects from [125m]Te. A matrix of PbTe irradiated with neutrons was found to have a chemical isomer shift of over $+0.1$ mm s^{-1} relative to a PbTe absorber [56]. Furthermore this shift difference decreased exponentially with a time constant of about ten days, the implication being that the difference is due to radiation damage causing a defect structure which anneals at room temperature. It was proposed that a radiation-induced distortion of the band structure in the PbTe semiconductor alters the *s*-electron density at the nucleus.

Independent measurements on neutron-irradiated PbTe, Te metal, and TeO$_2$ (before and after thermal annealing) showed no evidence at all for anomalous effects [57]. The conditions used in this case generated only displacements by the thermal neutron capture process, suggesting that this was not the cause of the earlier observations which may have been made under conditions where fast neutron elastic scattering can also take place.

Metallic Phases and Tellurides

Tellurium metal shows a substantial quadrupole splitting [48]. The structure contains spiral chains of Te atoms, and a crude estimation of the electric field gradient at the tellurium nucleus leads to a value for the nuclear quadrupole moment of $|Q| = 0.20$ barn [48]. This agrees well with an earlier value derived by similar means of $|Q| = 0.17$ barn [58]. Single-crystal measurements have also been made, showing that e^2qQ is negative [46]. An anisotropy of the recoil-free fraction was also claimed.

The quadrupole splitting decreases from 7·68(6) mm s^{-1} at 4·8 K to 7·36(12) mm s^{-1} at 77·9 K, and can be correlated with molecular torsional motions of the spiral chains [49]. A fuller lattice-dynamical theory has also been given [59]. Tellurium metal is a semiconductor at room temperature and pressure, but becomes metallic under high pressure. This is seen in the [125m]Te Mössbauer spectrum as a disappearance of the quadrupole splitting in the metallic phase [60].

The tellurides ZnTe, CdTe, and HgTe have the zinc-blende structure in which the tellurium atoms are tetrahedrally coordinated to the metal. Accordingly there is no electric field gradient at the tellurium and the resonances comprise single lines [49, 61]. PbTe is also cubic. MnTe has a hexagonal structure, and although considered to be nearly ionic with a closed-shell Te^{2-} configuration [58], has been shown to have a small electric field gradient [49]. CrTe shows magnetic broadening at 80 K [58].

FeTe and FeTe$_2$ have tetragonal and orthorhombic structures respectively, and both show a quadrupole splitting (see Table 15.2) [61]. The cubic form

[*Refs.* on p. 489]

of AuTe$_2$ gives an unsplit resonance, but monoclinic AuTe$_2$ shows a greater linewidth from a small unresolved quadrupole interaction [62]. Te$_{70}$Cu$_{25}$Au$_5$ shows a substantial quadrupole splitting [7·4(2) mm s^{-1}] which is very similar to the value for Te metal and is consistent with the belief that both have the same covalently bonded tellurium chain structure [62].

The Debye approximation is not valid for a diatomic cubic lattice, and both the recoil-free fraction and the effective Debye temperature for each atom are strongly dependent on the mass-ratio of the two atoms. However, if the mass difference is small the difference in Debye temperature should also be small. This has been verified for SnTe using ^{119m}Sn (^{119}Sn) and ^{125}I (^{125}Te) sources and an SnTe absorber, as well as an Sn^{129}Te source (decaying

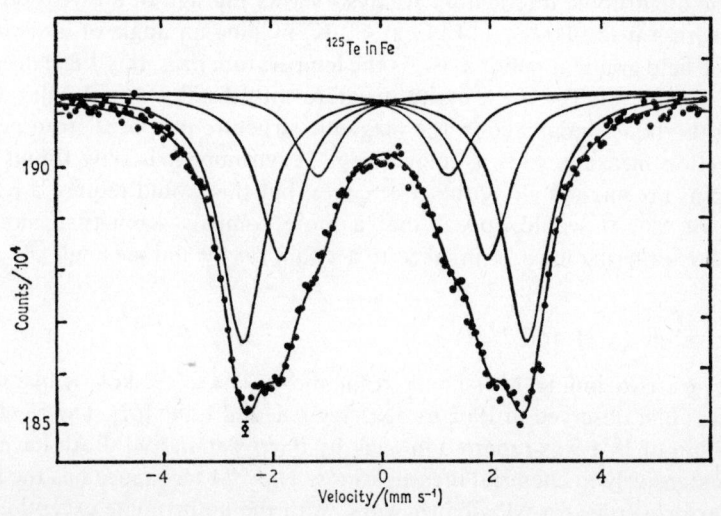

Fig. 15.16 The magnetic hyperfine splitting at a ^{125}Te nucleus in iron metal (produced by decay of ^{125}Sb) showing the six lines appropriate to an $\frac{3}{2} \to \frac{1}{2}$ M1 decay. [Ref. 65, Fig. 1]

to an iodine impurity atom) in an ^{129}I experiment [63]. The effective Debye temperatures deduced from the temperature dependence of the three resonances were $\theta(^{119}$Sn$) = 132(3)$ K, $\theta(^{125}$Te$) = 141(5)$ K, and $\theta(^{129}$I$) = 139(3)$ K.

Magnetic splitting of the ^{125}Te resonance is less helpful than in ^{57}Fe and ^{119}Sn because of the comparatively poor resolution obtainable. A field was first reported of about 600 kG at ^{125}Te nuclei in iron metal foil [58], and the measurements have been progressively improved [64] to give the spectrum at 4·2 K shown in Fig. 15.16 [65]. The ground-state nuclear moment is known to be -0.8872 n.m., and the final computer analysis of the data derives a value for the first excited-state moment of $\mu_e = +0.60(2)$ n.m. The magnetic fields at ^{125}Te daughter nuclei in ^{125}Sb-doped iron, cobalt, and nickel are then $+657(20)$, $|505(20)|$, and $+170(10)$ kG. These fields are produced by a transferred spin-polarisation mechanism in an analagous way to fields at ^{119}Sn.

A less accurate value for μ_e of $+0.74(7)$ had been derived previously from the spectra at 80 K of $CuCr_2Te_4$, which is a ferromagnetic spinel [66]. The apparent doublet spectrum corresponds to an unresolved magnetic splitting of 148(5) kG. Again the field is produced by spin polarisation, this time by the magnetic chromium cations.

$MnTe_2$ is antiferromagnetic below 83·8 K. Above this temperature it shows a simple quadrupole split spectrum with $\Delta = 7.7$ mm s^{-1} [44]. Below the Néel temperature the spectrum shape becomes broad and asymmetric due to the combined effects of a transferred hyperfine magnetic and an electric quadrupole interaction. Analysis shows the sign of e^2qQ to be negative, with a field of $H = 114$ kG at 4·2 K, making an angle of 30° with the electric field gradient major axis. As the temperature rises, this angle decreases substantially. $MnTe_2$ has a pyrite structure with the Te_2^{2-} molecular anions along the body diagonals. In the magnetic structure proposed from neutron diffraction measurements a unique Te_2^{2-} environment is only found when the spins are aligned along the cube edges, but this would require a relating angle of 55°. It would appear that a more complex screw-type model of magnetic ordering must be invoked to account for the unique angle observed.

15.6 Iodine (^{127}I and ^{129}I)

There are two iodine Mössbauer resonances. The 27·72-keV transition of ^{129}I was first observed in 1962 by Jha, Segnan, and Lang [67]. The 57·60-keV transition of ^{127}I was reported in 1964 by Barros et al. [68]. Both have been used extensively in chemical investigations. The ^{129}I resonance has the better nuclear properties for Mössbauer work, with the unfortunate exception that the ground state is radioactive with a half-life of 1.7×10^7 y. Because of this, ^{129}I absorbers must be specially prepared and handled, whereas natural iodine comprises the stable ^{127}I in 100% abundance.

Nuclear Properties

The first excited state of ^{129}I can be populated by decay of 33-day ^{129m}Te or 70-minute ^{129}Te. Both parents are conveniently produced by the ^{128}Te(n, γ) reaction. The decay scheme (shown in simplified form in Fig. 15.17a) is very complex, and there is some divergence of opinion regarding the details. We have adapted the recent work of Berzins et al. [69]. Early ^{129}I measurements were made using inaccurate values for some of the relevant nuclear constants. The currently accepted value for the ^{129}I Mössbauer γ-ray energy is $^{129}E_\gamma = 27.72(6)$ keV [70]. The excited-state lifetime is $t_\frac{1}{2} = 16.8(2)$ ns, giving a natural width of 0·59 mm s^{-1} [70]. The γ-ray has nearly pure M1 multipolarity, and the nuclear spin states are $^{129}I_e = \frac{5}{2}$ and $^{129}I_g = \frac{7}{2}$.

Better-quality spectra can be obtained from the ^{129}Te parent rather than ^{129m}Te, but the very short half-life (70 m) is a serious disadvantage in the

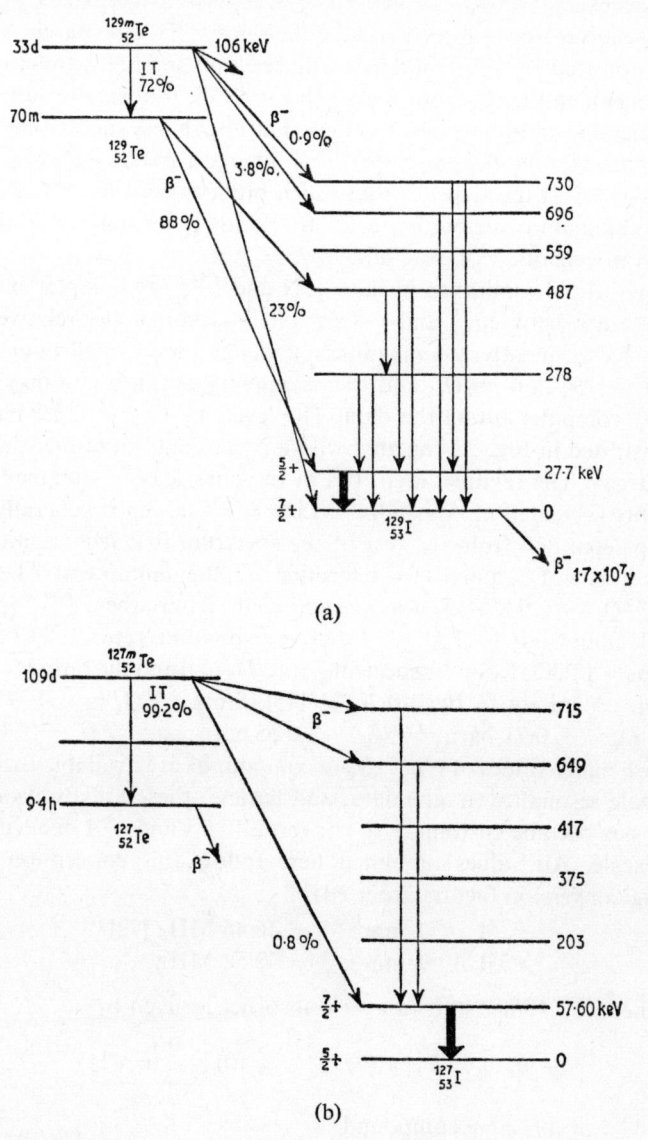

Fig. 15.17 (a) The decay scheme of ^{129}I; (b) the decay scheme of ^{127}I.

absence of nearby reactor facilities [70]. The source matrix universally adopted for both ^{129}I and ^{127}I is ZnTe. The usual preparation of ^{129m}Te involves irradiation of isotopically enriched ^{66}Zn^{128}Te to reduce unwanted

[*Refs. on p. 489*]

by-products, although for ^{129}Te it is better to first irradiate ^{128}Te and then to combine this with zinc.

The energy of the first excited state in ^{127}I is 57·60(2) keV [71], which results in lower recoil-free fractions than in ^{129}I. The parent isotope is ^{127m}Te produced by a ^{126}Te$(n, \gamma)^{127m}$Te reaction and has a 109-day half-life (decay scheme in Fig. 15.17b). Although Zn^{127m}Te is normally used, H$_6$TeO$_6$ and Te metal sources have also been used although less successfully [72]. The excited-state lifetime is $t_{\frac{1}{2}} = 1·86(11)$ ns, giving a natural linewidth of 2·54 mm s^{-1}. The ^{127}I resonance is thus much broader than its ^{129}I counterpart, which is the main contributing factor to the lower resolution obtained from ^{127}I data in general. $^{127}I_e = \frac{7}{2}$ and $^{127}I_g = \frac{5}{2}$.

The quadrupole splittings in both ^{127}I and ^{129}I are complex because the transitions are between $\frac{5}{2}$ and $\frac{7}{2}$ spin-states, although the relative order is reversed. Consequently, the magnitude and sign of the quadrupole coupling constants e^2qQ_g and e^2qQ_e, and the asymmetry parameter η may be determined by computer fitting the data. The level splitting for ^{129}I has already been illustrated in Fig. 3.3 together with a representation of the spectrum for an M1 decay. The relative intensities of the lines may be obtained from the coefficients tabulated in Appendix 2. Line 8 in Fig. 3.3 is generally so weak and is so separated from the rest of the spectrum that it is usually ignored because it is not required to determine all the unknowns. The ratio of $^{127}Q_e/^{127}Q_g = +0·896(2)$ was determined from the ^{127}I spectra of KICl$_4$.H$_2$O and KICl$_2$.H$_2$O [73]. The corresponding ratio $^{129}Q_e/^{129}Q_g$ was derived as $+1·23(2)$ from the spectrum of KIO$_3$ [74] and the ratio $^{129}Q_g/^{127}Q_g$ is known accurately to be $+0·70121$ [75]. Since $^{127}Q_g = -0·79$ barn, we have $^{127}Q_e = -0·71$ barn; $^{129}Q_g = -0·55$ barn; and $^{129}Q_e = -0·68$ barn.

Because many values of $e^2q^{127}Q_g$ in compounds are available from nuclear quadrupole resonance (n.q.r.) data, and because these are always quoted in MHz, it has become customary to convert all ^{127}I and ^{129}I quadrupole data to this scale. All values tabulated here follow this convention, and the following conversion factors are used:

^{127}I   1 mm s^{-1} = 46·46 MHz [72]
^{129}I   1 mm s^{-1} = 32·58 MHz

The chemical isomer shift in a ^{129}I absorber is given by

$$\delta_{129} = K\{|\psi_s(0)|_A^2 - |\psi_s(0)|_S^2\}\frac{\delta R}{R}(^{129}\text{I})$$

and for ^{127}I in the same compound

$$\delta_{127} = K\{|\psi_s(0)|_A^2 - |\psi_s(0)|_S^2\}\frac{\delta R}{R}(^{127}\text{I})$$

Thus $\dfrac{\delta_{127}}{\delta_{129}} = \dfrac{\delta R/R(^{127}\text{I})}{\delta R/R(^{129}\text{I})} = \rho(R)$

[Refs. on p. 489]

[Note that the δ values must be given in energy units, i.e. if δ is given in velocity units then $\rho(R) = \delta_{127}E_\gamma^{127}/(\delta_{129}E_\gamma^{129})$]. The chemical isomer shifts of ^{127}I and ^{129}I show opposing trends because the ratio $\rho(R)$ is negative in sign. Data for both isotopes in $Na_3H_2IO_6$ and KIO_4 relative to ZnTe sources gave $\rho(R) = -0.78(4)$ [76], but this calculation used a value for E_γ^{129} since shown to be in error. A more accurate value can be obtained from the data replotted in Fig. 15.18 which gives $\rho(R) = -0.65$. The chemical isomer shift of ^{127}I is smaller in magnitude and opposite in sign to that of ^{129}I and, as Fig. 15.18 shows, the direct proportionality is valid. The line drawn should

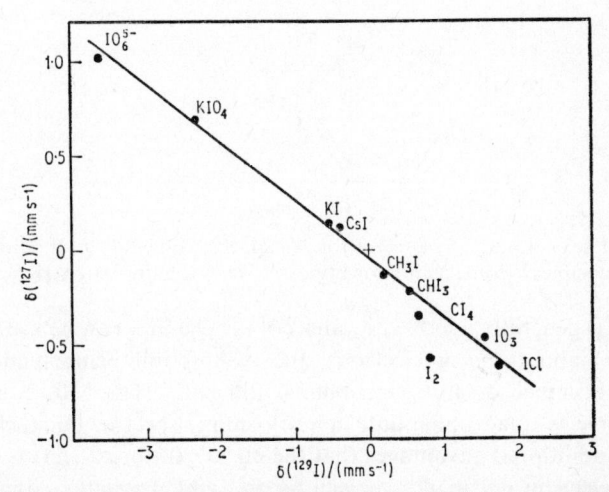

Fig. 15.18 The chemical isomer shifts for ^{127}I and ^{129}I in a number of iodine compounds. The solid line is for $\{\delta R/R(^{127}I)\}/\{\delta R/R(^{129}I)\} = -0.65$. All data are with respect to ZnTe sources.

pass through the origin corresponding to identical sources. The opposite signs of the shifts and the larger linewidth of the ^{127}I resonance are clearly illustrated by the spectra in Fig. 15.19.

Comparatively little interest has been shown in the numerical value of $\delta\langle R^2\rangle/\langle R^2\rangle$ ($= 2\,\delta R/R$). Derived values are $\delta\langle R^2\rangle/\langle R^2\rangle(^{129}I) = 6 \times 10^{-5}$ [74], subsequently updated to 10×10^{-5} [82]; $\delta\langle R^2\rangle/\langle R^2\rangle(^{127}I) = -5.6 \times 10^{-5}$ obtained from IO_3^- and IO_4^- data [83]. More recent comparison of ^{119}Sn, ^{121}Sb, ^{125}Te, ^{127,129}I, and ^{129}Xe data has given the substantially larger values of $\delta\langle R^2\rangle/\langle R^2\rangle(^{127}I) = -4.8 \times 10^{-4}$ and $\delta\langle R^2\rangle/\langle R^2\rangle(^{129}I) = 6.2 \times 10^{-4}$ [31].

Chemical Interpretation of Iodine Spectra

The quadrupole coupling constants obtained from the ^{127}I and ^{129}I quadrupole interactions are exactly analogous to those measured by the nuclear quadrupole resonance method, and where figures are available for both, the

[*Refs.* on p. 489]

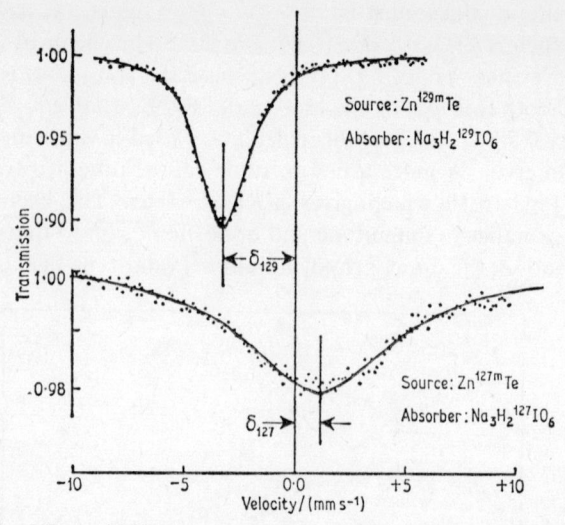

Fig. 15.19 The ^{127}I and ^{129}I spectra for $Na_3H_2IO_6$ showing the smaller linewidth, and larger chemical isomer shift (of opposite sign) for the latter. [Ref. 76, Fig. 1]

agreement is generally good. The values of e^2qQ and η can be used to obtain information about the 5p-electrons, the assumption being made that 5d-orbitals if occupied do not contribute significantly. The Mössbauer spectra give less precise quadrupole data than do n.q.r. spectra, but they have the significant additional advantages that the chemical isomer shift is dependent on the 5s-electron occupancy as well as 5p-, and that lattice-dynamical information can be derived. Mössbauer spectroscopy is therefore a more powerful technique than n.q.r. in this context.

The chemical bonding in iodine compounds is much simpler to describe than that in tin, antimony, or tellurium which precede it in the Periodic Table. This is particularly true where the iodine forms only one bond to another atom. As a result it is possible to develop a quantitative interpretation of the Mössbauer parameters. The equations given here were first formulated by Hafemeister *et al.* [74] and subsequently revised to a more elegant form by Perlow and Perlow [72].

The chemical isomer shifts for ^{127}I and ^{129}I can be expressed quantitatively in terms of the departure of the iodine electronic configuration from the $5s^25p^6$ closed shell of the I$^-$ anion. If we define the number of 5s- and 5p-electron 'holes' in the closed I$^-$ shell as h_s and h_p respectively, it is possible to write the shift with respect to I$^-$ as

$$\delta_A = K\{-h_s + \gamma(h_p + h_s)(2 - h_s)\} \qquad 15.1$$

K and γ are constants, K being numerically different for ^{127}I and ^{129}I since it contains the appropriate $\delta R/R$ value. The term $-Kh_s$ represents the change

[*Refs. on p. 489*]

in shift due to a loss of 5s-electrons. The $(2 - h_s)$ 5s-electrons remaining are then deshielded by a total decrease in the number of electrons of $h_p + h_s$. This assumes that the 5s- and 5p-electrons produce an equivalent shielding effect. In the event that the direct change in the number of s-electrons is negligible,

$$\delta_A = 2K\gamma h_p \qquad 15.2$$

Hafemeister studied the change in ^{129}I chemical isomer shift in the alkali iodides for which h_s is zero, and deduced a value of $\gamma = 0.097$. The detailed calculations used Slater's shielding rules and the Fermi–Segré formula. A more recent revision has given $\gamma = 0.07$, and this is the value we adopt [77]. The derivation of a numerical value for K requires a more circumspect approach.

The relative occupation of the 5s- and 5p-orbitals is not immediately given by the chemical isomer shift, but may be derived in conjunction with the quadrupole splitting to which h_s does not contribute. The relevant theory is directly adopted from n.q.r. spectroscopy [78]. The Townes and Dailey theory says that the principal value of the molecular field gradient (eq_{mol}) is related to the atomic electric field gradient from a 5p-hole in the $5s^2 5p^6$ configuration (eq_{at}) by

$$eq_{mol} = -eq_{at} U_p \qquad 15.3$$

where U_p is the 5p-electron imbalance. U_p is defined in terms of the 5p-electron populations in the x, y, and z directions, U_x, U_y, and U_z, by

$$U_p = -U_z + \frac{U_x + U_y}{2} \qquad 15.4$$

The asymmetry parameter is given by

$$\eta = \frac{q_{xx} - q_{yy}}{q_{zz}} = \frac{3}{2}\left(\frac{U_x - U_y}{U_p}\right) \qquad 15.5$$

For axial symmetry $U_x = U_y$ and $\eta = 0$. The value of $e^2 q_{at}{}^{127}Q$ is known accurately to be $+2293$ MHz [79], so that U_p is obtained directly from $e^2 q_{mol}{}^{127}Q$ by proportionality.

Having calculated U_z, U_x, and U_y from $e^2 q^{127}Q$ and η, h_p can be derived from

$$h_p = 6 - (U_x + U_y + U_z) \qquad 15.6$$

Using these equations and experimental data for ^{127}I, Perlow [72] was able to obtain a value for $2K\gamma$ of -0.56 mm s^{-1} (i.e. the ^{127}I shift per 5p-electron hole). Thence, using $\gamma = 0.07$, one obtains $-K = 4.0$ mm s^{-1} (the *direct* term in the shift per 5s-electron hole) and $-K + 2K\gamma = +3.4$ mm s^{-1} (the *total* shift per 5s-electron hole when $h_s \ll 1$). Note that the loss of a 5s-electron causes a significantly larger shift than loss of a 5p-electron, as indeed one would expect.

[Refs. on p. 489]

The chemical isomer shift in ^{127}I (in mm s^{-1} relative to the source standard, ZnTe, rather than to the isolated I$^-$ ion) is then given by

$$^{127}\delta_{ZnTe} = -4\cdot0\{-h_s + 0\cdot07(h_p + h_s)(2 - h_s)\} + 0\cdot16$$
$$\simeq +3\cdot44h_s - 0\cdot56h_p + 0\cdot16 \qquad 15.7$$

which for no 5s-participation in the bonding becomes

$$^{127}\delta_{ZnTe} = -0\cdot56h_p + 0\cdot16 \qquad 15.8$$

The factor 0·16 was estimated from experimental data.

Equivalent expressions for ^{129}I derived independently [80] are

$$^{129}\delta_{ZnTe} = -8\cdot2h_s + 1\cdot36h_p - 0\cdot54 \qquad 15.9$$

and
$$^{129}\delta_{ZnTe} = 1\cdot36h_p - 0\cdot54 \qquad 15.10$$

(all in units of mm s^{-1}). Equations 15.7–15.8 and 15.9–15.10 are not entirely consistent, e.g. 1·36 in equation 15.10 corresponds to a factor of $-0\cdot48$ in equation 15.8. However, the ^{129}I data were derived on the primary assumption that solid iodine (I$_2$) has an h_p value of 1·00. Recent data on frozen solutions of iodine [81] have shown that these provide a better description of the I$_2$ molecule than solid iodine, and that new equations can be formulated as

$$^{129}\delta_{ZnTe} = -8\cdot2h_s + 1\cdot5h_p - 0\cdot54 \text{ mm s}^{-1} \qquad 15.11$$

$$^{129}\delta_{ZnTe} = 1\cdot5h_p - 0\cdot54 \text{ mm s}^{-1} \qquad 15.12$$

These are in better accord with those for ^{127}I, and where possible all ^{129}I data listed here have been recalculated using equations 15.11–15.12. [NOTE: the coefficient of 8·2 for h_s has not been recalculated from Ref. 70 using the later data because probable error is large in any case.] The range of validity of these equations has not been fully established. It seems certain that they remain valid to h_p values of up to 1·5. Recent data on the fluorine complexes of iodine can be naively interpreted using the same equations, and it may be that the whole range of iodine compounds are encompassed.

Once U_z, U_x, and U_y have been determined for an iodine atom with a single bond, it is comparatively easy to express [77] the amount of π-character by

$$U_x = 2 - \pi_x, \quad U_y = 2 - \pi_y$$

The degree of 5s-participation, σ- and π-character, and bond ionicity can all be determined in favourable circumstances.

Additional discussion on many aspects of iodine work including interpretation has been given in recent reviews [84, 85].

In presenting the detailed results which follow, we have attempted to standardise the interpretation using equations 15.7–15.8 and 15.11–15.12 as already mentioned. Quadrupole splittings for both transitions are given in

[Refs. on p. 489]

Table 15.3 Mössbauer data for ^{127}I

Compound	T/K	$^{127}\delta$(ZnTe) /(mm s^{-1})	$e^2q^{127}Q_g$ /MHz	η	U_p	h_p	h_s	Reference
NaI	HE	+0.14(2)	—	—	—	0.04	—	72
KI	HE	+0.14(2)	—	—	—	0.04	—	72
CsI	HE	+0.12(2)	—	—	—	0.07	—	72
HI(aqueous)	HE	+0.16(2)	—	—	—	0.00	—	72
HI	HE	−0.51(14)	−1640(40)	—	+0.72	0.72	—	72
I$_2$	HE	−0.58(7)	−2238(20)	0.12(2)	+0.98	1.11	—	72
ICl	HE	−0.62(4)	−2868(20)	—	+1.26	1.39	—	72
KICl$_2$.H$_2$O	HE	−0.58(4)	−3189(20)	—	+1.39	1.39	—	72
KICl$_4$.H$_2$O	HE	−1.39(5)	+3094(20)	—	−1.35	2.70	—	72
CI$_4$	4.2	−0.35(3)	−2160	—	+0.94	0.94	0.004	77
CHI$_3$	4.2	−0.21(3)	−2060	—	+0.90	0.90	0.034	77
CH$_2$I$_2$	4.2	−0.14(4)	−1920	—	+0.84	0.84	0.046	77
CH$_3$I	4.2	−0.01(7)	−1775	—	+0.77	0.77	0.071	77
NaIO$_3$	HE	−0.44(5)	+1092	0	+0.48	—	—	83
NaIO$_3$.H$_2$O	HE	−0.41(5)	+1108	0	+0.48	—	—	83
Na$_2$Mn(IO$_3$)$_6$	HE	−0.55(4)	+1080	0	+0.47	—	—	83
KIO$_4$	HE	+0.70(2)	—	—	—	—	—	72
	20	+0.68(9)	—	—	—	—	—	76
	HE	+0.85(7)	—	—	—	—	—	83
Na$_3$H$_2$IO$_6$	HE	+1.02(1)	—	—	—	—	—	72
	20	+1.19(5)	—	—	—	—	—	76
	HE	+1.02(4)	—	—	—	—	—	83

MHz on the ^{127}I scale. Numerical values for ^{127}I are collected in Table 15.3 and for ^{129}I in Table 15.4. In several cases the revised calibrations cause large changes in the calculated bonding parameters, and some of the original interpretations become suspect.

Alkali-metal Iodides (I$^-$)

The alkali-metal iodides provided a convenient starting point for calibration of the ^{129}I chemical isomer shift [74] because independent values for h_p were available from dynamic quadrupole-coupling measurements. The corresponding values for h_p derived from the chemical isomer shift are given in Table 15.4. It may be assumed that there is no 5s-character in the very weak covalent bonding. There is no direct correlation with electronegativity of the cation, and detailed calculations show that it is necessary to consider the overlap deformation of the free-ion wavefunctions [82].

Less accurate values are available for ^{127}I in NaI, KI, and CsI [72]. A number of lattice-dynamical calculations have been made [86–89].

Hydrogen Iodide (HI)

The ^{127}I spectrum of anhydrous HI is totally different to that of a frozen aqueous solution [72]. The chemical isomer shift of the latter (Table 15.3) is characteristic of ionic I$^-$, but the bonding of the anhydrous form which shows unresolved quadrupole splitting has not been fully analysed.

Table 15.4 Mössbauer data for ^{129}I

Compound	T/K	$^{129}\delta$ (ZnTe) /(mm s^{-1})	$e^2q^{127}Q_g$ /MHz	η	U_p	h_p	h_s	Reference
LiI	80	−0·380(25)	0	—	—	0·11	—	74
NaI	80	−0·460(25)	0	—	—	0·05	—	74
KI	80	−0·510(25)	0	—	—	0·02	—	74
RbI	80	−0·430(25)	0	—	—	0·07	—	74
CsI	80	−0·37(25)	0	—	—	0·11	—	74
I$_2$	100	+0·82(1)	−2085(20)	0·16	+0·91	—	—	80
	80	+0·83(1)	−2156(10)	0·16(3)	+0·94	—	—	90
I$_2$ (hexane)	88	+0·98(5)	−2263(20)	0	+0·99	—	—	81
I$_2$ (CCl$_4$)	88	+0·91(5)	−2273(20)	0	+0·99	—	—	81
I$_2$ (argon)	22	+0·93(5)	−2231(20)	0	+0·98	—	—	81
I$_2$ (benzene)	88	+0·76(5)	−2412(20)	0	+1·05	—	—	81
IBr	LN	+1·23(2)	−2892(10)	0·06(2)	+1·26	1·18	—	90
ICl	LN	+1·73(5)	−3131(10)	0·06(3)	+1·38	1·48	—	90
I$_2$Cl$_6$	LN	+3·50(2)	+3060(10)	0·06(2)	−1·35	2·70	—	90
I$_2$Cl$_4$Br$_2$	LN	+3·48(2)	+3040(10)	0·06(2)	−1·33	2·68	—	90
	LN	+2·82(2)	+2916(10)	0·06(2)	−1·27	2·24	—	90
ICN	LN	+1·19(2)	−2640(20)	0·00(2)	+1·15	1·15	—	92
2IBr-2,2′-bipyridine	4·2	+1·35(9)	−2040(10)	0·15	+1·27	1·26	—	93
2IBr-4,4′-bipyridine	4·2	+1·48(2)	−2180(6)	<0·03	+1·36	1·34	—	93
IBr-pyridine	4·2	+1·61(3)	−2260(9)	<0·03	+1·41	1·43	—	93
2ICl-2,2′-bipyridine	4·2	+1·90(3)	−2400(9)	<0·03	+1·49	1·63	—	93
2ICl-4,4′-bipyridine	4·2	+1·43(2)	−2403(10)	<0·03	+1·49	1·31	—	93
ICl-pyridine	4·2	+1·74(2)	−2320(10)	<0·03	+1·44	1·52	—	93
ICl-pentamethylene tetrazole	4·2	+1·90(2)	−2380(10)	<0·03	+1·48	1·63	—	93

Table 15.4 (continued)

Compound	T/K	$^{129}\delta$(ZnTe) /(mm s^{-1})	$e^2q^{127}Qg$ /MHz	η	U_p	h_p	h_s	Reference
CsI$_3$	4·2	+1·40	$\begin{cases}-2500\\-1460\\-830\end{cases}$	—	+1·09 +0·64 +0·36	— — —	$\Big\}$	94
benzamide-HI$_3$	4·2	+1·33	$\begin{cases}-2460\\-1180\end{cases}$	— —	+1·07 +0·51	— —	$\Big\}$	94
amylose-I$_3$	4·2	+1·31	$\begin{cases}-2450\\-930\end{cases}$	— —	+1·07 +0·41	— —	$\Big\}$	94
IF$_5$	90	+3·00(1)	+1073(2)	0	−0·47	4·25	0·35	95
IF$_7$	90	−4·56(1)	−148(4)	0	+0·13	5·87	1·56	95
IF$_6^+$AsF$_6^-$	90	−4·68	0	—	0	6	1·60	96
IF$_6^-$Cs$^+$	90	+2·45	−1414	0·98	+0·62	4·76	0·50	96
CI$_4$	85	+0·65(3)	−2102(10)	0	+0·91	0·91	0·02	97
CHI$_3$	85	+0·53(3)	−2029(10)	0	+0·88	0·88	0·03	97
	4·2	+0·50	−2068	0	+0·90	0·90	0·03	77
CH$_3$I	85	+0·20(3)	−1739(10)	0	+0·76	0·76	0·04	97
SnI$_4$	85	+0·43(3)	−1364	0·00(3)	+0·59	0·64	—	98
GeI$_4$	85	+0·48(3)	−1500(10)	0	+0·65	0·68	—	101
SiI$_4$	85	+0·26(3)	−1335(10)	0	+0·58	0·53	—	101
CrI$_3$	78	+0·23(5)	+662(8)	0·35(5)	−0·28	—	—	105
	21	(H=25(5)kG)	+727(10)	0·35	−0·32	—	—	105
KIO$_3$	80	+1·56(20)	+997*	—	−0·43	—	—	74
NH$_4$IO$_3$	80	+1·31(20)	?	—	—	—	—	74
Ba(IO$_3$)$_2$	80	+1·11(20)	+1030*	—	−0·45	—	—	74
KIO$_4$	80	−2·34(6)	0	—	—	—	—	74
IO$_6^{5-}$	?	−3·10	0	—	—	—	—	90
I$_2$O$_4$	80	$\begin{cases}+2·65(24)\\+2·66(29)\end{cases}$	+3016(77) +1125(25)	0·252(4) 0·46(2)	−1·31 −0·49	2·35 2·37	$\Big\}$	108
^{129}I/PbTe	80	+0·22(5)	0	—	—	—	—	74
^{129}I/Te metal	80	+0·71(2)	−532(10)	0·80(5)	+0·23	—	—	110
^{129}I/TeO$_2$	80	+1·52(1)	+1121(10)	0·55(5)	−0·49	1·37	—	110
^{129}I/Te(NO$_3$)$_4$	80	+2·52(1)	+1139(10)	0	−0·50	2·04	—	110

* From n.q.r. data.

Molecular Iodine (I₂)

The ^{129}I spectrum of solid iodine [80, 90] is shown in Fig. 15.20 and illustrates the high resolution which can be obtained with this resonance. It was originally used to calibrate the ^{129}I chemical isomer shift under the assumption that the bonds are of pure p-character, i.e. the value of h_p is 1·00. The presence of an asymmetry parameter, however, argues against this, and an admixture of an s^2p^4d configuration into the s^2p^5 state leading to a higher value for h_p has been suggested [72]. ^{127}I data are also available.

Frozen solutions of molecular iodine can give information about the solvent–solute interaction [81]. The spectra of iodine in hexane, CCl₄, and solid argon are very similar and differ from solid iodine in showing no asymmetry parameter. It therefore seems likely that the species observed is a 'free' iodine molecule. These values were used to derive the currently accepted calibration of the ^{129}I chemical isomer shift. The spectrum in benzene is considerably different because of charge transfer from the benzene to the iodine.

Iodine Monobromide (IBr)

The ^{129}I resonance has been recorded [90]. The $e^2q^{127}Q_g$ value of $-2892(10)$ MHz yields a value for U_p of 1·26, and the asymmetry parameter is close to zero. Assuming no 5s-participation in the bonding gives $h_p = 1\cdot26$ while the chemical isomer shift and equation 15·12 give $h_p = 1\cdot18$. The discrepancy could be interpreted as a value for h_s of only 0·01. Thus 0·18e^- is transferred to the bromine from iodine.

ICl and I₂Cl₆

The ^{129}I data [90] for ICl give a U_p value of 1·38 from $e^2q^{127}Q_g$ and $h_p = 1\cdot48$ from the chemical isomer shift assuming only 5p-character in the bonding. The difference could require a π-bonding character of 3%, but the original calculations derived a π-character of 10% which was thought to be unlikely and an explanation was given in terms of intermolecular bonding. It is significant to note that ICl, ICl₂⁻, I₂Cl₆, and ICl₄⁻ which have the iodine atom in different coordination numbers and different ionic states all give an $e^2q^{127}Q_g$ value in the range 3060–3110 MHz, i.e. the value is characteristic of an I–Cl bond.

I₂Cl₆ has a planar bridged structure with two identical 4-coordinated iodine atoms (a):

[Refs. on p. 489]

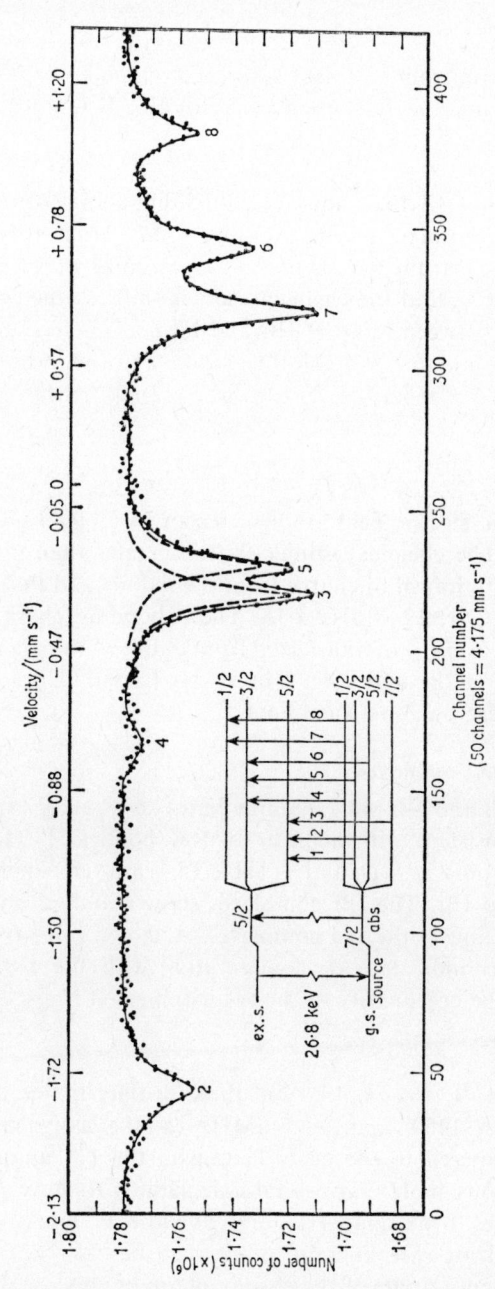

Fig. 15.20 The ^{129}I resonance in solid ^{129}I$_2$ at 100 K. The line numbered 1 on the decay scheme is off the scale of the spectrum. It is usually ignored because of its low intensity. [Ref. 80, Fig. 3]

[Refs. on p. 489]

The large positive chemical isomer shift indicates a very low 5s-participation and gives a value for h_p of 2·70, i.e. each chlorine removes effectively $0·42e^-$ from the iodine atom to give a $5s^2 5p^{3·03}$ configuration.

Lower resolution data from ^{127}I have also been obtained for ICl [72], and lattice dynamical calculations have been made for IBr, ICl, and I_2Cl_6 [91].

$I_2Cl_4Br_2$

The ^{129}I spectrum of $I_2Cl_4Br_2$ shows two different iodine environments, confirming that both bromines are on the same iodine atom as in structure (b) above [90]. The spectrum parameters for one iodine atom correspond closely to those in I_2Cl_6, and the chemical isomer shift for the other iodine atom is intermediate between those of IBr and I_2Cl_6. This second atom has $1·24e^-$ less than in iodine (for two chlorine–iodine and two bromine–iodine bonds the transfer is $2 \times 0·42e^- + 2 \times 0·18e^- = 1·20e^-$, which is in good agreement).

ICN

The axially symmetric electric field gradient tensor in ^{129}ICN gives a value for U_p of 1·15 [92]. The chemical isomer shift is greater than in molecular iodine, implying a withdrawal of charge from the iodine, and the value of h_p derived from equation 15.12 is also 1·15. The π-bonding character is obviously very low, and $0·15e^-$ are transferred from iodine to the cyanide group, compared to $0·18e^-$ in IBr and $0·42e^-$ in an I–Cl bond. No evidence for intermolecular covalent bonding was found.

ICl– and IBr–Pyridine Complexes

The complexes of ICl and IBr with pyridine bases show a very small asymmetry parameter consistent with a linear N–I–X bond [93]. The bonding parameters derived from $e^2q^{127}Q_g$ and δ (Table 15.4) are very similar to those of the parent ICl and IBr. The IBr complexes show mainly σ-bonding, but there is more π-bonding in the ICl complexes. A larger value of η in 2,2'-bipyridine $(IBr)_2$ may indicate a *cis* configuration with the two IBr units possibly destroying the coplanarity of the two conjugated rings.

The I_3^- Anion

The spectrum of $Cs^{129}I_3$ is complex, and three distinct iodine atoms with $e^2q^{127}Q_g = -2500, -1460,$ and -830 MHz can be assigned [94]. The highest value is attributed to the central atom in the I_3^- anion which is expected from previous molecular-orbital calculations to have h_p close to unity. The complexes benzamide–HI_3 and amylose–I_3 give very similar spectra in which at least two different iodines can be discerned. This confirms that I_3^- is present as part of the chromophore in amylose–I_3, which is the familiar blue starch–iodine complex. A full analysis of the bonding was prevented by the difficulty of determining all the chemical isomer shifts accurately.

KICl$_2$.H$_2$O and KICl$_4$.H$_2$O

The ^{127}I spectra of KICl$_2$.H$_2$O and KICl$_4$.H$_2$O show $e^2q^{127}Q_g$ values of opposite sign as expected from their molecular structures [72, 73]. In square-planar ICl$_4^-$ the loss of electronic charge from the $5p_x$- and $5p_y$-orbitals gives a positive value of $e^2q^{127}Q_g$, and in linear ICl$_2^-$, the loss is from the $5p_z$-orbitals, giving a negative value of $e^2q^{127}Q_g$. The reversal in sign is clearly illustrated in Fig. 15.21. Assuming no 5s-participation in the bonding, the

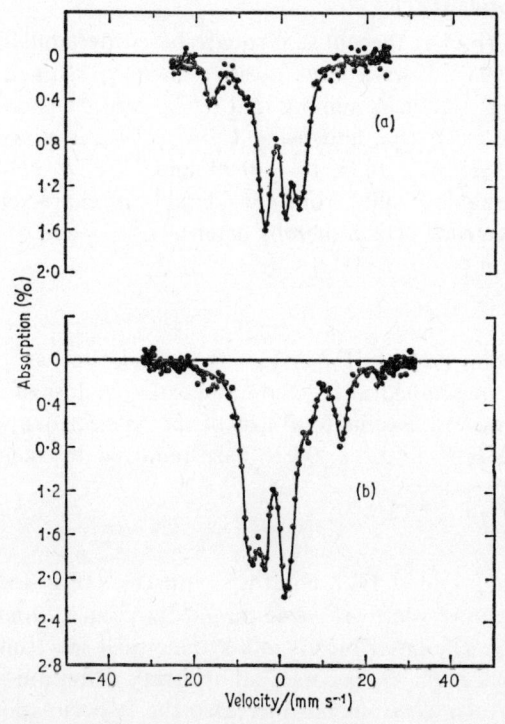

Fig. 15.21 ^{127}I spectra of (a) KICl$_4$.H$_2$O and (b) KICl$_2$.H$_2$O illustrating the reversal of the sign of the quadrupole coupling constant. [Ref. 72, Fig. 3]

U_p values correspond to h_p values of 2·70 in ICl$_4^-$ and 1·39 in ICl$_2^-$, a loss of about 0·42e^- and 0·19e^- from the iodine per chlorine atom respectively.

Iodine Heptafluoride (IF$_7$)

IF$_7$ has a pentagonal bipyramidal geometry. The ^{129}I resonance [95] shows a large negative chemical isomer shift (4·56 mm s^{-1}) and an unusually small coupling constant (−148 MHz). As the 5s-electrons participate in the bonding of both IF$_7$ and IF$_5$ (see following section), it is more difficult to obtain quantitative information. Since the I–F bond is highly ionic, one can assume that the value of $U_p = +0.13$ is derived from a lower limit to $U_z + U_x + U_y$, i.e. $U_z = 0$, $U_x + U_y = 0.13$. These figures correspond to $h_p = 5.87$, from which in conjunction with the chemical isomer shift one obtains $h_s = 1.56$. Thus $6.43e^-$ are transferred from iodine to fluorine, i.e. $0.92e^-$ per I–F bond. Any involvement of 5d- or 5f-orbitals on iodine is likely to be equivalent in effect on both $e^2q^{127}Q_g$ and $^{129}\delta$ to a complete transfer of these electrons to the fluorine because of the small electric field gradient and shielding effects they produce.

Iodine Pentafluoride (IF$_5$)

The compound IF$_5$ has the unusual square-based pyramidal geometry. The ^{129}I resonance [95] shows a large positive chemical isomer shift of $+3.00$ mm s^{-1} (see Fig. 15.22). Assuming that $0.92e^-$ are transferred from iodine to each fluorine as in IF$_7$, and using $U_p = -0.47$, it is possible to derive values of $h_p = 4.25$, $h_s = 0.35$, $U_z = 0.90$, and $U_x = U_y = 0.85$. The larger value of U_z probably results from considerable p_z character in the pair of non-bonding electrons, which thereby determines the sign of the quadrupole coupling constant ($+1073$ MHz).

IF$_6$$^+AsF_6$$^-$

The ^{129}I spectrum [96] of IF$_6$$^+AsF_6$$^-$ is a single line at -4.68 mm s^{-1}, consistent with an octahedral IF$_6$$^+$ ion and extensive loss of 5s-electrons. As an upper limit we can assume total loss of the 5p-electrons, i.e. $h_p = 6$, and can then derive $h_s = 1.60$, i.e. $0.93e^-$ are removed by each fluorine atom from I$^+$.

Cs$^+$IF$_6$$^-$

The ^{129}I spectrum [96] of IF$_6$$^-$ contrasts with IF$_6$$^+$ (Fig. 15.23) in having a large chemical isomer shift of $+2.45$ mm s^{-1} and an asymmetry parameter of close to unity. The latter results in a symmetrical spectrum because when $\eta \to 1$ the sign of $e^2q^{127}Q_g$ becomes an arbitrary definition as $V_{zz} = -V_{yy}$. The addition of two electrons has increased the 5s-occupation considerably. $U_p = +0.62$, so that with $\eta = 0.98$ we have $U_x - U_y = +0.41$. A limiting value for electron removal is obtained by setting $U_y = 0$ so $U_x = 0.41$ and $U_z = 0.82$, whence $h_p = 4.76$ and $h_s = 0.50$. These figures correspond to a removal of $0.88e^-$ per fluorine atom from I$^-$. It can be seen that these crudely estimated figures for IF$_7$, IF$_5$, IF$_6$$^+$, and IF$_6$$^-$ are reasonably self-consistent despite the fact that the chemical isomer shift calibration is being extrapolated to large values of h_p. A possible structure for IF$_6$$^-$ is based on that of IF$_7$ but with a lone-pair of electrons replacing one of the equatorial fluorine atoms.

[Refs. on p. 489]

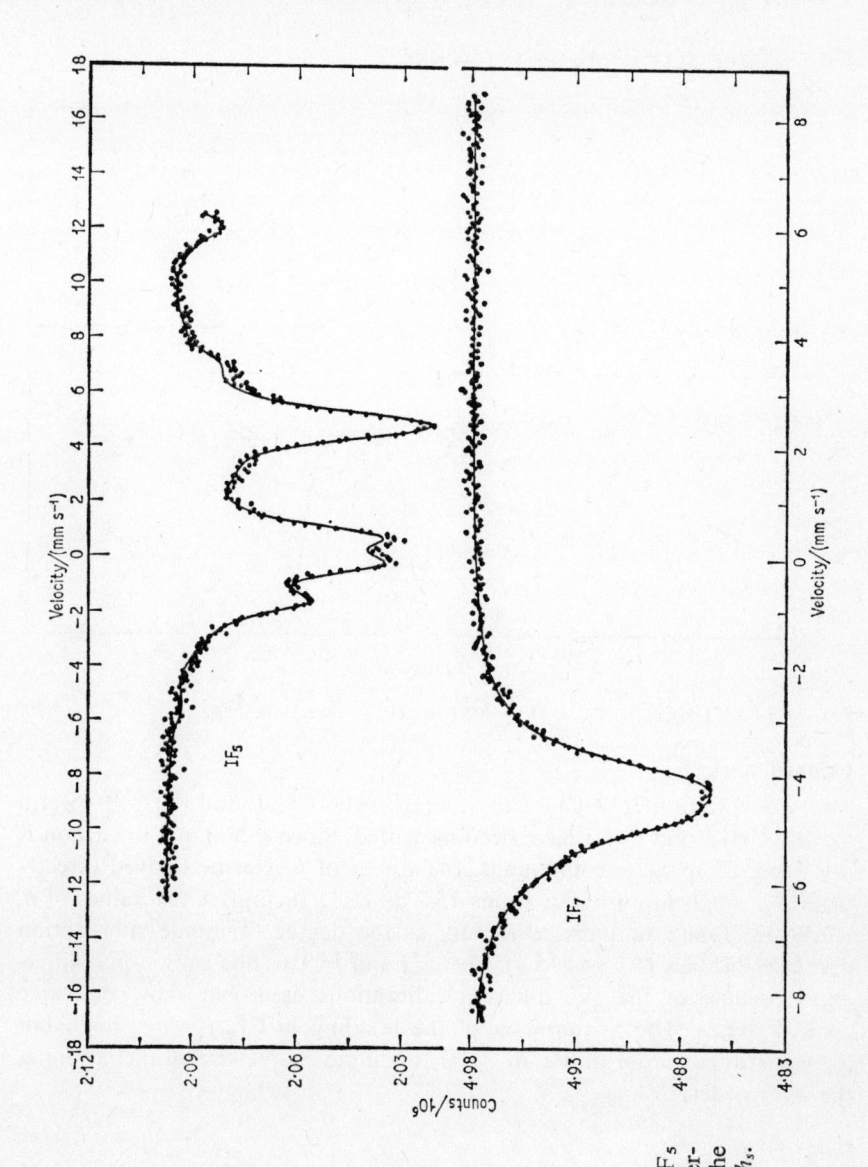

Fig. 15.22 ^{129}I spectra of IF$_5$ and IF$_7$. Note the large difference in shift caused by the alteration in the value of h_s. [Ref. 95, Fig. 2]

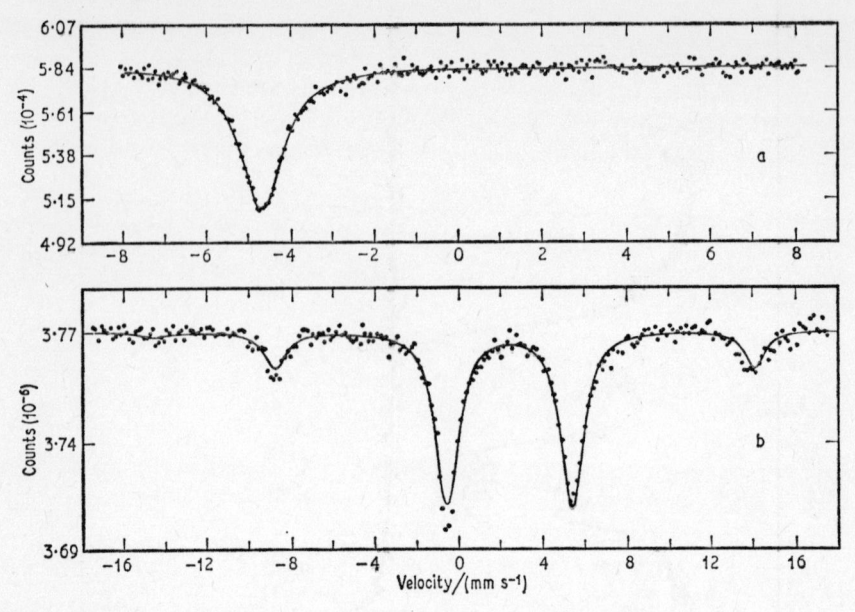

Fig. 15.23 ^{129}I spectra of (a) IF$_6^+$ and (b) IF$_6^-$. [Ref. 96, Fig. 1]

Iodomethanes

The ^{127}I spectra [97] of CI$_4$, CHI$_3$, CH$_2$I$_2$, and CH$_3$I, and the ^{129}I spectra of CI$_4$, CHI$_3$, and CH$_3$I have been measured. Since π-bonding to carbon is not expected in these compounds, the values of h_p can be derived directly from U_p. Application of equations 15.7 or 15.11 then gives the values of h_s which are found to increase rapidly as the degree of iodine substitution decreases (Tables 15.3 and 15.4). The ^{127}I and ^{129}I results show some differences because of the two different calibrations used, but show the same general trends. The 5s-character of the bonding in CI$_4$ is very small, but increases to the order of 8% in CH$_3$I, the ionic character also increasing in the same order.

SnI$_4$

Data for both ^{119}Sn and ^{129}I in SnI$_4$ are available [98]. The U_p value of $+0.59$ and $h_p = 0.64$ from equation 15.12 have been held to give $U_x = U_y = 1.98$ and $U_z = 1.39$, i.e. there is little π-electron interaction and 0·41e^- is transferred to iodine in each Sn–I bond. However, a more detailed interpretation of both the ^{119}Sn and ^{129}I data disagrees with this [99]. The new interpretation invokes a 23% π-character in the Sn–I bond, with 5–6% s-hybridisation in the iodine σ-bonding orbital. The iodine atom is essentially neutral with a charge of only $-0.015e^-$. This picture is more consistent with available data on the tin tetrahalides.

[*Refs.* on p. 489]

The temperature-dependence data give effective lattice temperatures of $\theta(^{119}Sn) = 166$ K and $\theta(^{129}I) = 85$ K, showing how markedly molecular solids differ from simple lattice theories. The four vibrational frequencies of SnI_4 are known from I.R./Raman data to be 47, 63, 149, and 216 cm^{-1}, the tin participating only in the 216 cm^{-1} mode. Thus in the temperature range used for the measurements (~80–200 K) the iodine 47 cm^{-1} (68 K) vibration is almost fully excited whereas the tin 216 cm^{-1} (311 K) mode is not. This accounts for the lower lattice temperature of iodine. More detailed molecular-dynamical calculations for SnI_4 have introduced the intermolecular translational and rotational vibrations [100].

GeI_4 and SiI_4

Similar studies on ^{129}I in GeI_4 and SiI_4 have been made [101]. The values of U_p and h_p (Table 15.4) show that there is very little 5s- or π-character in the bonding. The number of electrons transferred to iodine is $0.32e^-$ in GeI_4 and $0.47e^-$ in SiI_4, compared to $0.41e^-$ in SnI_4 and $0.11e^-$ in CI_4. The temperature-dependence data give $\theta(GeI_4) = 87$ K and $\theta(SiI_4) = 93$ K for the iodine atoms only, and a lattice-dynamical treatment was given. A small Karyagin effect was claimed.

CrI_3

We have not as yet mentioned magnetic hyperfine splitting in iodine. This is because there are few instances where it is known and in most of these the overall result is mainly one of line broadening. For example a 54.4-kG external field applied to a $K^{129}I$ absorber resulted in only broadening of the single resonance line into an apparent overlapping doublet, although computer analysis did give a value for $^{129}\mu_e$ of $+2.84(5)$ nuclear magnetons [102]. The only iodine compound for which a magnetic interaction has been observed, albeit produced by a transferred exchange interaction, is ferromagnetic CrI_3. The ^{127}I resonance proved to have inadequate resolution [103]. The value of the excited-state magnetic moment was not known, but a value of $^{127}\mu_e = 2.02(15)$ n.m. has been recently measured by a perturbed angular correlation method [104]. More satisfactory data have been obtained with ^{129}I [105]. The quadrupole spectrum at 78 K ($T_C = 70$ K) has an asymmetry parameter of 0.35. At 21 K the basic spectrum is retained but with broadening because of the additional magnetic splitting produced by a field of 25 kG as shown in Fig. 15.24. Computer analysis suggests that the angle β between V_{zz} and H is defined by $\cos \beta = 0.45$. The actual structure of CrI_3 is not yet fully known.

IO_3^-, IO_4^-, and IO_6^{5-}

The chemical isomer shift values for $^{129}IO_3^-$ (+1.56 mm s^{-1}), $^{129}IO_4^-$ (−2.34 mm s^{-1}), and $^{129}IO_6^{5-}$ (−3.10 mm s^{-1}) highlight the sensitivity of this parameter to the relative occupation of the 5s- and 5p-orbitals, rather than to formal oxidation state [74, 90, 106]. Assuming that KIO_3 has no 5s-character in the bonding a value of $h_p = 1·40$ is derived from equation 15.12, i.e. $0·46e^-$ is removed from the I^- configuration per I–O bond. IO_4^- and

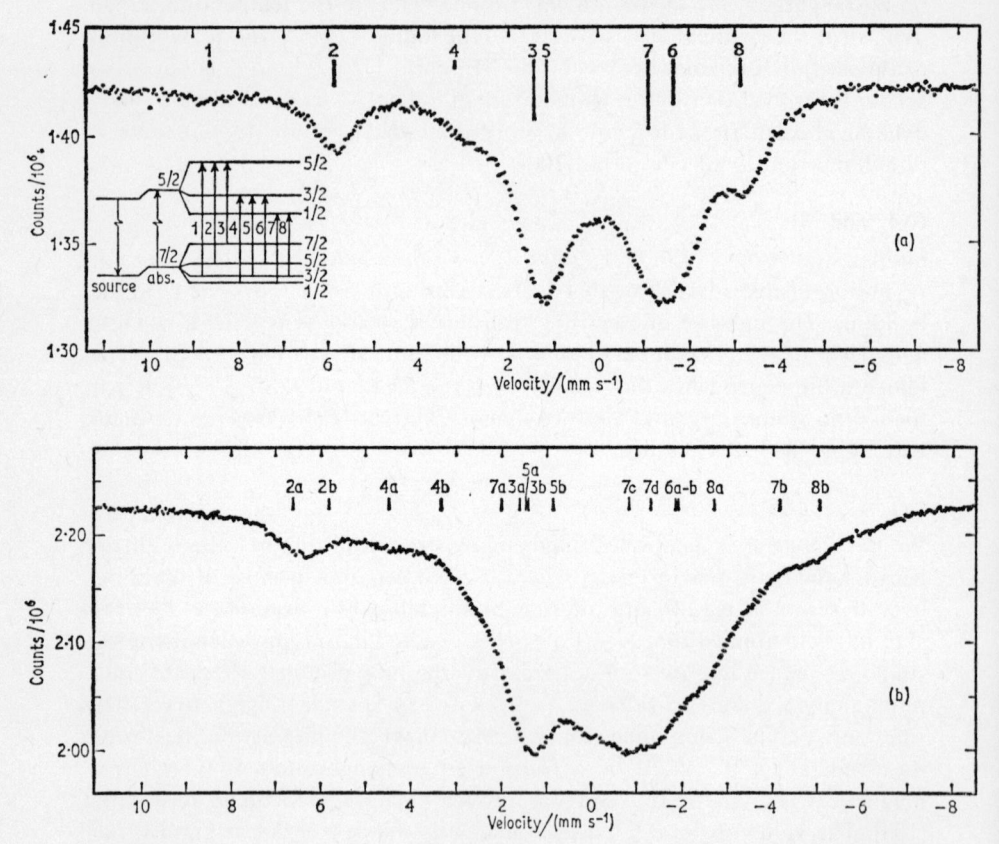

Fig. 15.24 ^{129}I spectra of CrI_3 at (a) 78 K and (b) 21 K. The magnetic interaction raises the remaining degeneracies and the resultant proliferation in transitions is seen as line broadening. [Ref. 105, Fig. 1]

IO_6^{5-} are more difficult to discuss because 5s-electrons do participate in the bonding of these compounds.

^{127}I data were used to derive a value for $\delta R/R$ (^{127}I) and thence to calibrate ^{125}Te data (see Chapter 15.5) [25].

I_2O_4

The oxide I_2O_4 contains two inequivalent iodine atoms [107, 108], parameters being given in Table 15.4. One of the atoms is similar to that in IO_3^-

[*Refs.* on p. 489]

and a structure has been proposed in which an IO$^+$ group is bonded covalently to an IO$_3^-$ unit via oxygen bridging atoms.

Impurity Studies

The high-spin states of the iodine isotopes allow definition of many of the parameters of an iodine impurity atom in a host lattice, and some interesting results are beginning to emerge.

^{129m}Te nuclei have been implanted in an iron foil magnetised perpendicular to the direction of observation to reduce the number of hyperfine lines with non-zero intensity [109]. A magnetic field of 1·13(4) MG at 100 K was observed at the daughter ^{129}I nucleus (1 MG = 10^6 gauss).

A source of ^{129m}Te(OH)$_6$ gives the identical single-line spectrum to that of Na$_3$H$_2$IO$_6$. and it appears that the daughter iodine atom retains the octahedral coordination to oxygen [72].

The decay of ^{129m}Te in TeO$_2$ and Te(NO$_3$)$_4$ has been used to study their structures [110]. It was assumed that the values of U_p in TeO$_2$ and Te(NO$_3$)$_4$ are not changed by the nuclear transformation. The β-decay does not have the same catastrophic effect on the electronic environment as electron-capture decay. The ratio of the ^{125}Te quadrupole splitting $\{e^2q^{125}Q(1+\frac{1}{3}\eta^2)^{\frac{1}{2}}\}$ to the ^{129}I quadrupole coupling constant $\{e^2q^{127}Q_g\}$ is 0·28 in both cases. However, the values of eq_{at} for tellurium and iodine are not identical (eq_{at}(Te)/eq_{at}(I) = 0·85) implying that a 5p-hole in tellurium has a greater radial extent than in iodine. Whether the value of U_p will be affected or not by the possible change in orbital overlap because of this contraction has not been satisfactorily answered. Certainly in tellurium metal considerable electronic reorganisation takes place and the ratio of the quadrupole couplings does not agree with the other data. Correlation of the^{129}I parameters with the structures of TeO$_2$ and Te(NO$_3$)$_4$ was attempted.

The spinel CuCr$_2$Te$_4$ is ferromagnetic (T_C = 365 K) and the daughter of (^{129m}Te) shows a magnetic field of | 48(2) kG | as a broadening at liquid nitrogen temperature [111]. The width of the resonance decreases in a 13-kG external field, showing that $H = -48$ kG. This is significantly smaller than the field of 148 kG already described for the ^{125}Te resonance (Chapter 15.5) [66]. Covalent mixing of the 5p_π-orbitals with the 3d_π-orbitals on the chromium is lower in I$^-$ than in Te^{2-} so that the exchange polarisation is weaker in iodine.

The use of ^{129}I in determining the lattice temperatures of three nuclei in an SnTe matrix has alrcady been referred to under ^{125}Te [63].

A Cr^{129}Te alloy has been shown to give a magnetically broadened ^{129}I resonance with a field of about 70 kG [74].

Mn^{129m}Te$_2$ also shows magnetic splitting below its Néel temperature of 84 K [112], the magnetic field at 4·2 K being 215 kG. The sublattice magnetisation is increased near the impurity, indicating an increase in the Mn–ligand–Mn exchange on substituting I$^-$ for Te$^-$.

[*Refs. on p. 489*]

15.7 Xenon (^{129}Xe, ^{131}Xe)

There are two xenon Mössbauer resonances known, the 39·58-keV level of ^{129}Xe, and the 80·16-keV level of ^{131}Xe. The former has been used exclusively for all chemical applications. Its excited-state lifetime is $t_{\frac{1}{2}} = 1\cdot01(4)$ ns [113], giving a natural linewidth of 6·85 mm s^{-1}. The decay from the $I_e = \frac{3}{2}$ first excited state to the $I_g = \frac{1}{2}$ ground state is of M1 character, and is highly converted. The precursor is ^{129}I, which decays by β-emission with a lifetime of $1\cdot7 \times 10^7$ y directly to the Mössbauer level (Fig. 15.25). Good source

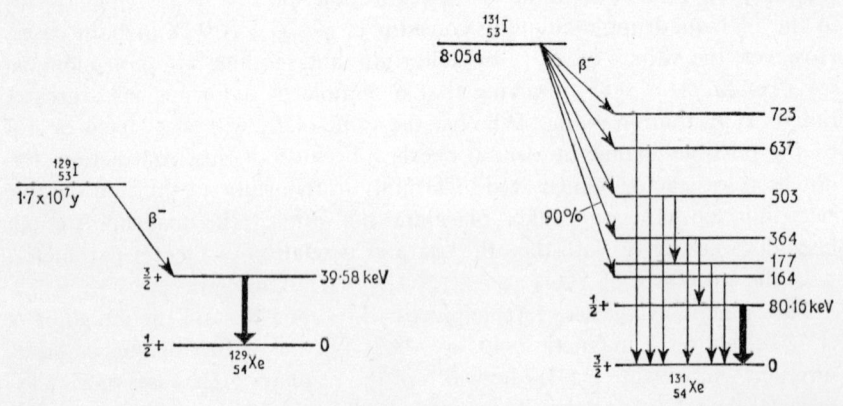

Fig. 15.25 The decay schemes of ^{129}Xe and ^{131}Xe populated by iodine precursors.

matrices are Na^{129}I, K^{129}IO$_4$, and Na$_3$H$_2$^{129}IO$_6$ which give narrow single lines, the last named giving the best recoil-free fraction. The ^{131}Xe resonance is from the $I_e = \frac{1}{2}$ first excited state to the $I_g = \frac{3}{2}$ ground state and is populated by the 8·05-d β-active ^{131}I precursor. The excited-state lifetime of $t_{\frac{1}{2}} = 0\cdot496(21)$ ns [114] gives a natural linewidth of 6·8 mm s^{-1}. A source of Na$_2$H$_3$^{131}IO$_6$ has been used successfully [115].

The published data for ^{129}Xe and ^{131}Xe stem as a series of continuing experiments from the first observation by G. J. Perlow and co-workers in 1963 [116]. Early work established the characteristics of the ^{129}Xe resonance in xenon hydroquinone clathrate, Na$_4$XeO$_6$.H$_2$O, XeF$_2$, and XeF$_4$ [116,

[Refs. on p. 489]

117], but the ^{131}Xe resonance was found to be more difficult to detect, and has only been reported in XeF$_4$ [115]. The work has been reviewed in detail [85]. The spectra consist of either a simple line as in xenon hydroquinone clathrate and Na$_4$XeO$_6$.H$_2$O, or a quadrupole split doublet as in XeF$_2$ and XeF$_4$.

The ^{129}Xe and ^{131}Xe spectra of XeF$_4$ (Fig. 15.26) are both quadrupole split [115]. This provides a means of calculating the excited-state quadrupole moment of ^{129}Xe from the ground-state moment of ^{131}Xe which is -0.12 barn. The ratio $|e^2q^{129}Q_e/e^2q^{131}Q_g| = 3.45(9)$ gives $|^{129}Q_e| = 0.41$ barn.

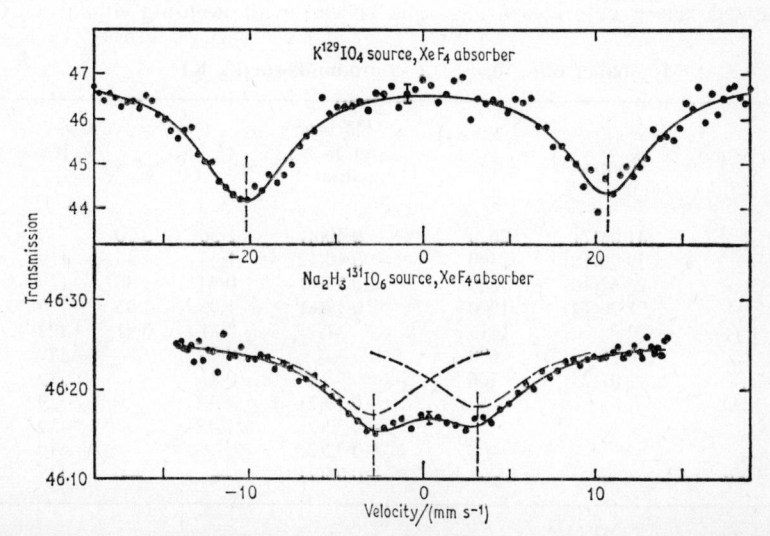

Fig. 15.26 The ^{129}Xe and ^{131}Xe spectra of XeF$_4$ showing the change in quadrupole splitting produced by the different magnitudes of the quadrupole moments of the $\frac{3}{2}$ states. [Ref. 115, Fig. 1]

The sign was determined from experiments with oriented KICl$_4$ sources to be negative, so that $^{129}Q_e = -0.41$ barn.

Special interest attaches to the chemical state of the ^{129}Xe daughter atom of the ^{129}I β-decay. The single-line sources, K^{129}IO$_4$ and Na$_3$H$_2$^{129}IO$_6$, show small but significant chemical isomer shifts when compared with a xenon clathrate absorber [118, 119]. Numerical values are collected in Table 15.5. This proves that there is some degree of chemical binding. One can postulate the reactions

$$IO_4^- \rightarrow \{XeO_4\}$$
$$IO_6^{5-} \rightarrow \{XeO_6^{4-}\}$$

where { } denotes a molecular species trapped in a host solid. Furthermore, the compound XeO$_3$ shows a quadrupole splitting of 10.95 mm s^{-1} while on the other hand the source Na^{129}IO$_3$ with a xenon clathrate single-line absorber shows a splitting of 11·07(25) mm s⁻¹. This is good evidence for the formation of an {XeO₃} molecule in the KIO₃ lattice by the β-decay. Additional support comes from the chemical isomer shift which we discuss shortly.

Very similar experiments using K^{129}ICl₄.H₂O and K^{129}ICl₂.H₂O appear to give spectra from the hitherto unknown molecules, {XeCl₄} and {XeCl₂} [119, 120]. The spectra, which are among the more dramatic results obtained by Mössbauer spectroscopy, are illustrated in Fig. 15.27 together with those of XeF₄ and XeF₂. The well-resolved quadrupole splittings show that the assumed xenon chlorides are unique species with bonding similar to the

Table 15.5 Mössbauer data for ^{129}Xe compounds (at 4.2 K)

Compound	$^{129}\Delta$ /(mm s⁻¹)*	$e^2q^{129}Q_e$ /MHz*	$^{129}\delta$/(mm s⁻¹) (rel. to Xe clathrate)	U_p	h_p	Reference
XeF₄†	41·04(7)	2620	0·40(4)	1·50	3·00	119
XeF₂	39·00(10)	2490	0·10(12)	1·43	1·43	119
{XeCl₄}	25·62(10)	1640	0·25(8)	0·94	1·88	119
{XeCl₂}	28·20(14)	1800	0·17(8)	1·03	1·03	119
{XeBr₂}	22·2	1415	—	0·81	0·81	122
XeO₃	10·95	698	—	0·40	—	119
{XeO₃}	11·07(25)	706	—	0·41	—	119
Na₄XeO₆	—	—	−0·19(2)	—	—	119
{XeO₆⁴⁻}	—	—	−0·22(2)	—	—	119
{XeO₄}	—	—	−0·22(2)	—	—	119
solid Xe	—	—	−0·05(7)	—	—	119

* 1 mm s⁻¹ = 31·93 MHz.
† Corresponding values for ^{131}Xe are $^{131}\Delta$ = 5·97(16) mm s⁻¹ and $e^2q^{131}Q_g$ = 772(21) MHz).

fluorides. There is a suggestion of other products in the XeCl₂ spectrum, but these have not been identified.

The theory of the quadrupole splitting and chemical isomer shift of ^{127}I (see preceding section) may be applied to xenon, and values of U_p, the imbalance in the 5p-orbitals, are calculated [120] directly from $e^2q^{129}Q_e$ using the known value for the quadrupole coupling constant in ^{131}Xe due to a $5p_z$ hole (−505 MHz, i.e. −1740 MHz for a $5p_z$ hole in ^{129}Xe). The values of h_p are derived assuming that there is no 5s- or π-character in the bonding and that $\eta = 0$. The h_p values for ICl₄⁻ and ICl₂⁻ are 2·70 and 1·39 respectively. By comparison with the h_p values from the ^{127}I data, one can see that changing iodine to xenon causes transfer of 0·82e^- and 0·36e^- respectively towards the xenon, being 0·20e^- and 0·18e^- respectively per chlorine atom. As might be expected fluorine withdraws more electron density from xenon than chlorine (see Table 15.5).

[*Refs. on p. 489*]

XENON | 485

Stable bonds are not always formed during the β-decay. A source of $^{129}I_2$ showed only a single-line resonance without any quadrupole splitting. Apparently atomic xenon is formed which does not interact with the immediate neighbour undecayed iodine whose chemical state is unknown [119].

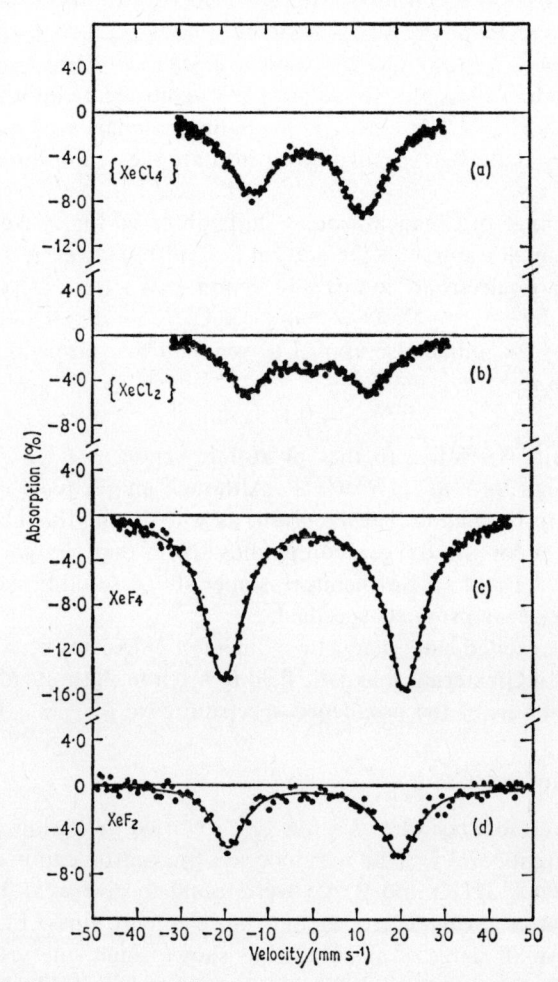

Fig. 15.27 Mössbauer spectra (^{129}Xe) of the xenon halides. [Ref. 119, Fig. 1]

It should be pointed out that although entities such as {XeCl$_4$} and {XeCl$_2$} were not known as stable compounds at the time of these experiments, the observation of their Mössbauer spectrum requires only that each molecule is stable for a little longer than the nanosecond required for emission of the γ-ray whose energy characterises it. Chemical change after this event (if it occurs) is not registered in the Mössbauer spectrum. The infrared spectrum

[*Refs. on p. 489*]

of XeCl$_2$ was reported independently at about the same time, although again the compound was not isolated [121].

A possible border-line case is found in the decay of ^{129}IBr$_2^-$. A source of KIBr$_2$ shows evidence of a quadrupole splitting of 22·2 mm s^{-1} (1415 MHz), although the overall spectrum is strongly dependent on the individual source preparation and there are signs of other species [122]. Assuming that the major product is a linear {XeBr$_2$} molecule, one can deduce $h_p = 0·81$, i.e. bromine withdraws less electron density than chlorine or fluorine.

The decay of Cs^{129}IBr$_2$ gives a single-line resonance of which detailed study suggests that IBr$_2^-$ breaks down into atomic xenon during the decay [122].

The total range of chemical isomer shifts observed for ^{129}Xe (Table 15.5) fall within a small fraction of the natural linewidth. However, comparison of the shifts for isoelectronic iodine and xenon pairs, ICl$_4^-$/{XeCl$_4$}, ICl$_2^-$/{XeCl$_2$}, I$^-$/Xe^0, IO$_4^-$/{XeO$_4$}, IO$_6^{5-}$/XeO$_6^{4-}$ does show that the same general theory for iodine also applies to xenon. The expression equivalent to equation 15.8 is

$$^{129}\delta_{Xe} = 0·13 h_p \text{ mm s}^{-1} \qquad 15.13$$

where the shift is relative to that of atomic xenon. $\delta\langle R^2 \rangle / \langle R^2 \rangle (^{129}\text{Xe})$ is opposite in sign to $\delta\langle R^2 \rangle / \langle R^2 \rangle (^{127}\text{I})$. Although an adequate description of the bonding in the halides can be given, as with iodine this has not proved possible as yet for the oxygen compounds. Note that the good agreement between the ^{127}I and ^{129}Xe chemical isomer shifts is additional evidence in favour of the decay products specified.

Finally, the excited-state magnetic moment of ^{129}Xe has been measured by applying a 78-kG external magnetic field to a xenon clathrate absorber [123]. Computer analysis of the broadened spectrum gave $\mu_e^{129} = +0·68(30)$ n.m.

15.8 Caesium (^{133}Cs)

The 81·0-keV resonance of ^{133}Cs was first reported by Perlow et al. in 1965 [124]. The parent is 7·2 y ^{133}Ba which decays by electron capture (Fig. 15.28). Sources of BaCl$_2$·2H$_2$O and BaAl$_4$ were found to give weak resonances at 4·2 K, the latter giving narrower and more intense lines. Extended work [125] with a small range of absorbers has shown single-line resonances only with small chemical isomer shifts which are listed in Table 15.6. A linear correlation between the shifts of the caesium halides and the ^{133}Cs nuclear magnetic resonance chemical shifts was found. Detailed calculations of the effects of orbital overlap in these compounds gave an approximate value for $\delta\langle R^2 \rangle / \langle R^2 \rangle$ of $+17·8 \times 10^{-5}$.

Independent work used ^{133}Ba in CaF$_2$ as the source [126]. Where identical compounds were studied the agreement in values is within the quoted errors. A second value for $\delta\langle R^2 \rangle / \langle R^2 \rangle$ of $+4 \times 10^{-4}$ was calculated on the basis

[Refs. on p. 489]

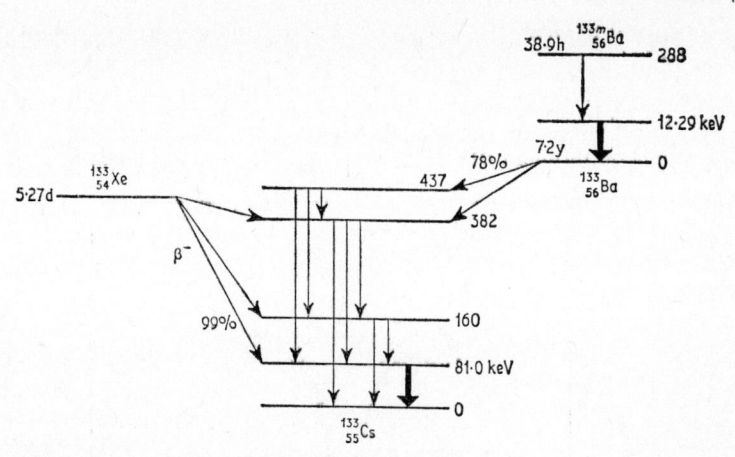

Fig. 15.28 The decay schemes of ^{133}Cs and ^{133}Ba.

Table 15.6 Chemical isomer shifts in ^{133}Cs (at 4.2 K)

Compound	δ (rel. to BaAl$_4$)* /(mm s^{-1})	Reference
Cs	−0·164(57)	125
CsI	−0·247(4)	125
	−0·239(11)	126
CsBr	−0·269(4)	125
	−0·264(9)	126
CsCl	−0·269(3)	125
	−0·269(4)	126
CsF	−0·278(2)	125
	−0·260(14)	126
CsMnF$_3$	−0·313(6)	125
CsN$_3$	−0·26(1)	125
Cs$_2$CO$_3$	−0·32(1)	125
CsNO$_3$	−0·320(10)	126
Cs$_2$SO$_4$	−0·311(7)	126
Cs$_2$Cr$_2$O$_7$	−0·309(8)	126
CsClO$_3$	−0·305(6)	126
CsBi$_2$	−0·209(20)	126
^{133}Ba/BaCl$_2$.H$_2$O	−0·31(2)	125
^{133}Ba/BaF$_2$	−0·28(2)	125
^{133}Ba/CaF$_2$	−0·270(4)	126

* Values from ref. 125 have been corrected for the zero-point motion in the second-order Doppler shift by up to 0·006 mm s^{-1}. Data from ref. 126 were quoted relative to ^{133}Ba/CaF$_2$ and have been converted using CsCl as the reference, but the errors quoted are the original values.

[*Refs. on p. 489*]

of the chemical isomer shift values for $CsNO_3$ and $CsBi_2$, the mean of the two results being approximately $+3\cdot4 \times 10^{-4}$.

An external magnetic field of 77·65 kG applied to $CsMnF_3$ produces magnetic broadening as illustrated in Fig. 15.29 [127]. Analysis gives a value for the magnetic moment of the $I_e = \frac{5}{2}$ state of $\mu_e = +3\cdot44(2)$ n.m.

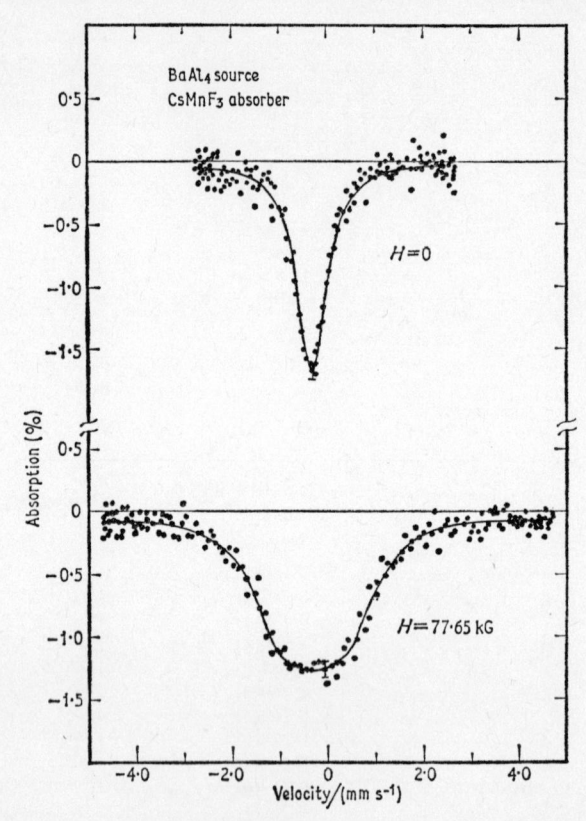

Fig. 15.29 Mössbauer spectra of $^{133}CsMnF_3$ in zero applied field and in a field of 77·65 kG. [Ref. 127, Fig. 1]

^{133}Cs impurity atoms in an iron foil have been studied by implantation from ^{133}Xe atoms which β-decay to caesium [128]. Partially resolved magnetic hyperfine structure is seen, giving a value for the field at the ^{133}Cs nucleus of $+273(10)$ kG.

15.9 Barium (^{133}Ba)

Only one reference has been made to the 12·29-keV resonance of ^{133}Ba [129]. This resonance is unusual in that the ground state is radioactive with a half-life of 7·2 y, but is populated via the short-lived 39 h isotope ^{133m}Ba, which

[*Refs. on p. 489*]

effectively suppresses the natural radioactive background in the actual experiment. Sources of BaO and BaSO$_4$ with BaO absorbers gave an effect of up to 1% in the temperature range 4·2–300 K, but the linewidths obtained were 8·6 mm s^{-1} as opposed to a natural width of 2·7 mm s^{-1}, derived from the excited-state lifetime of 8·1(2·0) ns [130]. No explanation for this could be found.

REFERENCES

[1] S. L. Ruby and R. E. Holland, *Phys. Rev. Letters*, 1965, **14,** 591.
[2] F. J. Lynch and R. E. Holland, *Phys. Rev.*, 1959, **114,** 825.
[3] D. W. Hafemeister and E. Brooks Shera, *Phys. Rev. Letters*, 1965, **14,** 593.
[4] P. K. Tseng, S. L. Ruby, and D. H. Vincent, *Phys. Rev.*, 1968, **172,** 249.
[5] D. Raj and S. P. Puri, *Phys. Stat. Sol.*, 1969, **34,** K13.
[6] G. Czjzek, J. L. C. Ford, F. E. Obenshain, and D. Seyboth, *Phys. Letters*, 1966, **19,** 673.
[7] G. Czjzek, J. L. C. Ford, J. C. Love, F. E. Obenshain, and H. H. F. Wegener, *Phys. Rev. Letters*, 1967, **18,** 529.
[8] G. Czjzek, J. L. C. Ford, J. C. Love, F. E. Obenshain, and H. H. F. Wegener, *Phys. Rev.*, 1968, **174,** 331.
[9] R. E. Holland and F. J. Lynch, *Phys. Rev.*, 1961, **121,** 1464.
[10] B. H. Zimmermann, H. Jena, G. Ischenko, H. Kilian, and D. Seyboth, *Phys. Stat. Sol.*, 1968, **27,** 639.
[11] Y. P. Varshni and R. Blanchard, *Phys. Letters*, 1969, **30A,** 238.
[12] Y. Hazony, P. Hillman, M. Pasternak, and S. Ruby, *Phys. Letters*, 1962, **2,** 337.
[13] S. L. Ruby and H. Selig, *Phys. Rev.*, 1966, **147,** 348.
[14] M. Pasternak and T. Sonnino, *Phys. Rev.*, 1967, **164,** 384.
[15] V. M. Krasnoperov, A. N. Murin, N. K. Cherezov, and I. A. Yutlandov, *Doklady Akad. Nauk S.S.S.R.*, 1969, **186,** 296.
[16] S. Bukshpan, C. Goldstein, and T. Sonnino, *Phys. Letters*, 1968, **27A,** 372.
[17] B. Barnett and Y. Hazony, *J. Chem. Phys.*, 1965, **43,** 3462.
[18] Y. Hazony and S. L. Ruby, *J. Chem. Phys.*, 1968, **49,** 1478.
[19] M. Pasternak, A. Simopoulos, S. Bukshpan, and T. Sonnino, *Phys. Letters*, 1966, **22,** 52.
[20] K. Gilbert and C. E. Violet, *Phys. Letters*, 1968, **28A,** 285.
[21] J. S. Brown, *Phys. Rev.*, 1969, **187,** 401.
[22] L. E. Campbell, G. J. Perlow, and M. A. Grace, *Phys. Rev.*, 1969, **178,** 1728.
[23] M. Greenshpan. D. Treves, S. Bukshpan, and T. Sonnino, *Phys. Rev.*, 1969, **178,** 1802.
[24] R. E. Snyder and G. B. Beard, *Phys. Letters*, 1965, **15,** 264.
[25] S. E. Gukasyan and V. S. Shpinel, *Phys. Stat. Sol.*, 1968, **29,** 49.
[26] V. S. Shpinel, V. A. Bryukhanov, V. Kothekar, B. Z. Iofa, and S. I. Semenov, *Symposia Faraday Soc. No. 1*, 1968, 69.
[27] S. L. Ruby and C. E. Johnson, *Phys. Letters*, 1967, **26A,** 60.
[28] S. L. Ruby, B. J. Evans, and S. S. Hafner, *Solid State Commun.*, 1968, **6,** 277.
[29] S. L. Ruby, G. M. Kalvius, R. E. Snyder, and G. B. Beard, *Phys. Rev.*, 1966, **148,** 176.

[30] S. L. Ruby, G. M. Kalvius, G. B. Beard, and R. E. Snyder, *Phys. Rev.* 1967, **159**, 239.
[31] S. L. Ruby and G. K. Shenoy, *Phys. Rev.*, 1969, **186**, 326.
[32] S. L. Ruby and G. M. Kalvius, *Phys. Rev.*, 1967, **155**, 353.
[33] V. A. Bryukhanov, B. Z. Iofa, V. Kothekar, S. I. Semenov, and V. S. Shpinel, *Zhur. eksp. teor. Fiz.*, 1967, **53**, 1582.
[34] V. Kothekar, B. Z. Iofa, S. I. Semenov, and V. S. Shpinel, *Zhur. eksp. teor. Fiz.*, 1968, **55**, 160.
[35] L. H. Bowen, J. G. Stevens, and G. G. Long, *J. Chem. Phys.*, 1969, **51**, 2010.
[36] G. G. Long, J. G. Stevens, and L. H. Bowen, *Inorg. Nuclear Chem. Letters*, 1969, **5**, 799.
[37] T. Birchall and B. Della Valle, *Chem. Comm.*, 1970, 675.
[38] G. G. Long, J. G. Stevens, H. Lawrence, and S. L. Ruby, *Inorg. Nuclear Chem. Letters*, 1969, **5**, 21.
[39] V. A. Bryukhanov, B. Z. Iofa, and S. I. Semenov, *Radiokhimiya*, 1969, **11**, 362.
[40] Pham Zuy Hien, V. G. Shapiro, and V. S. Shpinel, *Zhur. eksp. teor. Fiz.*, 1962, **42**, 703 (*Soviet Physics – JETP*, 1962, **15**, 489).
[41] N. Shikazono, T. Shoji, H. Takekoshi, and P. Tseng, *J. Phys. Soc. Japan*, 1962, **17**, 1205.
[42] H. Voorthius, W. Beens, and H. Verheul, *Physica*, 1967, **33**, 695.
[43] E. P. Stepanov, K. P. Aleshin, R. A. Manapov, B. N. Samoilov, V. V. Sklyarevskii, and V. G. Stankevich, *Phys. Letters*, 1963, **6**, 155.
[44] M. Pasternak and A. L. Spijkervet, *Phys. Rev.*, 1969, **181**, 574.
[45] N. E. Erickson and A. G. Maddock, *J. Chem. Soc. (A)*, 1970, 1665.
[46] R. N. Kuzmin, A. A. Opalenko, V. S. Shpinel, and I. A. Avenarius, *Zhur. eksp. teor. Fiz.*, 1969, **56**, 167 (*Soviet Physics – JETP*, 1969, **29**, 94).
[47] V. A. Lebedev, R. A. Lebedev, A. M. Babeshkin, and A. N. Nesmeyanov, *Vestnik Moskov. Univ., Khim.*, 1969, **24**, 128.
[48] C. E. Violet, R. Booth, and F. Wooton, *Phys. Letters*, 1963, **5**, 230.
[49] C. E. Violet and R. Booth, *Phys. Rev.*, 1966, **144**, 225.
[50] T. C. Gibb, R. Greatrex, N. N. Greenwood, and A. C. Sarma, *J. Chem. Soc. (A)*, 1970, 212.
[51] P. Jung and W. Triftshäuser, *Phys. Rev.*, 1968, **175**, 512.
[52] M. L. Unland, *J. Chem. Phys.*, 1968, **49**, 4514.
[53] V. S. Shpinel, V. A. Bryukhanov, V. Kothekar and B. Z. Iofa, *Zhur. eksp. teor. Fiz.*, 1967, **53**, 23 (*Soviet Physics – JETP*, 1968, **26**, 16).
[54] V. A. Bryukhanov, B. Z. Iofa, A. A. Opalenko, and V. S. Shpinel, *Zhur. Neorg. Khim.*, 1967, **12**, 1985 (*Russian J. Inorg. Chem.*, 1967, 1044).
[55] M. Pasternak and S. Bukshpan, *Phys. Rev.*, 1967, **163**, 297.
[56] E. P. Stepanov and A. Yu. Aleksandrov, *ZETF Letters*, 1967, **5**, 101 (*JETP Letters*, 1967, **5**, 83).
[57] J. F. Ullrich and D. H. Vincent, *J. Phys. Chem. Solids*, 1969, **30**, 1189.
[58] N. Shikazono, *J. Phys. Soc. Japan*, 1963, **18**, 925.
[59] J. Baijal and U. Baijal, *J. Phys. Soc. Japan*, 1967, **22**, 1507.
[60] I. V. Berman, N. B. Brandt, R. I. Kuzmin, A. A. Opalenko, and S. S. Slobodchikov, *ZETF Letters*, 1969, **10**, 373.
[61] G. Albanese, C. L. Lamborizio, and I. Ortalli, *Nuovo Cimento*, 1967, **50B**, 65.
[62] C. C. Tsuei and E. E. Kankeleit, *Phys. Rev.*, 1967, **162**, 312.
[63] S. Bukshpan, *Solid State Commun.*, 1968, **6**, 477.
[64] R. B. Frankel, J. Huntzicker, E. Matthias, S. S. Rosenblum, D. A. Shirley, and N. J. Stone, *Phys. Letters*, 1965, **15**, 163.

[65] R. B. Frankel, J. J. Huntzicker, D. A. Shirley, and N. J. Stone, *Phys. Letters*, 1968, **26A**, 452.
[66] J. F. Ullrich and D. H. Vincent, *Phys. Letters*, 1967, **25A**, 731.
[67] S. Jha, R. Segnan, and G. Lang, *Phys. Rev.*, 1962, **128**, 1160.
[68] F. de S. Barros, N. Ivantchev, S. Jha, and K. R. Reddy, *Phys. Letters*, 1964, **13**, 142.
[69] G. Berzins, L. M. Berger, W. H. Kelly, W. B. Walters, and G. E. Gordon, *Nuclear Phys.*, 1967, **93**, 456.
[70] R. Sanders and H. de Waard, *Phys. Rev.*, 1966, **146**, 907.
[71] J. S. Geiger, R. L. Graham, I. Bergstrom, and F. Brown, *Nuclear Phys.*, 1965, **68**, 358.
[72] G. J. Perlow and M. R. Perlow, *J. Chem. Phys.*, 1966, **45**, 2193.
[73] G. J. Perlow and S. L. Ruby, *Phys. Letters*, 1964, **13**, 198.
[74] D. W. Hafemeister, G. de Pasquali, and H. de Waard, *Phys. Rev.*, 1964, **135**, B1089.
[75] R. Livingstone and H. Zeldes, *Phys. Rev.*, 1953, **90**, 609.
[76] K. R. Reddy, F. de S. Barros, and S. DeBenedetti, *Phys. Letters*, 1966, **20**, 297.
[77] B. S. Ehrlich and M. Kaplan, *J. Chem. Phys.*, 1969, **50**, 2041.
[78] T. P. Das and E. L. Hahn, 'Nuclear Quadrupole Resonance Spectroscopy', Supplement I of *Solid State Physics*, Academic Press Inc., New York, 1958.
[79] R. Livingstone and H. Zeldes, *Phys. Rev.*, 1953, **90**, 609.
[80] M. Pasternak, A. Simopoulos, and Y. Hazony, *Phys. Rev.*, 1965, **140**, A1892.
[81] S. Bukshpan, C. Goldstein, and T. Sonnino, *J. Chem. Phys.*, 1968, **49**, 5477.
[82] W. H. Flygare and D. W. Hafemeister, *J. Chem. Phys.*, 1965, **43**, 789.
[83] P. Jung and W. Triftshäuser, *Phys. Rev.*, 1968, **175**, 512.
[84] M. Pasternak, *Symposia Faraday Soc. No. 1*, 1967, p. 119.
[85] G. J. Perlow, 'Chemical Applications of Mössbauer Spectroscopy', Ed. V. I. Goldanskii and R. H. Herber, Academic Press Inc., New York, 1968, p. 378.
[86] S. S. Jaswal, *Phys. Letters*, 1965, **19**, 369.
[87] S. S. Jaswal, *Phys. Rev.*, 1966, **144**, 353.
[88] R. Kamal, R. G. Mendiratta, S. B. Raju and L. M. Tiwari, *Phys. Letters*, 1967, **25A**, 503.
[89] K. Mahesh and N. D. Sharma, *Phys. Letters*, 1968, **28A**, 377.
[90] M. Pasternak and T. Sonnino, *J. Chem. Phys.*, 1968, **48**, 1997.
[91] M. Pasternak and T. Sonnino, *J. Chem. Phys.*, 1968, **48**, 2004.
[92] M. Pasternak and T. Sonnino, *J. Chem. Phys.*, 1968, **48**, 2009.
[93] C. I. Wynter, J. Hill, W. Bledsoe, G. K. Shenoy, and S. L. Ruby, *J. Chem. Phys.*, 1969, **50**, 3872.
[94] B. S. Ehrlich and M. Kaplan, *J. Chem. Phys.*, 1969, **51**, 603.
[95] S. Bukshpan, C. Goldstein, and J. Soriano, *J. Chem. Phys.*, 1969, **51**, 3976.
[96] S. Bukshpan, J. Soriano, and J. Shamir, *Chem. Phys. Letters*, 1969, **4**, 241.
[97] S. Bukshpan and T. Sonnino, *J. Chem. Phys.*, 1968, **48**, 4442.
[98] S. Bukshpan and R. H. Herber, *J. Chem. Phys.*, 1967, **46**, 3375.
[99] B. S. Ehrlich and M. Kaplan, *Chem. Phys. Letters*, 1969, **3**, 161.
[100] Y. Hazony, *J. Chem. Phys.*, 1968, **49**, 159.
[101] S. Bukshpan, *J. Chem. Phys.*, 1968, **48**, 4242.
[102] H. de Waard and J. Heberle, *Phys. Rev.*, 1964, **136**, B1615.
[103] G. M. Kalvius, L. D. Oppliger, and S. L. Ruby, *Phys. Letters*, 1965, **18**, 241.

[104] A. G. Svensson, R. W. Sommerfeldt, L. O. Norlin, and P. N. Tandon, *Nuclear Phys.*, 1967, **A95**, 653.
[105] C. Goldstein and M. Pasternak, *Phys. Rev.*, 1969, **177**, 481.
[106] H. de Waard, G. de Pasquali, and D. Hafemeister, *Phys. Letters*, 1963, **5**, 217.
[107] Yu. S. Grushko, A. N. Murin, B. G. Lure, and A. V. Motornyi, *Fiz. Tverd. Tela*, 1968, **10**, 3704.
[108] Yu. S. Grushko, B. G. Lure, and A. N. Murin, *Fiz. Tverd. Tela*, 1969, **11**, 2144.
[109] H. de Waard and S. A. Drentje, *Phys. Letters*, 1966, **20**, 38.
[110] M. Pasternak and S. Bukshpan, *Phys. Rev.*, 1967, **163**, 297.
[111] M. Pasternak and H. de Waard, *Phys. Letters*, 1968, **28A**, 298.
[112] M. Pasternak, *Phys. Rev.*, 1969, **184**, 523.
[113] J. S. Geiger, R. L. Graham, I. Bergstrom, and F. Brown, *Nuclear Phys.*, 1965, **68**, 352.
[114] R. S. Weaver, *Canad. J. Phys.*, 1962, **40**, 1684.
[115] G. J. Perlow, *Phys. Rev.*, 1964, **135**, B1102.
[116] C. L. Chernick, C. E. Johnson, J. G. Malm, G. J. Perlow, and M. R. Perlow, *Phys. Letters*, 1963, **5**, 103.
[117] G. J. Perlow and M. R. Perlow, *Rev. Mod. Phys.*, 1964, **36**, 353.
[118] G. J. Perlow and M. R. Perlow, 'Chemical Effects of Nuclear Transformations', Vol. II, p. 443, International Atomic Energy Agency, Vienna, 1965.
[119] G. J. Perlow and M. R. Perlow, *J. Chem. Phys.*, 1968, **48**, 955.
[120] G. J. Perlow and M. R. Perlow, *J. Chem. Phys.*, 1964, **41**, 1157.
[121] L. Y. Nelson and G. C. Pimentel, *Inorg. Chem.*, 1967, **6**, 1758.
[122] G. J. Perlow and H. Yoshida, *J. Chem. Phys.*, 1968, **49**, 1474.
[123] L. E. Campbell, G. J. Perlow, and N. C. Sandstrom, p. 161 of 'Hyperfine Structure and Nuclear Radiations'. Ed. E. Matthias and D. A. Shirley, North Holland, 1968.
[124] G. J. Perlow, A. J. F. Boyle, J. H. Marshall, and S. L. Ruby, *Phys. Letters*, 1965, **17**, 219.
[125] A. J. F. Boyle and G. J. Perlow, *Phys. Rev.*, 1966, **149**, 165.
[126] W. Henning, D. Quitmann, E. Steichele, S. Hufner, and P. Kienle, *Z. Physik*, 1968, **209**, 33.
[127] L. E. Campbell and G. J. Perlow, *Nuclear Phys.*, 1968, **A109**, 59.
[128] H. de Waard and S. R. Reintsema, *Phys. Letters*, 1969, **29A**, 290.
[129] A. J. F. Boyle and G. J. Perlow, *Phys. Rev.*, 1969, **180**, 625.
[130] J. E. Thun, S. Törnkvist, F. Falk, and H. Snellman, *Nuclear Phys.*, 1965, **67**, 625.

16 | Other Transition-metal Elements

A Mössbauer resonance is known in at least one isotope of fourteen transition metals in addition to iron. However, none has been extensively used up to the present time. Several of them present extreme difficulties in measurement, but as this chapter will show, sufficient background information has been collected to assess the feasibility of chemical application.

The elements will be discussed in the order of increasing atomic number in the Periodic Table, i.e. nickel, zinc, technetium, ruthenium, silver, hafnium, tantalum, tungsten, rhenium, osmium, iridium, platinum, gold, and mercury. Full numerical data of the relevant nuclear properties are summarized, as for other elements, in Appendix 1.

16.1 Nickel (^{61}Ni)

The 67·4-keV resonance of ^{61}Ni was first reported in 1961 by Obenshain and Wegener [1], who used the β-decay of 99-minute ^{61}Co to populate the excited level (see Fig. 16.1). The source in these initial experiments at 80 K was a

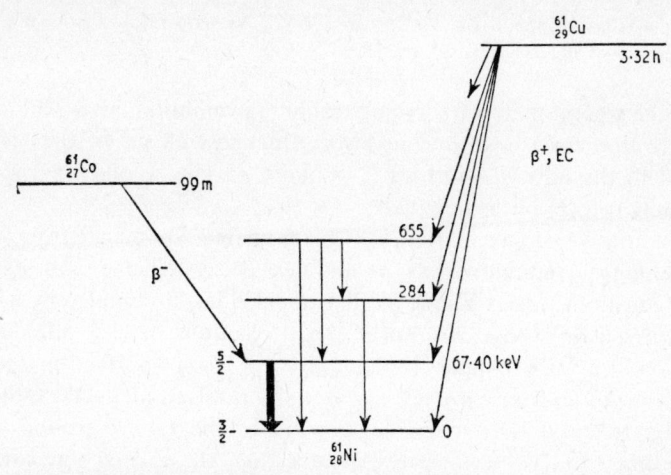

Fig. 16.1 Decay scheme of ^{61}Ni.

nickel foil enriched in ^{64}Ni so that the parent radioisotope could be made by the ^{64}Ni(p, α)^{61}Co reaction [1, 2], and the absorbers were nickel metal. The apparent single-line resonance was in fact broadened by magnetic hyperfine splitting in both source and absorber, and analysis of the spectrum shape in an external magnetic field gave the ratio of the magnetic moments $\mu_e/\mu_g = -0.47(8)$.

Very similar results were obtained using Coulomb excitation of a nickel foil target by 25-MeV oxygen ions [3], and the Coulomb-recoil implantation technique has also been demonstrated [4].

A second ^{61}Co preparation is ^{62}Ni(γ, p)^{61}Co, and this gives good results [4, 5]. A 15% chromium-nickel alloy is cubic, non-magnetic at 80 K, and

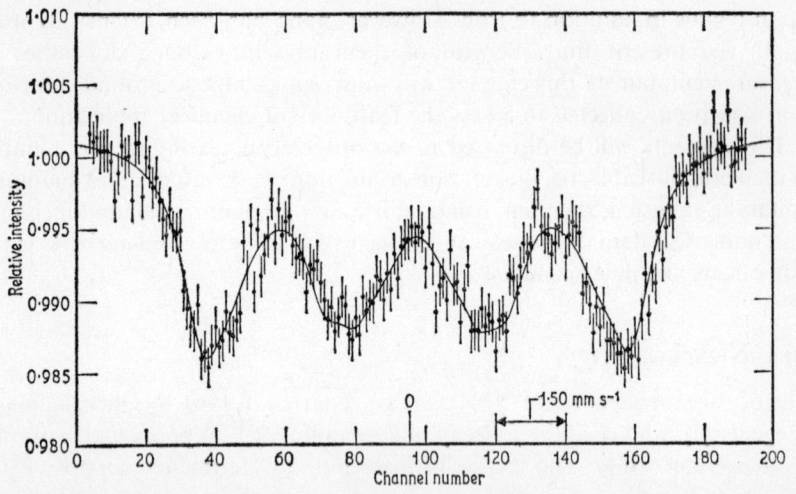

Fig. 16.2 Mössbauer spectrum of ^{61}Ni in a 1·5% Ni–iron alloy absorber at 80 K. The source was 15% Ni–chromium. [Ref. 5, Fig. 5]

proves to be a good matrix for bremsstrahlung irradiation, giving a linewidth after correction for source and absorber thickness effects of 0·97 mm s^{-1} compared to the natural width of 0·78 mm s^{-1}. The recoil-free fraction at 80 K is only 0·10(1) but is adequate.

The alternative decay parent, ^{61}Cu, prepared by a ^{63}Cu(γ, $2n$)^{61}Cu bremsstrahlung irradiation, has a complex decay scheme and its γ-ray energy spectrum is not as simple as that of ^{61}Co [5]. It can also be made by ^{58}Ni(α, p)^{61}Cu or ^{58}Ni(α, n)^{61}Zn(EC)^{61}Cu reactions, and a radiochemical preparation of a ^{61}Cu/copper source matrix has been described in detail [6].

A 1·5% nickel-in-iron alloy gives a partially resolved magnetic splitting at 80 K (Fig. 16.2) [5]. The magnetic moment of the $I_g = \frac{3}{2}$ ground state is −0·7487 n.m., and is considerably greater than the excited-state moment. The level splitting is illustrated in Fig. 16.3 and shows how the lines

[*Refs. on p. 532*]

cluster into four groups. Analysis of the spectrum in Fig. 16.2 gave $\mu_e/\mu_g = -0.559(12)$ with an internal field of 241 kG. Independent measurements on a 4% nickel–iron alloy at 4·2 K gave $\mu_e/\mu_g = -0.577(25)$ [7],

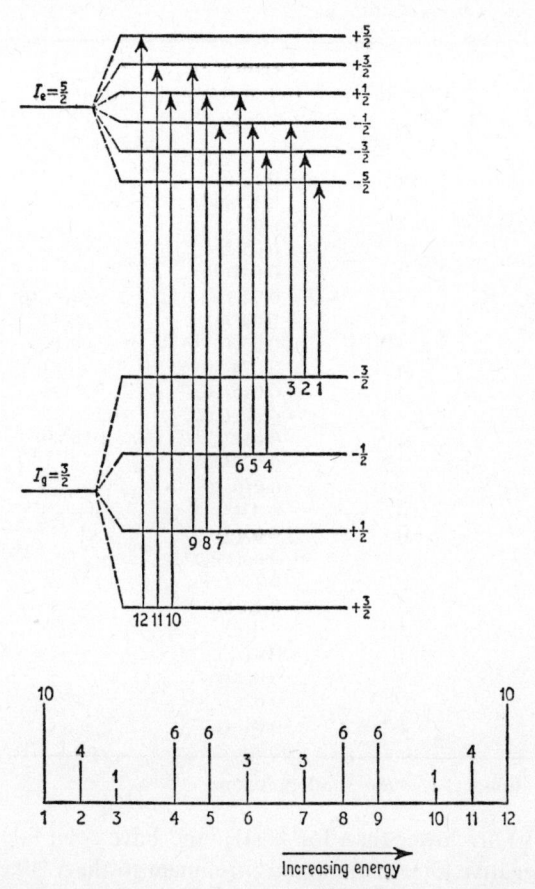

Fig. 16.3 Energy level diagram drawn using $\mu(\frac{5}{2})/\mu(\frac{3}{2}) = -0.637$ to show the magnetic fine structure of Fig. 16.2. The numbers above the lines in the bar diagram represent the relative intensities of the transitions and may be obtained from the coefficients in Appendix 2.

later updated to $-0.637(42)$ after additional experiments including polarisation in an external field [8]. The latter value gives $\mu_e = +0.477(31)$ n.m. and is the most recent value for this parameter.

The much smaller internal field in nickel metal results in only a broad line, which has given $\mu_e/\mu_g = -0.551(38)$ [4]. The field is 97 kG at 4·2 K [7]. NiO also gives an unresolved magnetic splitting with a field of 98 kG at 4·2 K [7] and 96 kG at 80 K [4].

The chemical isomer shifts in [61]Ni are extremely small compared with the natural linewidth (0·78 mm s^{-1}), the range of known values being 0·46 mm s^{-1} (Ni^{II}F$_2$ and [Ni^{IV}Mo$_9$O$_{32}$]$^{6-}$). Available data are given in Table 16.1.

Table 16.1 Chemical isomer shifts in ^{61}Ni

Compound	T/K	δ/(mm s^{-1}) (rel. to 15% Cr/Ni)	e^2qQ_e /(mm s^{-1})	Reference
K$_2$NiF$_4$	80	+0·13(5)	—	5
NiF$_2$	80	+0·126(40)	—	5
NiCl$_2$.6H$_2$O	4·2	+0·07(5)	—	10
NiO	80	+0·05(2)	—	5
(Et$_4$N)$_2$[NiCl$_4$]	4·2	+0·041(20)	—	8
(Et$_4$N)[Ph$_3$PNiBr$_3$]	4·2	+0·04(1)	+1·00(3·20)	10
[Ph$_3$MeAs]$_2$[NiCl$_4$]	4·2	+0·03(1)	−0·88(40)	10
*(diap)$_2$Ni	4·2	+0·03(1)	−2·00(1·2)	10
Ni(en)$_2$Cl$_2$	4·2	+0·024(30)	—	8
Ni(en)$_3$Cl$_2$.6H$_2$O	4·2	+0·015(10)	—	10
Ni(CO)$_4$	4·2	−0·002(10)	—	8
(Et$_4$N)$_2$[NiBr$_4$]	4·2	−0·01(2)	−1·16(60)	10
Ni(PCl$_3$)$_4$	4·2	−0·023(20)	—	8
K$_2$[Ni(CN)$_4$]	4·2	−0·050(20)	—	8
(NH$_4$)$_6$[NiMo$_9$O$_{32}$]	80	−0·33(8)	—	5
Ni	80	+0·02(2)	—	5
CuNi (2%)	80	−0·03(1)	—	5
CuNi (20%)	80	−0·10(3)	—	5
FeNi (1·5%)	80	+0·10(1)	—	5
GdNi$_2$	4·2	−0·01(2)	—	8
HoNi$_2$	4·2	−0·01(2)	—	8
ErNi$_2$	4·2	+0·05(3)	—	8
TmNi$_2$	4·2	+0·02(2)	—	8
YbNi$_2$	4·2	+0·04(4)	—	8

* diap = N,N'-diphenyl-1-amino-3-iminopropene

Shifts for Ni(IV) are lower than for Ni(II), and have been taken to indicate that $\delta R/R$ is negative [5]. An analogous argument to the Walker–Wertheim–Jaccarino calibration of ^{57}Fe gives an approximate value of $\delta R/R = -2·5 \times 10^{-4}$. An independent estimate of -3×10^{-4} comes from the difference of $-0·064$ mm s^{-1} between Ni(PCl$_3$)$_4$ and (Et$_4$N)$_2$NiCl$_4$ [8].

Quadrupole splitting of the $I_g = \frac{3}{2}$ ground state is expected to be small because the quadrupole moment Q_g is only $+0·162(15)$ barn [9]. The presence of an electric field gradient can be expected to give essentially three transitions to the $|\pm\frac{5}{2}\rangle$, $|\pm\frac{3}{2}\rangle$, and $|\pm\frac{1}{2}\rangle$ excited-state sublevels if Q_e is large. A slight asymmetry of the resonance in the [NiMo$_9$O$_{32}$]$^{6-}$ complex has been attributed to a positive value of e^2qQ_e [5]. Small values of e^2qQ_e have also been found in four other nickel(II) compounds as given in Table 16.1 [10], and have been taken to support a negative sign for Q_e.

The boracite Ni$_3$B$_7$O$_{13}$I gives a combined magnetic/quadrupole interaction at 4·2 K with a magnetic field of 210 kG [8]. In this case the value of e^2qQ_g was estimated to be $-0·6$ mm s^{-1} from the analogous ^{57}Fe data, assuming the same electric field gradient and the appropriate quadrupole moments. It proved impossible to fit the experimental spectrum, although Q_e/Q_g was considered to be negative, and there seems to be a possibility of more than one site being present in the ordered phase.

The magnetic field at ^{61}Ni at 4·2 K in iron–nickel alloys falls smoothly from 235 kG at 4 at. % Ni to 77 kG at 100 at. % Ni [8, 11]. The main contribution to the field is from conduction-electron polarisation. The rare-earth/nickel alloys (see table) do not show a significant field at 4·2 K [8].

Some lattice-dynamical calculations for nickel metal have been made [12].

16.2 Zinc (^{67}Zn)

The 93·26-keV excited state of ^{67}Zn has a lifetime of 9400 ns. This gives an unusually narrow Mössbauer linewidth of $\Gamma_r = 3·12 \times 10^{-4}$ mm s^{-1}. The potentially high precision which this offers promised spectacular results in measurements of, for example, the gravitational red-shift. However, the normal level of acoustic vibrations in the laboratory is greater than the natural linewidth, and consequently the first attempts to observe the resonance were unsuccessful [13].

The existence of resonance absorption for this transition was first demonstrated in 1960 using an on–off type of experiment [14, 15]. The decay scheme is shown in Fig. 16.4. A source of 78-hour ^{67}Ga made by the ^{66}Zn$(d, n)^{67}$Ga reaction in a matrix of ZnO, and subsequently annealed was rigidly clamped to a ZnO absorber. Both were maintained at 2 K, but with the absorber in a

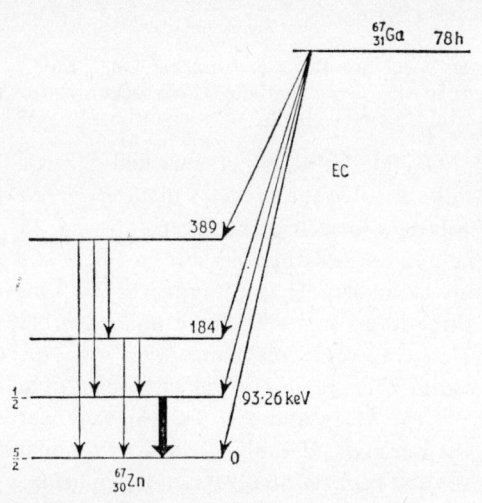

Fig. 16.4 Decay scheme of ^{67}Zn.

magnetic field varying from zero to 700 G produced by a small solenoid. The hyperfine splitting of the absorber caused a variation in the transmission of the 93-keV γ-rays as the magnetic field increased, and indicated a resonant absorption of about 0·3% maximum. Very similar results were obtained independently for zinc metal [16].

An attempt was made to produce conventional velocity-scan spectra by using a piezoelectric quartz oscillator rigidly fastened between source and absorber to provide the relative motion [17]. However, the results obtained for ZnO were not consistent.

Recent experiments have been more successful [18]. Once again the source

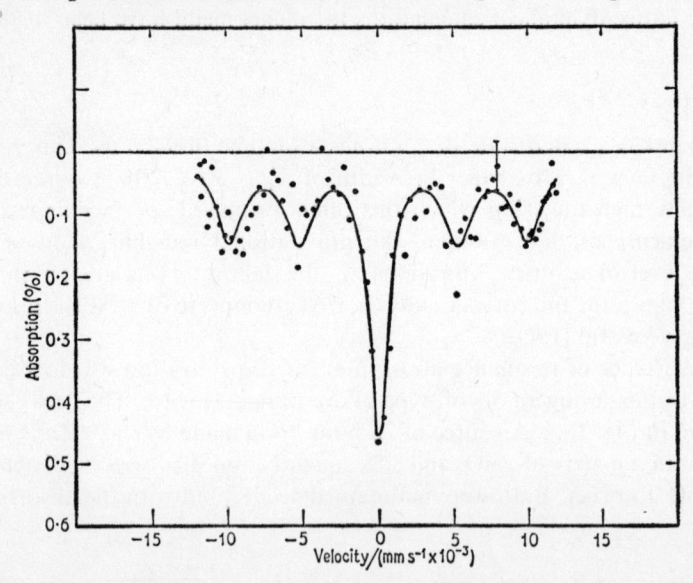

Fig. 16.5 Mössbauer spectrum at 4·2 K from a ^{67}Ga/^{66}ZnO source and a ^{67}ZnO absorber. Only five of the seven predicted lines occur within the velocity range scanned. [Ref. 18, Fig. 1]

was ^{67}Ga in ^{66}ZnO and the absorber enriched ^{67}ZnO. The piezoelectric drive fastened rigidly between the two was made from ten quartz rings so as to increase the total piezoelectric displacement, and the whole unit was cooled to liquid helium temperature. As seen in Fig. 16.5 hyperfine structure was recorded from a quadrupole splitting in the ZnO matrices. The $I_g = \frac{5}{2}$ state splits into three levels in both source and absorber, while the $I_e = \frac{1}{2}$ state is unsplit. Therefore seven resonance lines are expected of which five were observed within the maximum velocity scan obtainable. The parameters $e^2qQ_g = 2·47(3)$ MHz and $\eta = 0·23(6)$ were derived. The line intensities and widths were considerably affected by the manner in which both source and absorber had been thermally treated. Application of pressure to the source matrix caused a small reduction in the quadrupole coupling constant.

[*Refs. on p. 532*]

16.3 Technetium (^{99}Tc)

There has been only one series of measurements to date on the 140·5-keV ^{99}Tc resonance, and these were published in 1968 [19]. The Mössbauer level is populated by a complex β-decay from ^{99}Mo, the principal details of which are illustrated in Fig. 16.6. The experiments were made using a metallic

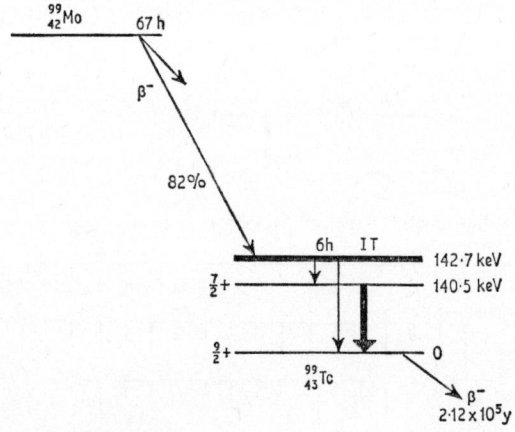

Fig. 16.6 Decay scheme of ^{99}Tc.

molybdenum source and a metallic technetium absorber. Extrapolation of the linewidth to zero absorber thickness led to an estimate for the excited-state lifetime of $t_{\frac{1}{2}} = 0\cdot192(10)$ ns. No information is available on hyperfine interactions other than that the electric field gradient in hexagonal technetium metal was estimated to contribute only 2% broadening to the observed line.

16.4 Ruthenium (^{99}Ru)

The 90-keV resonance in ^{99}Ru was first reported in 1963, by Kistner, Monaro, and Segnan [20]. The first excited level is populated by a complicated EC decay of 16-day ^{99}Rh (see Fig. 16.7). The latter is prepared by a ^{99}Ru(p, n)^{99}Rh cyclotron irradiation [20, 21], or by the alternative ^{99}Ru(d, $2n$)^{99}Rh reaction [22], both of which are expensive in relation to the usable lifetime of the source. Matrices of natural or enriched ruthenium metal can be irradiated and used directly, although a process for chemical separation of the ^{99}Rh activity followed by re-incorporation into ruthenium metal has been described [23].

Ruthenium metal has a hexagonal symmetry, but the source/absorber combination gives a linewidth of about 0·24 mm s^{-1}, which is not substantially greater than the natural width of 0·15 mm s^{-1} [21].

[*Refs. on p. 532*]

500 | OTHER TRANSITION-METAL ELEMENTS

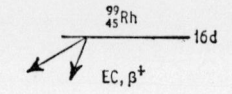

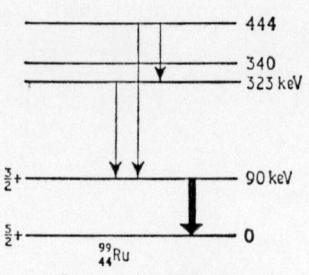

Fig. 16.7 Decay scheme of ^{99}Ru (simplified).

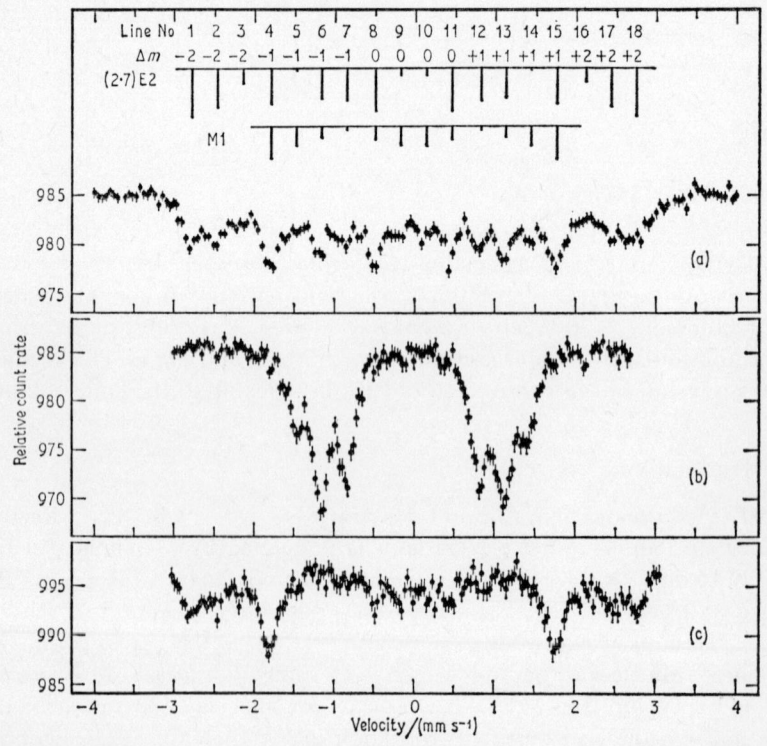

Fig. 16.8 ^{99}Ru magnetic splitting in an absorber of 2·3 at. % ^{99}Ru in metallic iron at 4·2 K. (a) unpolarised; (b) absorber magnetised parallel, (c) magnetised perpendicular to the gamma-ray axis. The line positions and intensities for the E2 and M1 components of the unpolarised spectrum are shown separately as a bar diagram. [Ref. 21, Fig. 2]

[*Refs. on p. 532*]

The Laves phase GdRu$_2$, although ferromagnetic at 84 K, shows no significant magnetic broadening at 4·2 K, but the alloy Ru$_{0.3}$Cu$_{0.7}$ gives a broad resonance with a width of ~0·96 mm s^{-1} [21]. Significant magnetic splitting is found in iron metal doped with 2·3 at. % ^{99}Ru [21]. The unpolarised spectrum for such a foil at 4·2 K is shown in Fig. 16.8a, together with those for it magnetised parallel (8b) and perpendicular (8c) to the γ-ray direction. The 18 lines observed confirm the assignment of $I = \frac{3}{2}$ to the excited state. The intensity data showed that there is considerable E2/M1 admixture in this transition. The value of E2/M1 = δ^2 is 2·7(6), and the appropriate coefficients for E2 and M1 radiation are given in Appendix 2. The line positions and relative intensities are indicated for the unpolarised foil in Fig. 16.8a. The derived ratio $\mu_e/\mu_g = 0.456(10)$ taken in conjunction with the known value for $\mu_g = -0.63(15)$ n.m. gives $\mu_e = -0.29(7)$ n.m. The magnetic field at the ^{99}Ru nucleus is 500(120) kG at 4·2 K.

The use of polarised magnetic spectra from ^{99}Ru in iron metal for studies on time-reversal invariance, although extremely interesting, is beyond our present scope [21, 24, 25].

The early work by Kistner showed partially resolved quadrupole splittings and large chemical isomer shifts in RuO$_2$ and Ru(π-C$_5$H$_5$)$_2$ [21]. More

Table 16.2 Mössbauer parameters for ^{99}Ru spectra at 4·2 K

Compound	Oxidation state	δ/(mm s^{-1}) (rel. to Ru metal)	$\frac{1}{2}e^2qQ_e$ /(mm s^{-1})	Reference
RuO$_4$	VIII ($4d^0$)	+1·06(1)	—	22
KRuO$_4$	VII ($4d^1$)	+0·82(2)	0·37(2)	22
BaRuO$_4$.H$_2$O	VI ($4d^2$)	+0·38(1)	0·44(2)	22
RuO$_2$	IV ($4d^4$)	−0·26(1)	0·50(1)	22
		−0·24(5)	?	21
		−0·22		26
K$_2$[RuCl$_6$]	IV ($4d^4$)	−0·31(1)	0·41(1)	22
		−0·26	0·23	26
[Ru(NH$_3$)$_4$OHCl]Cl.2H$_2$O	III ($4d^5$)	−0·39(1)	—	22
K$_3$[RuCl$_6$]	III ($4d^5$)	−0·35	0·41	26
(Bu$_4$N)$_3$[Ru(SCN)$_6$]	III ($4d^5$)	−0·49(4)	0·53(5)	22
[Ru(NH$_3$)$_6$]Cl$_3$	III ($4d^5$)	−0·49(1)	—	22
[Ru(NH$_3$)$_5$Cl]Cl$_2$	III ($4d^5$)	−0·53(1)	0·38(1)	22
[Ru(bipy)$_3$](ClO$_4$)$_3$	III ($4d^5$)	−0·54(1)	—	22
RuBr$_3$	III ($4d^5$)	−0·75(2)	0·65(3)	22
[Ru(bipy)$_2$ox].4H$_2$O	II ($4d^6$)	−0·63(2)	0·30(2)	22
[Ru(NH$_3$)$_4$(HSO$_3$)$_2$]	II ($4d^6$)	−0·66(1)	—	22
K$_4$[Ru(CN)$_6$].3H$_2$O	II ($4d^6$)	−0·22(1)	—	22
K$_4$[Ru(CN)$_6$]	II ($4d^6$)	−0·22(1)	—	22
	II ($4d^6$)	−0·25	—	26
Ru(CO)$_2$Cl$_2$	II ($4d^6$)	−0·23(3)	—	22
Ru(π-C$_5$H$_5$)$_2$	II ($4d^6$)	−0·75(2)	0·43(2)	22
		−0·76(4)	0·46(4)	21

extensive studies [22, 26] have established a strong similarity to ^{57}Fe in that the chemical isomer shift is dependent on the oxidation state. Typical spectra are given in Fig. 16.9, and the detailed values are in Table 16.2. The scale of

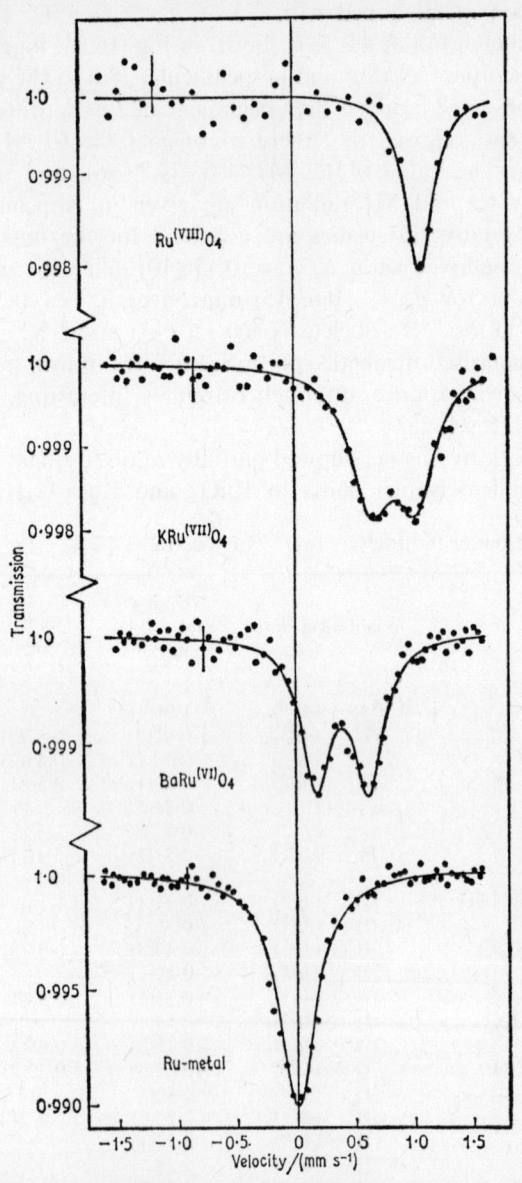

Fig. 16.9 ^{99}Ru spectra in RuO$_4$, KRuO$_4$, BaRuO$_4$, and Ru metal. Note the dependence of the chemical isomer shift on oxidation state and the quadrupole splitting from RuO$_4^-$ and RuO$_4^{2-}$. [Ref. 22, Fig. 1]

[*Refs. on p. 532*]

chemical isomer shifts is illustrated in Fig. 16.10. The shift increases as the oxidation state increases, a reversal to the established trend for ^{57}Fe which implies by analogy that $\delta R/R$ is probably positive in ruthenium.

The quadrupole moment of the excited state is at least a factor of 3 greater than that for the ground state. This results in an apparent doublet spectrum

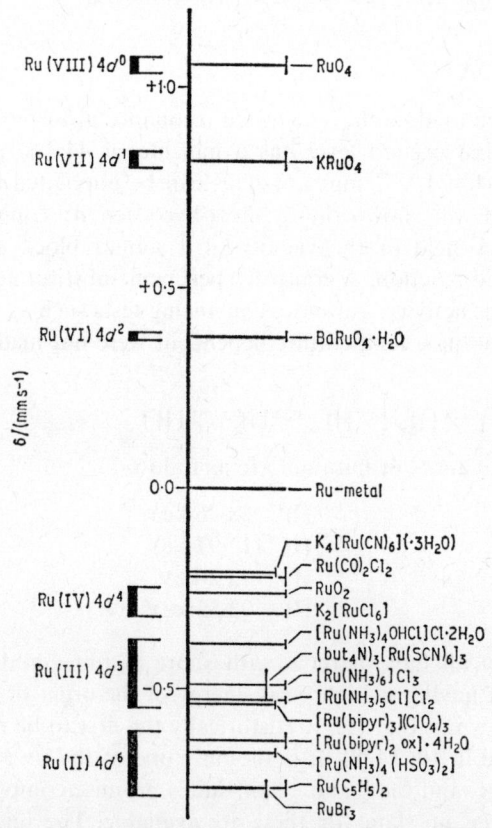

Fig. 16.10 Schematic presentation of the chemical isomer shifts in ^{99}Ru. [Ref. 22, Fig. 4]

(Fig. 16.9) produced by the quadrupole splitting of the $I = \frac{3}{2}$ excited state, each component of the doublet being in fact an unresolved triplet.

As with iron, the ruthenium oxidation states are largely determined by the d-electron configuration. By analogy with K$_4$[Fe(CN)$_6$] the higher shift found in K$_4$[Ru(CN)$_6$] and its trihydrate than in other Ru(II) compounds results from the back-donation to the ligands. The absence of splitting in RuO$_4$ (spherical d^0) contrasts with that found in RuO$_4^-$(d^1) and RuO$_4^{2-}$(d^2) which have a distorted tetrahedral environment and partially filled electronic levels. Splittings in [RuCl$_6$]$^{2-}$(d^4) and [Ru(SCN)$_6$]$^{3-}$(d^5) would be predicted

[*Refs. on p. 532*]

by analogy with similar non-spherical t_{2g} configurations in iron, as would the lack of splitting in the $[Ru(CN)_6]^{4-}(d^6)$ ion.

Detailed correlations on the scale of ^{57}Fe will not be possible until more data are available, but clearly similar trends will be found. Unavoidable difficulties are the very weak resonance effects even at helium temperature and the short lifetime of the cyclotron-irradiated source.

16.5 Silver (^{107}Ag)

A claim has been made to have detected resonance absorption in ^{107}Ag [27]. The 93·1-keV first excited level has a half-life of 44·3 s, giving a natural linewidth of $6·64 \times 10^{-11}$ mm s^{-1}. This can be populated by EC-β^+-decay in ^{107}Cd, and it was shown that a short-lived activity could be transferred to a silver block held in the vicinity of a similar block activated by the ^{107}Ag$(p, n)^{107}$Cd reaction. A control experiment substituting a copper block did not show this activity, but more convincing tests such as to show that the resonant absorption is temperature dependent were not made.

16.6 Hafnium (^{176}Hf, ^{177}Hf, ^{178}Hf, ^{180}Hf)

The known resonances of hafnium are as follows:

^{176}Hf: 88·36-keV
^{177}Hf: 112·97-keV
^{178}Hf: 93·2-keV
^{180}Hf: 93·33-keV

Unfortunately all have parents with short lifetimes and also suffer the consequences of having a high γ-ray energy of the order of 100 keV. Decay schemes are shown in Fig. 16.11. Historically the first to be reported was the ^{177}Hf resonance in 1963 [28]. It is the only one of the four not involving a 2+ → 0+ decay, and the $\frac{9}{2}$ and $\frac{7}{2}$ spin-states promise complicated hyperfine spectra. However, no data for these are available. The initial experiments were made using sources of 6·7-d ^{177}Lu prepared in Lu$_2$O$_3$ or Lu metal matrices by neutron irradiation. Resonances were obtained for Hf metal and HfO$_2$ absorbers of various thicknesses, leading to an estimate for the excited-state lifetime of 0·43(4) ns, in good agreement with other values from several delayed coincidence measurements which average about 0·5 ns.

The ^{176}Hf and ^{180}Hf resonances were described in 1966 [29], followed by ^{178}Hf in 1968 [30]. The parent for ^{176}Hf is 3·7 h ^{176}Lu made by neutron irradiation of ^{175}Lu. The source matrix ^{176}Lu$_2$O$_3$ is unsatisfactory because it contains two distinct sites [29], both of which show a quadrupole splitting. Better results have since been obtained using ^{176}LuRh$_2$ [31]. Good sources for ^{178}Hf can be made by a $(p, 4n)$ reaction on a tantalum foil to give

[Refs. on p. 532]

^{178}W [31], or by incorporating ^{178}W into molybdenum metal [30]. A ^{181}Ta$(d, 5n)^{178}$W preparation has also been used [33]. The ^{180}Hf resonance has been studied using a quadrupole split source of ^{180m}HfO$_2$ [29, 31].

The available data contain several conflicts: for example in one case an axially symmetric electric field gradient has been assumed in HfO$_2$ [30],

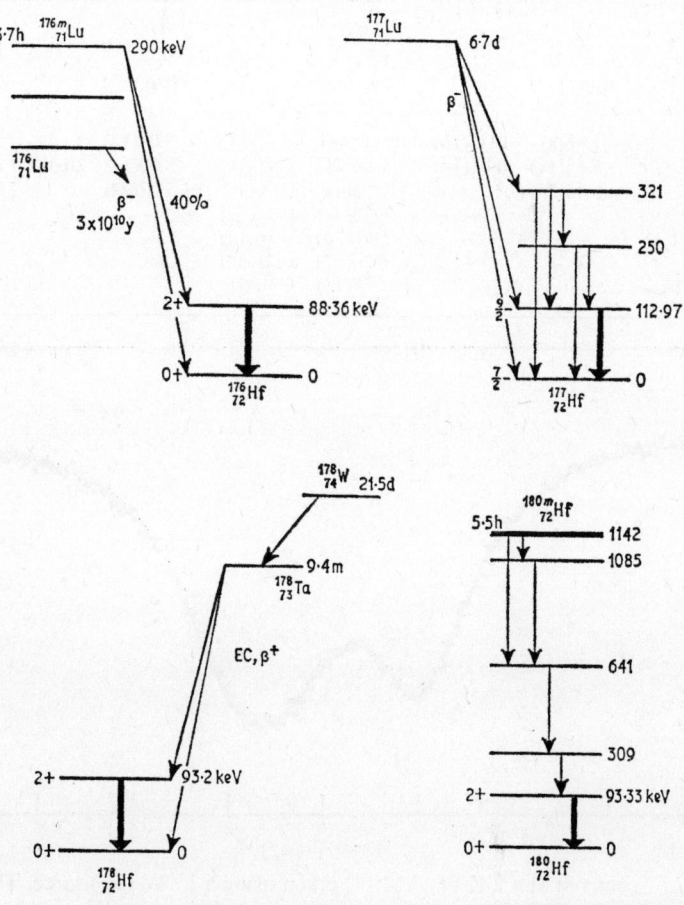

Fig. 16.11 Decay schemes of ^{176}Hf, ^{177}Hf, ^{178}Hf, and ^{180}Hf.

whereas other workers have found this not to be the case [31]. In Table 16.3 are collected numerical values which are probably the best available. A typical spectrum of ^{178}HfO$_2$ is shown in Fig. 16.12. Some of these figures were used to derive the following ratios of the 2+ state quadrupole moments [31]: $Q^{176}/Q^{178} = 1·040(8)$, $Q^{178}/Q^{180} = 1·014(13)$, $Q^{176}/Q^{180} = 1·053(15)$.

A large value of η and a Karyagin effect (anisotropy of f) have been found in (NH$_4$)$_2$HfF$_6$ [30]. The alloys HfCo$_2$, Zr$_{0·5}$Hf$_{0·5}$Fe$_2$, and Co$_2$HfAl give small unresolved magnetic splittings.

[*Refs. on p. 532*]

506 | OTHER TRANSITION-METAL ELEMENTS

The chemical isomer shifts between HfO_2 and Hf metal are less than 0·3 mm s^{-1} for ^{176,178,180}Hf and have not been measured accurately, but approximate values for $\delta R/R$ of 0·28, 0·12, and 0·14 ($\times 10^{-4}$) respectively

Table 16.3 Quadrupole splitting data for hafnium at 4·2 K

Compound	^{176}Hf e^2qQ /(mm s^{-1})	η	^{178}Hf e^2qQ /(mm s^{-1})	η	^{180}Hf e^2qQ /(mm s^{-1})	η	Reference
HfB$_2$	−7·96(8)	0·37(16)	−7·18(3)	0·42(5)	−7·01(12)	(0·42)	31
HfO$_2$	−8·65(11)	0·44(14)	−8·00(3)	0·48(4)	−8·00(16)	(0·48)	31
Hf	−6·40(24)	0	−5·94(4)	0	−5·77(20)	0	31
Hf(NO$_3$)$_4$	—	—	8·18(31)	0·57(15)	—	—	31
HfOCl$_2$.8H$_2$O	—	—	5·96(20)	0·86(12)	—	—	31
HfCl$_4$	—	—	6·26(57)	0·71(36)	—	—	31
(NH$_4$)$_2$HfF$_6$	—	—	+8·50(15)	0·90(5)	—	—	21

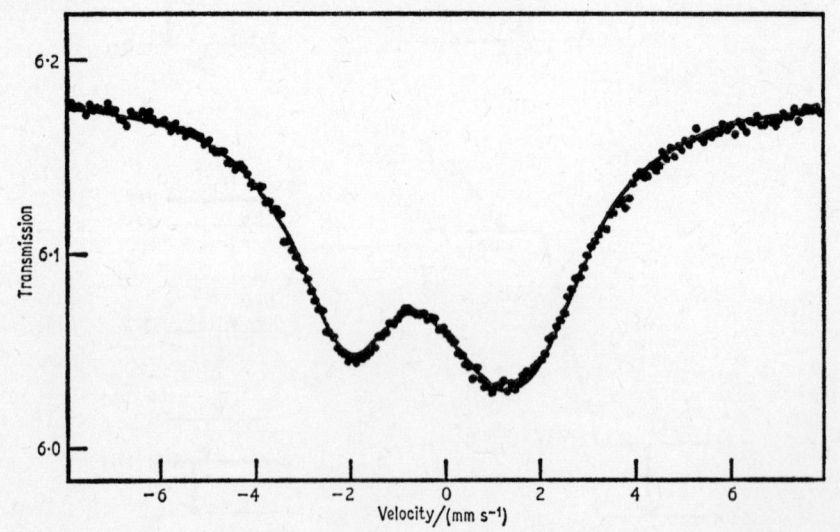

Fig. 16.12 Spectrum at 4·2 K of ^{178}HfO$_2$ taken using a ^{178}W/Ta source. The solid line corresponds to $e^2qQ = -8\cdot00$ mm s^{-1}, $\eta = 0\cdot475$, and a linewidth of 2·78 mm s^{-1}. [Ref. 31, Fig. 2]

have been estimated [31]. Single-crystal studies of hafnium metal agree with the powder data [32].

A 1 at. % alloy of ^{178}W in iron (source) with a HfN single-line absorber shows a partially resolved five-line magnetic spectrum [33]. The known excited-state magnetic moment for ^{178}Hf of $\mu_e = 0\cdot71$ n.m. leads to values for the internal magnetic field of 606 kG at 4·2 K and 547 kG at 78 K. By contrast, a 1 at. % alloy of ^{180m}Hf in iron for which $\mu_e = 0\cdot74$ n.m. gives a

[Refs. on p. 532]

field of only 334 kG at 4·2 K. The reason for this anomaly is not yet clear, but it seems likely that the ^{180}Hf is clustering or forming HfFe$_2$ so that the chemical environment is not identical in the two sources.

Coulombic excitation of ^{178}Hf and ^{180}Hf by 6-MeV α-particles in targets of HfC and HfN at 78 K has shown considerable line broadening which is not a feature of these materials used as absorbers [34]. It may be presumed that the effect is a result of radiation damage associated with the recoil of the excited nuclei. The latter come to rest at lattice sites with one or more lattice vacancies nearby which generate an electric field gradient at the nucleus.

16.7 Tantalum (^{181}Ta)

Resonances at 6·25-keV and 136·25-keV are known in ^{181}Ta. The 6·25-keV resonance of ^{181}Ta suffers from the same disadvantage as ^{67}Zn, namely a very narrow natural linewidth which in this instance is $6·5 \times 10^{-3}$ mm s^{-1}. The first experiments in 1964 used a piezoelectric drive to produce the small velocities required [35]; the absorbers were TaC and KTaO$_3$ and the sources were ^{181}W in tungsten metal or Ba$_2$CaWO$_6$ ($t_{\frac{1}{2}} = 140$ d, see Fig. 16.13). All these matrices have cubic symmetry. A small effect was found only

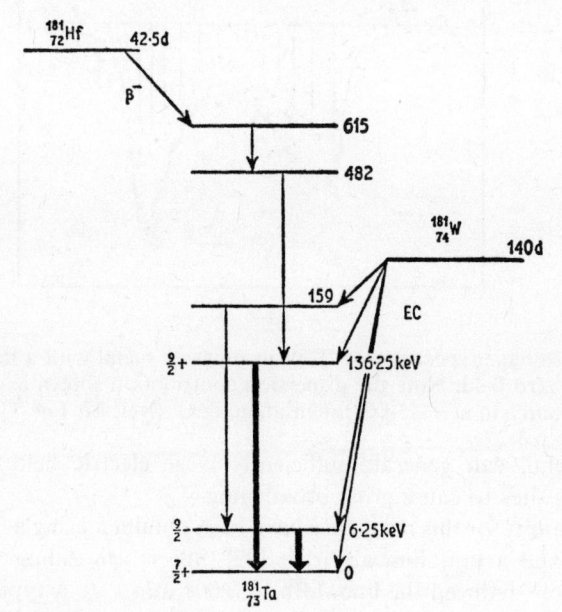

Fig. 16.13 Decay scheme of ^{181}Ta.

for the combination of an annealed tungsten source and TaC absorber, the width of the line being 18 times greater than expected.

[Refs. on p. 532]

508 | OTHER TRANSITION-METAL ELEMENTS

Independent measurements using ^{181}W in tantalum or tungsten as source, and tantalum metal as absorber also showed gross broadening [36]. Attempts to use the ^{181}Hf parent in an HfC matrix gave no effect. The large quadrupole moment of ^{181}Ta (+3·9 barn in the ground state) causes any electric field gradient to be seen in the spectrum as a splitting much greater than the natural width. Any chemical isomer shift is also large. In consequence, the presence of interstitial impurities, such as oxygen, nitrogen, or carbon in

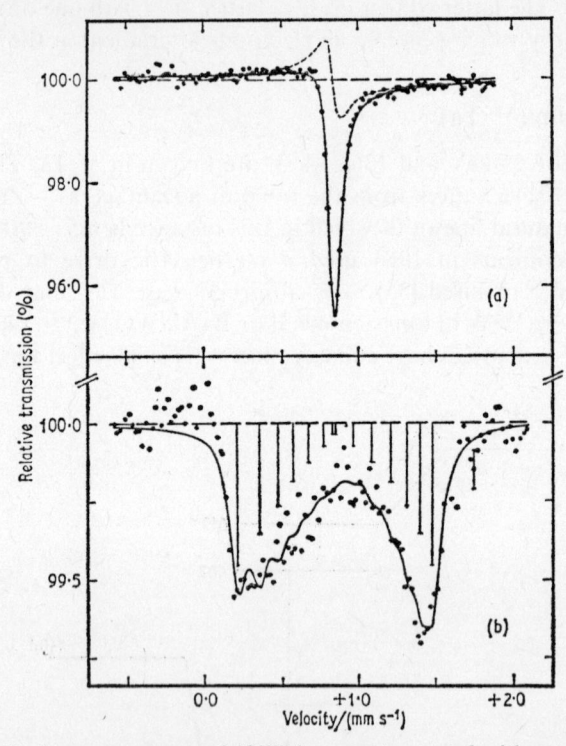

Fig. 16.14 Mössbauer spectrum of ^{181}W in tungsten metal with a tantalum metal absorber: (a) zero field. Note the dispersion contribution shown as a dotted line. (b) with the source in a 1·445-kG longitudinal field. [Ref. 37, Fig. 1]

tantalum metal, can generate sufficiently large electric field gradients at neighbouring sites to cause gross broadening.

The best results for this resonance have been obtained using a ^{181}W activity in tungsten and a tantalum absorber [37]. Special annealing treatment of both metal foils reduced the linewidth to 0·091 mm s^{-1}. A typical spectrum is shown in Fig. 16.14a. A characteristic feature of the resonance (to be discussed more fully below) is the superposition of a dispersion component as shown by the dotted line. The chemical isomer shift between tungsten sources and tantalum absorbers varied between 0·83 and 0·94 mm s^{-1}, imply-

[Refs. on p. 532]

ing a large value of $\delta R/R$. ^{181}W/Ta sources and tantalum absorbers gave shifts ranging from 0·04 to 0·15 mm s^{-1}, and there appears to be a strong correlation between the shift and the impurity content of the metals.

A ^{181}W/W source in a longitudinal field of 1·445 kG gives the magnetic spectrum shown in Fig. 16.14b. The $\frac{9}{2} \to \frac{7}{2}$ E1 decay gives 8 pairs of hyperfine lines in the ratios 36 : 28 : 21 : 15 : 10 : 6 : 3 : 1. The asymmetry of the spectrum confirms the suspected quadrupole effect in the source, and computer analysis gave the excited-state magnetic moment as $\mu_e = +5·20(15)$ n.m. (using $\mu_g = +2·35(1)$ n.m.) and the excited-state quadrupole moment as $Q_e = +2·9(1·2)$ barn (using $Q_g = 3·9(4)$ barn). μ_e was also estimated from a zero relative velocity experiment in which the transmission was measured as a function of applied magnetic field, giving a finally adopted value for μ_e of +5·14(15) n.m.

The dispersion term has since been shown to arise from interference between nuclear absorption followed by internal conversion and photoelectric absorption [38, 39]. Such an interference is significant because the nuclear transition has E1 multipolarity rather than M1 or E2. The only other known examples are the E1 decays in ^{161}Dy and ^{153}Eu. The dispersion term does not appear in the emission spectrum of the source, but is a property of the absorption.

The 136·25-keV transition has only been briefly investigated using a ^{181}W/tungsten source and a tantalum absorber at 4·2 K [19]. The linewidth of the unsplit resonance extrapolated to zero absorber thickness gave an estimate for the excited-state lifetime of 0·035(2) ns.

16.8 Tungsten (^{182}W, ^{183}W, ^{184}W, ^{186}W)

There are five known resonances for tungsten:

^{182}W: 100·10-keV
^{183}W: 46·48-keV and 99·08-keV
^{184}W: 111·2-keV
^{186}W: 122·6-keV

and the decay schemes for these are shown in Fig. 16.15. The 100-keV resonance in ^{182}W was observed in 1959 in the first confirmation of Mössbauer's original work on an element other than iridium [40]. Independent work was reported in 1960 involving an 'on–off' experiment with a tantalum source and a tungsten absorber at 78 K and room temperature [41]. A conversion-electron detection system was used in 1961 to enhance the very weak effect from the 100-keV transition [42]. The cubic lattices of the tantalum source and tungsten absorber resulted in a single line from which the excited-state lifetime was estimated to be $t_{\frac{1}{2}} = 1·34(13)$ ns, in good agreement with 1·37(1) ns from electronic measurements. γ–γ coincidence counting methods have also been used [43].

[Refs. on p. 532]

510 | OTHER TRANSITION-METAL ELEMENTS

Neutron irradiation of ^{181}Ta can be used to generate two Mössbauer parents by the ^{181}Ta$(n, \gamma)^{182}$Ta$(n,\gamma)^{183}$Ta reactions, thereby populating the 100-keV level of ^{182}W and the 46·5- and 99-keV levels of ^{183}W. All three resonances were found in tungsten metal in 1962 [44]. The unknown $t_{\frac{1}{2}}$ values for the ^{183}W resonances were estimated from the linewidths as

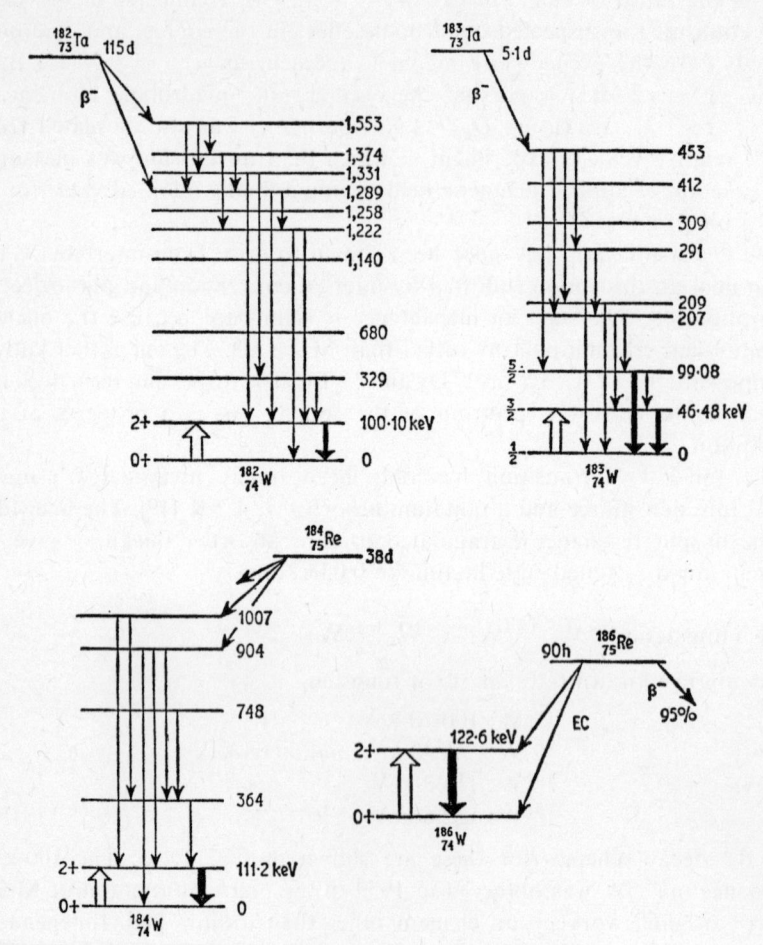

Fig. 16.15 Decay schemes for ^{182}W, ^{183}W, ^{184}W, and ^{186}W.

0·15 ns and 0·57 ns respectively. Hyperfine effects were first recorded in 1965 with the observation of a ^{182}W quadrupole splitting in WO$_3$ [45].

In an extended series of experiments the ^{182}W, ^{183}W (99 keV), ^{184}W, and ^{186}W resonances have been obtained at 4·2 K for WO$_3$, WS$_2$, and a tungsten–iron alloy [46–48]. The short lifetime of ^{183}Ta (5 days) and the longer lifetime of ^{182}Ta (115 days) allow the respective 99-keV and 100-keV

[Refs. on p. 532]

tungsten resonances to be studied using the same ^{181}Ta(n, γ) source preparation. The ^{183}W resonance can be measured immediately after irradiation, while the ^{182}W resonance is studied when the ^{183}Ta activity has effectively died. Enrichment of absorbers is advisable for ^{183}W, the excited-state lifetimes of which were determined to be $t_\frac{1}{2}$ (46·5 keV) = 0·184(5) ns, $t_\frac{1}{2}$ (99 keV) $\geqslant$ 0·7 ns [46].

The parents for the ^{184}W and ^{186}W resonances are made by the ^{185}Re$(p, pn)^{184}$Re [47] and ^{185}Re$(n, \gamma)^{186}$Re [48] reactions. The rhenium source matrix has a hexagonal lattice, and consequently a small quadrupole interaction. The values of e^2qQ_e (c/E_γ) in these sources are 3·0(5) mm s^{-1} (^{184}W) and 2·6(5) mm s^{-1} (^{186}W), and this has to be taken into account in analysis of the spectra.

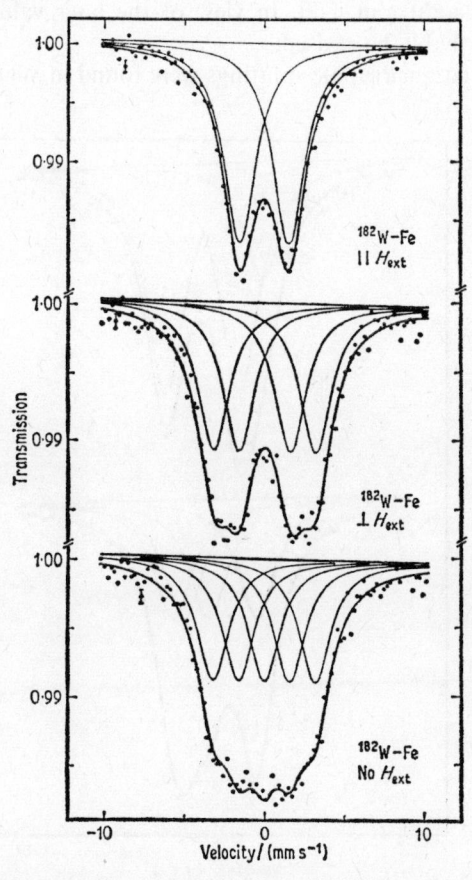

Fig. 16.16 Mössbauer spectra of ^{182}W in a 1·8% W-in-iron alloy. The unpolarised absorber shows five lines of equal intensity from the 0+ to 2+ transition as shown in the bottom spectrum. Polarisation in an external field gives the simpler spectra shown above and facilitates interpretation. [Ref. 46, Fig. 2]

512 | OTHER TRANSITION-METAL ELEMENTS

A 1·8 at. % W/iron metal absorber polarised in applied magnetic fields shows a partially resolved magnetic hyperfine spectrum (Fig. 16.16). N.m.r. data have given a value for the internal field of 630(13) kG, and this was used to determine the excited-state magnetic moment as $^{182}\mu_e = 0.532(18)$ n.m. [48]. Smaller magnetic fields were found in cobalt- and nickel-containing tungsten. More recent experiments have used applied fields of up to 120 kG on ^{182}W in iron metal at 4·2 K [49]. Samples with 0·5, 1·5, 3·3, and 5·0 at. % tungsten gave essentially similar spectra, and analysis of the data gave a field of -708 kG and $^{182}\mu_e = 0.44(4)$ n.m.

The ^{183}W (99 keV), ^{184}W, and ^{186}W resonances give lower resolution, but equivalent spectra were analysed with the help of the ^{182}W data ($^{182}\mu_e = 0.532$) to give $^{183}\mu$ (99) $= 0.930(42)$ n.m., $^{184}\mu_e = 0.590(20)$ n.m., and $^{186}\mu_e = 0.624(22)$ n.m. [48]. In view of the later value for $^{182}\mu_e$, the figures should probably be revised.

Partially resolved quadrupole splittings were found in WO$_3$ and WS$_2$ (Fig.

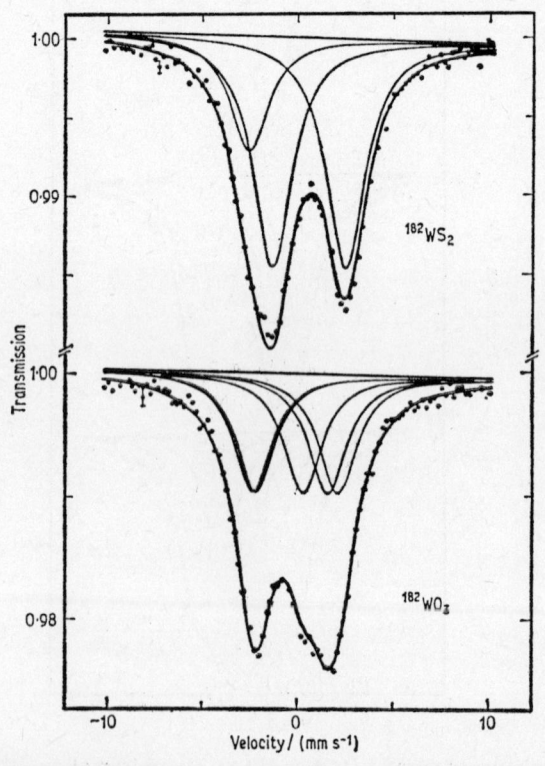

Fig. 16.17 Quadrupole splittings in ^{182}WS$_2$ and ^{182}WO$_3$. The increased number of lines in WO$_3$ is due to the presence of a finite asymmetry parameter, $\eta = 0.63$. [Ref. 46, Fig. 5]

[*Refs.* on p. 532]

16.17) with the coupling constants given in Table 16.4. From these can be derived the ratios of the quadrupole moments which are [48]

$$^{182}Q : {}^{183}Q \ (99) : {}^{184}Q : {}^{186}Q = 1 : 0 \cdot 94(4) : 0 \cdot 94(2) : 0 \cdot 88(2).$$

Table 16.4 Quadrupole coupling constants for tungsten at 4.2 K

	$e^2qQ(cE_\gamma^{-1})/$(mm s^{-1})			Reference
	WS$_2$	WO$_3$	WSe$_2$	
^{182}W (100)	10·00(7)	−8·35(8)	9·19(10)	48, 55
^{183}W (99)	9·55(38)	−7·92(37)	—	46
^{184}W (111)	8·46(10)	−7·37(14)	—	48
^{186}W (122)	7·21(11)	−5·93(27)	6·74(23)	48, 55
		$\eta = 0 \cdot 63(2)$		

The magnetic moments of ^{182}W and ^{186}W have also been measured by using a scattering technique in which the Mössbauer angular scattering distribution as shown by the γ-ray energy spectrum of the scattered radiation is measured as a function of a magnetic field which can be reversed in direction [50]. The values obtained were $^{182}\mu_e = 0 \cdot 466(54)$ n.m. and $^{186}\mu_e = 0 \cdot 68(6)$ n.m.

Chemical isomer shifts for the tungsten isotopes are generally within experimental error of the observed line position. The only unambiguous example is for ^{182}W where a shift of $-0 \cdot 25(5)$ mm s^{-1} was found between a tungsten absorber and a WCl$_6$ absorber [51]. An approximate estimate of $\delta R/R = 0 \cdot 65 \times 10^{-4}$ was derived. Equivalent shift values for ^{184}W and ^{186}W were $0 \cdot 17(46)$ and $-0 \cdot 10(22)$ mm s^{-1}, well within experimental error.

Little is known about hyperfine interactions of the ^{183}W 46·48-keV resonance because of the large linewidth associated with this transition. Early single-crystal data for WO$_3$ suggested the presence of a quadrupole splitting, giving $^{183}Q(44 \cdot 5)/^{182}Q_e = 0 \cdot 88(14)$ [52]. This resonance has also been used to show that the recoil-free fraction and hence the Debye temperature of microcrystalline tungsten metal are less than in large crystals [53].

Coulomb excitation of the 44·5-keV ^{183}W level has been achieved using a 3·3-MeV proton beam and a tungsten target foil at 77 K [54]. The lifetime obtained was $t_{\frac{1}{2}} = 0 \cdot 194(10)$ ns. The target recoilless fraction was lower than expected, either because of radiation damage or localised heating effects. The ^{182}W, ^{184}W, and ^{186}W levels have been simultaneously Coulomb-excited by 8-MeV He^{2+} ions on a tungsten target [55]. The values for $e^2qQ(c/E_\gamma)$ obtained from single crystals of WS$_2$ agree within experimental error with those from more conventional experiments. ^{182}Ta and ^{186}Re sources were also used, and parameters for WSe$_2$ are included in Table 16.4. The quadrupole moment ratios were derived as

$$^{182}\text{W} : {}^{184}\text{W} : {}^{186}\text{W} = 1 : 0 \cdot 930(16) : 0 \cdot 908(24)$$

in excellent agreement with the previous values.

[Refs. on p. 532]

514 | OTHER TRANSITION-METAL ELEMENTS

16.9 Rhenium (^{187}Re)

The 134·24-keV resonance of ^{187}Re was reported in 1960 [56]. In addition to the inherent disadvantages of a high γ-ray energy, the excited-state lifetime is also shorter than usual, although this parameter was unknown until determined from the Mössbauer resonance. A tungsten source (see simplified decay scheme in Fig. 16.18) and rhenium absorber were used at 20 K, and

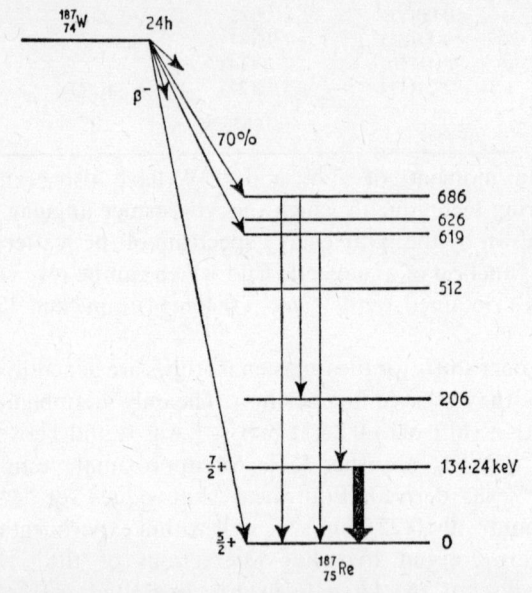

Fig. 16.18 Decay scheme of ^{187}Re.

the experimental linewidth after thickness correction gave $t_{\frac{1}{2}} = 0.0104(14)$ ns. This compares well with the later value from other methods of $t_{\frac{1}{2}} = 0.0101(13)$ ns. ($\Gamma_r = 202$ mm s^{-1}) [57]. No other data are available.

16.10 Osmium (^{186}Os, ^{188}Os, ^{189}Os)

Of the five known resonances in osmium:

^{186}Os: 137·16-keV
^{188}Os: 155·0-keV
^{189}Os: 36·22-keV, 69·59-keV, and 95·3-keV

(see Fig. 16.19), two, ^{186}Os and ^{188}Os, have very high energies and are populated by short-lived parent isotopes. The only data for these have therefore come from scattering experiments [50, 58, 59]. The ^{186}Re and ^{188}Re

[*Refs.* on p. 532]

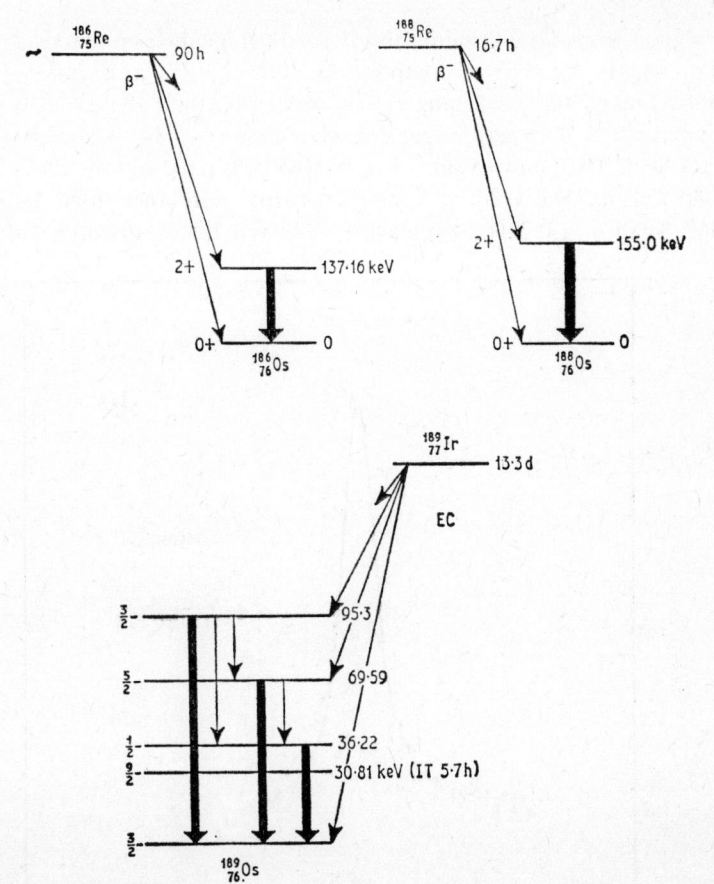

Fig. 16.19 Decay schemes of ^{186}Os, ^{188}Os, and ^{189}Os.

parents can be activated simultaneously by neutron irradiation of natural rhenium. An osmium metal scatterer was used, and a typical spectrum for ^{186}Os is shown in Fig. 16.20 [58]. The ^{186}Os recoil-free fraction is only 0·043 at 26 K, and that of ^{188}Os is even lower at 0·021. Both values decrease rapidly with rising temperature. Line broadening above the natural width may be attributed to an unresolved quadrupole interaction. The type of specialised scattering experiment used to determine the excited-state magnetic moments of ^{182}W and ^{186}W also gave the values $\mu_e(^{186}\text{Os}) = 0\cdot64(3)$ n.m. and $\mu_e(^{188}\text{Os}) = 0\cdot620(54)$ n.m. [50]. Similarly, angular correlation of the scattered radiation was used to study the quadrupole splitting in osmium metal [59].

The 36·22-keV ^{189}Os resonance was described in 1969 [60]. A source of ^{189}Ir chemically separated from rhenium and incorporated into cubic iridium metal was used. The parent isotope of ^{189}Ir is made by the

$^{187}(\alpha, 2n)^{189}$Ir reaction. Osmium metal gave a single-line resonance at 4·2 K (see Fig. 16.20). An alloy of composition ^{189}Ir$_{0.01}$Fe$_{0.99}$ as a source showed unresolved magnetic broadening. As shown in Fig. 16.21 the $\Delta m = 0$ transitions of the $\frac{1}{2} \to \frac{3}{2}$ hyperfine pattern were eliminated by polarisation in a 29·7-kG field. The known value of $\mu_g = 0.6565(3)$ n.m. and the n.m.r. value of 1130 kG for the field at osmium in iron were then used to derive μ_e (36·2 keV) = +0·226(29) n.m. The observed linewidths after thickness

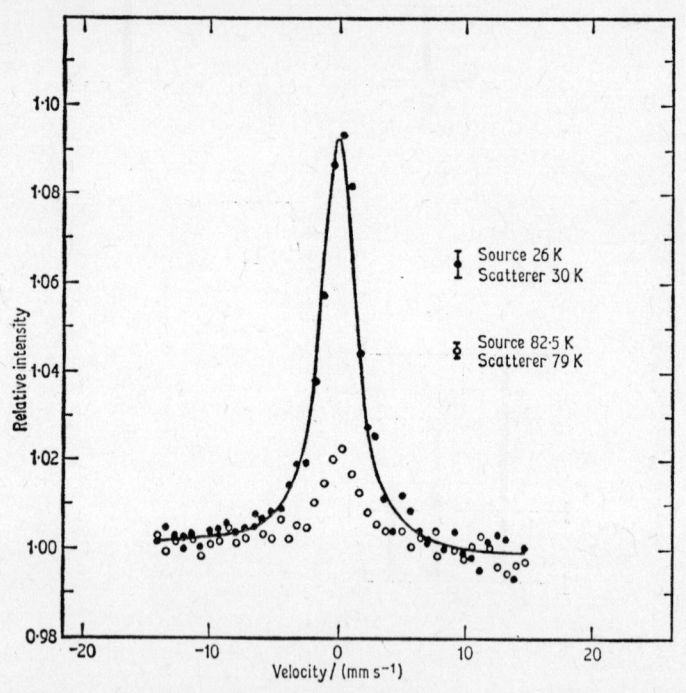

Fig. 16.20 The Mössbauer spectrum of the 137-keV ^{186}Os resonance obtained using an osmium metal scatterer. [Ref. 58, Fig. 4]

correction led to an estimate for the excited-state lifetime of $t_{\frac{1}{2}} = 0.72(4)$ ns ($\Gamma_r = 10.5$ mm s^{-1}).

The 69·6-keV third excited-state resonance of ^{189}Os was first reported in 1967 [61]. The γ-ray cannot be resolved from the 71·3-keV K_β X-ray which is more intense by a factor of approximately 10. The first measurements using a rhenium source matrix and a natural osmium absorber (^{189}Os = 16·1%) at 78 K gave a surprisingly large effect, although hyperfine structure was not resolved.

In subsequent work the ^{189}Ir was separated from the rhenium and incorporated into cubic iridium metal [62]. The single lines observed for K$_2$OsCl$_6$ and K$_2$Os(CN)$_6$ at 4·2 K were used to derive an estimate for $t_{\frac{1}{2}}$ of

[Refs. on p. 532]

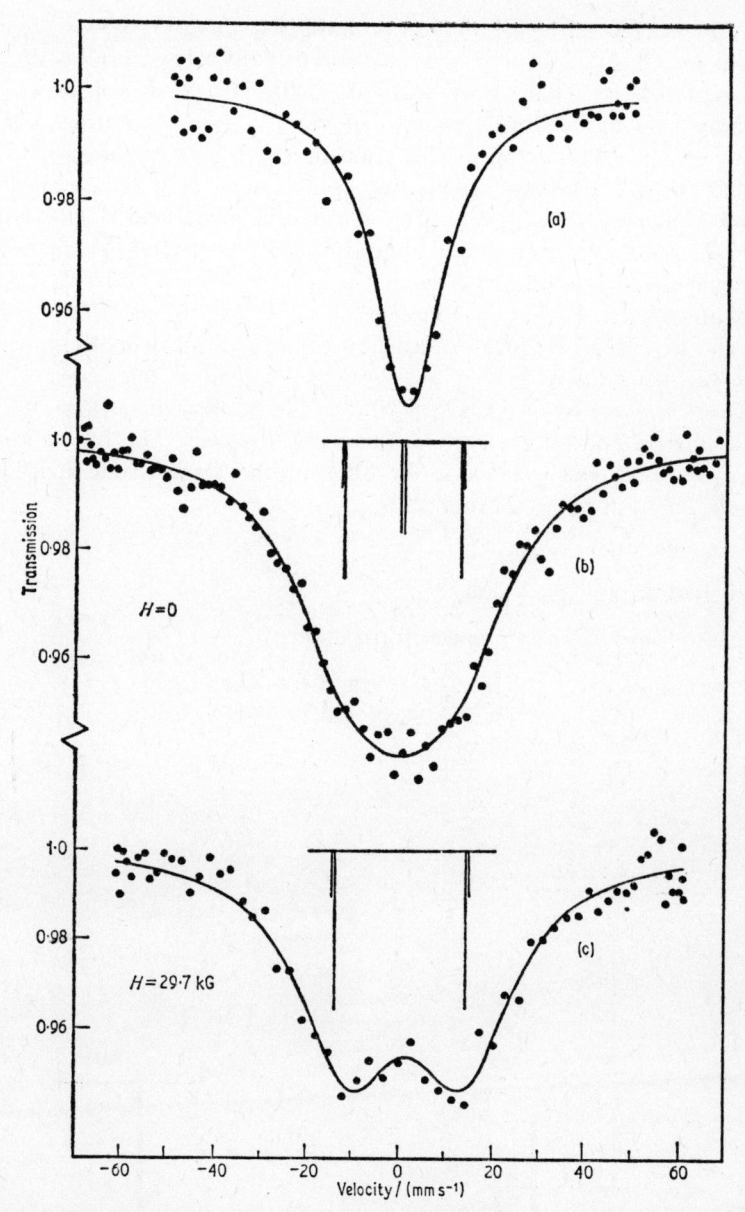

Fig. 16.21 The Mössbauer spectrum of the 36·2-keV ^{189}Os transition in (a) osmium metal using an iridium metal source, (b) and (c) osmium metal using a ^{189}Ir/iron metal source in zero and non-zero applied magnetic fields. The bars indicate the component lines and their intensities. [Ref. 60, Fig. 1]

[*Refs. on p. 532*]

518 | OTHER TRANSITION-METAL ELEMENTS

1·83(20) ns ($\Gamma_r = 2\cdot15(25)$ mm s^{-1}). Later experiments showed $t_{\frac{1}{2}}$ to be 1·63(4) ns [63]. The magnetic hyperfine spectrum of a 1 at. % alloy of ^{189}Os in iron metal is partially resolved, and polarisation studies in an external magnetic field have allowed analysis of the $\frac{5}{2} \to \frac{3}{2}$ level splitting. There is substantial E2/M1 mixing in this transition (cf. ^{99}Ru). The ratio of the g-values $g_e/g_g = 0\cdot905(9)$ gives the excited-state magnetic moment as μ_e (69·6 keV) = 0·988(10) n.m. The internal magnetic field is 1100(20) kG at 4·2 K, and the E2/M1 mixing parameter is $\delta = +0\cdot71(8)$ [$\delta^2 = 0\cdot50(11)$].

Independent data for the 69·6-keV transition using a 3 at. % Os–iron metal absorber at 4·2 K gave μ_e (69·6 keV) = 0·965(20) n.m., $H = 1084(52)$ kG, and $\delta^2 = 0\cdot57(21)$ [63]. An attempt to observe quadrupole splitting in OsS$_2$ was inconclusive.

The 95·3-keV resonance has been observed at 4·2 K using a rhenium metal source matrix and an enriched osmium-189 absorber. The resonance was extremely weak but gave a lower limit to the excited-state half-life of $t_{\frac{1}{2}} = 0\cdot132(15)$ ns ($\Gamma_r = 22$ mm s^{-1}).

16.11 Iridium (^{191}Ir, ^{193}Ir)

There are four known resonances for iridium:

^{191}Ir: 82·3-keV and 129·4-keV
^{193}Ir: 73·0-keV and 138·9-keV

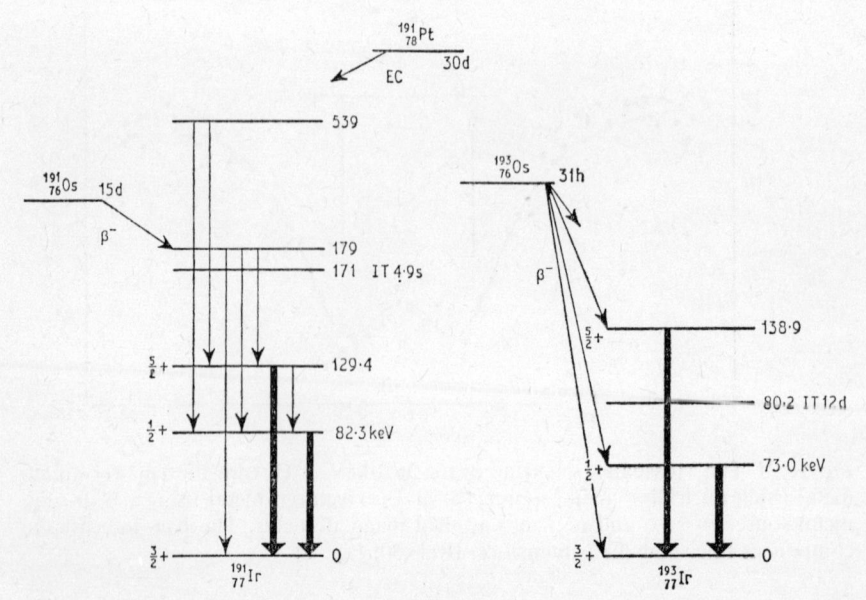

Fig. 16.22 Decay scheme for ^{191}Ir and ^{193}Ir.

[Refs. on p. 532]

It will hardly be necessary to recall that it was with the 129·4-keV γ-transition in ^{191}Ir that R. L. Mössbauer first demonstrated nuclear resonance absorption [64]. The source used was 16-day ^{191}Os (see Fig. 16.22 for decay scheme) and the absorber was iridium metal. The transmission of the γ-rays decreased unexpectedly as the temperature was lowered from 370 K to 90 K. Subsequently he initiated the use of velocity scanning [65, 66], and derived the excited-state half-life as $t_{\frac{1}{2}} = 0·099$ ns.

These results were verified by several groups independently [40, 41, 67]. There is some divergence of opinion as to the half-life of the excited state. At least seven values are available [68] ranging from 0·080 to 0·132 ns, the most recent Mössbauer values, which give a lower limit to $t_{\frac{1}{2}}$, being 0·100(7) ns [68] and 0·090(2) ns [19]. It is this latter value which we have adopted.

The angular dependence of the scattering of 129-keV γ-rays has been used to determine the E2/M1 mixing ratio as

$$\text{E2/M1} = \delta^2 = 0·13 \ (\delta = -0·36^{+0·04}_{-0·01}) \ [69].$$

Little is known about hyperfine effects in this resonance, but chemical isomer shifts have been given for IrO_2, $IrCl_3$, and Ir metal [70], and these are given in Table 16.5 together with values for the other transitions. Discussion of their significance is deferred for the moment.

The ^{191}Ir isotope shows a second Mössbauer transition at 82·3 keV. This level is not populated to any significant extent by the decay of ^{191}Os but the complicated EC-decay of ^{191}Pt does show a significant proportion of these γ-rays. A Mössbauer resonance was reported in 1967 for a source of ^{191}Pt in iridium metal made by the ^{191}Ir$(p, n)^{191}$Pt reaction and an iridium absorber, both at 80 K [71]. Another method of preparing the activity is by ^{191}Ir$(d, 2n)^{191}$Pt [72].

The half-life of the 82·3-keV excited state as determined by two delayed coincidence methods and from the Mössbauer linewidth is $t_{\frac{1}{2}} = 3·8(3)$ ns [68]. A ^{191}Pt/iridium source and an iridium absorber were used for the latter measurements because both have cubic lattices.

Magnetic hyperfine splitting has been observed in a source of ^{191}Pt in iron metal [68, 72]. All eight lines of the $\frac{3}{2} \rightarrow \frac{1}{2}$ transition are allowed because of an appreciable E2/M1 mixing, and the magnetic moment of the ground state is known to be $+0·1453(6)$ n.m. In one analysis the field at the ^{191}Ir was assumed to be 1500 kG, the value found for ^{193}Ir in a ^{193}Os/Fe source [72]. The spectrum (Fig. 16.23) obtained then gave μ_e (82 keV) $= +0·541(5)$ n.m. and the E2/M1 ratio, $\delta^2 = 0·64(10)$. Independent data gave μ_e (82 keV) $= +0·515(25)$ n.m. and $H = 1557(52)$ kG [68]. Data for iridium metal, IrO_2, and $IrCl_3$ showed large chemical isomer shifts, and a well-resolved quadrupole splitting in IrO_2. Comparison with the quadrupole splitting from the 73-keV ^{193}Ir transition gave the quadrupole moment ratio of the $I = \frac{3}{2}$ ground states as $^{191}Q_g/^{193}Q_g = 1·03(3)$.

[Refs. on p. 532]

The 73·0-keV transition in ^{193}Ir was first mentioned briefly in an 'on–off' type of experiment in 1960 [41], but more detailed data were not published until 1967 [73]. Since then it has been shown to be the best transition of the four for studying hyperfine interactions particularly because of its low natural

Table 16.5 Chemical isomer shifts, δ, in iridium at 4·2 K

	$\delta(^{191}$Ir$)/($mm s$^{-1})$		$\delta(^{193}$Ir$)/($mm s$^{-1})$		
	82 keV (^{191}Pt/Pt)	129 keV (^{191}Os/Os)	73 keV (^{193}Os/Os)	139 keV (^{193}Os/Os)	Reference
Iridium(VI)					
IrF$_6$			+0·89(2)		74
Iridium(IV)					
H$_2$IrCl$_6$			−1·471(10)		70
(NH$_4$)$_2$IrCl$_6$			−1·478(10)		70
			−1·45(5)		73
Na$_2$IrCl$_6$			−1·634(20)		70
K$_2$IrCl$_6$			−1·45(5)		73
IrCl$_4$			−1·4(3)		73
IrO$_2$	−0·269(73)		−1·466(10)		70
			−1·2(1)		73
Iridium(III)					
K$_3$IrCl$_6$			−2·697(33)		70
			−2·85(10)		73
IrCl$_3$	−1·458(40)	−0·318(44)	−2·490(17)	+0·142(113)	70
IrCl$_3$.4H$_2$O			−2·284(30)		70
IrBr$_3$.H$_2$O			−2·597(50)		70
IrBr$_2$OH.H$_2$O			−2·666(50)		70
IrI$_3$			−2·20(5)		73
Metals and Alloys					
Ir-metal	+0·654(24)	−0·041(25)	−0·542(6)	+0·019(64)	70
10% Ir–Fe		+0·098(36)	+0·413(9)	−0·004(82)	70
rare-earth Ir$_2$			+0·3(1)		73

linewidth, 0·625 mm s^{-1}. The parent isotope is 31-hour ^{193}Os prepared by a ^{192}Os$(n, \gamma)^{193}$Os reaction.

An absorber of 1 at. % Ir in iron metal gives eight lines appropriate to $\frac{1}{2} \to \frac{3}{2}$ transition with E2/M1 mixing, thereby confirming the excited-state spin to be $I = \frac{1}{2}$ [73]. Analysis of the spectrum gave $\delta^2 = 0.37(6)$, μ_e (73 keV) $= +0.48$ n.m., and $H_{\text{eff}} = 1430(80)$ kG. A similar spectrum obtained independently is shown in Fig. 16.22 [72], and in conjunction with a polarised spectrum gave $\delta^2 = 0.312$, $\delta = −0.558(5)$, and μ_e (73 keV) $= +0.470(1)$ n.m.

The measured ratio of the nuclear magnetic moments, μ_e/μ_g, is normally considered to be a constant for a given Mössbauer transition. However, this need not be the case. The hyperfine interaction is produced by an inter-

[Refs. on p. 532]

action of the nuclear magnetic moment (which has a radial distribution produced by the orbital and spin motions of the unpaired nucleons) and a magnetic field. The latter can be an applied field which is uniform over the nuclear radius, or a contact hyperfine field produced by electron density

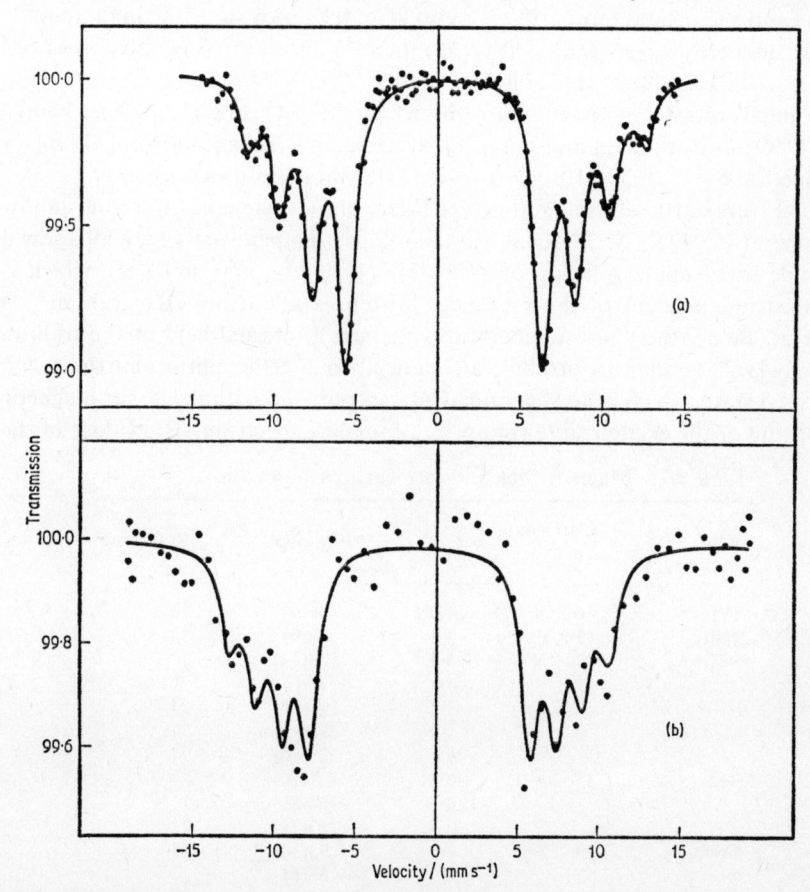

Fig. 16.23 Mössbauer spectra of (a) 73-keV transition of ^{193}Ir using an osmium source and a 2·7 at. % Ir/Fe alloy, (b) 82-keV transition of ^{191}Ir using an iridium absorber and a ^{191}Pt/Fe Source. Eight lines are seen from the $\frac{1}{2} \rightarrow \frac{3}{2}$ transition because of substantial E2/M1 mixing. [Ref. 72, Fig. 1]

which can also vary radially. The interactions in these two cases may differ slightly, and the apparent variation in the ratio μ_e/μ_g so produced is referred to as a 'hyperfine anomaly'. Such an effect has been found in the 73-keV transition of ^{193}Ir which has favourable nuclear characteristics for this type of investigation [74]. A 73-kG field applied externally to ^{193}Ir metal produces a ratio of $\mu_e/\mu_g = +3\cdot17(1)$ [74]. The equivalent value from a hyperfine field at ^{193}Ir in iron metal is $\mu_e/\mu_g = +2\cdot958(6)$ [72]. Similarly, the value for the

[*Refs. on p. 532*]

ratio in IrF$_6$ at 4·2 K, at which temperature the compound is antiferromagnetic, is +3·015(13) [74]. In the latter case the anomaly results mainly from the effects of the orbital contribution to the field. The total magnetic field of −1850 kG in IrF$_6$ comes mainly from the Fermi term ($H_S = -1650$ kG) and the orbital term ($H_L = -200$ kG). IrF$_6$ also shows a small quadrupole interaction ($\frac{1}{2}e^2qQ = -0.42(3)$ mm s^{-1}) and a large positive chemical isomer shift as shown in Table 16.5.

Time-reversal experiments have been carried out using a magnetic source of ^{193}Os in iron metal and magnetic absorbers of iridium in iron [75, 76]. A value for $\delta = +0.556(10)$ was obtained during these measurements.

The rare-earth alloys of the type MIr$_2$ show magnetic ordering at low temperatures [73]. At 78 K, above their Curie temperatures, they all show a simple quadrupole splitting of 212 MHz ($\frac{1}{2}e^2qQ = 3.60$ mm s^{-1}), which is almost independent of the rare earth within experimental error (about ±4 MHz). Below the Curie temperatures the small internal field at the iridium is not large enough to produce a significant magnetic splitting in the $I = \frac{3}{2}$ ground state, which is however quadrupole split, and with the larger magnetic splitting of the excited state results in a four-line spectrum. The values of the

Table 16.6 Magnetic fields in rare-earth/iridium alloys

Alloy	Curie temp. T_C/K	T/K	H_{eff}/kG	Reference
PrIr$_2$	16	4·2	46(9)	73
NdIr$_2$	11·8	4·2	48(9)	73
SmIr$_2$	37	4·2	89(13)	73
		4·2	+93(6)	77
GdIr$_2$	88	4·2	192(14)	73
		4·2	238(13)	77
		20	229(10)	77
TbIr$_2$	45	4·2	167(10)	73
		4·2	164(8)	77
		20	153(7)	77
DyIr$_2$	23	4·2	93(10)	73
		4·2	−126(6)	77
		20	−73(5)	77
HoIr$_2$	12	4·2	40(8)	73
		4·2	64(5)	77
ErI$_2$	>4	1·5	34(5)	77

fields deduced are given in Table 16.6 together with values obtained independently for essentially the same series [77]. The temperature dependence of the field was measured in DyIr$_2$ [77] and GdIr$_2$ [73], and the signs of the fields in DyIr$_2$ and SmIr$_2$ obtained by applying a 57-kG external field.

The extrapolated saturation field at the iridium is a linear function of the projection of the spin of the rare-earth ion, $(g_J - 1)J$, as shown in Fig. 16.24 [73, 77]. This shows that the induced field at the iridium arises by spin polarisation from the magnetic 4f-shell of the rare earth via the conduction electrons. However, the fact that the field is not zero when $(g_J - 1)J$ is zero shows that the interrelationship is not as simple as might at first be thought [73].

The range of chemical isomer shifts for 73-keV ^{193}Ir are greater than the

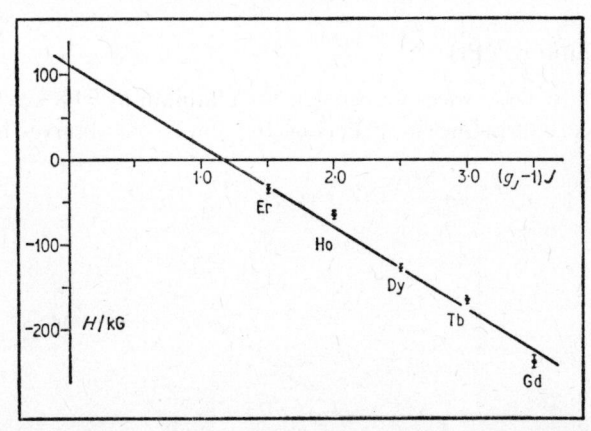

Fig. 16.24 The saturation field at the ^{193}Ir nucleus in rare-earth–Ir$_2$ alloys as a function of the projection of the spin of the rare earth, $(g_J - 1)J$. [Ref. 77, Fig. 6]

natural linewidth, and available values are given in Table 16.5. The Ir(III) results fall in the range -2.2 to -2.9 mm s^{-1} relative to ^{193}Os/Os metal, while Ir(IV) is in the range -1.2 to -1.6 mm s^{-1}. The only value for Ir(VI) is $+0.9$ mm s^{-1}. The respective formal electronic configurations are $5d^6$, $5d^5$, and $5d^3$ respectively. Under the assumption that the shift is determined primarily by changes in d-shielding, one can infer that $\delta R/R$ is positive in sign.

Clear quadrupole splitting is found in IrO$_2$ ($\frac{1}{2}e^2qQ = 2.7$ mm s^{-1}) consistent with its t_{2g}^5 configuration, but the IrCl$_6^{2-}$ salts have cubic symmetry. Small splittings were inferred in IrI$_3$ ($\frac{1}{2}e^2qQ = 0.7$ mm s^{-1}) and K$_3$IrCl$_6$ (0.6 mm s^{-1}) from line broadening only [73].

The 73-keV ^{193}Ir spectra of sources of ^{193}Os in OsO$_4$, K$_2$OsO$_4$.2H$_2$O, and Os(C$_5$H$_5$)$_2$ have indicated that the Ir β-decay daughter nucleus is isoelectronic with the osmium parent [78]. However, in the Os(IV) hexahalides indications were found of Ir(IV), Ir(III), and Ir(II) states.

The 138.9-keV resonance in ^{193}Ir was first reported in 1967 when chemical isomer shifts were given for Ir metal, IrCl$_3$, and a 10 wt % Ir–Fe alloy [70]. The values recorded were almost identical within experimental error. The excited-state half-life has been measured using an osmium source and an iridium absorber to be $t_{\frac{1}{2}} = 0.080(2)$ ns ($\Gamma_r = 24.6$ mm s^{-1}).

As might be anticipated there is a linear relationship between the chemical isomer shifts of any two isotopes for a series of compounds [70]. The values in Table 16.5 have been used to deduce the $\langle R_e^2 \rangle - \langle R_g^2 \rangle$ ($= \omega$) ratios

524 | OTHER TRANSITION-METAL ELEMENTS

$^{191}_{82}\omega/^{193}_{73}\omega = +1\cdot22(2)$, $^{191}_{129}\omega/^{191}_{82}\omega = +0\cdot21(2)$, $^{193}_{139}\omega/^{193}_{73}\omega = -0\cdot10(9)$. (In these ratios the superscript numerals refer to the mass number of the iridium isotope and the subscript numerals to the transition energy.) An estimate of $^{193}_{73}\omega$ of $+6 \times 10^{-3}$ fm^2 from isotope shift data was then used to derive via equation 3.11 some values for $\delta\langle R^2\rangle/\langle R^2\rangle$.

16.12 Platinum (^{195}Pt)

Two Mössbauer resonances are known for platinum at 99·8 keV and 129·4 keV. The 99-keV resonance in ^{195}Pt (Fig. 16.25) was first observed in platinum

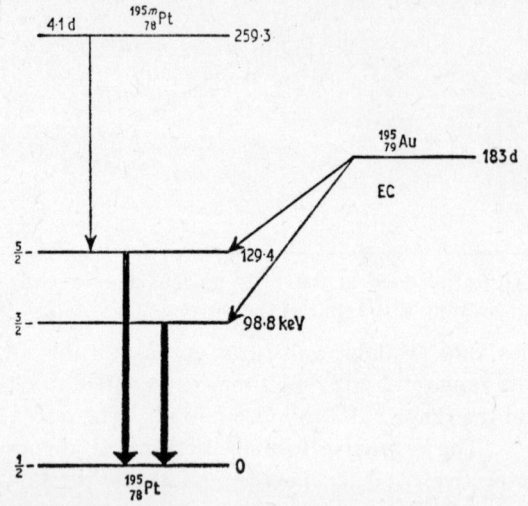

Fig. 16.25 Decay scheme of ^{195}Pt.

metal in 1965 [79]. The source used in early measurements was 183-day ^{195}Au diffused into platinum metal. This has the disadvantage of being a resonant matrix. More recently, ^{195}Au in copper or iridium has been found to give narrow lines [80, 81]. The iridium matrix can be activated directly by α-bombardment. The alternative parent of ^{195m}Pt prepared by neutron capture in ^{194}Pt has also been tried but is less successful [80].

Careful thickness studies with platinum-foil absorbers and a ^{195}Au/platinum source in the temperature range 20 K–77 K gave accurate values for the recoilless fraction [79]. At 20 K the values were $f_s = 0\cdot089(2)$ and $f_a = 0\cdot129(8)$, and at 77 K $f_s = 0\cdot021(3)$ and $f_a = 0\cdot043(11)$. Extrapolation to zero thickness gave $t_{\frac{1}{2}} = 0\cdot17$ ns, a later independent value being $0\cdot152(14)$ ns [80].

Magnetic hyperfine interactions are seen in appropriate alloys. Early studies on Pt$_{0\cdot1}$Fe$_{0\cdot9}$, Pt$_{0\cdot3}$Fe$_{0\cdot7}$, Pt$_{0\cdot5}$Fe$_{0\cdot5}$, and Pt$_{0\cdot75}$Fe$_{0\cdot25}$ showed that the large natural linewidth of the 98·8-keV transition (16 mm s^{-1}) obscured

[Refs. on p. 532]

much of the fine structure and a unique analysis of the data was not obtained [82]. Better spectra were obtained subsequently using a scattering

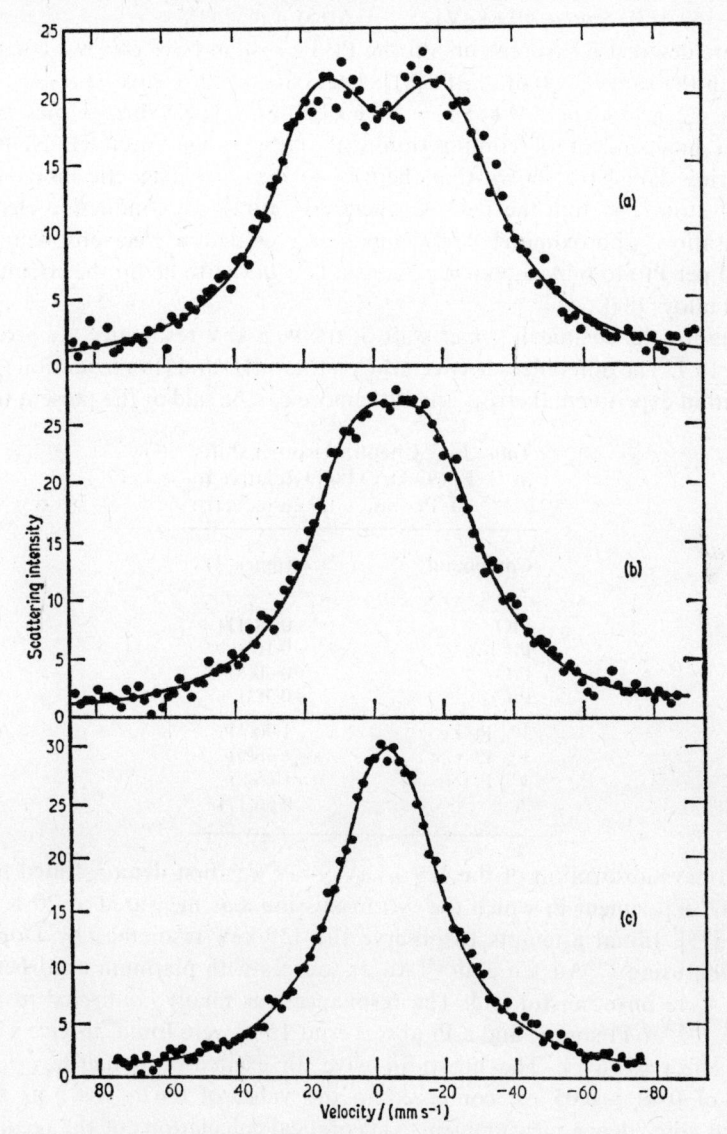

Fig. 16.26 Mössbauer spectra at 29 K for (a) Pt$_{0.3}$Fe$_{0.7}$, (b) Pt$_{0.07}$Co$_{0.93}$, and (c) Pt$_{0.07}$Ni$_{0.93}$, using a scattering geometry. [Ref. 83, Fig. 2]

technique [83] and are illustrated in Fig. 16.26. The internal magnetic fields are large: Pt$_{0.3}$Fe$_{0.7}$ ($H_{eff} = -1240$ kG), Pt$_{0.07}$Co$_{0.93}$ ($H_{eff} = -770$ kG), Pt$_{0.07}$Ni$_{0.93}$ ($H_{eff} = -340$ kG). Computer analysis using the known value for μ_g of $+0.6060$ n.m. gave μ_e (99 keV) $= -0.64(15)$ n.m. Independent measurements on 3 at. % Pt alloys gave Fe (-1190 kG), Co (-860 kG), Ni (-360 kG), and μ_e (99 keV) $= -0.61(5)$ n.m. [84].

More detailed measurements on the Pt/Fe system have covered compositions in the range 3–50 at. % Pt [81]. The 3 at. % alloy gave $H_{\text{eff}} = -1260$ kG at 4·2 K and μ_e (99 keV) $= -0.60(15)$ n.m. The value of the field is almost independent of composition within the range specified. Neutron-scattering data have shown that there is no localised magnetic moment on the Pt atoms, so that the field is generated entirely by conduction-electron polarisation, approximately 0·07 unpaired conduction electrons being required per Pt atom. A similar mechanism is believed to act in the cobalt and nickel alloys [84].

Data for the chemical isomer shift of the 98·8-keV resonance are given in Table 16.7. The differences between the platinum(II) and (IV) oxidation states are within experimental error, and little more can be said at the present time.

Table 16.7 Chemical isomer shifts in ^{195}Pt (99 keV) [84] (Relative to a ^{195}Au/Pt source taken as zero)

Compound	δ/(mm s^{-1})
PtO	$-0.34(11)$
PtCl$_2$	$-0.1(2)$
PtO$_2$	$-0.40(8)$
PtCl$_4$	$-0.3(3)$
3% Pt–Fe	$-1.90(11)$
3% Pt–Co	$-1.96(9)$
3% Pt–Ni	$-1.65(8)$
Pt	$+0.06(17)$

Resonant absorption of the 129·4-keV γ-ray was first demonstrated in an 'on–off' experiment in which the γ-transmission was measured at 20 K and 65 K [85]. Initial attempts to observe the 129-keV resonance by Doppler scanning using ^{195}Au/Cu and ^{195}Au/Ir sources with platinum absorbers at 4·2 K were unsuccessful [80]. The resonance was finally confirmed in 1969 [86]. A ^{195}Au/Pt source and a Pt absorber at 15 K were found to give a very weak effect (0·16%). The linewidth gave an excited-state half-life (lower limit) of 0.49 ± 0.05 ns, compared to the value of 0.62 ± 0.07 ns from delayed coincidence measurements. Theoretical calculations of the recoilless fraction for both transitions have been made [87].

16.13 Gold (^{197}Au)

The 77·34-keV ^{197}Au resonance was first reported in 1960 in gold-foil absorbers using both ^{197}Pt/platinum and ^{197}Hg/mercury sources at 4·2 K

[88]. The latter gave only a weak effect, although it could be enhanced by conversion to a gold amalgam. All subsequent work has used the platinum matrix. The appropriate decay schemes are shown in Fig. 16.27. Estimates of the excited-state half-life made by extrapolating the linewidth of a ^{197}Pt/Pt source and a gold-foil absorber to zero absorber thickness, are $t_{\frac{1}{2}} = 1\cdot93(10)$ ns [89] and $1\cdot892(14)$ ns [19].

The majority of the initial ^{197}Au work has centred on the gold alloys and gold impurity atoms in metals. Interpretation of the data has proved difficult, and for this reason only the more important features are summarised here.

A number of recoil-free fraction and lattice-dynamical studies have been made on gold metal [90, 91]. The resonance in gold microcrystals of mean diameter 20 and 6 nm shows a greater recoil-free fraction in the smaller crystals corresponding to an increase in the effective Debye temperature from

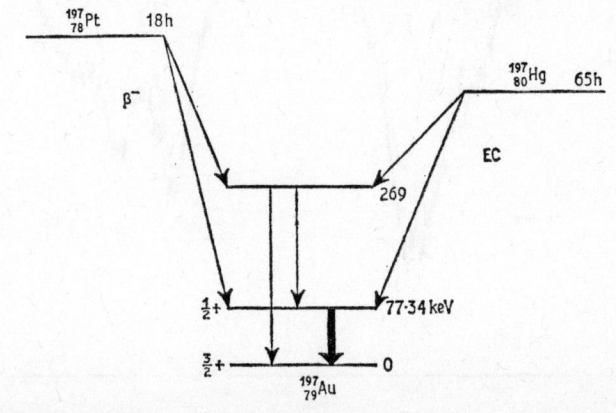

Fig. 16.27 Decay scheme of ^{197}Au.

163 K to 173 K [92]. This has been shown to be compatible with an observed concomitant decrease in the crystal lattice spacings [93].

Magnetic hyperfine splitting is seen at nominally diamagnetic ^{197}Au in ferromagnetic alloys, presumably as a result of conduction-electron polarisation. The 6s-contribution to the conduction band becomes partially spin-unpaired by interaction with the magnetic atoms. The lines are usually only partly resolved, and the first magnetic data were not analysed successfully [94], but later studies of a 0·5% Au in iron alloy gave $H_{\text{eff}} = 1460$ kG with a value for the excited-state magnetic moment of $\mu_e = +0\cdot38(8)$ n.m. [89]. The fields at 1% Au in cobalt and nickel were 1180 and 420 kG respectively. Independent results using 1% ^{197}Pt in the ferromagnetic metals as sources and a gold absorber gave $H_{\text{eff}} = -1420$ kG (Fe), -990 kG (Co cubic), -980 kG (Co hexagonal), and -340 kG (Ni) [95], with $\mu_e = +0\cdot37(4)$ n.m. The signs of the fields were obtained from applied field measurements.

528 | OTHER TRANSITION-METAL ELEMENTS

The E2 admixture into the otherwise M1 radiation is not negligible ($\delta^2 = 0.11$), and has a significant effect on line intensities. This has to be taken into account in accurate analysis of the magnetic splittings. A dilute solid solution of gold in iron (0·5 or 1 at. %) has been fully analysed and shows a field of 1280(25) kG at 21 K [96]. With the known ground-state moment of +0·1449 n.m. the excited-state moment was calculated as $\mu_e = +0.419(5)$ n.m.

Magnetic broadening has also been found in 85–95 at. % Au/iron alloys,

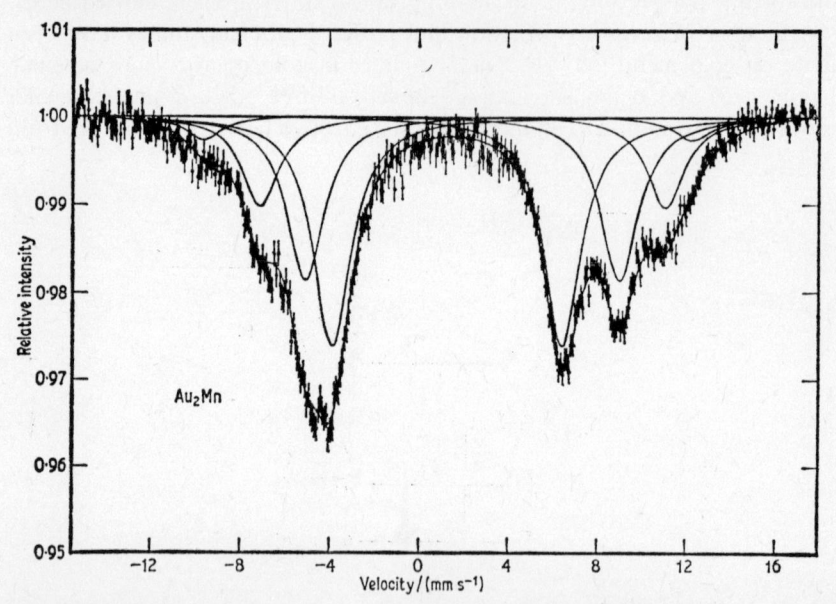

Fig. 16.28 Magnetic splitting of ^{197}Au$_2$Mn showing the eight lines for a $\frac{1}{2} \to \frac{3}{2}$ E2/M1 transition. The asymmetry is due to a quadrupole interaction. [Ref. 98, Fig. 2]

the spectra being asymmetric because of several distinct environments contributing to the total envelope, but no detailed interpretation has been given [97].

One of the best resolved ^{197}Au magnetic spectra is that of Au$_2$Mn at 4·2 K, which is illustrated in Fig. 16.28 [98]. The eight allowed lines for a $\frac{3}{2} \to \frac{1}{2}$ transition with E2 admixture are clearly shown, and the asymmetry in the spectrum is due to the presence of a quadrupole interaction ($e^2qQ = 2.705(70)$ mm s^{-1}), which is directed perpendicular to the magnetic axis ($H_{\text{eff}} = 1571$ kG). An unusually large pressure dependence of the magnetic field was interpreted as due to the uncoiling of the spiral spin structure of the alloy.

The intermetallic compound Au$_4$V is one of the few alloys of non-magnetic

[Refs. on p. 532]

metals to show magnetic ordering (T_C = 55 K). The magnetic broadening of the line at 4·2 K corresponds to a field of 230 kG [99]. Independent data gave 185 kG [100]. The small localised moment at the vanadium causes spin polarisation of the 6s-electrons, and there is no localised atomic magnetic moment as such on the gold atoms. The known moment of 4·15 B.M. at manganese in Au$_4$Mn and the ^{197}Au internal field of 847 kG allow a naive estimate of the vanadium moment to be set at 0·92 B.M. per atom.

An early analysis of the large chemical isomer shifts (Table 16.8) found for ^{197}Pt impurity atoms in Fe, Co, and Ni confirmed that $\delta R/R$ is positive [101]. The shift of ^{197}Au in nickel alloys at 4·2 K decrease linearly by a total of

Table 16.8 Chemical isomer shifts in ^{197}Au relative to Au metal at 4·2 K

Compound	δ/(mm s^{-1})	Reference
^{197}Pt doped metals*		
Li	7·4(5)	104
Be	5·6(5)	104
Mg	6·3(3)	104
Al	7·6(4)	104
Si	3·3(3)	104
Ca	8·7(5)	104
Fe	5·4(2)	104
	5·1(5)	94
Co	5·3(2)	104
	4·3(4)	94
Ni	4·7(2)	104
	3·8(4)	94
	5·4(2)	105
Cu	4·2(4)	104
	4·4(2)	105
Zn	3·4(5)	104
Ge	4·5(4)	104
Y	7·6(3)	104
Pd	2·4(3)	104
	2·4(2)	105
Ag	1·3(2)	104
	2·1(2)	105
Sn	4·3(4)	104
Te	1·9(3)	104
Pt	1·2(1)	104
	1·4(2)	105
Au	0	104
Se	1·8(4)	104
Alloys		
Al$_2$Au	7·1(4)	104
Au$_2$Mn	2·79(2)	98

* The sign of δ has been reversed for all ^{197}Pt doped shift values so that a positive shift indicates a higher energy state.

5·5 mm s^{-1} with increasing gold concentration from 0 to 100%. However, an attempt to relate this observation to the electronic band structure was only partially successful [89].

Several workers have attempted to rationalise the chemical isomer shifts observed at ^{197}Au impurity nuclei in various metals with limited success [102, 103]. The most comprehensive set of data comes from Barrett et al. [104], who doped ^{197}Pt into 20 metals. The shift correlates very approximately with the electronegativity of the host, and a crude interpretation is that electrons are transferred in varying degrees to the 6s-shell of the gold. An estimate of $\delta R/R = 1.9(6) \times 10^{-3}$ was obtained. Additional evidence in favour of an increased 6s-population has come from a comparison of the shift and residual electrical resistivity of ^{197}Au alloys with Cu, Ag, Pd, and Pt [105]. The pressure dependence up to 70·6 kbar of the chemical isomer shift in a gold foil at 4·2 K has been obtained and with detailed analysis leads to a value for $\delta R/R$ of $\sim +1.5 \times 10^{-4}$ [106].

In a study of the superconducting alloys Au$_2$Bi and AuNb$_3$, no significant change in the recoil-free fraction was found at the superconducting transition temperature [107]. Preliminary investigations on copper–gold alloys have revealed a difference in chemical isomer shift between ordered and disordered alloys at each particular composition [108].

Early data for gold compounds such as AuI, AuBr, AuCl, AuCl$_3$, AuF$_3$, KAuCl$_4$, KAuBr$_4$, KAuF$_4$, and KAu(CN)$_2$ have only been referred to obliquely in discussing other subjects [109, 110]. Recent data for 30 gold complexes are tabulated in Table 16.9 [111]. The range of chemical isomer shifts is larger for the gold(I) complexes than for gold(III), but whether this is a reflection of greater s-character in the bonding in the former or to the small variety of coordinating ligands in the latter is not certain. The quadrupole splitting in the gold(I) compounds is invariably larger than in gold(III), although the explanation for this is also obscure. This feature may well serve to characterise the oxidation state, which cannot be determined from the chemical isomer shift alone. Systematic variation of the quadrupole splitting with the chemical isomer shift was found in several series of related complexes as for example in LAuCl (L = Me$_2$S, py, Ph$_3$As, (C$_6$F$_5$)Ph$_2$P, and Ph$_3$P). Discussions in terms of the chemical bonding are however only tentative at the present time.

HAuCl$_4$ and KAu(CN)$_2$ have been decomposed on MgO and η-Al$_2$O$_3$ surfaces to give supported gold catalysts [112]. KAu(CN)$_2$ has a higher decomposition temperature than HAuCl$_4$ on the η-Al$_2$O$_3$ surface.

16.14 Mercury (^{201}Hg)

The 32·19-keV ^{201}Hg resonance which was first recorded in 1969 has several undesirable characteristics [113]. The short excited-state lifetime gives a large natural linewidth of 43 mm s⁻¹, and the γ-ray is highly internally converted with $\alpha_T = 39$. The ^{201}Tl precursor is therefore very inefficient. No Doppler-scan spectra have been recorded, and the resonance was recorded as a change

Table 16.9 Mössbauer parameters in gold compounds at 4·2 K [111]

Compound	Δ/(mm s⁻¹)	δ†/(mm s⁻¹)
Gold(I)		
Ph₃PAuMe	10·35(6)	4·93(4)
Ph₃PAuCN	10·5(2)	3·9(2)
Ph₃PAuOCOMe	7·6(2)	3·3(2)
Ph₃PAuN₃	8·4(2)	3·3(2)
(C₆F₅)Ph₂PAuCl	7·87(4)	2·93(3)
Ph₃PAuCl	7·47(13)	2·96(7)
Ph₃PAuBr	7·40(10)	2·76(6)
Ph₃AsAuCl	7·00(7)	1·92(3)
C₅H₅NAuCl	6·4(2)	1·7(1)
Me₂SAuCl	6·42(3)	1·26(2)
Ph₃PAuI	8·3(1)	1·24(6)
AuCl	4·6(4)	−1·2(2)
AuI	4·4(1)	−1·32(3)
AuCN	8·39(2)	2·37(2)
Gold(III)		
Ph₃PAuBr₃	3·37(5)	2·11(3)
Ph₃PAuCl₃	3·25(8)	2·06(4)
C₅H₅NAuCl₃	*	1·45(5)
Me₂SAuCl₃	2·20(8)	1·26(5)
p-MeC₆H₄NCAuCl₃	2·00(9)	0·75(5)
AuCl₃	*	0·83(9)
AuBr₃	1·8(2)	0·18(10)
KAuF₄	*	−0·04(2)
Ph₄AsAuCl₄	1·88(7)	1·09(4)
KAuCl₄.2H₂O	1·4(1)	0·87(8)
NaAuCl₄.xH₂O	1·4(2)	0·81(4)
NH₄AuCl₄.xH₂O	1·7(1)	0·86(5)
HAuCl₄.xH₂O	*	0·60(8)
Ph₄AsAuBr₄	1·5(2)	0·85(11)
KAuBr₄.2H₂O	*	0·56(7)
KAuI₄.xH₂O	2·1(2)	0·27(8)
Au₂O₃.xH₂O	2·3(2)	0·63(8)
gold metal		−1·18(3)

* No observed splitting.
† Relative to ^{197}Pt/Pt source.

in transmission of 32-keV γ-rays between 300 K and 77 K at zero velocity. The source was prepared by the ^{201}Hg(p, n)^{201}Tl reaction on frozen mercury, and the self-resonance in the source was further enhanced by clamping it to an absorber of HgO. The observed absorption cross-section favours an $I_e = \frac{1}{2}$ excited state.

[*Refs. on p. 532*]

REFERENCES

[1] F. E. Obenshain and H. H. F. Wegener, *Phys. Rev.*, 1961, **121**, 1344.
[2] H. H. F. Wegener and F. E. Obenshain, *Z. Physik*, 1961, **163**, 17.
[3] D. Seyboth, F. E. Obenshain, and G. Czjzek, *Phys. Rev. Letters*, 1965, **14**, 954.
[4] J. C. Love, G. Czjzek, J. J. Spijkerman, and D. K. Snediker, p. 124 of 'Hyperfine Structure and Nuclear Radiations', Ed. E. Matthias and D. A. Shirley, North-Holland, Amsterdam, 1968.
[5] J. J. Spijkerman, *Symposia Faraday Soc. No. 1*, 1968, 134.
[6] F. Ambe, S. Ambe, M. Takeda, H. H. Wei, K. Ohki, and N. Saito, *Radiochem. Radioanal. Letters*, 1969, **1**, 341.
[7] U. Erich and D. Quitmann, p. 130 of 'Hyperfine Structure and Nuclear Radiations', Ed. E. Matthias and D. A. Shirley, North-Holland, Amsterdam, 1968.
[8] U. Erich, *Z. Physik*, 1969, **227**, 25.
[9] W. S. Childs and L. S. Goodman, *Phys. Rev.*, 1968, **170**, 136.
[10] U. Erich, K. Frölich, P. Gütlich, and G. A. Webb, *Inorg. Nuclear Chem. Letters*, 1969, **5**, 855.
[11] U. Erich, E. Kankeleit, H. Prange, and S. Hufner, *J. Appl. Phys.*, 1969, **40**, 1491.
[12] D. Raj and S. P. Puri, *J. Phys. Soc. Japan*, 1969, **27**, 788.
[13] R. V. Pound and G. A. Rebka, *Phys. Rev. Letters*, 1960, **4**, 397.
[14] D. E. Nagle, P. P. Craig, and W. E. Keller, *Nature*, 1960, **186**, 707.
[15] P. P. Craig, D. E. Nagle, and D. R. F. Cochran, *Phys. Rev. Letters*, 1960, **4**, 561.
[16] S. I. Aksenov, V. P. Alfimenkov, V. I. Lushchikov, Yu. M. Ostanevich, F. L. Shapiro, and W. K. Yen, *Zhur. eksp. teor. Fiz.*, 1961, **40**, 88 (*Soviet Physics – JETP*, 1961, **13**, 62).
[17] V. P. Alfimenkov, Yu. M. Ostanevich, T. Ruskov, A. V. Strelkov, F. L. Shapiro, and W. K. Yen, *Zhur. eksp. teor. Fiz.*, 1962, **42**, 1029 (*Soviet Physics – JETP*, 1962, **15**, 713).
[18] H. de Waard and G. J. Perlow, *Phys. Rev. Letters*, 1970, **24**, 566.
[19] P. Steiner, E. Gerdau, W. Hautsch, and D. Steenken, p. 364 of 'Hyperfine Structure and Nuclear Radiations', Ed. E. Matthias and D. A. Shirley, North Holland, 1968; *Z. Physik*, 1969, **221**, 281.
[20] O. C. Kistner, S. Monaro, and R. Segnan, *Phys. Letters*, 1963, **5**, 299.
[21] O. C. Kistner, *Phys. Rev.*, 1966, **144**, 1022.
[22] G. Kaindl, W. Potzel, F. Wagner, U. Zahn, and R. L. Mössbauer, *Z. Physik*, 1969, **226**, 103.
[23] O. C. Kistner, 'Mössbauer Effect Methodology', Vol. 3, p. 217, Ed. I. J. Gruverman, Plenum Press, New York, 1967.
[24] O. C. Kistner, *Phys. Rev. Letters*, 1967, **19**, 872.
[25] O. C. Kistner, p. 295 of 'Hyperfine Structure and Nuclear Radiations', Ed. E. Matthias and D. A. Shirley, North-Holland, Amsterdam, 1968.
[26] C. A. Clausen, R. A. Prados, and M. L. Good, *Chem. Comm.*, 1969, 1188.
[27] G. E. Bizina, A. G. Beda, N. A. Burgov, and A. V. Davydov, *Zhur. eksp. teor. Fiz.*, 1963, **45**, 1408 (*Soviet Physics – JETP*, 1969, **18**, 973).
[28] W. Wiedemann, P. Kienle, and F. Stanek, *Z. angew. Physik*, 1963, **15**, S.7.
[29] E. Gerdau, H. J. Korner, J. Lerch, and P. Steiner, *Z. Naturforsch.*, 1966, **21a**, 941.
[30] E. Gerdau, P. Steiner, and D. Steenken, p. 261 of 'Hyperfine Structure and

Nuclear Radiations', Ed. E. Matthias and D. A. Shirley, North-Holland, Amsterdam, 1968.
[31] R. E. Snyder, J. W. Ross, and D. St P. Bunbury, *J. Phys. C*, 1968, Ser. 2, **1**, 1662.
[32] P. Boolchand, B. L. Robinson, and S. Jha, *Phys. Rev.*, 1969, **187**, 475.
[33] P. Steiner, E. Gerdau, and D. Steenken, *Proc. Roy. Soc.*, 1969, **311A**, 177.
[34] C. G. Jacobs, N. Hershkowitz, and J. B. Jeffries, *Phys. Letters*, 1969, **28A**, 498.
[35] S. G. Cohen, A. Marinov, and J. I. Budnick, *Phys. Letters*, 1964, **12**, 38.
[36] W. A. Steyert, R. D. Taylor, and E. K. Storms, *Phys. Rev. Letters*, 1965, **14**, 739.
[37] C. Sauer, E. Matthias, and R. L. Mössbauer, *Phys. Rev. Letters*, 1968, **21**, 961; C. Sauer, *Z. Physik*, 1969, **222**, 439.
[38] G. T. Trammell and J. P. Hannon, *Phys. Rev.*, 1969, **180**, 337.
[39] Yu. M. Kagan, A. M. Afanasev, and V. K. Voitovetskii, *ZETF Letters*, 1969, **9**, 155 (*JETP Letters*, 1969, **9**, 91).
[40] L. L. Lee, L. Meyer-Schutzmeister, J. P. Schiffer, and D. Vincent, *Phys. Rev. Letters*, 1959, **3**, 223.
[41] A. Bussiere de Nercy, M. Langevin, and M. Spighel, *Compt. rend.*, 1960, **250**, 1031; *J. Phys. Radium*, 1960, **21**, 288.
[42] E. Kankeleit, *Z. Physik*, 1961, **164**, 442.
[43] C. Bonchev, L. Mitrani, S. Ormandjiev, B. Skorchev, and I. Ouzounov, *Compt. rend., Acad. Bulgare Sci.*, 1963, **16**, 15.
[44] O. I. Sumbaev, A. I. Smirnov, and V. S. Zykov, *Zhur. eksp. teor. Fiz.*, 1962, **42**, 115 (*Soviet Physics – JETP*, 1962, **15**, 82).
[45] N. Shikazono, H. Takekoshi, and T. Shinjo, *J. Phys. Soc. Japan*, 1965, **20**, 271.
[46] D. Agresti, E. Kankeleit, and B. Persson, *Phys. Rev.*, 1967, **155**, 1342.
[47] B. Persson, H. Blumberg, and D. Agresti, *Phys. Letters*, 1967, **24B**, 522.
[48] B. Persson, H. Blumberg, and D. Agresti, *Phys. Rev.*, 1968, **170**, 1066.
[49] R. B. Frankel, Y. Chow, L. Grodzins, and J. Wulff, *Phys. Rev.*, 1969, **186**, 381.
[50] Y. W. Chow, L. Grodzins, and P. H. Barrett, *Phys. Letters*, 1965, **15**, 369.
[51] S. G. Cohen, N. A. Blum, Y. W. Chow, R. B. Frankel, and L. Grodzins, *Phys. Rev. Letters*, 1966, **16**, 322.
[52] N. Shikazono, H. Takekoshi, and T. Shoji, *J. Phys. Soc. Japan*, 1966, **21**, 829.
[53] S. Roth and E. M. Horl, *Phys. Letters*, 1967, **25A**, 299.
[54] K. A. Hardy, D. C. Russell, and R. M. Wilenzick, *Phys. Letters*, 1968, **27A**, 422.
[55] Y. W. Chow, E. S. Greenbaum, R. H. Howes, F. H. H. Hsu, P. H. Swerdlow, and C. S. Wu, *Phys. Letters*, 1969, **30B**, 171.
[56] R. L. Mössbauer and W. H. Wiedemann, *Z. Physik*, 1960, **159**, 33.
[57] A. E. Blaugrund, Y. Dar, G. Goldring, and E. Z. Skurnik, *Nuclear Phys.*, 1963, **45**, 54.
[58] R. J. Morrison, M. Atac, P. Debrunner, and H. Frauenfelder, *Phys. Letters*, 1964, **12**, 35.
[59] L. Grodzins and Y. W. Chow, *Phys. Rev.*, 1966, **142**, 86.
[60] F. Wagner, G. Kaindl, H. Bohn, U. Biebl, H. Schaller, and P. Kienle, *Phys. Letters*, 1969, **28B**, 548.
[61] S. Jha, W. R. Owens, M. C. Gregory, and B. L. Robinson, *Phys. Letters*, 1967, **25B**, 115.
[62] B. Persson, H. Blumberg, and M. Bent, *Phys. Rev.*, 1968, **174**, 1509.
[63] M. C. Gregory, B. L. Robinson and S. Jha, *Phys. Rev.*, 1969, **180**, 1158.
[64] R. L. Mössbauer, *Z. Physik*, 1958, **151**, 124.

[65] R. L. Mössbauer, *Naturwiss.*, 1958, **45**, 538.
[66] R. L. Mössbauer, *Z. Naturforsch.*, 1959, **14A**, 211.
[67] P. P. Craig, J. G. Dash, A. D. McGuire, D. Nagle, and R. D. Reiswig, *Phys. Rev. Letters*, 1959, **3**, 221.
[68] W. R. Owens, B. L. Robinson, and S. Jha, *Phys. Rev.*, 1969, **185**, 1555.
[69] F. Wittmann, *Z. Naturforsch.*, 1964, **19A**, 1409.
[70] F. Wagner, J. Klockner, H. J. Korner, H. Schaller, and P. Kienle, *Phys. Letters*, 1967, **25B**, 253.
[71] S. Jha, W. R. Owens, M. C. Gregory, and B. L. Robinson, *Phys. Letters*, 1967, **25B**, 115.
[72] F. Wagner, G. Kaindl, P. Kienle, and H. J. Korner, *Z. Physik*, 1967, **207**, 500.
[73] U. Atzmony, E. R. Bauminger, D. Lebenbaum, A. Mustachi, S. Ofer, and J. H. Wernick, *Phys. Rev.*, 1967, **163**, 314.
[74] G. J. Perlow, W. Henning, D. Olson, and G. L. Goodman, *Phys. Rev. Letters*, 1969, **23**, 680.
[75] M. Atac, B. Chrisman, P. Debrunner, and H. Frauenfelder, *Phys. Rev. Letters*, 1968, **20**, 691.
[76] E. Zech, F. Wagner, H. J. Korner, and P. Kienle, p. 314 of 'Hyperfine Structure and Nuclear Radiations', Ed. E. Matthias and D. A. Shirley, North-Holland, Amsterdam, 1968.
[77] A. Heuberger, F. Pobell, and P. Kienle, *Z. Physik*, 1967, **205**, 503.
[78] P. Rother, F. Wagner, and U. Zahn, *Radiochim. Acta*, 1969, **11**, 203.
[79] J. R. Harris, N. Benczer-Koller, and G. M. Rothberg, *Phys. Rev.*, 1965, **137**, A1101.
[80] A. B. Buyrn and L. Grodzins, *Phys. Letters*, 1966, **21**, 389.
[81] A. B. Buyrn, L. Grodzins, N.A.Blum, and J. Wulff, *Phys. Rev.*, 1967, **163**, 286.
[82] N. Benczer-Koller, J. R. Harris, and G. M. Rothberg, *Phys. Rev.*, 1965, **140**, B547.
[83] M. Atac, P. Debrunner, and H. Frauenfelder, *Phys. Letters*, 1966, **21**, 699.
[84] D. Agresti, E. Kankeleit, and B. Persson, *Phys. Rev.*, 1967, **155**, 1339.
[85] J. R. Harris, G. M. Rothberg, and N. Benczer-Koller, *Phys. Rev.*, 1965, **138**, B554.
[86] R. M. Wilenzick, K. A. Hardy, J. A. Hicks, and W. R. Owens, *Phys. Letters*, 1969, **29A**, 678.
[87] N. Malathi and V. K. Garg, *Phys. Letters*, 1969, **30A**, 219.
[88] D. Nagle, P. P. Craig, J. G. Dash, and R. D. Reiswig, *Phys. Rev. Letters*, 1960, **4**, 237.
[89] L. D. Roberts and J. O. Thomson, *Phys. Rev.*, 1963, **129**, 664.
[90] H. J. Andra, C. M. H. Hashmi, P. Kienle, and F. W. Stanek, *Z. Naturforsch.*, 1963, **18A**, 687.
[91] J. Speth and F. W. Stanek, *Z. Naturforsch.*, 1965, **20A**, 1175.
[92] S. W. Marshall and R. M. Wilenzick, *Phys. Rev. Letters*, 1966, **16**, 219.
[93] D. Schrocer, *Phys. Letters*, 1966, **21**, 123.
[94] D. A. Shirley, M. Kaplan, and P. Axel, *Phys. Rev.*, 1961, **123**, 816.
[95] R. W. Grant, M. Kaplan, D. A. Keller, and D. A. Shirley, *Phys. Rev.*, 1964, **133**, A1062.
[96] R. L. Cohen, *Phys. Rev.*, 1968, **171**, 343.
[97] R. J. Borg and D. N. Pipkorn, *J. Appl. Phys.*, 1969, **40**, 1483.
[98] J. O. Thompson, P. G. Huray, D. O. Patterson, and L. D. Roberts, p. 557 of 'Hyperfine Structure and Nuclear Radiations', Ed. E. Matthias and D. A. Shirley, North-Holland, Amsterdam, 1968.

[99] B. D. Dunlap, J. B. Darby, and C. W. Kimball, *Phys. Letters*, 1967, **25A**, 431.
[100] R. L. Cohen, R. C. Sherwood, and J. H. Wernick, *Phys. Letters*, 1968, **26A**, 462.
[101] D. A. Shirley, *Phys. Rev.*, 1961, **124**, 354.
[102] N. N. Delyagin, *Fiz. Tverd. Tela*, 1966, **8**, 3426 (*Soviet Physics – Solid State*, 1967, **8**, 2748).
[103] V. V. Chekin, *Zhur. eksp. teor. Fiz.*, 1968, **54**, 1829.
[104] P. H. Barrett, R. W. Grant, M. Kaplan, D. A. Keller, and D. A. Shirley, *J. Chem. Phys.*, 1963, **39**, 1035.
[105] L. D. Roberts, R. L. Becker, F. E. Obenshain, and J. O. Thomson, *Phys. Rev.*, 1965, **137**, A895.
[106] L. D. Roberts, D. O. Patterson, J. O. Thomson, and P. R. Levey, *Phys. Rev.*, 1969, **179**, 656.
[107] P. P. Craig, D. E. Nagle, and R. D. Reiswig, *J. Phys. and Chem. Solids*, 1960, **17**, 168.
[108] P. G. Huray, L. D. Roberts and J. O. Thomson, p. 596 of 'Hyperfine Structure and Nuclear Radiations', Ed. E. Matthias and D. A. Shirley, North-Holland, Amsterdam, 1968.
[109] D. A. Shirley, *Rev. Mod. Phys.*, 1964, **36**, 339.
[110] D. A. Shirley, Chap. 9 in 'Chemical Applications of Mössbauer Spectroscopy', Ed. V. I. Goldanskii and R. H. Herber, Academic Press, New York, 1968.
[111] J. S. Charlton and D. I. Nichols, *J. Chem. Soc. (A)*, 1970, 1484.
[112] W. N. Delgass, M. Boudart, and G. Parravano, *J. Phys. Chem.*, 1968, **72**, 3563.
[113] D. E. Carlson and A. A. Temperley, *Phys. Letters*, 1969, **30B**, 322.

17 | The Rare-earth Elements

Mössbauer spectroscopy of the rare-earth elements as a group has gained considerable attention from physicists. Successive addition of 4f-electrons along the series has little influence on many of the chemical properties, which are mainly those of a tripositive cation. However, such uniformity does not extend to properties derived from the electronic spin moment or from the atomic nucleus and therefore to the Mössbauer spectra. A resonance is known in at least one isotope of every rare-earth element except the first (cerium) and the last (lutetium). Furthermore, a resonance is often known for several isotopes, and in some instances for several transitions within a given isotope. It has therefore been possible to measure such properties as the nuclear magnetic moment for a large number of nuclear states within a small range of masses, and much of the work done has been directed to this end. The significance of the results to an understanding of the structure of the nucleus will not be considered here. Rather we shall describe the experimental phenomena and any derived nuclear parameters, together with the smaller proportion of data accrued which are chemically interesting. Numerical data for all transitions are summarised in Appendix 1.

In addition to the large hyperfine fields recorded in magnetically ordered materials such as the metals themselves and the rare-earth iron garnets, it is not uncommon to find magnetic fields in paramagnetic compounds. Electronic relaxation times are frequently found to be quite long at low temperatures, and the temperature dependence of the relaxation behaviour can give a detailed picture of the electronic properties of the ion. However, such paramagnetic broadening can cause difficulties in finding a satisfactory unbroadened source matrix for use at very low temperatures.

Many of the nuclei in the rare-earth region are appreciably deformed from spherical symmetry by rotational effects. Consequently the nuclear quadrupole moments are large and can give sizeable quadrupole splittings. Usually the electric field gradient is that from the 4f-electrons, but it should be noted that for Eu^{2+}, Gd^{3+}, and Yb^{2+} ions the electronic configurations have spherical symmetry (4f^7 or 4f^{14}), while for Eu^{3+} (4f^6) the total angular momentum is zero ($^7F_0$ state). In these instances any quadrupole splitting must arise from the much smaller lattice contribution.

[*Refs. on p. 590*]

NEODYMIUM | 537

Large chemical isomer shifts are also observed in some instances, but less correlation has been done in this area.

The detailed Hamiltonians appropriate to the electronic and nuclear properties of the rare earths in general have been excellently summarised elsewhere [1]. They are not given here explicitly because of the more limited depth of treatment. Note, however, that the vectorial addition of the total orbital, L, and spin angular momentum, S, denoted by J is the most useful quantum number for describing electronic states. Any crystalline field potential then acts as a perturbation to the appropriate J state. This is opposite to the situation found in ^{57}Fe, where the crystalline field is the dominant term.

17.1 Praseodymium (^{141}Pr)

The high energy of the 145·43-keV transition in ^{141}Pr is not conducive to a strong Mössbauer resonance and indeed the first attempts to observe it

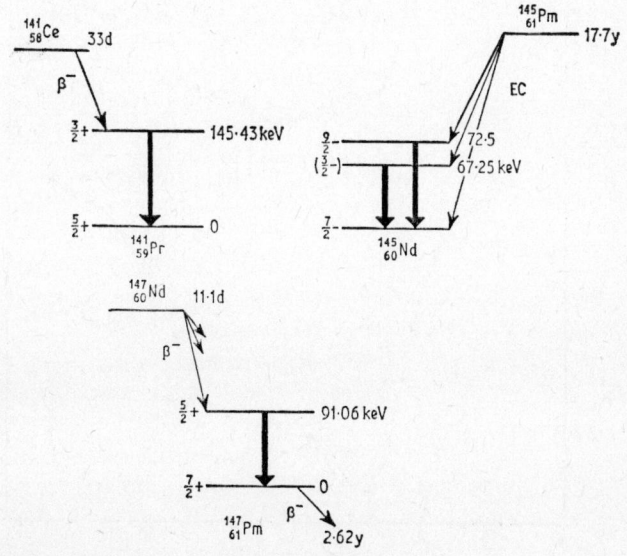

Fig. 17.1 Decay schemes for ^{141}Pr, ^{145}Nd, and ^{147}Pm.

using a ^{141}CeO$_2$ source and a Pr$_6$O$_{11}$ absorber or scatterer were not successful [2]. Later scattering experiments at lower temperatures with CeO$_2$ and CeF$_3$ sources and a Pr$_6$O$_{11}$ scatterer did however give a weak effect [3]. The simple decay scheme of the ^{141}Ce parent is shown in Fig. 17.1, but no information on hyperfine interactions is available.

17.2 Neodymium (^{145}Nd)

The 67·25-keV and 72·5-keV resonances in ^{145}Nd were first observed by Kaindl and Mössbauer in 1968 [4]. Both levels are populated by the decay of

[*Refs. on p. 590*]

538 | THE RARE-EARTH ELEMENTS

17·7-y ^{145}Pm (Fg. 17.1). This can be prepared by the reaction sequence ^{144}Sm$(n, \gamma)^{145}$Sm (EC 340 d)^{145}Pm with ion-exchange separation of the ^{145}Pm from the samarium after a suitable lapse of time (~6 months). A satisfactory source matrix can be made from ^{144}Nd$_2$O$_3$ doped with the ^{145}Pm, and all data have been obtained at 4·2 K.

The similarity in energy of the two γ-rays makes resolution difficult even with a Ge(Li) detection system. Although the broad 72-keV resonance

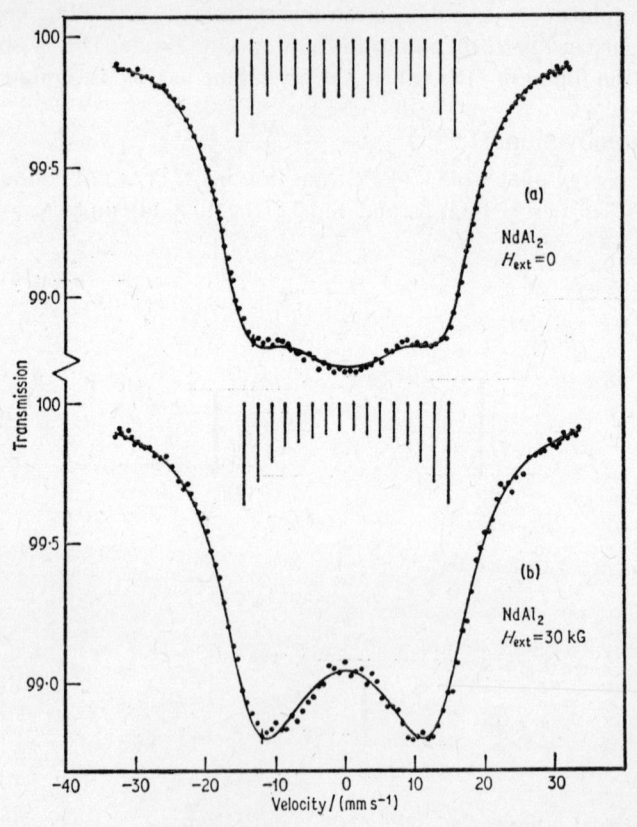

Fig. 17.2 Magnetic hyperfine splitting of the 72·5-keV resonance of ^{145}Nd in NdAl$_2$: (a) in zero applied field, (b) in a 30-kG longitudinal field. The solid lines are computer fits using the component lines indicated by bars. [Ref. 5, Fig. 1]

($\Gamma_r = 5\cdot 4$ mm s^{-1}) can be detected uniquely, the narrower 67-keV resonance ($\Gamma_r = 0\cdot 12$ mm s^{-1}) is weaker and is seen superimposed on the other from which it can be subtracted numerically. The initial results with a ^{145}Nd$_2$O$_3$ absorber showed single lines which appeared to be slightly broadened by an unresolved qudrupole splitting in the oxide.

In later measurements magnetic splitting was obtained in antiferromagnetic NdSb and ferromagnetic NdAl$_2$ [5]. Spectra for the 72-keV reso-

[Refs. on p. 590]

nance in the latter in a zero and a 30-kG external field are shown in Fig. 17.2. The excited-state spin quantum number was originally considered to be $I_e(72) = \frac{5}{2}$, but the only satisfactory interpretation of the data is that indicated by the solid lines in Fig. 17.2 which use $I_e(72) = \frac{9}{2}$. The bar diagrams are the component transitions for a $\frac{7}{2} \to \frac{9}{2}$ transition. Use of the known ground-state moment of $\mu_g(72) = -0.654$ (4) n.m. gave $\mu_e(72) = -1.121$ (13) n.m. The internal fields in NdSb, NdAl$_2$, and NdCo$_2$ are 3090, 2450, and 2560 kG respectively.

17.3 Promethium (^{147}Pm)

Promethium is another poor candidate for chemical Mössbauer spectroscopy as it is not a naturally occurring element. A resonance has been observed for the ^{147}Pm isotope (half-life 2·62 y) using sources of ^{147}Nd prepared by an (n, γ) reaction on ^{146}Nd$_2$O$_3$ [6]. Nine compounds as absorbers all gave a 91·06-keV single resonance line at 4·2 K with no detectable chemical isomer shift. No further details are available.

17.4 Samarium (^{149}Sm, ^{152}Sm, ^{154}Sm)

The known resonances in samarium are:

^{149}Sm: 22·5-keV
^{152}Sm: 121·78-keV
^{154}Sm: 81·99-keV

The 22·5-keV resonance in ^{149}Sm was first observed in 1962 independently by Jha, Segnan, and Lang [7], and by Alfimenkov et al. [8], thereby showing that this transition is to the ground state. Both groups used a source of ^{149}Eu (decay scheme in Fig. 17.3) in Sm$_2$O$_3$ with an absorber of ^{149}Sm$_2$O$_3$, and observed a single-line resonance. These measurements were later repeated for several absorber thicknesses [9]. The source activity can be generated by the ^{149}Sm$(p, n)^{149}$Eu reaction in a matrix of ^{149}Sm$_2$O$_3$, and can be separated by ion-exchange techniques and incorporated into Eu$_2$O$_3$ [10, 11]. An alternative to this is ^{150}Sm$(p, 2n)^{149}$Eu [10], or a spallation reaction by irradiation of tantalum with protons [9]. The recoil-free fraction of the Eu$_2$O$_3$ matrix is large enough to allow its use at room temperature. A ^{149}Eu$_2$O$_3$ source and a ^{149}Sm$_2$O$_3$ absorber show no resolved hyperfine effects from 20 to 300 K, although there is some degree of broadening [10].

Spectra of ferromagnetic samarium iron garnet (SmIG) were originally obtained to study the magnetic hyperfine splitting of ^{149}Sm [10]. However, there are three magnetically inequivalent rare-earth sites with approximately cubic symmetry in this material. Satisfactory interpretation was not possible until the magnetic-moment ratio had been determined accurately from data on SmCl$_3$.6H$_2$O (see below), and the original value for $^{149}\mu_e/^{149}\mu_g$ of $+1\cdot26(4)$ was found to be in error. Subsequently the temperature dependence of the magnetic field and the magnetisation were re-analysed [12]. The magnetic field is essentially the same for all three sites. Of particular importance is the

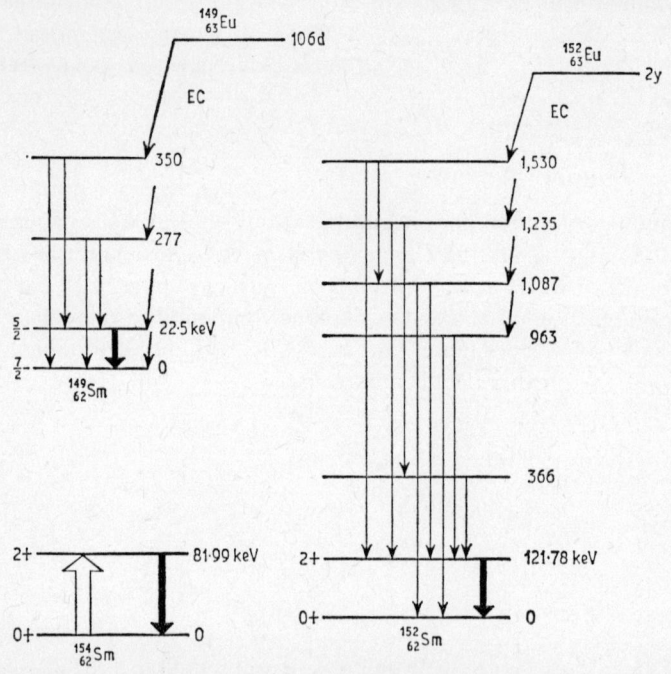

Fig. 17.3 Decay schemes for ^{149}Sm, ^{152}Sm, and ^{154}Sm.

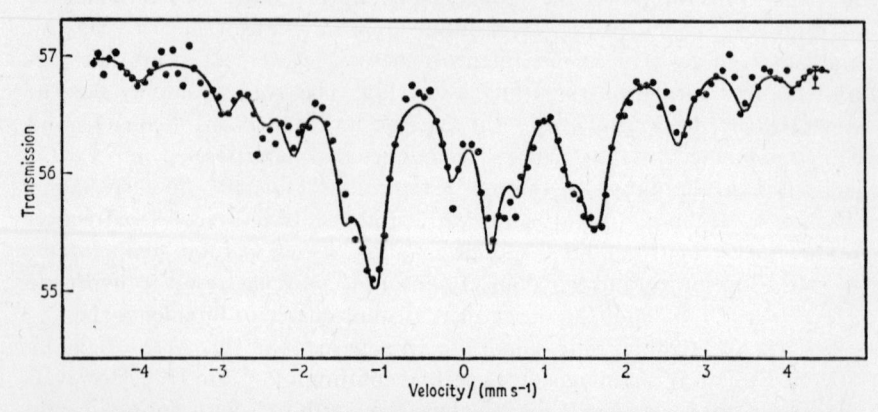

Fig. 17.4 The spectrum of SmCl$_3$.6H$_2$O at 4·2 K for the 22·5-keV ^{149}Sm transition. The solid line represents a constructed spectrum with $^{149}\mu_e/^{149}\mu_g = 0\cdot93$, $H_{eff} = 3450$ kG, and $e^2qQ_e = 14\cdot7$ mm s^{-1}. [Ref. 11]

[*Refs.* on p. 590]

first excited electronic state with $J = \frac{7}{2}$, which mixes into the ground-state $J = \frac{5}{2}$ manifold because the separation of 1100 cm^{-1} is small enough to allow significant population at room temperature.

An unambiguous value for the ^{149}Sm excited-state moment was obtained from the spectrum of SmCl$_3$.6H$_2$O at 4.2 K [11]. This is illustrated in Fig. 17.4, and features many lines because of the high spin-states involved ($I_g = \frac{7}{2}$, $I_e = \frac{5}{2}$). The compound contains only one site and shows a well-resolved magnetic interaction from slow paramagnetic relaxation. The computed reconstruction of the spectrum shown as a solid line gave $^{149}\mu_e/^{149}\mu_g = +0.93(1)$ and, with $^{149}\mu_g = -0.665$ n.m., one obtains $^{149}\mu_e = -0.62$ n.m. The parameters for the magnetic field and quadrupole interaction are given in Table 17.1. The quadrupole splitting of the ground state is smaller than that of the excited state by at least a factor of 8 and is therefore ignored.

Table 17.1 Magnetic and quadrupole parameters for ^{149}Sm and ^{154}Sm

Compound	T/K	H_{eff}/(kG)[a]	$e^2q^{149}Q_e$ /(mm s^{-1})[b]	Reference
Samarium-149				
SmCl$_3$.6H$_2$O	4.2	3450(105)	−14.7(1.8)	11
Sm(NO$_3$)$_3$.6H$_2$O	4.2	1900(280)		11
SmIG	16	2900(175)	−11.0(3.3)	11
SmAl$_2$	20	3520(175)		11
SmFe$_2$	20	3250(175)		11
SmNi	20	3250(345)		11
Sm metal	4.2	3450(175)	−13.2(3.3)	11
Samarium-154				
SmCl$_3$.6H$_2$O	4.2	3450[c]	13.0(7)[d]	17
	30	3444	11.0(1.0)[d]	17
SmIG	4.2	3040(120)	3.1(1)[d]	17

[a] converted from MHz using 1 MHz = 6.9064 kG.
[b] converted from MHz using 1 MHz = 0.0551 mm s^{-1}.
[c] assumed from ^{149}Sm data.
[d] $e^2q^{154}Q_e$ (derived using 1 MHz = 0.01505 mm s^{-1}).

The magnetic field in SmCl$_3$.6H$_2$O corresponds closely to that predicted for the free-ion Sm^{3+} in the fully magnetised state $J_z = J$ and, by assuming that this configuration also produces the electric field gradient tensor, a value of +0.40(6) barn was derived for the quantity $(1 - R)^{149}Q_e$ [11].

A similar paramagnetic hyperfine splitting was found in Sm(NO$_3$)$_3$.6H$_2$O, and more conventional magnetic ordering in several alloys (see Table 17.1) [11].

Initial chemical isomer shift measurements on ^{149}Sm were inconclusive [10], but more recent data have shown values in the order SmCl$_2$ < Sm$_2$O$_3$

[*Refs. on p. 590*]

< Sm metal [13]. The difference between Sm^{2+} and Sm^{3+} is caused by the different shielding of the closed-shell s-electrons by the $4f^6$ and $4f^5$ configurations respectively. The addition of an f-electron reduces the s-electron density at the nucleus, and therefore it can be inferred that $\delta\langle R^2\rangle$ $[=\frac{2}{3}R^2(\delta R/R)]$ is positive. The difference in electron density $|\psi(0)|^2(Sm^{3+}) - |\psi(0)|^2(Sm^{2+})$ at the nucleus was estimated to be $+3.4 \times 10^{26}$ cm^{-3}. The conduction-electron density at samarium metal was then deduced to be $+1.4 \times 10^{26}$ cm^{-3} from the observed shifts.

The second samarium Mössbauer resonance, involving the 121·78-keV transition of ^{152}Sm, was reported by two groups independently in 1967 [14, 15] and is difficult to observe because of the high energy of the γ-ray causing the resonance to be weak even at 4·2 K. A convenient source is ^{152}Eu in Gd_2O_3, and the decay scheme is shown in Fig. 17.3.

The magnetic hyperfine spectrum of ^{152}Sm in samarium iron garnet is shown in Fig. 17.5 and comprises five lines as expected for a $0^+ \to 2^+$

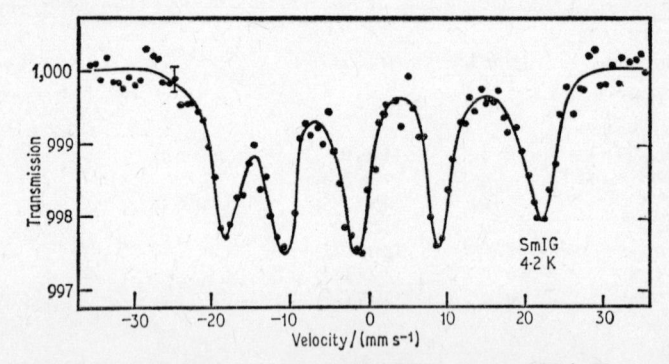

Fig. 17.5 The ^{152}Sm spectrum of samarium iron garnet (122-keV transition) at 4·2 K. The five lines correspond to a $0^+ \to 2^+$ magnetic hyperfine splitting. [Ref. 16, Fig. 1]

transition [16]. As with the ^{149}Sm resonance the three fields are not distinguished. A value of H_{eff} at 20 K of 3080(150) kG was used to derive a value for $^{152}\mu_e$ of $+0.832(50)$ n.m., but it should be noted that this field differs slightly from other quoted values.

A source of ^{152}Eu in CaF_2 decays to the Sm^{2+} ion with which an Sm_2O_3 absorber gives a chemical isomer shift of $+1.65(15)$ mm s^{-1} at 4·2 K (i.e. the transition in Sm^{2+} is lower in energy by 1·65 mm s^{-1}) [15]. Although the $I_g = 0$ ground state and $I_e = 2$ excited state are different rotational levels of the nucleus, the radius is altered by centrifugal stretching. As with ^{149}Sm the sign of $\delta\langle R^2\rangle$ is positive, and calculations gave $\delta\langle R^2\rangle/\langle R^2\rangle = +10(3) \times 10^{-4}$.

Independent comparison of $^{152}Eu/CaF_2$ and $^{152}Eu/Gd_2O_3$ sources at 4·2 K with an Sm_2O_3 absorber gave apparently different spectra [14]. The two

[*Refs. on p. 590*]

sites in Sm_2O_3 showed different relaxation times at 4·2 K so that the observed spectrum was a singlet superimposed on a quintet splitting. At 1·8 K two magnetic splittings pertain. However, the shift change of 1·64(6) mm s^{-1} between Sm^{3+} and Sm^{2+} is in good agreement with the previously mentioned value. Confirmation of the Sm^{2+} spin-state in CaF_2 was obtained from an absorber of CaF_2 doped with samarium which gave a five-line Sm^{3+} contribution and a sharp Sm^{2+} line. The derived value for $\delta\langle R^2\rangle$ was 19×10^{-3} fm^2, giving $\delta\langle R^2\rangle/\langle R^2\rangle = 7·7 \times 10^{-4}$.

The third samarium Mössbauer transition, stemming from the 82-keV level of ^{154}Sm, has no radioactive precursor but measurements have been made using Coulomb excitation [17, 18]. The target was Sm_2O_3 at 30–35 K. The absorption spectrum of $SmCl_3.6H_2O$ was a five-line hyperfine spectrum from the $0^+ \to 2^+$ levels. Taking the field in $SmCl_3.6H_2O$ as being 3450 kG (from ^{149}Sm data) gave $^{154}\mu_e = +0·778(36)$ n.m. The ^{154}Sm data for samarium iron garnet gave the internal field as 3040 kG. The quadrupole splitting values led to an estimate of $(1 - R)^{154}Q_e = -1·33(46)$ barn.

The chemical isomer shift of ^{154}Sm in $SmCl_2$ is 0·15 mm s^{-1} less than in Sm_2O_3 [18], from whence $\delta\langle R^2\rangle/\langle R^2\rangle = 0·46(34) \times 10^{-4}$.

17.5 Europium (^{151}Eu, ^{153}Eu)

The four known resonances in europium are:

^{151}Eu: 21·6-keV,
^{153}Eu: 83·37-keV, 97·43-keV, and 103·179-keV

The dominance of the 21·6-keV ^{151}Eu transition makes it convenient to discuss this before the three ^{153}Eu transitions, pp. 555 ff.

Europium-151 Parameters

The 21·6-keV transition in ^{151}Eu was first recorded in 1962 by Shirley et al. [19]. They used the ^{151}Gd parent (decay scheme shown in Fig. 17.6) in matrices of Nd_2O_3 and Eu_2O_3 with absorbers of Eu_2O_3. The oxides gave a single-line resonance without significant hyperfine interactions. The low energy of the transition lessens the problems associated with nuclear recoil, and high f-factors can be obtained in many compounds. The ease with which the resonance can be observed has led to more chemical studies with europium than for any other rare-earth element.

Both the ^{151}Sm and ^{151}Gd parents are long-lived, although the latter is more efficient in populating the 21·6-keV level. The most popular source matrices have been $^{151}Sm_2O_3$ and $^{151}Gd_2O_3$, although $^{151}SmF_3$ has also been used [20]. A detailed description has been given for the preparation of ^{151}Gd activity by the ^{151}Eu$(d, 2n)^{151}$Gd process, followed by its incorporation into a $^{153}Eu_2O_3$ non-resonant matrix [21]. All these sources give a high recoil-free fraction at room temperature and a single line with no appreciable hyperfine broadening.

Magnetic hyperfine splitting was first recorded for europium metal at 4·2 K in 1963 [22]. This confirmed the excited-state spin quantum number as $I_e = \frac{7}{2}$. The magnetic pattern is generally simpler than the $\frac{5}{2}, \frac{7}{2}$ spin-states would imply because many component lines overlap, and a recent spectrum

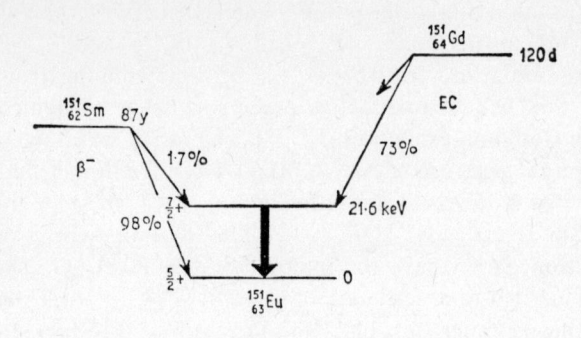

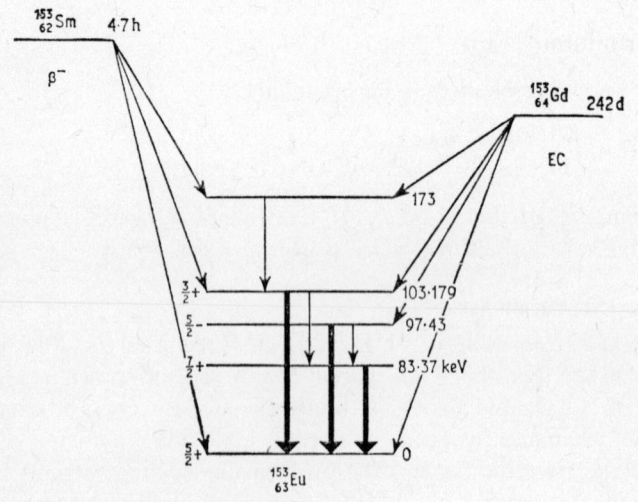

Fig. 17.6 Decay schemes for ^{151}Eu and ^{153}Eu.

for EuS is shown in Fig. 17.7 [23]. The original europium-metal data were analysed [22] to give a value for $^{151}\mu_e/^{151}\mu_g = +0.739(9)$. The result was confirmed independently using europium iron garnet (EuIG), giving $^{151}\mu_e/^{151}\mu_g = +0.74(5)$ [24]. The most accurate value for the ratio of the magnetic moments comes from europium iron garnet data [25] for which $^{151}\mu_e/^{151}\mu_g = +0.7465(7)$. Taking $^{151}\mu_g = +3.465(1)$ n.m. gives $^{151}\mu_g = +2.587(3)$ n.m. The possibility of the existence of a 'hyperfine structure anomaly' has already been discussed in connection with ^{193}Ir (p. 520). The hyperfine splitting in EuIG arises from the orbital momentum of the 4f-electrons in the Eu^{3+} ion, and the field is therefore constant over the nuclear radius. In EuO and EuS the field is generated by a contact interaction via core polarisation in the Eu^{2+} ion, and is not constant. The experimentally observed [23] ratios of $^{151}\mu_e/^{151}\mu_g$ at 4·2 K have been given as: EuIG 0·7465(7); EuO 0·7523(7); EuS 0·7525(7); which clearly illustrate the phenomenon.

It is difficult to measure a quadrupole splitting in europium compounds because of the electronic configurations of Eu^{2+} and Eu^{3+}. The Eu^{2+} ion has a $4f^7$ half-filled shell with an electronic ground state of $^8S_{\frac{7}{2}}$, which cannot

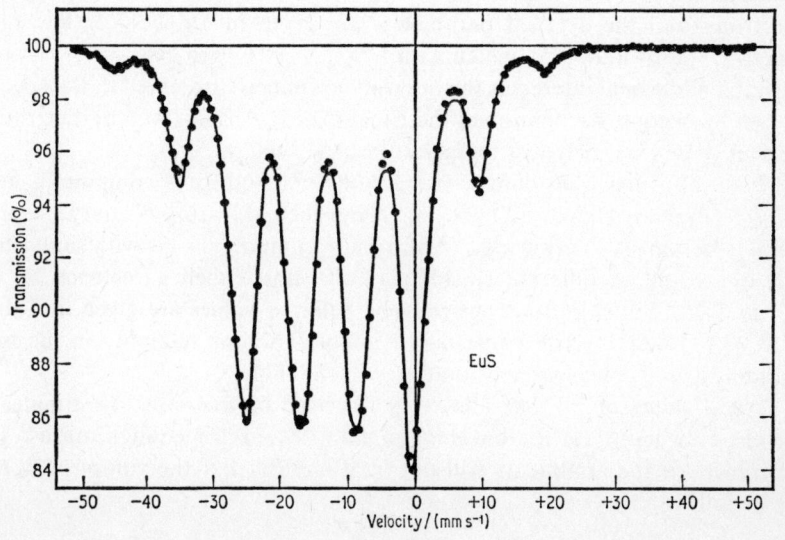

Fig. 17.7 The ^{151}Eu resonance in EuS at 4·2 K. The number of lines from the $\frac{5}{2}$, $\frac{7}{2}$ transition is reduced by accidental coincidences. Note the large negative chemical isomer shift for Eu^{2+} with respect to Eu^{3+} (i.e. the source of Eu$_2$O$_3$). [Ref. 23, Fig. 1]

produce a valence-electron contribution to the electric field gradient. Similarly the Eu^{3+} $4f^6$ configuration has a ground state of $^7F_0$, which has zero total angular momentum. Therefore any quadrupole splitting in Eu^{2+} and Eu^{3+} compounds can only arise from the lattice contributions to the electric field gradient, which are comparatively small. Such quadrupole splitting as does occur is generally manifest as a slight broadening of the resonance line, and analysis cannot be made because of the large number of lines from the $\frac{5}{2}$, $\frac{7}{2}$ spin-states. This can have the effect of causing small errors to be registered in the chemical isomer shift if the envelope is computed as a single Lorentzian, because of the asymmetric nature of the unresolved splitting [26]. Small quadrupole interactions are more readily detected in a magnetic spectrum, but the $^7F_0$ Eu^{3+} state is non-magnetic, and those Eu^{2+} compounds which are magnetic above 4·2 K have cubic lattices.

Initial attempts to measure $^{151}Q_e$ used europium iron garnet in which the Eu^{3+} is magnetically ordered by exchange polarisation effects from the iron [24]. However, failure to take into account the two Eu^{3+} lattice sites led to an inaccurate analysis, and careful reinvestigation gave a value for $^{151}Q_e/^{151}Q_g$ of +1·28(5) [25, 27].

A more direct measurement has used the non-cubic Eu^{2+} compounds EuSO$_4$, EuCl$_2$, EuC$_2$O$_4$, and EuCO$_3$ [28]. They are all magnetically ordered at 0·054 K. This extremely low temperature was achieved using a ^{3}He/^{4}He dilution refrigerator. The quadrupole interaction perturbs the magnetic spectrum, and the detailed parameters are given in Table 17.2. The ratio $^{151}Q_e/^{151}Q_g = +1\cdot30(5)$, which with $^{151}Q_g = 1\cdot16$ barn gives $^{151}Q_e = 1\cdot51$ barn. Of additional interest is the intensity asymmetry observed in the spectra caused by unequal occupation of these magnetic hyperfine levels of the ground state at these extremely low temperatures (see ^{57}Fe, p. 307).

Chemical isomer shift data for ranges of Eu^{2+} and Eu^{3+} compounds have come from many sources [20, 29–33]. The principal feature is a large separation ($\sim$15 mm s^{-1}) between Eu^{2+} and Eu^{3+} compounds. As with samarium, this is a result of different shielding of the closed-shell s-electrons by the $4f^6$ and $4f^5$ configurations respectively. Selected values are given in Tables 17.2–17.4. Where several values are available, the first reference in the final columns denotes the source of that used in the table.

Several values of $^{151}\delta\langle R^2\rangle$ have been derived on the basis of estimates of the electron density at the nucleus for the $4f^6$ and $4f^7$ configurations. The differences in the arguments will not be discussed, but the values of $\delta\langle R^2\rangle$ are as follows:

+0·030(10) fm^2 ($\delta R/R = 1\cdot22 \times 10^{-3}$) [20, 32]
+0·011(3) fm^2 ($\delta R/R = 0\cdot45 \times 10^{-3}$) [30]
+0·014(2) fm^2 ($\delta R/R = 0\cdot57 \times 10^{-3}$) [30]

Both Eu$_2$O$_3$ and EuF$_3$ have been used as reference standards for the chemical isomer shift. However, there is some confusion as to their inter-relationship. Careful measurement [34] has shown that a typical sample of oxide has a shift of +1·060(11) mm s^{-1} with respect to EuF$_3$.2H$_2$O. The anhydrous EuF$_3$ gives a shift of −0·052 mm s^{-1} w.r.t. EuF$_3$.2H$_2$O. Although there is no shift between EuF$_3$.2H$_2$O and a ^{151}SmF$_3$ source, the shift of the same oxide sample w.r.t. a ^{151}Sm$_2$O$_3$ source was −0·045 mm s^{-1}. Measurements of several oxide samples showed variations of the order of 0·2 mm s^{-1} in the shift. The situation is further aggravated by the existence of two forms of Eu$_2$O$_3$. The C form as usually produced is converted irreversibly to the monoclinic B modification at above 1400 K. Whether there is a difference in the two chemical isomer shifts is not clear from the literature, and in only

Table 17.2 ^{151}Eu Mössbauer parameters for Eu^{2+} compounds

Compound	T/K	δ (Eu$_2$O$_3$) /(mm s^{-1})	H_{eff}/kG	(e^2qQ_g/h) /MHz	Reference
EuF$_2$	RT	−14·17(9)			30, 35, 37
EuCl$_2$	RT	−14·33(9)			30, 32, 20
	4·2	−13·98(11)			29
	0·054		315(3)	−145(15)	28
EuBr$_2$	RT	−14·02(10)			30
EuI$_2$	RT	−14·04(11)			30
EuSO$_4$	RT	−14·57(10)			30, 32, 20
	4·2	−14·64(10)			29
	0·054		324(2)	+200(20)	28
EuCO$_3$	RT	−14·03(9)			30, 20
	4·2	−13·97(11)			29
	0·054		319(3)	−58(8)	28
EuC$_2$O$_4$	RT	−13·47(9)			30
	0·054		337(3)	+70(8)	28
EuMoO$_4$	RT	−13·98(16)			30
EuHPO$_4$	RT	−13·71(8)			30
Eu$_2$SiO$_4$	RT	−11·56(6)			30
EuB$_6$	4·2	−13·56(11)			29
Eu(C$_5$H$_5$)$_2$	RT	−13·2(5)			32
Eu(OH)$_2$	RT	−14·9(5)			32, 20
EuO	RT	−11·87(11)			30, 29, 32
	4	−12·1(2)	300(8)		36
Eu$_3$O$_4$	RT	{−12·6(1)			31
		{+ 0·6(2)	305(8) at 1·2 K		36
EuTiO$_3$	RT	−13·5(1)			31
EuZrO$_3$	RT	−14·1(2)			31
EuS	RT	−12·52(8)			30, 20, 32, 31
	80	−12·51(10)			29
	HE	−12·6(3)	331		37
EuGd$_2$S$_4$	RT	−11·9(2)			31
Eu$_3$S$_4$	RT	−12·6(2)			31
EuSm$_2$S$_4$	RT	−12·6(2)			31
EuLa$_2$S$_4$	RT	−12·9(2)			31
EuSc$_2$S$_4$	RT	−13·0(2)			31
EuYb$_2$S$_4$	RT	−13·8(2)			31
EuY$_2$S$_4$	RT	−13·6(2)			31
EuSe	RT	−12·65(8)			30, 20, 32, 31
	80	−12·08(10)			29
	HE	−12·6(3)	331		37
Eu$_{0.6}$Gd$_{0.4}$Se	RT	−12·2(2)			31
EuNd$_2$Se$_4$	RT	−12·9(2)			31
EuLa$_2$Se$_4$	RT	−13·1(2)			31
EuSc$_2$Se$_4$	RT	−13·4(2)			31
EuLu$_2$Se$_4$	RT	−13·6(2)			31
EuHo$_2$Se$_4$	RT	−13·8(2)			31
EuY$_2$Se$_4$	RT	−13·8(2)			31
EuTe	RT	−12·87(9)			30, 20, 32
	80	−12·71(10)			29
	HE		255		43

[*Refs. on p. 590*]

Table 17.3 ^{151}Eu Mössbauer parameters for Eu^{3+} compounds

Compound	T/K	δ (Eu$_2$O$_3$)/(mm s^{-1})	H_{eff}/kG	(e^2qQ_g/h)/MHz	Reference
EuF$_3$	RT	−0·59(2)			30
EuCl$_3$	RT	−0·29(9)			30, 33
EuBr$_3$	RT	−0·06(5)			30
EuI$_3$	RT	−0·06(4)			30
EuOF	RT	−0·72(2)			30
EuOCl	RT	−0·43(4)			30
EuOBr	RT	−0·29(3)			30
EuOI	RT	+0·02(9)			30
Eu$_2$(SO$_4$)$_3$	RT	−0·54(2)			30, 33
Eu$_2$(CO$_3$)$_3$	RT	−0·62(10)			30
Eu$_2$(MoO$_4$)$_3$	RT	−0·46(6)			30
Eu$_2$(HPO$_4$)$_3$	RT	−0·34(5)			30
Eu$_2$(C$_2$O$_4$)$_3$	RT	−0·54(1)			30, 33
Eu(ClO$_4$)$_3$	RT	−0·81(3)			33
EuPO$_3$	RT	−0·56(8)			33
Eu(OCOMe)$_3$	RT	−0·43(4)			33
Eu(EtSO$_4$)$_3$	RT	+0·25(5)			32, 20
Eu$_2$O$_3$	RT	+0·22(1)*			30
Eu$_2$S$_3$	RT	+0·19(7)			30, 20, 32
Eu$_2$Se$_3$	RT	+0·06(20)			30
Eu$_2$Te$_3$	RT	−0·14(8)			30, 20, 32
EuIG	(0)		{630, 570}	{−25(6), +15(6)}	25
	RT	+0·50(5)			20, 32

* This value was quoted relative to the monoclinic B form of Eu$_2$O$_3$ and possibly refers to the C form.

Table 17.4 ^{151}Eu Mössbauer parameters for europium alloys

Compound	T/K	δ (Eu$_2$O$_3$)/(mm s^{-1})	H_{eff}/kG	Reference
Eu	100	−8·70(8)		29
	4·2	−8·2(3)	265(10)	37, 22
Eu/Yb	4·2	−8·2(3)		53
Eu/Ba	4·2	−8·3(3)		53
EuCu$_2$	4·2	−8·8(2)	196(10)	54
EuZn$_2$	4·2	−9·4(2)	238(10)	54
EuPt$_2$	4·2	−9·9(2)	80(20)	54
EuPd$_2$	4·2	−9·5(2)	⩽20	54
EuPd$_3$	4·2	+3·6(2)		54
EuRh$_2$	4·2	+2·0(2)		54
EuAl$_2$	77	−9·7(3)		37
	1·8		278(10)	37
EuAl$_4$	77	−11·4(3)		37
	1·8		290(10)	37
EuSi$_2$	83	−10·6(3)		55
	4·2		286(10)	55
EuGe$_2$	295	−11·5(2)		55
	4·2		229(10)	55
EuSn$_3$	4·2	−10·9(5)	280(8)	56
EuSn	4·2	−12·0(5)	222(7)	56

one paper has the form of Eu_2O_3 used as a reference been clearly stated [30]. Consequently the adopted standard is not a good one, and the data tabulated are probably only accurate to within this uncertainty. We have quoted many of the values from the largest single set of data on the ground that these values will be self-consistent.

Correlation of the ionicity of the bonding with the chemical isomer shift in a number of similar compounds has revealed systematic relationships [30]. In series such as EuX_3, EuOX, and EuX_2 (X = I, Br, Cl, F) the shift decreases as the ionicity increases. The reverse trend is found in Eu_2Y_3 (Y = Te, Se, S, O) and this was taken to indicate a participation of the 5p-electrons in the

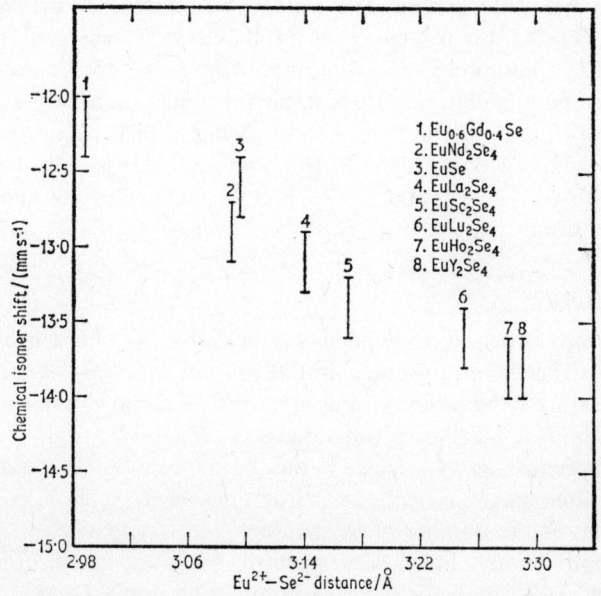

Fig. 17.8 Chemical isomer shifts of divalent europium selenides correlated with the mean Eu^{2+}–Se^{2-} distance. [Ref. 31, Fig. 3]

bonding. The electronic configurations of $4f^7 6s$ and $4f^7$ were deduced to differ by 14·8(4) mm s^{-1}, and $4f^6$ and $4f^7$ by 13·1(3) mm s^{-1}, leading to the values for $\delta\langle R^2\rangle$ given earlier.

Another useful systematic study has been made of a number of oxides, sulphides, and selenides by comparison of the Eu^{2+} chemical isomer shift with the mean Eu^{2+}–anion distance as estimated from the crystallographic lattice constants [31]. The correlation for the selenides is illustrated in Fig. 17.8. In all three cases there is an increase in the shift and s-electron density as the mean ionic separation decreases. Although the changes were suggested to be basically a pressure effect, this may well be a simplification in view of the results given later in this section for Eu^{2+} ions doped into CaF_2 and CaS.

[*Refs. on p. 590*]

The observed range of chemical isomer shifts for Eu^{3+} compounds is about -0.8 to $+0.3$ mm s^{-1} w.r.t. Eu_2O_3. This implies a considerable deviation from the ionic $4f^6 5s^2 5p^6$ configuration. Covalent bonding involving one or both of the 4f- and 6p-orbital types has been proposed [33], but no satisfactory quantitative interpretation has yet been given.

We shall now consider some of the groups of chemical compounds in more detail.

Europium Halides

The low-temperature magnetic spectrum of $EuCl_2$ has already been described but has not been interpreted. The EuF_2-EuF_3 system is particularly interesting [35]. The ^{151}Eu resonance at 77 K shows a superposition of both Eu^{2+} and Eu^{3+} components for the intermediate compositions. This means that the europium oxidation state is stable for longer than 10^{-8} s and there is no fast electron-hopping process. The Eu^{2+} line, which is at -14.2 mm s^{-1} in EuF_2, becomes a doublet for $EuF_{2.4}$ and higher stoichiometric ratios of fluorine, with $\delta = -15.0$ and -13.9 mm s^{-1}. Intermediate phases are believed to be produced, e.g. EuF_2-$EuF_{2.25}$ and $EuF_{2.25}$-$EuF_{2.43}$.

Europium Oxides

The oxide EuO orders ferromagnetically below 67 K with a field at 4 K of 300 kG [36]. The chemical isomer shift at this temperature is -12.1 mm s^{-1}. The temperature dependence of the magnetic field shows reasonable agreement with the $J = \frac{7}{2}$ Brillouin function. If the Eu^{2+} electronic configuration can be represented as $4f^7 5s^2 6s^x$, it has been proposed that the hyperfine field be written as $H_{\text{eff}} = H_{\text{core}} + H(6s^x)$ where H_{core} is a constant and $H(6s^x)$ varies as the number of 6s electrons and is opposite in sign to H_{core} [37]. An approximately linear relationship between the chemical isomer shift and the field was found for the compounds EuS, EuO, $EuAl_4$, $EuAl_2$, and Eu metal, but in view of later data for other compounds it seems that the concept is not general.

An attempt to measure the second-order Doppler shift of Eu_2O_3 between 100 K and 1000 K has revealed considerable deviations from the expected behaviour [38]. Temperature hysteresis effects are also found. The thermal properties of Eu_2O_3 are not well characterised, but it probably loses oxygen to become nonstoichiometric, the exact composition being dependent on the previous history of the sample.

The different magnetic behaviour of Eu^{2+} and Eu^{3+} ions is clearly seen in Eu_3O_4 [36]. This oxide becomes antiferromagnetic below 6.2 K. Above the Néel temperature there are two peaks with chemical isomer shifts of $+0.6$ mm s^{-1} (Eu^{3+}) and -12.5 mm s^{-1} (Eu^{2+}). In the magnetic phase only the Eu^{2+} ions order, with a field at 1.2 K of 305 kG. The temperature dependence of the field lies somewhat higher than the $J = \frac{7}{2}$ Brillouin function, although the reason for this is unknown.

In europium iron garnet both Eu^{3+} sites are magnetically ordered as well as the iron as a result of exchange and crystal-field interactions mixing the ^7F_J excited states into the non-magnetic $^7F_0$ ground state. There are two inequivalent sites, and parameters are given in Table 17.3. Early measurements [24] failed to distinguish the two sites which have similar parameters and this led to an erroneous measurement of the quadrupole effect (see earlier). The most recent experimental investigation has given accurate values for the magnetic field and quadrupole splitting at both sites between 1·5 K and 456 K as well as the nuclear constants detailed earlier [25]. At 0 K the two magnetic fields are 630 and 570 kG. A detailed crystal-field theoretical treatment has been given of the temperature dependence of the hyperfine field and quadrupole splitting [39].

In the gallium-substituted europium iron garnet the gallium goes almost entirely ($x < 1·6$) to the tetrahedral d site to give $\{Eu_3\}[Fe_2](Ga_xFe_{3-x})O_{12}$ [40]. An increase in gallium content changes the relative intensities of the two ^{57}Fe magnetic splittings, but also has the more dramatic result of causing a complete collapse of the ^{151}Eu magnetic spectrum at 4 K as x increases to 3·0. The europium field is produced mainly by an exchange interaction with the d sites. As the Fe atoms at these sites are progressively replaced by gallium the observed spectrum can be assumed to be a superposition of three components : (a) Eu^{3+} with two Fe^{3+} neighbours on d sites (statistical weight X^2 where X is the concentration of iron on d sites, which can be determined from the ^{57}Fe spectrum). (b) Eu^{3+} with one Fe^{3+} neighbour on d sites (statistical weight $2X[1 - X]$). (c) Eu^{3+} with no Fe neighbours on d sites (statistical weight $[1 - X]^2$). This simulation may be further improved by including a 12% contribution to the exchange field from the four third-nearest iron neighbours in tetrahedral sites. The actual ^{151}Eu spectra observed are shown in Fig. 17.9, and these can be simulated very accurately using the above model.

Further detailed information on exchange interactions can be obtained from the system $\{Eu_3\}[Sc_xFe_{2-x}](Fe_3)O_{12}$ [41]. The scandium substitutes exclusively at the octahedral a sites. Since the Eu–Fe exchange takes place via the tetrahedral d sites only, and the ^{57}Fe hyperfine field is constant for values of x up to 1·5, one expects to see no change in the ^{151}Eu spectrum. Unexpectedly, however, the increasing scandium content is seen to result in a partial collapse of the ^{151}Eu magnetic spectrum. An interpretation can be given by assuming that the iron d site spins are canted from the [111] direction. Each d spin with only one a site iron neighbour is canted by an average angle α, and each d spin with no a site Fe neighbours is canted by an average angle γ. It is then possible to calculate a statistical spectrum which simulates the data accurately, although the model is not necessarily the only possibility.

[Refs. on p. 590]

In europium iron garnet the magnetisation is in the [111] direction, but in samarium iron garnet at low temperatures it is close to the [110] direction [42]. Accordingly there is a significant difference between the spectra of a ^{151}EuIG absorber and a ^{151}SmIG source. The latter shows three fields of 634, 484, and

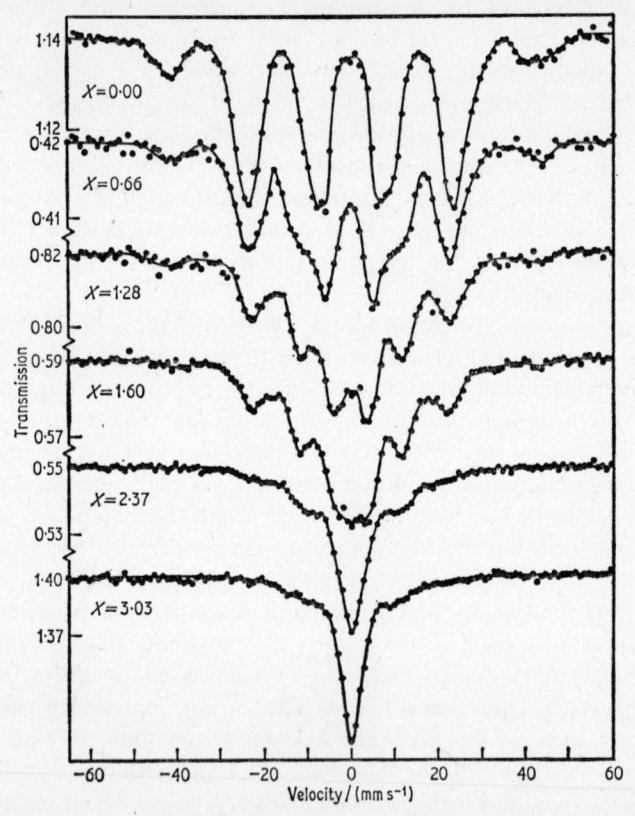

Fig. 17.9 ^{151}Eu spectra of gallium-substituted europium iron garnets of the type Eu$_3$Fe$_{5-x}$Ga$_x$O$_{12}$ showing the reduction in the exchange interaction at the Eu^{3+} ion with increasing gallium substitution. [Ref. 40, Fig. 3]

438 kG at 4·2 K. Mixed phases of the type {Eu$_x$Sm$_{1-x}$}IG are most closely related to the samarium garnet, but there is considerable spin canting, and indeed in SmIG itself the magnetisation is close to the [111] direction at 85 K.

Chalcogenides

The dependence of the chemical isomer shift on the interionic distance in Eu^{2+} sulphides and selenides has already been discussed. The shift and hyperfine fields in EuO, EuS, EuSe, and EuTe can be correlated empirically with the *d*- and *f*-character of the electron-conduction bands [43].

The sulphide Eu$_3$S$_4$ can be formulated as (Eu^{2+}Eu$_2^{3+}$)S$_4$ with all europium

[Refs. on p. 590]

cations occupying equivalent sites. Below 210 K the Mössbauer spectrum shows separate Eu^{2+} and Eu^{3+} resonances [44], but above this temperature motional narrowing occurs until a single line is found at about 300 K. This is illustrated in Fig. 17.10. Electron hopping takes place with an activation

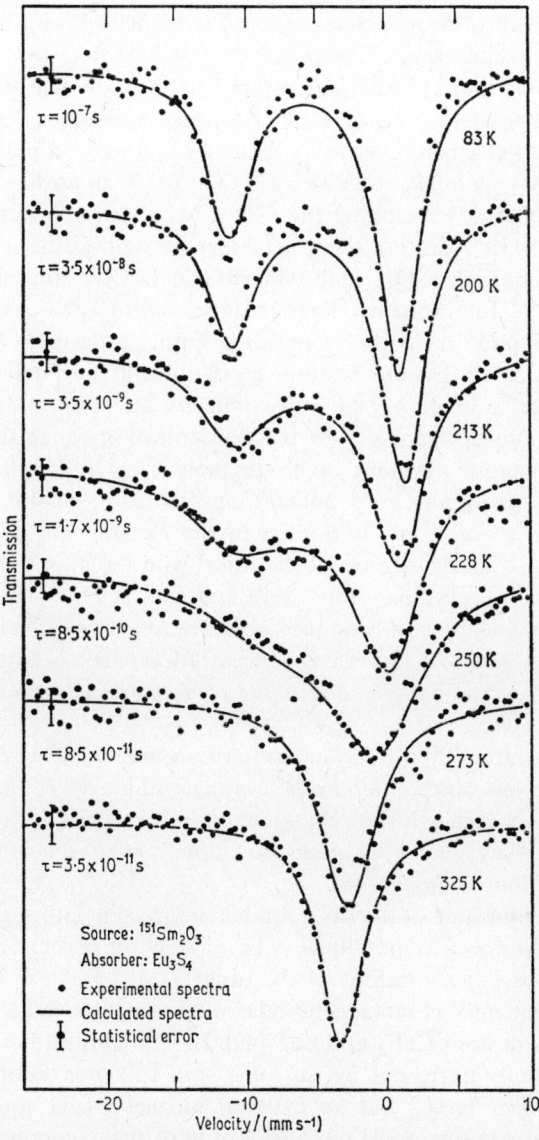

Fig. 17.10 ^{151}Eu spectra of Eu_3S_4. The solid lines are curves computed with a relaxation model whose time constant is τ. Fast electron hopping occurs at high temperature causing an average spectrum to be seen. [Ref. 44, Fig. 1]

energy of 0·24 eV, so that electronic exchange increases rapidly with rising temperature. This phenomenon is not found in Eu_3O_4 because it has a structure in which the Eu^{2+} and Eu^{3+} sites are not equivalent.

Other Europium Compounds

The ^{151}Eu chemical isomer shifts of five Eu^{3+} chelates with ligands such as ethylenediaminetetraacetic acid are in the range $-0·52$ to $-0·31$ mm s^{-1} [45]. All show an unusually large resonant effect at room temperature, a result of the lattice dynamics associated with this type of complex. A small participation of the 4f-electrons in covalent bonding has been proposed.

The thermal decomposition of $Eu_2(C_2O_4)_3.6H_2O$ in air has been followed by several techniques including the ^{151}Eu Mössbauer spectrum [46]. After straightforward dehydration there is further decomposition to a phase with empirical formula $EuCO_3$, and then to $Eu_2O_2CO_3$ and finally Eu_2O_3. However, the ^{151}Eu resonance shows that the '$EuCO_3$' phase is unequivocally a Eu^{3+} compound and is therefore not a simple carbonate. Eu^{2+} ions can only be produced if decomposition is carried out in a reducing atmosphere.

The chemical isomer shift of Eu^{2+} impurity atoms in a CaF_2 lattice decreases by about 1·0 mm s^{-1} in the concentration range 0·5–3 mole %, thereafter remaining constant up to 10 mole % [47]. The lattice constant increases with increasing Eu^{2+} content, and it was originally proposed that the effect was a result of a decrease in the electrostatic pressure on the Eu^{2+} ions. EuF_2 and CaF_2 are isostructural with the fluorite lattice, but the latter has smaller cell dimensions. EuS and CaS are also isostructural, the calcium compound again having the smaller lattice. However, in this case the Eu^{2+} shift *increases* with increasing concentration which is opposite to expectation if electrostatic compression is the correct explanation for both cases [48].

The Eu^{2+} coordination is octahedral in CaS and eightfold cubic in CaF_2. This difference in coordination leads to a reversal in sign of the fourth-order and some higher terms in the cubic potential. The opposing trends observed can then be interpreted by a π-electron donation to the otherwise vacant 5d-orbitals of the europium [48].

The $^8S_{\frac{7}{2}}$ ground state of the Eu^{2+} ion has a low spin–lattice relaxation rate, and the spin–spin relaxation frequency becomes of the order of the Mössbauer state lifetime at concentrations of the order of 5% Eu^{2+} in a diamagnetic lattice. Consequently paramagnetic relaxation phenomena have been found for Eu^{2+} ions in both CaF_2 and CaS [49]. In zero applied field at 4·2 K the lines are partially narrowed by isotropic spin–spin relaxation among electronic or nuclear levels, but an external magnetic field causes electronic polarisation and an increased resolution of hyperfine components.

[Refs. on p. 590]

Metals and Alloys

Europium metal is antiferromagnetic and its hyperfine spectrum was originally used to deduce the excited-state spin and magnetic moment (see earlier) [22]. Specific heat data for europium metal had shown indications of a transition at 16·1 K. However, the Mössbauer spectrum featured an additional paramagnetic component with a shift characteristic of Eu^{2+} at all temperatures above about 14 K [50]. Apparently the metal can become contaminated with the hydride EuH_2 during preparation, and it is the magnetic ordering of this phase at 16 K which was detected in the specific heat measurements.

The temperature dependence of the internal magnetic field in europium metal shows a sudden collapse at 88·5 K where a first-order phase transition causes the field to fall from 40% of the saturation value to zero [51].

A systematic study of the Eu/Yb and Eu/Ba alloys has been made [52, 53]. In the ytterbium system, the Curie temperature falls from 90 to 5 K and the saturation field also falls from 265 to 160 kG as the ytterbium content increases from 0 to 92 at. %. The relationships are linear apart from a discontinuity at 50 at. % where there is a phase change. Similarly for barium the Curie temperature falls from 90 to 40 K and the field from 265 to 206 kG as the barium content rises to 50 at. %. However, the chemical isomer shift is not significantly altered. The sign of the magnetic field is known to be negative from neutron diffraction data. Calculations suggest that a contribution of -340 kG to the field in europium metal arises from core polarisation, that $+190$ kG comes from conduction-electron polarisation by the atoms own 4f-electrons, and that -115 kG comes from conduction-electron polarisation, overlap, and covalency effects from neighbouring atoms.

Of alloys with transition metals, $EuCu_2$, $EuZn_2$, $EuPt_2$, and $EuPd_2$ give chemical isomer shifts in the range -9.5 to -8.8 mm s^{-1}, which corresponds roughly to a Eu^{2+} configuration (see Table 17.4), while $EuPd_3$ and $EuRh_2$ show positive shifts more consistent with Eu^{3+} [54]. Although the similar shifts of the first four alloys suggest a similarity in the 6s-electron charge density, the wide variation in hyperfine field points to different spin densities from conduction-electron polarisation.

The intermetallic compounds $EuSi_2$ and $EuGe_2$ are both magnetic at 4·2 K and are generally equivalent to Eu^{2+} compounds [55].

$EuSn_3$ and $EuSn$ both show a magnetic interaction at 4·2 K, but more information comes from the ^{119}Sn spectra which show two distinct tin sites in $EuSn_3$ [56].

Europium-153

On–off resonance experiments with the 103·2-keV transition of ^{153}Eu were made in 1960 [57], but the first Mössbauer spectrum was not reported until 1964 [58]; the first 97·4-keV resonance was reported in 1965 [59], and the

83·4-keV in 1968 [60]. The 83·4- and 103·2-keV levels of ^{153}Eu are conveniently populated by the ^{153}Sm parent, and the decay scheme is shown in Fig. 17.6. The 97·4-keV level is populated by the EC decay of ^{153}Gd. In this case the 103-keV γ-ray is also emitted, and the similarity in energies means that the observed spectrum is usually a superposition of the two components which have to be separated by computation. The source matrices adopted are the same as for ^{151}Eu.

The interference of the photoelectric effect and nuclear resonance absorption which gave a dispersion term to the line-shape of ^{181}Ta (see p. 508) also occurs in the 97·4-keV E1 transition of ^{153}Eu. A small asymmetry has been

Table 17.5 ^{153}Eu Mössbauer parameters

Compound	T/K	$\delta(Eu_2O_3)$ /(mm s^{-1})	H/kG	$(e^2q^{153}Q_g/h)$ /MHz	Reference
103-keV					
EuSO$_4$	4·2	+17·53(13)			29, 54, 65
EuO	4·2	+14·50(12)			29
EuB$_6$	4·2	+16·20(11)			29, 65
EuS	4·2	+15·0(1·5)			63
EuCl$_2$	4·2	+17·0(1·0)			63
Eu$_2$O$_3$	4·2	−0·245(26)			65
EuF$_3$	4·2	+0·94(14)			65
Eu	4·2	+10·30(26)			29
EuAl$_3$	4·2	+13·5(4)			65
EuAl$_2$	4·2	+10·3(3)			62
EuAl$_4$	4·2	+13·3(4)			62
EuCu$_2$	4·2	+9·8(4)			62
EuPt$_2$	4·2	+10·4(8)			62
EuRh$_2$	4·2	−2·1(5)			62
EuPd$_3$	4·2	−4·0(3)			62
EuIG	4·2	+0·42(7)	{636, 562}	{−57, +38}	67
97-keV					
EuSO$_4$	20	+15·1(1·5)			29, 59
EuS	4·2	+15·0(1·5)			29, 63
EuCl$_2$	4·2	+17·0(1·0)			29, 63
Eu	20	+11·0(1·5)			29
EuAl$_2$	4·2	+9·3(5)			62
EuAl$_4$	4·2	+12·7(6)			62
EuPt$_2$	4·2	+9·4(6)			62
EuRh$_2$	4·2	−1·8(6)			62
EuPd$_3$	4·2	−3·1(4)			62
83-keV					
EuSO$_4$	4·2	+0·65(22)			65
EuB$_6$	4·2	+0·35(14)			65
Eu$_2$O$_3$	4·2	+0·00(10)			65
EuF$_3$	4·2	−0·01(16)			65
EuAl$_3$	4·2	+0·38(22)			65

[Refs. on p. 590]

detected for a ^{153}GdF$_3$ source with an Eu$_2$O$_3$ absorber, but is absent as expected for the 103·2-keV transition [61].

Chemical isomer shifts for all four europium resonances have been measured in a number of compounds. The ^{153}Eu values are given in Table 17·5. Comparison of pairs of values for two transitions should show a linear relationship. An early attempt to verify this for the ^{151}Eu and ^{153}Eu (97 and 103 keV) transitions in Eu$_2$O$_3$, EuSO$_4$, and Eu metal disclosed a large deviation [59], but this was later shown to be a result of impurity in the metal [62].

The first value for $\delta \langle R^2 \rangle_{103}/\delta \langle R^2 \rangle_{97}$ was derived from data for EuCl$_2$, EuS, and Eu$_2$O$_3$ and gave a value of 1·05(10) [63]. A good linear relationship has also been shown in seven compounds for ^{151}Eu$_{21·6}$, ^{153}Eu$_{97}$ and ^{153}Eu$_{103}$ [62]. Independent data [29, 64] gave $\delta \langle R^2 \rangle_{103}/\delta \langle R^2 \rangle_{21·6} = -5·67(3)$ and $\delta \langle R^2 \rangle_{103}/\delta \langle R^2 \rangle_{97} = +1·09(6)$. By contrast chemical isomer shifts are small for the 83·4-keV resonance because the transition is a rotational one; however, they were observed for Eu$_2$O$_3$ and EuSO$_4$ in 1968 [60], and comparison with the 103-keV resonance has given a value of $\delta \langle R^2 \rangle_{83}/\delta \langle R^2 \rangle_{103} = +0·025(8)$ [65].

The magnetic splitting of the 83-keV transition in EuIG is badly resolved, but, by using some of the parameters from the 103-keV data, the magnetic-moment ratio has been determined as $^{153}\mu_{83}/^{153}\mu_g = +1·18(4)$, which with $^{153}\mu_g = +1·529(8)$ n.m. gives $^{153}\mu_{83} = +1·80(3)$ n.m. [65].

The broad linewidth (10·7 mm s^{-1}) of the 97-keV resonance also prevents resolution of the magnetic structure in europium iron garnet, but again using the known parameters it is possible to analyse the broadened envelope numerically [66]. The ratio $^{153}\mu_{97}/^{153}\mu_g$ is 2·10(15) from which $^{153}\mu_{97} = +3·21(22)$ n.m.

Better resolution can be achieved with the 103-keV transition. The first analysis gave a field of 600 kG and $^{153}\mu_{103} = +2·01(4)$ n.m. [58]. A more detailed study has taken into account the two europium sites [67]. The parameters derived at 4·2 K are given in Table 17·5. The ratios

$$^{153}\mu_{103}/^{153}\mu_g = 1·332 \text{ and } ^{153}Q_{103}/^{153}Q_g = 0·522$$

give values of $^{153}\mu_{103} = 2·04$ n.m. and $^{153}Q_{103} = +1·5$ barn. A third analysis gave $H_{\text{eff}} = 632$ kG and $^{153}\mu_{103}/^{153}\mu_g = +1·25(4)$ [65].

The magnetic field at Eu^{3+} ions measured by the 103-keV resonance in rare-earth garnets doped with 2·5% ^{151}Sm (M$_3$Fe$_5$O$_{12}$, M = Gd, Tb, Dy, Ho, Er, Eu, Tm, Yb, Lu, Y) increases by only 8% in that order [68]. Identical fields in YIG and LuIG in which the rare earth is diamagnetic show that the rare-earth–iron exchange interaction is insensitive to the lattice parameters which differ by 1%. However, in the series as a whole, the Eu^{3+}/rare-earth exchange interaction does alter and the field shows a linear dependence with the spin moment of the host, $\langle S \rangle = (g_J - 1)\langle J \rangle$. As already seen for the

^{151}Eu resonance the SmIG spectrum is different because it is the only one to magnetise in the [110] direction instead of [111].

A source of the alloy SmFe$_7$ has shown a very large hyperfine splitting of the 103-keV resonance as shown in Fig. 17.11 [69]. The magnetic field is

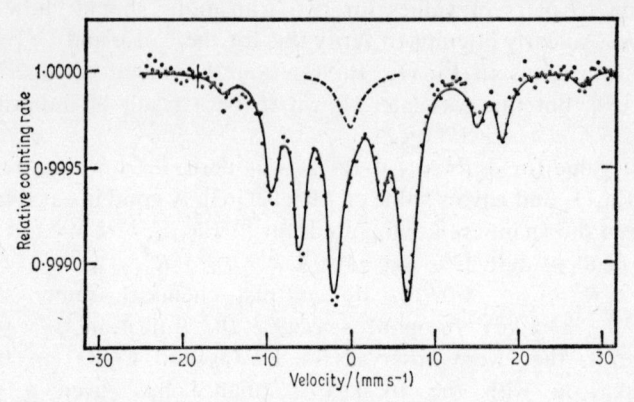

Fig. 17.11 The 103-keV ^{153}Eu resonance from a source of ^{153}SmFe$_7$ at 4·2 K. The contribution to the spectrum by Sm$_2$O$_3$ impurity is shown as a dotted line, and the solid curve is a theoretical fit using the parameters described in the text. [Ref. 69, Fig. 1]

1337(10) kG with an electric field gradient of $e^2q/4h = -96(8)$ MHz/barn. The ratio of the magnetic moments is $^{153}\mu_{103}/^{153}\mu_g = 1\cdot335(7)$ with $^{153}Q_{103}/^{153}Q_g = 0\cdot524(25)$. The large field was considered to arise from the exchange interaction (−1635 kG), core polarisation (−190 kG) and conduction electrons (+490 kG). Large fields were also found in Sm$_2$Co$_{17}$ (1544 kG) and Sm$_2$Ni$_{17}$ (360 kG).

17.6 Gadolinium (^{154}Gd, ^{155}Gd, ^{156}Gd, ^{157}Gd, ^{158}Gd, ^{160}Gd)

Gadolinium is unique in having eight known Mössbauer resonances in six isotopes:

^{154}Gd: 123·07-keV
^{155}Gd: 60·00-keV, 86·54-keV, and 105·32-keV
^{156}Gd: 88·97-keV
^{157}Gd: 64·0-keV
^{158}Gd: 79·51-keV
^{160}Gd: 75·3-keV

none of which unfortunately is easy to observe. Most require difficult and expensive experimentation, or do not show resolved hyperfine effects. Consequently little chemistry has been done beyond determining some of the nuclear parameters of interest. The chemistry of gadolinium centres around

[*Refs.* on p. 590]

the Gd^{3+} ion, which being in the $^8S_{\frac{7}{2}}$ configuration does not show a valence-electron contribution to the electric field gradient, and consequently quadrupole interactions are small.

For the four lighter isotopes ^{154,155,156,157}Gd the radioactive precursor is

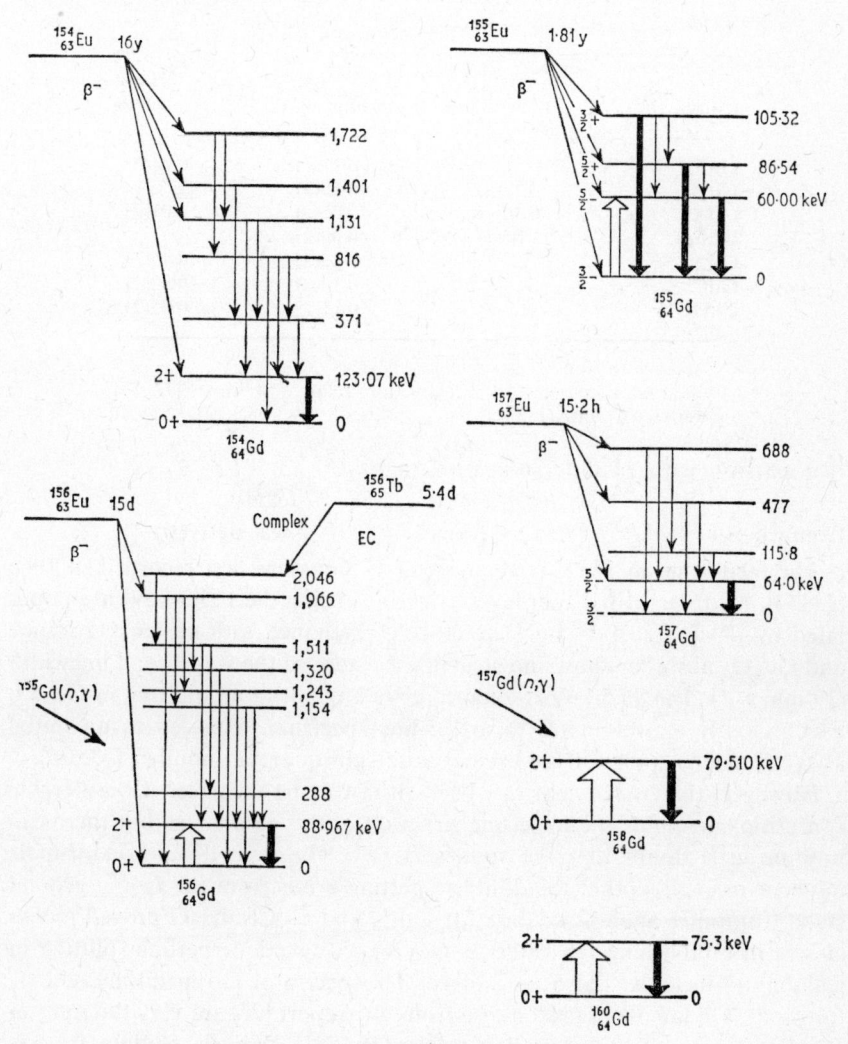

Fig. 17.12 Decay schemes for the six gadolinium isotopes.

usually a europium isotope. The Mössbauer level of ^{156}Gd can also be populated by EC decay of ^{156}Tb [70]. The decay schemes are complex and are shown in simplified form in Fig. 17.12. Coulomb excitation has been used for ^{156,158,160}Gd, and *in situ* Gd(n, γ) reactions for ^{156,158}Gd.

[*Refs. on p. 590*]

The 123·0-keV resonance of ^{154}Gd is very weak (~0·02%) even at 4·2 K, but chemical isomer shifts have been recorded. Initial data using a ^{154}Eu$_2$O$_3$ source and absorbers of Gd$_2$O$_3$ and GdAl$_2$ were inconclusive [71], but later results from a ^{154}EuF$_3$ source gave the values shown in Table 17.6 [72].

Table 17.6 Chemical isomer shifts in gadolinium at 4·2 K

Compound	δ (^{154}Gd)[a] /(mm s^{-1})	δ (^{155}Gd)[b] /(mm s^{-1})	δ (^{156}Gd)[c] /(mm s^{-1})
GdF$_3$		+0·11(2)	−0·04(2)
Gd$_2$O$_3$	+0·12(7)	+0·10(3)	
GdAl$_3$	+0·30(16)	−0·01(2)	+0·01(2)
GdFe$_2$	+1·02(19)	−0·39(2)	
Gd		−0·43(2)	+0·16(3)
GdCl$_3$		+0·12(2)	−0·02(2)
GdOOH		+0·11(2)	−0·05(2)

[a] relative to EuF$_3$ [72].
[b] data refer to the 86·5-keV transition, relative to SmAl$_3$ [72, 83].
[c] relative to SmAl$_3$ [83].

Comparison with ^{155}Gd (86-keV) data gave
$$^{154}\delta\langle R^2\rangle/^{155}\delta\langle R^2\rangle_{86} = -2·62(60)$$
from whence $^{154}\delta\langle R^2\rangle/\langle R^2\rangle = +7·5(2·3) \times 10^{-4}$ was derived.

The 60·0- and 86·5-keV resonances of ^{155}Gd were first recorded in 1966 [73, 74]. All three Mössbauer levels (the third being the 105·3-keV) are populated by β^--decay of ^{155}Eu. The 60-keV resonance with an Eu$_2$O$_3$ source and Gd$_2$O$_3$ absorber shows no structure because of the very broad linewidth (19 mm s^{-1}). The 86·5-keV resonance gives a quadrupole splitting in Gd$_2$O$_3$ which closely approximates to a two-line spectrum. This led to an initial assignment of $I_e = \frac{1}{2}$ for the excited-state spin quantum number [73, 75].

However, this assignment has been shown to be incorrect. Considerable confusion surrounds the hyperfine structure of this resonance, but there can now be little doubt that the spin is $I_e = \frac{5}{2}$. The excited-state quadrupole moment is small so that the doublet splitting arises from the $I_g = \frac{3}{2}$ ground state. Computer analysis of data for GdF$_3$ and GdCl$_3$.6H$_2$O proved inconclusive in establishing the value of Q_e [76]. Magnetic hyperfine splitting in gadolinium metal was also unresolved. The spectra of Gd metal and GdAl$_2$ for the ^{155}Gd (86·5-keV) resonance from one report [77] are very different in character to those in succeeding works [78, 79]. Analysis of data from a (^{155}Eu)Sm$_2$O$_3$ source and a GdFe$_2$ absorber has given $^{155}\mu_{86} = -0·53(5)$ n.m. [79]. Probably the most accurate data available are those shown in Fig. 17.13. The spectrum of ^{155}Eu in an Sm$_2$O$_3$ source matrix with a single-line GdRh$_2$ absorber clearly shows the quadrupole splitting of the excited state [78]. A better source matrix is SmH$_2$ which is, however, still split, but

[Refs. on p. 590]

it did allow analysis of the quadrupole splitting in GdF$_3 \cdot \tfrac{1}{2}$H$_2$O. The quadrupole moment ratio is $^{155}Q_{86}/^{155}Q_g = 0.12(1)$. Similarly the magnetic spectrum of GdFe$_2$ gave $^{155}\mu_{86}/^{155}\mu_g = 2.23(2)$ which with $^{155}\mu_g = -0.254(3)$ n.m. gives $^{155}\mu_{86} = -0.515(1)$ n.m.

Although it was originally reported that large magnetic fields could be

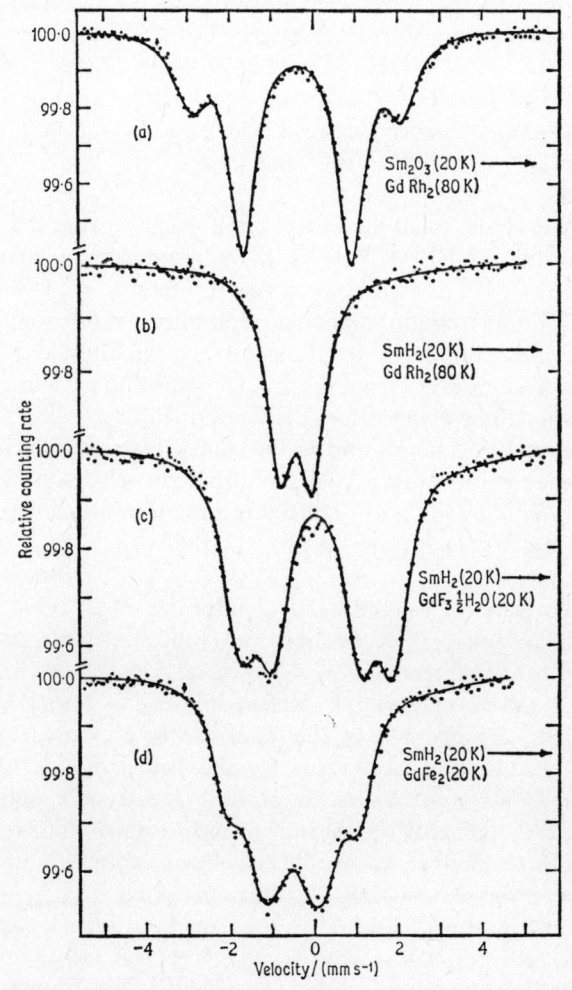

Fig. 17.13 The 86·5-keV ^{155}Gd resonance showing quadrupole and magnetic splitting. In all cases the source is also split by quadrupole interaction. [Ref. 78, Fig. 1]

induced at gadolinium nuclei in Gd metal or GdFe$_2$ by applying an external magnetic field [80], this has since been shown to be in error [81]. The internal field in GdFe$_2$ is 430 kG and in Gd metal is 305 kG. The origins of the hyperfine field have been discussed [82].

[*Refs. on p. 590*]

562 | THE RARE-EARTH ELEMENTS

Less data are available for the 105·3-keV ^{155}Gd resonance, although the excited-state spin of $I_e = \frac{3}{2}$ has been confirmed [79]. Quadrupole splitting in GdAl$_2$ gives $^{155}Q_{105}/^{155}Q_g \sim 1\cdot 0$. The magnetic spectrum of GdFe$_2$ has not been analysed unambiguously, and values for $^{155}\mu_{105}$ of $+0\cdot13(4)$ n.m. and $-0\cdot38(6)$ n.m. are both feasible on present evidence.

Chemical isomer shifts have been measured in five compounds using the 86·5-keV ^{155}Gd and the ^{156}Gd resonance with a source of SmAl$_3$ containing ^{155,156}Eu [83]. Values are given in Table 17.6. The linear interrelationship gives $^{156}\delta\langle R^2\rangle/^{155}\delta\langle R^2\rangle_{86} = -0\cdot36(6)$. From estimates of the s-electron density difference between Gd^{3+} and Gd metal the values $^{155}\delta\langle R^2\rangle/\langle R^2\rangle_{86} = -3\cdot0(8) \times 10^{-4}$ and $^{156}\delta\langle R^2\rangle/\langle R^2\rangle = +1\cdot0(3) \times 10^{-4}$ were deduced.

The 64-keV level of ^{157}Gd has a lifetime of 460 ns, giving it a very narrow natural linewidth of only $0\cdot0093$ mm s^{-1}; the resonance was first reported in 1966 [84]. The ^{157}Eu parent has a short lifetime of 15·4 h and the ^{158}Gd$(\gamma,p)^{157}$Eu preparation requires separation of the required source activity from other contaminants. The combined quadrupole splitting of the Gd$_2$O$_3$ source and absorber from the $\frac{3}{2}, \frac{5}{2}$ E1 transition gave line separations an order of magnitude greater than the linewidth.

Later work has used EuF$_2$ and CeO$_2$ source matrices which give single albeit broadened emission lines [76], but although well-resolved quadrupole splitting was found in GdF$_3$, it is difficult to obtain statistically good spectra. The ratios $^{157}Q_e/^{157}Q_g = 1\cdot78(4)$ and $^{155}Q_g$ (86 keV)$/^{157}Q_g = 0\cdot78(6)$ were found, from which $^{157}Q_g = 1\cdot67(27)$ barn. A small chemical isomer shift was found between GdF$_3$ and an EuF$_2$ source.

The ^{158}Gd resonance can be observed following neutron capture in ^{157}Gd at 21 K. The targets used were Gd metal, Gd$_2$O$_3$, or Gd$_{0\cdot03}$Y$_{0\cdot97}$Al$_2$ [85, 86]. Unresolved magnetic hyperfine splitting is found in Gd metal and GdN. The magnetic field at the latter can be estimated to be 359 kG from other data, leading to a value for the field in Gd of 312 kG and $^{158}\mu_e = +0\cdot770(44)$ n.m. Small unresolved quadrupole splittings were seen in Gd, GdF$_3$, and GdCl$_3$, and the chemical isomer shifts relative to Gd metal were within experimental error. The neutron capture technique has also been used for ^{156}Gd using Gd or Gd$_2$O$_3$ targets [86].

Coulombic excitation by 3-MeV protons has been used for several of the gadolinium levels [75, 87]. In each case the target was Gd$_2$O$_3$ at 4·2 or 77 K enriched in the appropriate isotope. The ^{156,158,160}Gd resonances showed unresolved quadrupole splitting which was analysed to give the ratios $^{156}Q_e/^{155}Q_g = 1\cdot04(2)$, $^{158}Q_e/^{155}Q_g = 1\cdot14(2)$, and $^{160}Q_e/^{155}Q_g = 1\cdot18(2)$. The 60-keV ^{155}Gd resonance in Gd$_2$O$_3$ showed no hyperfine structure because of the large linewidth.

[*Refs. on p. 590*

17.7 Terbium (^{159}Tb)

The 58·0-keV resonance in ^{159}Tb suffers from the disadvantage of a very large natural linewidth (36 mm s^{-1}). Initial measurements in 1966 used both the ^{159}Gd and ^{159}Dy parents (see Fig. 17.14) in the form of the respective sesquioxides, the gadolinium isotope giving the better results [88]. A resonance was found in Tb$_4$O$_7$, Tb$_2$(CO$_3$)$_3$, and Tb metal at 80 K. However, as

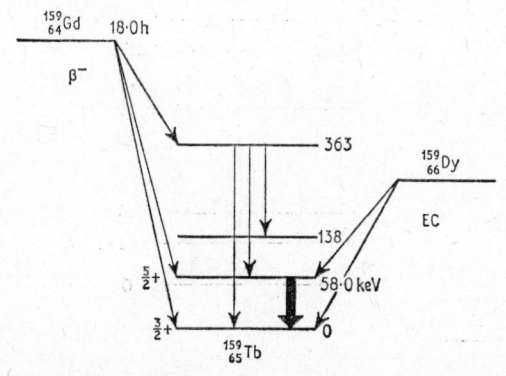

Fig. 17.14 Decay scheme for ^{159}Tb.

might be anticipated no hyperfine effects were resolved. Essentially similar data were given independently for Tb$_2$O$_3$, TbAl$_2$, TbFe$_2$, and Tb metal [89]. The broad envelopes of the TbFe$_2$ and Tb spectra were attributed to ferromagnetic hyperfine structure. Analysis by computer suggested a value for μ_e of +1·50 n.m., but this was not considered to be unambiguous. Interest in the resonance appears to have lapsed.

17.8 Dysprosium (^{160}Dy, ^{161}Dy, ^{162}Dy, ^{164}Dy)

Dysprosium has six known Mössbauer resonances:

^{160}Dy: 86·79-keV
^{161}Dy: 25·65-keV, 43·81-keV, and 74·57-keV
^{162}Dy: 80·7-keV
^{164}Dy: 73·392-keV

but only one of these, the 25·65-keV resonance in ^{161}Dy, has been extensively studied. It was first reported in 1960 for a ^{161}Gd$_2$O$_3$ source and a Dy$_2$O$_3$ absorber, the observed linewidth being two orders of magnitude greater than the natural width [90]. This early work was repeated and confirmed by several laboratories [91–93], and the presence of paramagnetic relaxation in the oxide matrices even at room temperature was realised. The interpretation of some early data at higher temperatures, where relaxation is expected to be less important [94], in terms of resolved magnetic splitting of the source appears inconsistent with later work on dysprosium relaxation.

The decay schemes for the various isotopes are given in Fig. 17.15. The 25·65-keV ^{161}Dy resonance will be discussed first. It has proved very popular

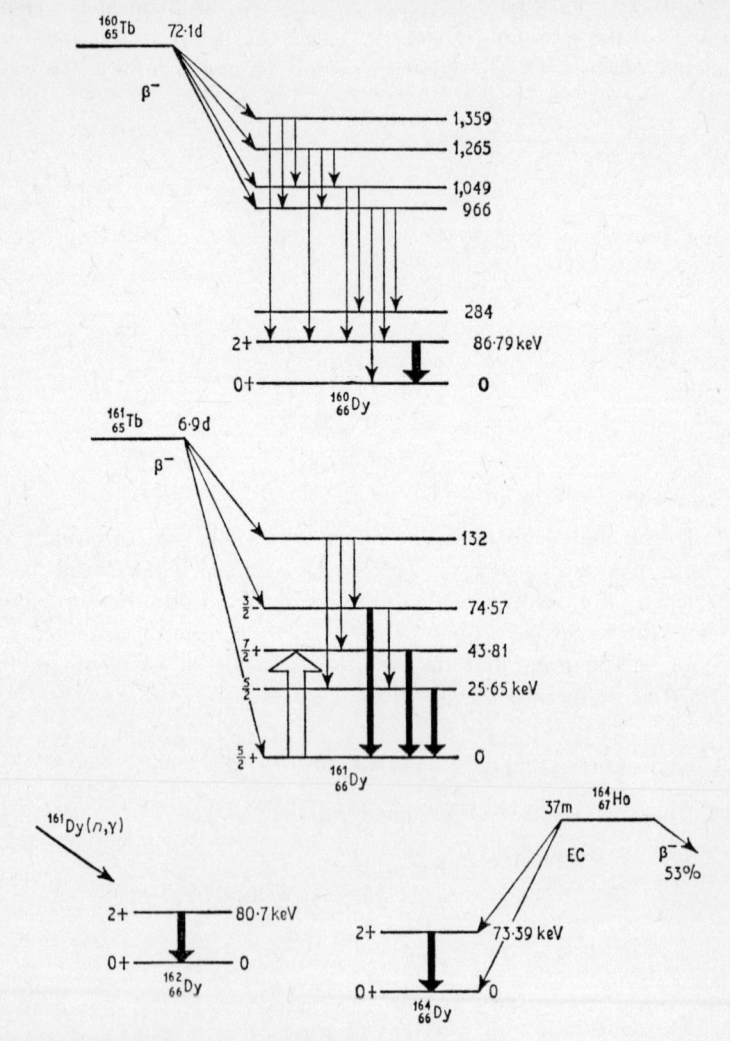

Fig. 17.15 Decay schemes for dysprosium.

for several reasons. The recoil-free fraction in most absorbers is adequate for experimentation at room temperature, which means that sources can be used under these conditions. Hyperfine splittings can be as much as three orders of magnitude greater than the natural linewidth, and the spin-states of $I_g = \frac{5}{2}$, $I_e = \frac{5}{2}$ produce enough lines to allow detailed interpretation. Para-

[*Refs. on p. 590*]

magnetic relaxation effects are often seen at low temperatures, although these can cause other problems such as unwanted broadening of the source emission line. However, it is experimentally easier to record a large hyperfine splitting using a partially broadened source. Relaxation broadening is found for example in a Gd_2O_3 matrix at room temperature activated by the $^{160}Gd(n, \gamma)^{161}Gd(\beta^-\ 3\cdot 7\ m)^{161}Tb$ reaction. DyF_3 as an absorber gives a narrow line, but unfortunately activated $^{161}GdF_3$ as a source does not. However, the compromise of a mixed fluoride $DyGdF_6$ is very successful, and a detailed preparation procedure has been given [95]. Other source matrices which have been used include ^{161}Gd in magnesium [96] and copper [97].

The dispersion term in the resonance line of an E1 transition first detected in ^{181}Ta has also been reported in the 25·65-keV ^{161}Dy resonance, although in this case the effect is much smaller [98, 99].

The ground state of the Dy^{3+} ion is $^6H_{\frac{15}{2}}$, and in many paramagnetic compounds it is the $|J_z = \pm\frac{15}{2}\rangle$ Kramers' doublet which lies lowest in the $J = \frac{15}{2}$ manifold. Substantial data on this system are already available from the electron-spin resonance technique. The maximum principal value that an anisotropic axially symmetrical $g =$ tensor can have in the free ion is $g_z = 19\cdot 6, g_x = g_y = 0$, which gives a maximum magnetic field contribution of $^{161}\mu_g\mu_N H_{eff}/(Ih) = -826$ MHz. This corresponds to a field of 5740 kG, but because of the close correlations with e.s.r. data it has become customary to quote internal magnetic fields in MHz (the expression $^{161}\mu_g\mu_N H_{eff}/(Ih)$ is often written as $g_0\beta_N H/h$ or in abbreviated form as $g_0\beta_N H$). Similarly the quadrupole coupling constant $e^2q^{161}Q_g$ is given in MHz (100 MHz = 4·836 mm s^{-1} for the 25·65-keV transition in ^{161}Dy).

The free-ion value of H_{eff} is frequently found in compounds with slow relaxation rates where the magnetic interactions involve dysprosium only, but variations are found for example where the compound contains a 3d-transition metal such as iron. The highly anisotropic g-tensor of the Dy^{3+} results in slow relaxation such as is usually only expected for magnetically diluted solids. Increase in temperature causes population of the excited-state Kramers' doublets and a collapse of hyperfine structure, detailed theory for which has been given [100]. Magnetic ordering of the Dy^{3+} ions generally splits the ground-state $|\pm\frac{15}{2}\rangle$ doublet, but causes little mixing in of the excited-state electronic levels. Consequently the hyperfine spectra below and immediately above the ordering temperature are often identical.

Several values for the ratios of the excited- and ground-state magnetic and quadrupole moments have been measured, and these are listed in chronological order in Table 17.7. Taking the value of $^{161}\mu_g = -0\cdot 472(13)$ n.m. and $^{161}\mu_{26}/^{161}\mu_g = -1\cdot 25(3)$, one obtains $^{161}\mu_{26} = +0\cdot 59(3)$ n.m.

In presenting results for particular compounds it is more convenient to use the order: oxide systems; other Dy^{3+} compounds; and alloys.

Table 17.7 Nuclear constants for dysprosium-161

Compound	μ_e/μ_g	Q_e/Q_g	Reference
25·65-keV			
Dy$_3$Fe$_5$O$_{12}$(DyIG)	−1·14(15)	0·75(40)	97
Dy, DyFe$_2$, DyAl$_2$	−1·19(5)	0·95(10)	101
Dy$_3$Al$_5$O$_{12}$(DyAlG)	−1·21(2)	0·98(3)	100
^{161}Dy/Gd	−1·2(1)	0·85(1)	102
74·57-keV			
Dy	+0·84(5)	0·48(4)	96
Dy	+0·852(12)	0·60(5)	103
Dy(NO$_3$)$_3$.5H$_2$O	+0·828(12)	0·58(4)	104
Dy$_3$Al$_5$O$_{12}$(DyAlG)	+0·84(3)	0·56(4)	105

Dysprosium Oxides

In the orthoferrite DyFeO$_3$ the iron sublattice orders magnetically at 653 K, but the dysprosium lattice remains paramagnetic above about 4·5 K at which temperature antiferromagnetic ordering occurs. Between 5 and 50 K the ^{161}Dy spectrum shows a well-resolved magnetic spectrum, the parameters of which are given in Table 17.8 [106]. Using the effective field approximation

Table 17.8 ^{161}Dy (26-keV) parameters in dysprosium compounds

Compound	T/K	g_z	$\{^{161}\mu_g\mu_N H_{eff}/(Ih)\}$ /MHz	$\frac{1}{4}e^2q^{161}Q_g$ /MHz	Reference
DyFeO$_3$	5	19·7	−830(20)	+435(20)	106
DyCrO$_3$	4·2		−806(20)	+420(20)	107
DyIG$^{(a)}$	85		−400(40)	+120(30)	97
	300		−84(2)	<20	97
DyAlG$^{(b)}$	4·2	18·2	−769(15)	+370(15)	100
Dy$_2$O$_3$	4·2	17·2	−727(10)	+290(30)	113
Dy$_2$(MoO$_4$)$_3$	4·2	18·9	−799(10)	+500(30)	113
DyMn$_2$O$_5$	4·2	19·7	−830(10)	+436(30)	113
DyES$^{(c)}$	4·2	10·75	−446(12)	−158(6)	111
Dy-oxalate	4·2	18·6	−787(10)	+575(30)	113
Dy-acetate.4H$_2$O	4·2	18·9	−796(10)	+625(30)	113
DyF$_3$.5H$_2$O	4·2	18·6	−785(10)	+525(30)	113
Dy$_2$(CO)$_3$	4·2	18·6	−780(10)	+550(30)	113
DyCl$_3$.6H$_2$O	4·2	16·0	−675(10)	+375(30)	113
Dy(NO$_3$)$_3$.6H$_2$O	4·2	17·4	−734(10)	+500(30)	113
DyPO$_4$.5H$_2$O	4·2	18·8	−795(10)	+575(30)	113

(a) dysprosium iron garnet Dy$_3$Fe$_5$O$_{12}$.
(b) dysprosium aluminium garnet Dy$_3$Al$_5$O$_{12}$.
(c) dysprosium ethyl sulphate.

for the energy levels one expects the ground state for the Dy^{3+} ion to be the $|J_z = \pm\frac{15}{2}\rangle$ Kramers' doublet at low temperature. The electric field gradient

($\frac{1}{4}e^2q^{161}Q_g$) from such a state has been estimated to be 630 MHz, so that the observed value of 435 MHz implies a contribution from the crystal lattice of −195 MHz. The values for g_z and the magnetic interaction of 19·7 and −830 MHz are very close to those for the free ion with $J_z = \pm\frac{15}{2}$. Above 50 K the increase in the thermal population of excited electronic levels causes a rapid decrease in the spin relaxation time and a collapse of the spectrum to a single line. The quadrupole splitting is only detected in the magnetically split spectrum.

The orthochromite, $DyCrO_3$, is paramagnetic above the Néel temperature of 2·16 K, but at 4·2 K a well-resolved magnetic hyperfine spectrum is seen because of a long paramagnetic relaxation time [107]. The parameters (Table 17.8) correspond closely to those for $DyFeO_3$. The spin relaxation time of the Dy^{3+} decreases rapidly with rising temperature, falling from 7 ns at 20 K to 0·05 ns at 78 K. The result is a complete collapse of the magnetic hyperfine spectrum so that the room-temperature spectrum is seen as a single line. This behaviour is clearly illustrated in Fig. 17.16.

Dysprosium iron garnet, $Dy_3Fe_5O_{12}$ (or DyIG), is ferromagnetic and gives a magnetic hyperfine splitting (see Table 17.8) [97]. Although the internal magnetic field is expected to decrease with rising temperature, the quadrupole interaction also diminishes because of a differing thermal population of the various J_z states split by the magnetic field which, in DyIG is well below the free-ion value, and may be taken to indicate the crystal-field interaction to be stronger in this case.

The spectra of dysprosium aluminium garnet (DyAlG) show similar effects to $DyCrO_3$, with a resolved magnetic spectrum below 20 K which does not change on passing through the Néel point at 2·49 K [100]. The relaxation time decreases from 10 ns at 4·2 K to 3 ns at 20 K.

Dysprosium gallium garnet is unique in that it shows a single line resonance even at 4·2 K because of an unusually fast spin–spin relaxation time [108].

Dy_2O_3 contains two distinct Dy sites, but these are not clearly distinguished in the Mössbauer spectrum. The paramagnetic hyperfine spectrum features only slight motional narrowing at 20 K, which implies an unusually long relaxation time [109].

Residual broadening is still present at room temperature. A theoretical treatment has given a satisfactory explanation of the long relaxation times [110].

Other Dysprosium Compounds

Dysprosium ethyl sulphate [$Dy(C_2H_5SO_4)_3.9H_2O$, or DyES] is unusual in that it shows a paramagnetic hyperfine interaction considerably less than that of the free ion (Table 17.8) [111]. Single-crystal measurements confirm that the magnetic field coincides with the crystal c axis and is collinear with the

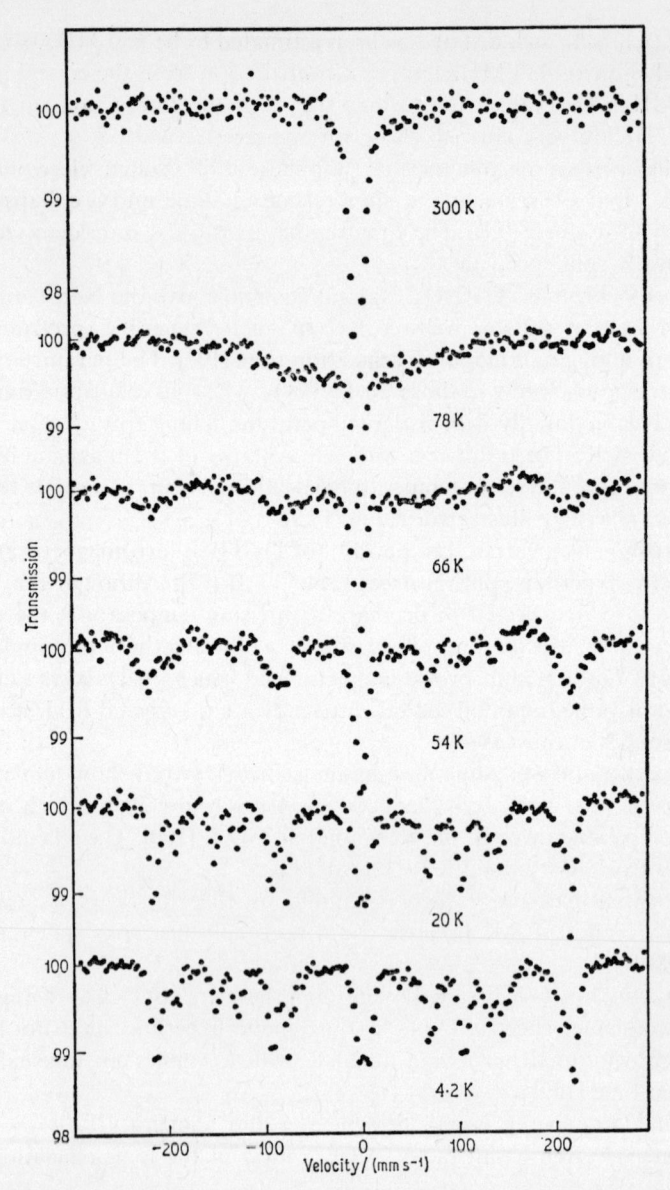

Fig. 17.16 The 25·65-keV ^{161}Dy resonance in DyCrO$_3$. Increasing temperature causes a reduction in the spin relaxation time and a complete collapse of the spectrum. [Ref. 107, Fig. 1]

[*Refs. on p. 590*]

axially symmetric quadrupole interaction. The magnetic field at 4·2 K of −446 MHz agrees with the g_z value of 10·75. The ground-state 4f-wavefunctions for DyES are accurately known, and contribute only −38 MHz to the observed quadrupole effect. The remaining −120 MHz derives from 'lattice' contributions. Detailed calculations of the relaxation phenomena below 14 K have been made and reproduce accurately the spectra observed in practice [112].

The paramagnetic hyperfine spectra of dysprosium oxalate $Dy_2(C_2O_4)_3$, dysprosium acetate $Dy(CH_3CO_2)_3.4H_2O$, $DyF_3.5H_2O$, $Dy(NO_3)_3.6H_2O$, $DyPO_4.5H_2O$, $Dy_2(MoO_4)_3$, and $DyMn_2O_5$ are comparable with $DyFeO_3$ [113]. All show a highly anisotropic g-tensor with g_z close to the limit of 19·6 for a $|J_z = \pm\frac{15}{2}\rangle$ doublet. The electric field gradient is axially symmetrical and collinear with the hyperfine field in each case. Dy_2O_3 and $DyCl_3.6H_2O$ show substantial differences from the others and presumably have more complicated ground states as found in DyES. The spectrum for the acetate is not that predicted from previous e.s.r. results which are now held to pertain to a minority constituent in the acetate, the ground state for the principal species being non-resonant.

The Dy^{2+} ion has been detected in CaF_2 doped with Dy^{3+} and reduced electrolytically [13]. The chemical isomer shift is about −7 mm s⁻¹ relative to Dy^{3+} in DyF_3, and this value was used to estimate the conduction-electron density in Dy metal.

Chemical isomer shifts are rarely reported for Dy^{3+} compounds because they are small compared to any magnetic splitting. However, a small number have been measured at room temperature where the resonances are narrower

Table 17.9 Chemical isomer shifts for ¹⁶¹Dy (26-keV) at room temperature relative to DyF_3 [114]

Compound	δ/(mm s⁻¹)	Compound	δ/(mm s⁻¹)
Dy	3·05(8)	$DyH_{2·08}$	0·50
DyN	0·85	DyF_3	0·00
Dy_2O_3	0·56	$DyF_3.\frac{1}{2}H_2O$	−0·04
$DyH_{2·90}$	0·55	$Dy_2(SO_4)_3.8H_2O$	−0·46

and these are given in Table 17.9 [114]. They were interpreted in terms of partial covalent bonding, but no serious systematic study has been made.

Frozen aqueous solutions of $Dy(ClO_4)_3$ show a drastic reduction in the recoil-free fraction at 183 K, similar to the effects seen with ⁵⁷Fe and ¹¹⁹Sn (pp. 140 and 393) as a result of structural change in the water [115].

Metallic Systems

Dysprosium metal is ferromagnetic below 85 K, antiferromagnetic between 85 and 178·5 K, and paramagnetic above this. The low-temperature magnetic

[Refs. on p. 590]

spectrum is typical of the Dy^{3+} free ion in a $J_z = \frac{15}{2}$ state and a typical spectrum measured at 77 K is shown in Fig. 17.17 [96]. The observed temperature dependence of the hyperfine field shown in reduced form in

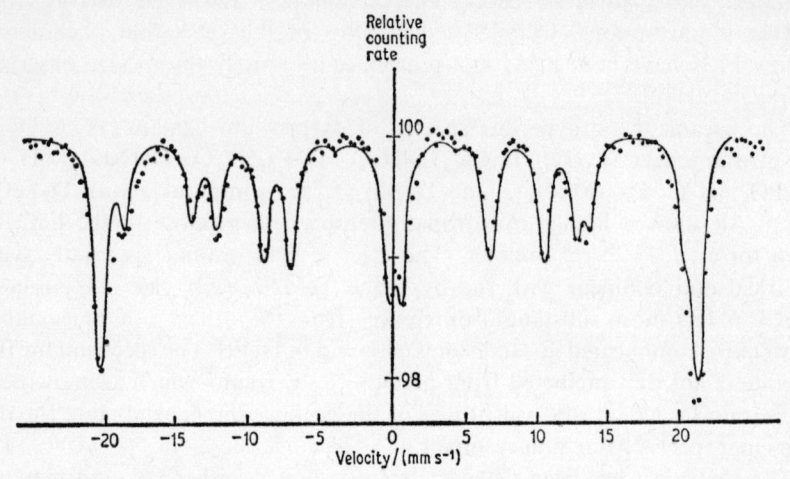

Fig. 17.17 The 25·65-keV ^{161}Dy resonance in Dy metal at 77 K. The solid line is a computed curve. [Ref. 96, Fig. 2]

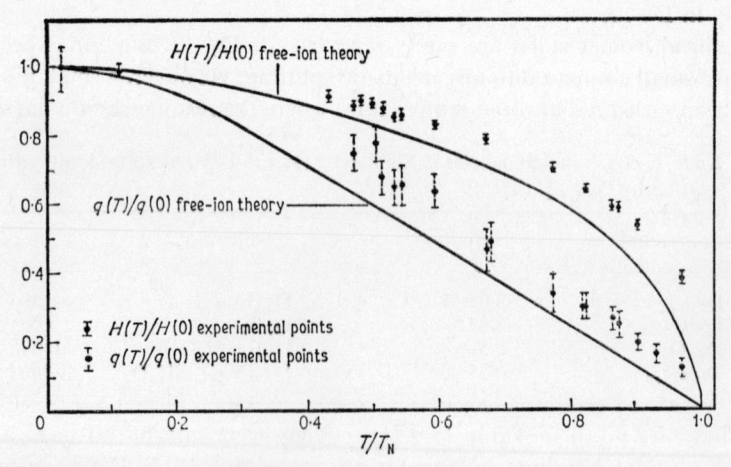

Fig. 17.18 The reduced values of the magnetic dipole, $H(T)$, and electric quadrupole splittings, $q(T)$, in Dy metal as a function of temperature, T. The solid lines represent predictions based on a free-ion model, and T_N is the Néel temperature. [Ref. 96, Fig. 3]

Fig. 17.18 agrees to a first approximation with the prediction of this model. Similarly the temperature dependence of the quadrupole splitting which is produced by thermal excitation within the J_z states split by the exchange

[*Refs. on p. 590*]

field can be reproduced. No discontinuity occurs at the ferromagnetic–antiferromagnetic transition at 85 K.

A large number of dysprosium intermetallic compounds are magnetically ordered at 4·2 K and show a magnetic splitting very close to the free-ion value (see Table 17.10) [101, 116, 117]. The quadrupole interaction, $\frac{1}{4}e^2qQ$, is

Table 17.10 ^{161}Dy (26-keV) parameters in metallic systems

Compound	$\{^{161}\mu_g\mu_N H_{eff}/(Ih)\}^*$ /MHz	δ/(mm s^{-1})†	Reference
Dy	−826	1·9	116
DyAl$_2$	−838	0·55	116
DyMn$_2$	−826	1·65	116
DyFe$_2$	−942	1·8	116
DyCo$_2$	−859	1·7	116
DyNi$_2$	−814	1·9	116
DyCu$_2$	−826	−0·2	116
DyGa$_2$	−834	−0·4	116
DyRu$_2$	−826	1·0	116
DyRh$_2$	−814	1·2	116
DyIr$_2$	−842	0·3	116
DyPt$_2$	−834	0·45	116
DyFe$_5$	−892	0·8	116
DyCo$_5$	−867	0·5	116
DyNi$_5$	−790	0·65	116, 118
DyNi	−842	1·45	116
Dy$_5$Si$_3$	−846	0·86	120
Dy$_5$Ge$_3$	−850	0·58	120
DySi$_2$	−838	−0·73	120
DyGe$_2$	−854	−0·98	120

* At 4·2 K.
† Measured at higher temperature; relative to Dy$_2$O$_3$.

in the range 560 to 680 MHz in all cases, close to the value of 630 MHz for the free ion. The spectra at 77 K and 300 K are all single lines and allow measurement of the chemical isomer shifts. These show distinct differences according to whether the other metal is a 3d-, 4d-, or 5d-transition metal, and it may be presumed that this is a result of a basically different behaviour in the conduction band.

More extensive temperature-dependence studies have been made on DyCo$_5$ and DyNi$_5$ [118]. The effective field in DyNi$_5$ is higher than that predicted from magnetic susceptibility measurements and shows that the nickel sublattice contributes to the total magnetisation by antiferromagnetic coupling. The temperature dependence of the relaxation in DyAl$_2$, DyNi$_2$, and DyCo$_2$ has also been studied in more detail [119].

The intermetallic compounds Dy$_5$Ge$_3$, Dy$_5$Si$_3$, DyGe$_2$, and DySi$_2$ also show parameters close to the free-ion values at 4·2 K [120].

Although the gadolinium Mössbauer resonances are not easy to study, it is possible to gain information from ^{161}Dy impurity atoms in gadolinium compounds. A source of ^{161}Gd in gadolinium metal at 5 K gives a ^{161}Dy magnetic splitting [121]. Part of the spectrum collapses to a single line at 30 K, but the greater part remains magnetic until the Curie point of the gadolinium. It was suggested that the decay of the ^{161}Tb intermediate nucleus results in two distinct electronic configurations about the ^{161}Dy daughter atom, which is unusual for a metallic matrix. The data for the temperature dependence of the hyperfine field have recently been analysed in more detail [122]. Each Dy impurity atom appears to reside on a regular Gd lattice site and the exchange field between Dy and Gd is effectively the same as that between Gd and Gd. With these assumptions a satisfactory simulation of the experimental data can be given.

Other Dysprosium Isotopes

Initial work on the ^{161}Dy 74·57-keV resonance in Dy_2O_3 did not give unambiguous data for the magnetic hyperfine structure [92]. An improvement in the 'signal-to-noise' ratio of the detection system was obtained by counting in coincidence the two γ-rays emitted in cascade in a proportion of the total decays by an *absorbing* nucleus *after* excitation [123]. The method suffers from the obvious disadvantage of requiring a high-intensity source to achieve a satisfactory counting rate.

More successful data accumulation was obtained using a ^{161}Tb/magnesium

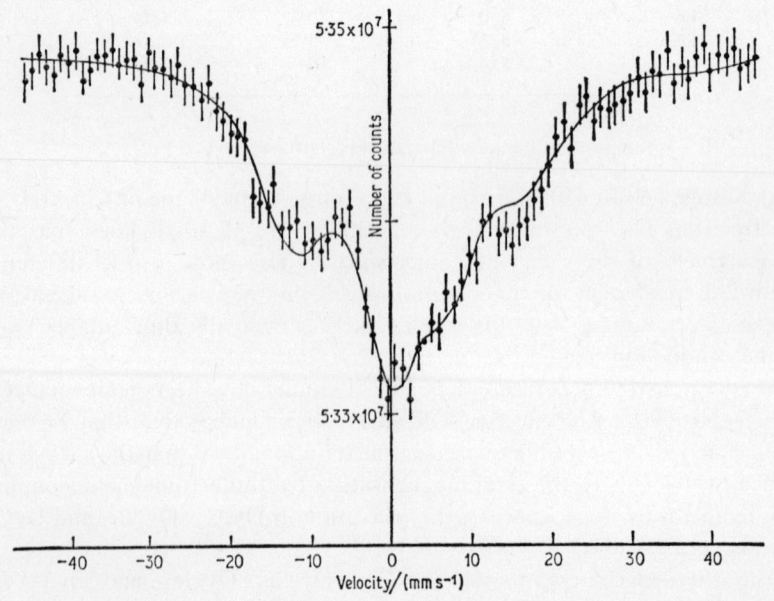

Fig. 17.19 The 74·57-keV ^{161}Dy resonance in Dy metal at 4·2 K. [Ref. 96, Fig. 4]

[*Refs. on p. 590*]

source and dysprosium metal at 4·2 K [96]. As seen in Fig. 17.19 the magnetic hyperfine structure is still unresolved, but with the ground-state parameters accurately known from the 26-keV data it proved possible to perform a computer analysis. The nuclear constants are given in Table 17.7. Basically similar independent data for Dy metal [103] Dy(NO$_3$)$_3$.5H$_2$O [104], DyAlG and Dy metal [105] confirmed this analysis. DyF$_3$.$\frac{1}{2}$H$_2$O and DyNi$_2$ gave single line resonances with a relative shift of 0·9 mm s^{-1} [104]. This is energetically equivalent to the shift in the 26-keV resonance, so that $\delta\langle R^2\rangle$ is approximately the same for both transitions.

The 43·8-keV ^{161}Dy transition was not recorded until 1969 when it was observed following Coulomb excitation of ^{161}Dy in a Dy$_2$O$_3$ target at 80 K with 3·3-MeV α-particles [124]. Dy$_2$O$_3$ and DyIG absorbers showed paramagnetic relaxation broadening at low temperatures, but no analysis for the excited-state nuclear parameters was attempted on the preliminary data.

The first major report of the ^{160}Dy resonance in 1965 [125] introduced the source matrix Tb$_{0.2}$Y$_{0.8}$Al$_2$, which is close to a single-line source at 25 K. DyF$_3$ as absorber at 20 K showed a single line. Dy$_2$O$_3$, (^{160}Dy)Fe$_2$Tb, and DyIG all showed five-line hyperfine patterns ($I_e = 2$, $I_g = 0$), the last named having two inequivalent sites which had not been distinguished earlier in ^{161}Dy. Analysis by comparison with ^{161}Dy data gave a value for $^{160}\mu_e$ of +0·74(7) n.m. Independent work with a Tb$_{0.05}$La$_{0.95}$Al$_2$ source at 20 K and a Dy$_2$O$_3$ absorber also gave $^{160}\mu_e = 0.74(8)$ n.m. [126].

The 80·7-keV level of ^{162}Dy has no radioactive parent, but can be studied by an *in situ* ^{161}Dy(n, γ)^{162}Dy reaction [103]. A DyF$_3$ target and a DyCl$_3$.6H$_2$O absorber gave a partially resolved five-line spectrum, and analysis gave $^{162}\mu_e = 0.74(8)$ n.m.

Coulombic excitation of a ^{164}Dy$_2$O$_3$ target gave well-resolved hyperfine structure in Dy$_2$O$_3$ at 4·2 K [127, 128], but as the velocity range scanned did not include all the component lines no calculation of the magnetic moment was made.

An alternative method for observing this resonance is to use the 37-m decay of ^{164}Ho [129, 130]. This can be prepared in a matrix of HoAl$_2$ by the ^{165}Ho(γ, n)^{164}Ho reaction, the emission line being unsplit. Analysis of the five-line spectrum of DyIG gave the values $^{164}\mu_e/^{161}\mu_g = -1.78(8)$ and $^{164}Q_e/^{161}Q_g = -0.83(7)$.

17.9 Holmium (^{165}Ho)

A report of a 94·70-keV resonance in ^{165}Ho appeared in 1966 [131]. The decay scheme is comparatively simple (Fig. 17.20), but the source lifetime is short and the natural linewidth of 96 mm s^{-1} is too large to expect resolved hyperfine effects. Consequently the resonance in ^{165}Ho$_2$O$_3$ is a single line, and no further attempt has been made to develop the use of this resonance.

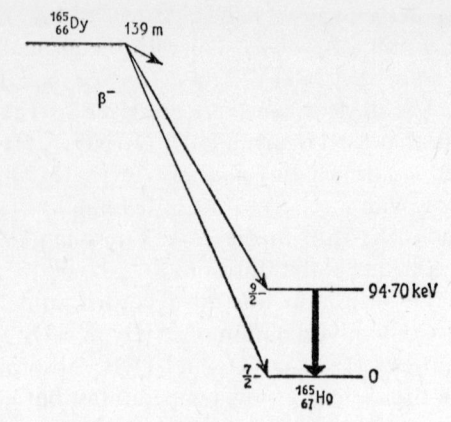

Fig. 17.20 The decay scheme for ^{165}Ho.

17.10 Erbium (^{164}Er, ^{166}Er, ^{167}Er, ^{168}Er, ^{170}Er)

There are five erbium Mössbauer resonances:

^{164}Er: 91·5-keV
^{166}Er: 80·56-keV
^{167}Er: 79·32-keV
^{168}Er: 79·8-keV
^{170}Er: 79·3-keV

and the appropriate isotope decay schemes are shown in Fig. 17.21. The short-lived ^{164}Ho isotope can be used to populate the 91·5-keV level of ^{164}Er. The compound HoAl$_2$ which has cubic symmetry provides a satisfactory, narrow emission line provided that it is kept above its Curie temperature of 25 K [130, 132]. At 4·2 K an absorber of ErCl$_3$.6H$_2$O shows a well-resolved five-line spectrum from paramagnetic hyperfine relaxation. The 0+ → 2+ transition has a magnetically split excited state because of a long spin–spin relaxation time. Comparison with similar spectra for ^{166}Er (see later) allows a direct determination of the ratio of the excited-state magnetic moments $^{164}\mu_e/^{166}\mu_e = 1·103(15)$, from which $^{164}\mu_e = 0·700(24)$ n.m.

The ^{166}Er resonance was first detected in an 'on–off' experiment [57]. Early Mössbauer spectra obtained at 20 K using an Ho$_2$O$_3$ source and an Er$_2$O$_3$ absorber were of poor quality [133, 134], and in the light of later data were probably misinterpreted [135]. The ^{166}Ho$_2$O$_3$ source suffers from line broadening, and better results can be obtained with ^{166}HoAl$_2$, although the high energy of the γ-ray necessitates cooling below room temperature to obtain an adequate recoilless fraction.

The paramagnetic hyperfine spectrum at 4·2 K of ErCl$_3$.6H$_2$O has five well-resolved lines for both ^{166}Er and ^{168}Er. This is illustrated in Fig. 17.22;

[*Refs. on p. 590*]

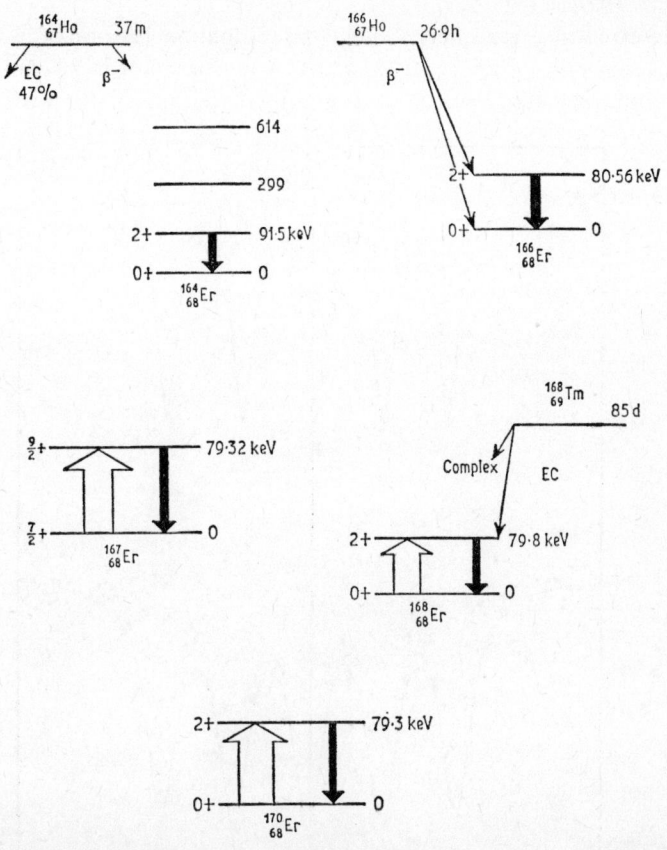

Fig. 17.21 The decay schemes for erbium.

the sources used were ^{166}HoAl$_2$ and ^{168}TmAl$_2$ respectively [136]. Comparison of a ^{166}Er spectrum with data from e.s.r. measurements on ^{167}Er in ErCl$_3$.6H$_2$O had already given a direct evaluation of the excited-state moment of ^{166}Er as $^{166}\mu_e = 0.624(20)$ n.m. [137]. Comparison with the $^{168}\mu_e$ data then gave $^{168}\mu_e/^{166}\mu_e = 1.042(8)$, from which $^{168}\mu_e = 0.662(20)$ n.m.

In nearly all cases of rare-earth paramagnetic hyperfine structure the long relaxation time necessary for observation is associated with a highly anisotropic hyperfine tensor. For general symmetry this is of the form

$$\mathcal{H} = A_Z S_Z I_Z + \tfrac{1}{2} A_X S_+ I_- + \tfrac{1}{2} A_Y S_- I_+$$

and when $A_Z \gg A_X, A_Y$ all but the first term may be omitted. Under these conditions, which also apply to magnetically ordered systems because the magnetisation is highly anisotropic, the level with spin I splits into $2I + 1$ equi-spaced levels. Thus in ErCl$_3$.6H$_2$O the $I_e = 2$ state splits into five levels with equal separation (any quadrupole interaction being small). This is

[Refs. on p. 590]

usually referred to as the effective-field approximation because of its analogy to the magnetically ordered system. If the A tensor is not highly anisotropic the spin–spin relaxation time is short and the hyperfine splitting is not seen.

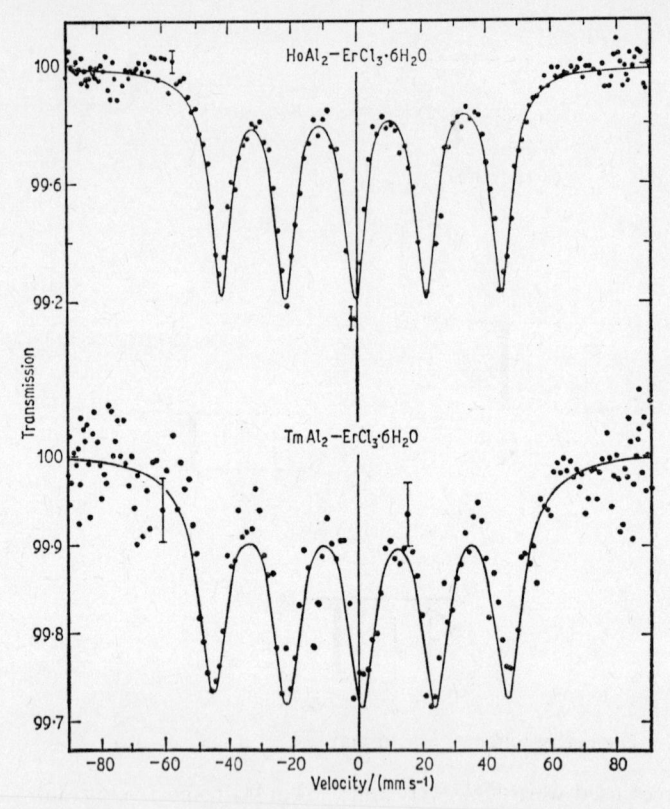

Fig. 17.22 Mössbauer spectra for $ErCl_3.6H_2O$ using the 80·56-keV ^{166}Er transition (source $HoAl_2$) and the 79·8-keV ^{168}Er transition (source $TmAl_2$). [Ref. 136, Fig. 1]

However, in a magnetically dilute system it is possible in principle to slow the spin–spin relaxation to the point where hyperfine structure is seen, and in this case the matrix elements involving S_+ and S_- (i.e. a mixing of different electronic states) cause an unequal spacing of the hyperfine levels.

This phenomenon is appropriate to the paramagnetic hyperfine spectra of magnetically dilute iron phases, and contrasts with the magnetically concentrated rare-earth paramagnetic hyperfine spectra where the effective-field approximation holds. However, it is possible to revert to the former situation by magnetic dilution. This has been shown in erbium ethyl sulphate diluted with the corresponding yttrium salt [138]. The hyperfine tensor was already known from e.s.r. data to be less anisotropic than usual, and the spec-

[*Refs. on p. 590*]

trum of a 2·4 at. % ^{166}Er sample is shown in Fig. 17.23. The five component lines are unequally spaced, and the possibility of a quadrupole interaction as an alternative explanation was eliminated by calculations.

ErFeO$_3$ is antiferromagnetic below 4·3 K and ferromagnetic below 640 K. The five-line ^{166}Er spectrum observed at 1·5 K is narrowed by relaxation effects at 3·4 K and becomes a single line by 4·2 K [139]. Spin relaxation

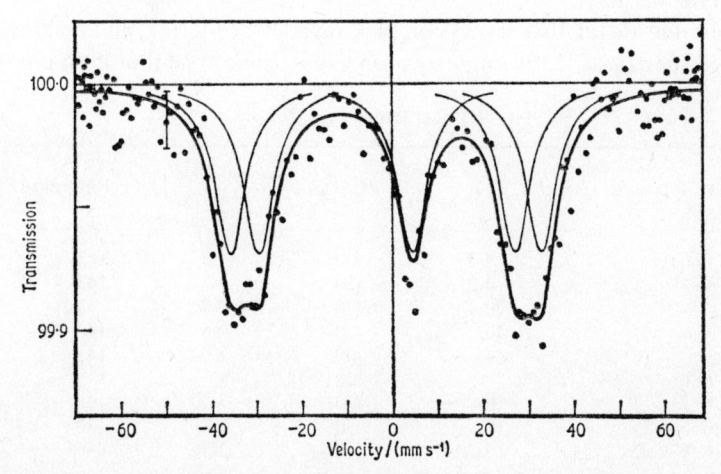

Fig. 17.23 The spectrum at 4·2 K of a sample of (Er$_{2\cdot4}$Y$_{97\cdot6}$) (C$_2$H$_5$SO$_4$)$_3$.9H$_2$O showing the unequally spaced lines of the paramagnetic hyperfine spectrum caused by a breakdown of the effective-field approximation. [Ref. 138, Fig. 1]

in a magnetically ordered system is less common than paramagnetic relaxation, and a model has been described which simulates the observed behaviour accurately [140].

A temperature-dependence study of ErCrO$_3$ between 4 and 40 K also showed relaxation narrowing [141]. Although the compound is anti-ferromagnetic below 133 K, the Er^{3+} spins do not order until below 16·8 K. This results in a distinct change in the nature of the relaxation above and below this temperature as seen in the temperature dependence of the spin-relaxation time. The quadrupole coupling is opposite in sign to that of Er metal (see later) and presumably is dominated by the lattice contribution. The magnetic field of 5300 kG (Table 17.11) is substantially less than the 7700 kG calculated for a $4f^{11}$ ($^4I_{\frac{15}{2}}$) free-ion configuration.

The temperature dependence of the magnetic field and quadrupole interaction in erbium metal have been followed between 4·2 and 40 K and analysed [142]. An estimate of $-1\cdot9(4)$ barn was suggested for the excited-state quadrupole moment. In a more detailed study, the line intensities of the hyperfine spectrum in a single crystal of Er metal have been correlated with the magnetic structure previously determined by neutron diffraction methods [143].

[*Refs. on p. 590*]

578 | THE RARE-EARTH ELEMENTS

For a single crystal at 55 K with the c axis parallel to the axis of observation, the spectrum consists of two lines only because all the spin moments lie along this axis and the other three lines for the $0+ \to 2+$ transition have zero probability. At 4·2 K the spins form a spiral arrangement leading to a finite probability for all five lines. Both types of ordering are complex, but a statistical calculation of the transition probabilities agreed well with the experimental data.

Holmium metal also shows complex magnetic ordering, and the line intensities in spectra of ^{166}Er impurity atoms in a single crystal of holmium reflect

Table 17.11 Mössbauer parameters in ^{166}Er

Compound	T/K	H_{eff}/kG	$\frac{1}{4}e^2qQ$ /(mm s^{-1})	Reference
ErCrO$_3$	4·2	5300	−1·2	141
Er	4·2	7550	3·5	143
	20·4	6990	2·7	143
	40	6170	0·6	143
	44	5400	1·5	143
ErFe$_2$	20	8400	—	135
ErAg	(0)	6480	2·50	146
ErNi$_2$	(0)	7440	2·88	146
ErNiCo	(0)	7830	3·86	146
ErNi$_{0\cdot5}$Co$_{1\cdot5}$	(0)	7940	4·08	146
ErCo$_2$	(0)	8180	3·82	146
ErAl$_2$	(0)	7700	—	147
cubic ErAl$_3$	(0)	4050	—	148
hexagaonal ErAl$_3$	(0)	5100 ⎫ 6450 ⎭	—	149

the changes in magnetic structure with increasing temperature [144]. A spiral spin arrangement is preserved throughout, but the alignment with respect to the c axis changes with temperature.

Paramagnetic hyperfine splitting has been recorded for ^{166}Er impurity atoms in zirconium metal at 4·2 K [145]. It is most unusual to observe such an effect in a metallic host, and the relaxation processes which cause narrowing at higher temperatures are rather different to those found in insulating materials.

The alloy ErFe$_2$ shows a five-line magnetic spectrum at 20 K with an estimated field of 8400 kG [135]. ErMnFe$_3$ also shows magnetic splitting.

ErCo$_{1\cdot5}$Ni$_{0\cdot5}$, ErCoNi, ErNi$_2$, and ErAg all show a low-temperature magnetic spectrum with electronic relaxation [146]. ErCo$_2$ differs in that it gives a first-order phase transition at 35 K resulting in a single-line spectrum above this temperature. ErAg and ErNi$_2$ show fields of less than the free-ion value because of an interaction with the crystalline field which causes a partial quenching of the magnetic moment.

[Refs. on p. 590]

Similar relaxation effects are found in ErAl$_2$ [147] and in both the cubic and hexagonal forms of ErAl$_3$ below their respective Curie points [148]. In all cases the motional narrowing allows a temperature-dependence study of the spin-relaxation time constant. Hexagonal ErAl$_3$ shows two erbium sites.

The 79·3-keV resonance in ^{167}Er was the last to be reported [149]. Coulomb excitation of an ^{167}Er$_2$O$_3$ target by 3-meV protons populates the first excited level, and the spectrum of an Er$_2$O$_3$ absorber at 30 K is a single line. The linewidth of 33·4 mm s^{-1} corresponds to a lower limit to the excited-state lifetime of 0·103 ns. No hyperfine effects have been reported.

The use of ^{168}Tm to populate ^{168}Er has already been mentioned. Coulomb excitation of the 79·8-keV resonance has also been used [150, 151]. An Er$_2$O$_3$ target gave a single emission line, and enabled the five-line magnetic spectrum to be resolved in Er metal at 4·2 K. Assuming a field of 7460 kG the data gave $^{168}\mu_e = 0.66(4)$ n.m.

The 79·3-keV resonance in ^{170}Er can also be observed by Coulomb excitation techniques [152]. A target of ErAl$_2$ gave a narrow line with an ErAl$_2$ absorber above the ordering temperature of 12·5 K [153]. Comparison of ^{166}Er and ^{170}Er spectra in ErFe$_2$ at 30 K gave the following parameters from the combined magnetic and quadrupole interactions: $^{170}\mu_e/^{166}\mu_e = 1.002(13)$, $^{170}Q_e/^{166}Q_e = 1.05(16)$, and $^{170}\mu_e = 0.638(22)$ n.m.

17.11 Thulium (^{169}Tm)

The decay of erbium-169 as shown in Fig. 17.24 populates the 8·40-keV level of ^{169}Tm with high efficiency, and the low energy of the latter is compatible with a good Mössbauer resonance. Unfortunately the 8·40-keV γ-ray has a very large internal conversion coefficient ($\alpha = 220 \pm 50$) [154], and the difficulties associated with detecting such a weak and comparatively 'soft' γ-ray have the cumulative effect of reducing the resonance absorption to about the 1% level. However, the low recoil energy allows measurements to be made at temperatures of up to 1000 K quite easily. Although the natural linewidth is large (9·3 mm s^{-1}) it is possible to resolve magnetic and electric quadrupole interactions. The first report in 1961 was of a spectrum from Tm$_2$O$_3$ with a ^{169}Er$_2$O$_3$ source [155]. Hyperfine structure was seen and later measured in more detail [156, 157].

Although ^{169}ErF$_3$ and ^{169}Er$_2$O$_3$ have been widely used as sources, they both give narrow linewidths without hyperfine splitting only at temperatures considerably above 300 K. More recently the material Na$_3$^{169}ErF$_6$ has been found to give a narrow emission line at room temperature, and a detailed preparation has been given [158].

The degree of internal conversion in the P electronic shell (principal quantum number $n = 6$) of a ^{169}Tm atom has been found to be lower in

[Refs. on p. 590]

580 | THE RARE-EARTH ELEMENTS

matrices of Tm_2O_3 and WO_3 than in W metal, although there is no difference in the M, N, and O shells ($n = 3-5$) [159]. This implies a decrease in the 6s-electron density at the nucleus. Wavefunction calculations were also made, but not correlated with Mössbauer chemical isomer shifts as had already been done for ^{119}Sn.

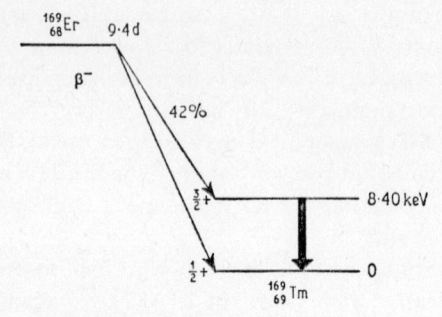

Fig. 17.24 The decay scheme for ^{169}Tm.

The nature of the pseudoquadrupole shift in a Mössbauer resonance has already been described in detail in Section 3.10. Such a phenomenon has been observed in $TmCl_3.6H_2O$ and $(Tm_{0.36}Y_{1.64})(SO_4)_3.8H_2O$ [160, 161]. Both these compounds have an excited electronic state within 2 cm^{-1} of the ground state, and the shifts observed between 1·2 and 4·2 K are illustrated in Fig. 17.25. Relaxation between the two electronic states causes a partial 'narrowing' of the four predicted lines to produce an asymmetric doublet. The temperature dependence of the shift in $TmCl_3.6H_2O$ below 4·2 K leads to a value for the separation of the electronic states of 1·11 cm^{-1}.

The crystalline field parameters for thulium ethyl sulphate are known from optical absorption studies. The quadrupole splitting between 6 and 250 K shows strong temperature dependence, and detailed analysis in terms of the 4f-wavefunctions and thermal population of the excited electronic levels gave a value for $^{169}Q_e = -1·1(1)$ barn [162]. In more detailed measurements a source of $^{169}ErF_3$ was maintained at 550 K at which temperature it shows its narrowest emission line [163, 164]. Although the electric field gradient in thulium ethyl sulphate has axial symmetry, all the energy levels in the $J = 6$ manifold (the ground state for the Tm^{3+} ion is $^3H_6$) have a complicated form, e.g. the singlet ground state is $0·119 | -6\rangle + 0·986 | 0\rangle + 0·119 | +6\rangle$ and can produce a quadrupole splitting of 79 mm s^{-1} at zero temperature. Detailed crystalline electric field interpretation of the quadrupole splitting data from 9·6 to 300 K gave an accurate simulation of the data.

The quadrupole splitting of one of the non-cubic sites in C-type Tm_2O_3 decreases to near zero between 11 and 700 K [164]. Similar data for $^{169}Er_2O_3$ above 700 K show an increase in quadrupole splitting with rising temperature,

[Refs. on p. 590]

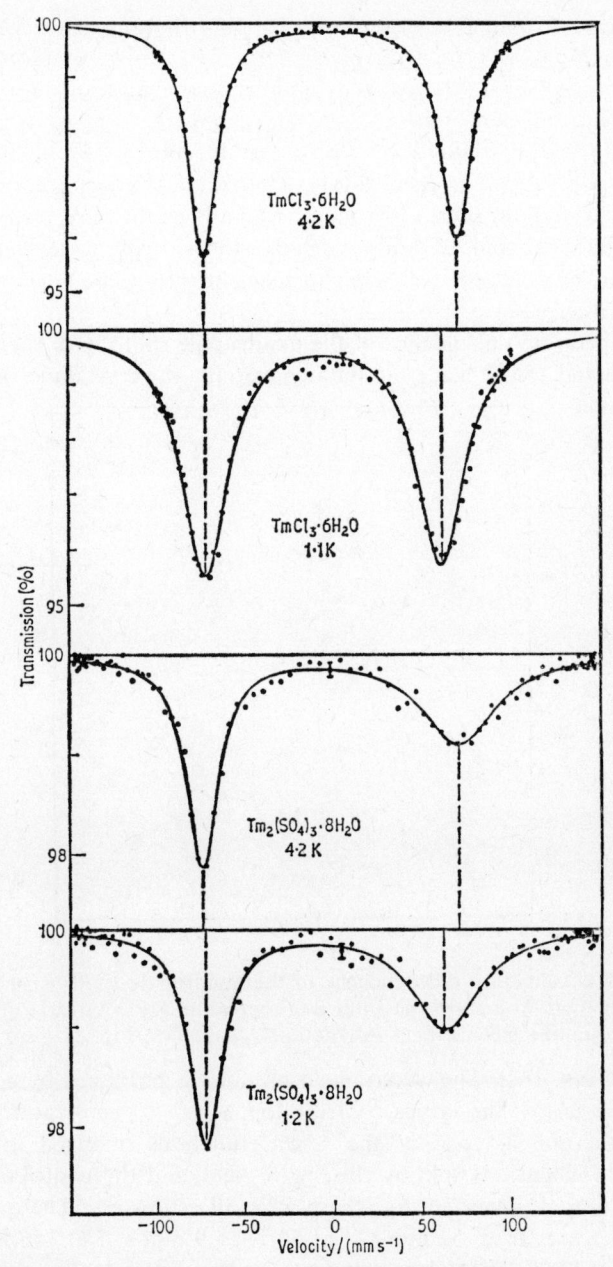

Fig. 17.25 Spectra of TmCl$_3$.6H$_2$O and Tm$_2$(SO$_4$)$_3$.8H$_2$O at low temperatures. Note the shift in the centroid of each spectrum with change in temperature. [Ref. 160, Fig. 1]

[*Refs. on p. 590*]

which suggest that in this case the lattice contribution to the electric field gradient is now greater than the valence term. The problems of estimating the electronic shielding factors for the closed-shell configurations were discussed in detail.

Some preliminary data have been given for other thulium compounds [165]. C-type Tm_2O_3, $Tm(benzoate)_3$, $Tm(oxinate)_3$, and $Tm(acac)_3.3H_2O$ all show two thulium sites, only one of which is quadrupole split. Chemical isomer shifts of 12 and 37 mm s^{-1} relative to the oxide were found only in TmF_3 and $TmCl_3$ respectively, but insufficient information is available for systematic interpretation.

The temperature dependence of the quadrupole splitting in thulium metal between 59 and 156 K has given data concerning the crystalline electric field

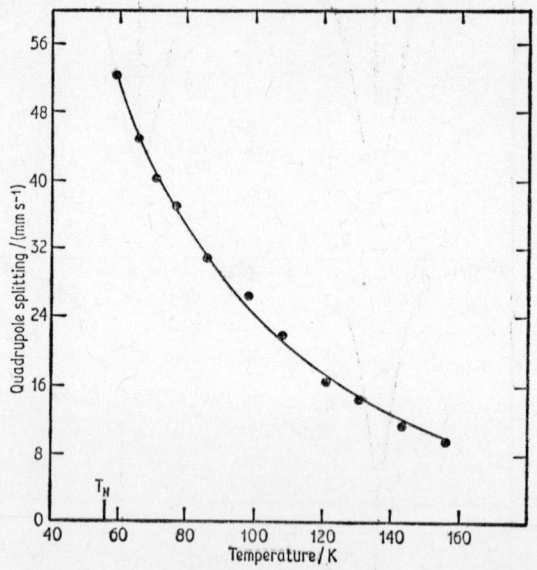

Fig. 17.26 The temperature dependence of the quadrupole splitting of the ^{169}Tm resonance in thulium metal. The solid line represents a theoretical curve derived from the crystalline electric field interactions. [Ref. 166, Fig. 1]

interaction [166, 167]. The electric field gradient is produced by a combination of three terms: the lattice contribution, which is temperature independent, the thermal average of the 4f-wavefunctions involved in the $^3H_6$ ground term when it is split by the crystal field, and the contribution from the conduction electron density; these were all estimated. The quadrupole splitting decreases from 52 mm s^{-1} at 59 K to 9 mm s^{-1} at 156 K as illustrated in Fig. 17.26. The valence-electron contribution is larger in magnitude but opposite in sign to the lattice term at low temperature, but decreases with increasing temperature. In principle the sign of the electric field gradient may reverse at some temperature above 156 K. The quadrupole splitting extrapolated to 0 K from the calculated crystal-field parameters is 151 mm s^{-1}, corresponding to a $|J_z = \pm 6\rangle$ ground state. This value is in good agreement with 146 mm s^{-1} found in the magnetically ordered phase which also has a $|\pm 6\rangle$ ground state [168]. The first excited level, $|J_z = \pm 5\rangle$, is 23·3 cm^{-1} above the ground state with a total splitting in the $J = 6$ manifold of 76·1 cm^{-1}.

The metal is magnetically ordered below 56 K and at 5 K shows a six-line hyperfine spectrum. Initial data were analysed to give a ratio for μ_e/μ_g of −2·33(4) and a field of 7000 kG [168]. Spectra closer to the Néel temperature are more complicated and have only been fully analysed in recent work embodying data on the magnetic structure from neutron-scattering measurements [169, 170]. The thulium spins lie along the crystal c axis and are collinear with the axial electric field gradient. The magnetic moment for any given Tm atom can be represented by a Fourier series such that the magnetic unit cell contains seven atoms in four equivalent sites. At low temperature all the sites become identical, and new data for the six-line spectrum at 5 K gave $\mu_e/\mu_g = -2·22(7)$ [170]. At higher temperatures no analysis of data is possible using a four-site model with a repeating distance of seven units. The assumption must be made that the magnetic and lattice unit cells are not commensurate. The magnetic structure repeats at intervals of $7 + \varepsilon$ where $|\varepsilon| \ll |$, and in this way an additional averaging is included which is insensitive to the value of ε and generates an infinite number of inequivalent sites. The theoretical spectra computed on this model agree very well with the experimental data, as can be seen in Fig. 17.27.

Initial measurements on thulium iron garnet TmIG, failed to observe all peaks in the spectrum [171], but later data gave values for the internal magnetic fields at the two thulium sites of 1700 and 810 kG at 20 K [172]. The fields are reduced from the free-ion value and that in Tm metal by anisotropic crystal-field and exchange interactions.

A crystal-field treatment, similar to that used for the metal, of the quadrupole splittings in TmRu$_2$, TmRe$_2$, and TmMn$_2$ below 150 K has given the total splittings of the $J = 6$ manifold as 94·0, 87·5, and 41·5 cm^{-1} respectively [173]. All three alloys showed a strong temperature dependence in the quadrupole splitting as already found in Tm metal, although the lattice term is substantially larger in TmRu$_2$ and TmRe$_2$ and the conduction-electron contributions have the opposite sign.

Thulium is magnetically ordered in the alloy Fe$_2$Tm, and the internal magnetic field decreases from 7200 kG at 4 K to 1800 kG at 400 K [174]. Detailed analysis of the magnetic and quadrupole interactions gave $\mu_e/\mu_g = -2·17(10)$ and $Q_e = -1·3$ barn. The field at the Tm is produced by an exchange interaction between the 4f-electrons and the iron atoms. This dominates the crystal-field interaction, prevents substantial mixing of the J_z states, and allows a comparatively simple analysis of the detailed data to be made.

[*Refs. on p. 590*]

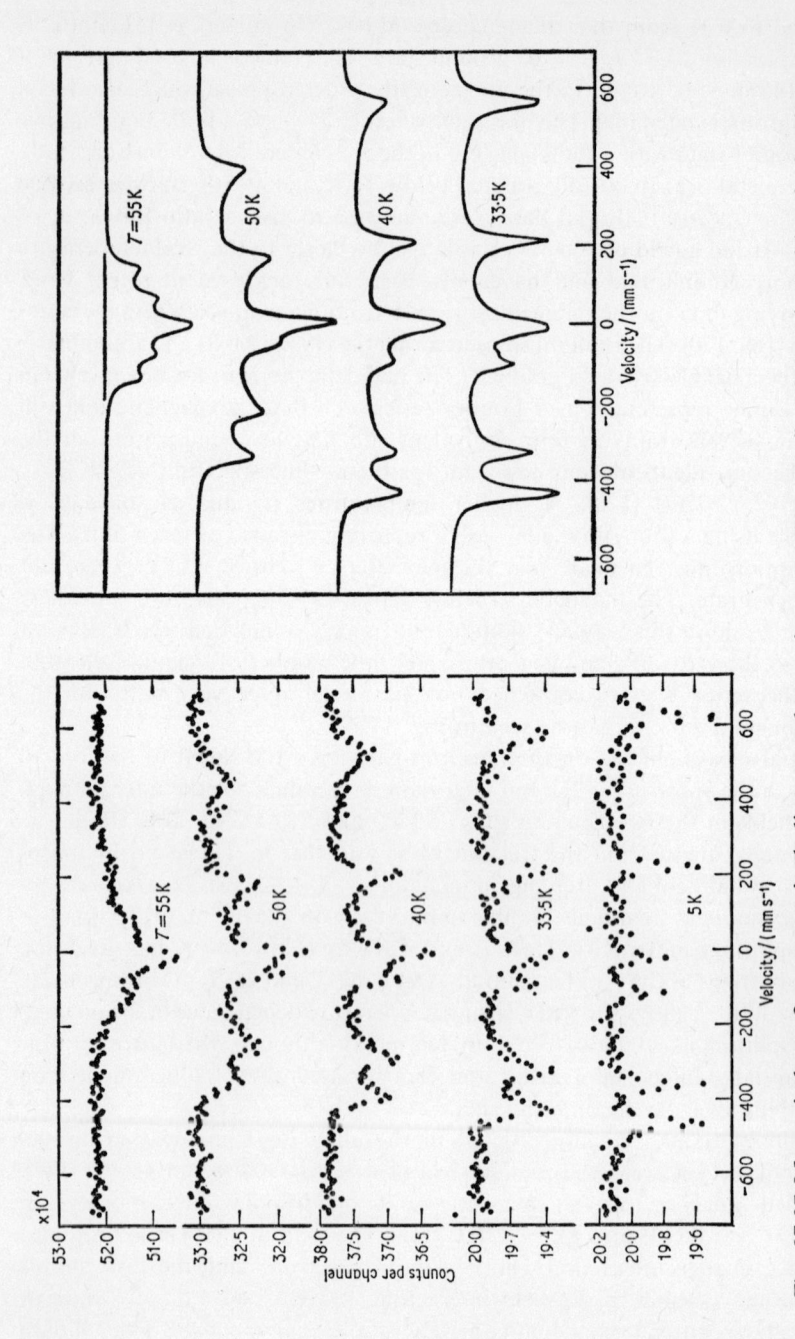

Fig. 17.27 Magnetic hyperfine spectra in thulium metal at various temperatures (in K) together with calculated spectra using a model in which the magnetic and crystal lattices are incommensurate. [Ref. 170, Fig. 2]

[Refs. on p. 590]

Preliminary temperature-dependence data on thulium-doped soda-silica glasses have shown the feasibility of determining at least some of the crystalline electric field parameters in unknown environments [175].

Recoilless scattering experiments have been made using thulium metal, but are beyond the scope of this book [154].

17.12 Ytterbium (^{170}Yb, ^{171}Yb, ^{172}Yb, ^{174}Yb, ^{176}Yb)

There are six known Mössbauer resonances in ytterbium:

^{170}Yb: 84·26-keV
^{171}Yb: 66·74-keV and 75·89-keV
^{172}Yb: 78·74-keV
^{174}Yb: 76·5-keV
^{176}Yb: 82·1-keV

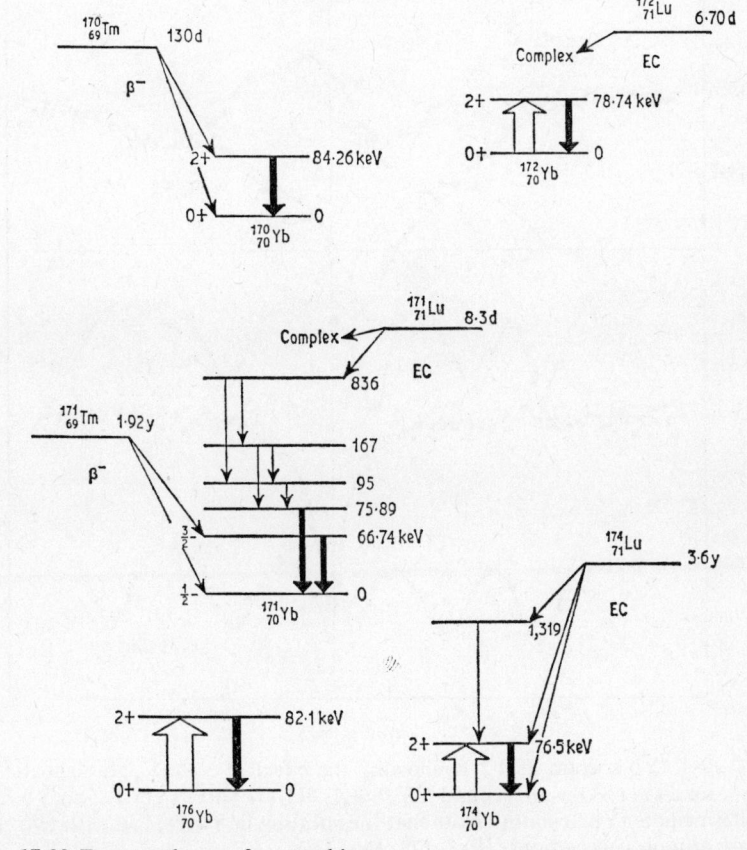

Fig. 17.28 Decay schemes for ytterbium.

All involve rotational excited states of the appropriate isotope, with a γ-ray energy of about 70–80 keV. In terms of experimental difficulty, all require low temperatures to achieve an adequate recoilless fraction, and only ^{170}Yb, ^{171}Yb (66·74-keV), and ^{174}Yb have convenient precursors. The decay schemes are shown in Fig. 17.28. The Yb^{2+} oxidation state has a $4f^{14}$ configuration and is diamagnetic. As it is this electronic configuration which effectively exists in ytterbium metal, no magnetic effects are found in this matrix. The Yb^{3+} ion usually has a $^2F_{\frac{7}{2}}$ ground state, and has magnetic properties analogous to those of Er^{3+}.

The 84·26-keV resonance in ^{170}Yb was first recorded in 1962 using a Tm metal source and a Yb$_2$O$_3$ absorber at 20 K [176]. A partially resolved quadrupole splitting was found in the oxide, the source being unsplit because the metal has cubic symmetry. The ^{170}Yb spectrum of YbCl$_3$.6H$_2$O at

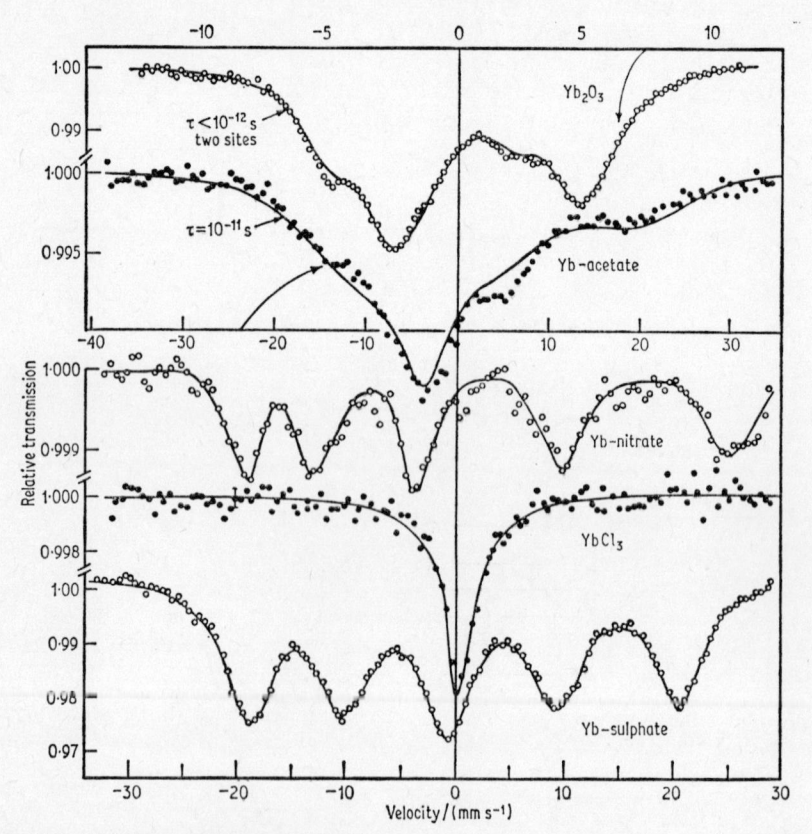

Fig. 17.29 ^{170}Yb spectra at 4·2 K showing the effect of a long paramagnetic relaxation time (Yb(NO$_3$)$_3$.6H$_2$O and Yb$_2$(SO$_4$)$_3$.8H$_2$O), short (YbCl$_3$ and Yb$_2$O$_3$), and intermediate (Yb acetate). Note that the splitting in Yb$_2$O$_3$ is due to two sites with quadrupole interactions. [Ref. 178, Fig. 1]

[*Refs.* on p. 590]

4·2 K shows five lines from a paramagnetic hyperfine splitting with a long relaxation time of this $0+ \rightarrow 2+$ transition [177]. The magnetic properties of the $^2F_{\frac{7}{2}}$ ground state are similar to those found in $ErCl_3.6H_2O$. Both compounds have an anisotropic hyperfine interaction. Analysis of the ^{170}Yb data gave $^{170}\mu_e = 0.668(10)$ n.m.

Of other ytterbium compounds studied by the ^{170}Yb resonance, $Yb(NO_3)_3.6H_2O$ and $Yb_2(SO_4)_3.8H_2O$ also show paramagnetic hyperfine splitting like that in the chloride, and presumably have a similar anisotropic hyperfine interaction [178]. As may be seen in Fig. 17.29 the five lines are unequally spaced because of a large quadrupole interaction. Yb_2O_3 and $YbCl_3$ show no paramagnetic broadening at 4·2 K, while YbF_3, the garnet $Yb_3Ga_5O_{12}$ and Yb acetate are intermediate cases with substantial broadening. The lower magnetic field in the acetate (2390 kG) reflects the increased crystalline field interaction in this compound compared to the sulphate (2620 kG) and the nitrate (3000 kG). The largest quadrupole interaction recorded was $e^2q^{170}Q_e = 25.4$ mm s^{-1} in YbF_3.

Chemical isomer shifts recorded for ^{170}Yb in $YbCl_2$, $YbSO_4$, Yb, $YbAl_2$, $YbSi_2$, $YbCl_3$ and YbGa garnet, although small, showed a distinct difference between Yb^{2+} and Yb^{3+} [179]. The maximum shift of $+0.63(10)$ mm s^{-1} between $YbSO_4$ and YbGa garnet, when taken in conjunction with estimates of electron densities at the ytterbium nucleus, showed $\delta\langle R^2 \rangle/\langle R^2 \rangle$ to be positive.

Independent data compared the shifts of the ^{170}Yb and 67-keV ^{171}Yb resonances in a short series of compounds [83]. As expected there is an approximately linear correlation which gives the ratio

$$^{170}\delta\langle R^2 \rangle/^{171}\delta\langle R^2 \rangle_{67} = +1.12(14).$$

This is illustrated in Fig. 17.30. The separation between Yb^{2+} and Yb^{3+} in the $^{171}Yb_{67}$ resonance is $+0.50(5)$ mm s^{-1}. Estimates for $\delta\langle R^2 \rangle/\langle R^2 \rangle$ derived were $+4.8(1.2) \times 10^{-5}$ and $+4.3(1.1) \times 10^{-5}$ for ^{170}Yb and ^{171}Yb respectively. Line splittings were recorded at 4·2 K in $YbSO_4$ and YbOOH in addition to those already mentioned, but were not interpreted in detail.

Ytterbium iron garnet is magnetically ordered at low temperature. At 4·2 K the ^{170}Yb spectrum is a five-line magnetic one with a field of 1830 kG [180]. As the temperature is raised the thermal population of the higher electronic levels allows a faster electronic relaxation and motional narrowing occurs in an analogous manner to that found in $ErFeO_3$. Paramagnetic relaxation has been observed in the intermetallic compounds $YbNi_5$ and $YbPd_3$ at 4·2 K [181]. $YbNi_2$ shows complex magnetic behaviour which is not fully understood.

The 66·74-keV resonance in ^{171}Yb was reported by Kalvius in 1965 [182]. The source of $^{171}Tm_2O_3$ was quadrupole split. A Yb_2O_3 absorber showed a quadrupole splitting of 10·7 mm s^{-1} ($= 2.4$ μeV) which by comparison with

^{170}Yb data gave $^{171}Q_{67}/^{170}Q_e = 0.73$. The resonance in YbCl$_3$.6H$_2$O at 4·2 K showed eight lines because of paramagnetic hyperfine splitting. One set of measurements used an Er$_2$(^{171}Tm)O$_3$ source which showed quadrupole splitting and complicated the experimental data [183]. The $\frac{1}{2} \to \frac{3}{2}$ transition features considerable E2/M1 (= δ^2) mixing which results in all eight magnetic lines having a finite intensity. Detailed analysis gave the values $^{171}\mu_{67} = 0.351(3)$ n.m., $^{171}Q_e/^{170}Q_e = 0.89$, and $\delta^2 = 0.45(3)$. The internal magnetic field was 2970 kG. Independent data published simultaneously for

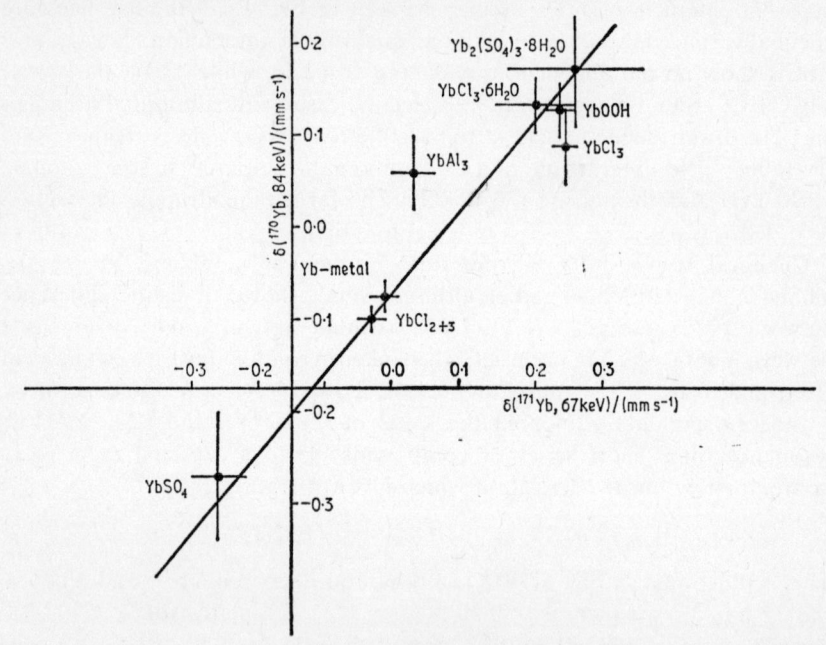

Fig. 17.30 Chemical isomer shifts for the 84-keV ^{170}Yb resonance as a function of the 67-keV ^{171}Yb resonance in the same absorber. [Ref. 83, Fig. 3]

YbCl$_3$.6H$_2$O were obtained using a cubic alloy Er$_{10}$(^{171}Tm)Al$_{90}$, giving the simpler spectra shown in Fig. 17.31. The ratio $^{171}\mu_{67}/^{171}\mu_g = 0.2355(15)$ results in $^{171}\mu_{67} = 0.349(3)$ n.m. in good agreement with the other value. The E2/M1 mixing ratio was $\delta^2 = 0.49(6)$. Interference effects were observed between the multipoles using an oriented single crystal of YbCl$_3$.6H$_2$O. This is analogous to the ^{99}Ru work, and the measured phase angle was in agreement with time-reversal invariance.

The ^{171}Tm parent used for the 66·74-keV resonance does not populate the 75·89-keV level, and the shorter-lived ^{171}Lu parent must be used [185]. This has a very complex decay, and a lithium-drifted germanium detector is required to detect the γ-ray with sufficient discrimination. The ^{169}Tm($\alpha, 2n$)^{171}Lu reaction provides a convenient route so that cubic

[*Refs. on p. 590*]

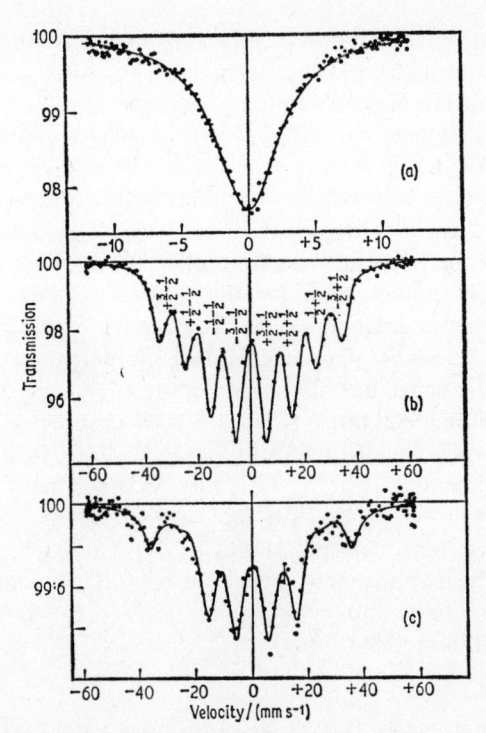

Fig. 17.31 ^{171}Yb 66·7-keV spectra of (a) Yb metal, (b) YbCl$_3$.6H$_2$O polycrystalline, (c) YbCl$_3$.6H$_2$O single crystal with magnetic axis perpendicular to direction of observation. [Ref. 184, Fig. 1]

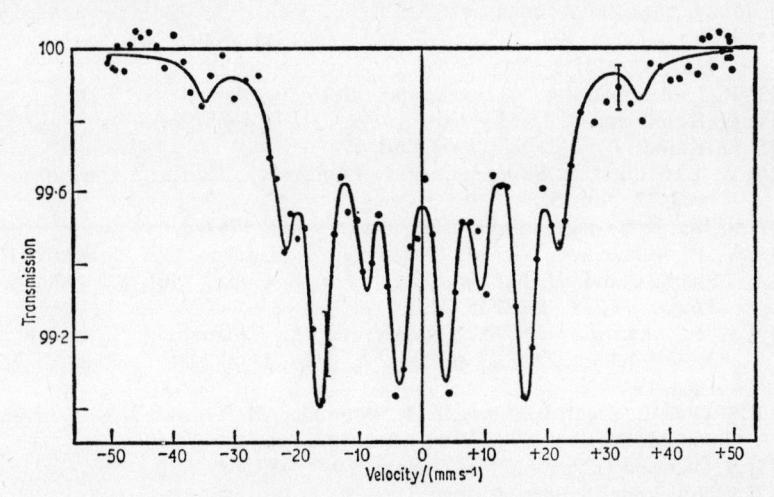

Fig. 17.32 ^{171}Yb 75·9-keV spectrum of YbCl$_3$.6H$_2$O at 4·2 K. [Ref. 186, Fig. 1]

[*Refs. on p. 590*]

thulium metal or $TmAl_2$ can be used as a source. Yb_2O_3 at 4·2 K gave a partially resolved quadrupole splitting from which $e^2q^{171}Q_{76} = 25·6(4)$ mm s^{-1} (= 6·5 μeV) [185]. The 66·7-keV resonance gave $e^2q^{171}Q_{67} = 20·7(4)$ mm s^{-1} (= 4·6 μeV), and thus $^{171}Q_{76}/^{171}Q_{67} = 1·41$. The paramagnetic hyperfine spectrum of $YbCl_3.6H_2O$ is shown in Fig. 17.32, and can be analysed in a similar manner to give $^{171}\mu_{76}/^{171}\mu_g = 2·055(10)$, from which a value can be derived for $^{171}\mu_{76} = 1·01(1)$ n.m. [186].

The paramagnetic hyperfine spectrum of ^{174}Yb in $YbCl_3.6H_2O$ at 4·2 K is a well-resolved quintet from the $0+ \rightarrow 2+$ 76·5-keV transition [187]. Comparison with the analogous ^{170}Yb data gave $^{170}\mu_e/^{174}\mu_e = 0·994(15)$ whence $^{174}\mu_e = 0·680(20)$ n.m. In this case the parent isotope was ^{174}Lu, but the ^{174}Yb resonance has also been detected following Coulomb excitation of an ytterbium metal target [188]. A Yb_2O_3 absorber gave a quadrupole split resonance which yielded the ratio $^{174}Q_e/^{170}Q_e = 1·04(2)$.

Parallel experiments on the ^{172}Yb and ^{176}Yb resonances with a Yb_2O_3 target gave $^{176}Q_e/^{170}Q_e = 1·045(2)$ and $^{172}Q_e/^{170}Q_e = 1·01(2)$. The ^{172}Yb resonance has also been detected in $YbCl_3.6H_2O$ at 4·2 K using a lutetium metal source. The paramagnetic hyperfine splitting is comparable to that for ^{170}Yb and ^{174}Yb, and gives values of $^{170}\mu_e/^{172}\mu_e = 1·009(17)$ and $^{172}\mu_e = 0·670(20)$ n.m. [189].

REFERENCES

[1] S. Ofer, I. Nowik, and S. G. Cohen, p. 427 of 'Chemical Applications of Mössbauer Spectroscopy', Ed. V. I. Goldanskii and R. H. Herber, Academic Press, N.Y., 1968.
[2] V. A. Bukarev, *Zhur. eksp. teor. Fiz.*, 1963, **44**, 852 (*Soviet Physics – JETP*, 1963, **17**, 579).
[3] R. J. Morrison, see *Nuclear Science Abstracts*, 1965, **19**, No. 6638.
[4] G. Kaindl and R. L. Mössbauer, *Phys. Letters*, 1968, **26B**, 386.
[5] G. Kaindl, *Phys. Letters*, 1968, **28B**, 171.
[6] W. L. Croft, J. A. Stone, and W. L. Pillinger, see *Nuclear Science Abstracts*, 1968, **22**, No. 35289.
[7] S. Jha, R. Segnan, and G. Lang, *Phys. Letters*, 1962, **2**, 117.
[8] V. P. Alfimenkov, Yu. M. Ostanevich, T. Ruskov, A. V. Strelkov, F. L. Shapiro, and W. K. Yen, *Zhur. eksp. teor. Fiz.*, 1962, **42**, 1036 (*Soviet Physics – JETP*, 1962, **15**, 718).
[9] V. P. Alfimenkov, N. A. Lebedev, Yu. M. Ostanevich, T. Ruskov, and A. V. Strelkov, *Zhur. eksp. teor. Fiz.*, 1964, **46**, 482 (*Soviet Physics – JETP*, 1964, **19**, 326).
[10] S. Ofer, E. Segal, I. Nowik, E. R. Bauminger, L. Grodzins, A. J. Freeman, and M. Schieber, *Phys. Rev.*, 1965, **137**, A627.
[11] S. Ofer and I. Nowik, *Nuclear Phys.*, 1967, **A93**, 689.
[12] I. Nowik and S. Ofer, *J. Appl. Phys.*, 1968, **39**, 1252.

REFERENCES | 591

[13] W. Henning, G. Kaindl, P. Kienle, H. J. Korner, H. Kulzer, K. E. Rehm, and N. Edelstein, *Phys. Letters*, 1968, **28A**, 209.
[14] P. Steiner, E. Gerdau, P. Kienle, and H. J. Korner, *Phys. Letters*, 1967, **24B**, 515.
[15] D. Yeboah-Amankwah, L. Grodzins, and R. B. Frankel, *Phys. Rev. Letters*, 1967, **18**, 791.
[16] U. Atzmony, E. R. Bauminger, D. Froindlich, and S. Ofer, *Phys. Letters*, 1967, **26B**, 81.
[17] R. M. Wheeler, U. Atzmony, and J. C. Walker, *Phys. Rev.*, 1969, **186**, 1280.
[18] R. M. Wheeler, U. Atzmony, K. A. Hardy, and J. C. Walker, *Phys. Letters*, 1970, **31B**, 206.
[19] D. A. Shirley, M. Kaplan, R. W. Grant, and D. A. Keller, *Phys. Rev.*, 1962, **127**, 2097.
[20] S. Hufner, P. Kienle, D. Quitmann, and P. Brix, *Z. Physik*, 1965, **187**, 67.
[21] R. J. Bullock, N. R. Large, I. L. Jenkins, A. G. Wain, P. Glentworth, and D. A. Newton, *J. Inorg. Nuclear Chem.*, 1969, **31**, 1929.
[22] P. H. Barrett and D. A. Shirley, *Phys. Rev.*, 1963, **131**, 123.
[23] G. Crecelius and S. Hufner, *Phys. Letters*, 1969, **30A**, 124.
[24] I. Nowik and S. Ofer, *Phys. Rev.*, 1963, **132**, 241.
[25] M. Stachel, S. Hufner, G. Crecelius, and D. Quitmann, *Phys. Rev.*, 1969, **186**, 355.
[26] B. A. Goodman, N. N. Greenwood, and G. E. Turner, *Chem. Phys. Letters*, 1970, **5**, 181.
[27] M. Stachel, S. Hufner, G. Crecelius, and D. Quitmann, *Phys. Letters*, 1968, **28A**, 188.
[28] G. M. Kalvius, G. K. Shenoy, G. J. Ehnholm, T. E. Katila, O. V. Lounasmaa, and P. Reivari, *Phys. Rev.*, 1969, **187**, 1503.
[29] E. Steichele, *Z. Physik*, 1967, **201**, 331.
[30] G. Gerth, P. Kienle, and K. Luchner, *Phys. Letters*, 1968, **27A**, 557.
[31] O. Berkooz, *J. Phys. and Chem. Solids*, 1969, **30**, 1763.
[32] P. Brix, S. Hufner, P. Kienle, and D. Quitmann, *Phys. Letters*, 1964, **13**, 140.
[33] F. A. Deeney, J. A. Delaney, and V. P. Ruddy, *Phys. Letters*, 1967, **25A**, 370.
[34] N. R. Large, R. J. Bullock, P. Glentworth, and D. A. Newton, *Phys. Letters*, 1969, **29A**, 352.
[35] E. Catalano, R. G. Bedford, V. G. Silveira, and H. H. Wickman, *J. Phys. and Chem. Solids*, 1969, **30**, 1613.
[36] H. H. Wickman and E. Catalano, *J. Appl. Phys.*, 1968, **39**, 1248.
[37] H. H. Wickman, I. Nowik, J. H. Wernick, D. A. Shirley, and R. B. Frankel, *J. Appl. Phys.*, 1966, **37**, 1246.
[38] F. A. Deeney, J. A. Delaney, and V. P. Ruddy, *Phys. Letters*, 1968, **27A**, 571.
[39] H. Eicher, *Z. Physik*, 1964, **179**, 264.
[40] I. Nowik and S. Ofer, *Phys. Rev.*, 1967, **153**, 409.
[41] E. R. Bauminger, I. Nowik, and S. Ofer, *Phys. Letters*, 1969, **29A**, 328.
[42] U. Atzmony, E. R. Bauminger, A. Mustachi, I. Nowik, S. Ofer, and M. Tassa, *Phys. Rev.*, 1969, **179**, 514.
[43] A. A. Gomes, R. M. Xavier, and J. Danon, *Chem. Phys. Letters*, 1969, **4**, 239.
[44] O. Berkooz, M. Malamud, and S. Shtrikman, *Solid State Commun.*, 1968, **6**, 185.
[45] F. A. Deeney, J. A. Delaney, and V. P. Ruddy, *J. Inorg. Nuclear Chem.*, 1968, **30**, 1175.
[46] P. K. Gallagher, F. Schrey, and B. Prescott, *Inorg. Chem.*, 1970, **9**, 215.

[47] H. Maletta, W. Heidrich, and R. L. Mössbauer, *Phys. Letters*, 1967, **25A**, 295.
[48] H. H. Wickman, M. Robbins, E. Buehler, and E. Catalano, *Phys. Letters*, 1970, **31A**, 59.
[49] H. H. Wickman, *Phys. Letters*, 1970, **31A**, 29.
[50] O. V. Lounasmaa and G. M. Kalvius, *Phys. Letters*, 1967, **26A**, 21.
[51] R. L. Cohen, S. Hufner, and K. W. West, *Phys. Letters*, 1969, **28A**, 582; *J. Appl. Phys.*, 1969, **40**, 1366; *Phys. Rev.*, 1969, **184**, 263.
[52] S. Hufner, *Phys. Rev.*, 1967, **19**, 1034.
[53] S. Hufner and J. H. Wiedersich, *Phys. Rev.*, 1968, **173**, 448.
[54] H. H. Wickman, J. H. Wernick, R. C. Sherwood, and C. F. Wagner, *J. Phys. and Chem. Solids*, 1968, **29**, 181.
[55] I. Shidlovsky and I. Mayer, *J. Phys. and Chem. Solids*, 1969, **30**, 1207.
[56] M. Loewenhaupt and S. Hufner, *Phys. Letters*, 1969, **30A**, 309.
[57] A. Bussiere de Nercy, M. Langevin, and M. Spighel, *Compt. rend.*, 1960, **250**, 1031.
[58] U. Atzmony, A. Mualem, and S. Ofer, *Phys. Rev.*, 1964, **136**, B1237.
[59] U. Atzmony and S. Ofer, *Phys. Letters*, 1965, **14**, 284.
[60] U. Atzmony, E. R. Bauminger, D. Froindlich, J. Hess, and S. Ofer, *Phys. Letters*, 1968, **26B**, 613.
[61] W. Henning, G. Baehre, and P. Kienle, *Phys. Letters*, 1970, **31B**, 203.
[62] U. Atzmony, E. R. Bauminger, I. Nowik, S. Ofer, and J. H. Wernick, *Phys. Rev.*, 1967, **156**, 262.
[63] E. Steichele, S. Hufner, and P. Kienle, *Phys. Letters*, 1965, **14**, 321.
[64] E. Steichele, S. Hufner, and P. Kienle, *Phys. Letters*, 1966, **21**, 220.
[65] M. Richter, W. Henning, and P. Kienle, *Z. Physik*, 1969, **218**, 223.
[66] U. Atzmony and S. Ofer, *Phys. Rev.*, 1966, **145**, 915.
[67] E. R. Bauminger, I. Nowik, and S. Ofer, *Phys. Letters*, 1969, **29A**, 199.
[68] U. Atzmony, E. R. Bauminger, B. Einhorn, J. Hess, A. Mustachi, and S. Ofer, *J. Appl. Phys.*, 1968, **39**, 1250.
[69] H. Armon, E. R. Bauminger, J. Hess, A. Mustachi, and S. Ofer, *Phys. Letters*, 1969, **28A**, 528.
[70] F. F. Tomblin and P. H. Barrett, p. 245 of 'Hyperfine Structure and Nuclear Radiations', Ed. E. Matthias and D. A. Shirley, North-Holland, Amsterdam, 1968.
[71] J. Gal, E. R. Bauminger, and S. Ofer, *Phys. Letters*, 1968, **27B**, 552.
[72] K. E. Rehm, W. Henning, and P. Kienle, *Phys. Rev. Letters*, 1969, **22**, 790.
[73] R. R. Stevens, Y. K. Lee, and J. C. Walker, *Phys. Letters*, 1966, **21**, 401.
[74] A. E. Balabanov, N. N. Delyagin, and H. El Sayes, *Yadern Fiz.*, 1966, **3**, 209.
[75] R. R. Stevens, J. S. Eck, E. T. Ritter, Y. K. Lee, and J. C. Walker, *Phys. Rev.*, 1967, **158**, 1118.
[76] H. Prange, *Z. Physik*, 1968, **212**, 415.
[77] N. Y. Delyagin, H. El Sayes, and V. S. Shpinel, *Zhur. eksp. teor. Fiz.*, 1966, **51**, 95 (*Soviet Physics – JETP*, 1967, **24**, 64).
[78] E. R. Bauminger, D. Froindlich, A. Mustachi, I. Nowik, S. Ofer, and S. Samuelov, *Phys. Letters*, 1969, **30B**, 530.
[79] H. Blumberg, B. Persson, and M. Bent, *Phys. Rev.*, 1968, **170**, 1076.
[80] B. Persson, H. Blumberg, and M. Bent, *Phys. Letters*, 1968, **27A**, 189.
[81] H. Maletta, R. B. Frankel, W. Henning, and R. L. Mössbauer, *Phys. Letters*, 1969, **28A**, 557.
[82] R. B. Frankel, *Phys. Letters*, 1969, **30A**, 269.
[83] W. Henning, *Z. Physik*, 1968, **217**, 438.

[84] H. Prange and P. Kienle, *Phys. Letters*, 1966, **23**, 681.
[85] J. Fink, *Z. Physik*, 1967, **207**, 225.
[86] J. Fink and P. Kienle, *Phys. Letters*, 1965, **17**, 326.
[87] J. Eck, Y. K. Lee, E. T. Ritter, R. R. Stevens, and J. C. Walker, *Phys. Rev. Letters*, 1966, **17**, 120.
[88] J. C. Woolum and A. J. Bearden, *Phys. Rev.*, 1966, **142**, 143.
[89] U. Atzmony, E. R. Bauminger, and S. Ofer, *Nuclear Phys.*, 1966, **89**, 433.
[90] S. Ofer, P. Avivi, R. Bauminger, A. Marinov, and S. G. Cohen, *Phys. Rev.*, 1960, **120**, 406.
[91] S. Jha, R. K. Gupta, H. G. Devare, G. C. Pramila, and R. S. Raghavan, *Nuovo Cimento*, 1961, **19**, 682.
[92] V. V. Sklyarevskii, B. N. Samoilov, and E. P. Stepanov, *Zhur. eksp. teor. Fiz.*, 1961, **40**, 1874 (*Soviet Physics-JETP*, 1961, **13**, 1316).
[93] P. K. Tseng, N. Shikazono, H. Takekoshi, and T. Shoji, *J. Phys. Soc. Japan*, 1961, **16**, 1790.
[94] V. V. Sklyarevskii, K. P. Aleshin, V. D. Gorobchenko, I. I. Lukashevich, B. N. Samoilov, and E. P. Stepanov, *Phys. Letters*, 1963, **6**, 157.
[95] R. L. Cohen and H. J. Guggenheim, *Nuclear Instr. Methods*, 1969, **71**, 27.
[96] G. J. Bowden, D. St P. Bunbury, and J. M. Williams, *Proc. Phys. Soc.*, 1967, **91**, 612.
[97] R. Bauminger, S. G. Cohen, A. Marinov, and S. Ofer, *Phys. Rev. Letters*, 1961, **6**, 467.
[98] I. K. Lukashevich, V. D. Gorobchenko, V. V. Sklyarevskii, and N. I. Filippov, *Phys. Letters*, 1970, **31A**, 112.
[99] V. D. Gorobchenko, I. I. Lukashevich, V. V. Sklyarevskii, and N. I. Filippov, *ZETF Letters*, 1969, **9**, 237.
[100] I. Nowik and H. H. Wickman, *Phys. Rev.*, 1965, **140**, A869.
[101] S. Ofer, M. Rakavy, E. Segal, and B. Khurgin, *Phys. Rev.*, 1965, **138**, A241.
[102] I. I. Lykashevich, V. V. Sklyarevskii, K. P. Aleshin, B. N. Samoilov, E. P. Stepanov, and N. I. Filippov, *ZETF Letters*, 1966, **3**, 81 (*JETP Letters*, 1966, **3**, 50).
[103] W. Henning, D. Heunemann, W. Weber, P. Kienle, and H. J. Korner, *Z. Physik*, 1967, **207**, 505.
[104] B. Khurgin, S. Ofer, and M. Rakavy, *Nuclear Phys.*, 1968, **110A**, 577.
[105] G. Crecelius and D. Quitmann, p. 172 of 'Hyperfine Structure and Nuclear Radiations', Ed. E. Matthias and D. A. Shirley, North-Holland, Amsterdam, 1968.
[106] I. Nowik and H. J. Williams, *Phys. Letters*, 1966, **20**, 154.
[107] M. Eibschutz and L. G. Van Uitert, *Phys. Rev.*, 1969, **177**, 502.
[108] S. Hufner, H. H. Wickman, and C. F. Wagner, p. 952 of 'Hyperfine Structure and Nuclear Radiations', Ed. E. Matthias and D. A. Shirley, North-Holland, Amsterdam, 1968.
[109] S. Ofer, B. Khurgin, M. Rakavy, and I. Nowik, *Phys. Letters*, 1964, **11**, 205.
[110] I. Nowik, *Phys. Letters*, 1965, **15**, 219.
[111] H. H. Wickman and I. Nowik, *Phys. Rev.*, 1966, **142**, 115.
[112] S. Hufner, H. H. Wickman, and C. F. Wagner, *Phys. Rev.*, 1968, **169**, 247.
[113] H. H. Wickman and I. Nowik, *J. Phys. and Chem. Solids*, 1967, **28**, 2099.
[114] T. P. Abeles, W. G. Bos, and P. J. Ouseph, *J. Phys. and Chem. Solids*, 1969, **30**, 2159.
[115] I. Dezsi, N. A. Eissa, L. Keszthelyi, B. Molnar, and D. L. Nagy, *Phys. Stat. Sol.*, 1968, **30**, 215.

[116] I. Nowik, S. Ofer, and J. H. Wernick, *Phys. Letters*, 1966, **20**, 232.
[117] S. Ofer and E. Segal, *Phys. Rev.*, 1966, **141**, 448.
[118] I. Nowik and J. H. Wernick, *Phys. Rev.*, 1965, **140**, A131.
[119] B. Khurgin, I. Nowik, M. Rakavy and S. Ofer, *J. Phys. and Chem. Solids*, 1970, **31**, 49.
[120] I. Shidlovsky and I. Mayer, *J. Phys. and Chem. Solids*, 1969, **30**, 1207.
[121] I. I. Lykashevich, V. V. Sklyarevskii, K. P. Aleshin, B. N. Samoilov, E. P. Stepanov, and N. I. Filippov, *ZETF Letters*, 1966, **3**, 81 (*JETP Letters*, 1966, **3**, 50).
[122] I. Nowik, *J. Appl. Phys.*, 1969, **40**, 414.
[123] W. J. Nicholson, *Phys. Letters*, 1964, **10**, 184.
[124] S. Sylvester, J. E. McQueen, and D. Schroeer, *Solid State Commun.*, 1969, **7**, 673.
[125] R. L. Cohen, *Phys. Rev.*, 1965, **137**, A1809.
[126] S. Ofer, M. Rakavy, and E. Segal, *Nuclear Phys.*, 1965, **69**, 173.
[127] R. R. Stevens, Jr, J. S. Eck, E. T. Ritter, Y. K. Lee, and J. C. Walker, *Phys. Rev.*, 1967, **158**, 1118.
[128] J. Eck, Y. K. Lee, E. T. Ritter, R. R. Stevens, Jr, and J. C. Walker, *Phys. Rev. Letters*, 1966, **17**, 120.
[129] E. Munck, D. Quitmann, and S. Hufner, *Z. Naturforsch.*, 1966, **21A**, 2120.
[130] E. Munck, *Z. Physik*, 1968, **208**, 164.
[131] T. Ruskov, T. Tomov, and H. Popov, *Compt. rend. Bulg. Sci.*, 1966, **19**, 701.
[132] E. Munck, D. Quitmann, and S. Hufner, *Phys. Letters*, 1967, **24B**, 392.
[133] R. L. Mössbauer, F. W. Stanek, and W. H. Wiedemann, *Z. Physik*, 1961, **161**, 388.
[134] F. W. Stanek, *Z. Physik*, 1962, **166**, 6.
[135] R. L. Cohen and J. H. Wernick, *Phys. Rev.*, 1964, **134**, B503.
[136] E. Munck, D. Quitmann, and S. Hufner, *Z. Naturforsch.*, 1966, **21A**, 847.
[137] H. Dobler, G. Petrich, S. Hufner, P. Kienle, W. Wiedemann, and H. Eicher, *Phys. Letters*, 1964, **10**, 319.
[138] E. R. Seidel, G. Kaindl, M. J. Clauser, and R. L. Mössbauer, *Phys. Letters*, 1967, **25A**, 328.
[139] W. Wiedemann and W. Zinn, *Z. angew. Phys.*, 1966, **20**, 327.
[140] I. Nowik and H. H. Wickman, *Phys. Rev. Letters*, 1966, **17**, 949.
[141] M. Eibschutz, R. L. Cohen, and K. W. West, *Phys. Rev.*, 1969, **178**, 572.
[142] S. Hufner, P. Kienle, W. Wiedemann, and H. Eicher, *Z. Physik*, 1965, **182**, 499.
[143] R. A. Reese and R. G. Barnes, *Phys. Rev.*, 1967, **163**, 465.
[144] R. A. Reese and R. G. Barnes, *J. Appl. Phys.*, 1969, **40**, 1493.
[145] L. L. Hirst, E. R. Seidel, and R. L. Mössbauer, *Phys. Letters*, 1969, **29A**, 673.
[146] G. Petrich, *Z. Physik*, 1969, **221**, 431.
[147] W. Wiedemann and W. Zinn, *Phys. Letters*, 1967, **24A**, 506.
[148] W. Zinn and W. Wiedemann, *J. Appl. Phys.*, 1968, **39**, 839.
[149] R. M. Wilenzick, K. A. Hardy, J. A. Hicks, and W. R. Owens, *Phys. Letters*, 1969, **30B**, 167.
[150] R. R. Stevens, J. S. Eck, E. T. Ritter, Y. K. Lee, and J. C. Walker, *Phys. Rev.*, 1967, **158**, 1118.
[151] J. Eck, Y. K. Lee, E. T. Ritter, R. R. Stevens, and J. C. Walker, *Phys. Rev. Letters*, 1966, **17**, 120.
[152] J. W. Wiggins, J. R. Oleson, Y. K. Lee, and J. C. Walker, *Rev. Sci. Instr.*, 1968, **39**, 995.

[153] J. W. Wiggins and J. C. Walker, *Phys. Rev.*, 1969, **177,** 1786.
[154] F. E. Wagner, *Z. Physik*, 1968, **210,** 361.
[155] M. Kalvius, P. Kienle, K. Bockmann, and H. Eicher, *Z. Physik*, 1961, **163,** 87.
[156] M. Kalvius, *Z. Naturforsch.*, 1962, **17A,** 248.
[157] M. Kalvius, W. Wiedemann, R. Koch, P. Kienle, and H. Eicher, *Z. Physik*, 1962, **170,** 267.
[158] C. I. Wynter, J. J. Spijkerman, H. H. Stadelmaier, and C. H. Cheek, *Nature*, 1969, **223,** 1055.
[159] T. A. Carlson, P. Erman, and K. Fransson, *Nuclear Phys.*, 1968, **111A,** 371.
[160] M. J. Clauser, E. Kankeleit, and R. L. Mössbauer, *Phys. Rev. Letters*, 1966, **17,** 5.
[161] M. J. Clauser and R. L. Mössbauer, *Phys. Rev.*, 1969, **178,** 559.
[162] S. Hufner, M. Kalvius, P. Kienle, W. Wiedemann, and H. Eicher, *Z. Physik*, 1963, **175,** 416.
[163] R. G. Barnes, E. Kankeleit, R. L. Mössbauer, and J. M. Poindexter, *Phys. Rev. Letters*, 1963, **11,** 253.
[164] R. G. Barnes, R. L. Mössbauer, E. Kankeleit, and J. M. Poindexter, *Phys. Rev.*, 1964, **136,** A175.
[165] C. I. Wynter, C. H. Cheek, M. D. Taylor, and J. J. Spijkerman, *Nature*, 1968, **218,** 1047.
[166] D. L. Uhrich, D. J. Genin, and R. G. Barnes, *Phys. Letters*, 1967, **24A,** 338.
[167] D. L. Uhrich and R. G. Barnes, *Phys., Rev.*, 1967, **164,** 428.
[168] M. Kalvius, P. Kienle, H. Eicher, W. Wiedemann, and C. Schuler, *Z. Physik*, 1963, **172,** 231.
[169] R. L. Cohen, *Phys. Letters*, 1967, **24A,** 674.
[170] R. L. Cohen, *Phys. Rev.*, 1968, **169,** 432.
[171] I. Nowik and S. Ofer, *Phys. Letters*, 1963, **3,** 192.
[172] R. L. Cohen, *Phys. Letters*, 1963, **5,** 177.
[173] D. L. Uhrich, D. J. Genin, and R. G. Barnes, *Phys. Rev.*, 1968, **166,** 261.
[174] R. L. Cohen, *Phys. Rev.*, 1964, **134,** A94.
[175] D. L. Uhrich and R. G. Barnes, *Phys. Chem. Glasses*, 1968, **9,** 184.
[176] F. E. Wagner, F. W. Stanek, P. Kienle, and H. Eicher, *Z. Physik*, 1962, **166,** 1.
[177] A. Huller, W. Wiedemann, P. Kienle, and S. Hufner, *Phys. Letters*, 1965, **15,** 269.
[178] I. Nowik and S. Ofer, *J. Phys. and Chem. Solids*, 1968, **29,** 2117.
[179] U. Atzmony, E. R. Bauminger, J. Hess, A. Mustachi, and S. Ofer, *Phys. Rev. Letters*, 1967, **18,** 1061.
[180] S. Ofer and I. Nowik, *Phys. Letters*, 1967, **24A,** 88.
[181] I. Nowik, S. Ofer, and J. H. Wernick, *Phys. Letters*, 1967, **24A,** 89.
[182] G. M. Kalvius, *Phys. Rev.*, 1965, **137,** B1441.
[183] C. Gunther and E. Kankeleit, *Phys. Letters*, 1966, **22,** 443.
[184] W. Henning, P. Kienle, E. Steichele, and F. Wagner, *Phys. Letters*, 1966, **22,** 446.
[185] G. M. Kalvius and J. K. Tison, *Phys. Rev.*, 1966, **152,** 829.
[186] W. Henning, P. Kienle, and H. J. Korner, *Z. Physik*, 1967, **199,** 207.
[187] E. Munck, S. Hufner, H. Prange, and D. Quitmann, *Z. Naturforsch.*, 1966, **A21,** 1507.
[188] J. S. Eck, Y. K. Lee, J. C. Walker, and R. R. Stevens, *Phys. Rev.*, 1967, **156,** 246.
[189] E. Munck, D. Quitmann, H. Prange, and S. Hufner, *Z. Naturforsch.*, 1966, **21A,** 1318.

18 | The Actinide Elements

Interest in the actinides stems from the importance of measuring the physical nuclear constants required for the study of the nuclear structure in this region of comparative instability. Few low-energy levels are conveniently populated by a radioactive parent, but the successful and detailed studies with ^{237}Np using both ^{237}U and ^{241}Am parents have stimulated the use of more difficult techniques such as Coulomb excitation. Consequently the ^{232}Th, ^{231}Pa, ^{238}U, and ^{243}Am resonances have now also been detected. Full details of the known nuclear parameters of these nuclides are tabulated in Appendix I.

18.1 Thorium (^{232}Th)

The 49·8-keV resonance of thorium-232 was reported in 1968 following experiments using Coulomb excitation of a thorium metal target by 4·5-MeV α-particles at 80 K [1]. The transition is highly internally converted ($\alpha_T = 260$), which reduces the effective excitation rate considerably. The linewidth obtained from an absorber of ThO$_2$ at 110 K was 16·7 mm s^{-1}, which is close to the natural width of 15·7 mm s^{-1} thereby implying negligible quadrupole interaction, consistent with the cubic lattices of the metal and ThO$_2$. No hyperfine effects have yet been reported.

18.2 Protactinium (^{231}Pa)

The 84·2-keV resonance of ^{231}Pa was first reported in 1968 and also requires complicated experimentation [2]. The 25·5-hour precursor of ^{231}Th is troublesome to prepare and has to be separated from impurities and fission products following an (n, γ) reaction on separated ^{230}Th. The decay scheme is complex and the relevant details are shown in simplified form in Fig. 18.1. The 84·2-keV level is probably the *third* excited state of ^{231}Pa. This isotope is itself radioactive and decays by α-emission with a half-life of $3·25 \times 10^4$ y. It is only recently that quantities of this isotope adequate for preparing chemical compounds have become available.

[*Refs. on p. 604*]

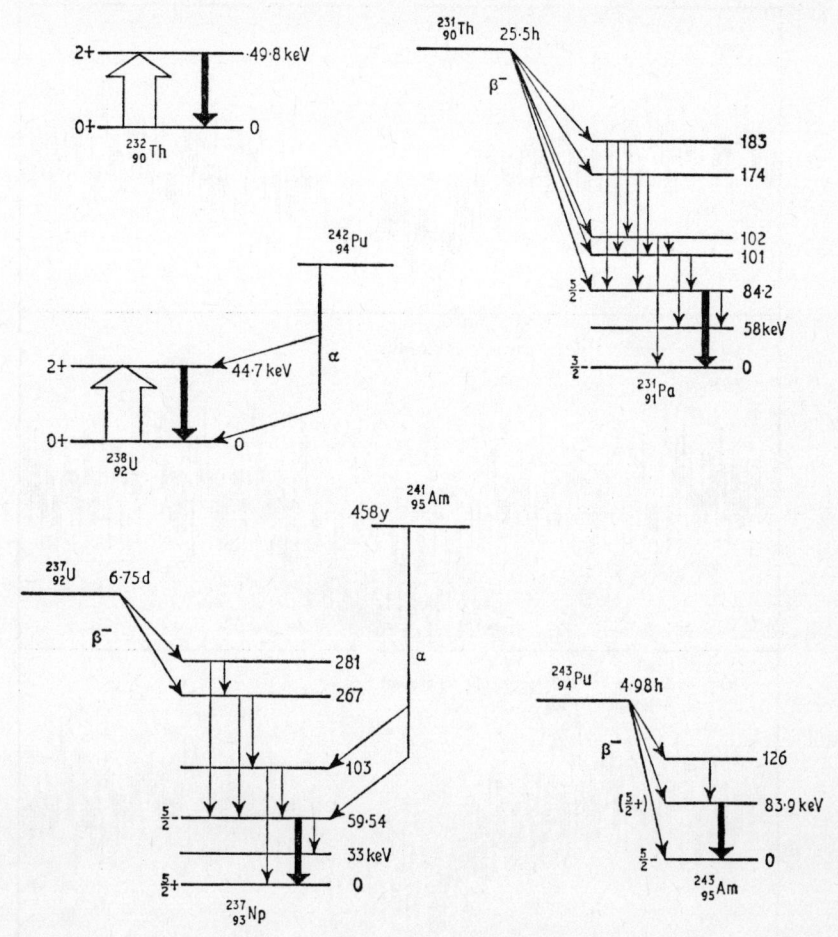

Fig. 18.1 Decay schemes for ^{232}Th, ^{231}Pa, ^{238}U, ^{237}Np, and ^{243}Am.

Spectra with a source of ThO_2 and absorbers of Pa_2O_5 and PaO_2 at 4·2 K are shown in Fig. 18.2. Although there is some suggestion of hyperfine interactions, the lines are far broader than the natural width of 0·079 mm s^{-1}, which makes analysis difficult. If the hyperfine interactions are large compared to the linewidth, as appears likely, the effects of radiation damage on the environment of the Mössbauer atoms could be a significant factor.

18.3 Uranium (^{238}U)

The 44·7-keV transition in ^{238}U suffers from an extremely high internal conversion coefficient α_T of about 625. A resonance was first described in 1967 following Coulomb excitation of a metal target at 80 K by 3-MeV

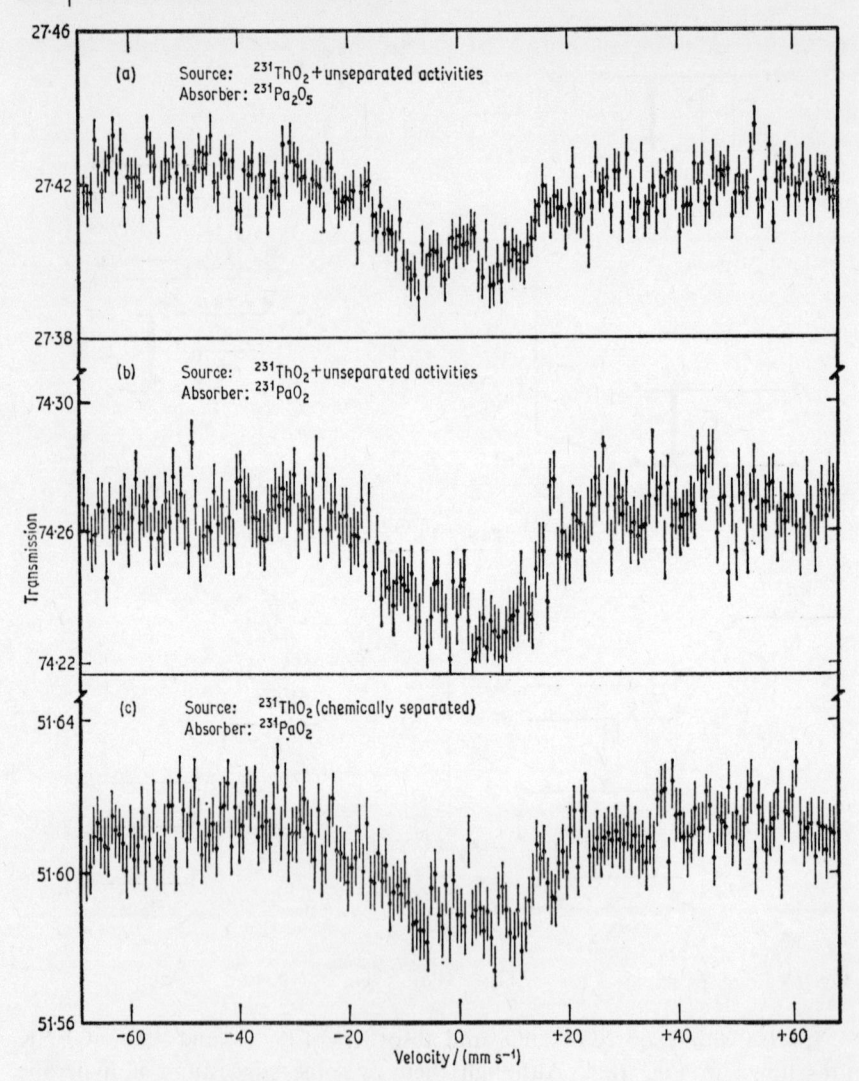

Fig. 18.2 Spectra of $^{231}Pa_2O_5$ and $^{231}PaO_2$ at 4·2 K. Note the broad resonance lines ($\Gamma_r = 0.079$ mm s^{-1}). [Ref. 2, Fig. 3]

α-particles [3]. Although the spectra were of comparatively poor quality, suggestions of a partially resolved hyperfine structure in a U_3O_8 absorber were found.

More satisfactory results have come from a ^{242}Pu source which decays by α-emission directly to the Mössbauer level [4]. An α-decay parent had been previously used in ^{237}Np work (see below) but the change in the lifetime of the Mössbauer level from 63 ns in ^{237}Np to 0·23 ns in ^{238}U gave rise to fears

[*Refs. on p. 604*]

that the recoil after-effects and local kinetic heating might be a more serious problem in the study of the latter. Fortunately this has not proved to be the case. The long lifetime of the ^{242}Pu source and the high internal conversion result in very low counting rates and approximately equal cross-sections for photoelectric and Mössbauer processes. The result is that the absorptions are weak even at 4·2 K. As normally obtained, ^{242}Pu is contaminated with the shorter-lived ^{241}Pu isotope which decays to ^{241}Am. The 44·7-keV γ-rays can only be detected satisfactorily after a chemical separation from the americium impurity, a treatment which has to be repeated at least every three months. PuO$_2$ is cubic and provides a suitable matrix.

An absorber of UO$_2$ at 77 K gave a linewidth of 48 mm s^{-1} which compares favourably with the natural width of 27 mm s^{-1} and shows that the α-decay

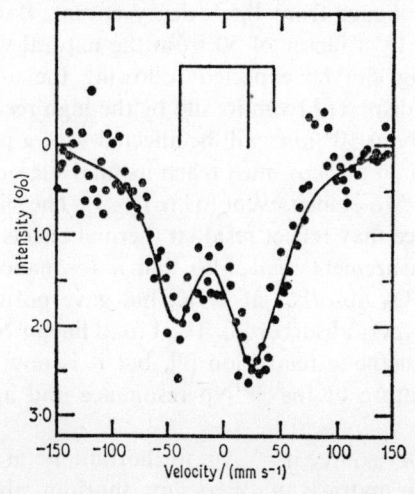

Fig. 18.3 The ^{238}U resonance in (UO$_2$)(NO$_3$)$_2$.6H$_2$O at 4·2 K. The bar diagram is a quadrupole splitting of the $0+ \to 2+$ transition. [Ref. 4, Fig. 4]

induces no serious after-effects. UO$_2$ is antiferromagnetic below 30·8 K, and the U^{4+} ion has a $5f^2$ configuration. Accordingly, the resonance at 4·2 K was broadened by a magnetic hyperfine splitting, and, using an assumed value for μ_e of 0·5 n.m., a value of $H_{\text{eff}} = 2700$ kG was derived.

UF$_4$ has a non-cubic symmetry, and shows a quadrupole interaction of $e^2qQ = 80$ mm s^{-1} at 4·2 K (from the $I = 2$ excited state). The U^{6+} ion has a $5f^0$ configuration and magnetic ordering is not found, but quadrupole splittings of $e^2qQ = -169$ mm s^{-1} in (UO$_2$)(NO$_3$)$_2$.6H$_2$O and $+160$ mm s^{-1} in UO$_3$ were found. The former is illustrated in Fig. 18.3. The excited-state quadrupole moment is unknown but is probably about -3 barn. Chemical isomer shifts for U^{4+} and U^{6+} were the same within experimental error (~ 2 mm s^{-1}), which places an upper limit of 10^{-5} on $\delta \langle R^2 \rangle / \langle R^2 \rangle$. This is not unexpected for a pure rotational nuclear transition.

[Refs. on p. 604]

α-uranium metal shows a quadrupole splitting of $e^2qQ = -84$ mm s^{-1} at 4·2 K, but no magnetic field greater than the upper limit of resolution of 300 kG. The cubic Laves phase UFe$_2$ is ferromagnetic below 195 K, but again the field at the uranium is below significant detection limits at 4 K.

18.4 Neptunium (^{237}Np)

The first experiments with ^{237}Np (59·54-keV) in 1964 used precursors of ^{241}Am (α-decay) and ^{237}U (β-decay) [5]. The source matrices were 1% solid solutions of AmO$_2$ and UO$_2$ in NpO$_2$ respectively, and an absorber of NpO$_2$ was used at 77 and 4·2 K. Although the detection of a resonance from the ^{237}U source was not unexpected, it was more surprising to find a good (if slightly weaker) resonance from the α-decay parent. Both resonances were however broadened by a factor of 50 from the natural width. Considerable local kinetic heating can be expected following the α-decay because the decaying nucleus is displaced from its site by the high recoil energy. Calculations suggest that about 10^5 ions will be affected with a peak temperature of 1000 K, but the excited nucleus must reach its final site on a time-scale much less than that of the Mössbauer event (63 ns) [6, 7]. The slightly lower f factor for the ^{241}Am source may reflect residual thermal effects.

Independent measurements failed to find a resonance using an AmCl$_3$ source and an Np$_3$O$_8$ absorber at 77 K, but gave positive results with an AmO$_2$ source and NpO$_2$ absorber [8]. The broad line in NpO$_2$ was originally attributed to paramagnetic relaxation [9], but it is now known that broad linewidths are a feature of the ^{237}Np resonance and are not of magnetic origin.

A more satisfactory source is ^{241}Am in thorium metal [10]. Detailed comparison of ^{241}Am in matrices of americium, thorium, aluminium, gold, and copper metals with NpO$_2$ or NpAl$_2$ absorbers has shown that the thorium source gives the narrowest linewidths [11]. A ^{241}Am/Th source at 4·2 K and an NpAl$_2$ absorber at 78 K gave a width of 1·1 mm s^{-1} which is, however, still considerably greater than the natural width of 0·073 mm s^{-1}. A source of ^{241}AmO$_2$ showed evidence for multiple charge states of the daughter neptunium ion.

The $\frac{5}{2}+ \to \frac{5}{2}-$ spin-states of the E1 transition result in a large number of hyperfine lines, and although the resonance linewidths are invariably broad, they have proved to be much less than the separations between hyperfine components in many cases. Typical hyperfine spectra are shown in Fig. 18.4.

The alloy NpAl$_2$ has a cubic lattice and at 77 K gives only a single resonance line [12], but at 4·2 K a well-resolved magnetic hyperfine splitting is seen. Although a unique value for μ_e/μ_g cannot be deduced if the positions only of the lines are considered, the unusually large splitting of the energy levels at 4·2 K results in nuclear polarisation from a Boltzmann occupation favour-

[Refs. on p. 604]

ing the lower levels. This causes a slight asymmetry in the line intensities similar to that already discussed for ^{57}Co metal (p. 307) which eliminates all values with the exception of $\mu_e/\mu_g = +0.537(5)$.

NpAl$_2$ orders ferromagnetically below 55.8 K. The chemical isomer shift is typical of Np^{4+} (see later) and implies a $^4I_{\frac{9}{2}}$ ($5f^3$) ground state [13]. However, the temperature dependence of the magnetic splitting deviates considerably from a simple molecular field dependence model with $J = \frac{9}{2}$, and

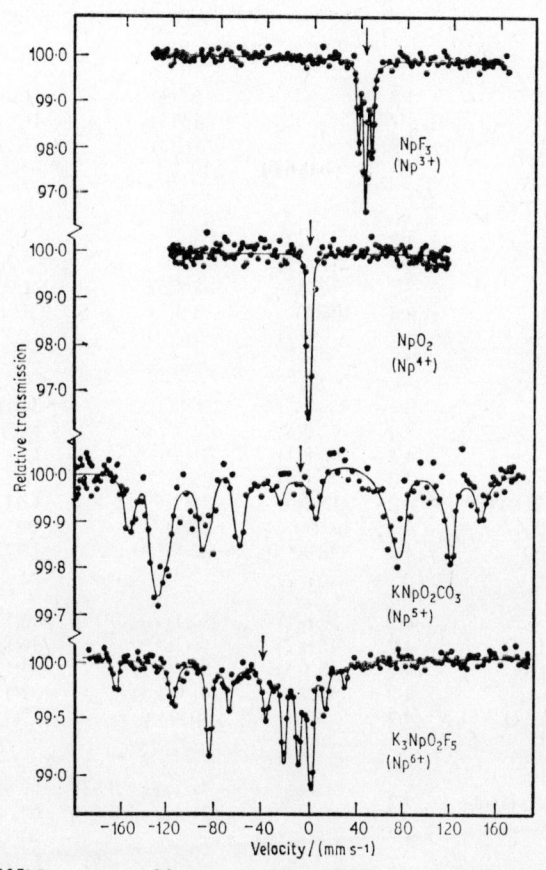

Fig. 18.4 The ^{237}Np spectra of four neptunium oxidation states showing quadrupole splitting (NpF$_3$), combined quadrupole/magnetic interaction (K$_3$NpO$_2$F$_5$ and KNpO$_2$CO$_3$), and the large chemical isomer shifts. [Ref. 11, Fig. 4]

a biquadratic exchange interaction must be invoked. Although the alloy is cubic, the unquenched orbital angular momentum gives rise to a small quadrupole splitting in the magnetically ordered state of $e^2qQ = -0.56(20)$ mm s^{-1} at 4.2 K in an analogous manner to equivalent rare-earth systems.

NpCl$_4$ shows a quadrupole splitting at 77 K which gives $Q_e/Q_g = +1.0(1)$

[*Refs. on p. 604*]

[12], and a combined magnetic/quadrupole interaction at 4·2 K. Parameters are given in Table 18.1. The exact value of μ_0 is uncertain but is believed to be of the order of 2·8 n.m., which implies that the fields in $NpAl_2$ and $NpCl_4$ are about 3000 kG. The value of Q_g has also not been determined directly,

Table 18.1 Mössbauer parameters in neptunium compounds

Compound	T/K	$\{\mu_g\mu_N H_{eff}/I\}$ /(mm s^{-1})	$\frac{1}{4}e^2qQ_g$ /(mm s^{-1})	δ (NpO$_2$) /(mm s^{-1})	Reference
Neptunium(III)					
NpF$_3$	4·2	—	5·5(3)	+41·0(5)	11
NpCl$_3$	4·2	—	4·9(5)	+41(1)	15
NpBr$_3$	4·2	—	5·5(3)	+41(1)	15
Np$_2$S$_3$	77	60·0(6·0)	10·4(3)	+35(1)	15
Neptunium (IV)					
NpO$_2$	4·2	0·7(2)	—	0	11
NpCl$_4$	4·2	47·4(5)	—	+2·2(5)	15
	77	—	+8·8(5)	+1·7(5)	12
NpBr$_4$	4·2	104(1)	−3·0(5)	+3·3(1·3)	15
Np(Et$_2$NCS$_2$)$_4$	4·2	74·6(5)	−7·2(1·0)	+1·3(5)	15
Neptunium(V)					
Np$_3$O$_8$	4·2	{ — , 93·0(5) }	{ 25·8(3), +12·9(1·0) }	{ −24(1), −15 5(7) }	15
KNpO$_2$CO$_3$	4·2	108·8(1)	−31(5)	−12(1)	11
	4·2	98·1(5)	+25·5(5)	−6(1)	15
(NpO$_2$)C$_2$O$_4$H.2H$_2$O	4·2	115·2(5)	+28(2)	−17(1)	11
	4·2	101(1)	+23(1)	−14(1)	15
(NpO$_2$)OH.xH$_2$O	4·2	98·4(1·0)	+22·5(2·5)	−18(1)	15
Neptunium(VI)					
K$_3$NpO$_2$F$_5$	4·2	39·6(1)	+50(1) (η=0·15)	−46(2)	11
NpO$_2$(NO$_3$)$_2$.xH$_2$O	4·2	46·0(2)	+18(10)	−36(2)	11
Na(NpO$_2$)(C$_2$H$_3$O$_2$)$_3$	4·2	48·1(5)	+59·6(1·5)	−34(1)	15
Rb(NpO$_2$)(NO$_3$)$_3$	4·2	52·9(5)	+61·9(1·5)	−32(1)	15
(NH$_4$)$_2$Np$_2$O$_7$.H$_2$O	4·2	—	40·4(1·0)	−39(1)	15
Neptunium(VII)					
Co(NH$_3$)$_6$NpO$_5$.xH$_2$O	4·2	{ —, — }	{ +31(1) (η=0·83), +21(1) (η=0·69) }	{ −62·8(8), −62·8(8) }	17
Np^{VII}/NaOH	4·2	—	—	−60(1)	17
ozonised BaNp$_2$O$_7$.xH$_2$O	4·2	{ —, — }	{ +25(2) (η=0·6), +28(2) (η=0·6) }	{ −62(1), −42(1) }	17
Metals					
Neptunium	4·2	{ —, — }	{ 22·3(4), 7·0(1) }	{ −1·3(1), +1·3(1) }	15
NpAl$_2$	4·2	53·4(5)	—	+5·7(5)	12
	4·2	54·0(5)	—	+6·3(5)	11
NpAl$_4$	4·2	49·0(5)	−4·7(1·0)	+14(1)	15
NpC	4·2	84·0(5)	—	−12(1)	11
	4·2	84·6(1·0)	—	−12(2)	15

[*Refs.* on p. 604]

but a value of +4·1 barn has been proposed by drawing an analogy between the ^{241}Am nucleus and the first excited state of ^{237}Np which have identical quantum states [14].

Np and Am metals showed no magnetic ordering down to 1·7 K but do give a quadrupole interaction [13].

Chemical isomer shifts, quadrupole splittings, and magnetic hyperfine splittings have been measured in a number of compounds of neptunium with different formal oxidation states [11, 15]. Typical spectra are shown in Fig. 18.4 and values are given in Table 18.1. The pure magnetic spectra in NpAl$_2$ and NpC gave $\mu_e/\mu_g = 0.533(5)$. A quadrupole interaction in NpF$_3$ gave $Q_e/Q_g = 1.0(1)$.

The chemical isomer shifts decrease considerably as the formal oxidation state of the neptunium increases from +3 to +6. Comparison of the shifts with electron charge densities at the nucleus derived from non-relativistic Hartree–Fock wavefunctions for $5f^4$ to $5f^1$ configurations showed a linear interrelationship. An approximate value for $\delta\langle R^2\rangle/\langle R^2\rangle$ of -3.5×10^{-4} was derived.

NpO$_2$ has the CaF$_2$ structure and is antiferromagnetic below 22 K, resulting in a slight broadening of the Mössbauer resonance [16]. The internal magnetic field at 4·2 K is of the order of 40 kG in NpO$_{2.03}$ and only slightly greater in NpO$_{1.92}$. Although the latter is highly non-stoichiometric no evidence was found for the Np^{3+} ion. The magnetic moment in NpO$_2$ at each neptunium ion is only of the order of 0·01 B.M., which indicates an unusually severe quenching of the moment by crystalline electric field and exchange splitting interactions.

The unusual oxidation state Np^{7+}, which has a $5f^0$ configuration, has been observed in the compound [Co(NH$_3$)$_6$]NpO$_5$.xH$_2$O (which contains two distinct Np sites); in an NaOH matrix; and in ozonised BaNp$_2$O$_7$.xH$_2$O [17]. As seen in Table 18.1 and Fig. 18.5 the chemical isomer shift values are the

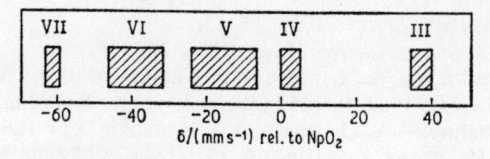

Fig. 18.5 The known ranges for the chemical isomer shift in neptunium compounds correlated with the formal oxidation state.

most negative known for neptunium. The effective range of shifts between Np^{7+} and Np^{3+} is 100 mm s^{-1}. The significance of covalent bonding cannot yet be adequately assessed. NpO$_2^+$ and NpO$_2^{2+}$ are known to feature some degree of covalency and the same may be implied for NpO$_5^{3-}$. Accordingly, electron-density calculations for the free ions yield only an approximate calibration.

[*Refs. on p. 604*]

The extreme sensitivity of the chemical isomer shift to environment may be the cause of the line broadening in ^{237}Np spectra particularly in view of the inescapable presence of radiation damage.

18.5 Americium (^{243}Am)

The 83·9-keV resonance of ^{243}Am is difficult to observe [18]. The short half-life of the americium ground state (7·95 × 10^3 y) makes all the absorbers extremely radioactive, with additional activity from the daughter ^{239}Np. A Ge(Li) detector can resolve the required γ-ray, but the experimental limits on quantities of material are severe. A ^{243}PuO$_2$ source can be made by the ^{242}PuO$_2(n, \gamma)$ reaction. Single resonance lines in AmO$_2$ and AmF$_3$ showed a relative shift of +53(1) mm s^{-1}, which is an extremely large shift between neighbouring oxidation states. With electronic configurations of $5f^5$ and $5f^6$ respectively and estimated electron densities, a value of $\delta\langle R^2\rangle/\langle R^2\rangle = -9(3) \times 10^{-4}$ was derived. The AmO$_2$ resonance narrows considerably as the temperature is raised from 4·2 to 77 K, possibly as a result of a hyperfine interaction.

REFERENCES

[1] N. Hershkowitz, C. G. Jacobs, and K. A. Murphy, *Phys. Letters*, 1968, **27B**, 563.
[2] W. L. Croft, J. A. Stone, and W. L. Pillinger, *J. Inorg. Nuclear Chem.*, 1968, **30**, 3203.
[3] J. R. Oleson, Y. K. Lee, J. C. Walker, and J. W. Wiggins, *Phys. Letters*, 1967, **25B**, 258.
[4] S. L. Ruby, G. M. Kalvius, B. D. Dunlap, G. K. Shenoy, D. Cohen, M. B. Brodsky, and D. J. Lam, *Phys. Rev.*, 1969, **184**, 374.
[5] J. A. Stone and W. L. Pillinger, *Phys. Rev. Letters*, 1964, **13**, 200.
[6] J. G. Mullen, *Phys. Letters*, 1965, **15**, 15.
[7] M. Kaplan, *J. Inorg. Nuclear Chem.*, 1966, **28**, 331.
[8] G. N. Belozerskii, Yu. A. Nemilov, A. A. Chaikhorskii, and A. V. Shvedchikov, *Fiz. Tverd. Tela*, 1967, **9**, 1252 (*Soviet Physics – Solid State*, 1967, **9**, 978).
[9] V. A. Bryukhanov, V. V. Ovechkin, A. I. Peryshkin, E. I. Rzhekhina, and V. S. Shpinel, *Fiz. Tverd. Tela*, 1967, **9**, 1519 (*Soviet Physics – Solid State*, 1967, **9**, 1189).
[10] B. M. Aleksandrov, A. V. Kalyamin, A. S. Krivokhatskii, B. G. Lure, A. N. Murin, and Yu. F. Romanov, *Fiz. Tverd. Tela*, 1968, **10**, 1896.
[11] B. D. Dunlap, G. M. Kalvius, S. L. Ruby, M. B. Brodsky, and D. Cohen, *Phys. Rev.*, 1968, **171**, 316.
[12] J. A. Stone and W. L. Pillinger, *Phys. Rev.*, 1968, **165**, 1319.
[13] B. D. Dunlap, M. B. Brodsky, G. M. Kalvius, G. K. Shenoy, and D. J. Lam, *J. Appl. Phys.*, 1969, **40**, 1495.
[14] B. D. Dunlap and G. M. Kalvius, *Phys. Rev.*, 1969, **186**, 1296.

[15] J. A. Stone and W. L. Pillinger, *Symposia Faraday Soc. No. 1*, 1968, 77.
[16] B. D. Dunlap, G. M. Kalvius, D. J. Lam, and M. B. Brodsky, *J. Phys. and Chem. Solids*, 1968, **29**, 1365.
[17] J. A. Stone, W. L. Pillinger, and D. G. Karraker, *Inorg. Chem.*, 1969, **8**, 2519.
[18] G. M. Kalvius, S. L. Ruby, B. D. Dunlap, G. K. Shenoy, D. Cohen, and M. B. Brodsky, *Phys. Letters*, 1969, **29B**, 489.

Appendix 1 | Table of Nuclear Data for Mössbauer Transitions

Data are given in the following table for all known Mössbauer transitions. The figures given are a personal selection from available data. The following derivations have been used.

$$\Gamma_r = \frac{273 \cdot 8}{E_\gamma t_{\frac{1}{2}}} \text{ mm s}^{-1} \; (E_\gamma \text{ in keV}, t_{\frac{1}{2}} \text{ in ns})$$

$$E_R = 5 \cdot 36942 \times 10^{-4} \frac{(E_\gamma)^2}{m} \text{ eV} \; (E_\gamma \text{ in keV}, m \text{ in a.m.u.})$$

$$\sigma_0 = \frac{2 \cdot 446 \times 10^{-15}}{(E_\gamma^2)} \frac{2I_e + 1}{2I_g + 1} \frac{1}{1 + \alpha_T} \text{ cm}^2 \; (E_\gamma \text{ in keV})$$

The parameters given are:

- E_γ energy of γ-ray transition in keV
- Γ_r natural linewidth ($= 2\Gamma$) in mm s^{-1}
- $I_g, I_e (\pm)$ ground- and excited-state nuclear spin quantum numbers (parity)
- a percentage natural abundance of resonant isotope
- $t_{\frac{1}{2}}$ excited-state half-life in ns
- α_T total internal conversion coefficient
- E_R recoil energy of nucleus (in eV $\times 10^{-2}$)
- σ_0 resonant absorption cross-section (in cm$^2 \times 10^{-18}$)
- E2/M1 the multipolarity of the radiation
- μ_g, μ_e the ground- and excited-state magnetic moments in n.m.
- Q_g, Q_e the ground- and excited-state quadrupole moments in barn
- *   data on the possible source reactions.

It should be noted that the degree of uncertainty in some of the parameters can be quite large, and more detailed comment together with important references may be found in the main text.

Isotope	E_γ/keV	Γ_r/(mm s^{-1})	I_g	I_e	a/%	$t_{\frac{1}{2}}$/ns	α_T	E_R /(eV×10^{-2})	σ_o /(cm^2×10^{-18})
^{40}K	29·4	2·4	4−	3−	0·012	3·9	0·35	1·16	1·6
^{57}Fe	14·412	0·192	$\frac{1}{2}$−	$\frac{3}{2}$−	2·17	99·3	8·17	0·195	2·57
^{57}Fe	136·32	0·23	$\frac{1}{2}$−	$\frac{5}{2}$−	2·17	8·9	0·14	17·5	0·346
^{61}Ni	67·40	0·78	$\frac{3}{2}$−	$\frac{5}{2}$−	1·25	5·2	0·12	4·00	0·72
^{67}Zn	93·26	3·12×10^{-4}	$\frac{5}{2}$−	$\frac{1}{2}$−	4·11	9400	0·54	6·98	0·061
^{73}Ge	67·03	2·2	$\frac{9}{2}$+	($\frac{7}{2}$+)	7·67	1·86	—	3·31	—
^{83}Kr	9·3	0·20	$\frac{9}{2}$+	$\frac{7}{2}$+	11·55	147	11	0·056	1·9
^{99}Tc	140·5	10·1	$\frac{9}{2}$+	$\frac{7}{2}$+	Nil	0·192	—	10·7	—
^{99}Ru	90	0·147	$\frac{5}{2}$+	$\frac{3}{2}$+	12·63	20·7	—	4·4	—
^{107}Ag	93·1	6·64×10^{-11}	$\frac{1}{2}$−	$\frac{7}{2}$+	51·35	44·3×10^9	20	4·34	0·053
^{119}Sn	23·875	0·626	$\frac{1}{2}$+	$\frac{3}{2}$+	8·58	18·3	5·12	0·258	1·40
^{121}Sb	37·15	2·1	$\frac{5}{2}$+	$\frac{7}{2}$+	57·25	3·5	∼10	0·612	0·21
^{125}Te	35·48	5·02	$\frac{1}{2}$+	$\frac{3}{2}$+	6·99	1·535	12·7	0·541	0·28
^{127}I	57·60	2·54	$\frac{5}{2}$+	$\frac{7}{2}$+	100	1·86	3·70	1·40	0·21
^{129}I	27·72	0·59	$\frac{7}{2}$+	$\frac{5}{2}$+	Nil	16·8	5·3	0·321	0·38
^{129}Xe	39·58	6·85	$\frac{1}{2}$+	$\frac{3}{2}$+	26·44	1·01	11·8	0·652	0·24
^{131}Xe	80·16	6·83	$\frac{3}{2}$+	$\frac{1}{2}$+	21·18	0·50	2·0	2·63	0·063
^{133}Cs	81·00	0·536	$\frac{7}{2}$+	$\frac{5}{2}$+	100	6·28	1·63	2·65	0·106
^{133}Ba	12·29	2·7	$\frac{1}{2}$+	$\frac{3}{2}$+	Nil	8·1	110	0·061	0·29
^{141}Pr	145·43	0·99	$\frac{5}{2}$+	$\frac{7}{2}$+	100	1·9	0·43	8·1	0·11
^{145}Nd	67·25	0·12	$\frac{7}{2}$−	($\frac{3}{2}$−)	8·29	33	6·6	1·67	—
^{145}Nd	72·5	5·4	$\frac{7}{2}$−	$\frac{9}{2}$−	8·29	0·7	—	1·95	—
^{147}Pm	91·06	1·16	$\frac{7}{2}$+	$\frac{5}{2}$+	Nil	2·59	1·9	3·03	0·076
^{149}Sm	22·5	1·60	$\frac{7}{2}$−	$\frac{5}{2}$−	13·9	7·6	(∼12)	0·182	—
^{152}Sm	121·78	1·61	0+	2+	26·6	1·4	0·10	5·27	0·75
^{154}Sm	81·99	1·1	0+	2+	22·6	3·0	5·0	2·34	—
^{151}Eu	21·6	1·44	$\frac{5}{2}$+	$\frac{7}{2}$+	47·8	8·8	29	0·166	0·23
^{153}Eu	83·37	2·93	$\frac{5}{2}$+	$\frac{7}{2}$+	52·2	1·12	5·5	2·44	0·072
^{153}Eu	97·43	10·7	$\frac{5}{2}$+	$\frac{5}{2}$−	52·2	0·26	0·42	3·32	0·182
^{153}Eu	103·12	0·68	$\frac{5}{2}$+	$\frac{5}{2}$+	52·2	3·9	1·55	3·74	0·060
^{154}Gd	123·07	1·88	0+	2+	2·23	1·18	1·1	5·29	0·38
^{155}Gd	60·00	19·0	$\frac{3}{2}$−	$\frac{5}{2}$−	15·0	0·24	7·5	1·25	0·12
^{155}Gd	86·54	0·514	$\frac{3}{2}$−	$\frac{5}{2}$+	15·0	6·15	0·49	2·60	0·33
^{155}Gd	105·32	2·6	$\frac{3}{2}$−	$\frac{3}{2}$+	15·0	1·0	—	3·84	—
^{156}Gd	88·97	1·40	0+	2+	20·6	2·2	4·1	2·72	0·30
^{157}Gd	64·0	0·0093	$\frac{3}{2}$−	$\frac{5}{2}$+	15·7	460	0·8	1·40	0·50
^{158}Gd	79·51	1·50	0+	2+	24·5	2·3	5·94	2·15	0·278
^{160}Gd	75·3	1·45	0+	2+	21·6	2·5	—	1·91	—
^{159}Tb	58·0	36·2	$\frac{3}{2}$+	$\frac{5}{2}$+	100	0·13	10	1·14	0·099
^{160}Dy	86·79	1·58	0+	2+	2·30	2·00	3·7	2·52	0·35
^{161}Dy	25·65	0·37	$\frac{5}{2}$+	$\frac{7}{2}$−	18·88	29	(2·5)	0·220	—
^{161}Dy	43·81	6·8	$\frac{5}{2}$+	$\frac{7}{2}$+	18·88	0·92	(5)	0·64	—
^{161}Dy	74·57	1·22	$\frac{5}{2}$+	$\frac{3}{2}$−	18·88	3·0	—	1·86	—
^{162}Dy	80·7	1·54	0+	2+	25·5	2·2	∼5	2·16	—
^{164}Dy	73·39	15·5	0+	2+	28·1	2·4	—	1·76	—
^{165}Ho	94·70	96	$\frac{7}{2}$−	$\frac{9}{2}$−	100	0·03	2·8	2·92	0·090
^{164}Er	91·5	1·97	0+	2+	1·56	1·52	2	2·74	0·487
^{166}Er	80·56	1·86	0+	2+	33·41	1·83	6·0	2·10	0·27
^{167}Er	79·32 >33		$\frac{7}{2}$+	$\frac{9}{2}$+	22·94	>0·103	(5·3)	2·02	—
^{168}Er	79·8	1·79	0+	2+	27·07	1·92	3·3	2·04	0·45
^{170}Er	79·3	∼0·2	0+	2+	14·88	—	—	1·99	—
^{169}Tm	8·40	9·3	$\frac{1}{2}$+	$\frac{3}{2}$+	100	3·5	220	0·024	0·31

Radi-ation	μ_g /n.m.	μ_e /n.m.	Q_g /barn	Q_e /barn	Source reactions β^- IT EC Coulomb Other	Popular parent
—	−1·298	—	−0·09	—	^{39}K$(^{n\,\gamma}_{d,p})$	
M1	+0·0902	−0·1547	0	0·2	* *	^{56}Fe$(^{n\,d}_{d\,p})$ ^{57}Co (EC 270 d)
E2	+0·0902	—	0	—	*	^{57}Co (EC 270 d)
M1	−0·7487	+0·47	+0·162	−0·3	* * *	^{61}Co (β^- 99 m)
E2	+0·8755	—	+0·16	—	*	^{67}Ga (EC 78 h)
—	−0·8788	—	−0·26	—	*	
M1	−0·967	−0·939	+0·2701	+0·459	* * *	^{83}Br (β^- 2·41 h)
M1 (+E2)	+5·6806	—	+0·3	—	*	^{99}Mo (β^- 67 h)
E2 /M1=2·7	−0·62	−0·29	±0·05	±0·15	*	^{99}Rh (EC 16 d)
—	−0·1135	—	—	—	*	^{107}Cd (EC 6·5 h)
M1	−1·041	+0·67	0	−0·08	* *	^{119m}Sn (IT 250 d)
M1	+3·359	+2·35	−0·26	−0·36	*	^{121m}Sn (β^- 76 y)
M1	−0·8872	+0·60	0	−0·2	* * *	^{125}I (EC 60 d)
M1	+2·809	+2·02	−0·79	−0·71	*	^{127m}Te (β^- 109 d)
M1	+2·617	+2·84	−0·55	−0·68	*	^{129m}Te (β^- 33 d)
M1	−0·7769	+0·68	0	−0·41	*	^{129}I (β^- 1·7×10^7 y)
M1	+0·6907	—	−0·12	0	*	^{131}I (β^- 8 d)
M1	+2·5789	+3·44	−0·003	—	* *	^{133}Ba (EC 7·2 y)
—	—	—	0	—	*	^{133m}Ba (IT 38·9 h)
M1	+4·3	—	−0·059	—	*	^{141}Ce (β^- 33 d)
E2	−0·654	—	−0·25	—	*	^{145}Pm (EC 17·7 y)
M1	−0·654	−1·121	−0·25	—	*	^{145}Pm (EC 17·7 y)
M1	+2·8	—	±0·9	—	*	^{147}Nd (β^- 11·1 d)
M1	−0·665	−0·620	≤0·06	0·40	*	^{149}Eu (EC 106 d)
E2	0	+0·832	0	—	*	^{152}Eu (EC 12 y)
E2	0	+0·778	0	—	*	
M1	+3·465	+2·587	+1·16	+1·51	* *	^{151}Gd (EC 120 d)
E2 /M1=0·56	+1·530	+1·80	+2·9	—	*	^{153}Sm (β^- 47 h)
E1	+1·530	+3·21	+2·9	—	*	^{153}Gd (EC 242 d)
M1	+1·530	+2·04	+2·9	+1·5	* *	^{153}Sm (β^- 47 h)
E2	0	+0·73	0	—	*	^{154}Eu (β^- 16 y)
M1	−0·254	—	+1·3	—	* *	^{155}Eu (β^- 1·81 y)
E1	−0·254	−0·515	+1·3	+0·16	*	^{155}Eu (β^- 1·81 y)
E1	−0·254	—	+1·3	+1·3	*	^{155}Eu (β^- 1·81 y)
E2	0	+0·60	0	+1·73	* * *	^{155}Gd(n,γ) ^{156}Eu (β^- 15 d)
E1	−0·339	—	+1·67	+2·97	*	^{157}Eu (β^- 15·2 h)
E2	0	+0·770	0	+1·5	*	^{157}Gd(n,γ)
E2	0	+0·61	0	+1·6	*	
M1	+1·9	+1·5	+2·0	—	* *	^{159}Gd (β^- 18·0 h)
E2	0	+0·74	0	—	*	^{160}Tb (β^- 72·1 d)
E1	−0·472	+0·59	+1·35	+1·36	*	^{161}Tb (β^- 6·9 d)
M1	−0·472	—	+1·35	—	*	
E1	−0·472	−0·396	+1·35	+0·75	*	^{161}Tb (β^- 6·9 d)
E2	0	+0·74	0	—		^{161}Dy(n,γ)
E2	0	+0·84	0	−1·12	* *	^{164}Ho (EC 37 m)
M1	+4·0	—	+2·82	—	*	^{165}Dy (β^- 139·2 m)
E2	0	+0·70	0	—	*	^{164}Ho (β^- 37 m)
E2	0	+0·62	0	−1·9	*	^{166}Ho (β^- 26·9 h)
M1+(E2)	−0·565	—	+2·83	—	*	
E2	0	+0·66	0	—	* *	^{168}Tm (EC 85 d)
E2	0	+0·62	0	—	*	
M1	−0·231	+0·50	0	−1·3	*	^{169}Er (β^- 9·4 d)

Isotope	E_γ/keV	Γ_r/(mm s^{-1})	I_g	I_e	a/%	$t_{\frac{1}{2}}$/ns	α_T	E_R/(eV×10^{-2})	σ_o/(cm^2×10^{-18})
^{170}Yb	84·26	2·07	0+	2+	3·1	1·57	6·7	2·24	0·22
^{171}Yb	66·74	4·8	$\frac{1}{2}-$	$\frac{3}{2}-$	14·3	0·86	11·3	1·40	0·088
^{171}Yb	75·89	2·1	$\frac{1}{2}-$	$\frac{5}{2}-$	14·3	1·7	7·2	1·81	0·155
^{172}Yb	78·74	2·21	0+	2+	21·8	1·57	—	1·94	—
^{174}Yb	76·5	1·88	0+	2+	31·6	1·9	—	1·81	—
^{176}Yb	82·1	1·67	0+	2+	12·7	2·0	—	2·06	—
^{176}Hf	88·36	2·22	0+	2+	5·21	1·39	5·5	2·38	0·24
^{177}Hf	112·97	4·8	$\frac{7}{2}-$	$\frac{9}{2}-$	18·56	0·5	2·2	3·86	0·075
^{178}Hf	93·2	1·96	0+	2+	27·1	1·50	1·0	2·62	0·70
^{180}Hf	93·33	1·96	0+	2+	35·22	1·50	4·2	2·60	0·27
^{181}Ta	6·25	0·0065	$\frac{7}{2}+$	$\frac{9}{2}-$	99·99	6800	45	0·0116	1·70
^{181}Ta	136·25	57	$\frac{7}{2}+$	$\frac{9}{2}+$	99·99	0·035	8	5·51	0·018
^{182}W	100·10	2·00	0+	2+	26·4	1·37	3·2	2·96	0·29
^{183}W	46·48	31	$\frac{1}{2}-$	$\frac{3}{2}-$	14·4	0·19	9·0	0·633	0·23
^{183}W	99·08	3·9	$\frac{1}{2}-$	$\frac{5}{2}-$	14·4	0·7	4·3	2·88	0·14
^{184}W	111·2	1·92	0+	2+	30·6	1·28	2·7	3·61	0·27
^{186}W	122·6	2·21	0+	2+	28·4	1·01	1·6	4·33	0·31
^{187}Re	134·24	202	$\frac{5}{2}+$	$\frac{7}{2}+$	62·93	0·0101	1·8	5·18	0·065
^{186}Os	137·16	2·38	0+	2+	1·59	0·84	1·2	5·44	0·29
^{188}Os	155·0	2·48	0+	2+	13·3	0·71	0·9	6·86	0·27
^{189}Os	36·22	10·5	$\frac{3}{2}-$	$\frac{1}{2}-$	16·1	0·72	—	0·372	—
^{189}Os	69·59	2·41	$\frac{3}{2}-$	$\frac{5}{2}-$	16·1	1·63	8·2	1·38	0·082
^{189}Os	95·3	22	$\frac{3}{2}-$	$\frac{3}{2}-$	16·1	$\geqslant 0·13$	—	2·58	—
^{191}Ir	82·33	0·85	$\frac{3}{2}+$	$\frac{1}{2}+$	38·5	3·9	—	1·91	—
^{191}Ir	129·39	23·5	$\frac{3}{2}+$	$\frac{5}{2}+$	38·5	0·090	2·4	4·71	0·065
^{193}Ir	73·0	0·60	$\frac{3}{2}+$	$\frac{1}{2}+$	61·5	6·2	~6	1·48	0·03
^{193}Ir	138·9	24·6	$\frac{3}{2}+$	$\frac{5}{2}+$	61·5	0·080	—	5·36	—
^{195}Pt	98·8	16	$\frac{1}{2}-$	$\frac{3}{2}-$	33·8	0·17	7·2	2·69	0·059
^{195}Pt	129·4	3·5	$\frac{1}{2}-$	$\frac{5}{2}-$	33·8	0·6	2	4·61	0·14
^{197}Au	77·34	1·87	$\frac{3}{2}+$	$\frac{1}{2}+$	100	1·892	4·0	1·63	0·041
^{201}Hg	32·19	43	$\frac{3}{2}-$	$\frac{1}{2}-$	13·22	0·2	39	0·276	0·03
^{232}Th	49·8	15·7	0+	2+	100	0·35	260	0·574	0·189
^{231}Pa	84·20	0·079	$\frac{3}{2}-$	$\frac{5}{2}+$	Nil	41	1·83	1·65	0·18
^{238}U	44·7	27	0+	2+	99·28	0·23	625	0·451	0·01
^{237}Np	59·54	0·073	$\frac{5}{2}+$	$\frac{5}{2}-$	Nil	63	1·06	0·803	0·33
^{243}Am	83·9	1·39	$\frac{5}{2}-$	$\frac{5}{2}+$	Nil	2·34	0·2	1·56	0·29

Radi-ation	μ_g/n.m.	μ_e/n.m.	Q_g/barn	Q_e/barn	Source reactions β^- IT EC Coulomb Other	Popular parent
E2	0	+0.67	0	$(1.0)^{170}$ *		^{170}Tm (β^- 130 d)
E2 /M1=0.47	+0.4930	+0.350	0	$(0.89)^{170}$ *		^{171}Tm (β^- 1.92 y)
E2	+0.4930	+1.01	0	$(1.03)^{170}$	*	^{171}Lu (EC 8.3 d)
E2	0	+0.67	0	$(1.01)^{170}$	* *	^{172}Lu (EC 6.70 d)
E2	0	+0.68	0	$(1.04)^{170}$	* *	^{174}Lu (EC 3.6 y)
E2	0	—	0	$(1.04)^{170}$	*	
E2	0	—	0	—	*	^{176m}Lu (β^- 3.7 h)
E2	+0.61	+1.0	—	—	*	^{177}Lu (β^- 6.7 d)
E2	0	+0.71	0	—	* *	^{178}W (2 EC 21.5 d)
E2	0	+0.74	0	—	* *	^{180m}Hf (IT 5.5 h)
E1	+2.35	+5.14	+3.9	+2.9	*	^{181}W (EC 140 d)
E2 /M1=0.19	+2.35	—	+3.9	—	*	^{181}W (EC 140 d)
E2	0	+0.532	0	$(1.0)^{182}$ *	*	^{182}Ta (β^- 115 d)
M1	+0.1172	−0.10	0	$(0.88)^{182}$ *	*	^{183}Ta (β^- 5.1 d)
E2	+0.1172	+0.93	0	$(0.94)^{182}$ *		^{183}Ta (β^- 5.1 d)
E2	0	+0.590	0	$(0.94)^{182}$	* *	^{184}Re (EC 38 d)
E2	0	+0.62	0	$(0.88)^{182}$	* *	^{186}Re (EC 90 d)
M1	+3.204	—	+2.6		*	^{187}W (β^- 23.9 h)
E2	0	+0.64	0	1.5	*	^{186}Re (β^- 90 h)
E2	0	+0.62	0	—	*	^{188}Re (β^- 16.7 h)
M1	+0.6565	+0.23	+0.91	0	*	^{189}Ir (EC 13.3 d)
E2 /M1=0.57	+0.6565	+0.988	+0.91	—	*	^{189}Ir (EC 13.3 d)
E2 /M1=0.09	+0.6565	—	+0.91	—	*	^{189}Ir (EC 13.3 d)
E2 /M1=0.64	+0.1453	+0.54	+1.5	—	*	^{191}Pt (EC 3.0 d)
E2 /M1=0.13	+0.1453	—	+1.5	—	*	^{191}Os (β^- 15 d)
E2 /M1=0.31	+0.1589	+0.470	+1.5	0	*	^{193}Os (β^- 31 h)
M1(+E2)	+0.1589	—	+1.5	—	*	^{193}Os (β^- 31 h)
M1	+0.6060	−0.61	0	—	* *	^{195}Au (EC 183 d)
E2	+0.6060	—	0	—	*	^{195}Au (EC 183 d)
E2 /M1=0.11	+0.1449	+0.419	+0.56	0	*	^{197}Pt (β^- 18 h)
M1	−0.5567	—	+0.50	0	*	^{201}Tl (EC 73 h)
E2	0	—	0	—	*	
E1	—	—	—	—	*	^{231}Th (β^- 25.5 h)
E2	0	—	0	—	*	^{242}Pu(α)
E1	+2.8	+1.5	+4.1	+4.1	*	^{241}Am(α)
E1	—	—	—	—	*	^{243}Pu (β^- 4.98 h)

Appendix 2 | The Relative Intensities of Hyperfine Lines

The Clebsch–Gordan coefficients and the angular dependence functions needed to calculate the relative intensities of the hyperfine lines for a $\frac{3}{2} \to \frac{1}{2}$ transition with M1 multipolarity have already been given in Table 3.2 (p. 67) and discussed in detail in Section 3.7. In this appendix are listed other useful sets of coefficients for the cases

(a) $\frac{3}{2} \to \frac{1}{2}$ E2
(b) $\frac{5}{2} \to \frac{3}{2}$ M1
(c) $\frac{5}{2} \to \frac{3}{2}$ E2
(d) $\frac{7}{2} \to \frac{5}{2}$ M1
(e) $2 \to 0$ E2
(f) $\frac{5}{2} \to \frac{5}{2}$ E1

The various quantities tabulated have been explained in Section 3.7, but the following additional notes referred to in the column headings are relevant:

(1) The Clebsch–Gordan coefficients $\langle \frac{1}{2}\,2\,-m_1 m\,|\,\frac{3}{2}\,m_2\rangle$ calculated using the formulae in ref. 29 of Chapter 3 for $\langle \frac{3}{2}\,\frac{1}{2}\,m_2 m_1\,|\,2m\rangle$ and converted using the relationship
$$\langle \tfrac{1}{2}\,2\,-m_1 m\,|\,\tfrac{3}{2} m_2\rangle = (-)^{\frac{1}{2}+m_1}\sqrt{\tfrac{4}{3}}\langle \tfrac{3}{2}\,\tfrac{1}{2}\,m_2 m_1\,|\,2m\rangle$$

(2) C^2 and Θ are the angular independent and dependent terms arbitrarily normalised.

(3) relative intensities observed at 0° and 90° to the principal axis. Normalisation arbitrary.

(4) the Clebsch–Gordan coefficients $\langle \frac{3}{2}\,1\,-m_1 m\,|\,\frac{5}{2} m_2\rangle$ calculated as in (1) using
$$\langle \tfrac{3}{2}\,1\,-m_1 m\,|\,\tfrac{5}{2} m_2\rangle = (-)^{\frac{3}{2}+m_1}\sqrt{2}\langle \tfrac{5}{2}\,\tfrac{3}{2}\,m_2 m_1\,|\,1m\rangle$$

(5) as (4) using
$$\langle \tfrac{3}{2}\,2\,-m_1 m\,|\,\tfrac{5}{2} m_2\rangle = (-)^{\frac{3}{2}+m_1}\sqrt{\tfrac{6}{5}}\langle \tfrac{5}{2}\,\tfrac{3}{2}\,m_2 m_1\,|\,2m\rangle$$

(6) the Clebsch–Gordan coefficients $\langle \frac{5}{2}\,1\,-m_1 m\,|\,\frac{7}{2} m_2\rangle$ calculated using the formulae in M. A. Melvin and N. V. V. J. Swamy, *Phys. Rev.*, 1957, **107**,

186, and the relationship
$$\langle \tfrac{5}{2} 1 -m_1 m \mid \tfrac{7}{2} m_2 \rangle = (-)^{\frac{5}{2}+m_1} \sqrt{\tfrac{8}{3}} \langle \tfrac{7}{2} \tfrac{5}{2} m_2 m_1 \mid 1m \rangle$$
(7) all the coefficients are unity
(8) as (6) using
$$\langle \tfrac{5}{2} 1 -m_1 m \mid \tfrac{5}{2} m_2 \rangle = (-)^{\frac{5}{2}+m_1} \sqrt{2} \langle \tfrac{5}{2} \tfrac{5}{2} m_2 m_1 \mid 1m \rangle$$

(a) A $\tfrac{3}{2} \to \tfrac{1}{2}$ transition with E2 multipolarity

Magnetic spectra (E2)

$-m_1$	m_2	m	C (1)	C^2 (2)	Θ (2)	$\theta = 90°$ (3)	$\theta = 0°$ (3)
$+\tfrac{1}{2}$	$+\tfrac{3}{2}$	$+1$	$+\sqrt{\tfrac{1}{5}}$	1	$\cos^2 \theta + \cos^2 2\theta$	1	2
$+\tfrac{1}{2}$	$+\tfrac{1}{2}$	0	$+\sqrt{\tfrac{2}{5}}$	2	$\tfrac{3}{2} \sin^2 2\theta$	0	0
$+\tfrac{1}{2}$	$-\tfrac{1}{2}$	-1	$+\sqrt{\tfrac{3}{5}}$	3	$\cos^2 \theta + \cos^2 2\theta$	3	6
$+\tfrac{1}{2}$	$-\tfrac{3}{2}$	-2	$+\sqrt{\tfrac{4}{5}}$	4	$\sin^2 \theta + \dfrac{\sin^2 2\theta}{4}$	4	0
$-\tfrac{1}{2}$	$+\tfrac{3}{2}$	$+2$	$-\sqrt{\tfrac{4}{5}}$	4	$\sin^2 \theta + \dfrac{\sin^2 2\theta}{4}$	4	0
$-\tfrac{1}{2}$	$+\tfrac{1}{2}$	$+1$	$-\sqrt{\tfrac{3}{5}}$	3	$\cos^2 \theta + \cos^2 2\theta$	3	6
$-\tfrac{1}{2}$	$-\tfrac{1}{2}$	0	$-\sqrt{\tfrac{2}{5}}$	2	$\tfrac{3}{2} \sin^2 2\theta$	0	0
$-\tfrac{1}{2}$	$-\tfrac{3}{2}$	-1	$-\sqrt{\tfrac{1}{5}}$	1	$\cos^2 \theta + \cos^2 2\theta$	1	2

Quadrupole spectra (E2)

Transition	C^2 (2)	Θ (2)	$\theta = 90°$ (3)	$\theta = 0°$ (3)
$\pm\tfrac{1}{2}, \pm\tfrac{3}{2}$	1	$2 + 3 \sin^2 \theta$	5	2
$\pm\tfrac{1}{2}, \pm\tfrac{1}{2}$	1	$3(1 + \cos^2 \theta)$	3	6

(b) A $\frac{5}{2} \to \frac{3}{2}$ transition with M1 multipolarity

Magnetic spectra (M1)

$-m_1$	m_2	m	C (4)	C^2 (2)	Θ (2)	$\theta = 90°$ (3)	$\theta = 0°$ (3)
$+\frac{3}{2}$	$+\frac{5}{2}$	$+1$	1	10		10	20
$+\frac{1}{2}$	$+\frac{3}{2}$	$+1$	$\sqrt{\frac{3}{5}}$	6	$1 + \cos^2\theta$	6	12
$-\frac{1}{2}$	$+\frac{1}{2}$	$+1$	$\sqrt{\frac{3}{10}}$	3		3	6
$-\frac{3}{2}$	$-\frac{1}{2}$	$+1$	$\sqrt{\frac{1}{10}}$	1		1	2
$+\frac{3}{2}$	$+\frac{3}{2}$	0	$\sqrt{\frac{2}{5}}$	4		8	0
$+\frac{1}{2}$	$+\frac{1}{2}$	0	$\sqrt{\frac{3}{5}}$	6	$2\sin^2\theta$	12	0
$-\frac{1}{2}$	$-\frac{1}{2}$	0	$\sqrt{\frac{3}{5}}$	6		12	0
$-\frac{3}{2}$	$-\frac{3}{2}$	0	$\sqrt{\frac{2}{5}}$	4		8	0
$+\frac{3}{2}$	$+\frac{1}{2}$	-1	$\sqrt{\frac{1}{10}}$	1		1	2
$+\frac{1}{2}$	$-\frac{1}{2}$	-1	$\sqrt{\frac{3}{10}}$	3	$1 + \cos^2\theta$	3	6
$-\frac{1}{2}$	$-\frac{3}{2}$	-1	$\sqrt{\frac{3}{5}}$	6		6	12
$-\frac{3}{2}$	$-\frac{5}{2}$	-1	1	10		10	20

Quadrupole spectra (M1)

Transitions	C^2 (2)	Θ (2)	$\theta = 90°$ (3)	$\theta = 0°$ (3)
$\pm\frac{3}{2}, \pm\frac{5}{2}$	10	$1 + \cos^2\theta$	10	20
$\pm\frac{3}{2}, \pm\frac{3}{2}$	4	$2\sin^2\theta$	8	0
$\pm\frac{3}{2}, \pm\frac{1}{2}$	1	$1 + \cos^2\theta$	1	2
$\pm\frac{1}{2}, \pm\frac{3}{2}$	6	$1 + \cos^2\theta$	6	12
$\pm\frac{1}{2}, \pm\frac{1}{2}$	9	$\frac{2}{3} + \sin^2\theta$	15	6

(c) A $\frac{5}{2} \to \frac{3}{2}$ transition with E2 multipolarity

Magnetic spectra (E2)

$-m_1$	m_2	m	C (5)	C^2 (2)	Θ (2)	$\theta = 90°$ (3)	$\theta = 0°$ (3)
$+\frac{1}{2}$	$+\frac{5}{2}$	$+2$	$-\sqrt{\frac{4}{7}}$	40 ⎱		40 ⎱	0
$-\frac{1}{2}$	$+\frac{3}{2}$	$+2$	$-\sqrt{\frac{16}{35}}$	32 ⎬ $\sin^2\theta + \frac{\sin^2 2\theta}{4}$	32 ⎬	0
$-\frac{3}{2}$	$+\frac{1}{2}$	$+2$	$-\sqrt{\frac{6}{35}}$	12 ⎰		12 ⎰	0
$+\frac{3}{2}$	$+\frac{5}{2}$	$+1$	$+\sqrt{\frac{3}{7}}$	30 ⎱		30 ⎱	60
$+\frac{1}{2}$	$+\frac{3}{2}$	$+1$	$-\sqrt{\frac{1}{35}}$	2 ⎬ $\cos^2\theta + \cos^2 2\theta$	2 ⎬	4
$-\frac{1}{2}$	$+\frac{1}{2}$	$+1$	$-\sqrt{\frac{15}{42}}$	25 ⎬	25 ⎬	50
$-\frac{3}{2}$	$-\frac{1}{2}$	$+1$	$-\sqrt{\frac{27}{70}}$	27 ⎰		27 ⎰	54
$+\frac{3}{2}$	$+\frac{3}{2}$	0	$+\sqrt{\frac{18}{35}}$	36 ⎱		0 ⎱	0
$+\frac{1}{2}$	$+\frac{1}{2}$	0	$+\sqrt{\frac{3}{35}}$	6 ⎬ $\frac{3}{2}\sin^2 2\theta$	0 ⎬	0
$-\frac{1}{2}$	$-\frac{1}{2}$	0	$-\sqrt{\frac{3}{35}}$	6 ⎬	0 ⎬	0
$-\frac{3}{2}$	$-\frac{3}{2}$	0	$-\sqrt{\frac{18}{35}}$	36 ⎰		0 ⎰	0
$+\frac{3}{2}$	$+\frac{1}{2}$	-1	$+\sqrt{\frac{27}{70}}$	27 ⎱		27 ⎱	54
$+\frac{1}{2}$	$-\frac{1}{2}$	-1	$+\sqrt{\frac{15}{42}}$	25 ⎬ $\cos^2\theta + \cos^2 2\theta$	25 ⎬	50
$-\frac{1}{2}$	$-\frac{3}{2}$	-1	$+\sqrt{\frac{1}{35}}$	2 ⎬	2 ⎬	4
$-\frac{3}{2}$	$-\frac{5}{2}$	-1	$-\sqrt{\frac{3}{7}}$	30 ⎰		30 ⎰	60
$+\frac{3}{2}$	$-\frac{1}{2}$	-2	$+\sqrt{\frac{6}{35}}$	12 ⎱		12 ⎱	0
$+\frac{1}{2}$	$-\frac{3}{2}$	-2	$+\sqrt{\frac{16}{35}}$	32 ⎬ $\sin^2\theta + \frac{\sin^2 2\theta}{4}$	32 ⎬	0
$-\frac{1}{2}$	$-\frac{5}{2}$	-2	$+\sqrt{\frac{4}{7}}$	40 ⎰		40 ⎰	0

Quadrupole spectra (E2)

Transitions	C^2 (2)	Θ (2)	$\theta = 90°$ (3)	$\theta = 0°$ (3)
$\pm\frac{3}{2}, \pm\frac{5}{2}$	30	$\cos^2\theta + \cos^2 2\theta$	30	60
$\pm\frac{3}{2}, \pm\frac{3}{2}$	36	$\frac{3}{2}\sin^2 2\theta$	0	0
$\pm\frac{3}{2}, \pm\frac{1}{2}$	39	$\frac{1}{13}(5 + 5\cos^2\theta + 8\cos^2 2\theta)$	39	54
$\pm\frac{1}{2}, \pm\frac{5}{2}$	40	$\sin^2\theta + \frac{\sin^2 2\theta}{4}$	40	0
$\pm\frac{1}{2}, \pm\frac{3}{2}$	34	$\frac{1}{17}(2 + 15\sin^2\theta + 3\sin^2 2\theta)$	34	4
$\pm\frac{1}{2}, \pm\frac{1}{2}$	31	$\frac{1}{31}(9 + 25\cos^2\theta + 16\cos^2 2\theta)$	25	50

(d) A $\frac{7}{2} \to \frac{5}{2}$ transition with M1 multipolarity

Magnetic spectra (M1)

$-m_1$	m_2	m	C (6)	C^2 (2)	Θ (2)	$\theta = 90°$ (3)	$\theta = 0°$ (3)
$+\frac{5}{2}$	$+\frac{7}{2}$	$+1$	1	21		21	42
$+\frac{3}{2}$	$+\frac{5}{2}$	$+1$	$\sqrt{\frac{5}{7}}$	15		15	30
$+\frac{1}{2}$	$+\frac{3}{2}$	$+1$	$\sqrt{\frac{10}{21}}$	10	$1 + \cos^2\theta$	10	20
$-\frac{1}{2}$	$+\frac{1}{2}$	$+1$	$\sqrt{\frac{2}{7}}$	6		6	12
$-\frac{3}{2}$	$-\frac{1}{2}$	$+1$	$\sqrt{\frac{1}{7}}$	3		3	6
$-\frac{5}{2}$	$-\frac{3}{2}$	$+1$	$\sqrt{\frac{1}{21}}$	1		1	2
$+\frac{5}{2}$	$+\frac{5}{2}$	0	$\sqrt{\frac{2}{7}}$	6		12	0
$+\frac{3}{2}$	$+\frac{3}{2}$	0	$\sqrt{\frac{10}{21}}$	10		20	0
$+\frac{1}{2}$	$+\frac{1}{2}$	0	$\sqrt{\frac{2}{7}}$	12	$2\sin^2\theta$	24	0
$-\frac{1}{2}$	$-\frac{1}{2}$	0	$\sqrt{\frac{4}{7}}$	12		24	0
$-\frac{3}{2}$	$-\frac{3}{2}$	0	$\sqrt{\frac{10}{21}}$	10		20	0
$-\frac{5}{2}$	$-\frac{5}{2}$	0	$\sqrt{\frac{2}{7}}$	6		12	0
$+\frac{5}{2}$	$+\frac{3}{2}$	-1	$\sqrt{\frac{1}{21}}$	1		1	2
$+\frac{3}{2}$	$+\frac{1}{2}$	-1	$\sqrt{\frac{1}{7}}$	3		3	6
$+\frac{1}{2}$	$-\frac{1}{2}$	-1	$\sqrt{\frac{2}{7}}$	6	$1 + \cos^2\theta$	6	12
$-\frac{1}{2}$	$-\frac{3}{2}$	-1	$\sqrt{\frac{10}{21}}$	10		10	20
$-\frac{3}{2}$	$-\frac{5}{2}$	-1	$\sqrt{\frac{5}{7}}$	15		15	30
$-\frac{5}{2}$	$-\frac{7}{2}$	-1	1	21		21	42

Quadrupole spectra (M1)

Transitions	C^2 (2)	Θ (2)	$\theta = 90°$ (3)	$\theta = 0°$ (3)
$\pm\frac{5}{2}, \pm\frac{7}{2}$	21	$1 + \cos^2\theta$	21	42
$\pm\frac{5}{2}, \pm\frac{5}{2}$	6	$2\sin^2\theta$	12	0
$\pm\frac{5}{2}, \pm\frac{3}{2}$	1	$1 + \cos^2\theta$	1	2
$\pm\frac{3}{2}, \pm\frac{5}{2}$	15	$1 + \cos^2\theta$	15	30
$\pm\frac{3}{2}, \pm\frac{3}{2}$	10	$2\sin^2\theta$	20	0
$\pm\frac{3}{2}, \pm\frac{1}{2}$	3	$1 + \cos^2\theta$	3	6
$\pm\frac{1}{2}, \pm\frac{3}{2}$	10	$1 + \cos^2\theta$	10	20
$\pm\frac{1}{2}, \pm\frac{1}{2}$	18	$\frac{2}{3} + \sin^2\theta$	30	12

(e) A 2 → 0 transition with E2 multipolarity

Magnetic spectra (E2)

$-m_1$	m_2	m	C^2 (7)	Θ (2)	$\theta = 90°$ (3)	$\theta = 0°$ (3)
0	+2	+2	1	$\sin^2\theta + \dfrac{\sin^2 2\theta}{4}$	1	0
0	+1	+1	1	$\cos^2\theta + \cos^2 2\theta$	1	2
0	0	0	1	$\tfrac{3}{2}\sin^2 2\theta$	0	0
0	−1	−1	1	$\cos^2\theta + \cos^2 2\theta$	1	2
0	−2	−2	1	$\sin^2\theta + \dfrac{\sin^2 2\theta}{4}$	1	0

Quadrupole spectra (E2)

Transitions	C^2 (2)	Θ (2)	$\theta = 90°$ (3)	$\theta = 0°$ (3)
0, ±2	2	$\sin^2\theta + \dfrac{\sin^2 2\theta}{4}$	2	0
0, ±1	2	$\cos^2\theta + \cos^2 2\theta$	2	4
0, 0	1	$\tfrac{3}{2}\sin^2 2\theta$	0	0

(f) A $\frac{5}{2} \to \frac{5}{2}$ transition with E1 multipolarity

Magnetic spectra (E1)

$-m_1$	m_2	m	C (8)	C^2 (2)	Θ (2)	$\theta = 90°$ (3)	$\theta = 0°$ (3)
$+\frac{5}{2}$	$+\frac{3}{2}$	-1	$\sqrt{\frac{2}{7}}$	10 ⎫		⎧ 5	10
$+\frac{3}{2}$	$+\frac{1}{2}$	-1	$\sqrt{\frac{16}{35}}$	16 ⎪		⎪ 8	16
$+\frac{1}{2}$	$-\frac{1}{2}$	-1	$\sqrt{\frac{18}{35}}$	18 ⎬	$1 + \cos^2 \theta$	⎨ 9	18
$-\frac{1}{2}$	$-\frac{3}{2}$	-1	$\sqrt{\frac{16}{35}}$	16 ⎪		⎪ 8	16
$-\frac{3}{2}$	$-\frac{5}{2}$	-1	$\sqrt{\frac{2}{7}}$	10 ⎭		⎩ 5	10
$+\frac{5}{2}$	$+\frac{5}{2}$	0	$\sqrt{\frac{5}{7}}$	25 ⎫		⎧ 25	0
$+\frac{3}{2}$	$+\frac{3}{2}$	0	$\sqrt{\frac{9}{35}}$	9 ⎪		⎪ 9	0
$+\frac{1}{2}$	$+\frac{1}{2}$	0	$\sqrt{\frac{1}{35}}$	1 ⎬	$2 \sin^2 \theta$	⎨ 1	0
$-\frac{1}{2}$	$-\frac{1}{2}$	0	$-\sqrt{\frac{1}{35}}$	1 ⎪		⎪ 1	0
$-\frac{3}{2}$	$-\frac{3}{2}$	0	$-\sqrt{\frac{9}{35}}$	9 ⎪		⎪ 9	0
$-\frac{5}{2}$	$-\frac{5}{2}$	0	$-\sqrt{\frac{5}{7}}$	25 ⎭		⎩ 25	0
$+\frac{3}{2}$	$+\frac{5}{2}$	$+1$	$-\sqrt{\frac{2}{7}}$	10 ⎫		⎧ 5	10
$+\frac{1}{2}$	$+\frac{3}{2}$	$+1$	$-\sqrt{\frac{16}{35}}$	16 ⎪		⎪ 8	16
$-\frac{1}{2}$	$+\frac{1}{2}$	$+1$	$-\sqrt{\frac{18}{35}}$	18 ⎬	$1 + \cos^2 \theta$	⎨ 9	18
$-\frac{3}{2}$	$-\frac{1}{2}$	$+1$	$-\sqrt{\frac{16}{35}}$	16 ⎪		⎪ 8	16
$-\frac{5}{2}$	$-\frac{3}{2}$	$+1$	$-\sqrt{\frac{2}{7}}$	10 ⎭		⎩ 5	10

Quadrupole spectra (E1)

Transitions	C^2 (2)	Θ (2)	$\theta = 90°$ (3)	$\theta = 0°$ (3)
$\pm\frac{1}{2}, \pm\frac{1}{2}$	19	$(10 + 8\cos^2 \theta)/19$	10	18
$\pm\frac{3}{2}, \pm\frac{3}{2}$	9	$\sin^2 \theta$	9	0
$\pm\frac{5}{2}, \pm\frac{5}{2}$	25	$\sin^2 \theta$	25	0
$\pm\frac{1}{2}, \pm\frac{3}{2}$	16 ⎫		⎧ 8	16
$\pm\frac{3}{2}, \pm\frac{1}{2}$	16 ⎬	$(1 + \cos^2 \theta)/2$	⎨ 8	16
$\pm\frac{3}{2}, \pm\frac{5}{2}$	10 ⎪		⎪ 5	10
$\pm\frac{5}{2}, \pm\frac{3}{2}$	10 ⎭		⎩ 5	10

Notes in the International System of Units (SI)

The greater part of the data included in this book was published before the change to the use of the SI system of units began to take effect. In practice this has had little effect on Mössbauer spectroscopy, and we have tended to use the originally quoted figures, giving conversions where appropriate. The following notes may be of assistance:

(1) The commonly used basic units are the metre (length), kilogramme (mass), second (time), and kelvin (thermodynamic temperature) with the symbols m, kg, s and K.
(2) Hyperfine interactions such as the chemical isomer shift and quadrupole splitting are usually referred to in units of mm s^{-1}, i.e. in velocity units, but the absolute energy of the interaction must be obtained from $\varepsilon = (v/c)E_\gamma$ as detailed in Section 1.6.
(3) Magnetic field flux density values are given in gauss (= 10^{-4} tesla), symbols G and T.
(4) Nuclear magnetic moments are in units of the nuclear magneton, μ_N or n.m. (= 5·04929 × 10^{-24} ergs per gauss or 5·04929 × 10^{-27} J T^{-1}). Similarly the Bohr magneton is μ_B = 0·92 × 10^{-20} ergs per gauss or 0·92 × 10^{-23} J T^{-1}.
(5) For crystallographic data the Ångstrom (Å) is 10^{-10} m = 10^{-1} nm.
(6) Nuclear quadrupole moments are in barn (10^{-28} m^2).
(7) Pressures are given in bar (= 10^5 N m^{-2}).
(8) Gamma-ray energies are given in keV (eV ≈ 1·6021 × 10^{-19}J).
(9) Some quadrupole interactions have been given in MHz because of the earlier use of this unit in nuclear quadrupole resonance spectroscopy. The appropriate conversion factors for the isotopes concerned are given in the text.

Author Index

Numbers in square parentheses are reference numbers appropriate to the page cited. Rounded parentheses are used where the reference is given in a footnote to a Table, rather than at the end of the chapter.

Abdel-Gawad, M., 295[239], 296[244], 296[239].
Abeledo, C. R., 202[13], 218[42], 344[98].
Abeles, T. P., 569[114].
Abidov, M. A., 419[221].
Ablov, A. V., 141[74], 142[74], 190[56][57].
Abragam, A., 59[20], 98(2), 116[4].
Adachi, K., 323[108].
Adloff, J. P., 141[70], 331[4][7][8], 332[4], 333[13].
Afanasev, A. M., 345[105], 509[39].
Agarwal, R. D., 333[21].
Agresti, D., 510[46][47][48], 511[46][47][48], 512[46][48], 513[48], 526[84].
Aiyama, Y., 262[65], 263[65], 265[74].
Aksenov, S. I., 498[16].
Alam, M., 338[58].
Albanese, G., 279[169][171], 340[71], 455[61], 456[61], 460[61].
Alcock, N. W., 411[188], 412[189].
Aleksandrov, A. Yu., 75[55], 402[147], 404[140][144][147], 405[144], 406[159], 407[162], 408[174], 409[174], 411[144][185], 412[159][185], 413[144][159], 460[56], 600[10].
Alekseev, L. A., 394[103], 421[242].
Alekseevskii, N. E., 417[204][206], 418[206].
Aleonard, R., 271[103].
Aleshin, K. P., 452[43], 563[94], 566[102], 572[121].
Alexander, S., 307[8].
Alff, G., 275[133].
Alfimenkov, V. P., 21[13], 498[16][17], 539[8][9].
Ali, K. M., 391[91], 393[91].
Aliev, L. A., 275[138].
Ambe, F., 158[125], 494[6].
Ambe, S., 494[6].

Amelinckx, S., 216[34], 288[217], 290[217], 291[217], 294[231], 335[34].
Anderson, T., 398[115].
Ando, E., 323[108].
Ando, K. J., 338[47].
Andra, H. J., 527[90].
Anisimov, K. N., 415[197].
Annersten, H., 294[233][234].
Aratini, M., 333[20].
Armon, H., 558[69].
Armstrong, R. J., 241[19], 248[19].
Arrott, A., 314[43].
Artmann, J. O., 95(11)(17).
Asakura, T., 365[24].
Asano, H., 315[49][54].
Asch, L., 141[70].
Asti, G., 279[169][171].
Atac, M., 70[40], 82[16], 514[58], 515[58], 516[58], 522[75], 525[83].
Atzmony, U., 520[73], 522[73], 523[73], 542[16], 543[17][18], 552[42], 555[58][59], 556[59][60][62], 557[58][59][60][62][66][68], 563[89], 587[179].
Aubke, F., 401[133], 413[133].
Avenarius, I. A., 452[46], 455[46], 460[46].
Avivi, P., 563[90].
Axel, P., 527[94], 529[94].
Axtmann, R. C., 122[21][22], 148[94], 150[94], 151[94].

Babeshkin, A. M., 387[67], 394[98], 398[118], 452[47].
Bacmann, M., 280[176].
Baehre, G., 557[61].
Baijal, J., 460[59].
Baijal, U., 460[59].
Bailey, R. E., 322[107].
Bai-shi, U., 346[116].

Baker, W. A., 143[76][79], 144[79], 145[86], 146[86], 161[137][139], 162[137], 172[13], 187[13][52][53], 188[13][52][53], 195[5], 196[5], 199[5].
Balabanov, A. E., 423[251], 560[74].
Ballhausen, C. F., 234[34].
Bancroft, G. M., 42[82], 65[30], 117[37], 129[37], 140[69], 141[69], 150[109], 154[109], 158[126], 159[126][129][131], 160[131], 161[138], 162[138], 183[42], 185[42][49], 186[48], 187[42][43][49], 189[42][49][55], 286[212][213], 288[212] [218], 289[212][218], 290[212][213][222] [223][224], 291[212][225], 292[212][213], 293[213][229], 295[241], 296[246].
Bando, Y., 241[49], 246[12], 254[37], 255[37][45], 256[49], 257[49], 265[73].
Banerjee, S. K., 103[33], 241[26], 250[24], 251[26], 261[24][26], 262[24][26][64].
Banfield, B., 366[28].
Banks, E., 283[184].
Bara, J., 341[85].
Baranovskii, V. I., 415[198].
Barloutaud, R., 29[44], 374[7][8][9].
Barnes, R. G., 39[72], 420[235], 577[143], 578[143][144], 580[163][164], 582[166] [167], 583[173], 585[175].
Barnett, B., 438[17].
Baros, F. de S., 134[59][60], 138[58][59] [60], 174[19], 462[68], 466[76], 469[76].
Baros, S. de S., 138[58].
Barrett, P. H., 513[50], 514[50], 515[50], 529[104], 530[104], 544[22], 548[22], 555[22], 559[70].
Bartenev, G. M., 398[120].
Bartsch, R. G., 365[23], 366[27].
Bashkirov, S. S., 99[25], 246[10], 271[107].
Bassi, G., 271[102][103].
Battey, M. H., 296[242].
Batti, P., 279[169].
Bauminger, E. R., 248[17], 251[17], 275[132], 520[73], 522[73], 523[73], 539[10], 541[10], 542[16], 551[41], 552[42], 556[60][62][67], 557[60][62][67] [68], 558[69], 560[71][78], 561[78], 563[89][90], 565[97], 566[97], 567[97], 587[179].
Beard, G. B., 248[18], 263[18], 441[24], 442[24][29], 443[29], 444[29][30], 445[30], 446[29], 447[30], 449[30].
Bearden, A. J., 19[7], 353[1], 354[1], 355[1], 356[7], 363[17], 365[23], 366[27], 389[73], 390[73], 563[88].
Bearden, J. A., 87[1].
Beasley, M. L., 179[30], 180[30], 181[30].
Beattie, I. R., 393[85].
Beaudreau, C. A., 353[1], 354[1], 355[1].

Becker, R. L., 529[105], 530[105].
Beda, A. G., 504[27].
Bedford, R. G., 547[35], 550[35].
Beens, W., 452[42].
Bekker, A. A., 387[67], 394[98], 398[118].
Belakhovsky, M., 279[173].
Bell, R. O., 272[117].
Belov, K. P., 394[104], 395[105], 396[105].
Belov, V. F., 266[80], 275[138][140], 279[170][172], 380[40][41], 395[40][41].
Belov, Yu. M., 369[34].
Belozerskii, G. N., 190[56], 275[146], 339[69], 340[72][73][74], 600[8].
Belyaev, L. M., 276[147], 394[97][104].
Belyustin, A. A., 346[116].
Bemski, G., 340[70], 344[97].
Benczer-Koller, N., 36[55], 38[55], 39[55], 524[79], 525[82], 526[85].
Benedek, G. B., 53[11], 54[11], 305[3], 307[12].
Bennett, L. H., 814[46][47], 315[48][58], 318[81].
Benshoshan, R., 231[23].
Benski, H. C., 87[5].
Bent, M., 516[62], 560[79], 561[80], 562[79].
Berger, L. M., 462[69].
Berger, W. G., 109[54][55].
Bergstrom, I., 464[71], 482[113].
Berkooz, O., 546[31], 547[31], 549[31], 553[44].
Berman, I. V., 460[60].
Bernas, H., 152[101], 316[70], 318[84][86].
Bernheim, R. A., 245[7].
Berrett, R. R., 161[136], 162[136], 184[46] [47], 185[47], 187[47][50], 188[47][50].
Bersohn, R., 98(1).
Bersuker, I. B., 190[57], 376[25].
Bertaut, E. F., 271[102][103], 272[111], 274[128], 278[163], 280[176], 284[190].
Bertelsen, U., 148[92].
Berzins, G., 462[69].
Bhasin, H. C., 339[65][66][67].
Bhide, V. G., 272[115], 333[24], 334[24], 338[45][56], 339[60][63][64][65][66][67], 379[33].
Biebi, U., 515[60], 517[60].
Birchall, T., 145[85], 146[85], 155[85], 213[31][32], 214[32], 234[40], 236[40], 278[164], 444[37], 451[37].
Bizina, G. E., 504[27].
Black, P. J., 30[45][46].
Blaise, A., 280[176].
Blanchard, R., 436[11].
Blander, M., 295[239], 296[239][244].
Blaugrund, A. E., 514[57].
Blayden, H. E., 393[85][86].

AUTHOR INDEX | 623

Bledsoe, W., 470[93], 474[93].
Bliznakov, G., 390[81], 391[81].
Blomstrom, D. C., 366[29].
Blount, J. F., 225[10].
Blow, S., 324[127][128].
Blum, N. A., 98(19), 216[36], 244[6], 270[99], 271[104], 306[6], 338[48][49][52], 342[88], 354[6], 513[51], 524[81], 526[81].
Blumberg, H., 510[47][48], 511[47][48], 512[48], 513[48], 516[62], 560[79], 561[80], 562[79].
Blume, M., 73[44], 74[44], 355[6].
Bockman, K., 579[155].
Bocquet, J. P., 371[1][2], 372[2], 376[1], 377[1].
Bodmer, A. R., 47[3].
Boehm, F., 88[8].
Böhmer, W. H., 197[7], 199[7].
Bohn, H., 515[60], 517[60].
Bokov, V. A., 275[143][144][145], 394[100], 398[111][112].
Bolduc, P. E., 307[19].
Bömmel, H., 246[13], 247[13].
Bonchev, C., 509[43].
Bonchev, Zw., 418[217].
Bondarev, D. A., 266[80].
Bontschev, Z, 393[93].
Boom, G., 39[68].
Booth, R., 19[4], 453[48][49], 454[49], 455[48][49], 456[49], 459[49], 460[48][49].
Bor, G., 233[29].
Borg, R. J., 317[74][75], 528[97].
Bornaz, M., 23[26], 265[75][76].
Borsa, F., 420[235].
Borshagovskii, B. V., 171[11], 172[11], 174[21], 177[21], 285[206].
Bos, W. G., 569[114].
Bosch, D., 424[257].
Both, E., 323[118], 324[118], 419[225].
Boudart, M., 345[107], 530[112].
Boutron, F., 98(2), 116[4].
Bowden, G. J., 324[124], 565[96], 570[96], 572[96], 573[96].
Bowen, L. H., 294[237], 444[35][36], 445[35], 450[35], 451[36].
Bowles, B. J., 418[216].
Bowman, J. D., 35[53].
Boyle, A. J. F., 11[7], 27[41], 74[46], 156[113], 254[35], 266[81], 374[16], 376[23], 379[23], 380[36], 382[23], 384[23], 418[24], 486[124][125], 487[125], 488[129].
Bradford, E., 74[50], 362[14].
Bradley, J. E. S., 26[38].
Brady, P. R., 333[16].

Brand, P., 390[75].
Branden, I., 393[87].
Brandt, N. B., 460[60].
Brauch, F., 206[23].
Bray, R. C., 367[30].
Bregadse, V. I., 402[147], 404[147].
Breit, G., 12[8], 41[1], 47[2].
Bressani, T., 27[39].
Britton, D., 410[177].
Brix, P., 543[20], 546[20][32], 547[20][32], 548[20][32].
Broadhurst, J. H., 26[29].
Brodsky, M. B., 598[4], 599[4], 600[11], 601[11], 602[11], 603[11][13][16], 604[18].
Brooks, Shera, E., 434[3].
Brovetto, P., 27[39].
Brown, D. B., 181[32], 182[33].
Brown, F., 464[71], 482[113].
Brown, J. S., 438[21].
Brown, M. G., 295[241], 296[246].
Brown, P. J., 323[116].
Bruckner, W., 151[116], 156[116].
Bryuchova, E. V., 406[160].
Bryukhanov, V. A., 375[17], 380[35][39], 390[80], 391[80], 406[158], 412[158], 420[237], 421[240][241][243][245], 442[26], 444[26], 445[26], 450[26][33][34], 452[39], 455[53][54], 456[53], 457[53][55], 600[9].
Buchanan, D. N. E., 113[3], 114[3], 115[3], 116[3], 117[3][11][17], 118[11], 119[11], 120[17], 121[17], 129[11], 148[93], 149[93], 150[93], 151[93], 280[178], 281[178][180][181], 282[181], 313[36], 314[36], 315[36], 322[105], 324[36], 331[5], 336[44], 343[94][96], 344[94], 346[115].
Buckley, A. N., 161[142], 163[142], 164[143].
Budnick, J. I., 507[35].
Buehler, E., 554[48].
Buisson, G., 271[102][103].
Bukarev, V. A., 374[15], 386[15], 391[15], 537[2].
Bukshpan, S., 26[33], 390[77], 391[77], 438[16][19], 439[16], 440[23], 454[55], 459[55], 461[63], 468[81], 470[81], 471[95][96][97][98][101][110], 472[81], 475[95], 476[95][96], 477[95], 478[96][97][98], 479[101], 481[63][110].
Bullock, R. J., 543[21], 546[34].
Bunbury, D. St P., 22[18], 231[24], 324[124], 345[101], 346[118][121], 347[118], 374[16], 376[23], 379[23], 380[36], 382[23], 384[23], 418[214], 505[31], 506[31], 565[96], 570[96], 572[96], 573[96].

Bunget, I., 265[72].
Burbridge, C. D., 117[26], 123[26], 141[72], 142[72], 143[72], 144[72][80], 145[80][84], 146[80][84].
Burewicz, A., 255[46].
Burger, K., 140[68], 147[88][90], 233[29].
Burgov, N. A., 82[18], 504[27].
Burin, K. L., 393[93].
Burlet, P., 284[190].
Burns, G., 98(3)(9), 275[136].
Burns, R. G., 286[212][213], 287[212], 288[212][218], 289[212][218], 290[212][213][222][224], 291[212], 292[212][213], 293[213][229].
Burton, J. W., 345[102].
Bussiere de Nercy, A., 509[41], 519[41], 520[41], 555[57], 574[57].
Butt, N. M., 38[59].
Buyrn, A. B., 524[80][81], 526[80][81].
Bykov, G. A., 401[135].

Cadeville, M. C., 318[82].
Camassei, F. D., 205[19].
Campbell, A. D., 345[100].
Campbell, I. A., 316[70], 318[84][86].
Campbell, L. E., 134[60], 138[60], 151[119][120], 156[119], 157[119][120], 438[22], 439[22], 486[123], 488[127].
Canner, J. P., 272[116].
Cape, J. A., 272[112][113], 288[215].
Carlow, J. S., 22[15].
Carlson, D. E., 530[113].
Carlson, T. A., 580[159].
Carter, W., 234[40], 236[40].
Carty, A. J., 234[40], 236[40].
Caspers, W. J., 74[48], 156[117].
Cassell, K., 19[3].
Catalano, E., 547[35][36], 550[35][36], 554[48].
Caughey, W. S., 353[1], 354[1], 355[1], 356[7], 363[17].
Cavanagh, J. F., 331[9].
Cervone, E., 205[19].
Chabanel, M., 322[100].
Chackett, G. A., 32[48], 89[10].
Chackett, K. F., 32[48], 89[10].
Chaikhorskii, A. A., 600[8].
Champeney, D. C., 81[8][11], 346[122], 348[122].
Champion, A. R., 39[75], 118[9], 119[9], 120[9], 148[9], 150[9], 155[9], 158[9], 171[7], 177[7], 181[7], 206[22].
Chandler, L., 134[45], 261[63], 262[63].
Chandra, G., 341[84].
Chandra, K., 133[43][44], 134[43][44], 171[10], 177[10], 285[204].
Chandra, S., 65[29], 76[60], 117[29], 124[29], 134[56], 138[56], 285[200], 338[58], 406[164], 407[164].
Chappert, J., 98(19), 216[36], 241[50], 257[50][51], 266[86], 267[86], 271[102][103], 274[127][128][130][131], 278[163], 284[190], 338[48][49][52].
Charlton, J. S., 530[111], 531[111].
Checherskii, V. D., 253[32].
Cheek, C. H., 354[4], 579[158], 582[165].
Chekin, V. V., 398[111][112], 418[218], 419[226], 420[231], 421[244], 530[103].
Chen, C. W., 316[64][65].
Cheng, H. S., 391[83], 392[83].
Cher, L., 346[116], 394[102].
Cherezov, N. K., 336[40], 438[15].
Chernick, C. L., 482[116].
Chevalier, R. R., 272[111], 279[173], 280[176].
Chiavassa, E., 27[39].
Childs, W. S., 496[9].
Chivers, T., 404[148].
Chohan, A. S., 38[59].
Chol, G., 266[78].
Chow, Y. W., 512[49], 513[50][51][55], 514[50][59], 515[50][59].
Chrisman, B., 70[40], 82[16], 522[75].
Christiansen, J., 109[53].
Christov, D., 393[93].
Christyakov, V. A., 246[10], 271[107].
Chu, Y. Y., 371[1][2], 372[2], 376[1], 377[1].
Cinader, C., 246[8].
Clark, H. C., 405[155].
Clark, M. G., 291[225][226].
Clark, P. E., 22[15].
Clarke, P. T., 261[62].
Clausen, C. A., 150[104], 153[104], 154[110], 502[26].
Clauser, M. J., 76[64][65], 171[6], 321[99], 576[138], 577[138], 580[160][161], 581[160].
Cocconi, C., 82[12].
Cochran, D. R. F., 38[62], 39[70], 70[34], 105[38], 308[20], 497[15].
Coey, J. M. D., 252[29], 274[129].
Cohen, D., 598[4], 599[4], 600[11], 601[11], 602[11], 603[11], 604[18].
Cohen, I. A., 356[8], 555[51].
Cohen, J., 39[75].
Cohen, R. L., 22[21][22], 528[96], 529[100], 573[125], 574[135], 577[141], 578[135][141], 583[169][170][171][174], 584[170].
Cohen, S. G., 76[61], 248[17], 251[17], 275[132], 507[35], 513[51], 537[1], 563[90], 565[95][97], 566[97], 567[97].
Coleman, C. F., 151[112], 156[112].

Collins, R. L., 65[31], 143[76], 172[13], 179 [30], 180[30], 181[30], 187[13], 188[13], 222[4], 226[13], 227[4][13][17], 228[4], 229[17], 231[25], 233[30], 345[104].
Collinson, D. W., 296[242].
Condon, E. U., 66[32].
Constabaris, G., 246[13], 247[13], 288[215], 338[47].
Cooke, R., 364[21], 367[21].
Cooper, A. R., 346[114].
Cooper, J. D., 23[27], 24[28], 317[80].
Cordey-Hayes, M., 382[46], 404[139][146], 405[146], 406[143][146], 409[146], 411 [146], 412[143], 420[228].
Corey, E. R., 415[193].
Costa, N. L., 150[100], 153[105].
Coston, C. J., 331[12], 334[30].
Cotton, E., 374[7].
Cox, D. E., 249[20], 269[20], 270[20], 271[20][105], 276[148], 279[20].
Cox, M., 160[132], 162[132], 164[132].
Craig, P. P., 308[20], 316[67][68][73], 330[1][2], 338[1][2], 342[87], 346[119], 497[14][15], 519[67], 527[88], 530[107].
Crangle, J., 316[72].
Cranshaw, T. E., 26[30], 41[79], 60[22], 80[1], 81[3][6], 103[3], 305[2], 313[38], 314[38][45], 315[38], 322[102], 338[54], 339[54], 380[43], 418[216], 422[248][249], 423[249][254].
Crasemann, B., 372[4].
Crecelius, G., 275[134], 544[23][25], 545[23], 546[25][27], 548[25], 551[25], 556[104], 573[105].
Croft, W. L., 539[6], 596[2], 598[2].
Cromar, D. T., 385[55].
Cser, L., 265[77], 309[25][26][27], 312[35].
Cullen, W. R., 227[16], 229[16], 230[26] [27], 232[26], 233[27].
Cunningham, D., 391[91], 393[91].
Curran, C., 391[89], 392[89], 402[89], 408[171], 409[171], 410[178], 411[178].
Cusanovich, M. A., 365[23], 366[27].
Czerlinsky, E. R., 275[142].
Czjzek, G., 435[6][7][8], 436[7][8], 494[3][4], 495[4].

Dahl, J. P., 234[34].
Dahl, L. F., 225[10][11].
Dähne, W., 390[74].
Daj, D., 286[210].
Dale, B. W., 190[58], 191[58], 205[20][21].
Daniels, J. M., 253[34].
Danilenko, L. E., 420[231].
Danon, J., 93[19], 94[19], 96[19], 141[70], 150[105], 153[105], 170[1], 182[36], 218[41], 344[97], 547[43], 552[43].
Dar, Y., 514[57].
Darby, J. B., 529[99].
Das, T. P., 54[14], 98(10), 467[78].
Dash, J. G., 38[66], 308[20], 341[81][82] [83], 519[67], 527[88].
da Silva, Z. A., 244[97].
Date, S. K., 338[56], 339[59].
Davies, A. G., 405[157].
Davies, C. G., 338[69], 389[69][72], 390 [72].
Davis, F. F., 369[33].
Davydov, A. V., 82[18], 504[27].
DeBenedetti, S., 21, 151[119][120], 155[111], 156[119], 157[119][120], 174[18], 175[23], 466[76], 469[76].
de Boer, E., 147[91].
Debrunner, P., 27[43], 33[52], 39[74][75], 54[13], 70[40], 82[16], 307[13], 336[35], 337[35], 364[21], 367[21], 514[58], 515[58], 516[58], 522[75], 525[83].
Decker, D. L., 307[16][18].
de Coster, M., 216[34], 288[217], 290 [217], 291[217], 294[231], 335[34], 339 [61].
Dederichs, P. H., 109[49].
Deeney, F. A., 546[33], 548[33], 550[33] [38], 554[45].
de Graaf, A. M., 344[97].
Dehn, J. T., 70[36], 105[39], 233[31], 279[174].
Dekhtiar, I. I., 315[52].
Dekker, A. J., 73[43], 74[48], 98(12), 156[114], 254[42].
Dekker, M., 231[20].
Delaney, J. A., 546[33], 548[33], 550[33] [38], 554[45].
Delapalme, A., 271[103].
Delgass, W. N., 125[32], 130[32], 133[32], 134[32], 135[32], 345[107], 530[112].
Delmelle, M., 346[117].
Delyagin, N. N., 375[17], 380[35][39], 406[158][159], 412[158][159], 413[159], 420[237], 421[240][241][243][245][246], 423[251], 530[102], 560[74][77].
Demyanets, L. N., 394[94].
Denno, W. S., 22[24].
Deo, R., 285[208].
de Pasquali, G., 39[74], 54[13], 280[179], 307[13], 331[6][12], 340[75], 341[79], 464[74], 465[74], 466[74], 469[74], 470[74], 471[74], 480[106], 481[74].
Devare, H. G., 563[91].
Devisheva, M. N., 275[140], 380[40][41], 395[40][41].
De Voe, J. R., 17, 26[31], 39[31], 41[78], 182[35].
de Vries, J. L. K. F., 147[91].

de Waard, H., 41[77], 151[99], 156[99], 462[70], 463[70], 464[74], 465[74], 466[74], 469[74], 470[74], 471[74], 479[102], 480[106], 481[74][111][189], 488[128], 498[18].
Dezsi, I., 137[53], 139[63][64], 143[78], 159[63][128], 195[4], 198[4], 241[47], 255[43][47], 256[47], 265[77], 266[82], 272[114], 309[25], 569[115].
Dharmawardena, K. G., 158[126], 159[126].
Dietl, J., 346 [110][111].
DiLorenzo, J. V., 140[66].
Djega-Mariadasson, C., 323[118], 324[118], 419[225].
Dmitrieva, T. V., 394[97].
Dobler, H., 575[137].
Do-Dinh, C., 278[163].
Donaldson, J. D., 382[48][49], 383[48], 384[51][53], 385[54], 386[49][51][53][56] [58][62], 387[49][56][58][62], 388[48] [51][62][69][71], 389[69][71][72], 390[72], 391[91], 393[91].
Dorfman, Ya. G., 408[174], 409[174].
Dosser, R. J., 196[8], 197[8], 199[8].
Drago, R. S., 400[128], 402[128], 406[128], 407[128], 408[128].
Drentje, S. A., 481[109].
Drever, R., 53[11], 54[11], 307[12].
Drevs, H., 206[23].
Drickamer, H. G., 39[74][75], 54[12][13], 118[9], 119[9], 120[9], 148[9], 150[9], 153[102], 155[9], 158[9], 171[7][8], 177[7], 181[7], 184[45], 206[22], 235[36], 246[9], 271[9], 280[179], 285[193], 307[13] [14], 331[12], 334[30], 340[75], 341[79].
Driker, G. Ya., 398[114].
Dryburgh, P. M., 89[12].
Dubery, J. M., 117[37], 129[37].
Dubovtsev, I. A., 323[113].
Duke, B. J., 41[80], 42[80].
Duncan, J. F., 141[71], 142[71], 143[71], 159[127], 178[29], 322[107], 333[16].
Dunlap, B. D., 56[16], 341[81][82], 529[99], 598[4], 599[4], 600[11], 601[11], 602[11], 603[11][13][14][16], 604[18].
Dunmyre, G. R., 423[250].
Dynamus, A., 270[100].
Dzevitskii, B. E., 390[79], 391[79], 393[92], 415[198].

Earls, D. E., 148[94], 150[94], 151[94], 171[4].
Eck, J. S., 38[61], 560[75], 562[75][87], 573[127][128], 579[150][151], 590[188].
Eckhause, M., 87[4].
Edelstein, N., 542[13], 569[13].
Edge, C. K., 39[74], 54[13], 307[13].

Edwards, C., 374[16], 376[23], 379[23], 380[36], 382[23], 384[23], 418[214].
Edwards, P. R., 117[36], 125[36], 126[36], 128[36], 105[106], 151[106], 153[106], 154[106], 155[106], 190[58], 191[58], 196[8], 197[8], 199[8], 205[21].
Efremov, E. N., 394[98].
Egelstaff, P. A., 81[6].
Egiazarov, B. G., 315[52].
Egyed, C. L., 147[90].
Ehnholm, G. J., 308[21], 546[28], 547[28].
Ehrlich, B. S., 467[77], 468[77], 469[77], 471[77][94], 474[94], 478[99].
Ehrman, J. R., 13[9], 33[51].
Eibschutz, M., 103(4), 117[16], 119[16], 260[61], 261[61], 263[61], 267[88], 269[97][98], 270[97], 273[120][122], 276[153], 280[177], 285[196][198][199], 286[198], 288[214], 397[109], 566[107], 567[107], 568[107], 577[141], 578[141].
Eicher, H., 98(6), 222[2], 223[2], 233[32], 363[16], 551[39], 575[137], 577[142], 579[155][157], 580[162], 583[168], 586[176].
Eilbeck, W. J., 196[8], 197[8], 199[8].
Einhorn, B., 557[68].
Einstein, F. W. B., 407[168].
Eissa, N. A., 241[47], 255[47], 256[47], 265[77], 569[115].
Elias, D. J., 249[21].
Elliott, J. A., 346[118][121], 347[118].
El Sayes, H., 560[74][77].
Elstner, E., 366[28].
Emerson, G. F., 231[25].
Emery, G. T., 371[1][2], 372[2], 376[1], 377[1].
Endoh, Y., 246[11], 254[37], 255[37], 314[41][42].
Epstein, L. M., 53[10], 161[135], 187[51], 188[51], 204[18], 212[30], 324[122], 353[2], 354[2], 417[202].
Eremenko, V. V., 253[32].
Erich, U., 145[87], 495[7], 496[8][10], 497[8][11].
Erickson, G. A., 38[66].
Erickson, N. E., 141[73], 142[73], 145[86], 146[86], 161[137], 162[137], 187[53], 188[53], 225[9], 402[260], 452[45], 454[45], 458[45].
Erlaki, G., 272[114].
Erman, P., 580[159].
Espinosa, G. P., 272[112][113], 276[152] [156][157], 277[156][157].
Evans, B. J., 252[30], 268[93][94], 283[187], 442[28], 444[28], 445[28], 446[28], 449[28].
Evans, D. E., 30[45].

AUTHOR INDEX | 627

Evans, M. C. W., 366[28].

Fabri, G., 340[71], 419[224].
Fackler, J. P., 154[108].
Fairhall, A. W., 225[9].
Falk, F., 489[130].
Fano, V., 340[76], 398[118].
Farmery, K., 222[6], 225[6][12].
Fatehally, R., 117[15], 119[15].
Fatseas, G. S. 268[90], 276[179], 323[110].
Feldman, J. L., 418[212].
Fenton, D. E., 415[196].
Fernandes, J. C., 340[70].
Fernandez-Moran, H., 295[240].
Filippov, N. I., 565[98][99], 596[102], 572 [121].
Filoti, G., 23[26], 98(22), 265[75][76].
Fine, M. E., 315[62].
Fink, J., 109[55], 562[85][86].
Fischer, E. O., 231[24].
Fisher, R. M., 312[34].
Fitzsimmons, B. W., 160[132], 161[136], 162[132][136], 164[132], 184[46][47], 185[47], 187[47][50], 188[47][50], 402[261], 408[169][170], 409[170], 417[203].
Flanders, P. J., 246[8].
Flannigan, W. T., 231[21].
Flinn, P. A., 20[9][10], 276[148], 309[28], 316[64][65], 320[87], 321[95], 376[27], 377[27], 383[50], 384[50].
Fluck, E., 171[15], 172[14][15], 174[15], 178[15], 183[14][15], 184[14], 206[23], 223[7], 286[211], 318[82].
Flygare, W. H., 465[82], 469[82].
Fodor, M., 255[43].
Folen, V. J., 248[18], 263[18].
Foner, S., 270[99], 271[104], 306[6].
Ford, B. F. E., 401[133], 412[186], 413[133] [186].
Ford, J. L. C., 435[6][7][8], 436[7][8].
Forder, R. A., 410[176].
Forester, D. W., 134[51], 136[51], 151[112], 156[112].
Forsyth, J. B., 137[55], 241[44], 255[44], 323[116].
Fox, R. A., 254[35].
Franck, C. W., 340[75].
Frank, E., 202[13], 218[42], 231[24].
Frankel, R. B., 98(19), 216[36], 266[86], 267[86], 270[99], 271[104], 306[6], 338[48][49][52], 342[88], 461[64][65], 512[49], 513[51], 542[15], 547[37], 548[37], 550[37], 561[81][82].
Franssen, P. J. M., 279[166].
Fransson, K., 580[159].
Fraser, M. J., 391[91], 393[91].
Frauenfelder, H., 8[5], 9[5], 27[40], 33[52], 38[62], 39[70][74], 54[13], 70[34][40], 82[13][16], 105[38], 307[13], 514[58], 515[58], 516[58], 522[75], 525[83].
Freeman, A. G., 292[227], 346[112] [113].
Freeman, A. J., 60[23], 97[24], 224[6], 271[104], 306[6], 539[10], 541[10].
Fridman, E. A., 309[29].
Friedberg, S. A., 138[58].
Friedt, J. M., 141[70], 331[4][7][8], 332[4], 333[13].
Fritchie, C. J., 415[192].
Fritz, R., 183[40].
Froindlich, D., 542[16], 556[60], 557[60], 560[78], 561[78].
Frölich, K., 496[10].
Fruchart, R., 318[86], 323[109].
Fujimoto, W. Y., 363[17].
Fujita, F. E., 320[89][90], 321[93].
Fujita, T., 117[34], 120[18], 122[18], 125[34], 129[18], 130[39], 331[10].
Fukase, M., 278[161].
Fullmer, L. D., 276[155], 278[155].
Fung, S. G., 171[8], 184[45].
Furlani, C., 205[19].

Gabriel, J. R., 65[26][27], 66[26], 74[46], 156[113].
Gal, J., 560[71].
Gallagher, P. K., 157[124], 159[124], 280[178], 281[178][180][181], 282[181], 283[183], 285[194], 554[46].
Ganiel, U., 116[7], 117[7], 118[7], 119[10], [13][14], 260[61], 261[61], 263[61], 269 [97][98], 270[97], 276[153], 288[214].
Garg, V. K., 526[87].
Garrod, R. E. B., 186[48].
Garten, R. L., 345[107].
Garwin, E. L., 82[13].
Gassenheimer, B., 410[179], 411[179].
Gastebois, J., 241[2], 242[2].
Gavrilova, G. Z., 323[113].
Gavron, A., 42[84].
Gay, P., 295[241], 296[246].
Geiger, J. S., 464[71], 482[113].
Gelberg, A., 23[26], 265[75][76].
Geller, S., 272[112][113], 274[126], 276 [152][155][156][157], 277[156][157], 278 [155].
Gen, M. Ya., 218[211].
Genin, D. J., 582[166], 583[173].
Genin, J. M., 320[87].
Gerard, A., 242[5], 269[96], 284[192], 346[117].
Gerber, W. D., 314[43].
Gerdau, E., 499[19], 504[29][30], 505[29] [30], 506[33], 509[19], 519[19], 527[19], 542[14].

AUTHOR INDEX

Germagnoli, E., 323[112], 419[224].
Gerson, R., 22[24], 272[116].
Gerth, G., 546[30], 547[30], 548[30], 549[30].
Gibb, T. C., 23[27], 41[80], 42[80], 72[41], 76[41], 100[29][30], 101[30], 103[33], 104[34], 126[35], 223[8], 224[8], 229[18], 230[18], 234[18], 236[18], 250[24], 261[24], 262[24], 271[108], 279[108], 292[228], 317[80], 381[45], 382[45], 383[45], 384[52], 385[52], 400[126], 401[126], 405[126], 453[50], 454[50], 455[50], 456[50], 457[50].
Gibson, J. F., 366[28].
Gielen, P. M., 76[61], 320[91], 322[91].
Gilbert, K., 438[20].
Ginsberg, A. P., 150[103], 153[103], 154[107].
Gittsovich, V. N., 275[146].
Gladkih, I., 265[77].
Gleason, T. G., 171[5].
Glentworth, P., 543[21], 546[34].
Godwin, R. P., 345[102].
Goldanskii, V. I., 75[52][54][57][59], 135[48], 136[50], 141[74], 142[74], 147[89], 171[11], 172[11], 174[21], 177[21], 190[56][57], 235[38], 275[140], 345[105][106], 369[32][34][35], 376[24][25][30], 377[31], 380[40][41], 391[76], 392[76], 393[92], 394[102], 395[40][41][106], 399[124], 402[137][147], 403[153], 404[137][145][147], 406[137][145][158][160], 407[161], 409[175], 411[137][184], 412[137][158][182][184], 413[137][184], 416[201], 417[137], 418[211].
Goldberg, D. A., 109[50].
Golding, R. M., 100[27][28], 119[12], 141[71], 142[71], 143[71], 174[17], 202[14], 204[14], 213[33], 216[33].
Goldring, G., 514[57].
Goldstein, C., 438[16], 439[16], 468[81], 470[81], 471[95][105], 472[81], 475[95], 476[95], 477[95], 479[105], 480[105].
Gotthardt, V., 424[256].
Gomes, A. A., 547[43], 552[43].
Gomolea, V., 98(22).
Gonser, U., 75[58], 90[14], 107[42], 125[32], 130[32], 133[32], 134[32][54], 135[32][47][49], 137[54], 182[37][38], 183[37], 263[70], 264[70], 265[70], 276[152][155][156][157], 277[156][157], 278[155], 307[17], 315[57], 317[57], 354[3], 356[3][9].
Good, M. L., 150[104], 153[104], 154[110], 502[26].
Goodenough, J. B., 104[35].
Goodgame, D. M. L., 117[26], 123[26], 141[72], 142[72], 143[72], 144[72][80], 145[80][84], 146[80][84], 197[9][10], 199[9][10].
Goodgame, M., 141[72], 142[72], 143[72], 144[72].
Goodings, D. A., 306[5].
Goodman, B. A., 384[52], 385[52], 401[134], 402[136][262][263], 545[26].
Goodman, G. L., 520[74], 521[74], 522[74].
Goodman, L. S., 496[9].
Goodman, R. H., 26[36][37].
Gopinathan, K. P., 88[9].
Gordon, G. E., 462[69].
Gorobchenko, V. D., 563[94], 565[98][99].
Gorodetsky, G., 273[122], 274[125].
Gorodinskii, G. M., 75[52], 376[24].
Goscinny, Y., 234[39], 236[39], 414[194], 415[194].
Gosselin, J. P., 346[114].
Gotlib, V. I., 387[67].
Grabari, V., 23[26].
Grace, M. A., 438[22], 439[22].
Graham, R. L., 464[71], 482[113].
Grant, R. W., 75[58], 90[14], 117[32], 125[32], 130[32], 133[32], 134[32][54], 135[32][47][49], 137[54], 182[37][38], 183[37], 272[112][113], 274[126], 276[152][155][156][157], 277[156][157], 278[155][160], 295[239], 296[239][244], 315[57], 317[57], 354[3], 356[3][9], 527[95], 529[104], 530[104], 543[19].
Grdenic, B., 387[60][61].
Greatrex, R., 72[41], 76[41], 222[5][6], 223[5][8], 224[5][8], 225[5][6][12], 229[18], 230[18], 231[22], 233[5][28], 234[5][18], 236[5][18], 402[263], 453[50], 454[50], 455[50], 456[50], 457[50].
Greenbaum, E. S., 513[55].
Greenshpan, M., 440[23].
Greenwood, N. N., 23[27], 24[28], 72[41], 76[41], 100[29], 103[33], 104[34], 126[35], 213[31][32], 214[32], 222[5][6], 223[5][8], 224[5][8], 225[5][6][12], 229[18], 230[18], 231[22], 233[5][28], 234[5][18], 236[5][18], 250[24], 261[24], 262[64], 271[108], 278[164], 279[108], 285[205], 286[205], 292[228], 296[242][247], 317[80], 376[28][29], 377[28], 378[28], 379[29], 384[52], 385[52], 390[82], 391[82], 392[82], 400[82][126], 401[82][126][134], 402[136][262][263], 403[28], 405[126], 414[82], 453[50], 454[50], 455[50], 456[50], 457[50], 545[26].
Gregory, M. C., 516[61], 518[63], 519[71].

AUTHOR INDEX | 629

Grodzins, L., 82[17], 244[6], 346[114], 512[49], 513[50][51], 514[50][59], 515[50][59], 524[80][81], 526[80][81], 539[10], 541[10], 542[15].
Gros, Y., 266[78], 315[55].
Grushko, Yu. S., 471[108], 480[107][108].
Grusin, P. L., 394[101][103], 421[242].
Gubin, S. P., 235[38].
Guggenheim, H. J., 117[11][16][17], 118[11], 119[11][17], 120[17], 121[17], 129[11], 148[93], 149[93], 150[93], 151[93], 331[11], 335[11], 336[11][44], 565[95].
Guimares, A. P., 324[124].
Gukasyan, S. E., 442[25], 445[26], 451[25].
Gunther, C., 588[183].
Gupta, M. P., 394[99].
Gupta, R. K., 563[91].
Gusev, I. A., 340[73].
Gütlich, P., 182[34], 496[10].

Haacke, G., 285[202], 321[98].
Hafemeister, D. W., 434[3], 464[74], 465[74][82], 466[74], 469[74][82], 470[74], 471[74], 480[106], 481[74].
Hafner, S. S., 98(21), 252[30], 268[93][94], 283[187], 290[219], 295[240], 296[244], 442[28], 444[28], 445[28], 446[28], 449[28].
Häggström, L., 294[233][234], 322[106], 323[115].
Hahn, E. L., 54[14], 467[78].
Hall, D. O., 366[26][28].
Hall, H. E., 11[7], 27[41] 346[118][121], 347[118], 374[16], 418[214].
Hall, L. H., 160[133][134], 161[133][134].
Ham, F. S., 338[50].
Hamaguchi, Y., 268[92].
Hamermesh, M., 104[36].
Hamilton, W. C., 405[156].
Hanna, S. S., 39[71], 103(1), 62[24], 104[36], 106[40], 109[51], 288[216], 305[1], 306[1][4], 380[37], 419[37][220].
Hannaford, P., 398[116][117].
Hannon, J. P., 509[38].
Hanzel, D., 22[16].
Harbourne, D. A., 227[16], 229[16], 230[26][27], 232[26], 233[27].
Hardy, K. A., 513[54], 526[86], 543[18], 578[149], 579[149].
Hargrove, R. S., 241[31], 252[28][31], 253[31].
Harmon, K. M., 391[84], 392[84].
Harris, C. B., 98(18), 235[35].
Harris, J. R., 524[79], 525[82], 526[85].
Harris, R. J., 87[4].
Harris, T. R., 420[228].
Harrison, P. G., 416[199].

MS—X

Hartmann-Boutron, F., 260[58][60], 263[60][67], 269[96].
Hashimoto, F., 333[15].
Hashmi, C. M. H., 527[90].
Hass, M., 248[18], 263[18].
Hasselbach, K. M., 182[34].
Hauser, M. G., 19[7].
Hautsch, W., 499[19], 509[19], 519[19], 527[19].
Hay, H. J., 81[6].
Hazony, Y., 22[17], 52[9], 53[9], 76[62], 87[3], 122[21][22][24], 148[94], 150[94], 151[94], 170[3], 171[3][4], 172[3], 330[3], 390[78], 437[12], 438[17][18], 468[80], 470[80], 472[80], 473[80], 479[100].
Heberle, J., 39[71][75], 62[24], 103(1), 106[40], 305[1], 306[1][4], 479[102].
Heberle, S., 104[36].
Hedley, I. G., 241[44], 255[44].
Heidrich, W., 554[47].
Heiman, N. D., 108[46].
Heine, V., 306[5].
Hembright, P., 354[4].
Hennig, K., 336[36][37][38][39], 393[95].
Henning, W., 486[126], 487[126], 520[74], 521[74], 522[74], 542[13], 556[65], 557[65], 557[61], 560[72][83], 561[81], 562[83], 566[103], 569[13], 573[103], 587[83], 588[83], 589[184][186], 590[186].
Herber, R. H., 36[55], 38[55], 39[55], 76[60], 217[38], 221[1], 222[3], 223[3], 228[14][15], 233[33], 234[1][33][39][41], 235[33], 236[39][41], 330[3], 333[17][18], 369[33], 374[12][14], 375[19][20], 390[77], 391[77][83], 392[83], 400[128][130][132], 402[128], 403[151], 404[141], 405[12][141][151], 406[128][130][141][164], 407[128][164], 408[128], 410[179], 411[130][179], 412[130], 413[130], 414[194], 415[194] [195], 471[98], 478[98].
Herbert, D., 364[22].
Hermon, E., 285[196][198], 286[198], 397[109].
Hermondsson, Y., 393[88].
Hershkowitz, N., 38[61][63], 109[56], 110[56], 318[83], 507[34], 596[1].
Herzenberg, C. L., 279[175], 285[207], 290[220], 294[232][238], 296[243].
Hess, F., 307[10].
Hess, J., 556[60], 557[60][68], 558[69], 587[179].
Hesse, L., 391[84], 392[84].
Heuberger, A., 522[77], 523[77].
Heunemann, D., 566[103], 573[103].
Hicks, J. A., 526[86], 578[149], 579[149].

Hicks, J. M., 33[49].
Hien, P. Z., 374[11], 394[96], 398[113][114], 417[204], 453[40].
Higashimura, T., 363[18], 364[18].
Hill, H. A. O., 202[15][16], 203[15], 204[15], 212[29], 218[43], 219[43].
Hill, J., 470[93], 474[93].
Hill, J. C., 400[128], 402[128], 406[128], 407[128], 408[128].
Hillman, P., 249[23], 437[12].
Hirahara, E., 283[185][186].
Hirai, A., 265[73].
Hirsch, A. A., 318[85], 319[85].
Hirst, L. L., 578[145].
Hobson, M. C., 345[99][100].
Hodgson, A. E. M., 254[36], 255[36].
Hoffmann, K. W., 122[19], 133[19].
Hofmann, W., 387[64].
Hohenemser, C., 87[5], 217[209], 372[3] 373[6], 418[6], 419[6].
Holland, R. E., 433[1][2], 435[9].
Hollander, J. M., 30[47], 31[47].
Holmes, L., 117[16], 119[16].
Holzapfel, W., 153[102].
Hoppe, R., 390[74].
Horl, E. M., 513[53].
Horton, G. K., 418[212].
Housley, R. M., 75[58], 88[6], 90[14], 107[43], 135[4][49], 151[99], 156[99][118], 157[118], 182[37][38], 183[37], 263[70], 264[70], 265[70], 295[239], 296[239][244], 307[10], 342[92].
Howard, C. J., 398[116].
Howard, D. G., 341[81][82][83], 342[91].
Howe, A. T., 296[247].
Howes, R. H., 513[55].
Howie, R. A., 288[218], 289[218].
Hoy, G. R., 65[29], 117[29], 124[29], 134[56][59], 138[56][58][59], 285[200][201], 338[58].
Hrynkiewicz, A. Z., 157[122], 174[22], 177[22][26], 178[27], 154[38][39], 155[39], 266[82], 341[85].
Hrynkiewicz, H. U., 341[85].
Hsu, F. H. H., 513[55].
Hudson, A., 268[95], 269[95], 278[159].
Huffman, G. P., 312[34], 423[250].
Hufner, S., 187[54], 188[54], 205[24], 206[24], 275[134], 279[168], 486[126], 487[126], 497[11], 543[20], 544[23][25], 545[23], 546[20][25][27][32], 547[20][32], 548[20][25][32][53][56], 551[25], 555[51][52][53][56], 556[63], 557[63][64], 567[108], 569[112], 573[129], 574[132], 575[136][137], 576[136], 577[142], 580[162], 587[177], 590[187][189].
Hull, G. W., 207[28].
Huller, A., 587[177].
Hulme, R., 407[165].
Huntzicker, J., 461[64][65].
Huray, P. G., 528[98], 529[98], 530[108].
Hurley, J. W., 122[21][22].

Iannarella, L., 182[36].
Ibraimov, N. S., 380[42], 420[42][227][236].
Ichiba, S., 392[90].
Ichida, T., 218[40].
Ichinose, N., 279[165].
Ilina, A. N., 369[32].
Illarionova, N. V., 38[60].
Imbert, P., 242[5], 260[59][60], 263[66][68], 269[96], 274[127], 284[192].
Ingalls, R., 39[73], 58[17], 94[20], 98[(5)(8)(20)], 100[26], 116[5], 118[8], 130[5], 133[5], 134[45], 138[5], 155[111], 306[7], 331[6][12], 334[30], 341[79].
Ino, H., 320[89][90], 321[93].
Inokuti, Y., 321[93].
Iofa, B. Z., 390[80], 391[80], 442[26], 444[26][34], 445[26], 450[26][33][34], 452[39], 455[53][54], 456[53], 457[53][55].
Isaacs, N. W., 407[167].
Isaak, G. R., 81[11], 88[7].
Isaak, U., 88[7].
Isakov, L. M., 315[52].
Ischenko, G., 435[10].
Ishigaki, A., 117[34], 125[34], 174[20], 175[20].
Ishikawa, Y., 251[25], 268[91], 283[186], 309[23], 314[41][42].
Ito, A., 117[34], 120[18], 122[18], 125[34], 129[18][42], 130[39], 133[42], 134[42], 135[42], 137[42], 138[42], 179[31], 180[31], 218[39], 242[4], 243[4], 251[25], 259[54], 260[54], 261[63], 262[63], 283[185][186], 309[23], 331[10].
Itoh, F., 317[79], 318[79].
Ivanov, Y. D., 369[34].
Ivantchev, N., 462[68].

Jaccarino, V., 91[16], 92[16], 313[36], 314[36], 315[36], 322[104], 324[36].
Jackson, F., 213[33], 216[33].
Jacobs, C. G., 507[34], 596[1].
Jagannathan, R., 333[19].
Jain, A. P., 380[43], 422[248], 423[254].
James, W. J., 272[116].
Janot, C., 317[77].
Jasal, S. S., 469[86][87].
Jaura, K. L., 402[262].
Jech, A. E., 344[98].
Jeffries, J. B., 318[83], 507[34].

Jelen, A., 388[71], 389[71].
Jena, H., 435[10].
Jenkins, I. L., 543[21].
Jesson, J. P., 197[12], 199[11][12], 200[11], 201[12].
Jha, S., 462[67][68], 516[61], 518[63], 519[68][71], 539[7], 563[91].
Jiggins, A. H., 19[3].
Johansson, A., 345[107].
Johnson, C. E., 19[5], 58[18], 60[21][22], 98(4)(16), 106[41], 117[36], 124[28][33], 125[33][36], 126[36], 128[36], 130[40], 131[41], 132[41], 137[55], 150[106], 151[106], 153[106], 154[106], 155[106], 190[58], 191[58], 196[8], 197[8], 198[8], 202[15][16], 203[15], 204[15], 205[20][21], 212[29], 218[43], 219[43], 241[26][44][48], 251[26], 255[44][48], 261[26], 262[26], 313[38], 314[38][45], 315[38][59], 322[102], 323[116], 338[54], 339[54], 341[80], 354[5], 366[26][28], 367[30], 368[31], 402[259], 442[27], 444[27], 445[27], 482[116].
Johnson, D. P., 38[66], 118[8], 241[22], 249[22], 250[22].
Jones, M. T., 414[191].
Jones, P., 393[85].
Jordanov, A., 418[217].
Josephson, B. D., 50[7].
Jung, P., 454[51], 455[51], 456[51], 458[51], 459[51], 465[83], 469[83].
Jura, G., 307[15].

Kachi, S., 338[57].
Kagan, Yu., 421[240][241].
Kagan, Yu. M., 509[39].
Kahn, U., 222[2], 223[2].
Kaindl, G., 499[22], 501[22], 502[22], 503[22], 515[60], 517[60], 519[72], 520[72], 521[72], 537[4], 538[5], 542[13], 569[13], 576[138], 577[138].
Kakabadse, G. J., 345[101].
Kalvius, G. M., 109[51], 268[93], 283[187], 290[219], 376[26], 378[26], 442[29], 443[29], 444[29][30], 445[30][32], 446[29], 447[30][32], 448[32], 449[30], 479[103], 546[28], 547[28], 555[50], 587[182], 588[185], 590[185], 598[4], 599[4], 600[11], 601[11], 602[11], 603[11][13][14][16], 604[18].
Kalvius, M., 38[65], 39[65], 222[2], 223[2], 279[168], 579[155][156][157], 580[162], 583[168].
Kalyamin, A. V., 600[10].
Kamal, R., 469[88].
Kamenar, B., 387[60][61].
Kammith, R. D. W., 404[139].

Kanamori, J., 116[6], 118[6], 129[6], 135[6], 271[106].
Kanekar, C. R., 159[127], 420[229].
Kankeleit, E., 26[32], 35[53], 38[64], 39[72], 76[64], 88[8], 497[11], 509[42], 510[46], 511[46], 512[46], 526[84], 580[160][163][164], 581[160], 588[183].
Kankeleit, E. E., 455[62], 456[62], 461[62].
Kanno, M., 373[5].
Kaplan, M., 39[67], 58[19], 98(14), 124[30], 125[30], 130[30], 133[30], 138[30], 139[65], 140[65][66], 467[77], 468[77], 469[77], 471[77][94], 474[94], 478[99], 527[94][95], 529[94][104], 530[104], 543[19], 600[7].
Kaplienko, A. I., 420[231].
Kaplow, R., 76[61], 320[91], 322[91].
Kappler, H. M., 19[8].
Karapandzic, M., 341[85].
Karasev, A. N., 398[110], 415[197].
Karchevskii, A. I., 324[119], 419[222].
Karraker, D. G., 602[17], 603[17].
Karyagin, S. V., 75[52][53], 376[24].
Kasai, N., 411[180].
Katila, T. E., 308[21], 546[28], 547[28].
Kaufmann, E. N., 35[53].
Keaton, P. W., 109[47][48][52].
Kedem, D., 42[84], 266[85].
Keller, D. A., 527[95], 529[104], 530[104], 543[19].
Keller, W. E., 308[20], 497[14].
Kelly, W. H., 248[18], 263[18], 462[69].
Kelsin, D., 22[16].
Kennard, C. H. L., 407[167].
Kerler, W., 134[46], 171[15], 172[12][15], 174[12][15], 178[15], 182[39], 183[15], 223[7], 286[211].
Kestigian, M., 119[10][14].
Keszthelyi, L., 137[53], 139[63][64], 159[63], 241[47], 255[47], 256[47], 265[77], 272[114], 309[25], 569[115].
Khan, A. M., 81[11].
Khimich, T. A., 266[80], 279[170][172].
Khrapov, V. V., 75[52][54], 190[56], 376[24], 391[76], 392[76], 393[92], 399[124], 402[137][147], 403[153], 404[137][147], 406[137][145][160], 407[161], 409[175], 411[137][184], 412[137][182][184], 413[137][184], 416[200][201], 417[137], 418[211].
Khurgin, B., 566[101][104], 567[109], 571[101][119], 573[104].
Kidron, A., 315[56], 320[92].
Kiener, V., 231[24].
Kienle, P., 222[2], 223[2][32], 279[168], 424[257], 486[126], 487[126], 504[28], 515[60], 517[60], 519[70][72], 520[70][72],

Kienle, P.—*contd.*
521[72], 522[76][77], 523[70][77], 527[90], 542[13][14], 543[20], 546[20][30], 547[20][30][32], 548[20][30][32], 549[30], 556[63][65], 557[61][63][64][65], 560[72], 562[84][86], 566[103], 569[13], 573[103], 575[137], 577[142], 579[155][157], 580[162], 583[168], 586[176], 587[177], 589[184][186], 590[186].
Kilian, H., 435[10].
Killean, R. C. G., 393[85].
Kilner, M., 222[6], 225[6][12].
Kim, D. J., 342[88].
Kimball, C. W., 314[40][43][44], 316[72], 529[99].
King, R. B., 221[1], 234[1].
Kingston, W. R., 222[3], 223[3], 333[17].
Kirkjan, C. R., 346[115].
Kiryanov, A. P., 417[206], 418[206].
Kistner, O. C., 70[39], 82[14][15], 87[2], 88[2], 90[15], 241[1], 371[1], 376[1], 377[1], 380[38], 381[38], 423[38], 499[20][21][23], 500[21], 501[21][24][25].
Kitchens, T. A., 342[86].
Kitching, W., 407[167].
Kiyama, M., 151[123], 157[123], 246[12], 254[37], 255[37][45], 265[73].
Klein, M. P., 365[25].
Kleinberger, R., 269[96].
Kleinstuck, K., 266[79].
Klemann, L. P., 391[84], 392[84].
Klockner, J., 519[70], 520[70], 523[70].
Klotz, I. M., 368[31].
Klumpp, W., 122[19], 133[19].
Knauer, R. C., 21[14], 315[60].
Knight, E., 366[29].
Knowles, P. F., 367[30].
Knox, G. R., 231[20][21].
Knudsen, J. M., 148[92], 282[182].
Koch, R., 579[157].
Kocher, C. W., 19[6], 148[95], 150[95], 151[95], 321[94], 391[84], 392[84].
Kodre, A., 22[16].
Kolobova, N. E., 415[197].
Komissarova, B. A., 417[207].
Komor, M., 159[128].
Komura, S., 324[126].
König, E., 143[77], 144[82], 145[82], 187[54], 188[54], 195[2][3][6], 196[3][6], 197[7], 199[7], 205[24], 206[24][25].
Konig, U., 266[78].
Kono, H., 117[38], 129[38].
Koon, N. C., 134[51], 136[51].
Kopfermann, H., 47[5].
Kordyuk, S. L., 258[52].
Korecz, L., 139[63], 147[88], 159[63], 233[29].

Korneev, V. P., 345[106].
Korner, H. J., 504[29], 505[29], 519[70][72], 520[70][72], 521[72], 522[76], 523[70], 542[13][14], 566[103], 569[13], 573[103], 589[186], 590[186].
Korneyev, E. V., 266[80], 279[170][172].
Korytko, L. A., 75[52], 376[24], 406[158], 412[158].
Kostiner, E., 229[19], 230[19].
Kostroun, V. A., 372[4].
Kostyanovsky, R. G., 399[124], 404[145], 406[145], 412[182].
Kosuge, K., 338[57].
Kotkhekar, V., 390[80], 391[80], 400[129], 442[26], 444[26][34], 445[26], 450[26][33][34], 455[53], 456[53], 457[53].
Kovats, T. A., 307[11].
Krasnoperov, V. M., 438[15].
Kravtsov, D. N., 416[200][201].
Kregzde, J., 356[9].
Krishman, R., 268[90].
Krivokhatskii, A. S., 600[10].
Krizhanskii, L. M., 75[52], 376[24], 379[34], 412[183], 413[183].
Krogh, H., 148[92].
Kubisz, J., 157[122].
Kubo, M., 54[15].
Kuhn, P., 171[15], 172[14][15], 174[15], 178[15], 183[14][15], 184[14], 206[23], 223[7], 286[211], 318[82].
Kulgawczuk, D. S., 157[122], 241[47], 254[38][39], 255[39][46][47], 256[47], 265[77], 266[82].
Kulzer, H., 542[13], 569[13].
Kundig, W., 65[28], 81[7], 241[31], 246[13], 247[13], 252[28][31], 253[31], 288[215], 338[47].
Kunitomi, N., 324[126].
Kupper, F. W., 145[87].
Kuprianova, E. J., 394[101].
Kurbatov, G. D., 271[107].
Kurichok, P. P., 266[80].
Kurkjian, C. R., 157[124], 159[124].
Kuzmin, R. N., 375[17], 380[42], 419[221], 420[42][227][236][237], 452[46], 455[46], 460 [46][60].

Lam, D. J., 598[4], 599[4], 603[13][16].
Lamb, W. E., 9[6].
Lambert, J. L., 160[133][134], 161[133][134].
Lamborizio, C. L., 279[171], 323[112], 340[71], 455[61], 456[61], 460[61].
Lamoreaux, R., 294[232].
Lang, G., 155[111], 174[18], 175[23][24], 176[24][25], 219[44], 356[10][11][12], 357[10], 358[10], 359[10], 360[10][13],

361[10][13], 362[10], 364[22], 365[24], 462[67], 539[7].
Langevin, M., 152[101], 509[41], 519[41], 520[41], 555[57], 574[57].
Large, N. R., 543[21], 546[34].
Larkindale, J. P., 346[113].
Larkworthy, L. F., 160[132], 162[132], 164[132].
Larson, A. C., 385[55].
Lawrence, H., 451[38], 452[38].
Lawrence, J. L., 393[85].
Lebedev, R. A., 452[47].
Lebedev, V. A., 452[47], 539[9].
Lebenbaum, D., 520[73], 522[73], 523[73].
le Caer, G., 317[77].
Lecocq, P., 323[110][118], 324[118], 419[225].
Lederer, C. M., 30[47], 31[47].
Lee, E. L., 307[19].
Lee, L. L., 509[40], 519[40].
Lee, Y. K., 109[47][48][50][52], 560[73][75], 562[75][87], 573[127][128], 579[150][151][152], 590[188], 598[3]
Lees, J. K., 376[27], 377[27], 383[50], 384[50].
Lefevre, H. W., 284[191].
Lefevre, S., 276[149].
Lefkowitz, I., 148[94], 150[94], 151[94], 171[4].
Lehman, G. W., 418[213].
Lehmann, C., 109[49].
Lehr, J., 87[5].
Leider, H. R., 338[51].
Lejeune, S., 407[166], 408[166].
Lependina, O. L., 408[174], 409[174].
Lerch, J., 504[29], 505[29].
Lesikar, A. V., 234[37], 235[37].
Letcher, J. H., 379[32].
Levey, P. R., 530[106].
Levinson, L. M., 274[123][125], 284[189].
Levinstein, H. J., 117[16][17], 119[16], 120[17], 121[17].
Levishiva, M. N., 395[106].
Levy, L., 37[57], 375[22].
Lewis, S. J., 321[95].
Liengme, B. V., 227[16], 229[16], 230[26][27], 232[26], 233[27], 345[107], 412[186], 413[186].
Likhtenstein, G. I., 369[35].
Lin, S. C. H., 321[96].
Lind, M. D., 151[97].
Lindgren, R. A., 342[93].
Lindley, D. H., 336[35], 337[35].
Lindley, P. F., 231[24].
Lindquist, R. H., 246[13], 247[13], 288[215], 338[47].
Linnett, J. W., 249[21].

Lipkin, H. J., 7[3], 9[3].
Lipkin, J., 21[11].
Lisichenko, V. I., 258[52].
Listner, J. O., 305[3].
Littlejohn, C., 104[36], 106[40].
Livingstone, R., 464[75], 467[79].
Locati, G. C., 419[224].
Loewenhaupt, M., 548[56], 555[56].
Long, G. G., 444[35][36], 445[35], 450[36], 451[36][38], 452[38].
Long, G. J., 161[139].
Longworth, G., 30[46], 316[66][69], 321[96], 418[215].
Lotgering, F. K., 285[203].
Lounasmaa, O. V., 308[21], 546[28], 547[28], 555[50].
Lovborg, L., 23[25].
Love, J. C., 435[7][8], 436[7][8], 494[4], 495[4].
Lubau, M., 274[123].
Luchner, K., 346[110][111], 546[30], 547[30], 548[30], 549[30].
Luciani, M. L., 205[19].
Ludwig, G. W., 102[31].
Lukashevich, I. I., 563[94], 565[98][99].
Lure, B. G., 335[31], 336[40][41][42][43], 471[108], 480[107][108], 600[10].
Lüscher, E., 82[13].
Lushchikov, V. I., 498[16].
Lykashevich, I. I., 566[102], 572[121].
Lynch, F. J., 433[2], 435[9].
Lyubutin, I. S., 275[141][147], 276[139], 394[97][104], 395[105], 396[105], 397[107][108].

MacChesney, J. B., 280[178], 281[178][180][181], 282[181], 283[183], 285[194].
McCleverty, J. A., 213[31].
McDonald, R. R., 385[55].
McDonald, W. S., 233[28].
McGuire, A. D., 519[67].
Machada, A. A. S. C., 197[9][10], 199[9][10].
Mackey, J. L., 345[104].
McKinley, S. V., 391[84], 392[84].
McMullin, P. G., 22[22].
McNab, T. K., 254[35], 266[81].
McNiff, E. J., 306[6].
McQueen, J. E., 573[124].
McWhinnie, W. R., 144[81], 408[172], 409[172].
Maddock, A. G., 42[82], 150[109], 154[109], 159[129][130][131], 160[131], 161[138], 162[138], 186[48], 286[212], 287[212], 288[212], 289[212], 290[212][222][224], 291[212], 292[212], 293[229], 452[45], 454[45], 458[45].

Madeja, K., 143[77], 144[82], 145[82], 187[54], 188[54], 195[2][3][6], 196[3][6], 197[7], 199[7], 205[24], 206[24][25].
Madsen, P. E., 314[45].
Maeda, T., 315[51], 320[89][90], 321[93].
Maeda, Y., 345[107], 363[18][19][20], 364[18][19][20].
Maer, K., 179[30], 180[30], 181[30].
Mahesh, K., 469[89].
Mahler, J. E., 231[25].
Major, J. K., 27[42].
Makarov, E. F., 75[52][54][57][59], 135[48], 136[50], 141[74], 142[74], 147[89], 190[56], 235[38], 266[79], 275[139][141], 285[195], 286[209], 345[105], 376[24][25][30], 377[31], 391[76], 392[76], 395[106], 397[108], 398[122], 406[158], 407[161], 411[184], 412[158][184], 413[184].
Makarov, V. A., 397[108].
Makhanov, G. T., 369[34].
Malamud, M., 553[44].
Malathi, N., 294[236], 526[87].
Malden, P. J., 294[235].
Maletta, H., 554[47], 561[81].
Maling, J. E., 98(15), 363[15].
Malm, J. G., 482[116].
Manapov, R. A., 271[107], 452[43].
Mandache, S., 98(22).
Mandzukov, Iv., 393[93].
Mannheim, P. D., 342[89].
Marcus, H. L., 315[61][62].
Mardanyan, S. S., 369[34].
Mareschal, J., 274[128].
Marfunin, A. S., 285[195][206], 286[209].
Margolis, L. Ya., 398[110].
Margulies, S., 13[9], 33[51][52], 82[13].
Maricondi, C., 353[2], 354[2].
Marini, J. L., 154[108].
Marinov, A., 248[17], 251[17], 275[132], 507[35], 563[90], 565[97], 566[97], 567[97].
Marks, J., 26[38].
Marsh, H. S., 389[73], 390[73].
Marshall, J. H., 486[124].
Marshall, S. W., 42[81], 527[92].
Marshall, W., 19[5], 60[21], 74[50], 98(4), 106[41], 130[40], 341[80], 356[10][11], 357[10], 358[10], 359[10], 360[10], 361[10], 362[10][14].
Martel, E., 263[68].
Martin, R. L., 194[1].
Marzolf, J. G., 70[36], 105[39].
Massey, A. G., 229[19], 230[19].
Matas, J., 171[9], 172[9].
Mather, H. B., 260[55][57], 263[57], 285[197], 333[19], 394[99].
Mathur, P. K., 161[140], 333[21][22].
Matlak, T., 341[85].
Mattern, P. L., 19[7].
Matthias, B. T., 39[70].
Matthias, E., 461[64], 508[37].
May, L., 403[152].
Mayer, A., 19[8].
Mayer, I., 548[55], 555[55], 571[120].
Mays, M. J., 140[69], 141[69], 183[42], 185[42][49], 186[48], 187[42][43][49], 189[42][49][55].
Mazak, R. A., 227[17], 229[17].
Meads, R. E., 294[235].
Medeiros, L. O., 159[129][130].
Meechan, C. J., 307[17], 315[57], 317[57].
Meisel, W., 336[36][37], 393[95].
Mekata, M., 316[63].
Mendiratta, R. G., 469[88].
Menth, A., 274[124].
Merritt, F. R., 207[27][28].
Mewissen, L., 339[61].
Meyer-Schutzmeister, L., 380[37], 419[37][220], 509[40], 519[40].
Michalski, M., 26[35].
Mikhalenkov, V. S., 315[52].
Mill, B. V., 276[147].
Milledge, H. J., 405[157].
Miller, C. A., 26[29].
Millett, L. E., 307[18].
Milligan, W. O., 179[30], 180[30], 181[30].
Mills, O. S., 231[24].
Milne, M., 393[85].
Minovka, A., 418[217].
Misetich, A., 98(19), 338[49][52].
Mishima, M., 392[90].
Misra, S. K., 70[36].
Mitina, L. P., 394[97].
Mitrani, L., 37[57], 375[22], 509[43].
Mitrofanov, K. P., 38[58][60], 374[13], 375[18][21], 398[114][119], 404[140], 406[159], 408[174], 409[174], 411[185], 412[159][185], 413[195], 417[13].
Mitui, T., 268[92].
Miyamoto, H., 241[49], 256[49], 257[49].
Mizoguchi, T., 265[74], 267[89], 268[89].
Mizushima, M., 283[184].
Mkrtchyan, A. R., 285[195][206], 286[209].
Moiner, B., 241[47], 255[47], 256[47].
Mok, K. F., 141[71], 142[71], 143[71], 159[127].
Moljk, A., 22[16].
Möller, H. S., 420[233][234], 424[256].
Molnar, B., 139[64], 143[78], 195[4], 198[4], 569[115].
Monaro, S., 499[20].
Montmory, M. C., 279[173].
Moon, P. B., 5[2], 81[8].

Moore, W. J., 387[63].
Mora, S., 323[112].
Moras, D., 399[125].
Morel, J. P., 267[87].
Morice, J. A., 283[188], 284[188], 285[188], 345[103].
Morimoto, T., 142[75].
Morita, Y., 363[18][20], 364[18][20].
Moriya, T., 320[89][90], 321[93].
Morrish, A. H., 104[34], 241[19], 248[19], 252[27][29], 253[33], 265[71], 266[83], 274[129].
Morrison, R. J., 514[58], 515[58], 516[58], 537[3].
Mosbaek, H., 164[144].
Moshkovskii, Yu. Sh., 369[32][34].
Moss, J. R., 233[28].
Moss, T. H., 353[1], 354[1], 355[1], 356[7], 363[17], 365[23], 366[27].
Mössbauer, R. L., 1[1], 17[1][2], 39[72], 76[64][65], 420[233], 424[256], 499[22], 501[22], 502[22], 503[22], 508[37], 514[56], 519[64][65][66], 537[4], 554[47], 561[81], 574[133], 576[138], 577[138], 578[145], 580[160][161][163][164], 581[160].
Motornyi, A. V., 480[107].
Moyzis, J. A., 39[75], 307[14].
Mozer, B., 316[67].
Mualem, A., 555[58], 557[58].
Muir, A. H., 88[8], 98(17), 125[32], 130[32], 133[32], 134[32], 135[32], 276[150], 295[239], 296[239][244], 307[17], 315[57], 317[57].
Mulay, L. N., 233[31], 279[174].
Mullen, J. G., 21[14], 216[35], 315[60], 333[25], 334[25][26][28], 335[32][33], 600[5].
Mullins, M. A., 391[89], 392[89], 402[89], 408[171], 409[171], 410[178], 411[178].
Multani, M. S., 272[115], 339[63][64], 379[33].
Munck, E., 573[129][130], 574[130][132], 575[136], 576[136], 590[187][189].
Mundt, W. A., 183[41].
Murakami, Y., 310[33].
Murin, A. N., 275[146], 335[31], 336[40][41][42][43], 340[73], 438[15], 471[108], 480[107][108], 600[10].
Murphy, K. A., 596[1].
Murray, G. A., 322[102].
Murray, K. S., 161[141], 163[142], 164[143].
Musci, M., 340[71], 419[224].
Mustachi, A., 89[11], 520[73], 522[73], 523[73], 552[42], 557[68], 558[69], 560[78], 561[78], 587[179].

Nadav, E., 26[34].
Nadzharyan, G. N., 285[195][206].
Nagarajan, R., 117[15], 119[15].
Nagle, D. E., 38[62], 39[70], 70[34], 105[38], 308[20], 342[87], 497[14][15], 519[67], 527[88], 530[107].
Nagy, D. L., 569[115].
Nakagwa, T., 273[119].
Nakamura, D., 54[15].
Nakamura, T., 246[11], 254[37], 255[37].
Nakamura, Y., 246[11], 315[50][53], 317[79], 318[79], 322[101], 338[57].
Nasielski, J., 407[166], 408[166].
Nasu, S., 310[33], 339[62].
Nath, A., 333[21].
Nathans, R., 316[64][65].
Naumov, V. G., 419[226], 421[244].
Nees, W. L., 342[91].
Negita, H., 392[90].
Nelson, J. A., 42[81].
Nelson, L. Y., 486[121].
Nemilov, Yu. A., 339[69], 340[72][73][74], 600[8].
Nesmeyanov, A. N., 235[38], 387[67], 394[98], 398[118], 416[200][201], 452[47].
Neuwirth, W., 171[12][15], 172[15], 174[12][15], 178[15], 183[15], 223[7], 286[211], 318[82].
Nevitt, M. V., 314[40][44].
Newnham, R., 279[173].
Newnham, R. E., 279[174].
Newton, D. A., 543[21], 546[34].
Ng, T. W., 234[40], 236[40].
Nichol, A. W., 22 [15].
Nichols, D. I., 530[111], 531[111].
Nicholson, D. G., 386[62], 387[62], 388[62].
Nicholson, W. J., 98(9), 275[136], 309[29], 572[123].
Nicol, M., 307[15].
Niculescu-Majewska, H., 98(22).
Niedzwiedz, S., 318[85], 319[85], 320[92].
Nielsen, J. W., 280[177].
Nikitina, S. V., 419[221].
Nikolaev, V. I., 323[113], 324[119][120], 419[222][223].
Nininger, R. C., 246[14].
Nishio, M., 310[33].
Nistor, C., 23[26].
Nizhankovskii, V. I., 417[206], 418[206].
Norem, P. C., 339[68].
Norlin, L. O., 479[104].
Normura, S., 273[119].
Novikov, G. V., 369[35], 394[100][102], 395[106].
Nowik, I., 74[47], 275[137], 537[1], 539[10][11], 540[11][12], 541[10][11],

Nowik, I.—contd.
544[24], 546[24], 547[37], 548[37], 550[37], 551[24][40][41], 552[40][42], 556[62][67], 557[62][67], 560[78], 561[78], 565[100], 566[100][106][113], 567[100], 569[109][110][111][113], 571[116][118][119], 572[122], 577[140], 583[171], 586[178], 587[178][180][181].
Nozik, A. J., 39[67], 58[19], 98(14), 124[30], 125[30], 130[30], 133[30], 138[30], 139[65], 140[65], 285[202], 321[98].
Nussbaum, R. H., 88[6], 342[91][92].
Nybakken, T. W., 22[23].

Oakzaki, A., 387[65].
Obenshain, F. E., 109[55], 151[112], 156[112], 435[6][7][8], 436[7][8], 493[1], 494[1][2][3], 529[105], 530[105].
O'Brien, R. J., 405[155].
O'Connor, D. A., 30[45][46], 33[50], 38[59].
O'Connor, J. E., 415[193].
Ofer, S., 248[17], 251[17], 275[132][137], 520[73], 522[73], 523[73], 537[1], 539[10][11], 540[11][12], 541[10][11], 542[16], 544[24], 546[24], 551[24][40][41], 552[40][42], 555[58][59], 556[59][60][62][67], 557[58][59][60][62][66][67], 558[69], 560[71][78], 561[78], 563[89][90], 565[97], 566[97][101][104], 567[97][109], 571[101][116][117][119], 573[104][126], 583[171], 586[178], 587[178][179][180][181].
Ohashi, S., 157[121], 290[221].
Ohki, K., 494[6].
Ohno, H., 312[63].
Ohta, K., 317[78].
Ohtsuki, M., 295[240].
Ok, H. N., 122[24], 136[52], 333[14][25], 334[25][26][28].
Okada, T., 268[92], 278[161].
Okamura, M.Y., 368[31].
Okawara, R., 411[180].
Okhlokystin, O. Yu., 75[55], 402[137][147], 404[137][144][147], 405[144], 406[137], 407[162], 409[175], 411[137][144][185], 412[137][183][185], 413[137][144][183], 417[137].
Okiji, A., 116[6], 118[6], 129[6], 135[6], 271[106].
Oleson, J. R., 579[152], 598[3].
Olsen, D., 65[27], 520[74], 521[74], 522[74].
Olsen, D. H., 403[154].
Ong, W. K., 42[82], 150[109], 154[109], 159[131], 160[131].
Ono, K., 117[34], 120[18], 122[18], 125[34], 129[18][42], 130[39], 133[42], 134[42][45], 135[42], 137[42], 138[42], 174[20], 175[20], 179[31], 180[31], 218[39], 242[4], 243[4], 251[25], 259[54], 260[54], 261[63], 262[63], 283[185][186], 309[23], 331[10].
Ono, Y., 321[93].
Oosterhuis, W. T., 174[18][19], 175[23][24], 176[24][25], 217[37], 219[44].
Opalenko, A. A., 452[46], 455[46][54], 460[46][60].
Oppliger, L. D., 479[103].
O'Reilly, W., 103[33], 241[26], 250[24], 251[26], 261[24][26], 262[24][26][64], 296[242].
Ormandjiev, S., 37[57], 375[22], 509[43].
Ortalli, I., 323[112], 340[71][76], 455[61], 456[61], 460[61].
Ostanevich, J., 309[25][26][27].
Ostanevich, Yu. M., 21[13], 346[116], 498[16][17], 539[8][9].
Ostergaard, P., 323[118], 324[118], 398[115], 419[225].
Oteng, R., 384[53], 385[54], 386[53].
Ouseph, P. J., 569[114].
Ouzounov, I., 509[43].
Ovechkin, V. V., 600[9].
Owen, D. B., 42[83].
Owens, W. R., 516[61], 519[68][71], 526[86], 578[149], 579[149].
Owusu, A., 161[136], 162[136], 187[50], 188[50].

Pachek, V. F., 369[35].
Pachevskaya, V. M., 416[201].
Packwood, R. H. 418[215].
Paez, E. A., 217[37].
Pahor, J., 22[16].
Pal, L., 309[25][26][27].
Palaith, D., 316[72].
Palenik, G. J., 234[40], 236[40].
Palmai, M., 26[34].
Panson, A. J., 387[66].
Panyushkin, V. N., 280[179], 387[67], 418[210].
Papp-Molnar, E., 147[90].
Parish, R. V., 23[27], 317[80], 400[127], 402[127][149][259].
Parisi, G. I., 403[151], 404[151].
Parravano, G., 530[112].
Partenova, V. P., 423[251].
Pasternak, M., 26[33], 76[62][63], 87[3], 437[12][14], 438[14][19], 452[44], 454[55], 456[44], 459[55], 462[44], 468[80][84], 470[80][90][92], 471[90][105][110], 472[80][90], 473[80], 474[90][91][92], 479[105], 480[90][105], 481[110][111][112].
Patterson, D. O., 528[98], 529[98], 530[106].

Pauling, L., 387[63].
Pauson, J., 231[21][22].
Pauthenet, R., 271[103].
Peacock, R. D., 404[139][146], 405[146], 406[146], 409[146], 411[146].
Peacock, R. N., 82[13].
Pebay-Peyroula, J. C., 315[55].
Pebler, J. W., 148[96].
Pecuil, T. E., 293[230].
Pelah, I., 344[95], 393[94].
Penfold, B. R., 407[168].
Penkov, I. N., 271[107].
Perisho, R. C., 316[68].
Perkins, H. K., 278[162].
Perkins, P. G., 376[28][29], 377[28], 378[28], 379[29], 403[28].
Perlman, I., 30[47], 31[47].
Perlman, M. L., 371[1][2], 372[2], 376[1], 377[1].
Perlow, G. J., 19[5], 62[24], 98(4), 103(2), 104[36], 106[40][41], 130[40], 306[4], 341[80], 438[22], 439[22], 464[72][73], 466[72], 467[72], 468[85], 469[72], 472[72], 474[72][73], 475[72], 482[115] [116], 483[85][115][117][118][119], 484[119][120][122], 485[119], 486[122] [123][124][125], 487[125], 488[127][129], 498[18], 520[74], 521[74], 522[74].
Perlow, M. R., 464[72], 466[72], 467[72], 469[72], 472[72], 474[72], 475[72], 482[116], 483[117][118][119], 484[119] [120], 485[119].
Persson, B., 35[53], 510[46][47][48], 511[46][47][48], 512[46][48], 513[48], 516[62], 526[84], 560[79], 561[80], 562[79].
Peryshkin, A. I., 600[9].
Petrich, G., 575[137], 578[146].
Petrov, K., 390[81], 391[81].
Petrovich, E. V., 374[10].
Pettit, R., 143[76], 172[13], 187[13], 188[13], 222[4], 226[13], 227[4][13], 228[4], 231[23][25].
Pfeiffer, L., 108[46].
Pfletschinger, E., 117[20], 122[20], 129[20], 130[20], 150[20], 151[20].
Philip, J., 391[89], 392[89], 402[89].
Phillips, E. A., 82[17].
Phillips, W. C., 314[44].
Phillips, W. D., 366[29].
Picou, J-L., 29[44], 374[7][9].
Piekoszewskii, J., 26[35].
Pierce, R. D., 280[177].
Pietro, A. S., 366[27].
Pillinger, W. L., 82[19], 539[6], 596[2], 598[2], 600[5][12], 602[12][15][17], 603[15][17].

Pimentel, G. C., 486[121].
Pinajian, J. J., 331[12].
Pipkorn, D. N., 39[74], 54[13], 307[13], 338[51], 528[97].
Pisarevskii, A. M., 346[116].
Plachinda, A. S., 345[105][106], 398[122].
Platt, R. H., 400[127], 402[127][149].
Plotnikova, M. V., 374[13], 375[18], 398[114], 408[174], 409[174], 417[13].
Pobell, F., 276[154], 278[154], 345[108] [109], 419[219], 420[230], 424[257], 522[77], 523[77].
Pocs, L., 139[63][64], 159[63].
Poder, C., 412[187], 413[187].
Podvalmykh, G. S., 266[80].
Poindexter, J. M., 39[72], 580[163][164].
Polak, L. S., 398[110], 404[140][144], 405[144], 406[159], 408[174], 409[174], 411[144][185], 412[159][185], 413[144] [159], 415[197].
Pollak, H., 288[217], 290[217], 291[217], 294[231].
Poller, R. C., 144[81], 408[172][173], 409[172][173], 414[173].
Polozova, I. P., 398[121].
Popov, A. V., 412[183], 413[183].
Popov, G. V., 275[143][144][145], 379[34].
Popov, H., 573[131].
Portier, J., 241[50], 257[50][51].
Potzel, W., 499[22], 501[22], 502[22], 503[22].
Poulson, K. G., 164[144].
Pound, R. V., 50[6], 53[11], 54[11], 80[2], 81[4][5], 307[12], 497[13].
Povitskii, V. A., 266[79], 275[139][141], 276[147], 285[195], 286[209], 397[108].
Pradaude, H. C., 306[6].
Prados, R. A., 502[26].
Pramila, G. C., 563[91].
Prange, H., 497[11], 560[76], 562[76][84], 590[187][189].
Prater, B. E., 140[69], 141[69], 183[42], 185[42][49], 186[48], 187[42][43][49], 189[42][49][55].
Prescott, B., 554[46].
Preston, R. S., 39[71], 103(1), 62[24], 104[36], 106[40], 305[1], 306[1], 307[9], 314[44], 316[72], 380[37], 419[37][220].
Price, D. C., 424[258].
Prince, R. H., 42[82], 150[109], 154[109], 159[131], 160[131].
Prokofev, A. K., 399[124], 404[145], 406[145], 412[182].
Proskuryakov, O. B., 394[100].
Puri, S. P., 133[43][44], 134[43][44], 171[10], 177[10], 285[204][208], 286[210], 294[236], 434[5], 497[12].

Puxley, D. C., 405[157].

Quaim, S. M., 340[77], 341[78].
Quidort, J., 241[2], 242[2], 374[7].
Quitmann, D., 275[134], 486[126], 487[126], 495[7], 543[20], 544[25], 546[20][25][27][32], 547[20][32], 548[20][25][32], 551[25], 556[105], 573[105][129], 574[132], 575[136], 576[136], 590[187] [189].

Rabinowitz, I. N., 369[33].
Radhakrishman, T. S., 341[84].
Raghaven, R. S., 563[91].
Raj, D., 133[44], 134[44], 171[10], 177[10], 285[204], 434[5], 497[12].
Raju, S. B., 469[88].
Rakavy, M., 566[101][104], 567[109], 571[101][119], 573[104][126].
Ramshesh, V., 161[140].
Randl, R. P., 161[138], 162[138].
Rao, K. R. P., 420[229].
Rao, V. J. S., 420[229].
Raper, G., 233[28].
Raymond, M., 98(21).
Rebka, G. A., 50[6], 80[2], 497[13].
Rebouillat, H. P., 271[103].
Rebouillat, J. P., 274[128].
Recknagel, E., 109[53].
Reddy, K. R., 462[68], 466[76], 469[77].
Reedijk, J., 145[83].
Rees, L. V. C., 283[188], 284[188], 285[188], 345[103].
Reese, R. A., 420[235], 577[143], 578[143] [144].
Reggev, Y., 26[33].
Rehm, K. E., 542[13], 560[72], 569[13].
Reichle, W. T., 400[130], 404[142], 406[130] [142], 411[130][142], 412[130][142], 413[130][142].
Reid, A. F., 278[162][164].
Reiff, W. M., 145[86], 146[86], 161[137] [139], 162[137], 187[53], 188[53].
Reintsema, S. R., 488[128].
Reiswig, R. D., 519[67], 527[88], 530[107].
Reivari, P., 41[79], 103(3), 305[2], 308[21] [22], 546[28], 547[28].
Remeika, J. P., 338[53].
Reno, R., 87[5].
Renovitch, G. A., 143[79], 144[79], 187[52], 188[52], 195[5], 196[5], 199[5].
Rensen, J. G., 279[167].
Rentzeperis, R. I., 389[70].
Richardson, J. E., 26[36].
Richter, F. W., 148[96].
Richter, M., 556[65], 557[65].
Rickard, D. T., 283[188], 284[188], 285[188].

Rickards, R., 202[15][16], 203[15], 204[15], 212[29], 218[43], 219[43].
Riddock, J., 345[101].
Ridout, M. S., 60[22], 124[33], 125[33], 313[38], 314[38][45], 315[38], 317[76], 322[102], 338[54], 339[54].
Riley, D. L., 279[175], 290[220], 294[232] [238], 296[243].
Rimmer, G. D., 404[139].
Ritter, E. T., 109[47], 109[48], 109[50], 109[52], 156[116], 560[75], 562[75][87], 573[127][128], 579[150][151].
Ritter, G., 151[116], 263[69].
Robbins, M., 274[124], 554[48].
Roberts, L. D., 151[112], 156[112], 527[89], 528[98], 529[98][105], 530[89][105][106] [108].
Robin, M. B., 150[103], 153[103], 154[107], 178[28].
Robinson, B. L., 516[61], 518[63], 519[68] [71].
Robinson, J. E., 346[120].
Rochev, V. Ya, 393[92], 399[124], 402[137] 403[153], 404[137], 406[137], 409[175], 411[137], 412[137][182], 413[137], 416[200][201], 417[137].
Roger, A., 323[109].
Rogers, K. A., 160[132], 162[132], 164[132].
Rogozev, B. I., 379[34], 412[183], 413[183].
Rokhlina, E. M., 416[200][201].
Romanov, V. P., 253[32], 398[111][112].
Romanov, Yu. F., 600[10].
Romasho, V. P., 315[52].
Ron, M., 249[23], 315[56], 318[85], 319[85], 320[92].
Rose, M. E., 66[33].
Rosenberg, M., 98(22), 265[75][76].
Rosenblum, S. S., 461[64].
Rosencuraig, A., 315[56].
Rosencwaig, A., 253[34].
Rosenthal, J. E., 47[1].
Ross, J. W., 505[31], 506[31].
Rosser, W. G. V., 51[8].
Rossiter, M. J., 254[36], 255[36], 259[53], 260[56], 261[62], 262[53], 263[53].
Rath, S., 513[53].
Rothberg, G. M., 524[79], 525[82], 526[85].
Rothem, T., 42[84], 266[85].
Rother, P., 523[78].
Roult, G., 272[111], 280[176].
Roy, R. B., 314[39].
Rubin, D., 22[20].
Rubinson, W., 88[9].
Rubinstein, M., 314[37].
Ruby, S. L., 33[49], 65[26], 66[26], 87[3], 249[20], 269[20], 270[20], 271[20][105], 276[148], 279[20], 309[28], 321[97],

376[26], 378[26], 393[94], 433[1], 434[4], 437[12][13], 438[18], 440[13], 441[13], 442[27][28][29], 443[29], 444[27][28][29][30][31], 445[27][28][30][32], 446[28][29], 447[30][32], 448[32], 449[28][30][31], 451[38], 452[38], 454[31], 455[31], 456[31], 458[31], 464[73], 465[81], 470[93], 474[73][93], 479[103], 480[4], 486[124], 598[4], 599[4], 600[11], 601[11], 602[11], 603[11], 604[18].
Ruddick, J. N. R., 390[82], 391[82], 392[82], 400[82], 401[82], 408[172][173], 409[172][173], 414[82][173].
Ruddy, V. P., 546[33], 548[33], 550[33][38], 554[45].
Ruderfer, M., 81[10].
Ruegg, F. C., 26[31], 39[31], 41[78], 182[35].
Rukhadze, E. G., 147[89].
Runcorn, S. K., 296[242].
Rundle, R. E., 403[154].
Rundqvist, S., 322[106].
Ruskov, T., 21[13], 498[17], 539[8][9], 573[131].
Russell, D. C., 513[54].
Ryasnyin, G. K., 75[55], 401[135], 407[162].
Rzhekhina, E. I., 600[9].

Sackmann, H., 390[75].
Sadykou, E. K., 99[25], 271[107].
Saito, N., 142[75], 158[125], 494[6].
Sakai, H., 392[90].
Saksonov, Yu. G., 394[100].
Salmon, J. F., 70[36], 105[39].
Salpeter, E. E., 82[12].
Samaraskii, Yu. A., 417[206], 418[206].
Samoilov, B. N., 452[43], 563[92][94], 571[121], 572[92].
Sams, J. R, 227[16], 229[16], 230[26][27], 232[26], 233[27], 345[107], 401[133], 409[148], 412[186][187], 413[133][186][187].
Samuelov, S., 560[78], 561[78], 566[102].
Sanders, R., 462[70], 463[70].
Sandstrom, N. C., 486[123].
Sano, H., 75[56], 117[38], 129[38], 158[125], 333[15][20], 373[5], 374[12], 375[19], 387[68], 400[131][132], 404[138], 405[12], 406[138], 407[163].
Saraswat, I. P., 294[236].
Sarma, A. C., 453[50], 454[50], 455[50], 456[50], 457[50].
Sastry, N. P., 117[15], 119[15].
Sato, K., 323[108].
Sauer, C., 508[37].
Sawatzky, G. A., 122[23], 241[19], 248[19], 251[271], 252[29], 253[33], 265[71], 266[83], 274[129].
Sawicka, B. D., 174[22], 177[22][26], 178[27].
Sawicki, A., 26[35].
Sawicki, J. A., 174[22], 177[22][26], 178[27].
Schaller, H., 515[60], 517[60], 519[70], 520[70], 523[70].
Schechter, B., 21[11].
Schideler, J. A., 138[57].
Schieber, M., 270[99], 271[104], 539[10], 541[10].
Schiffer, J. P., 80[1], 81[3][6], 509[40], 519[40].
Schlemper, E. O., 405[156], 410[177], 414[190].
Schmid, H., 139[61].
Schnorr, H., 336[36][37], 393[95].
Schreiner, W. N., 248[18], 263[18].
Schrey, F., 554[46].
Schroeer, D., 246[14][15], 334[27], 527[93], 573[124].
Schuele, W. J., 266[84], 267[84].
Schuler, C., 583[168].
Schunn, J., 415[192].
Schunn, R. A., 415[192].
Schwartz, B. B., 342[88].
Schwartz, L. H., 181[32], 315[61][62].
Schwerer, F. C., 423[250].
Scott, J. C., 345[107].
Seeley, N. J., 408[169][170], 409[170].
Segal, D. A., 26[33].
Segal, E., 248[17], 251[17], 539[10], 541[10], 566[101], 571[101][117], 573[126].
Segnan, R., 316[67][68][71], 462[67], 499[20], 539[7].
Seidel, E. R., 576[138], 577[138], 578[145].
Seifer, G. B., 171[11], 172[11], 174[21], 177[21].
Seitchik, J. A., 322[104].
Selig, H., 437[13], 440[13], 441[13].
Selyutin, G. J., 246[10].
Semenov, S. I., 442[26], 444[26][34], 445[26], 450[26][33][34], 452[39].
Semin, G. K., 406[160].
Senateur, J. P., 323[109].
Senior, B. J., 382[48][49], 383[48], 384[51][53], 386[49][51][53][56][58][62], 387[49][56][58][62], 388[48][51][62], 391[91], 393[91].
Seregin, P. P., 335[31], 336[40][41][42][43], 398[121].
Sergeev, V. P., 415[198].
Seyboth, D., 435[6][10], 494[3].
Shamir, J., 471[96], 476[96], 478[96].
Shaner, J. W., 244[6].

Shapiro, F. L., 21[13], 498[16][17], 539[8].
Shapiro, V. G., 417[204][205], 452[40].
Sharma, N. D., 469[89].
Sharma, R. R., 98(10).
Sharon, B., 39[69].
Shaw, B. L., 233[28].
Shcherbina, Yu. I., 324[119][120], 419[222][223].
Shechter, H., 249[23], 315[56], 318[85], 319[85], 320[92].
Sheldrick, G. M., 410[176].
Shenoy, G. K., 56[16], 117[15], 119[15], 272[109], 290[219], 333[24], 334[24], 338[45], 339[66], 444[31], 449[31], 454[31], 455[31], 456[31], 458[31], 465[31], 470[93], 474[93], 546[28], 547[28], 598[4], 599[4], 603[13], 604[18].
Sherwin, C. W., 81[9], 82[13].
Sherwood, R. C., 117[17], 120[17], 121[17], 274[124], 285[194], 322[104], 324[125], 529[100], 548[54], 555[54], 556[54].
Shidlovsky, I., 548[55], 555[55], 571[120].
Shier, J. S., 420[238][239].
Shiga, M., 246[11], 315[50][53], 338[57].
Shija, M., 254[37], 255[37].
Shikazono, N., 268[92], 353[53], 317[79], 318[79], 322[101], 324[126], 339[62], 452[41], 460[58], 461[58], 510[45], 513[52], 563[93].
Shimony, U., 282[182], 346[114].
Shinjo, T., 151[123], 157[123], 218[40], 241[49], 246[11][12], 254[37][40], 255[37][45], 256[49], 257[49], 265[73], 310[33], 317[79], 318[79], 322[101], 338[57], 339[59], 510[45], 513[52].
Shinohara, M., 117[34], 125[34], 174[20], 175[20].
Shipko, M. N., 279[170][172].
Shirane, G., 249[20], 269[20], 270[20], 271[20][105], 276[148], 279[20], 316[64][65], 321[97].
Shirley, D. A., 365[25], 461[64][65], 527[94][95], 529[94][101][104], 530[104][109][110], 543[19], 544[22], 547[37], 548[22][37], 550[37], 555[22].
Shlokov, G. N., 394[101][103].
Shoji, T., 452[41], 563[93].
Shortley, G. H., 66[32].
Shpinel, V. S., 38[58][60], 75[55], 374[11][13], 375[18][21], 380[35][39], 390[80], 391[80], 394[96], 398[113][114], 400[129], 401[135], 404[140][144], 405[144], 406[158][159], 407[162], 408[174], 409[174], 411[144][185], 412[158][159][185], 413[144][159], 415[197], 417[13][204][205][207], 420[237][243][245], 423[251], 442[25][26], 444[26][34], 445[25][26], 450[26][33][34], 451[25], 452[40][46], 455[46][53][54], 456[53], 457[53][55], 460[46], 560[77], 600[9].
Shriver, D. F., 181[32], 182[33].
Shtrikman, S., 21[11], 103(4), 107[44], 107[45], 116[7], 117[7], 118[7], 119[10][13][14], 246[8], 260[61], 261[61], 263[61], 266[84], 267[84][88], 269[98], 273[120][122], 274[123], 276[153], 285[196][198][199], 286[198], 397[109], 553[44].
Shuler, W. B., 87[4].
Shulman, R. G., 95[23], 170[2].
Shurcliff, W. A., 104[37].
Shvedchikov, A. V., 339[69], 340[72], 600[8].
Sidorov, T. A., 398[119].
Siegbahn, K., 36[54].
Siegwarth, J. D., 338[46].
Sienko, M. J., 278[162].
Silveira, V. G., 547[35], 550[35].
Šimánek, E., 94[21], 95[22].
Simkin, D. J., 117[25], 122[25], 129[25], 130[25], 245[7].
Simopoulos, A., 76[62], 342[89], 344[95], 438[19], 468[80], 470[80], 472[80], 473[80].
Simpson, W. B., 389[72], 390[72].
Singh, B., 32[48], 89[10].
Singh, K. P., 285[201].
Singwi, K. S., 346[120].
Sinha, A. P. B., 260[57], 263[57].
Sinn, E., 213[33], 216[33].
Sivardiere, J., 274[128].
Sklyarevskii, V. V., 452[43], 563[92][94], 565[98][99], 566[102], 572[92][121].
Skorchev, B., 509[43].
Skorchev, S., 336[39].
Skurnik, E. Z., 514[57].
Slobudchikov, S. S., 460[60].
Smirnov, A. I., 510[44].
Smirnov, Yu. P., 374[10].
Smith, A. W., 160[132], 162[132], 164[132], 408[169][170], 409[170].
Smith, D. L., 199[123].
Smith, P. J., 405[157].
Snediker, D. K., 494[4], 495[4].
Snellman, H., 489[130].
Snider, J. L., 81[4], 81[5].
Snyder, N. S., 417[208].
Snyder, R. E., 324[124], 376[26], 378[26], 441[24], 442[24][29], 443[29], 444[29][30], 445[30], 446[29], 447[30], 449[30], 505[31], 506[31].
Solly, B., 314[39].
Somekh, S., 107[45].
Sommerfeldt, R. W., 479[104].

Sonnino, T., 76[63], 183[41], 437[14], 438[14][16][19], 439[16], 440[23], 468[81], 470[81][90][92], 471[90][97], 472[81][90], 474[90][91][92], 478[97], 480[90].
Soriano, J., 471[95][96], 475[95], 476[95][96], 477[95], 478[96].
Sorokin, A. A., 417[207].
Southwall, W. H., 307[16].
Speth, J., 527[91].
Spighel, M., 509[41], 519[41], 520[41], 555[57], 574[57].
Spijkerman, J. J., 17, 26[31], 39[31], 41[78], 160[133][134], 161[133][134], 182[35], 354[4], 374[14], 375[20], 403[152], 494[4][5], 495[4], 496[5], 579[158], 582[165].
Spijkervet, A. L., 452[44], 456[44], 462[44].
Sprecher, N., 407[166], 408[166].
Sprenkel-Segal, E. L., 288[216].
Sprouse, G. D., 109[51].
Srivastava, J. K., 272[109][110].
Šroubek, Z., 94[21].
Stachel, M., 544[25], 546[25][27], 548[25], 551[25].
Stadelmaier, H. H., 579[158].
Stanek, F. W., 504[28], 527[90], 527[91], 574[133][134], 586[176].
Strauss, G. H., 314[37].
Stearns, M. B., 309[24][30], 310[30][31][32], 311[32], 312[30][32], 313[32], 314[37], 322[24][31][103].
Steen, C. F., 342[91].
Steenken, D., 499[19], 504[30], 505[30], 506[33], 509[19], 519[19], 527[19].
Steichele, E., 187[54], 188[54], 205[24], 206[24], 486[126], 487[126], 546[29], 547[29], 548[29], 556[29][63], 557[29][63][64], 589[184].
Steiner, P., 499[19], 504[29][30], 505[29][30], 506[33], 509[19], 519[19], 527[19], 542[14].
Stepanov, E. P., 452[43], 460[56], 563[92][94], 566[102], 571[121], 572[92].
Stephen, J., 89[13].
Stephenson, A., 296[242].
Sterk, E., 265[77].
Sternheimer, R. M., 98(7).
Stevens, J. G., 294[237], 444[35][36], 445[35], 450[35], 451[36][38], 452[38].
Stevens, R. R., 109[48][50], 560[73][75], 562[75][87], 573[127][128], 579[150][151], 590[188].
Steyert, W. A., 316[68][73], 342[86][87][90], 508[36].
Stöckler, H. A., 75[56], 333[20], 374[12], 387[68], 400[130][131][132], 404[138][141], 405[12][141], 406[130][138][141], 407[163], 411[130], 412[130], 413[130].
Stone, A. J., 42[82], 159[131], 286[213], 290[213], 291[225], 292[213], 293[213].
Stone, J. A., 83[19], 160[131], 539[6], 596[2], 598[2], 600[5][12], 602[12][15][17], 603[15][17].
Stone, N. J., 461[64][65].
Storms, E. K., 508[36].
Straub, D. K., 161[135], 204[18], 212[30], 353[2], 354[2], 417[202].
Street, R., 421[247], 424[258].
Strelkov, A. V., 21[13], 498[17], 539[8][9].
Stukan, R. A., 141[74], 142[74], 147[89], 171[11], 172[11], 174[21], 177[21], 235[38], 285[195][206], 286[209], 369[32][34], 377[31], 391[76], 392[76], 411[184], 412[184], 413[184].
Suba, L., 140[68].
Suenega, M., 179[31], 180[31].
Sugano, S., 95[23], 170[2].
Sukhovenkhov, V. F., 390[79], 391[79], 393[92].
Sumarokova, T. N., 391[76], 392[76].
Sumbaev, O. I., 510[44].
Sunyar, A. W., 87[2], 88[2], 90[15], 241[1], 380[38], 381[38].
Sutin, N., 141[73], 142[73], 346[119].
Suzdalev, I. P., 75[52][57][59], 135[48], 136[50], 258[52], 345[105][106], 376[24], 398[122], 406[158], 412[158], 418[211].
Svensson, A. G., 479[104].
Swan, J. B., 380[38], 381[38].
Swartzendruber, L. J., 314[46][47], 315[48][58], 318[81].
Sweet, R. M., 415[192].
Swerdlow, P. H., 513[55].
Sylvester, S., 573[124].
Syono, Y., 259[54], 260[54].
Syrtsova, L. A., 369[35].
Szytuza, A., 255[46].

Takada, T., 151[123], 157[123], 218[40], 241[49], 246[12], 254[37], 255[37][45], 256[49], 257[49], 265[73], 339[59].
Takaki, H., 246[12], 254[37], 255[37], 316[63], 317[79], 318[79], 338[57].
Takano, M., 151[123], 157[123].
Takashima, Y., 157[121], 290[221], 345[107].
Takeda, M., 142[75], 494[6].
Takeda, Y., 315[50].
Takei, W. J., 271[105], 276[148], 321[97].
Takekoshi, H., 324[126], 339[62], 452[41], 510[45], 513[52], 563[93].
Tamagawa, N., 339[59].
Tanaka, M., 262[65], 263[65], 265[74], 267[89], 268[89].

Tandon, P. N., 479[104].
Tarnoczi, T., 143[78], 147[90], 195[4], 198[4].
Tassa, M., 552[42].
Taube, R., 206[23].
Taylor, M. D., 582[165].
Taylor, R. D., 38[62], 39[70], 70[34], 105[38], 308[20], 342[86][87][90], 420[238] [239], 508[36].
Temkin, A. Ya., 408[174], 409[174].
Temperly, A. A., 284[191], 530[113].
Tenenbaum, Y., 285[199].
Terry, C., 138[57].
Thevarasa, M., 144[81], 408[172], 409 [172].
Thompson, D. T., 229[18], 230[18], 234[18], 236[18].
Thompson, J. O., 527[89], 528[98], 529[98] [105], 530[89][105][106][108].
Thorp, T. L., 205[20].
Thrane, N., 150[100], 151[100], 152[100].
Thum, J. E., 489[130].
Timms, R. E., 411[188], 412[189].
Tino, Y., 315[51].
Tison, J. K., 314[40], 588[185], 590[185].
Tiwari, L. M., 469[88].
Tokoro, T., 262[65], 263[65].
Tolkachev, S. S., 340[74].
Tomala, K., 254[38].
Tomblin, F. F., 559[70].
Tomilov, S. B., 339[69], 340[72], 346[116].
Tominaga, T., 142[75], 158[125].
Tomiyoshi, S., 323[114].
Tomov, T., 573[131].
Tompa, K., 143[78], 195[4], 198[4].
Tornkvist, S., 489[130].
Traff, J., 273[118], 323[118], 324[118], 419[225].
Trammell, G. T., 509[38].
Trautwein, A., 19[8], 363[16].
Treves, D., 21[11], 39[69], 103(4), 266[84], 267[84], 273[120][121][122], 284[189], 307[8], 440[23].
Triftshauser, W., 330[1][2], 334[27], 338[1] [2], 454[51], 455[51], 456[51], 458[51], 459[51], 465[83], 469[83].
Trofimenko, S., 197[12], 199[12], 201[12].
Trooster, J. M., 139[61][62], 147[91], 270[100], 271[101].
Trotter, J., 405[155].
Trousdale, W., 334[29], 342[93].
Trozzolo, A. M., 207[26][28], 208[26], 209[26].
Trukhtanov, V. A., 190[56], 275[140], 369[35], 380[40][41], 391[76], 392[76], 394[100][102], 395[40][41][106], 411[184], 412[184], 413[184].

Trumpy, G., 150[100], 151[100], 152[100], 323[118], 324[118], 419[225].
Tseng, P. K., 324[126], 434[4], 452[41], 480[4], 563[93].
Tsuchida, Y., 310[33].
Tsuei, C. C., 321[96], 455[62], 456[62], 461[62].
Tsyganov, A. D., 398[120].
Turner, G. E., 545[26].
Twist, W., 271[108], 279[108].
Tzara, C., 29[44], 374[8][9].

Ueda, I., 387[65].
Uhrich, D. L., 582[166][167], 538[173], 585[175].
Ullrich, J. F., 456[66], 460[57], 462[66], 481[66].
Umemoto, S., 345[107].
Underhill, A. E., 196[8], 197[8], 199[8].
Unland, M. L., 379[32], 454[52], 455[52], 456[52], 458[52].

Vali, V., 22[23].
Valle, B. D., 444[37], 451[37].
Van den Berg, J. M., 382[47], 387[57].
Van den Bergen, A., 161[141].
Van der Kraan, A. M., 145[83], 254[41].
Van der Steen, G. H. A. M., 285[203].
Van der Woude, F., 39[68], 73[43], 98(12), 103[32], 122[23], 156[114][117], 241[3], 242[3], 244[3], 245[3], 252[27], 253[33], 254[42], 265[71], 266[83].
Van Fleet, H. B., 307[16].
Van Loef, J. J., 254[41], 275[135], 279[166].
Van Stapele, R. P., 285 [203].
Van Uitert, L. G., 566[107], 567[107], 568[107].
Van Wieringen, J. S., 246[16], 279[167], 285[203].
Van Zorge, B. C., 74[48], 156[117].
Varret, F., 269[96].
Varshni, Y. P., 436[11].
Vaughan, R. W., 39[75], 54[12], 118[9], 119[9], 120[9], 148[9], 150[9], 155[9], 158[9], 235[36], 246[9], 271[9], 285[193].
Venevtsev, Yu, N., 394[96], 398[113] [114].
Venkateswarlu, K. S., 161[140].
Verheul, H., 452[42].
Verkin, B. I., 398[111], 418[218].
Verschueren, M., 339[61].
Vertes, A., 140[67][68], 147[90], 159[128].
Vincent, D. H., 62[24], 104[36], 106[40], 380[37], 419[37], 434[4], 456[66], 460[57], 462[66], 480[4], 481[66], 509[40], 519[40].
Vincze, I., 312[35].
Vinnikov, A. P., 418[218].

Vinogradov, I. A., 75[57][59], 135[48], 136[50].
Violet, C. E., 19[4], 37[56], 70[38], 307[19], 317[74][75], 438[20], 453[48][49], 454[49], 455[48][49], 456[49], 459[49], 460[48][49].
Virgo, D., 295[240], 296[244].
Viskov, A. S., 394[96], 398[113][114].
Visscher, W. M., 8[4], 70[34], 105[38].
Vogel, H., 19[8].
Voitovetskii, V. K., 509[39].
Vooght, J. de., 407[166], 408[166].
Voorthius, H., 452[42].
Voronov, F. F., 218[210].
Vucelic, M., 404[146], 405[146], 406[146] 409[146], 411[146].

Wagner, C. F., 202[17], 210[17], 212[17], 548[54], 555[54], 556[54], 567[108].
Wagner, F., 499[22], 501[22], 502[22], 503[22], 515[60], 517[60], 519[70], 520[70], 523[70], 519[72], 520[72], 521[72], 523[78], 589[184].
Wagner, F. E., 579[154], 585[154], 586[176].
Wain, A. G., 543[21].
Walker, J. C., 38[61][63], 108[46], 109[47][48][50][52][56], 171[5], 307[11], 543[17][18], 560[73][75], 562[75][87], 573[127][128], 579[150][151][152][153], 590[188], 598[3].
Walker, L. R., 91[16], 92[16].
Wall, D. H., 376[28][29], 377[28], 378[28], 379[29], 403[28].
Wallace, E. W., 324[121][122].
Walters, W. B., 462[69].
Wames, R. E., 418[213].
Wampler, J. M., 293[230].
Wäppling, R., 294[233][324], 314[39], 322[106], 323[115].
Watanabe, H., 278[161], 323[114].
Watson, K. J., 144[82], 145[82], 195[6], 196[6].
Watson, R. E., 60[23], 91[17][18], 92[18], 97[24], 314[47].
Weaver, C. E., 293[230].
Weaver, D. L., 217[37].
Weaver, R. S., 482[114].
Webb, G. A., 496[10].
Weber, W., 566[103], 573[103].
Webster, M., 393[85][86].
Wedd, R. W. J., 345[107].
Weed, S. B., 294[237].
Wegener, H., 74[49], 109[49], 151[116], 156[115][116].
Wegener, H. H. F., 435[7][8], 436[7][8], 493[1], 494[1][2].
Wei, C. H., 225[11].

Wei, H. H., 494[6].
Weiger, G., 109[53].
Weiher, J. F., 197[12], 199[11][12], 200[11], 201[12], 366[29].
Weisman, I. D., 318[81].
Weiss, R., 399[125].
Weissbluth, M., 98(15), 363[15].
Welsh, R. E., 87[4].
Wendling, R., 318[82].
Wernick, J. H., 313[36], 314[36], 315[36], 322[104][105], 324[36][123][125], 520[73], 522[73], 523[73], 529[100], 547[37], 548[37][54], 550[37], 555[54], 556[54][62], 557[62], 571[116][118], 574[135], 578[135], 587[181].
Wertheim, G. K., 22[22], 65[25], 73[42], 74[42], 84[20], 91[16], 92[16], 113[1][2][3], 114[3], 115[3], 116[1][3], 117[3][11][17], 118[11], 119[11], 120[17], 121[17], 129[11], 148[93], 149[93], 150[93], 151[93], 217[38], 221[1], 222[3], 223[3], 228[14], 233[33], 234[1][33], 235[33], 274[124], 275[133], 313[36], 314[36], 315[36], 322[104][105], 324[36][123][125], 331[5][11], 333[17][18][23], 335[11], 336[11][44], 338[53][55], 339[68], 343[94][96], 344[94].
West, B. O., 161[141], 555[51].
West, K. W., 577[141], 578[141].
Wheeler, R. M., 543[17][18].
White, A. H., 194[1].
Whitehead, A. B., 80[1].
Whitfield, H. J., 100[27], 202[14], 204[14], 268[95], 269[95], 276[151], 278[159], 285[205], 286[205], 292[227].
Wickman, H. H., 73[42], 74[42][45][51], 84[20], 202[17], 207[26][28], 208[26], 209[26], 210[17], 212[17], 338[55], 365[25], 547[35][36][37], 548[37][54], 550[35][36][37], 554[48][49], 555[54], 556[54], 565[100], 566[100][113], 567[100][108], 569[111][112][113], 577[140].
Wiedemann, W., 279[168], 345[109], 504[28], 514[56], 574[133], 575[137], 577[139][142], 578[147][148], 579[147][148][157], 580[162], 583[168], 587[177].
Wiedersich, H., 98(17), 117[32], 125[32], 130[32], 133[32], 134[32], 135[32], 276[150][152][155][156][157], 277[156][157], 278[155], 307[17], 315[57], 317[57].
Wiedersich, J. H., 548[53], 555[53].
Wieser, E., 266[79].
Wiggins, J. W., 579[152][153], 598[3].
Wigley, P. W.R., 178[29].
Wignall, J. W. G., 151[98], 155[98], 156[98], 157[98], 159[98], 398[116][117].

Wigner, E., 12[8].
Wilenzick, R. M., 42[81], 513[54], 526[86], 527[92], 578[149], 579[149].
Wilets, L., 47[4].
Wilkinson, C., 137[55].
Williams, H. J., 117[11], 118[11], 119[11], 129[11], 207[28], 322[104], 556[106].
Williams, I. R., 423[253].
Williams, J. M., 346[118], 380[44], 420[44][232], 565[96], 570[96], 572[96], 573[96].
Williams, P. G. L., 65[30].
Williams, R. J. P., 117[36], 125[36], 126[36], 128[36], 190[58], 191[58], 205[20][21], 367[31].
Wilson, G. V. H., 163[142], 164[143], 423[253].
Wilson, S. S., 310[32], 311[32], 312[32], 313[32].
Window, B., 315[59], 421[247], 423[252][253][255].
Winter, M. R. C., 368[31].
Winterberger, M., 284[192].
Wittmann, F., 98(13), 276[154], 278[154][158], 234[108], 345[109].
Wolfe, R., 280[177].
Wolfram, T., 418[213].
Wong, A. Y. C., 95[22].
Wong, M. K. F., 70[35].
Wood, J. C., 321[98].
Woodburg, H. H., 102[31].
Woodhams, F. W. D., 21[12], 26[12].
Woodruff, R. J., 154[108].
Woolum, J. C., 563[88].
Wooton, F., 453[48], 455[48], 460[48].
Wu, C. S., 513[55].
Wulff, J., 512[49], 524[81], 526[81].
Wynter, C. I., 354[4], 470[93], 474[93], 579[158], 582[165].

Xavier, R. M., 150[105], 153[105], 547[43], 552[43].

Yagnik, C. M., 260[55][57], 263[57], 272[116], 285[197].
Yakimov, S. S., 323[113], 324[120], 419[223].
Yamadaya, T., 268[92].
Yamamoto, H., 278[161], 323[111], 323[114], 323[117].
Yamamoto, N., 246[11][12], 254[37], 255[37][45].
Yaqub, M., 373[6], 418[6], 419[6].
Yasuda, K., 411[180].
Yasuoka, H., 265[73].
Yeats, P. A., 401[133], 413[133].
Yeboah-Amankwah, D., 542[15].
Yen, W. K., 498[16][17], 539[8].
Yerzinkian, A. L., 423[251].
Yonetani, T., 364[22], 365[24].
Yoshida, H., 484[122], 486[122].
Yoshikawa, K., 273[119].
Yoshioka, T., 279[165].
Young, A. E. 391[84], 392[84].
Yung, K., 336[39].
Yuschuk, S. I., 394[100][102].
Yushchuk, S. L., 275[143][144][145].
Yutlandov, I. A., 438[15].

Zahn, U., 233[32], 499[22], 501[22], 502[22], 503[22], 523[78].
Zakharkin, L. I., 402[147], 404[147].
Zane, R., 22[19].
Zech, E., 522[76].
Zeldes, H., 464[75], 467[79].
Zemcik, T., 171[9], 172[9], 320[88].
Zhdanov, G. S., 380[42], 420[42][236].
Ziebath, G., 117[27], 123[27], 151[27].
Zimmermann, B., 171[15], 172[15], 174[15], 178[15], 183[15], 223[7], 286[211], 435[10].
Zinn, V. W., 279[168].
Zinn, W., 577[139], 578[14][148], 579[147][148].
Zory, P., 124[31], 134[60], 138[60].
Zuckerman, J. J., 386[59], 389[73], 390[73], 399[123], 415[196], 416[199].
Zvenglinskii, B., 380[35].
Zykov, V. S., 374[10], 510[44].

Subject Index

absolute calibration of constant-acceleration drives, 40
absorber optimum thickness of, 33–5
absorber preparation, 30, 33–5
absorption cross section (*see* cross-section)
absorption lines, intensity of (*see* line intensities)
abundance (%) of resonant isotope, tabulation of, 607–11
acceleration shift, detection of, 81
actinide elements (*see* Chapter 18)
actinolite, 287, 290, 292
adenosine phosphate complexes, 369
adsorbed surface states, 86, 398
aether drift, investigation of, 81
alloys of iron, 308–17
almandite, 287, 288
alpha-emission source precursors, 30, 598–600, 604
aluminosilicates, 287, 291
americium (^{243}Am), 604
amosite, 292
amphiboles, 290–2
andradite, 287, 288
Ångstrom unit defined, 619
angular dependence of line intensities, 66–72
tabulations of, 612–18
angular momentum, conservation of, 56
anisotropic hyperfine tensor A in rare earths, 575, 576
anisotropy of: g-tensor in high-spin iron(III), 99
inertia, investigation of, 81, 82
mean-squared displacement, 342
recoil-free fraction (*see* Goldanskii–Karyagin effect)
anthophyllite, 287–90, 292
antiferromagnetic ordering in high-spin iron(II) compounds, 113–16, 118–25, 129 (*see also* magnetic interactions)
antiferromagnetism, 63 (*see also* Brillouin function, magnetic interactions, Néel temperature, etc.)
influence on relaxation times, 73
antimony (^{121}Sb), 441–52
decay scheme and sources, 441, 442
nuclear parameters, 442, 445, 449
antiphase boundaries, 312
antishielding factor (*see* Sternheimer antishielding factor)
Apollo 11 mission, 294
applications of the Mössbauer effect (*see* Chapter 4)
arandisite, 399
asymmetry due to Boltzmann occupation of nuclear hyperfine levels, 601
asymmetry of quadrupole doublet, 133, 153, 223 (*see also* Goldanskii–Karyagin effect, line intensities, etc.)
asymmetry parameter, definition of, 55, 58
effect in quadrupole splitting, 56
in cytochrome, 364
FeOF, 257
Fe_2O_3, 244
hafnium spectra, 505, 506
iodine spectra, 466, 467, 469–71, 474, 476
iron carbonyl complexes, 229
iron(II) halides, 117, 124
iron(II) oxyacid salts, 133, 134, 137, 138
$S = \frac{3}{2}$ iron(III) compounds, 208
tin-119 spectra, 381
zinc-61 spectra, 498
atomic beam measurements in KrF_2, 441
Auger cascades, 33, 84, 330, 373, 459
augite, 290
in lunar samples, 295
austenite, 320–2
austenitic steel, hydriding of, 344
Azotobacter vine landii, 369

bacteria, 365
bar (pressure unit) defined, 619

SUBJECT INDEX

barium (^{133}Ba), 488, 489
barium stannate as ^{119}Sn source, 375
barn (unit) defined, 619
beta-decay source precursors, 30, 31, 607–611
beta-tin (*see* tin metal)
biological compounds (*see* Chapter 13)
biotite, 294
Bohr magneton, units of, 619
Bohr nuclear magneton, 60, 619
Bokkeveld meteorite, 346
boron nitride intercallation compounds, 346
bremsstrahlung irradiation, 494
Breit–Wigner formula, 12
Brillouin function, 63
  for europium systems, 550, 551
  for iron compounds, 103, 104, 113, 218, 244, 245, 257, 261, 278, 305, 321
  for tin systems, 419, 424
brownmillerite, 280
Burn's screening rules, 378

caesium (^{133}Cs), 486–8
canfieldite (Ag$_8$SnS$_6$), 399
carborane derivatives: of iron, 236
  of tin, 402–4
cassiterite (mineral SnO$_2$), 399
catalysts studied by ^{57}Fe, 344, 345
cement (*see* portland cement)
cementite, 318–20
centre shift (*see* chemical isomer shift)
chalcogenides of iron, 283–6
chalcopyrite (CuFeS$_2$), 285
chelate complexes of low-spin iron, 187–91
chemical binding energies, 6, 7
chemical isomer shift (*see also* individual elements), 46–50
  absence of in ^{40}K, 434
  comparison between ^{127}I and ^{125}Te, 458
  dependence on coordination number, 240
  dependence on oxidation state, 91–4
  dependence on π-bonding, 95–6
  dependence on spin multiplicity, 91, 95
  discontinuity at α–γ phase transition, 307
  discontinuity at Curie point, 341
  effect of pressure on, 53, 54, 94, 118, 119, 148, 153, 171, 177, 235, 250, 251
  hysteresis in at phase transition, 309
  in americium compounds 604
  in antimony compounds using ^{121}Sb, 444–52
  in binary iron oxides and hydroxides (*see* iron oxides and hydroxides)
  in caesium compounds using ^{133}Cs, 487, 488
  in cyclopentadienyl iron derivatives, 233–6
  in cytochrome, 364, 365
  in dysprosium compounds using ^{161}Dy, 569, 571, 573
  in europium compounds using ^{151}Eu, 545–50, 552–5
  in europium compounds using ^{153}Eu, 556, 557
  in gadolinium compounds using gadolinium resonances, 560–2
  in germanium compounds using ^{73}Ge, 435, 436
  in gold alloys and compounds using ^{197}Au, 529–31
  in haeme derivatives (*see also* biological compounds), 354
  in haemoglobin derivatives, 357
  in hafnium compounds using ^{176}Hf, ^{178}Hf and ^{180}Hf, 506
  in impurity (source) studies, 329–36
  in iodates using ^{127}I and ^{129}I, 470, 471, 479–81
  in iodides using ^{127}I and ^{129}I, 469–79
  in iodine compounds, systematics of, 465–8
  in ^{127}I and ^{129}I compared, 464, 465
  in iridium compounds using ^{191}Ir and ^{193}Ir, 519–23
  in iron(I) high-spin systems, 216, 218, 219
  in iron(II) chelate complexes, 187–8
  in iron(II) halides, 116–19, 122, 127, 130
  in iron(II) nitrogen complexes, 141–7
  in iron(II) oxyacid salts, 134, 139
  in iron(II) $S = 1$ compounds, 205, 206
  in iron(III) chelate complexes, 159–64, 187–9
  in iron(III) halides, 148–54
  in iron(III) oxyacid salts, 155–9
  in iron(III) $S = \frac{3}{2}$ compounds, 206–8, 212, 213–15
  in iron(IV) complexes, 215–17
  in iron(IV) ternary oxides, 280–3
  in iron (VI) compounds, 217, 218
  in iron alloys, 309–24
  in iron carbonyl derivatives, 221–36
  in iron chalcogenides, 283–6
  in iron cyanide complexes, 169–87
  in iron dithiolate complexes, 212–16
  in iron hydroxides, 254–8
  in iron intermetallic compounds, 318, 321–4
  in iron metal, 304–7
  in iron oxides and hydroxides, 238–58 (*see also* ternary oxides, spinel oxides, silicate minerals, etc.)
  in krypton compounds using ^{83}Kr, 437–441
  in lunar samples, 294–6

SUBJECT INDEX | 647

in metalloproteins, 365–8
in metals doped with ^{57}Co, 340, 344
in neptunium compounds using ^{237}Np, 601–4
in organo-iron complexes, 221–36
in organotin compounds (see tin(IV) organic compounds)
in peroxidase, 363, 364
in platinum compounds using ^{195}Pt, 526
in ruthenium compounds using ^{99}Ru, 501–3
in samarium compounds using ^{149}Sm, ^{152}Sm and ^{154}Sm, 541–3
in silicate minerals, 286–94
in spin-crossover situations, 194–205
in spinel oxides, 258, 259
in tantalum using ^{181}Ta, 508, 509
in tellurium compounds using ^{125}Te, 453–9
in ternary oxides of iron, 270–80 (see also spinel oxides, iron(IV) ternary oxides, etc.)
in thulium compounds using ^{169}Tm, 582
in tin compounds, 375
in tin(II) compounds, 381–90
in tin(IV) inorganic compounds, 390–4, 398, 399, 478
in tin(IV) organic compounds, 399–417
in tin metal and alloys, 417–21
in tin sources, 374
in tungsten hexachloride using ^{182}W, 513
in uranium compounds using ^{238}U, 599
in xenon compounds, 484, 486
in ytterbium compounds using ytterbium resonances, 587, 588
relevance of second-order Doppler shift, 53
chemical reactions followed using ^{57}Fe, 329, 345
chemical shift (see chemical isomer shift)
chi-squared function, 42, 442
Chromatium strain D, 365, 366
cis and *trans* isomers and quadrupole splitting, 142, 184–6, 236
cis and *trans* stereochemistries in Sn(IV) complexes, 408–11, 413
clay minerals, 293, 294
Clebsch–Gordan coefficients, 66–72
tabulation of, 612–18
clinopyroxene (see monoclinic pyroxene)
'clock-paradox', 81
Clostridium pasteurianum, 366
cobalt-57 doped into cobalt compounds, 331
Compton scattering of γ-rays, 27, 35, 110
computer curve fitting, 41–3

computer simulation of Mössbauer spectrum, 65, 66
computers, on-line, 26
conduction electron polarisation, 61, 85, 104, 304, 309, 311–13, 316, 318, 329, 342, 527, 555
conduction electron spin density oscillations, 314
conservation: of angular momentum, 56
of energy, 2, 7
of momentum, 2
of parity, 56
constant-acceleration drives, 21–6, 39
constant-velocity drives, 19–21
advantages and disadvantages of, 21
calibration of, 39
controlled-temperature furnace, 39
conversion electrons, 38
coordination number, determination of, 84, 85
effect on chemical isomer shift, 240
core polarisation (see also Fermi contact term), 61, 85, 104, 300, 311–13, 318, 338, 555
cosine effect, correction for, 18, 21, 42
Coulomb excitation for sources, 30, 33, 84, 109, 434, 494, 507, 513, 543, 559, 562, 573, 579, 590, 597
Coulomb recoil implantation for ^{73}Ge, 435
Coulombic interactions (see also electric monopole interactions), 46, 47
counting statistics, 19, 41
covalency, influence on chemical isomer shift, 153, 154, 284
on magnetic fields, 150, 151, 202, 268
covalent compounds of iron (see Chapters 7, 8, 9)
crocidolite, 292
crossover complexes (see spin crossover)
crossover phenomena (see spin crossover)
cross-relaxation interactions, 72
cross-section for resonant absorption, 2, 11–16, 31, 34
tabulation of, 607–11
cryogenic equipment, 38, 39
cubanite (CuFe$_2$S$_3$), 286
cummingtonite, 287, 290, 292
Curie temperature, 63, 104, 249, 274, 279, 285, 305–7, 309, 315–18, 321, 322, 339, 341, 394, 395, 398, 522, 550, 555, 569, 572, 574, 579, 601
curve fitting by computer, 41–3
cyanide linkage isomerism, 181, 182
cyclopentadienyl derivatives (see ferrocene and its derivatives)
cyclopentadienyltin(II), 416
cylindrite (Pb$_3$Sn$_3$Sb$_2$S$_{14}$), 399

cytochrome, 364, 365

'dead-time' effects, 24, 25, 42
Debye model for lattice vibrations, 8, 10, 11, 52, 54
Debye temperature: using ^{57}Fe, 10. 52, 85, 253
  using ^{119}Sn, 372, 390, 400, 417, 419, 421, 429
  using ^{73}Ge, 435
  using ^{125}Te, 461
  using ^{129}I, 479
  using ^{183}W, 513
  using ^{197}Au, 527
Debye–Waller factor, 2, 9, 11, 348
decay after-effects, 32, 33, 84
  following α-particle emission, 599, 600, 614
  studied using ^{57}Co, 329–40
  studied using ^{119}Sn, 372, 373
  studied using ^{125}Te, 453, 459, 460
decay schemes (see individual elements)
deerite, 292
defect structure in Fe$_{1-x}$O, 249
delayed coincidence experiments, 33
demagnetising field, 60
derivative spectrometers, 26–8
detection equipment, 35–8
dicarbollide carborane derivatives, 236
dicyclopentadienyl tin(II), 416
diffusion broadening of Mössbauer resonance, 315
diffusion in liquids, 329, 346–8
diffusion in solids, 321, 346–8
dioctahedral micas, 294
diopside, 287, 290
dipolar (magnetic) contribution to hyperfine field, 150, 175, 206, 217, 244, 271
Dirac theory, 48
disorder phenomena in iron oxide phases, 239, 262
dispersion term in E1 spectrum: from ^{181}W/Ta, 509
  from ^{161}Dy, 565
  from ^{153}Eu, 556
dispersion theory, 9
dithiocarbamatoiron complexes, 202, 204, 206–12
dithiolate complexes: of iron, 212–16
  of tin, 416, 417
doped systems using ^{57}Co (see Chapter 12)
doping experiments with ^{119}Sn, 421–4
Doppler effect (externally applied), 15
  relativistic equation, 51
Doppler line broadening (see thermal broadening)
Doppler scanning (see Doppler velocity)

Doppler shift (see Doppler velocity, second order Doppler shift, etc.)
Doppler velocity, 15, 16–26 (see also chemical isomer shift)
drives, constant velocity, 19–21
dynamic Jahn–Teller effect, 141
dysprosium (^{160}Dy, ^{161}Dy, ^{162}Dy, ^{164}Dy), 563–73
  decay schemes, 563, 564
  hyperfine interactions in alloys and compounds, 566–73
  nuclear parameters, 563–6
dysprosium garnets, 567, 573
Dzyaloshinsky's theory, 246

Einstein model for lattice vibrations, 8, 10, 52, 423
electric dipole (E1) transitions, 66
electric field gradient tensor (see also quadrupole splitting), definition of, 54
  determination of sign of, 69
  lattice term, 58
  relation to electronic wave functions, 58, 59
  temperature dependence of, 59
  valence (ion) term, 58
electric monopole interactions (see also chemical isomer shift etc.), 46
electric quadrupole interactions (see also quadrupole splitting,) 54–9
  magnetic perturbation of, 65, 66
electric quadrupole (E2) transitions, 66
electrical conduction in Fe$_3$O$_4$, 253, 254
electromechanical drives (see constant-velocity drives, constant-acceleration drives)
electron-capture source precursors, 30, 31
electron configuration, determination of, 85
electronegativity and chemical isomer shift in
  iron(II) halides, 122
  tin(IV) halides, 390, 392
  using ^{129}I, 469
electron-hopping process, 85, 251 4, 262, 552–4
electronic spin–lattice interactions, 73
electronic spin relaxation, 63, 85
electronic spin–spin interactions, 73
electron spin resonance of biological compounds, 352, 360, 362, 365, 366
electron spin resonance data on ^{161}Dy, 565, 566, 569; on ^{167}Er, 575
energetic nuclear reactions, influence on ^{57}Fe resonance, 109
energy, conservation of, 2, 7
energy of γ-transitions, tabulation of, 607–611

SUBJECT INDEX | 649

energy transfer to lattice, 1, 6–9
energy units, 619
enzymes (*see* biological compounds)
epidote, 287, 291
equilibrium between spin states in iron compounds (*see* spin crossover)
erbium (^{164}Er, ^{166}Er, ^{167}Er, ^{168}Er, ^{170}Er), 574–9
escape peak, 37
Euglena green alga, 366
Eulerian angle transformation, 65
europium (^{151}Eu, ^{153}Eu), 543–58
  decay schemes, 543, 544
  hyperfine interactions, in ^{151}Eu, 544–55
  hyperfine interactions in ^{153}Eu, 556–8
  nuclear parameters for ^{151}Eu, 543–6
  nuclear parameters for ^{153}Eu, 555–8
europium iron garnet, 544–6, 551, 557
eV (electron volt) conversion to SI units, 619
exchange interaction, effect on chemical isomer shift, 246
exchange polarization effect, 85 (*see also* core polarization)
excited-state lifetime (*see* individual Mössbauer nuclides and tabulation on pp. 607–11)
experimental techniques, see Chapter 2, p. 17
E2/M1 mixing ratio δ^2, 70, 83, 105, 501, 518, 520, 528, 588, tabulation of pp. 607–11

Faraday effect in MgFe$_2$O$_4$, 263, 264
Fermi contact term, 60, 61, 103, 113, 150, 174, 175, 202, 206, 208, 217, 244, 248, 255, 271, 304, 384, 522
Fermi level in iron metal and alloys, 327, 341
Fermi radius in iron alloys, 314
Fermi–Segrè equation, 376, 378, 379, 467
Fermi–Segrè–Goudsmit formula, 92
Fermi surface, Fe$_2$B, 318
ferredoxin, 366, 367
ferrichrome A, 365, 366
ferricyanides (*see also* iron cyanide complexes), 173–8
ferrimagnetism (*see* magnetic interactions)
ferriprotoporphyrin chloride (*see* haemin)
ferriprotoporphyrin hydroxide (*see* haematin)
ferrocene and its derivatives, 233–6, 333, 414–16
ferrocyanides (*see also* iron cyanide complexes), 169–73
ferroelectric compounds, 339, 398
ferroelectric transition, influence on recoil-free fraction, 171

ferromagnetism (*see also* magnetic hyperfine interactions, Brillouin function, Curie temperature, etc.), 63
  in iron(II) halides, 122, 129
  induced by plastic deformation, 312
  influence on relaxation times, 73
ferroprotoporphyrin (*see* haeme)
forward–backward address scaling, 26
franckeite (Pb$_5$Sn$_3$Sb$_2$S$_{14}$), 399
frozen solutions, Mössbauer effect in, 86
  using ^{57}Fe, 139, 159, 175, 182, 222, 236, 364, 365
  using ^{119}Sn, 388, 393, 408
  using ^{121}Sb, 452
  using ^{125}Te, 455–7
  using ^{127}I or ^{129}I, 468, 469, 472
  using ^{161}Dy, 569
furnaces, 39

gadolinium (^{154}Gd, ^{155}Gd, ^{156}Gd, ^{157}Gd, ^{158}Gd, ^{160}Gd), 558–62
gamma function, 48
gamma-radiolysis of iron(III) oxalates, 158
gamma-ray energy, units defined, 619
gamma-ray mass absorption coefficient, 13, 14
  decay scheme (*see* individual nuclides)
  line, width, thickness broadening of, 14, 33–5
  velocity modulation of, 17–26 (*see also* Doppler velocity)
garnet group of silicates, 287, 288
garnets (M$_3$Fe$_5$O$_{12}$), 269, 274–6, 279, 280 (see also garnets, rare-earth iron)
garnets, rare-earth iron, 536, 539, 544–6, 551, 557, 558, 567, 573, 583, 587
garnets, studied by ^{121}Sb, 446–7
garnets (tin substituted), 394–7
gauss, conversion to SI units, 619
Gaussian distribution, 3
gear drives, 19, 20
general physics, application of Mössbauer effect in, 80–2
germanium (^{73}Ge), 434–6
gillespite, 287, 291
glassy materials, 346
glycerol doped with ^{57}Co/HCl, 346, 347
goethite (α-FeOOH), 254, 255
gold (^{197}Au), 526–31
Goldanskii effect (*see* Goldanskii–Karyagin effect)
Goldanskii–Karyagin effect, 74–6, 333
  in siderite, FeCO$_3$, 75, 135, 136
  in Me$_2$SnF$_2$, 76
  in BaFe$_{12}$O$_{19}$, 279
  in SnS, 387
  in adsorbed Sn^{2+}, 395

Goldanskii–Karyagin effect—*contd.*
  in organotin compounds, 407, 411
  absence from Sn monolayers, 418
  in iodides using ^{129}I, 479
  in $(NH_4)_2 HfF_6$, 505
graphite intercallation compounds (^{57}Fe), 346
gravitational red shift, detection of, 80
green alga *Euglena*, 366
Gruneisen's constant, 54
grunerite, 287, 290
g-factor, determination of, 83
g-tensor, anisotropy in high-spin Fe(III), 99
g-tensors in dysprosium compounds, 566–9
g-value, related to Mössbauer parameters, 278
g-values of biological compounds, 362, 364

haematin, 353
haematite (α-Fe_2O_3), 240–6
haeme, 353–65
haemeproteins, 353–65
haemin (^{57}Co-doped), 346
haemin derivatives, 353–6, 363
haemoglobin, 352, 353
haemoglobin (^{57}Co-doped), 346
haemoglobin derivatives, 356–63
hafnium (^{176}Hf, ^{177}Hf, ^{178}Hf, ^{180}Hf), 504–7
half-life of excited state, 5, 31 (*see also* lifetime of excited state and individual elements)
  tabulation of, 607–11
half-life of Mössbauer precursor, 30, 31
Hartree–Fock calculations: for ^{57}Fe, 91, 92, 306
  for ^{119}Sn, 377, 379
  for ^{119}Sn and ^{121}Sb, 448, 449
  for neptunium compounds, 603
hedenbergite, 287–90, 292
Heisenberg natural linewidth (*see also* tabulation for individual elements and Appendix I), 1, 5, 12, 13, 18, 47, 375
Heisenberg uncertainty principles, 5, 9
helium dilution refrigerator, 546
herzenbergite ([SnPb]SnS_2), 399
Heusler alloys of tin, 420
high-pressure cells, 39
high-spin iron complexes (*see* Chapter 6)
high-spin iron(I), 216, 217
high-spin iron(II) complexes, 113–47
high-spin iron(III) halides, 148–64
high-spin–low-spin crossover (*see* spin crossover)
histidine, 353
holmium (^{165}Ho), 573, 574

howieite, 292
hulsite, 399
hyperfine interactions (*see* Chapter 3, *see also* chemical isomer shift, magnetic interactions, and quadrupole splitting)
hyperfine magnetic field at ^{57}Fe, 103 (*see also* internal magnetic field, magnetic interactions, etc.*)
hyperfine magnetic field in FeF_2, temperature dependence of, 113–16
hyperfine structure anomaly, 521, 522, 544, 545
hysteresis effects in quadrupole spectra, 141
hysteresis in chemical isomer shift at phase transition, 309
hysteresis in pressure effects, 171
hysteresis in spin crossover effects, 200

ice, phase transitions in studied by Mössbauer effect, 139, 140
illite, 294
ilmenite in lunar samples, 294–6
ilvaite, 290
impurities and imperfections, influence on $FeCO_3$ spectra, 136
impurity atoms in β-tin, 418
impurity doping, 85
impurity studies with ^{57}Fe (*see* Chapter 12)
impurity studies in Mg_2SnO_4 following neutron irradiation, 398
insoluble Prussian blue, 178–180
instrumental drift, 24, 42
intensity of absorption lines (*see* line intensities)
intercallation compounds, 346
intermetallic iron compounds, 317–24
internal conversion coefficient (*see also* individual elements and Appendix 1), 12, 13, 31, 579, 597
  determination of, 83
internal conversion electrons, use of in ^{119}Sn β-spectrometers, 375
internal magnetic field (*see also* magnetic hyperfine interactions, antiferromagnetic ordering, ferromagnetic ordering, etc.)
  determination of sign of, 62
  effect of pressure on, 54
international system of units (SI), 619
Invar alloys, 315
inverse spin structure (*see* spinel oxides)
iodine (^{127}I, ^{129}I), 462–82
  compounds of, 469–81
  decay schemes, 462, 463
  impurity studies, 481, 482
  iodates, 479–81
  iodides, 469–79

nuclear parameters, 462–5, 479
quadrupole systematics, 464–9
sources, 462, 463
systematics of hyperfine interactions, 465–9
ionicity correlated with chemical isomer shift in ^{151}Eu, 549
iridium (^{191}Ir, ^{193}Ir), 518–24
  decay schemes, 518
  hyperfine interactions in compounds, 519–29
  initial discovery by R. L. Mössbauer, 519
  nuclear parameters, 518–21
iron-57 (*see also* Table of Contents for Chapters 5–13)
  chemical isomer shift systematics, 90–6
  coulombic excitation to 136 keV level, 109
  decay scheme, 87, 88
  enrichment studies of Prussian blue, 180, 181
  Hartree–Fock calculations for, 91, 92
  Heisenberg linewidth, 87
  impurity studies (*see* Chapter 12)
  influence of energetic nuclear reactions in resonance of, 109
  iron(I) in doped alkali halides, 335
  iron(I) systems, 216–19
  iron(II) carbonate, Goldanskii–Karyagin effect in, 75, 135, 136
  iron(II) complexes with nitrogen ligands, 140–7
  iron(II) compounds with $S = 1$ spin, 205, 206
  iron(II) fluoride, 113–17
  iron(II) halides, 113–30
  iron(II) oxyacid salts of, 130–40
  iron(III) chelate complexes, 159–64
  iron(III) compounds with $S = \frac{3}{2}$ spin, 206–15
  iron(III) halides, 148–55
  iron(III) hydroxide gels, 257, 258
  iron(III) oxide fluoride (FeOF), 257
  iron(III) oxide hydroxide (α, β, γ, δ-FeOOH), 254–6
  iron(III) salts of oxyacids, 155–9
  iron(IV) compounds, 215–17, 239
  iron(IV) ternary oxides, 280–3
  iron alloys, 308–17
  iron binary oxides and hydroxides, 240–58
  iron carbonyls, 222–6
  iron carbonyl anions, 222–6
  iron carbonyl derivatives, 221–36, 414, 415
  iron carbonyl hydride anions, 222–6
  iron chalcogenides, 283–6
  iron complexes (low-spin)—*see* Chapter 7
  iron cyanide complexes, 169–73 (*see also* Prussian Blue) molecular orbital analysis of, 95
  iron cyanide complexes (substituted), 182–7
  iron dithiolate complexes, 212–16
  iron foil, use as standard for ^{57}Fe chemical isomer shift, 90
  iron intermediate spin states, 205, 206
  iron intermetallic compounds, 317–24
  iron metal, 304–8
  iron oxides (*see* Chapter 10)
  iron phosphine complexes, 188, 189
  iron sulphides (*see* Chapter 10)
  isocyanide complexes of iron, 184, 185
  line-shapes in a fluctuating magnetic field, 74
  magnetic hyperfine interactions, 102–4
  natural abundance, 88
  nuclear parameters of, 87, 88, 97–9, 103
  pressure dependence of chemical isomer shift, 94
  quadrupole systematics, 96–102
  reference standards for chemical isomer shift, 89, 90
  source preparation, 89, 90
iron-57, 136·4-keV transition in, 110
isomer shift (*see* chemical isomer shift)
isomeric-transition source precursors, 30, 31

Jahn–Teller effects, 85, 141, 206, 265, 268
Japanese-radish peroxidase (JRP-*a*), 363, 364
jarosites, MFe$_3$(OH)$_6$(SO$_4$)$_2$, 157
jump diffusion, 321, 348

Karyagin effect (*see* Goldanskii–Karyagin effect)
Kinetics, studied by Mössbauer spectroscopy, 403
Kramers' doublets (*see also* electric quadrupole interactions), 56, 68, 72, 73, 104, 105, 151, 156, 157, 160, 173, 207, 208, 338, 354, 358, 365, 566
Krönecker delta function, 55
Krypton (^{83}Kr), 437–41

Lamor nuclear precession time, 72, 73
Landé splitting factor, 62
Laplace equation, 55
Lathe drives (*see* constant-velocity drives)
lattice defects, 33, 85
lattice dynamics, 10, 51, 85, 291, 342, 387, 438, 497, 527
lattice energies, 6, 7
lattice sum calculations, 246, 274, 276, 278

lattice vibrations, 7
  anharmonicity of, 11
Laves phases, 324, 501, 600
lepidocrocite (γ-FeOOH), 254–6
lifetime of excited state, determination of, 82 (see also individual nuclides and appendix 1)
ligand-filled crossover (see spin crossover)
ligand-field strengths, determination of, 85
  relation to ^{57}Fe chemical isomer shift, 187
line broadening, 14, 32–5
line intensities (see also Clebsch–Gordan coefficients, Goldanskii–Karyagin effect, saturation effects, recoil-free fraction, quadrupole spectra, partial orientation of crystallites, etc.)
  angular dependence of, 66–72
  in polarisation studies, 105
  saturation effects, 71
  tabulation of, 612–18
line-shapes of ^{57}Fe in fluctuating magnetic field, 74
linkage isomerism in cyanides, 181, 182
lithium drifted germanium detectors, 37
localised magnetic impurity moments, 342
löllingite (FeAs$_2$), 284
Lorenzian curves, computer fitting of, 41, 42
Lorenzian distribution, 12, 14, 16, 348, 459
Lorentz (magnetic field), 60
low-spin iron complexes (see Chapter 7)
low-spin iron(I), 218, 219
ludlamite (Fe$_3$(PO$_4$)$_2$.4H$_2$O), 138
lunar samples, 294–6

magnetic dilution (doping) technique, 175, 176, 304
magnetic dilution and spin relaxation time, 336 (see also spin relaxation)
magnetic dipole interactions, 46, 60, 62
magnetic dipole (M1) transitions, 66
magnetic field, sign of, 60–2
  sign of in iron metal, 306
magnetic field flux density, units of, 619
magnetic hyperfine interactions, 59–63 (see also antiferromagnetic, ferromagnetic, internal magnetic field, etc.)
  in ^{57}Fe spectra, 102–8
  in ^{119}Sn compounds, 394
  quadrupole perturbation of, 63–5
magnetic interactions in (see antiferromagnetism, Néel temperature, etc.)
antimony compounds using ^{121}Sb, 445–8
binary iron oxides and hydroxides (see iron oxides and hydroxides, ternary oxides, spinel oxides, silicate minerals, etc.)

caesium compounds using ^{133}Cs, 488
cytochrome, 364, 365
dysprosium compounds: using ^{161}Dy, 563, 565–72
  using ^{160}Dy, ^{164}Dy, 573
erbium compounds using erbium resonances, 574–9
europium compounds: using ^{151}Eu, 544–8, 550–5
  using ^{153}Eu, 556–8
[Fe(H$_2$O)$_6$]SiF$_6$, 130–2
ferricyanides, 174–7
gadolinium compounds using gadolinium resonances, 560–2
gold alloys using ^{197}Au, 527–9
haeme derivatives, 354, 355 (see also biological compounds)
haemoglobin derivatives, 357–62
hafnium alloys using ^{178}Hf and ^{180}Hf, 506, 507
ilmenite from the moon, 294–6
impurity (source) studies, 329, 331, 333–5, 338, 339
iodine compounds using ^{129}I, 479–81
iridium alloys and compounds using ^{191}Ir and ^{193}Ir, 519–23
iron(II) halides, 113–25, 127, 128
iron(II) oxyacid salts, 133, 135–8
iron(III) halides, 148–50, 153, 154
iron(III) oxyacid salts, 155–7
iron (IV) ternary oxides, 280–3
iron(VI) compounds, 217, 218
iron alloys, 309–17
iron chalcogenides, 283–6
iron cyanide complexes, 174–7, 180
iron intermetallic compounds, 317–24
iron metal, 304–8
  in lunar samples, 294–6
iron oxides and hydroxides, 238–58
krypton using ^{83}Kr, 438–9
low-spin iron(I) compounds, 218
metals doped with ^{57}Co, 341–4
metalloproteins, 365–8
nickel alloys and compounds using ^{61}Ni, 494, 495, 497
neodymium alloys using ^{145}Nd, 538, 539
neptunium compounds using ^{237}Np, 600–3
organotin compounds (see tin(IV) organic compounds)
osmium alloys using ^{189}Os, 516, 517
peroxidase, 364
platinum alloys using ^{195}Pt, 524–6
ruthenium alloys using ^{99}Ru, 500, 501
samarium compounds using ^{149}Sm, ^{152}Sm, ^{154}Sm, 539–43

silicate minerals, 288
spinel oxides, 258–69
$S = 1$ iron(II) compounds, 205, 206
$S = \frac{3}{2}$ iron(III) compounds, 207–12
tantalum using ^{181}W/Ta, 508, 509
tellurium compounds using ^{125}Te, 461, 462
ternary oxides of iron, 270–80 (*see also* spinel oxides; iron(IV) ternary oxides, etc.)
thulium compounds using ^{169}Tm, 583, 584
tin(II) compounds, 382–5
tin(IV) inorganic compounds, 394–7
tin(IV) organic compounds, 401
tin alloys, 419–24
tungsten alloys using ^{182}W, ^{183}W, ^{184}W, ^{186}W, 511–13
uranium compounds using ^{238}U, 599–600
xenon clathrate using ^{129}Xe, 486
ytterbium compounds, using ytterbium resonances, 586–90
magnetic moment and quadrupole splitting, relation between, 141
magnetic perturbation method for sign of e^2qQ, 130, 186, 190, 218, 233
magnetic perturbation of ^{119}Sn quadrupole interactions, 380–5, 394–7
magnetic properties: of alloys, 85, 103, 104
of spin-crossover complexes (*see* spin crossover)
magnetic spectra, angular dependence of, 67–9
appearance of more than six lines line in, 115, 116
influence of electronic relaxation time on, 73
magnetic susceptibility tensor, 131
magnetite (Fe$_3$O$_4$) (*see also* iron oxides), 251–4
hydrated, 345
in meteorite, 346
possible presence in lunar samples, 295
marcasite (FeS$_2$), 284, 285
martensite, 318–22
mechanical drives (*see* constant-velocity drives)
mercury (^{201}Hg), 530, 531
meridial geometric isomers, 199
mesoporphyrin, 353
metallic iron, 304–8
metalloproteins, 365–9
metal–metal bonded compounds, 403, 405, 414, 415
metals studied by ^{57}Co doping, 340–4
metamagnetism in FeCO$_3$, 136
meteorites, 346

methaemoglobin, 360
metmyoglobin, 363
Me$_2$SnF$_2$, Goldanskii–Karyagin effect in, 76
micas, 294
momentum, conservation of, 2
momentum transfer to lattice, 1, 6–9
monochromatic source of polarised radiation, 107
monoclinic pyroxenes in lunar samples, 295, 296
monolayers of tin, 418
montmorillonite, 294
moon rocks (*see* lunar samples)
Morin temperature, 240–6
Morin transition, 272, 284
Mössbauer effect, applications of (*see* Chapter 4)
in frozen solutions, 86
Mössbauer parameters, 84 (*see also* chemical isomer shift, quadrupole splitting, etc.)
Mössbauer sources (*see* individual nuclides)
desirable properties of, 32
Mössbauer spectrometers, 17–30
Mössbauer spectrum, 15–16
computer simulation of, 65, 66
Mössbauer transitions, tabulation of data, 607–11
motional narrowing (*see also* relaxation effects), 151
motional narrowing of magnetic interactions, 104
multiple twinning in Fe$_3$O$_4$, 252, 253
multipolarity of radiation, tabulation of, 607–11
multipolarity of γ-transition, 66
multiscalar-mode spectrometers, 25, 26
muscovite, 294
M1/E2 mixing ratio (*see* E2/M1 mixing ratio)

NaI/TL scintillation detector, 36
NASA Lunar Receiving Laboratory, 294
natural linewidths for Mössbauer transitions, tabulation of, 607–11
Néel temperature, 63
Néel temperature in
dysprosium compounds, 566–70
erbium compounds, 577
europium EuO, 55
high-spin iron(II) compounds, 113–16, 119–21, 124, 125, 129
ilmenite from moon, 295
impurity (source) studies, 331–4, 338
iron(II) oxyacid salts, 135–8
iron(III) halides, 148–50, 153, 154

Néel temperature in—*contd.*
  iron(IV) ternary oxides, 280–3
  iron(VI) compounds, 217, 218
  iron intermetallic compounds, 323, 324
  iron oxides and hydroxides, 244–7, 254–6
  lunar ilmenite, 295
  spinel oxides, 261–9
  tellurium compounds using ^{125}Te, 462
  tellurium compounds using ^{129}I (from ^{129m}Te), 481
  ternary oxides of iron, 272–80 (*see also* spinel oxides, iron(IV) ternary oxides, etc.)
  thulium compounds, 583
  tin alloys, 419, 421, 422
  tin doped into NiO and Cr_2O_3, 398
  uranium dioxide, 599
neodymium ^{145}Nd, 537–9
neptunite, 293
neptunium ^{237}Np, 600–4
neutron diffraction studies, 137
neutron irradiation for ^{83}Kr, 437
nickel (^{61}Ni), 493–7
nickel hydride studied by ^{57}Co doping, 342, 343
nitrogen fixation, 369
nitroprusside ion (*see* iron cyanide complexes, substituted; sodium nitroprusside)
non-Lorentzian line-shape, 249, 250, 268
  from unresolved asymmetric quadrupole interactions in ^{151}Eu, 545
non-resonant attenuation of γ-rays, 15, 34, 35
non-stoichiometry, 85, 239
  in Eu_2O_3, 550
  in $Fe_{1-x}O$, 248–51
  in Fe_3O_4, 253, 254
  in β-FeOOH, 254, 255
  in iron sulphides, 284
  in silicate minerals, 286–94
  in spinel oxides, 265
nordenskioldine ($CaSnB_2O_6$), 399
normal spinel structure (*see* spinel oxides)
nuclear Bohr magneton, 60
  units of, 619
nuclear data for Mössbauer transitions, tabulation of, 607–11
nuclear magnetic moments, 60
  determination of, 83
  of ^{57}Fe, 102, 103
  sign of, 61
  tabulation of, 607–11
nuclear magnetic resonance constants in tin(IV) compounds, 403
nuclear parameters (*see* individual elements)
nuclear physics, application of Mössbauer effect to, 82, 83

nuclear quadrupole moment, 54 (*see also* quadrupole moment)
  determination of, 83
  for ^{57}Fe, 97–9, 235
  tabulation of, 607–11
  units of, 619
nuclear quadrupole resonance, 54
  in antimony compounds, 442
  in iodine compounds, 465–7, 482
nuclear radius, change on excitation, 47–50, 83 (*see also under* individual elements; nuclear data)
nuclear reactions for source preparation, 30, 433, 434, 437
nuclear spin, determination of, 83
  quantum numbers, tabulation of, 607–11
nuclear volume change on excitation, 47–50
nuclear Zeeman effect (*see* magnetic hyperfine interaction)
nucleotides, 369

olivines, 287, 288, 292
on-line computers, 26
optical isomeric shift, 47
orbital contribution to magnetic hyperfine field, 150, 174, 206, 217, 244, 271, 522
orbital magnetic moment, 62
orbital reduction factor: for Fe^{2+}, 97
  for $[Fe^{III}(CN)_6]^{3-}$, 182
order–disorder phenomena, 85, 239, 240, 309
organometallic compounds of iron (*see* Chapter 9)
Orgueil meteorite, 346
orthoferrites ($MFeO_3$), 269, 273, 274
orthopyroxenes, 287–90, 292
osmium (^{186}Os, ^{188}Os, ^{189}Os), 514–18
ovens, 38–9
oxidation state, determination of, 84, 85, 91–5, 285, 286, 451, 452
  influence on ^{57}Fe quadrupole spectra, 99–102
oxime complexes of iron, 190, 191

palladium hydride system studied by ^{57}Co doping, 344
paraelectric region and recoil-free fraction, 398
parity, 66
  conservation of, 56
  of nuclear states, tabulation of, 607–11
partial chemical isomer shifts, 186–7, 221
partial orientation of crystallites, 68, 76
particle size, determination of, 85
  influence on recoil and diffusion process, 346
  influence on resonance, 35, 254
  influence on spin crossover, 200

SUBJECT INDEX | 655

Pauling electronegativity (see electronegativity)
pendulum drives, 20
pentacyanide complexes of iron, 188, 189
perovskites (MFeO₃), 269, 272, 273, 280, 283
perovskite stannates (MSnO₃), 394, 398
peroxidase, 363, 364
phase changes studied by Mössbauer effect, 86
phosphine complexes of iron, 188, 189
phosphors studied by ^{57}Fe, 345
photoelectric effect and Mössbauer dispersion term, 509, 556, 565
phthalocyanine complexes of iron, 190
piezoelectric drives, 21, 22, 498
pigeonite in lunar samples, 295
plastic-deformation induced ferro-magnetism, 312
platinum (^{195}Pt), 524-6
point-charge model for electric field gradient, 184-5
Poisson distribution, 41, 42
polarisation coefficients, 70
polarised iron metal-^{57}Co source and sign of e^2qQ, 256
polarised radiation studies with ^{57}Fe, 104-8
polarising power of cations, influence on chemical isomer shift of, 173, 174, 387
polycrystalline sample, effect of partial orientation, 68, 76
Pople-Segal-Santry self-consistent M.O. method, 378
porphyrin, 353
portland cement studied by ^{57}Fe, 345
potassium (^{40}K), 433, 434
praseodymium (^{141}Pr), 537
premelting anomaly in β-tin, 418
pressure, disproportionation of SnO under, 387
  effect on: chemical isomer shift, 53, 54, 94, 118, 119, 148, 153, 171, 177, 235, 250, 251, 285, 307, 420
  internal magnetic field, 54, 307, 334
  FeF₂ spectra, 118
  insoluble Prussian blue, 181
  Mössbauer parameters, 418
  quadrupole splitting, 54, 148, 153, 177, 206, 235, 246, 250, 251, 285, 498
  second-order Doppler shift, 53
  spin-density imbalance, 62
  generation of quadrupole splitting in SnO₂, 374
  units of, 619
pressure cells, 39
pressure effects, hysteresis in, 171

pressure-induced reduction of Fe³⁺ to Fe²⁺, 152, 153, 155-8, 331
pressure-induced reversal of sign of e^2qQ, 246
promethium (^{147}Pm), 539
protactinium (^{231}Pa), 596, 597
protein molecules (see biological compounds)
protoporphyrin, 353
Prussian blue, 178-82 (see also iron cyanide complexes)
pseudo-brookite, 279
pseudoquadrupole interaction, 76-8, 580
pulse-height analyser spectrometers, 23-5
pyrite (FeS₂), 284, 285
pyrope, 287, 288
pyroxene (see also monoclinic pyroxene, orthopyroxene), 295, 296
π-back donation (see π-bonding)
π-bonding character, determination of, 85
  in haemoglobin derivatives, 357-9
  in iodine compounds, 468, 472, 474, 478, 479
  influence on: ^{57}Fe chemical isomer shift, 95, 96
    Mössbauer parameters, 164, 169, 170, 183, 215, 226, 228, 229
    quadrupole interactions in Sn(IV) compounds, 400-2, 408
π-cyclopentadienyl derivatives (see ferrocene and its derivatives)
π-electron interaction in europium systems, 554

quadrupole coupling constant e^2qQ, sign of, 57, 58 (see also quadrupole splitting and individual elements)
  determination of sign by polarisation experiments, 106
  sign of in high-spin iron(II) halides and complexes, 106, 113, 119, 122, 124, 125, 127
quadrupole/dipole mixing ratio (see E2/M1 mixing ratio)
quadrupole moment, sign of (see also individual elements), 54
quadrupole perturbation of magnetic spectrum, 256
quadrupole spectra, angular dependence of, 67-9
  cause of asymmetric, 73, 129, 130
  hysteresis effects in, 141
  relative intensity of lines, 67-72
quadrupole splitting, effect of pressure on (see also electric quadrupole interactions, electric field gradient tensor, etc.), 54, 148, 153, 177, 206, 235, 246, 250, 251

quadrupole splitting in ^{57}Fe, 96–102
　dependence on oxidation state, 99–102
　effect of: covalency on, 99, 102
　　site symmetry on, 99, 100
　　spin multiplicity on, 99
　　spin–orbit coupling on, 100, 101
　linear relation with chemical isomer shift, 226
quadrupole splitting in
　antimony compounds using ^{121}Sb, 442–7
　binary iron oxides and hydroxides (see iron oxides and hydroxides)
　cyclopentadienyliron derivatives, 233–6
　cytochrome, 364–5
　dysprosium compounds using ^{161}Dy, 565–7, 569–71
　dysprosium compounds using ^{164}Dy, 573
　erbium compounds using erbium resonances, 577–9
　europium compounds using ^{151}Eu, 545–9
　europium compounds using ^{153}Eu, 556–8
　gadolinium compounds using gadolinium resonances, 560–2
　gold alloys and compounds using ^{197}Au, 528–31
　haeme derivatives, 353–6 (see also biological compounds)
　haemoglobin derivatives, 356–63
　hafnium compounds using ^{176}Hf, ^{178}Hf, ^{180}Hf, 505, 506
　high-spin iron(I) systems, 216, 218, 219
　impurity (source) studies, 333, 335–8
　iodates using ^{127}I and ^{129}I, 470–1, 479–81
　iodides using ^{127}I and ^{129}I, 469–79
　iodine, ^{127}I and ^{129}I compared, 364
　iodine compounds, systematics of, 467, 468
　iridium compounds using ^{191}Ir and ^{193}Ir, 519–23
　iron(II) chelate complexes, 187, 188
　iron(II) halides, 113, 116–20, 122–30
　irin(II) nitrogen complexes, 141–7
　iron(II) oxyacid salts, 130, 133, 139
　iron(III) chelate complexes, 159–64, 187, 189
　iron(III) halides, 148–55
　iron(III) oxyacid salts, 155–9
　iron(IV) complexes, 215–17
　iron(IV) ternary oxides, 280–3
　iron(VI) compounds, 217, 218
　iron alloys, 315, 317
　iron carbonyl derivatives, 221–36
　iron chalcogenides, 283–6
　iron cyanide complexes, 169–87
　iron dithiolate complexes, 212–16
　iron intermetallic compounds, 317, 320–4
　iron hydroxides, 254–8
　iron oxides and hydroxides, 238–58 (see also spinel oxides, silicate minerals, etc.)
　krypton compounds using ^{83}Kr, 437–41
　lunar samples, 294–6
　metalloproteins, 365–8
　metals doped with ^{57}Co, 341
　neptunium compounds using ^{237}Np, 601–3
　nickel compounds using ^{61}Ni, 496, 497
　organo-iron complexes, 221–36
　organotin compounds (see tin(IV) organic compounds)
　peroxidase, 363, 364
　ruthenium compounds using ^{99}Ru, 501–3
　samarium compounds using ^{149}Sm and ^{154}Sm, 540–2
　silicate minerals, 286–94
　SnO$_2$ under pressure, 374
　spin-crossover situations, 194–205
　spinel oxides, 258–69
　$S = 1$ iron(II) compounds, 205, 206
　$S = \frac{3}{2}$ iron(III) compounds, 206–8, 212, 213–15
　tantalum using ^{181}Ta, 508, 509
　tellurium compounds using ^{125}Te, 453–62
　ternary oxides of iron, 269–80 (see also spinel oxides, iron(IV), ternary oxides, etc.)
　thulium compounds using ^{169}Tm, 580–3
　tin(II) compounds, 382–90
　tin(IV) inorganic compounds, 390–5, 398, 399, 478
　tin(IV) organic compounds, 399–417
　tin metal and alloys, 417–20
　tungsten compounds using ^{67}Zn, 498
　uranium compounds using ^{238}U, 599, 600
　xenon compounds, 483–6
　ytterbium compounds using ytterbium resonances, 587, 588, 590
　zinc oxide using ^{67}Zn, 498
quadrupole systematics (see individual elements)
quadrupole–Zeeman interactions, 63–5
quality control by Mössbauer effect, 345
quaternary oxides of iron, 272–9

radiation damage, 604
radiolysis experiments (see also gamma-radiolysis), 331
rare-earth elements (see Chapter 17)
rare-earth iron garnets, 536, 539, 544–6, 551, 552, 557, 558, 583, 587
Rayleigh scattering of γ-rays, 27, 29, 30, 110, 374
recoil energy, 1–7, 9
　compensation of by ultracentrifuge, 5

SUBJECT INDEX | 657

implantation technique, 109
tabulation of, 607–11
recoil-free fraction, 2, 6, 8–16, 30, 34, 35, 85, 136, 140, 253, 342
  and diffusion processes, 348
  and paraelectric region, 398
  and polymer structure, 333
  anisotropy of (see Goldanskii–Karyagin effect)
  at ferroelectric transition, 171
  discontinuity at α–γ phase transition, 307
  effect of particle size, 246
  for ^{119}Sn sources, 374, 375, 394
  in Fe(OH)$_3$ gels, 257, 258
  in β-tin and its alloys, 417–20
  in krypton and krypton clathrate, 438
  of SnH$_4$, 402
  relation to Curie temperature, 394
  using ^{197}Au, 527
  using ^{161}Dy, 569
  using ^{40}K, 434
  using ^{61}Ni, 494
  using ^{237}Np, 600
  using ^{186}Os and ^{188}Os, 515
  using ^{195}Pt, 524, 526
  using ^{125}Te from ^{125}I, 459
recoil implantation for ^{73}Ge, 435
reduction factor F in iron compounds, 100, 101
refractive-index measurements by Mössbauer spectroscopy, 82
relativistic time-dilation, 417
relativity, application of Mössbauer effect to, 80–2
relaxation phenomena, 72–4 (see also biological compounds, spin–lattice relaxation, etc.)
repetitive velocity scan systems, 21–6
resolving power, intrinsic, 7
resonance absorption cross-section, 2, 11–16, 34
  tabulation of, 607–11
  of ultraviolet radiation, 4
resonance overlap, 4, 15, 18
resonance scintillation counter, 37, 38
  for ^{119}Sn, 375
rhenium (^{187}Re), 514
Rhodospirillum rubrum, 365
ribonucleic acid (RNA), 369
rotating cam drives, 19, 20
rotating disc drives, 20, 21
rotation matrices, 70
ruthenium (^{99}Ru), 499–504

samarium (^{149}Sm, ^{152}Sm, ^{154}Sm), 539–43
samarium iron garnet, 539
sandwich compounds (see ferrocene)

sapphirine, 292
saturation effects on line intensities, 71
scattering and interference of Mössbauer radiation, 82
scattering-geometry experiments, 27–30, 86, 514, 525
scintillation-crystal detectors, 36
screw-type magnetic ordering, 462
second-order Doppler shift, 50–3, 85, 171, 173, 246, 307, 310, 342
  in europium compounds, 550
  in iron compounds, 96
  in iron(II) halides, 127
  in tin compounds, 394
  using ^{40}K, 434
selection rules for γ-transitions, 56
s-electron density at the nucleus (see also chemical isomer shift), 48
  and shielding effects, 50
semiconductors studied by ^{57}Co doping, 338–9
shielding factor (see Sternheimer antishielding factor)
shift operators, 55
Shulman–Sugano treatment of bonding in iron cyanide complexes, 170
SI units, 619
siderite, 135, 136
silica gel, adsorption of Sn^{2+} on, 398
silicate minerals (see also lunar samples), 286–94, 346
silicon-29 n.m.r. data, 322
silver (^{107}Ag), 504
site occupancy in
  doped ferroelectrics, 339
  ferrite-type oxides, 85
  γ-Fe$_2$O$_3$, 248
  Fe$_4$N, 321
  silicate minerals (see silicate minerals)
  spinels, 258–69
  substituted europium iron garnet, 551, 552
  ternary iron oxides, 271–3, 278
site symmetry, determination of, 84, 85 (see also quadrupole interactions)
  effect on ^{57}Fe quadrupole spectra, 99, 100
Slater's shielding rules, 467
sodium nitroprusside
  Na$_2$[Fe(CN)$_5$NO].2H$_2$O, absolute value of quadrupole splitting in, 182, 183
  used as standard for ^{57}Fe chemical isomer shift, 90
solid-state diffusion, 346–8
solid-state physics and chemistry, application of Mössbauer effect to, 84–6
solid state reactions, studied by Mössbauer effect, 86

SUBJECT INDEX

soluble Prussian blue, 178–80
sources (*see* individual Mössbauer nuclides)
 desirable properties of, 32
source experiments with ^{57}Fe, 329–44
source motion, correction term for, 18, 19, 21
source preparation, 30–3 (*see also* individual elements)
spectrochemical series, relation to ^{57}Fe chemical isomer shift, 187
spin canting, 240
spin crossover, 85, 143–5
 in iron(II) compounds, 194–201
 in iron(III) compounds, 202–5
spin-density imbalance, 60–2
 effect of pressure on, 62
spin-flip in iron(II) halides, 122, 129
 in FeCO$_3$, 136
spin-flip process (*see also* Morin temperature), 244
spin–lattice relaxation time, 72, 73, 291
spin multiplicity in ^{57}Fe compounds, and chemical isomer shift, 91, 95, 168
 and quadrupole splitting, 99, 168
spin operators, 65
spin–orbit coupling, effects on ^{57}Fe quadrupole spectra, 100, 101
spin-relaxation effects in magnetic interactions (*see* superparamagnetism, motional narrowing, etc.)
spin–spin relaxation time, 72, 159, 163
 solvent dependence of, 159
spin waves, 73
spinach ferredoxin, 366, 367
spinel structure: of γ-Fe$_2$O$_3$, 246
 of Fe$_3$O$_4$, 251
 of CuCr$_2$Te$_4$ using ^{125}Te, 462; using ^{129}I (from ^{129m}Te), 481
 of oxides AB$_2$O$_4$, 258–69
 of oxides studied with ^{121}Sb, 446–7
 of sulphides (AB$_2$S$_4$), 285
 of tin (SnM$_2$O$_4$), 394
starch–iodine blue complex, 474
staurolite, 287–91
steel, 90, 319
Sternheimer, antishielding factor, 58, 99, 101
 shielding factor, 58, 97
structural assignments from Mössbauer spectra, 221, 223, 226, 229–33
sulphide spinels (*see* spinel sulphides)
superconducting magnet installation, 391
superconducting states in tin and its alloys, 419
superparamagnetism, 104, 157, 246, 254, 266, 268, 271, 338, 345
surface states, 85, 329, 344, 345

Taft inductive factor σ^*, 412
tantalum (^{181}Ta), 507–9
tealite (PbSnS$_2$), 399
technetium (^{99}Tc), 499
tellurium (^{125}Te), 452–62
 decay after-effects, 459, 460
 decay scheme, 453
 inorganic compounds, 453, 459
 sources, 452, 453
 tellurides, 460–2
tempering, effect on Mössbauer effect in martensite, 321
tephroite, 288
terbium (^{159}Tb), 563
ternary oxides of iron (*see also* spinel oxides, garnets, etc.), 269–83
tesla (SI unit), 619
tetrahedral FeX$_4^{2-}$ complexes, small distortion in, 125–8
thermal decomposition followed by Mössbauer effect, 142, 157, 554
thermal line broadening, 1–5, 7
thermal neutron beam to generate ^{40}K, 434
thermal red shift in β-tin, 417
thorium (^{232}Th), 596, 597
thulium (^{169}Tm), 579–85
thulium iron garnet, 583
thulium salts, pseudoquadrupole interactions in, 77
time-mode spectrometers, 25, 26
time-reversal invariance, 82, 501, 522
tin-119 (*see also* Chapter 14)
 decay scheme, 371, 372
 nuclear parameters, 371, 372, 375–80
 sign of $\delta R/R$, 375–9
 sources, 371–5
 tin(II) compounds, 381–90
 tin(IV) hydride derivatives, 402, 403, 405
 tin(IV) inorganic compounds, 390–9
 tin(IV) organic compounds, 399–417
 tin alloys, 417–24
 tin impurity atoms, 421–4
 tin metal, 417, 418
Torula utilis yeast strain, 364
Townes–Dailey theory of n.q.r., 467
trioctahedral micas, 294
troilite (FeS), 284
 in lunar samples, 295, 296
tungsten (^{182}W, ^{183}W, ^{184}W, ^{186}W), 509, 513
Turnbull's blue, 178–81
twinning in Fe$_3$O$_4$, 252, 253

units, note on, 619
uranium (^{238}U), 597–600

vacuum cryostats, 38

vacuum furnace, 39
valence-electron contribution to e^2qQ, 130
van der Graaff accelerator: for ^{40}K, 434
  for ^{73}Ge, 435
variable temperature control, 39
velocity calibration, 39–41
  by diffraction grating, 40
  by optical interferometer, 40, 41
  by quartz crystal modulation, 41
  international standards for, 39
velocity modulation of γ-rays, 17–26 (see also Doppler velocities, etc.)
Verwey transition in Fe$_3$O$_4$, 251–3
vibration frequencies (i.r./Raman) compared with Mössbauer data, 479
vibrational energy of a crystal, 6–8
vibrational modes (see Debye model, Einstein model, lattice vibrations, etc.)
vicinal geometric isomers, 199
viscous liquids, Mössbauer effect in, 6, 346, 347
vivianite, Fe$_3$(PO$_4$)$_2$.8H$_2$O, 137, 138

Walker–Wertheim–Jaccarino calibration of ^{57}Fe chemical isomer shift, 91–4
wave vector for γ-photon, 9
Weiss-field model, 266, 267
window materials, desirable properties of, 35
  for vacuum cryostats, 39
  iron impurity in, 39
Wustite (Fe$_{1-x}$O) (see also iron oxides), 248–51

WWJ model (see Walker–Wertheim–Jaccarino calibration)
WWJ-type calibration of ^{61}Ni data, 496

Xanthine oxidase, 367, 368
xenon compounds, synthesis by β-decay of ^{129}I compounds, 483–6
X-ray diffraction, compared with Mössbauer effect, 8, 9, 11

Yaffet–Kittel model, 266, 267
yeast, *Torula utilis*, 364
ytterbium (^{170}Yb, ^{171}Yb, ^{172}Yb, ^{174}Yb, ^{176}Yb), 585–90
ytterbium iron garnet, 587
yttrium iron garnet (YIG), use of in polarisation experiments, 107, 108
  substituted by tin, 394

Zeeman interaction (see magnetic hyperfine interactions, etc.)
Zeeman levels of ^{57}Co, 307, 308
Zeeman lines (see magnetic hyperfine interactions)
Zeeman precession frequency, 73
Zeeman–quadrupole Hamiltonian, 66
Zeeman–quadrupole interactions, 63–5
zeolites studied by ^{57}Fe, 345
  adsorption of Sn^{2+} on, 398
zero-phonon events (see recoil-free fraction)
zero-point motion, 50–3
zinc (^{67}Zn), 497–9
zinnwaldite, 294
zussmanite, 292